P9-DIY-428

ESSENTIALS OF THE DYNAMIC UNIVERSE

An Introduction to Astronomy

FOURTH EDITION

LOOKING NORTH

LOOKING EAST

LOOKING WEST

LOOKING SOUTH

This map represents the sky
at the following standard times
(for daylight saving time, add one hour):

JUNE 1 at 10 p.m.
JUNE 16 at 9 p.m.
JULY 1 at 8 p.m.

ESSENTIALS OF THE
DYNAMIC UNIVERSE
An Introduction to Astronomy

FOURTH EDITION

Theodore P. Snow
UNIVERSITY OF COLORADO, BOULDER

West Publishing Company
MINNEAPOLIS / ST. PAUL   NEW YORK   LOS ANGELES   SAN FRANCISCO

Illustrations: George V. Kelvin, Science Graphics;
Alexander Teshin Associates
Composition: Parkwood Composition, Inc.
Copy Editing: Patricia Lewis

Cover Images
The front cover features a photograph of the total
solar eclipse of July 11, 1991 from Mauna Kea,
Hawaii. Taken as part of the International
Multistation Coronal Experiment, this picture
shows a maximum activity corona. Photographer:
Serge Koutchmy, Paris Institut d'Astrophysique,
CNRS.

The back cover image was also taken at Mauna Kea,
Hawaii during the 1991 solar eclipse. Photographer:
Richard Wainscoat, Institute for Astronomy,
University of Hawaii.

West's Commitment to the Environment

In 1906, West Publishing Company began recycling materials
left over from the production of books. This began a tradition
of efficient and responsible use of resources. Today, up to 95
percent of our legal books and 70 percent of our college texts
are printed on recycled, acid-free stock. West also recycles
nearly 22 million pounds of scrap paper annually—the equivalent
of 181,717 trees. Since the 1960s, West has devised ways to
capture and recycle waste inks, solvents, oils, and vapors created
in the printing process. We also recycle plastics of all kinds,
wood, glass, corrugated cardboard, and batteries, and have
eliminated the use of styrofoam book packaging. We at West are
proud of the longevity and the scope of our commitment to our
environment.

Production, Prepress, Printing and Binding by West Publishing Company.

COPYRIGHT © 1984,
1987, 1990   By WEST PUBLISHING COMPANY
COPYRIGHT © 1993   By WEST PUBLISHING COMPANY
                             610 Opperman Drive
                             P.O. Box 64526
                             St. Paul, MN 55164-0526

All rights reserved

Printed in the United States of America

00 99 98 97 96 95 94 93   8 7 6 5 4 3 2 1

Library of Congress Cataloging-in-Publication Data

Snow, Theodore P. (Theodore Peck).
    Essentials of the dynamic universe / Theodore P. Snow.—4th ed.
        p.   cm.
    Includes index.
    ISBN 0-314-00876-4 (pbk.)
    1. Astronomy.   I. Title.
QB43.2.S663   1993   92-26738
520—dc20   CIP ∞

FOR
MAC, TYLER, AND REILLY

Contents

Chapter 7
The Other Terrestrials: Venus, Mars, and Mercury 147

Chapter 8
The Gas Giants and Pluto 175

Chapter 9
Space Debris 205

Chapter 10
Formation of the Solar System: Disks, Rings, and Moons 229

SECTION III
THE STARS   253

Chapter 11
The Sun   255

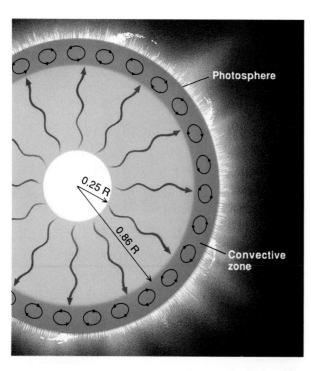

Photosphere

0.25 R

0.86 R

Convective
zone

Chapter 12
Observations and Basic Properties
of Stars   277

Chapter 13
Stellar Structure:
What Makes a Star Run?   303

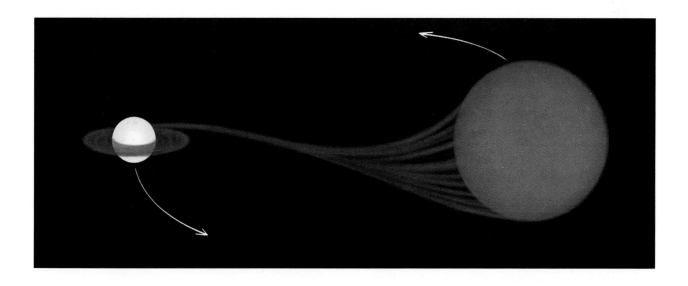

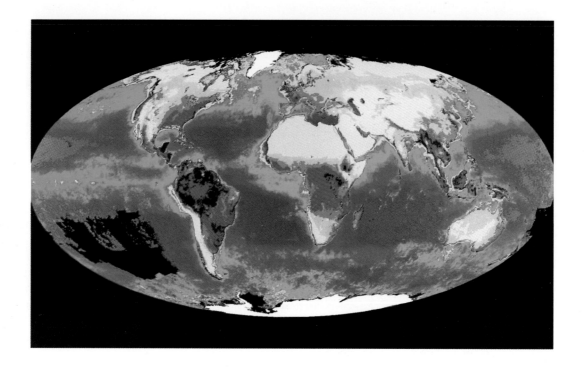

Preface

To study astronomy is, in a sense, the most human thing we can do. What distinguishes us from lower creatures, if not our curiosity, our compulsion to explore and discover? And what exemplifies this compulsion better than the study of the universe?

We probe the heavens (and the Earth) by all possible means, and we do so for no other reason than to learn whatever there is to be known. Astronomy has produced many useful by-products, of course, and could be (and often is) justified solely on that basis. That is not the real reason for astronomy, however.

This textbook represents an attempt by an astronomer to share both the knowledge and the intellectual gratification of our science. There is considerable beauty in the universe for the eye and mind to behold. Just as it is visually stimulating to gaze at a great glowing nebula or a colorful moon, it is intellectually pleasing to arrive at a new understanding of one of the grand themes of the cosmos. It is hoped that the reader of this book will gain by doing both.

This textbook is intended for the student who has not chosen science as his or her major area of study, but who needs an appreciation of science as a vital aspect of preparation for a career. It is as important for such a student to gain some perspective on the general nature of science as it is to learn a great deal of specific information about a particular discipline in the sciences. For that reason, this text stresses the philosophy and outlook of the scientist as well as the knowledge we have gathered about the physical universe we live in.

It is probably as important for the student to understand *how* we know what we know as it is to understand *what* we know. In this era of instantaneous communication and universal access to information, we need more than ever to be able to discriminate among competing hypotheses and to be able to judge the reasonableness of ideas that are advanced. An underlying theme of this text is that to know the workings of science is one of the most important tools we have for meeting the challenges of our technological society.

This edition, like its predecessors, covers the entire scope of astronomy, from the closest objects to the most distant, from the smallest to the largest. Many questions are left unanswered, because there is so much we do not know, and because it is important for students to be aware of that. Not knowing the answers, but knowing how to pursue them, is the essence of science.

This fourth edition has some new features and, of course, reflects all the exciting new developments in astronomy that have occurred since the third edition was published. Perhaps the most immediately obvious change is the nearly complete replacement of the photo and art program. Particularly noteworthy also are the major revision of the section on the solar system and the addition of Astronomical Activities at the ends of the chapters. Finally, at the end of each section is a new feature—Astronomical Updates. These brief updates on current events or controversies in astronomy were written as the book went to press.

One of the most appealing aspects of astronomy, for the professional and the non-science student alike, is the visual beauty of celestial objects. This beauty is enhanced in the new edition by the inclusion of many new photographs and drawings, printed in full color using modern reproduction techniques. Even in cases where the same photographs have been used, new originals were scanned using these modern techniques, thus bringing new life to even the older, more familiar photos. The new, more realistic drawings, which have been prepared by a widely respected scientific illustrator, will enhance the student's intuitive understanding.

Now that the age of *Voyager* is over, and eight of the nine planets have been visited by probes from Earth, a reorganization of the chapters on the solar system is timely. The *Voyager* spacecraft have largely revolutionized our knowledge of the outer planets, turning what was a collection of mysterious, faraway objects into a set of related bodies that seem a little closer to us, in mind if not in place. New themes and new understanding have emerged. In the revised solar system section (Chapters 5 through 10), we now start with a new general planetary science chapter, which gives an overview of the system along with explanations of the major physical processes that affect the planets. With this as background, the student then explores the Earth-Moon system, gaining further insight into general principles of planetary science that can be applied to the other planets. The following two chapters treat the terrestrial planets and the gas giant planets as distinct groupings. In each case, the chapter provides an overview of the general properties of the class of planets, followed by descriptions of the distinctive characteristics of the individual planets. One result of this reorganization is the reduction of the overall length of the book by two chapters. The section ends, as in the past, with a chapter summarizing the formation of the solar system, but now this chapter includes a unified discussion of rings and moons, which are known today to be general phenomena associated with the giant planets. It is hoped that this revised section will provide the student with a more modern, cohesive picture of our system of planets and of the modern planetary scientist's approach to its study.

Another major change in this edition is the addition of "hands-on" activities, provided at the end of each chapter, that the student can perform with little or no special equipment. These Astronomical Activities can be used as class assignments, extra credit work, or simply opportunities for interested students to learn a little more about astronomy and perhaps gain some familiarity with the sky in the process. In many cases, the activity requires the student to go outside and make simple naked-eye observations and record the results. Other activities involve using realistic (in some cases real) astronomical data to derive the properties of astronomical objects, thus allowing the student to gain some feeling for what it is like to conduct research and independently find answers to questions. A few activities require the use of a small telescope or binoculars, and for a small number, the student will need some materials such as paper or cardboard (including in one case two razor blades and a small piece of transmission grating, which is easily obtained through science supply stores or catalogs).

Astronomical Updates, brief articles that emphasize late-breaking developments or controversies in astronomy, are a new feature. These have been written as the book went to press and thus provide the student with some background on discoveries or issues that may appear in the popular literature while his or her astronomy course is still in progress.

In this edition special charts have been made, largely through the Astronomical Insights, to highlight important contributions of women to astronomical discoveries. This emphasis has been added because the many important breakthroughs by women represent major advances in the field, and it is hoped that by highlighting them here, we may help provide role models for young women taking astronomy courses who might have some interest in science as a career.

Other significant updates in this edition include current descriptions of new planetary probes such as *Magellan,* which has largely completed its mission to map the entire surface of Venus at high resolution, *Mars Observer,* on its way to the red planet as this

book goes to press, and *Galileo,* which has embarked on its five-year journey to Jupiter. The recent cosmic discoveries of the *COBE* mission are, of course, included, as are many new results from the *Hubble Space Telescope,* the *Compton Observatory* (formerly known as the *Gamma Ray Observatory*), the *Roentgen Satellite (ROSAT),* and the very recently launched *Extreme Ultraviolet Explorer.* Not only are new space missions covered in this edition, but extensive rewriting has been done (primarily in Chapter 4) to bring the student up-to-date on new developments in telescope technology on the ground.

Apart from the coverage of new space-based missions and ground-based telescopes, the new edition includes major revisions in the chapters on our galaxy (Chapters 16 and 17), the cosmic background radiation (Chapter 19), cosmology (Chapter 21), and the search for life in the universe (Chapter 22).

Comprehensive factual data are incorporated into the appendices, which have been updated with new information (as provided by the *Voyager* planetary encounters, for example). In addition, several appendices have been expanded, including those on stellar data, interstellar molecules, major telescopes, and groups and clusters of galaxies.

An important cultural change has been made in this edition, with a switch from centimeter-gram-second units to the *Système Internationale* (SI) units. Thus all quantitative references (with some exceptions mentioned below) are now given in SI units in the text, in tables, and in the appendices. The two significant exceptions are the retention of Angstrom units for wavelengths of light and the retention of units of gram/cm^3 in discussions of mass densities (which occur primarily in the solar system section). The choice of traditional Angstroms for wavelengths of light was based on comments from many users of the book, who remain more familiar and comfortable with this unit. Mass densities are expressed in grams/cm^3 because students intuitively understand this unit and find it easier to envision the density of a planet in terms of grams/cm^3 than in kg/m^3 due to the ready comparison with familiar substances such as water or rock.

The arrangement of the text remains traditional: an introductory section on the background of astronomy, both in history and in basic physics; a section on the solar system, presenting the planets as individuals before discussing interplanetary bodies and then the formation of the entire system; a section on stars and their lives and deaths; a section on galaxies, starting with two chapters on the Milky Way and then moving outward to the rest of the universe; and a final, brief section on the possibilities that life may exist elsewhere. At the beginning of each of these sections is an introduction that leads the student into the material, and at the end of each section is an Astronomical Update (described above).

The book is designed so that the sequence of sections may be easily altered. For example, to teach the sections on stars, the galaxy, and the universe before discussing the solar system, one need only skip directly from Chapter 4 to Chapter 11 and then go on to the end before returning to Chapter 5, where the solar system studies begin. The chapter on the Sun leads into the section on stars, so that skipping or delaying the solar system discussion will not prevent the student from learning about the nearest and best-understood star. The overview and summary chapters on the solar system (Chapters 5 and 10) include enough information on the Sun that the discussion of the system as a whole and its formation is complete as it stands.

The well-received Astronomical Insights have been carried over into this edition, with a substantial number of new ones added. These features, inserted at appropriate places within the chapters, describe people, discoveries, or current controversies or new hypotheses related to the subject matter of the text. They are meant to enhance the students' enjoyment of the material or understanding of complex topics, but above all they are designed to increase understanding of the scientific process.

Supplemental materials for this text include a revised edition of the *Instructor's Manual*, with new authors Thomas Hockey and Siobahn Morgan of the University of Northern Iowa, and an updated version of the *Study Guide*, prepared by Jeffrey O. Bennett and the undersigned (both of the University of Colorado). As before, the *Instructor's Manual* contains helpful discussions of teaching strategies, Learning Objectives, Key Terms (Concepts), Suggested Readings, and Suggestions for Audio Visual Aids. It provides a large number of exam questions (with answers) and gives complete answers to all the review questions from the main text. The *Study Guide*, intended to help the student obtain maximum benefit from the text, contains brief chapter summaries, lists of key words and phrases, self-tests, and complete bibliographies of articles on relevant topics, taken from a wide assortment of magazines and journals. In addition to the *Study Guide* and the *Instructor's Manual*, other teaching aids are offered to qualified adopters of the text. These include a set of color transparencies for use with overhead projectors and a set of color slides for use in the classroom, including many photographs from the book, as well as a number of spectacular images that do not appear there. In addition to these traditional instructional aids, two innovations are offered to qualified adopters as well. One is a laser disk, prepared by Optical Data Corporation, along with a booklet providing guidance for the instructor on find-

ing the images or movies that are relevant to material in the text. The other ground-breaking teaching aid is an interactive software package and instruction booklet (both prepared by Stephen Pompea of the University of Arizona). The software package allows students to reduce and analyze real astronomical images (obtained with visible-light CCD cameras and near-infrared cameras), in the process gaining insight into how computers are used to handle modern digital data.

At every step during the preparation of this text, vital assistance was provided by a number of people, whose help is acknowledged with gratitude (with apologies to anyone inadvertently omitted). The most important guidance and support were provided by my wife, Connie; by the West Publishing Company editor, Denise Simon; and by the production editor, Ann Rudrud of West. Further editorial support was provided by Bridget Neumayr of West. The new and improved drawings were prepared by George V. Kelvin of Science Graphics; some updated drawings were also provided by Alexander Teshin Associates. A very important contribution to the overall quality and accuracy of the book was made by Derek Wills of the University of Texas at Austin, who scrutinized the galley proofs, making many useful suggestions along the way, in addition to helping to catch the typographical errors that always creep in.

Much of the work of revising the text was done while the author was a Visiting Research Professor at the University of Sydney, Australia, where the support and assistance of Donald Melrose, director of the Research Centre for Theoretical Astrophysics, and of Barbara Dunn, administrative officer of the Centre, were very much appreciated.

Several of my colleagues at the University of Colorado and elsewhere have helped by reviewing sections of the text, providing new photographs, or updating data for tables and appendixes. In addition to the general reviewers listed below, J. M. Shull of the University of Colorado and Alan Stern of the Southwest Research Corporation read and commented on substantial sections of the book. New illustrations or photographs were provided through the courtesy and assistance of David Malin, David Allen, and Coral Cooksley of the Anglo-Australian Observatory; Kathy Hoyt of the U.S. Geological Survey; Karie Meyers and Emma Hardesty of the National Optical Astronomy Observatories; and Patricia Smiley of the National Radio Astronomy Observatory. Stephen Pompea of the University of Arizona provided a number of helpful suggestions for photographs and illustrations, in addition to making his instructional interactive software available as a unique learning aid for students. Others who graciously provided photos were Steve Albers (National Center for Atmospheric Research), Richard

Binzel (Massachusetts Institute of Technology), Gary Emerson (University of Colorado), Akiri Fujii (Japan), Leon Golub (Harvard-Smithsonian Center for Astrophysics), James Kaler (University of Illinois), Kevin Krisciunas (Joint Astronomy Center, Hawaii), Charles Lada (Harvard-Smithsonian Center for Astrophysics), Jeffrey Linsky (University of Colorado), John Mather (NASA Goddard Space Flight Center), and Richard Wainscoat (Institute for Astronomy, University of Hawaii). The cooperation and assistance of professional photographers Peter French, Rick Morley, and Jane and Roger Ressmeyer are also greatly appreciated.

Reviewers of the revised manuscript were:

Bill T. Adams, Jr.
Baylor University

Greg Bothun
University of Oregon

Thomas Hockey
University of Northern Iowa

Shiv S. Kumar
University of Virginia

Keung L. Luke
California State University, Longbeach

Gary D. Schmidt
University of Arizona

Derek Wills
University of Texas at Austin

To all of these people, and to the students whose responses to my teaching philosophies have also helped to shape this book, I am grateful. With their continued input, I trust that this book will continue to evolve, as does our understanding of the dynamic universe.

Theodore P. Snow
October 1992

THE NIGHTTIME SKY AND HISTORICAL ASTRONOMY

Our study of astronomy begins with an overview of the breadth, relevance, and excitement of the science that encompasses all of the universe. We begin, in Chapter 1, with a broad description of the nighttime sky and of human beings' fascination with its splendor. Next, in Chapter 2, we examine more closely the many phenomena that can be seen and appreciated with only our eyes as observing equipment. This arms us with all the knowledge our ancestors had as they attempted to develop a successful picture of the cosmos and their place in it. In Chapter 2 we go well beyond simply describing the sky; we also uncover the explanations for the observed phenomena. Thus we start with knowledge that took human beings millenia to develop.

We then trace the historical development of astronomy, beginning with the earliest recorded astronomical observations and hypotheses, and following the oft-times slow growth of science through the ages. The basis of our modern science lies in the Mediterranean region of Europe; therefore we emphasize developments there. We also explore the genesis of astronomy in other parts of the world, where sophisticated knowledge ultimately either led to dead ends or merged with the mainstream of development.

In Chapter 4 we discuss the groundwork for modern science that was laid during the Renaissance, when fresh ideas emerged in astronomy, as in all forms of human endeavor. We learn to appreciate the awesome breakthroughs made by giants such as Copernicus, Tycho, Kepler, and Galileo, who led the way toward a correct understanding of the universe and the place of people and the planet we inhabit.

The Essence of Astronomy

Telescope domes on Mauna Kea at sunset. (© Peter French)

PREVIEW

The principal point of this opening chapter is to give you a feeling for what astronomy is and what astronomers do. In the process, you will learn a bit about science: how we learn about things by seeking the best explanation for what we observe, and how we test our explanations by making predictions and comparing them with our observations. You will also learn a bit about the nighttime sky, as some of the most prominent objects are introduced. Along the way you will gain some feeling for the enormous range of sizes and distances in the universe and a better understanding of just how small our familiar planet is amid the vastness of the cosmos.

The oldest of all sciences is perhaps also the most beautiful. No artificial light show can rival the splendor of the heavens on a clear night, and few intellectual concepts can compare with the beauty of our modern understanding of the cosmos.

Today we study astronomy for a variety of reasons, some technological, some practical; but no one loses sight of the underlying majesty, of the human instinct for intellectual satisfaction. To study astronomy is to ask the grandest questions possible, and to find hints at their answers is to satisfy one of humankind's most deeply ingrained yearnings.

In this text we explore astronomy in the modern context, which is highly technical and sophisticated, but we will endeavor to retain the sense of wonder and beauty that has motivated the science since the beginning. Although some may argue that astronomy has little practical use, we will see that its origins are rooted in very practical requirements for methods for keeping time and maintaining calendars. In this chapter we begin our study by defining astronomy and introducing some simple terminology that will assist us in later chapters.

WHAT IS ASTRONOMY?

Astronomy is the science in which we consider the entire universe as our subject. It is the science in which we derive the properties of celestial objects and from these properties deduce the laws by which the universe operates. It is the science of everything.

Technically, we might say that astronomy is the science of everything except the Earth, or that it is the study of everything beyond the Earth's atmosphere, since the Earth and its atmosphere fall into the purview of other disciplines such as geophysics and atmospheric science. We will find, however, that the study of astronomy necessarily includes an examination of the properties and the evolution of the Earth and its atmosphere.

In the modern sense, astronomy is probably more aptly called **astrophysics.** Ever since the time of Sir Isaac Newton (the late seventeenth century), the universe has been explored through the application of the laws of physics—most of them derived from earthly experiments and observations—to celestial phenomena. Other scientific disciplines enter into our discussions as well: to study the planets, for example, we must know something of geology and geophysics; to analyze molecules in space, we must understand the principles of chemistry.

Astronomical observations were made and records kept at least as early as the time of the most ancient recorded history (Fig. 1.1), and we believe that the skies were studied and their cyclic motions pondered long before that. In the earliest times, astronomy had a practical motivation: knowledge of motions in the heavens made it possible to predict and plan for certain significant events, such as the changing of the seasons.

Along with the practical came the whimsical and the spiritual. In ancient times, events in the heavens were thought to exert some influence over the lives of people on Earth, and many early astronomers practiced what today we call astrology. Many a monarch retained the services of an astronomer, not only to foretell the seasons, but also to provide advice on strategies for war, love, politics, and business. In many cultures, religion and astronomy were intimately linked, so astronomers attained the importance that major religious leaders have today.

The rich and diverse Greek civilization that flourished for many centuries before and around the time of Christ developed several sciences—astronomy included—beyond the mystical and the spiritual. The Greeks had many preconceived notions about the nature of the universe, but they also made many rational advances in understanding the heavens, setting the stage for the development of modern science many centuries later. Following the era known as the Dark Ages, astronomy (like many other disciplines) experienced the pangs of rebirth during the Renaissance, becoming a rational and methodical science. During the Renaissance, battles were yet to be fought between religion and science, but the course was set. The work of pioneers such as Copernicus, Galileo, Kepler, and especially Newton placed astronomy on a firm physical basis.

Modern astronomy still has a practical aspect, but few who pursue the science do so for that reason. Today we study astronomy primarily for the sake of expanding our knowledge; it is research of the purest sort. Even so, many practical and concrete benefits

ASTRONOMICAL INSIGHT
The Philosophy of Science

Science progresses by hypothesis and test, by trial and error. Rather than adopting a hypothesis and then attempting to ignore or rationalize data that contradict this preconceived notion, a scientist is willing to revise a theory in the light of new information.

If a scientist adopts anything as an article of faith, it is that no theory is above the possibility of modification—that our knowledge always has room to grow. This willingness to change does not mean that scientific theories are wrong or invalid. A scientific theory should be regarded simply as the best explanation available that is consistent with the known facts. A theory is modified or replaced when more facts become known or when a better explanation is found.

A theory usually starts as a hypothesis, an initial suggestion made to explain some phenomenon that is

observed. As new information is gathered, the hypothesis develops into a theory, a framework allowing for prediction and test. Many theories fall by the wayside quickly because new information contradicts them, while others persist and become widely accepted because they are consistent with new data that become available. One feature of any theory is that it leads to specific predictions about what will happen in new situations. Part of the job of testing a new theory is to make predictions from it and then make observations or perform experiments to test those predictions.

One of the most important goals of this text is to impart to the student an understanding of how scientific progress is made—how theories are developed, tested, and revised. For this reason, many examples of discoveries are described, some in Astronomical Insights such as this one and others in the main text. These descriptions will help illustrate how scientists develop, test,

and modify theories in the light of new information. They will show that even the best scientists make errors, but that the common thread is their willingness to discard incorrect hypotheses when conflicting evidence is found or a better explanation appears. The reader will see that scientists do not always know what to look for, but that our knowledge of the universe grows because scientists seek the best explanation for what they observe, rather than trying to modify the observations to conform to existing theories. The hope is that these descriptions will help the student develop a sense of how science works and a better perspective on new theories and controversies that may be encountered outside the classroom.

Figure 1.1   An ancient astronomical site.   Stonehenge, a monument in England, was built according to astronomical alignments in prehistoric times. (C. D. McLoughlin)

derive from astronomy; witness, for example, the many technological spin-offs from the space program or the multitude of new physical and chemical processes, applicable here on Earth, that have been discovered through astronomical observation.

No matter how analytical we may be in modern astronomy, we never lose sight of the same basic human feelings that inspired our ancestors. The modern astronomer, who may spend observing time using a large telescope and a variety of complex electronic instruments (Fig. 1.2), still treasures the moments spent outside the telescope building, simply watching the skies with the same tools the ancients used.

A TYPICAL NIGHT OUTDOORS

Let us imagine that we are sitting outdoors on a fine, clear night, far from city lights and other distractions. This is easy to do and highly recommended; by sim-

ply getting out into the countryside, we can see many of the beautiful objects that we will be studying in more detail as we use this text.

The most obvious objects in the sky, assuming that the Moon is not in a bright phase, are the stars. They appear in profusion, scattered across the heavens, displaying a wide range of brightnesses and subtle variations in color. They twinkle, giving the appearance of vitality. Here and there we may see concentrations of stars in a cluster (Fig. 1.3) or possibly a dimly glowing gas cloud (Fig. 1.4).

If it is a moonless night, we see a broad, diffuse band of light across the sky. This is the Milky Way, our own galaxy of stars, seen edgewise from an interior position (Fig. 1.5). The Milky Way consists of billions of stars intermixed with patchy clouds of interstellar gas and dust.

Figure 1.3   A cluster of stars.   This group of stars is held together by mutual gravitational attraction. This cluster, called the Pleiades, is easily visible to the unaided eye in the autumn and winter. (© 1985 Anglo-Australian Telescope Board. Photo by D. Malin)

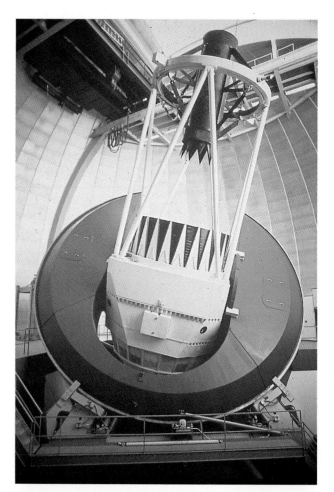

Figure 1.2   A large modern telescope.   This large reflecting telescope at the Cerro Tololo Inter-American Observatory, located in Chile and operated with U.S. government funds, has a diameter of 4 meters, or roughly 13 feet. (National Optical Astronomy Observatories)

Figure 1.4   A gaseous nebula.   Clouds of gas and tiny dust particles such as this are the birthplaces of stars. This is the Orion nebula, bright enough to be seen with only a modest telescope or pair of binoculars. (© 1981 Anglo-Australian Telescope Board. Photo by D. Malin)

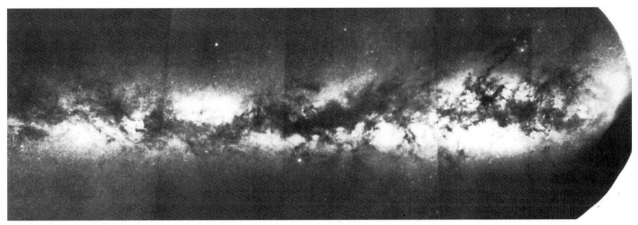

Figure 1.5   The Milky Way.   This composite photograph shows the hazy band of light across the sky known as the Milky Way. It is a cross-sectional view of the disk of our galaxy, which contains roughly 100 billion stars. (Mt. Wilson and Las Campanas Observatories, Carnegie Institution of Washington)

We may also see a few bright, steady objects that do not appear to twinkle. These are the planets, and as many as five of them may be visible on a given night, distributed along a great arc through the sky. Careful observation over several nights will reveal that the planets are all moving gradually along this arc, changing their positions relative to the fixed stars.

The most prominent object in the nighttime sky is usually the Moon (Fig. 1.6), which shines so brightly when near full that it drowns out the light of all but the brightest stars and planets. The Moon, about one-fourth the diameter of the Earth but some 387,000 km (or about 60 times the radius of the Earth) away, presents various appearances to us, depending on how much of its sunlit portion we see. The Moon is always found somewhere along the same east-west strip of the sky, called the **zodiac,** where the planets travel.

Occasionally we may see brief flashes or trails of light called **meteors.** These "shooting stars" can be spectacular events, particularly when they arrive with great frequency, as they do during a meteor shower.

Comets are occasional visitors to our sky (Fig. 1.7). Perhaps once or twice a year a comet is found that is bright enough to be seen with the unaided eye. These largely gaseous bodies orbit the Sun, as do the Earth and other planets, but in very elongated paths that bring them close enough to the Sun to heat up and glow visibly only for brief periods of days or weeks. In ancient times some of the more spectacular, bright comets were interpreted as harbingers of catastrophe.

The sky displays a complex pattern of motions, some of it evident to an alert watcher in an hour or so, other aspects requiring careful observations over hours, days, or weeks. The most obvious motion is

Figure 1.6   The Moon.   Astronauts on one of the Apollo missions made this photograph from space. The portion at the left in this image is not visible from Earth. (NASA).

the steady rotation of the entire sky; objects rise in the east and set in the west, as a reflection of the Earth's rotation. (The terms **rotation** and **revolution** are sometimes interchanged in everyday conversation, but here *rotation* means the same thing as *spin*, whereas *revolution* means *orbital motion*, such as the motion of the Earth around the Sun or of the Moon around the Earth.) Another motion that can be discerned readily is that of the Moon with respect to the stars. As it orbits the Earth, the Moon moves a distance in the sky equal to its own apparent diameter

ASTRONOMICAL INSIGHT
Astronomy and Astrology

There is an unfortunate tendency in modern society to confuse astrology and astronomy or, worse yet, to consider one a legitimate alternative to the other. Behind this tendency lies an even more unfortunate misunderstanding of what science is and of the distinction between scientific theory and beliefs adopted on faith.

Astrology, the belief system based on the premise that events and configurations in the sky influence human affairs and activities on Earth, arose at a primitive stage of human development, at a time when the Earth was still thought to be a flat disk under the dome of the heavens. Although ancient Greek astronomers did much to raise the study of the heavens to a scientific level, the astrological rationale for study of the heavens persisted through the subsequent Dark Ages. Belief in the predictions of astrology began to waver during the Renaissance, when the true nature of the heavens and the

motions of astronomical objects were untangled, but even so, substantial interest in astrology was maintained. In fact, some of the most eminent and forward-looking astronomers of the time practiced astrology. It is unfortunately true that, even today, some people continue to profess faith in astrology.

One of the basic lessons we have learned about the universe is that it is easy to make mistakes unless we are careful to be objective and to accept only conclusions that can be verified by repeated observations or experiments, or by making predictions that can be tested. Astrology fails to meet these criteria, utterly and abysmally. Researchers have made serious attempts to test astrological lore by statistical analysis of people born under different signs, but no trace of a correlation has ever been found.

Unfortunately, people who accept astrology usually do not demand

verification, but prefer instead to put their trust unquestioningly into the jumbled and vague predictions and aimless advice offered by astrologers. No real knowledge is to be gained by doing so; perhaps faith in astrology provides a sort of shelter from a world that is too complex for easy understanding and ready decision making.

Many phenomena defy the understanding of modern science, and many of these deserve more attention than they are getting. However, astrology is not one of them. As a phenomenon, astrology is worthy of study only by the sociological and psychological sciences, for its effects do not exist in the physical universe. Comparing astrological signs may be interesting party talk, but it is a good idea to bear in mind the difference between objective reality and subjective impressions. Failure to do so is failure to understand what science is.

Figure 1.7   A comet.   This is a view of Comet Kohoutek during its 1973 passage through the inner solar system. (NASA)

in just one hour, so it is possible to see its position change with respect to the background stars in a short time. Other cyclical motions, such as those of the planets as they gradually travel along their orbits about the Sun, require more patience and care to be observed. It is noteworthy, however, that ancient astronomers noticed many of the regular patterns, some of them quite subtle, in the motions of heavenly bodies. That they did so is a testimony to the care and diligence they applied to their studies of the skies.

THE VIEW FROM EARTH

When we look at the sky, we do not see it in three dimensions because there are no obvious clues to tell us the distances to the objects we see. Long ago this fact led to the concept of the **celestial sphere** (Fig. 1.8), in which the stars and other objects in the sky are envisioned as lying on the surface of a sphere that is centered on the Earth. Although we no longer

think of this as literal truth, the celestial sphere may still serve as a convenient device for discussing and visualizing the heavens.

Directions and separations of objects on the celestial sphere are measured in angular units because, lacking knowledge of distances to objects, we have no easy way to determine their actual separations in true distance units such as meters or kilometers. Thus we may specify where a star is by saying how many degrees, minutes, and seconds of angle it is from another star or from a reference direction.

We have noted that the stars rise and set with the daily rotation of the Earth. This gives us a natural basis for timekeeping, and our standard units of time are based on the Earth's rotation. The length of the day is equivalent to the rotational period of the Earth (a more specific definition is given in the next chapter).

Anyone who has traveled from one hemisphere to the other may have noticed that the visible stars are not the same in the Northern Hemisphere as in the Southern (Fig. 1.9). The portion of the sky that we can see depends on our latitude (our distance, in degrees, north or south of the equator). For those of us living in the Northern Hemisphere, a large region of the southern sky is hidden from view. We see different constellations as we travel north or south, a fact that was well known to early astronomers, who deduced from this and other evidence that the Earth is round. One consequence is that astronomers must

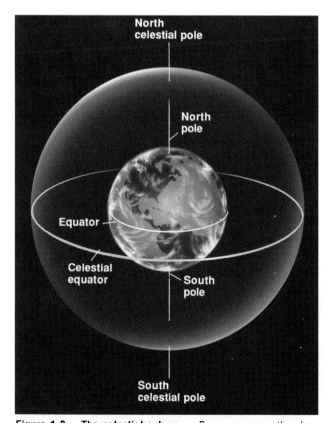

Figure 1.8 The celestial sphere. Because we see the sky in only two dimensions, it is useful and convenient to visualize it as a sphere centered on the Earth, with the stars and other bodies set on the surface of the sphere. We measure positions of objects on the celestial sphere in angular units because the actual distances are not directly known. (Throughout this text, we will learn about distance measurements in astronomy.)

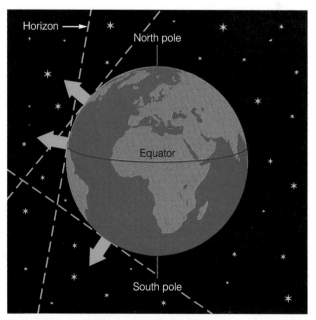

Figure 1.9 Latitude and our view of the sky. The portion of the sky that we see depends on where we are with respect to the Earth's equator. The arrows indicate the overhead direction from three different latitudes.

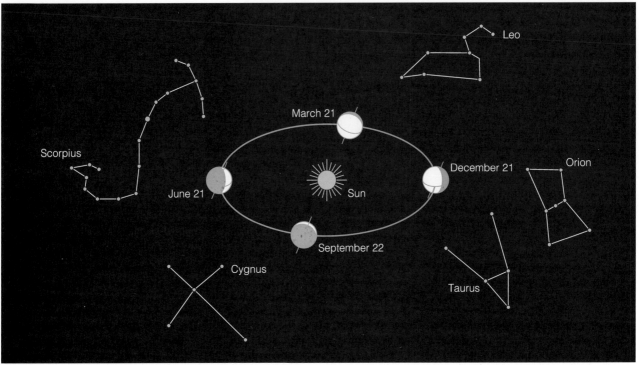

Figure 1.10   The changing view of the sky with the seasons.   As the Earth orbits the Sun, the portion of the sky that we can see at night changes. A few prominent constellations are shown here.

have telescopes in both hemispheres in order to study the entire sky.

We usually cannot observe celestial objects during daylight, so our view of the heavens at any particular time is limited to only the half of the sky that is visible at night. Because of the Earth's motion around the Sun, however, the part of the sky that is up during the night gradually changes (Fig. 1.10). Therefore, any part of the sky that is visible from a given location on the Earth can be observed at night at some time of the year; we need only wait until the appropriate time to observe a given object.

FROM THE EARTH TO THE UNIVERSE: THE SCALE OF THINGS

We have been describing the appearance of the sky to the unaided eye, which has necessarily limited us to nearby objects such as the Sun, the Moon, the planets, and the stars in our part of the local galaxy. It is interesting now to expand our horizons and to try to comprehend the scale of the universe beyond this local neighborhood. Some of the distance scales we discuss are listed in Table 1.1.

Even the nearest star is much farther from the Earth than any solar system object. If we take the average Sun-Earth distance as a unit of measure (we

TABLE 1.1   Size Scales in the Universe

Object or Phenomenon	Size[a]
Atomic nucleus	10^{-15} m
Atom	10^{-10}
Virus	5×10^{-8}
Interstellar dust grain	5×10^{-7}
Bacterium	10^{-6}
Human body cell	5×10^{-5}
Human	1.8
Planet Earth	1.3×10^{7}
Sun	1.4×10^{9}
Sun-Earth distance	1.5×10^{11}
Distance to nearest star	4.1×10^{16}
Milky Way galaxy	9×10^{20}
Distance to Andromeda Galaxy	2×10^{22}
Local group of galaxies	5×10^{22}
Rich cluster of galaxies	10^{23}
Supercluster	10^{24}
The universe	$>10^{26}$

[a]The sizes listed are meant to be typical, to illustrate the relative scales. For round objects, the diameter is used; for irregular objects, an approximate average dimension is given.

call this the **astronomical unit** or **AU**), the nearest star is nearly 300,000 of these units away from us. The most distant known planet, Pluto, is only about 40 AU from the Sun, so we see that the stars are much more widely dispersed in space than the objects within our solar system.

ASTRONOMICAL INSIGHT
What Is an Astronomer?

Throughout this text we will be referring to astronomers, scientists, astrophysicists, and physicists. In a book that endeavors to summarize all that we know about the universe and its contents, we can hardly omit a description of the people who devote their time to developing this knowledge.

There are thousands of people in the United States alone who study astronomy as either a vocation or an avocation. Representative of the latter are the amateur astronomers who engage in a variety of astronomical activities, often on their own, but in many cases through local and even nationwide organizations. Telescope making, astrophotography, long-term monitoring of variable stars, public programs, and just plain stargazing are included. If you wish to join such a group, to get advice on buying a telescope, or to learn techniques such as photography of celestial objects, your best bet is to get in touch with an amateur astronomy group. Local clubs exist in most major cities, and regional associations are everywhere. These groups may be difficult to find in the telephone book, but a telescope shop or planetarium is bound to know whom to contact.

The amateur astronomers in the United States outnumber the professionals. The American Astronomical

Society, the principal professional astronomy organization in the United States, has some 5,000 members. As a rule, these people fall into a limited number of categories: those who do research and teach at colleges and universities; those who do research and work at government-sponsored institutions, such as the national observatories and federal agencies like NASA; and those who perform research and related engineering functions in private industry, most often with companies involved in aerospace activities.

Most of the funding for research in astronomy, even for those not working directly for federal labs and observatories, comes from the government. A significant function of an astronomer on a university faculty is to write proposals, usually to the National Science Foundation or to NASA, asking for support for projected research programs.

The terms *astronomer* and *astrophysicist* have come to mean much the same thing, although historically, there was a difference. An astronomer studied the skies, gathering data but doing relatively little interpretation; an astrophysicist was primarily interested in understanding the physical nature of the universe and therefore carried out comprehensive analyses of astronomical data or did theoretical work, in both cases applying the laws of physics to phenomena in the heavens. Nearly all modern astronomers do astrophysics

to varying degrees, however, ranging from the observational astronomer at one end of the spectrum to the pure theorist at the other. The two terms for people engaged in these pursuits are therefore used interchangeably today. Many modern astronomers call on the fields of engineering (for instrument development), chemistry (in studying planetary and stellar atmospheres and the interstellar medium), geophysics (in probing interior conditions in planets and other solid bodies), and sometimes even biology, but always with an underlying foundation in physics.

If you want to become a professional astronomer, you should be aware from the outset that the field is small and job opportunities are both limited and often subject to the vagaries of federal funding. If you persist, the best course is to study physics, at least through the undergraduate level, and then to plan to attend graduate school in astronomy or physics. (The latter option may make you a bit more versatile and is never a handicap when entering astronomy later.) On the bright side, demographic studies have indicated a significant shortage of astronomers for a period beginning in the early 1990s and extending through the rest of this century. Perhaps the timing will be right for you.

If we could reduce the scale of the solar system, it might help us to visualize the relative distances. For example, if we let the Earth be the size of a basketball (let us say that its diameter is 0.3 m (about 12 in.)), we can convert the rest of the solar system to the same scale. The Sun would be 32.7 m (109 ft.) in diameter, and the distance from the Sun to the Earth would be 3.53 km (2.2 miles). Pluto would be an object the size of a tennis ball roughly 140 km (87 miles) away. The distance from the Sun to Alpha Centauri, the nearest star, would be over 1 million

km, or more than twice the actual Earth-Moon distance!

Now consider our galaxy, the vast collection of stars to which our Sun belongs (Fig. 1.11). The Milky Way galaxy contains roughly 100 billion (10^{11}) stars, arranged in a huge disklike structure having a diameter of about 100,000 (10^5) light-years. (A **light-year,** the distance light travels in a year at its speed of 300,000 km per second, is equal to about 9,500 billion km.) The distance to Alpha Centauri is about 4 light-years, and the most distant stars easily seen

Figure 1.11   A galaxy similar to the Milky Way.   We can not obtain an exterior view of our own galaxy, but this one is believed to resemble ours. Note that most of the stars lie in a disk, seen face-on in this photo. (© 1992 Anglo-Australian Telescope Board. Photo by D. Malin)

Figure 1.12   The Magellanic Clouds.   These two irregularly shaped, small galaxies lie just outside the Milky Way galaxy and orbit it, taking hundreds of millions of years for each complete orbit. (Fr. R. E. Royer)

with the unaided eye are several hundred light-years away. (The majority of the brightest stars in the sky are actually quite nearby by galactic standards; see Appendix 8.) Light from these stars has been traveling for hundreds of years when it reaches our eyes, and light from the far side of the galaxy takes about 100,000 years to reach us.

The nearest galaxies beyond the limits of the Milky Way are the Magellanic Clouds, two irregularly shaped fuzzy patches of light visible only from the Southern Hemisphere (Fig. 1.12). The Magellanic Clouds lie between 150,000 and 210,000 light-years from the Sun, so we see that they are not very far outside the galaxy. They are considered satellites of the Milky Way, orbiting it in a time of several hundred million years. Light from the Magellanic Clouds takes over 150,000 years to reach us. The most distant object visible to the unaided eye is another galaxy similar to the Milky Way, called the Andromeda galaxy, after its constellation, which is about 2.3 million light-years away (Fig. 1.13). When we look at the Andromeda galaxy, we are receiving light that has been traveling for more than 2 million years!

Even the distance to the Andromeda galaxy is insignificant compared with the scale of the universe

itself. The Milky Way, the Magellanic Clouds, the Andromeda galaxy, and a number of other galaxies all belong to a concentrated grouping, or cluster, of galaxies. Most of the other galaxies in the universe also belong to clusters (Fig. 1.14), whose diameters can be as large as tens of millions of light-years. Between clusters of galaxies, space is relatively empty. (Actually, this point is controversial, as we shall see; all we can say for certain is that there are relatively few visible galaxies between clusters.)

Clusters of galaxies are themselves grouped into larger conglomerates called **superclusters,** whose size scales are significant, even on the scale of the universe itself. A supercluster typically may have a diameter measured in the hundreds of millions of light-years. (*Diameter* is probably not a good word; superclusters seem to be sheetlike or filamentary structures, not rounded like clusters of galaxies.) It is difficult to imagine an organized object or collection of objects so large that it takes hundreds of millions of years for light to travel across it.

Beyond the size scale of the superclusters, we approach the scale of the universe itself. It is apparent from a variety of lines of evidence that the observable universe has an overall size scale measured in tens of billions of light-years. Light reaching us now from the farthest reaches of the universe has been traveling for many billions of years. This has the fascinating implication that we can see the universe as it was at that long-ago time when the light was emitted, a fact used by astronomers who study the origins of the universe.

Considering the sizes and distances of objects in the universe provides us a sobering perspective on ourselves and our tiny planet. It can be quite a revelation to see how much we presume to explain about the universe and how much we think we have learned about the origins, present state, and future evolution

Figure 1.13   The Andromeda galaxy.   At a distance of over 2 million light-years, this galaxy is the most distant object visible to the unaided eye. Without a telescope and a time-exposure photograph, the eye sees only an extended, fuzzy patch of light, rather than the detailed view shown here. The full extent of the galaxy, as revealed by a time exposure such as this, covers about 2½° on the sky, or five times the angular diameter of the Moon. (Palomar Observatory, California Institute of Technology)

of the cosmos and all it contains. It will be wise, however, to keep in mind that there is very much that we do not know.

PERSPECTIVE

The introduction to astronomy in this chapter prepares us for a plunge into more detailed discussions. We will begin these discussions in the next chapter by describing the nighttime sky as we see it without telescopes. We will describe in more detail the various bodies and motions observable to the unaided eye, and we will discuss the modern understanding of these phenomena. We will also see how the ancient philosophers fared in their attempts to understand what they could see in the nighttime sky.

Figure 1.14   A cluster of galaxies.   The faint, fuzzy objects in this photograph are galaxies belonging to a cluster that lies over a billion light-years from the Milky Way. Like our own, each galaxy contains billions of individual stars. (Palomar Observatory, California Institute of Technology)

SUMMARY

1. Astronomy is the science in which the entire universe is studied. The study of astronomy requires knowledge of several sciences, such as physics, chemistry, geology, and perhaps biology.

2. On a clear, dark night, the unaided eye can see up to five planets, the Moon, countless stars, and the occasional meteor or comet.

3. The Earth's rotation causes all objects in the sky to undergo daily motion, rising in the east and setting in the west.

4. Because we cannot directly see how far away objects are, we can only measure their positions or separations in angular units. For convenience, we visualize a celestial sphere, centered on the Earth, on which the astronomical bodies lie.

5. The part of the sky that can be seen depends on the latitude of the observer, and the portion visible at night depends on the time of year.

6. The most distant planet in our solar system is about 40 times farther from the Sun than the Earth is, whereas the nearest star is some 300,000 times the Sun-Earth distance, or 4 light-years, away. Our galaxy is about 100,000 light-years in diameter, the nearest galaxies are 150,000 to 2.3 million light-years distant, and clusters of galaxies are typically separated by millions of tens of millions of light-years. The size of the observable universe is measured in billions of light-years.

7. The light we receive from a distant object has been traveling toward us for as long as billions of years (in the case of the most distant galaxies), so we can observe the universe as it was long ago.

REVIEW QUESTIONS

1. Express in your own words the nature of scientific theory, and explain the differences between theory, speculation, and faith.

2. Discuss how you might test the validity of astrology; that is, describe an experiment or observation that would show whether predictions made by astrologers are correct or incorrect.

3. Why do you think it is best to get away from the city to view the nighttime sky?

4. We have noted that the planets appear to travel across the sky along the same strip. What does this tell us about the orbits of the planets around the Sun?

5. Explain why positions of objects in the sky are measured in angles rather than in units of linear distance such as meters or kilometers.

6. Explain why it is necessary to have observatories in both the Southern and Northern Hemispheres. At what latitude is the largest possible fraction of the sky visible?

7. An angle of one degree contains 60 minutes of arc, and a minute of arc contains 60 seconds of arc. How many arcseconds are in a full circle (360°)? The Sun's angular diameter is 30 arcminutes. What is it in degrees and in arcseconds?

8. Suppose the scale of the solar system were changed so that the Earth was the size of a marble (with a diameter of 1 centimeter). On this scale, what would be the diameter of the Sun? What would be the Earth-Moon distance? What would be the Sun-Earth distance? How far from the Sun would Pluto be? How far away would the nearest star be? (Note: to answer these questions, you need to know the true sizes and distances involved. Most are given in the text of this chapter, but many can also be found easily in Appendix 7).

9. Using information in Appendix 7, calculate the light-travel time from the Sun to Earth, to Jupiter, and to Pluto. (You will find it easiest to use the speed of light in metric units, which is 300,000 kilometers per second.)

10. If galaxies formed 10 billion years ago, how far away must we look in order to see the universe at a time before any galaxies existed?

ADDITIONAL READINGS

We can further appreciate the essence of astronomy, as well as its beauty, by reading a wide range of books and periodicals. Many bookstores contain volumes of astronomical photographs, as well as numerous books on astronomy written for the layperson.

Periodicals that are particularly well suited to students using this text include *Sky and Telescope, Mercury,* and *Astronomy. Sky and Telescope* is especially recommended for those wishing to carry out projects such as telescope building or astrophotography; it includes monthly charts showing the positions of the stars and planets as they change throughout the year.

Bahcall, J. N. 1991. U.S. Astronomy's next decade. *Sky and Telescope* 81(6):584.

Culver, R. 1984. *Astrology, true or false: A scientific evaluation.* New York: Prometheus Books.

Dean, G. 1987. Does astrology need to be true? Parts 1 and 2. *Skeptical Inquirer* 11(2):166, 11(3):257.

Dobson, A. K. and Bracher, K. 1992. Urania's heritage: a historical introduction to women in astronomy. *Mercury* 21(1):4.

Fraknoi, A. 1990, Scientific responses to pseudoscience related to astronomy. *Mercury* 19:144.

Radner, D., and M. Radner, 1982. *Science and unreason.* Belmont, Calif.: Wadsworth. (Discusses the difference between science and pseudoscience.)

Trimble, V. and Elson, R. 1991. Astronomy as a national asset. *Sky and Telescope* 82(5):485.

ASTRONOMICAL ACTIVITY
Becoming Familiar with the Sky

Many students taking astronomy for the first time may not be familiar with the nighttime sky. As a starting exercise, then, it is a good idea to go outdoors on a clear night and make some simple observations. Doing this will help acquaint you with the sky and will impart to you some of the sense of awe and wonder that inspires astronomers to pursue careers in this field.

To begin, you must have a clear night, preferably with no Moon (that is, you should choose a time when the Moon is between its third quarter and new phases, so that it will not rise until very late at night, leaving the sky dark during the early evening). You will be able to see a lot more if you can get away from city lights—bright lights "night-blind" your eyes, and scattered light from street lights, glowing neon signs, and the like obscures the sky. Take a sky chart with you (the ones inside the front and back covers of this book will do, or you can find monthly charts in magazines like *Sky and Telescope*). If you bring a flashlight to see the chart, tape red cellophane or plastic over it; red light does not ruin your night vision the way white light does.

Start by just looking at the sky and allowing your eyes to become comfortable with the grand view of thousands of stars. In time you will begin to notice patterns of relatively bright stars; you can then start to match these with the chart you have brought. These patterns are the same as those seen by the earliest humans, and various cultures have attached a great deal of significance to them. Learning the constellations, as the patterns are called, and some of the mythology associated with them is fun. The particular constellations you will see depend on the time of the year when you look. During the year, you should go out every month or two and identify the new constellations that have shifted into view during evening hours.

Once your eyes have adapted to the darkness, on a spring or summer night you should be able to make out a hazy, bright band stretching across the entire sky. This is the Milky Way, our home galaxy. You are seeing a cross-sectional view of a great disk; the solar system is located within the disk, about two-thirds of the way out from the center. Later you will learn that the Milky Way contains about 100 billion stars, and that this great pinwheel of stars is just one of hundreds of billions in the universe. At this point in your nighttime viewing session, you may begin to gain some appreciation for how small and insignificant the Earth is on the grand scale of things.

At almost any time of the year, you will see at least one or two planets. These will be obvious because they do not "twinkle" the way stars do, and they will be among the brightest objects you see. Venus and Jupiter are usually brighter than any star, as is Mars at its most favorable position; Saturn also ranks with the brightest stars. Mercury, the other planet that can be seen without binoculars or a telescope, is also quite bright, but is always very close to the Sun (the Astronomical Activity in Chapter 7 provides some guidance for viewing Mercury). You will notice that the planets fall along a line that stretches across the sky from east to west; this is the **ecliptic,** and it represents an edge-on view of the disk of the solar system (the specific definition is given in Chapter 2.)

If the Moon is near third quarter when you are observing, it will rise around midnight. As it does, you will see how its scattered light makes the entire sky glow faintly. It soon becomes impossible to see the faintest stars, and the Milky Way is drowned out as well. The Moon itself is a fascinating body to look at; even without binoculars or a telescope, you can make out surface markings. You can also see the Moon's motion relative to the fixed stars; in a time of only one hour, the Moon moves a distance approximately equal to its own diameter. You should be able to observe this by noting the Moon's position relative to stars near it on the sky (the planets also move with respect to the fixed stars, but much more slowly, so that days or weeks are required to detect the motion; see the Astronomical Activity in Chapter 8).

During your evening outdoors, you may be lucky enough to see other celestial phenomena, such as meteors. It is not unusual to see several over a few hours. Meteors are caused by small particles of rock (often no bigger than grains of sand) that enter the Earth's atmosphere from space and are burned up by the heat of friction as they collide with air molecules.

During your observing session, you will see the same view of the sky as your ancient ancestors, but with a couple of differences. No matter how far you go away from city lights, your view is affected, at least a little, by human influences. Even if scattered light is minimized, the air overhead is less clear than it was before the industrial revolution due to global air pollution (this effect is actually very small in remote areas, but can be very significant near cities). In addition, you very likely may see artificial satellites moving across the sky; these are quite obvious because they do not twinkle and move rapidly (most will move in the west-to-east direction, but you may see some in polar orbits, which move from north to south). And almost certainly you will see the lights of aircraft as they carry people across the sky.

In modern times, as in the past, spending time viewing the sky on a clear night is a refreshing experience and one that you may enjoy long after your astronomy course is over. Perhaps a lasting appreciation for the pleasure of doing so will be one of the most valuable lessons you learn from your course in astronomy.

Cycles and Seasons: Motions in the Sky

The Moon moving across the Sun during the 1991 solar eclipse at La Paz, Mexico
(A. Fujii, H. Tomioka, and Y. Shiono)

PREVIEW

In this chapter you will learn about the nighttime sky—its most prominent objects and the many regular motions that they undergo. There are daily motions, caused by the rotation of our observing platform, the Earth; yearly cycles, caused by the Earth's orbital motion about the Sun; monthly phenomena, due to the Moon's orbit about the Earth; and longer-term motions of the planets, as they follow their own paths around the Sun. The ancient astronomers observed and contemplated all of these motions, so in learning about them here, we will be setting the stage for our discussions of historical astronomy in later sections of the chapter. There you will see how the ancients fared in their attempts to explain what they saw. From these discussions you should gain an appreciation for both the sophistication of the observations made by the early scientists and the difficulty of unraveling the true nature of the heavens from our vantage point on a moving, spinning planet.

We view the heavens from a moving platform. The Earth spins while it travels around the Sun in a nearly circular path, and these motions create both daily and yearly cycles of celestial events as seen from the Earth's surface. The other eight planets all behave in similar fashion, and most, including the Earth, have one or more satellites orbiting them. Because we are viewing the skies from a moving vantage point and because all celestial objects have motions of their own, our impression is that the heavens are very complex. It took human beings many centuries of thought, observation, and technological development to arrive at the simple explanation summarized in this paragraph, an explanation that most of us, in these modern times, learn from early childhood.

In this chapter, we will learn about the nighttime sky as it appears to the unaided eye. We will develop a modern understanding of the objects that can be seen without a telescope and how their simple motions create complex paths through the sky as seen from the Earth. In the process, we will develop an appreciation for the task of the ancient philosophers who strove to comprehend the workings of the universe, for the view of the heavens described here is the sum of all the evidence available to pretechnological cultures.

RHYTHMS OF THE COSMOS

The observed motions of celestial bodies result from a combination of rotational and orbital motions, including the spin of the Earth and its annual movement around the Sun, as well as the individual mo-

tions of the Moon and the planets. We can best understand the overall machinery of the solar system by examining the individual parts.

Many of the motions in the heavens are cyclical; that is, they repeat regularly, in a well-defined period. Table 2.1 lists the periods for several motions in the solar system (unfamiliar terms in the table are explained later in the chapter).

DAILY MOTIONS

The most obvious of the many motions that affect our view of the universe are the daily cycles of all celestial objects caused by the rotation of the Earth. The Earth spins on its axis in about 24 hours, so we, on its surface, see a continuously changing view of the heavens. We see the Sun rise and set, along with the Moon, the planets, and most of the stars. Even though we understand that these daily, or **diurnal,** motions are the result of the Earth's spin, we still refer to them as though the objects in the sky themselves were moving. The rotation of the Earth means that a person standing on the equator covers the entire circumference (about 25,000 miles) in 24 hours; yet our senses give us no feeling of motion.

In science in general, and especially in astronomy, we must always be careful to note that what we ob-

TABLE 2.1   Periods of Significant Motions[a]

Motion	Period
Sidereal day	$23^h56^m4.098^s$
Mean solar day	$24^h00^m00^s$
Tropical year (equinox to equinox)	$365^d5^h48^m45^s$
Sidereal year (fixed stars)	$365^d6^h9^m10^s$
Synodic month	$29^d12^h44^m3^s$
Sidereal month	$27^d7^h43^m11^s$
Mercury: Sidereal period	87.969^d
Synodic period	115.88^d
Venus: Sidereal period	224.701^d
Synodic period	583.92^d
Mars: Sidereal period	1.88089^y
Synodic period	2.1354^y
Jupiter: Sidereal period	11.86223^y
Synodic period	1.0921^y
Saturn: Sidereal period	29.4577^y
Synodic period	1.0352^y
Uranus: Sidereal period	84.0139^y
Synodic period	1.0121^y
Neptune: Sidereal period	164.793^y
Synodic period	1.00615^y
Pluto: Sidereal period	248.5^y
Synodic period	1.0041^y

[a]Units for lunar and planetary motions are mean solar days or tropical years.

serve depends on the frame of reference in which our observations are made. Our view of the sky from a location on the Earth's surface is strongly affected by our reference frame and its own motions. Physicists have learned that there is no absolute reference frame, that in any situation we must define the frame in which we observe and make conclusions. In astronomy we can use the reference frame of the Earth, as we are doing now in discussing our view of the nighttime sky, but we use more general frames in other contexts, such as the framework established by the (approximately) fixed stars or perhaps the reference frame of distant galaxies.

In the reference frame of the Earth, the Earth's rotation forms the basis for our timekeeping system, since the length of the day is a natural unit of time on which to base our lives—one to which, indeed, nearly all earthly species have adapted. The day is divided into 24 hours, each containing 60 minutes, each of which in turn consists of 60 seconds. These divisions are based on the numbering system developed several thousand years ago, largely by the Babylonians (see discussion later in this chapter).

Careful observation shows that, in the reference frame of the Earth, the Sun and the Moon take longer to complete their daily cycles than do the stars. Each has its own motion that carries it in the same direction as the Earth's rotation, opposite to the daily rising and setting of the stars (Fig. 2.1). The Sun and the Moon, therefore, shift to the east with respect to the stars a little each day, so it takes a little longer for each to return to the same position from our point of view. Thus, the Sun rises about 4 minutes later each day, by comparison with the stars, and the Moon rises almost an hour later each day. The planets also move, but so slowly that their diurnal motions are not easily distinguished by eye from those of the stars.

The day that forms the basis for our timekeeping system is the **solar day** rather than the **sidereal day** (Fig. 2.2), which is the rotation period of the Earth in the reference frame of the fixed stars. (The term *sidereal* means "with respect to the stars.") Because the Earth's orbital speed is not precisely constant, the length of the solar day varies a little throughout the year. It would be inconvenient to allow the hour, minute, and second to vary along with it, so the average length of the solar day, called the **mean solar day,** has been adopted as our timekeeping standard. The mean solar day is 3 minutes and 56 seconds longer than the sidereal day.

Today time is measured precisely with atomic clocks but is still kept synchronized with time based on the Earth's rotation. To measure the Earth's rotation period with respect to the stars (that is, the sidereal day), astronomers refer to the **meridian,** the imaginary north-south line passing overhead at an observer's position. As the Earth rotates, each star crosses the meridian once every sidereal day. Special instruments (Fig. 2.3) are used to measure the precise

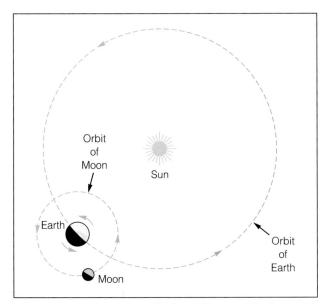

Figure 2.1   Motions of the Earth and Moon.   The Earth and Moon spin and move along their orbits at the same time. The combination of motions affects the apparent path of the Sun and the Moon as seen by an observer on the spinning Earth. (Not to scale.)

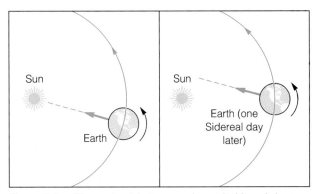

Figure 2.2   The contrast between solar and sidereal days. The blue arrow indicates the overhead direction from a fixed point on the Earth. From noon one day (left), it takes one sidereal day for the arrow to point again in the same direction, as seen by a distant observer. Because the Earth has moved, however, it will be about 4 minutes later when the arrow points directly at the Sun again; hence the solar day is nearly 4 minutes longer than the sidereal day. (Not to scale.)

ASTRONOMICAL INSIGHT
Keeping in Touch with the Stars

As mentioned in the text, modern timekeeping is done with atomic clocks, which have several advantages over the use of transit telescopes to monitor sidereal time. For example, any physics laboratory can maintain its own atomic clock and make precise observations with it, without the need to have a transit telescope. Also, the signals from atomic clocks can be transmitted by radio from laboratory to laboratory, so that experiments requiring accurate relative timing can be performed. It must be stressed, however, that the atomic clocks are used only as measures of relative time, the time from some adopted reference time. For this purpose they are very precise, but they cannot adequately maintain correct time relative to the Earth's rotation, because the Earth's rotation is not perfectly constant.

All methods of timekeeping are still based on the rotation of the Earth, as measured through the observations of stars. Hence, atomic clocks must occasionally be updated to account for small changes that occur in the Earth's rotation rate, when it may speed up or slow down. These changes occur for many reasons, not all of them well understood. Some appear to be random variations in the rotation rate, and others seem to be long-term, gradual changes.

One persistent change that always occurs in the same direction is a very gradual slowing of the Earth's rotation rate, probably because of internal frictional forces exerted by the tidal force of the Moon (this is discussed more fully in Chapters 3 and 6, where the effect on the Earth-Moon orbit is described). The combined effect of all the changes in the past century has increased the length of the day by about 0.0014 seconds. Given enough time, this slowing of the Earth's rotation rate reduces the number of days in a year, since each day is longer than it used to be. Fossil evidence indicates that the year once contained about 400 days, instead of the current 365¼.

The short-term random changes in the Earth's rotation rate can either increase or decrease the length of the day. Researchers think that some of these changes are caused by the flow of molten material deep in the Earth's core, but the causes of most of them are not known. Some of them are probably related to a slow shrinking of the Earth as its interior gradually cools; these minor effects act to speed up the rotation temporarily.

Over long periods of time, however, the overall trend is toward a decrease in the rotation rate and a lengthening of the day. To keep atomic clocks in synchronization with the Earth's spin, therefore, it is occasionally necessary to add some time to the clocks. Hence, every few years, a "leap second" is added to all official atomic clocks to bring their timekeeping into accord with the Earth's rotation, that is, with sidereal time. This is distinct from the leap year, when a full day is added every fourth year. This added day is inserted because the standard calendar has exactly 365 days in a year, whereas the actual length of the year is closer to 365¼ days. Thus, a full day must be added every 4 years to make up for the one-quarter days that have been dropped. Interestingly, if the length of the day keeps increasing as expected, in a few million years the year will contain exactly 365 days, and for a while there will be no need for leap years.

You may occasionally notice small announcements in newspapers about the addition of "leap seconds" to official clocks. These small changes in official time make little practical difference in our daily lives, but now you understand that they are part of a large-scale astronomical process, with important long-term effects.

moment when a star crosses the meridian, thus measuring sidereal time. Allowance must be made in determining the mean solar time from these measurements.

Atomic clocks measure intervals of time very accurately because they are based on the vibration frequencies of certain kinds of atoms, which are very constant. These clocks have several advantages over the measurement of sidereal time based on observations of stars. For example, atomic clocks keep a continuous record of the passage of time so they can be referred to whenever needed. Also, identical clocks can be placed in many locations and kept coordi-

nated, a practice that allows many laboratories to measure the time precisely.

We say that atomic clocks measure true physical time instead of sidereal time, but these clocks are adjusted as needed so that they stay in agreement with sidereal time. Such adjustments are needed because occasionally the Earth's rotation period changes very slightly.

Let us return now to the diurnal motions of celestial objects. We have mentioned that the stars rise and set as the Earth rotates, but in fact not all stars do this. This is because, from a given latitude on the Earth (except right at the equator), it is possible to

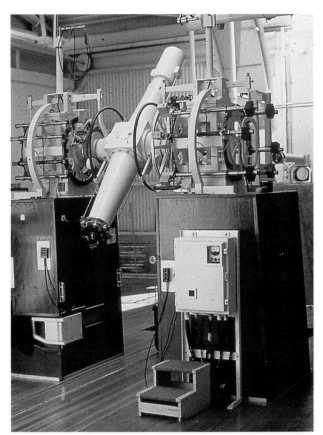

Figure 2.4　Star trails illustrating the Earth's rotation. The curved streaks shown in this ten-hour exposure are the trails left by stars on the photographic film as the Earth rotated. This is a view of Mt. Rainier in Washington state, as seen from the south. (© Rick Morley Photography)

Figure 2.3　A transit circle. Such a device normally points straight up. It is used to record the times when certain reference stars pass over the meridian (the north-south line that passes directly overhead) and is therefore helpful in measuring the sidereal day. (U.S. Naval Observatory)

see an area of the sky, around one of the poles, that is never obscured by the Earth. From a latitude of 30°N, for example, we can see the sky to an angular distance of 30° beyond the North Pole. Therefore, we can always see the whole part of the sky that lies within 30° of the North Pole (but we can never see the sky within 30° of the South Pole), and stars in this part of the sky circle the pole but do not set (Fig. 2.4) The same thing happens around the South Pole for observers in the Southern Hemisphere.

The point about which the stars circle, directly over the Earth's North Pole, is the **north celestial pole,** the point where the earth's rotational pole is projected onto the celestial sphere. Similarly, the projection of the South Pole onto the celestial sphere is the **south celestial pole.** A bright star called Polaris lies very close to the position of the North celestial pole and is therefore almost stationary (in the reference frame of the Earth) throughout the night. The projection of the Earth's equator onto the sky is

called, naturally enough, the **celestial equator.** We will now examine the usefulness of the celestial poles and the equator as reference points for measuring star positions.

Astronomical Coordinate Systems

Our discussion of diurnal motions, latitude, and the changing appearance of the sky as we move over the surface of the Earth has led us to a point where it is convenient to introduce the notion of coordinate systems for the skies. Just as it is useful to be able to describe the position of a place on the surface of the Earth, it is also important for astronomers to be able to record the position of a particular object in the sky so that they can return to it later or communicate to other astronomers where it is. The establishment of a coordinate system naturally requires us to specify the reference frame in which the coordinates are measured. Astronomers use different reference frames for different purposes, but the one most often used is based on the Earth (Fig. 2.5).

In this coordinate system, called the **equatorial coordinate system,** one reference direction is the celestial equator. To specify a star's position in this system, two coordinates are needed. One of them is the angular distance of the star north or south of the celestial equator; this is called the **declination** of the

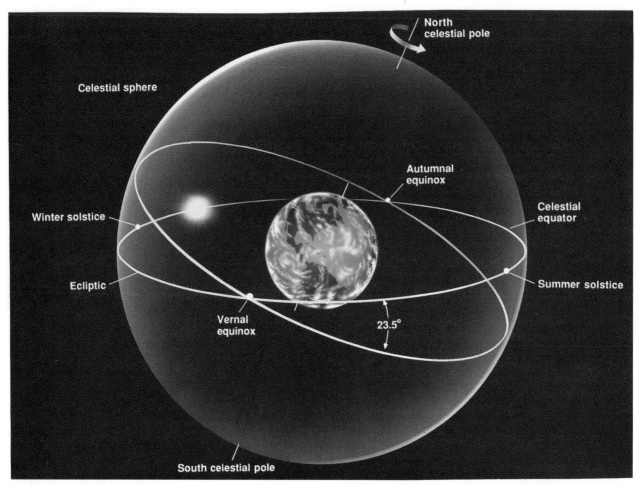

Figure 2.5   Equatorial coordinates.   Here are illustrated the measurements of star position in declination and right ascension. Declination is the distance (in degrees, minutes, and seconds of arc) of a star north (+) or south (−) of the celestial equator. Right ascension is the angular distance (in hours, minutes, and seconds of time) of the star to the east of the direction of the vernal equinox, a fixed direction in space.

star. The declination is analogous to the latitude of a position on the Earth's surface, except that here we speak of angular distance on the sky. North declinations are assigned positive values (such as $+23°14'$ for a star $23°14'$ north of the celestial equator), and south declinations are assigned negative values (the Magellanic Clouds, for example, lie at declinations near $-70°$).

The second coordinate needed to specify an object's position is called **right ascension.** The right ascension measures the angular distance of the object to the east of a fixed direction on the sky and is therefore analogous to longitude on the Earth's surface. Because the Earth is rotating, however, we cannot measure right ascension from some fixed point defined on the Earth's surface; we must instead specify a reference direction on the sky. This direction is defined by the line of intersection of the Earth's equatorial plane and the plane of its orbit. This is the direction toward the Sun at the time of the vernal

equinox (defined in the next section). When this coordinate system was established, the line of intersection of these planes pointed toward the constellation Aries, but because of gradual changes in the Earth's orientation (described later), it now points toward Pisces and will soon enter Aquarius; this is why popular astrology refers to the "dawning of the age of Aquarius." Nevertheless, the position toward which this line points is still called the *First Point of Aries.*

Because it takes 24 sidereal hours for a given star to make a complete circuit of the sky as the Earth rotates, its position in the east-west direction is measured in units of sidereal hours, minutes, and seconds rather than in angular units. Thus, a star located in the direction of our reference point has a right ascension of $0^h0^m0^s$, whereas a star one-quarter of the way around the sky to the east of that point has a right ascension of $6^h0^m0^s$.

There is one complication in using the equatorial coordinate system. The Earth's rotational axis slowly

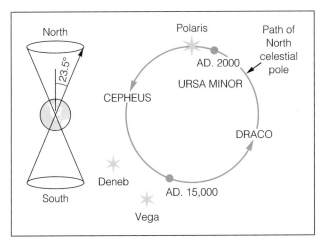

Figure 2.6   Precession.   The Earth's axis is tilted 23½° away from perpendicular to the orbital plane, and it wobbles on its axis, so that an extension of the axis describes a conical pattern (left) in a time of about 26,000 years. The north celestial pole therefore follows a circular path on the sky (right).

wobbles (Fig. 2.6), in a motion called **precession,** and this causes the celestial equator to move slowly through the sky. It takes some 26,000 years for the Earth to complete one cycle of this motion, which is exactly like the wobbling of a play top or gyroscope. The motion of our coordinate system is therefore very gradual, causing star positions to shift extremely slowly. Despite the small magnitude of the effect, observers noticed it more than 2,000 years ago, and modern astronomers must allow for it when planning observations. Typically, one preparation for observing is to "precess" the coordinates of all the target stars (that is, to calculate their current locations in the Earth's reference frame). Catalogs of star positions always specify the date for which they are valid so that astronomers know how much precession to allow for.

Other coordinate systems (that is, other reference frames) sometimes are used in astronomy, although equatorial coordinates are by far the most widely employed, being the most practical for use in observations. Solar system astronomers sometimes use a coordinate system based on the plane of the **ecliptic** (the plane defined by the Earth's orbit around the Sun) and the direction toward the Sun, whereas those studying the Milky Way may use coordinates based on the plane of the galaxy's disk and the direction toward its center. Ecliptic and galactic coordinates are particularly useful in studies of the structure of the solar system or of the galaxy, because an object's position immediately tells the astronomer where it lies with respect to the overall organization of the solar system or galaxy.

ANNUAL MOTIONS: THE SEASONS

We turn our attention now to celestial phenomena that are caused by the Earth's motion as it orbits the Sun. One aspect of this, the daily eastward motion of the Sun with respect to the stars, has already been mentioned in connection with the difference between the solar and the sidereal days. In the reference frame of the fixed stars, this is only an apparent motion caused by our changing angle of view as we move with the Earth in its orbit. As the Earth moves about the Sun, it travels in a fixed plane, so the apparent annual path of the Sun through the constellations is the same each year (Fig. 2.7). This apparent path of the Sun, defined by the plane of the Earth's orbital motion about the Sun, is the **ecliptic.** The sequence of constellations through which the Sun passes is called the **zodiac.** There are twelve principal constellations of the zodiac, identified since antiquity and once thought by astronomers to have significance for our daily lives (see the comments on astrology in Chapter 1).

Because we can see stars only in the nighttime sky, the constellations most easily visible to us at any given time are the ones at least a few hours of right ascension to the east or west of the Sun. This means that most stars, except those near the poles, cannot be observed year-round, as discussed in Chapter 1.

The orbital planes of the other planets and of the Moon are closely aligned with that of the Earth, so the planets and the Moon are always seen near the ecliptic. Hence, most of the major naked-eye objects in the solar system pass through the same sequence of constellations, the zodiac, so it is not surprising that ancient astronomers attached great significance to this sequence.

Besides causing the apparent annual motions of the Sun and planets, the Earth's orbital motion has a second, and far more significant, effect on us: It creates our seasons. The Earth's spin axis is tilted with respect to the perpendicular to its orbital plane, so during the course of a year, portions of the Earth away from the equator are exposed to varying amounts of sunlight. Summer in the Northern Hemisphere occurs when the North Pole is tipped toward the Sun; winter occurs during the opposite part of the Earth's orbit, as the pole, which remains fixed in orientation, is tilted away from the Sun (Fig. 2.8). The tremendous seasonal variations in climate at intermediate latitudes are caused by a combination of two effects: (1) the length of time the Sun is up varies so that in summer, for example, the Sun has more time to heat the Earth's surface; and (2) the Sun's rays strike the ground at a more nearly perpendicular angle in the

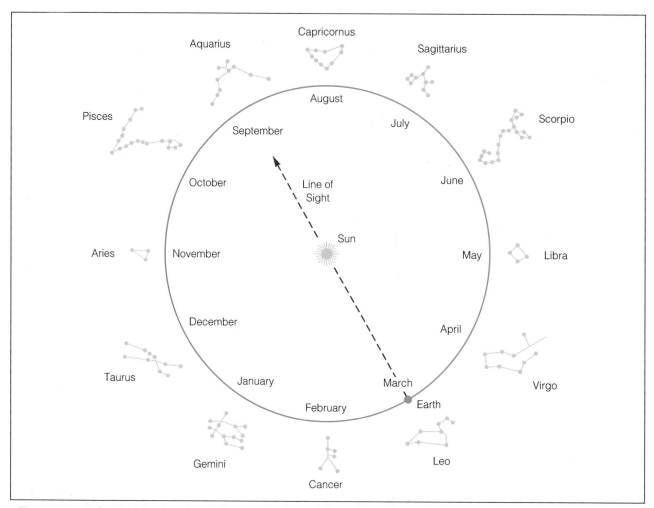

Figure 2.7   The path of the Sun through the constellations of the zodiac.   The dates refer to the position of the Earth each month. To see which constellation the Sun is in during a given month, imagine a line drawn from the Earth's position through the Sun; that line will extend to the Sun's constellation. For example, in March the Sun is in Aquarius.

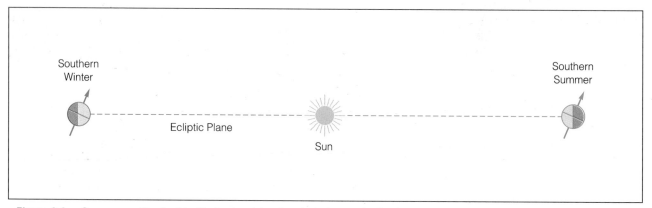

Figure 2.8   Seasons.   The Earth's tilted axis retains its orientation as the Earth orbits the Sun. Thus, at opposite points in the orbit, each hemisphere has winter and summer, depending on whether that hemisphere is tipped toward the Sun or away from it.

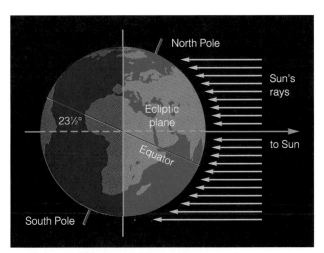

Figure 2.9   Effect of the Earth's tilted axis. The Sun's rays strike the ground at varying angles, depending on the latitude and the time of the year. Therefore, the solar heating, which depends on how directly the rays reach the ground, varies with the season. Here it is summer in the Northern Hemisphere; the Sun's rays are nearly perpendicular to the Earth's surface at low to moderate northern latitudes, but they strike the surface very obliquely at southern latitudes.

summer (Fig. 2.9), so the Sun's intensity is much greater, heating the surface more efficiently.

The Earth's axis is tilted 23½° from the perpendicular to the orbital plane. Therefore, during the year, the Sun, as seen from the Earth's surface, can appear directly overhead (that is, at the **zenith**) as far north and south of the equator as 23½°, defining a region called the **tropical zone** (Fig. 2.10). For people who live outside the tropics, the Sun can never be directly overhead. When the North Pole is tilted most nearly in the direction toward the Sun, an occasion occurring around June 21 and called the **summer solstice,** the Sun passes directly overhead at 23½° N latitude at local noon, but it passes to the south of the zenith for anyone at more northerly latitudes. On this occasion, sunlight covers the entire north polar region to a latitude as far as 23½° south of the pole. This defines the **Arctic Circle** (see Fig. 2.10), and at the time of the solstice the entire circle has daylight for all 24 hours of the Earth's rotation. At the pole itself there is constant daylight for 6 months.

During winter in the Northern Hemisphere, the Sun does not rise as high above the southern horizon at local noon as it does during the summer, because the South Pole is now tilted toward the Sun. At the **winter solstice,** when the South Pole is pointed most

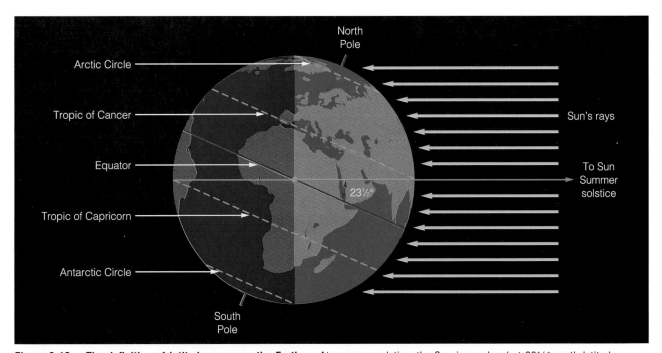

Figure 2.10   The definition of latitude zones on the Earth. At summer solstice, the Sun is overhead at 23½° north latitude, its northernmost point. This defines the Tropic of Cancer, the northern limit of the tropical zone. At the same time the entire area within 23½° of the North Pole is in daylight throughout the Earth's rotation; the boundary of this region is the Arctic Circle. Similarly, the Antarctic Circle receives no sunlight at all during a complete rotation of the Earth. Six months later, the Sun is overhead at the Tropic of Capricorn.

nearly in the direction of the Sun, the Sun's midday height above the horizon, as viewed from the Northern Hemisphere, is the lowest of the year.

If we follow the Sun's motion north and south of the equator throughout the year, we find that it follows a graceful curve as it traverses its range from $+23\frac{1}{2}°$ (north) declination to $-23\frac{1}{2}°$ (south) declination (Fig. 2.11). The Sun crosses the equator twice in its yearly excursion, at the times when the Earth's North Pole is pointed in a direction 90° from the Earth-Sun line. At these times, the lengths of day and night in both hemispheres are equal. These occasions, which are referred to as the **vernal** (spring) and **autumnal** (fall) **equinoxes,** occur on about March 21 and September 23, respectively. The direction to the Sun at the time of the vernal equinox coincides exactly with the direction of 0^h right ascension. (Remember, the definition of this direction is that it lies along the line of intersection of the Earth's equatorial plane and the orbital plane; see Fig. 2.5.)

Calendars

We have referred to the association of astronomy and timekeeping, particularly in ancient times, when astronomers were responsible for predicting major natural events such as changes of season. Today everyone knows when these things will occur and generally keeps track of dates for all purposes with a modern calendar (Fig. 2.12), which can be purchased well in advance for any given year. It is interesting to realize, however, that even today our calendars, and hence the annual pattern of our daily lives, are determined by the motions of bodies in the heavens.

Various ancient cultures established a number of calendars, all of which were based to some degree on the length of the year. In several cases it was deemed unlikely or unacceptable that the year should not be evenly divisible into a round number of lunar months or even days, so some of the old calendars adopted lengths of the year that did not agree with the actual period of the Earth's orbit of the Sun. All such calendars had to be adjusted every so often or fall badly out of agreement with the seasons.

The length of the year that is important in calendar making is not the sidereal period of the Earth's orbit, as we might expect. The reason is that, because of precession, the seasons will gradually shift with respect to the sidereal year. Consider the length of time from one vernal equinox to the next. As the Earth travels around the Sun, the orientation of the Earth's axis does not stay precisely fixed, but instead changes just slightly because of precession. As a result, the next vernal equinox, which occurs when the Earth's axis returns to precisely the same orientation (with respect to the stars) it had at the previous equinox, takes place just a little sooner than it would have if precession did not occur. The difference between the sidereal year and the **tropical year** (the length of time for a full cycle of seasons to occur) is just 20 minutes. Hence, calendars today are based on the tropical year. Precise values for both the sidereal and the tropical year are included in Appendix 2.

Until the time of Julius Caesar, most calendars were based on the assumption that the year included a whole number of days, even though it was well known that this was not so. The errors that built up were allowed to accumulate until extra days would be inserted to bring the calendar back into agreement with the seasons. Caesar commissioned the development of a new calendar, now called the **Julian calendar,** in which the length of the year was taken to be $365\frac{1}{4}$ days, only a few minutes in error. Every fourth year an extra day was inserted, a practice that we still follow. The year was divided into 12 months that did not correspond exactly to lunar months, so that there would at least be a whole number of months in a year. Caesar was also responsible for beginning the year on January 1; it had begun in March before his reform, which took place in 46 B.C.

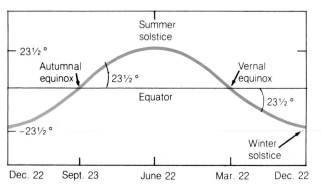

Figure 2.11   The path of the Sun through the sky.   Because of the Earth's orbital motion and the tilt of its axis, the Sun's annual path through the sky has the shape illustrated here.

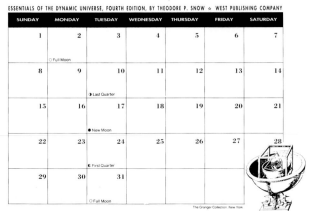

Figure 2.12   Modern calendar showing the Moon's phases.

In 1582, to correct for the few minutes' error that crept into the Julian calendar because the tropical year is not precisely 365 days long, Pope Gregory XIII ordered additional reform. As a result, in the modern calendar (called the **Gregorian calendar**), leap year is occasionally not observed. Leap year is ignored (the extra day is not inserted after February 28) in century years not divisible by 400. Thus, there will be a leap year in the year 2000, which is divisible by 400, but not in the years 2100, 2200, and 2300. With this refinement, the Gregorian calendar builds up an error of only 1 day in 3,300 years, a sufficient degree of accuracy for most people's appointment books!

THE MOON AND ITS PHASES

Except for the Sun, the Moon is the brightest object in the sky, and it, too, has its own complex motions as seen from the Earth. The Moon's motion with re-

spect to the stars is much more rapid than that of the Sun and is therefore more easily noticed. It is not difficult, in fact, to observe this lunar motion during the course of an evening. The Moon moves about 13° across the sky every day, or 1° every 2 hours. Its angular diameter is about 1/2°, which means that the Moon moves a distance in the sky about equal to its own diameter every hour.

The plane of the Moon's orbit is closely aligned with that of the Earth's orbit, so the Moon stays near the ecliptic, as seen from the Earth. In the reference frame of the fixed stars, it takes the Moon a little over 27 days ($27^d7^h43^m11^s.5$) to make one trip around the Earth. Because the Earth moves at the same time, however, it appears to us in the Earth's reference frame that the Moon takes longer than 27 days to make one complete circuit; from our point of view, the Moon takes $29^d12^h44^m2^s.8$ to complete its full cycle of phases (Fig. 2.13). The shorter orbital period is called the **sidereal period,** because it is the time

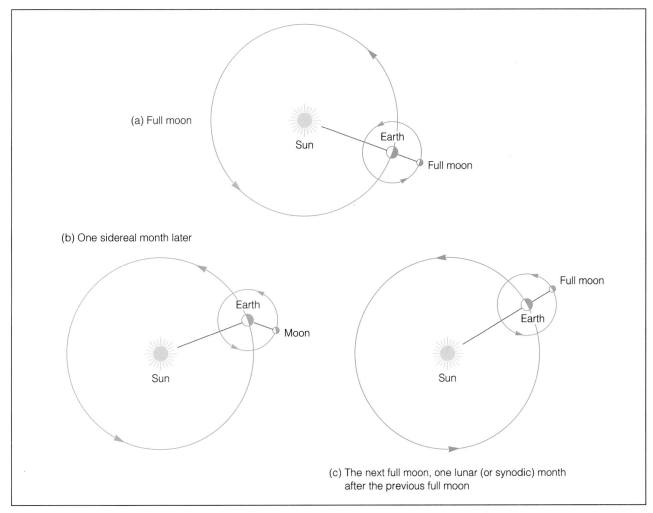

(a) Full moon

Sun

Earth

Full moon

(b) One sidereal month later

Earth

Moon

Sun

Full moon

Earth

Sun

(c) The next full moon, one lunar (or synodic) month after the previous full moon

Figure 2.13   Sidereal and synodic periods of the Moon.   Because the Earth moves in its orbit while the Moon orbits it, the Moon goes through more than one full circle (as seen by an outside observer) to go from one full moon to the next. Hence the lunar (or synodic) month is about two days longer than the Moon's sidereal period.

required for the Moon to go around the Earth as seen in a fixed reference frame with respect to the stars. The observed cycle of phases, which is the time required for the Moon to return to a given alignment with respect to the Sun, is called the **synodic period,** or the **lunar month**. The difference is akin to the distinction between the solar and sidereal days discussed in the preceding section. In both cases, it is the Earth's motion about the Sun that lengthens the time it takes to complete a full cycle as we see it.

The Moon always keeps the same side facing the Earth, because its rotation period is equal to its orbital period (Fig. 2.14). It is a common misconception that the Moon does not rotate. It should be clear that the Moon is rotating in the reference frame of the stars, for if it were not, we would see all sides of the Moon as it circled the Earth. The fact that the orbital and spin periods are equal is not a coincidence, but is a result of the tidal forces exerted on the Moon and the Earth by their mutual gravitational pull (this is discussed in Chapters 3 and 6). The phenomenon of matching orbital and spin periods is called **synchronous rotation,** and it is common in the universe, both within the solar system and in double stars.

Since the Moon does not emit light of its own but shines instead by reflected sunlight, we easily see only those portions of its surface that are sunlit. As the Moon orbits the Earth and our viewpoint relative to the Sun's direction changes, we see varying fractions of the daylit half of the Moon. This causes the Moon's apparent shape to change drastically during the month, the sequence of shapes being referred to as the **phases** of the Moon (Fig. 2.15). The full cycle of phases is completed during one synodic period, or lunar month, of about 29½ days.

The extremes of the cycle are represented by the **full moon** (occurring when it is directly opposite the Sun, so we see its entire sunlit hemisphere) and the **new moon** (when the Moon is between the Earth and the Sun, with its dark side facing us). The new moon cannot be observed because it is very dim and is only up during daytime. Hence, our nighttime sky is moonless at the time of the new moon.

Just as we speak of phases of the Moon, which really refer to its apparent shape as seen from the Earth, we also can speak of its **configurations,** which describe its position with respect to the Earth-Sun direction. For example, when a full moon occurs, the Moon is at **opposition** (it is in the direction opposite to that of the Sun), and a new moon occurs at **conjunction** (when the Moon lies in the same direction as the Sun). We can follow the Moon through its phases as we trace its configurations, beginning with the new moon, which takes place at conjunction (refer to Fig. 2.15). During the first week following conjunction, as the Moon moves toward **quadrature** (the position 90° from the Earth-Sun line), it appears to us to have a crescent shape that becomes thicker each night. This is called the **waxing crescent** phase; when the Moon reaches quadrature, in which we see exactly one-half of the sunlit hemisphere, it has reached the phase called **first quarter.** For the next week, as the Moon goes from quadrature toward opposition and we see more and more of the daylit side, its phase is said to be **waxing gibbous.** After the full moon, as it again approaches quadrature, the phase is **waning gibbous,** and this time at quadrature the phase is called **third quarter.** First quarter can easily be distinguished from third quarter by noting the time of night: If the Moon is already up when the Sun sets, it is first quarter; but if the Moon does not rise until midnight, it is third quarter. After third quarter, as the Moon moves closer to the Sun, we see a diminishing slice of its sunlit side, and it is in the **waning crescent** phase.

ECLIPSES OF THE SUN AND MOON

An eclipse of the Sun, or **solar eclipse,** occurs when the Moon passes directly in front of the Sun, as seen from the Earth. A **lunar eclipse,** by contrast, occurs when the Moon passes through the Earth's shadow, so that for a brief period the Moon is not directly illuminated (Fig. 2.16).

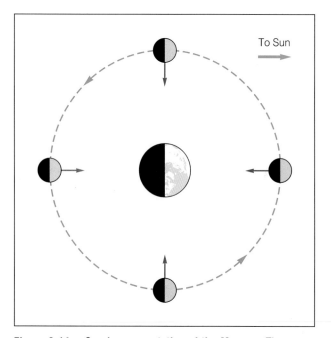

Figure 2.14   Synchronous rotation of the Moon.   The arrow, fixed to a specific point on the Moon, illustrates that the Moon spins once during each orbit of the Earth. Thus the Moon keeps the same side facing the Earth at all times. (Not to scale.)

To Sun

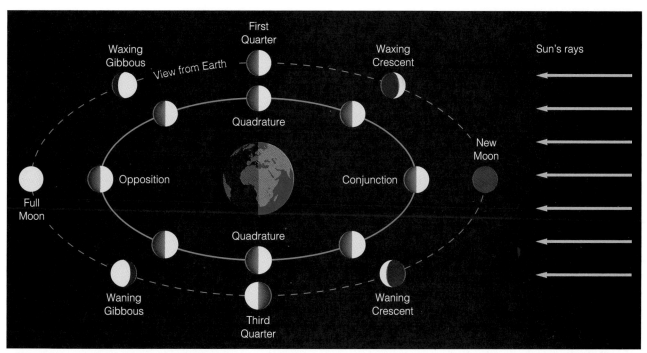

Figure 2.15   Lunar phases and configurations.   As the Moon orbits the Earth, we see varying portions of its sunlit side. The phases sketched here (outside the circle representing the Moon's orbit) show the Moon as it appears to an observer in the Northern Hemisphere.

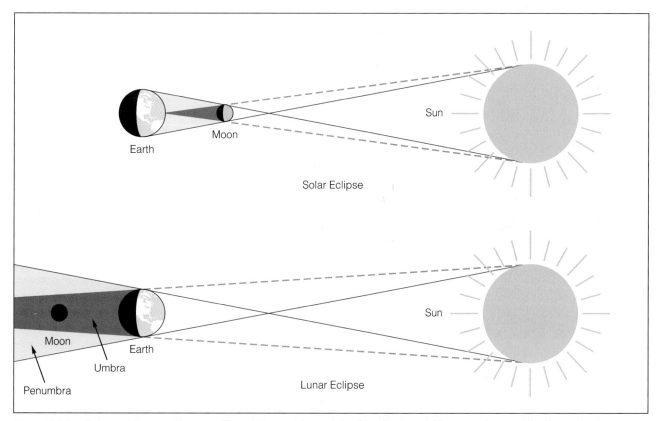

Figure 2.16   Solar and lunar eclipses.   The relative positions of the Sun, Earth, and Moon are shown at the times of solar (upper) and lunar (lower) eclipses. Only the dark portion of the shadow (the umbra) is shown in each case. (Not to scale.)

TABLE 2.2   Total Solar Eclipses of the Future

Date of Eclipse	Duration (min)	Location
November 3, 1994	4.6	South America
October 24, 1995	2.4	South Asia
March 9, 1997	2.8	Siberia, Arctic regions
February 26, 1998	4.4	Central America
August 11, 1999	2.6	Central Europe, central Asia
June 21, 2001		South Atlantic, South Africa
December 4, 2002		South Africa, Indian Ocean
March 29, 2006		South Atlantic, Africa, Middle East
August 1, 2008		Greenland, North Atlantic, Russia
July 22, 2009		Indonesia, South Pacific
July 11, 2010		South Pacific
November 13, 2012		South Pacific
March 20, 2015		North Atlantic
August 21, 2017		United States, Atlantic Ocean

From what has been stated so far, it may seem that the Moon should pass directly in front of the Sun on each trip around the Earth, and through the Earth's shadow at each opposition, producing alternating solar and lunar eclipses at 2-week intervals. However, this is not the case; the reason is that the Moon's orbital plane does not lie exactly in the ecliptic but is tilted by about 5°. Therefore, the Moon usually passes just above or below the Sun as it goes through conjunction and similarly misses the Earth's shadow at opposition. On each trip around the Earth, the Moon passes through the ecliptic at only the two points where the planes of the Earth's and the Moon's orbits intersect. Because the Moon's orbital plane wobbles slowly, in a precessional motion similar to that of the Earth's spin axis, the line of intersection with the Earth's orbital plane slowly moves around. The combination of this motion, the Moon's orbital motion, and the movement of the Earth around the Sun creates a cycle of eclipses, with the same pattern recurring every 18 years (see Table 2.2 for a list of future solar eclipses). This cycle of eclipses, called the **saros**, was recognized in antiquity.

It is purely coincidental that the Moon and the Sun have nearly equal angular diameters, so that the Moon neatly blocks out the disk of the Sun during a solar eclipse (Figs. 2.17 and 2.18). The angular diameter of an object is inversely proportional to its distance, meaning that the farther away it is, the smaller it looks. The Sun is much larger than the Moon, but it is also much more distant. The two objects have almost exactly the same angular diameter, because the ratio of the Sun's diameter to that of the Moon just happens to be almost the same as the ratio of the Sun's distance to that of the Moon (Fig. 2.19).

If a total solar eclipse occurs at the time when the Moon is farthest from the Earth in its slightly non-circular orbit, it does not quite block all of the Sun's disk, leaving instead an outer ring of the Sun visible. This is called an **annular eclipse** (Fig. 2.20). Because a total (or annular) solar eclipse requires precise alignment of the Sun and the Moon, an eclipse will appear total only along a well-defined, narrow path of the Earth's surface (Fig. 2.21). In a wider zone outside the path, the Moon appears to block only a portion of the Sun's disk; people in this zone see a partial solar eclipse.

During a lunar eclipse, when the Moon passes through the Earth's shadow, observers everywhere on the nighttime side of the Earth see the same portion of the Moon eclipsed. If the Moon passes through

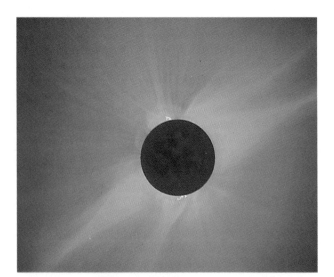

Figure 2.17   A total solar eclipse.   This occurs when the Moon entirely blocks our view of the Sun. At this time we see the Sun's corona. This photo is a sum of seven individual exposures made from Baja California during the eclipse of July 11, 1991. Special processing was used to bring out details of coronal structure. (E. E. Barnard Observatory; photo by Steven Albers, Dennis DiCicco, Gary Emerson, and David Sime)

the **umbra** (the dark inner portion of the Earth's shadow), the eclipse is total, as no part of the Moon's surface is exposed to direct sunlight. If the Moon misses being fully immersed in the umbra, it passes through the **penumbra** and undergoes a partial eclipse.

PLANETARY MOTIONS

Just as in the case of the Sun and the Moon, careful observation reveals that the planets, too, move with respect to the background stars. Indeed, this fact lent the planets their generic name, since **planet** is the Greek word for "wanderer." The planets all orbit the Sun in the same direction as the Earth does and, as mentioned earlier, in nearly the same plane, so they appear to move nearly in the ecliptic, through the constellations of the zodiac.

The two planets lying within the orbit of the Earth, Mercury and Venus, are called **inferior planets** and can never appear far from the Sun in our sky (Fig. 2.22). For Mercury, the greatest angular distance from the Sun, called the **greatest elongation,** is about 28°, whereas Venus can be seen as far as 47° from the Sun. Like the Moon, the planets have specific configurations, referring to their positions with respect to the Sun-Earth line. An inferior planet is said to be at **inferior conjunction** when it lies between the Earth and the Sun (or in **transit** if the planet crosses directly in front of the Sun's disk) and at **superior conjunction** when it is aligned with the Sun but on the far side.

The outer planets, or **superior planets,** can be seen in any direction with respect to the Sun, including opposition, when they are in the opposite direction from the Sun (Fig. 2.22). Conjunction for a superior planet can occur only when the planet is

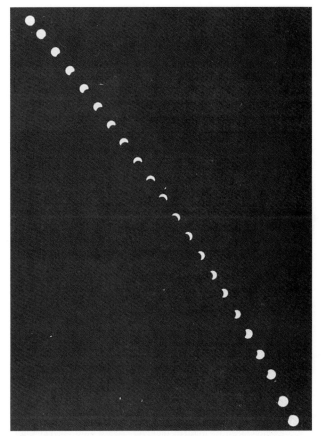

Figure 2.18   A solar eclipse sequence.   This series of photographs illustrates the Moon's progression across the disk of the Sun during a partial solar eclipse. (NASA)

aligned with the Sun but on the far side of it, a configuration analogous to superior conjunction for an inferior planet. **Quadrature** occurs when a superior planet is seen 90° from the direction of the Sun.

Each planet has a sidereal period and a synodic period, the former being the orbital period as seen in the fixed framework of the stars and the latter being

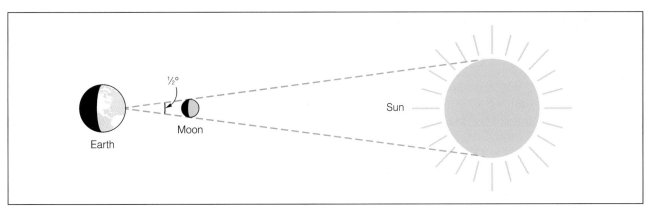

Figure 2.19   Angular diameters of the Moon and Sun.   The actual diameters and the relative distances of the Sun and Moon compensate one another so that the angular diameters are equal. (Not to scale.)

Figure 2.20   An annular eclipse.   This photograph shows an eclipse that was seen from southern California in 1992. Because the Moon was relatively distant from the Earth at the time of the eclipse, its angular diameter was a little smaller than that of the Sun. (National Optical Astronomy Observatories)

Figure 2.21   The Moon's shadow during a solar eclipse. This photograph, taken from space, shows the shadow of the Moon on the Earth during a solar eclipse. The eclipse appeared total only to observers on the Earth who were located directly in the center of the shadow's path (that is, in the umbra). (NASA)

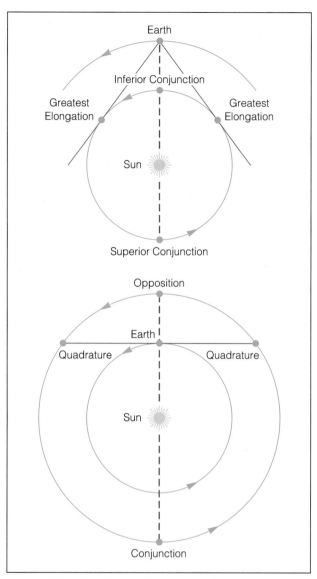

Figure 2.22   Planetary configurations.   The upper drawing illustrates the configurations (planetary positions relative to the Sun-Earth line) for an inferior planet; the lower drawing shows the configurations for a superior planet.

the length of time it takes the planet to pass through the full sequence of configurations, as from one conjunction or one opposition to the next (Fig. 2.23). The situation is much like that of two runners on a track. The time it takes the faster runner to overtake the slower one is analogous to the synodic period, whereas the time it takes to simply circle the track corresponds to the sidereal period.

The motion of the Earth has one very important effect on planetary motions. As we go outward from the Sun, each successive planet has a slower speed in its orbit (see the discussion of Kepler's laws of planetary motion in Chapter 3). This means that the Earth, moving faster than the superior planets, periodically passes each of them; this occurs once every synodic period. As the Earth overtakes one of the superior planets, there is an interval of time during which our line of sight to that planet sweeps backward with respect to the background stars, making it appear that the planet is moving backward. The same thing happens when we, in a rapidly moving automobile, pass a slowly moving vehicle on a curved road; for a brief moment the other vehicle appears to move backward with respect to the fixed background. Ancient astronomers thought this apparent backward movement, called **retrograde motion,** was a real motion of the superior planets rather than merely a reflection of the Earth's motion (Fig. 2.24). This mistaken belief greatly complicated many of the early models of the universe and gave rise to a whole class of theories in which the planets were thought to move in small circles called **epicycles,** which in turn orbited the Earth (see the discussions of Hipparchus and Ptolemy later in this chapter).

Even the inferior planets undergo retrograde motion as they overtake the Earth, but this is difficult to observe, since it occurs near the time of inferior conjunction, when they lie near the direction of the Sun.

HISTORICAL DEVELOPMENTS

Now that we have discussed most of the phenomena that can be observed without telescopes and are thus aware of nearly all the data available to ancient watchers of the sky, it is appropriate to review the development of human understanding of these phenomena.

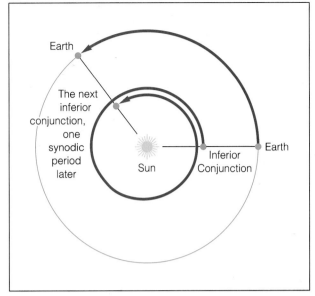

Figure 2.23   The synodic period of an inferior planet.   The inner planets travel faster than the Earth in their orbits and therefore "lap" the Earth, much as a fast runner laps a slower runner on a track. This illustration shows approximately the situation for Mercury, which has a synodic period of about 116 days, or roughly one-third of a year.

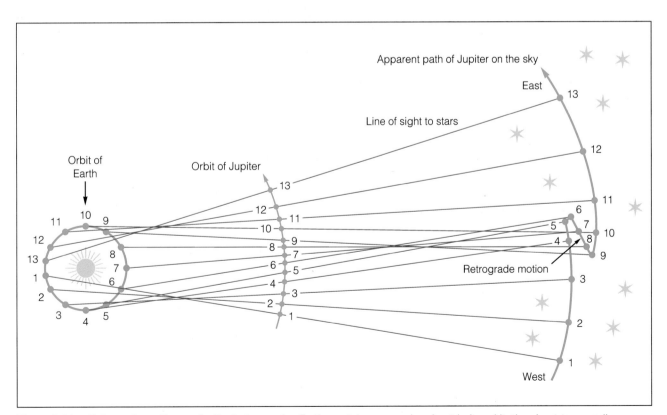

Figure 2.24   Retrograde motion.   As the faster-moving Earth overtakes a superior planet in its orbit, the planet temporarily appears to move backward with respect to the fixed stars. This sketch illustrates the modern explanation of something that took ancient astronomers a long time to understand correctly.

It is likely that people were preoccupied with the heavens from the time they first became aware of their environment. The speculative mood that we, in these modern times, can conjure up only by disregarding our daily pressures and escaping into the countryside on a clear night to look at the stars, must have dominated the nighttimes of the earliest cultures. The following discussion emphasizes developments that occurred around the shores of the Mediterranean, because these achievements laid the foundation for the modern understanding of the universe.

In this brief overview of early astronomy, we confine our discussion to developments that led, more or less directly, to modern understandings. Many important and, in some cases, rather sophisticated discoveries occurred in other parts of the world, but these discoveries did not play direct roles in the development of modern astronomy. Among those not discussed, but important in their own right, are the ancient astronomies of China and India and the early American cultures.

The Earliest Astronomy

The earliest records of astronomical lore have been found in the region east of the Mediterranean now known as Iraq (Fig. 2.25), where the Babylonian culture flourished for many centuries, beginning around 2000 B.C. The Babylonians attained an accurate knowledge of the length of the year, establishing the basis for the modern 12-month calendar, and bequeathed us our timekeeping and angular measures that are based on the number 60.

Parallel with the developments in Babylonia, an early Greek civilization arose on the shores of the Mediterranean (Fig. 2.25), at some unknown time coming in contact with the Babylonian culture. The ancient Greek traditions, known to us largely through the writings of Homer (Fig. 2.26), gave birth to our modern constellation names and other astronomical lore and laid the foundation for the first rational scientific inquiry. (Table 2.3 lists some achievements of the ancient Greeks.)

The first formal scientific thought is associated with the philosopher **Thales** (624–547 B.C.) and his followers, who formed the so-called Ionian school of thought. The principal contribution of Thales was the idea that rational inquiry can go beyond *describing* the universe to *understanding* it. The Ionians did develop a primitive **cosmology**, or theory of the universe, in which all the basic elements of the universe were formed from water, the primeval substance.

Another major school of thought centered around

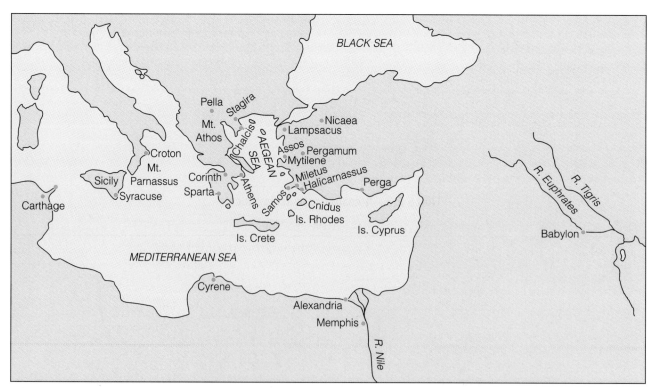

Figure 2.25   The ancient Mediterranean.   This map shows the locations of many of the sites mentioned in the text. To the right is the region of Babylonia, now Iraq. On the northern shores of the Mediterranean Sea lay most of the Greek empire, in what is now Italy, Greece, and Turkey. (C. Ronan, *The Astronomers* [New York: Hill and Wang] 1964)

Pythagoras (c. 570–500 B.C.) and his followers, who believed that natural phenomena could be described mathematically, a belief that is at the heart of all modern science. Pythagoras himself is credited with being the first to assert that the Earth is round and that all heavenly bodies move in circles, ideas that never thereafter lost favor in ancient times.

Plato and Aristotle

One of the most influential figures in the development of Greek philosophy was **Plato** (428–347 B.C.; Fig. 2.27), who established an academy in Athens in about the year 387 B.C., where he taught his ideas of natural philosophy. Plato's fundamental precept was that what we see of the material world is only an imperfect representation of the ideal creation. This doctrine had the corollary that one can learn more about the universe by reason than by observation, since observation can only present us with an incomplete picture. Hence Plato's ideas of the universe, described in his *Republic,* were based on certain idealized assumptions that he found reasonable. Among the most important of these assumptions was the idea that all motions in the universe are perfectly circular and that all astronomical bodies are spherical. Thus he adopted the Pythagorean view that the Sun, Moon, and all the planets moved in combinations of circular motions about the Earth. Plato evidently thought these objects were fixed to clear, ethereal spheres that rotated.

Plato's most renowned student was **Aristotle** (384–322 B.C.), who was the first to adopt physical laws and show why, in the context of those laws, the universe works as it does. He taught that circular motions are the only natural motions and that the center of the Earth is the center of the universe. He also

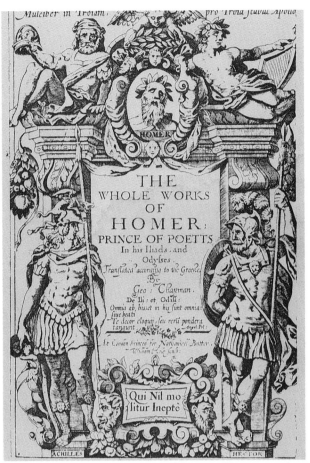

Figure 2.26   Homer's *Iliad* and *Odyssey*. These two epic poems by Homer incorporate much more ancient legends and astronomical lore, including myths from the Minoan civilization that had flourished earlier on the isle of Crete. (The Granger Collection)

TABLE 2.3   Notable Greek Achievements

Date	Name	Discovery or Achievement
c. 900–800 B.C.	Homer	*Iliad* and *Odyssey*; summaries of legends
c. 624–547 B.C.	Thales	Rational inquiry leads to knowledge of universe
c. 611–546 B.C.	Anaximander	Universal medium; primitive cosmology
c. 570–500 B.C.	Pythagoras	Mathematical representation of universe; round Earth
c. 500–400 B.C.	Philolaus	Earth orbits central fire
c. 500–428 B.C.	Anaxagoras	Moon reflects sunlight; correct explanation of eclipses
c. 428–347 B.C.	Plato	Material world imperfect; deduce properties of universe by reason
c. 408–356 B.C.	Eudoxus	First mathematical cosmology; nested spheres
c. 384–322 B.C.	Aristotle	Concept of physical laws; proof that Earth is round
c. 310–230 B.C.	Aristarchus	Relative sizes, distances of Sun and Moon; first heliocentric theory
c. 273–? B.C.	Eratosthenes	Accurate size of Earth
c. 265–190 B.C.	Apollonius	Introduction of the epicycle
c. 200–100 B.C.	Hipparchus	Many astronomical developments; full mathematical epicyclic cosmology
c. A.D. 100–200	Ptolemy	*Almagest*; elaborate epicyclic model

ASTRONOMICAL INSIGHT
The Mythology of the Constellations

As noted in the text, Homer's *Iliad* and *Odyssey*, which appeared around 900–800 B.C., contain descriptions of the constellations and their meanings. These descriptions were based on legends from the Minoan civilization that were already ancient in Homer's time.

Little precise information is available as to when and how the mythology of the constellations arose. However, it has been possible to deduce roughly the era that gave birth to them. Careful examination of the ancient constellations shows them to be distributed symmetrically about a point in the sky that was at the north celestial pole around 2600 B.C., so it is probable that the constellations were invented at about that time. (The pole has since shifted away from the center of these constellations because of precession.) There are few ancient constellations near the southern pole, in regions not visible from the latitude of Crete, a fact that gives independent support to the supposition that the legends arose in the Minoan culture.

It is likely that ancient peoples did not take the constellations as literally as is usually supposed. Rather than being considered faithful depictions of the people and events with which they are associated, the constellations should more properly be viewed as symbolic representations. They probably did not originally derive their names from their imagined resemblance to certain characters; instead, areas of the sky were dedicated in honor of prominent figures of mythology, and the familiar pictures of these figures were then fitted to the patterns of bright stars. This explanation helps account for the lack of obvious resemblance between the star patterns and the figures and events they supposedly represent.

The constellation names were translated from the Greek of Homer to Latin when the Roman Empire rose to dominance, and for the most part the names we use today are the Latin ones. Interestingly, the names of prominent stars went through another transformation, being translated from Greek into Arabic. Today's star names are Arabic designations that usually are literal indications of the place these stars hold in their constellations. Betelgeuse, for example, the name of the prominent red star in the shoulder of Orion, is translated as "the armpit of the giant" or "the armpit of the central one." Thus, in today's standard usage, we often adopt Arabic names for stars in constellations with Latin designations, although both translations are based on Greek descriptions of the ancient Minoan legends.

Ancient and nonscientific though they are, the constellations have a significant impact on the nomenclature of modern astronomy. The modern constellations, many of which correspond to ancient ones, refer to very specific regions of the sky with well-defined boundaries that have been agreed on by the international community of astronomers. The "official" constellations are based on those of legend, containing within them the ancient figures of mythology, but their boundaries have been extended so that every part of the sky falls within one constellation or another. A map of the constellations looks a bit like a map of the western United States, where the boundaries are generally straight lines, but the shapes are irregular. As an alternative to the Arabic names for the brightest stars, modern astronomers often use designations based on a star's rank within its constellation, with letters of the Greek alphabet used to indicate specific stars. Thus, Betelgeuse, the brightest star in Orion, is also called α Orionis.

It can be fun and informative to learn the prominent constellations, and this is taught in many general astronomy classes. It is worth keeping in mind, however, that memorizing star patterns on the sky is not the same as scientific inquiry, but should be regarded instead as a convenient method for remembering the locations of objects that are interesting to view.

---

believed that the world is composed of four elements: earth, air, fire, and water. He could demonstrate, in the context of his adopted physical laws, that the universe was spherical and that the Earth was also spherical. He had three proofs of the latter:

1. Only at the surface of a sphere do all falling objects seek the center by falling straight down. (It was another premise of his that falling objects were following their natural inclination to reach the center of the universe.)

2. The view of the constellations changes as one travels north or south.

3. During lunar eclipses it can be seen that the shadow of the Earth is curved (Fig. 2.28).

By relating his theories to observation in this manner, Aristotle broke with the tradition of Plato to some extent, although he approached the problem in the same manner, letting reason rather than observation guide the way.

Other tenets of Aristotle included the conclusion

Figure 2.27   Plato.   Plato's beliefs that the universe and all celestial objects were perfect and immutable had great influence on later philosophers. (The Granger Collection)

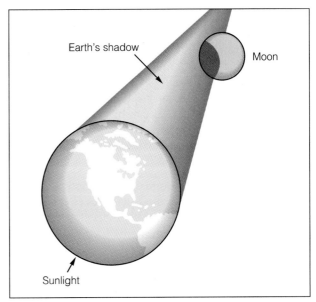

Figure 2.28   Curvature of the Earth's shadow on the Moon. Only a spherical body can cast a circular shadow for all alignments of the Sun, Moon, and Earth.

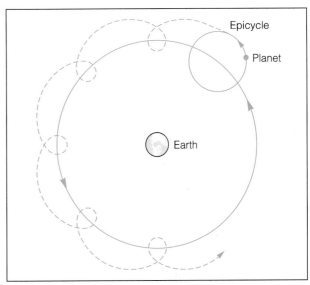

Figure 2.29   The epicycle.   Ancient astronomers realized that planetary motions could be represented by a combination of motions involving an epicycle, which carries a planet as it spins while orbiting the Earth.

that the universe is finite in size (this led directly to his belief that the heavenly bodies can follow only circular motions, because otherwise they might encounter the edge of the universe) and that the heavenly bodies were made of a fifth fundamental substance, which is called the aether.

The Later Greeks: Sophisticated Cosmologies

After Aristotle's period, the center of Greek scientific thought moved across the Mediterranean Sea to Alexandria, the capital city established in 332 B.C. by Alexander the Great near the site of the present city of Cairo. The first prominent astronomer of this era was **Aristarchus** (c. 310–230 B.C.), the first scientist to adopt the idea that the Sun, not the Earth, is at the center of the universe. His conclusion was based on geometrical arguments showing that the Sun is much larger than the Earth; therefore it seemed natural for the Sun to be at the center of the universe. The heliocentric hypothesis of Aristarchus failed to attract many followers at the time, due largely to a lack of any concrete evidence that the Earth was in motion and also to a general satisfaction with the Aristotelian viewpoint, which had no recognized flaws.

By the third century B.C. the need for more precise mathematical models of the universe became apparent as better observing techniques were developed. A mathematical concept that provides the needed precision while preserving the precepts of Aristotle was the **epicycle,** a small circle on which a planet moves; the center of the epicycle in turn orbits the Earth following a large circle called a **deferent** (Fig. 2.29). An important advantage of the epicycle was that it could explain the retrograde motions of the superior planets.

The epicyclic motions of the celestial bodies were refined further by **Hipparchus** (Fig. 2.30), who was active during the middle of the second century B.C. (Very little is known about his life, not even the dates of his birth and death.) Hipparchus, who did most of his work at his observatory on the island of Rhodes, was one of the greatest astronomers of antiquity. Among his major contributions are the following:

1. The first use of trigonometry in astronomical work (in fact, he is largely credited with its invention, although many of the concepts were developed earlier).
2. The refinement of instruments for measuring star positions, along with the first known use of a celestial coordinate system akin to our modern equatorial coordinates, which enabled him to compile a catalog of some 850 stars.
3. Refinement of the methods of Aristarchus for measuring the relative sizes of the Earth, Moon, and Sun.
4. The invention of the stellar magnitude system for estimating star brightnesses, a system still in use today (with minor modifications).

5. Perhaps most impressive, the discovery of the precession of star positions, which he accomplished by comparing his observations with some that were made 160 years earlier.

Although Hipparchus did not add much that was new to the then-current models of planetary motions, he did compile an extensive series of observations of solar motion. This led him to a refinement of the epicyclic theory, namely, that the Earth was off-center in the large circle on which the Sun moved. This idea of an **eccentric,** or off-center, orbit accounted rather accurately for the observed variations in the speed of the Sun in its annual motion.

Despite the prior suggestions that the Sun was the central body in the universe, Hipparchus clung to the Earth-centered view. Apparently his main reasons for doing so were that the geocentric picture was quite capable of explaining the observations and that there was no evidence that the Earth moved. There was, in fact, an observational argument that the Earth did *not* move. Hipparchus knew that nearby stars should appear to move back and forth on the sky as a reflection of the Earth's orbital motion (Fig. 2.31), but no such **stellar parallax** could be detected. (The shifts do occur, but they are far too small to have been measured with the crude instruments of the Greek astronomers.)

After the great work of Hipparchus, almost three hundred years passed before any significant new astronomical developments occurred. Claudius Ptolemaeus, or simply **Ptolemy** (Fig. 2.32), lived in the middle of the second century A.D. and undertook to summarize all the world's knowledge of astronomy.

Figure 2.30   A fanciful rendition of Hipparchus at his observatory.   Hipparchus made important discoveries based on his observations and on his advanced methods for analyzing data. (The Bettmann Archive)

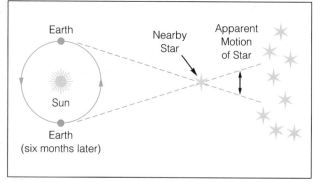

Figure 2.31   Stellar parallax.   As the Earth orbits the Sun, our line of sight toward a nearby star varies, causing the star's position (with respect to more distant stars) to change. The change of position is greatly exaggerated in this sketch; in reality, even the largest stellar parallax displacements are too small to have been measured by ancient astronomers.

The Renaissance
and the Laws of Motion

Space shuttle astronauts demonstrating weightlessness. (NASA)

PREVIEW

In this chapter you will begin to see how modern scientists attacked the celestial mysteries that confounded the ancients. We start by following the progress made during the Renaissance, when intellectual giants such as Copernicus, Brahe, Kepler, and Galileo began to discover the underlying mechanisms and the true relationships of the Earth, Sun, Moon, and planets. Here you will see how complex phenomena can be explained by a few simple principles, and you will get a feeling for the way in which science progresses through the comparison of prediction and observation. The latter portion of the chapter is devoted to the laws of gravity and motion, many of which were revealed by Newton. From these discussions you will gain an understanding of the mechanisms that underlie many of the astronomical phenomena to come in the rest of this text, so the lessons learned here will be important throughout the remainder of your study of astronomy.

Figure 3.1 Nicolaus Copernicus. The view of Copernicus that the Sun, rather than the Earth, occupied the central place in the cosmos had great influence on later scientists. (The Granger Collection)

The fifteenth century saw the beginnings of a reawakening of intellectual spirit in Europe. Some scientific studies began at the major universities; increased maritime explorations brought demands for better means of celestial navigation; and the art of printing was developed, opening the way for widespread dissemination of information.

Major advances in all the sciences accompanied the new developments in other fields of human endeavor. In this chapter we discuss the principal achievements in Renaissance science that led to the development of modern astronomy. In doing so, we discuss the accomplishments of five major figures; Copernicus, Brahe, Kepler, Galileo, and Newton.

COPERNICUS: THE HELIOCENTRIC VIEW REVISITED

Some nineteen years before the epic voyage of Columbus, Niklas Koppernigk (Fig. 3.1) was born in Torun, in the northern part of Poland. As a young man he attended the University of Cracow; here his fondness for Latin, the universal language of scholars, led him to change his surname to Copernicus. At Cracow he developed an avid interest in astronomy, becoming fully acquainted with the Aristotelian view as well as the Ptolemaic model of planetary motions. He persisted in his study of astronomy, and by 1514 he had developed some doubts about the validity of the accepted system.

His reasons for doing so have been the subject of some uncertainty and misconception. It was long assumed that Copernicus was encouraged to adopt the Sun-centered view of the universe because he recognized shortcomings in Ptolemy's geocentric model. There is, however, no evidence of widespread dissatisfaction with the Ptolemaic system or any records indicating that Copernicus himself found serious inaccuracies in it. His reasons for adopting the heliocentric viewpoint were more subtle.

The basis for his conversion was primarily philosophical. Copernicus believed that his new system presented a pleasing and unifying model of the universe and its motions. While he was no doubt encouraged by the climate of change and cultural revolution that was sweeping Europe with the advent of the Renaissance, he did not adopt the heliocentric view just to be different or to improve the accuracy or reduce the complexity of the accepted view. The Copernican model was, in fact, no more accurate in predicting planetary motions than the Ptolemaic system, and it was just as complex. Copernicus adhered to the notion of perfect circular motions and was obliged to include small epicyles in order to match the observed planetary positions. Apparently one of

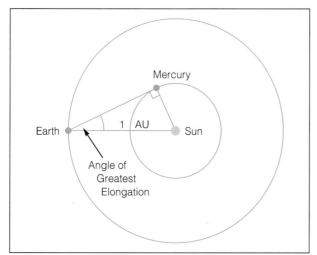

Figure 3.2 Copernicus's method for finding relative planetary distances from the Sun. For an inferior planet such as Mercury or Venus, Copernicus knew the angle of greatest elongation and was therefore able to reconstruct the triangle shown, which provided the Sun-Mercury (or Sun-Venus) distance relative to the Sun-Earth distance (that is, the astronomical unit). For superior planets, similar but slightly more complicated considerations provided the same information.

Figure 3.3 Tycho Brahe. Tycho's principal contribution, a massive collection of accurate observations of planetary positions, was crucial to subsequent advances in understanding planetary motions. (The Granger Collection)

the most pleasing aspects of the Sun-centered system to Copernicus was the fact that the relative distances of the planets could be deduced (Fig. 3.2) and were found to have a certain regularity; that is, the spacings between planets increase systematically with distance from the Sun. Copernicus was also able to determine the relative speeds of the planets in their orbits, finding that each planet moves more slowly than the next one closer to the Sun. Like Aristarchus long before him, Copernicus also recognized that the Sun is the largest body in the solar system, which he considered a strong argument for placing the Sun at the center.

Copernicus first circulated his ideas informally sometime before 1514 in a manuscript called *Commentariolis*, which drew increasing attention over the next several years. The Church voiced no opposition, despite the fact that the ideas expressed in the work were in strong contradiction to commonly accepted Church doctrine. Copernicus was apparently reluctant to publish his findings in a more formal way for fear of raising controversy and was continually rechecking his calculations. Publication of his findings in a book called *De Revolutionibus Orbium Coelestium (On the Revolution of the Celestial Sphere)*, or simply *De Revolutionibus*, finally took place in 1543, when Copernicus was near death. He did not live to see the profound impact of his work.

TYCHO BRAHE: ADVANCED OBSERVATIONS

Although Copernicus had contributed little observationally and had not demanded a close match between theory and observation as long as general agreement was found, there was in fact a strong need for improved precision in astronomical measurements. The next influential character in the historical sequence met this need; if he had not, further progress would have been seriously delayed, as would the eventual acceptance of the Copernican doctrine.

Tycho Brahe (Fig. 3.3) was born in 1546, some three years after the death of Copernicus, in the extreme southern portion of modern Sweden (the region was part of Denmark at the time). Of noble descent, Tycho spent his youth in comfortable surroundings and was well educated, first at Copenhagen University, then at Leipzig, where he insisted on studying mathematics and astronomy despite his family's wish that he pursue a law career. As a result of some notable observations and their interpretation, Tycho eventually developed a strong reputation as an astronomer (and astrologer—the distinction was still scarcely recognized) and attracted the attention of the Danish king, Frederick II. In 1575 the king ceded to Tycho the island of Hveen, about fourteen miles north of Copenhagen, along with enough servant support and financial assistance to allow him to build and maintain his own observatory.

Even before this time, Tycho had shown an acute interest in astronomical instruments, and with the grant to build his observatory, this interest bore fruit. He devised a variety of instruments (Fig. 3.4), which, although they did not encompass any new principles, were capable of more accurate readings than any before his time.

His observational contributions were significant for both the unprecedented accuracy of his data and the completeness of his records. Until his time, the general practice of astronomers had been to record the positions of the planets only at notable points in their

travels, such as when a superior planet comes to a halt just before beginning retrograde motion. Tycho made much more systematic observations, recording planetary positions at times other than just the significant turning points in their motions. He also made multiple observations in many instances, allowing the results to be averaged to improve their accuracy. Tycho himself did not attempt any extensive analysis of his data, but the vast collection of measurements that he gathered over the years contained the information needed to reveal the basis of the planetary motions.

Tycho was unable to accept the heliocentric view, primarily because he could find no evidence that the Earth was moving. He tried and failed to detect stellar parallax, which he thought he should see with his accurate observations if the Earth really moved. Furthermore, as a strict Protestant he found it philosophically difficult to accept a moving Earth when the Scriptures stated that the Earth is fixed at the center of the universe. On the other hand, he realized that the Copernican system had advantages of mathematical simplicity over the Ptolemaic model, and in the end he was ingenious enough to devise a model that satisfied all of his criteria. He imagined that the Earth was fixed with the Sun orbiting it, but that all the other planets orbited the Sun (Fig. 3.5). Mathematically, this model is equivalent to the Copernican system in terms of accounting for the motions of the planets as seen from the Earth. The idea never received much acceptance, however, and Tycho is remembered primarily for his fine observations.

Tycho Brahe died in 1601. The task of seeking the secrets contained in Tycho's data was left to those who followed him, particularly a young astronomer named Johannes Kepler.

Figure 3.4   The Great Mural Quadrant at Hveen. This instrument was used to measure the angular positions of stars and planets with respect to the horizon. Its large size made the angle markings easy to read accurately. (Photo Researchers, Inc.)

JOHANNES KEPLER AND THE LAWS OF PLANETARY MOTION

Until now it has been possible to follow developments in a straight sequence, but here we must begin to cover events that occurred at nearly the same time in different locations. To complete the thread begun with our discussion of Tycho Brahe, this section is concerned with Johannes Kepler (Fig. 3.6), who worked briefly with Tycho himself and then spent many years analyzing the great wealth of observational data that Tycho had accumulated. We must keep in mind, however, that during the same period, Galileo Galilei was at work in Italy, and each man was aware of the other's accomplishments.

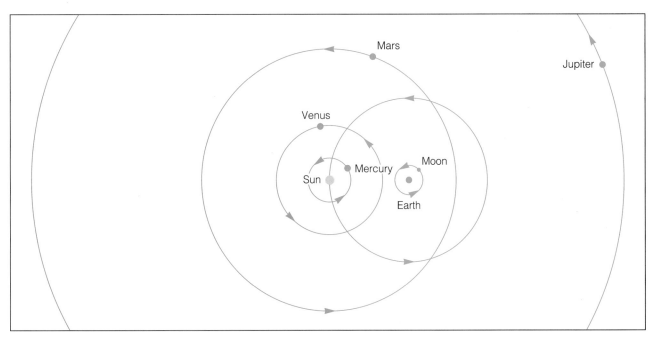

Figure 3.5   Tycho's model of the universe.   Tycho held that the Earth was fixed and that the Sun orbited the Earth. The planets in turn orbited the Sun. This system was never worked out in mathematical detail, but it successfully preserved both the advantages of the Copernican system and the spirit of the ancient teachings. It also accounted for the lack of observed stellar parallax, since the Earth was fixed.

Born in 1571, Kepler was a sickly youngster, seemingly headed for a career in theology. While attending university, however, he encountered a professor of astronomy who inspired in him a strong interest in the Copernican system, which he adopted wholeheartedly. Kepler thereafter devoted his life to seeking the underlying harmony of the cosmos. He devoted great effort to searching for simple numerical relationships among the planets and, in doing so, embarked on several false turns and erroneous paths. He was a true scientist, however, in that he was always willing to discard his ideas if the data did not support them.

Kepler went to work as Tycho's assistant in 1600, and a year later found himself the beneficiary of the massive collection of data left behind at Tycho's death. Kepler's main mission, due both to the wishes of Tycho and to his own interests, was to develop a refined understanding of the planetary motions and to upgrade the tables used to predict their positions. He set to work first on the planet Mars; the data were particularly extensive, and its motions were among the most difficult to explain in the established Ptolemaic system (or in the Copernican system with its requirement of circular motions only). By a very complex process, Kepler was able to separate the effects of the Earth's motion from those of Mars itself, so

Figure 3.6   Johannes Kepler.   Kepler's fascination with numerical relationships led him to discover the true nature of planetary motions. (The Bettmann Archive)

that he could map out the path that Mars followed with respect to the Sun.

By 1604 Kepler had determined that the orbit of Mars was some kind of oval, and further experimentation revealed that it was fitted precisely by a simple geometric figure called an **ellipse** (Fig. 3.7). This is a closed curve defined by a fixed total distance from two points called **foci** (singular: **focus**), and indeed Kepler found that the Sun was at the precise location of one focus of the ellipse. He later generalized this discovery to apply to all of the planets, and it became known as Kepler's first law of planetary motion: The orbit of each planet is an ellipse, with the Sun at one focus.

Further analysis of the motion of Mars revealed a second characteristic: The planet moves fastest in its orbit when it is nearest the Sun, and slowest when it is on the opposite side of its orbit, farthest from the Sun. Mathematically, Kepler's second discovery was that a line connecting Mars to the Sun sweeps out equal areas of space in equal intervals of time (Fig. 3.7). Again, Kepler later stated that this law too, applies to all of the planets.

Kepler's results on the orbit of Mars were published in 1609 in a book entitled *The New Astronomy: Commentaries on the Motions of Mars.* The book received a great deal of attention. In 1619 Kepler published a book entitled *The Harmony of the World,* in which he reported his discovery of a simple relationship between the orbital periods of the planets and their average distances from the Sun. Now known as Kepler's third law or simply the harmonic law, it

states that the square of the period of a planet is proportional to the cube of the semimajor axis (which is half of the long axis of an ellipse). Put in other terms, this says that $P^2 = a^3$, where P is the sidereal period of a planet in years, and a is the semimajor axis in terms of the Sun-Earth distance (i.e., in terms of the **astronomical unit** or **AU**) (Table 3.1).

In another major work, the *Epitome of the Copernican Astronomy*—published in parts in 1618, 1620, and 1621—Kepler presented a summary of the state of astronomy at that time, including Galileo's discoveries. In this book Kepler generalized his laws, explicitly stating that all of the planets behaved similarly to Mars, something that had clearly been his belief all along.

By the time the *Epitome* was published, the Roman Catholic church was in a very intolerant frame of mind, in contrast with the situation at the time of Copernicus nearly a hundred years before, and Kepler's treatise soon found itself on the *Index of Prohibited Books,* along with *De Revolutionibus.*

In 1627, Kepler published his last significant astronomical work, a table of planetary positions based on his laws of motion, which could be used to predict planetary motions accurately. These tables, which he called the *Rudolphine Tables* in honor of a former benefactor, were used for the next several years. The *Rudolphine Tables* represented an improvement in accuracy over any previous tables by a factor of nearly 100, a resounding and remarkable confirmation of the validity of Kepler's laws. In a very real sense the *Rudolphine Tables* represented Kepler's life's work, since

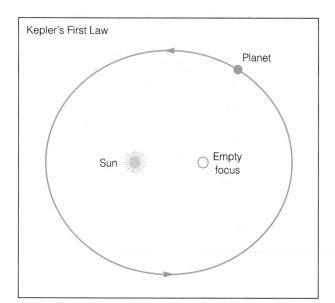

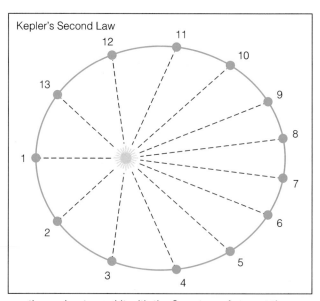

Figure 3.7   The ellipse.   At the left is an exaggerated ellipse representing a planetary orbit with the Sun at one focus; at the right is a similarly exaggerated ellipse with lines drawn to illustrate Kepler's second law. If the numbers represent the planet's position at equal time intervals, the areas of the triangular segments are equal.

TABLE 3.1 Testing Kepler's Third Law*

Planet	a (AU)	P (years)	a³	P²
Mercury	0.387	0.241	0.058	0.058
Venus	0.723	0.615	0.378	0.378
Earth	1.000	1.000	1.000	1.000
Mars	1.523	1.881	3.533	3.538
Jupiter	5.203	11.86	140.85	140.66
Saturn	9.539	29.46	867.98	867.89
Uranus	19.18	84.01	7,055.79	7,057.68
Neptune	30.06	164.8	27,162.32	27,159.04
Pluto	39.44	248.4	61,349.46	61,762.56

*The minor disagreement between P^2 and a^3 seen here for the outermost planets do not indicate failures of Kepler's third law; instead they reflect inaccuracies in the measured values of P and a.

with their publication he completed the task set before him when he first went to work for Tycho. Kepler died in 1630 at the age of 59.

GALILEO, EXPERIMENTAL PHYSICS, AND THE TELESCOPE

Very strong contrasts can be drawn between Kepler and his great contemporary, Galileo Galilei (Fig. 3.8), who was born in Pisa in northern Italy in 1564. Where Kepler was fascinated with universal harmony and therefore with the underlying principles on which the universe operates, Galileo was primarily concerned with the nature of physical phenomena and was less devoted to finding fundamental causes. Galileo wanted to know how the laws of nature operated, whereas Kepler sought the reason for their existence.

Galileo's approach was level-headed and rational in the extreme. He used simple experiment and deduction in advancing his perception of the universe and has frequently been cited as the first truly modern scientist, although others of his time probably deserve a share of that recognition. A follower of Plato and Aristotle, whose works still dominated in Galileo's time, would proceed by rational thought from standard unproven assumptions; Galileo found it much more sensible to begin with experiment or observation and work toward a recognition of the underlying principles. In doing this, Galileo founded an entirely new basis for scientific inquiry, an achievement in many ways more profound than his contributions to astronomy, which were considerable.

Galileo's discoveries in physics, having to do with the motions of objects, were published in his later years, after his astronomical career had been forcibly

Figure 3.8 Galileo Galilei. From his numerous experiments and observations, Galileo deduced the nature of the cosmos. His methodical attacks on traditional beliefs caused him personal troubles but greatly influenced contemporary thinking. (The Granger Collection)

ended by Church decree. It was in fact an early interest in *mechanics*, the science of the laws of motion, that lured Galileo away from a career in medicine, the subject of his first studies. Galileo's contributions

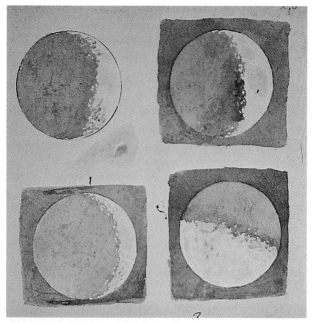

Figure 3.9   Galileo's sketches of the Moon.   Galileo's observation of mountains, craters, and "seas" led him to believe that the Moon was not the perfect heavenly body envisioned in the ancient Greek teachings. (The Granger Collection)

to physics will be described briefly later in this chapter.

Galileo's astronomical discoveries were quite sufficient to earn him a major place in history, and his flair for debate and his habit of ridiculing those whose arguments he disproved made him famous in his own time, though not universally loved.

In 1609 Galileo learned of the invention of the telescope and devised one of his own, which he soon put to use in systematic observations of the heavens. Despite the poor quality of the instrument, Galileo made a number of important discoveries almost at once and reported them in 1610 in a publication called *The Starry Messenger*. Here Galileo showed that the Moon was not a perfect sphere but was covered with craters and mountains (Fig. 3.9). He also reported that the broad band of the Milky Way consisted of countless stars and, most significant of all, that Jupiter was attended by four satellites, whose motions he observed long enough to establish that they orbited the parent planet (Fig. 3.10). All of these discoveries, but especially the latter, violated the ancient philosophies of an idealized universe centered on the Earth (Table 3.2). The satellites of Jupiter helped convince Galileo that there were centers of motion other than the Earth.

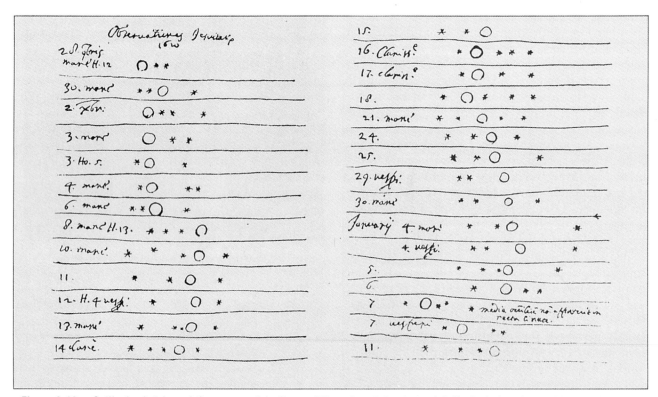

Figure 3.10   Galileo's sketches of the moons of Jupiter.   This series of drawings of Galileo's Jovian observations is often attributed to The Starry Messenger, but in fact was made some years later. Galileo's discovery of moons orbiting Jupiter showed that there are heavenly bodies that do not orbit the Earth. (Yerkes Observatory)

TABLE 3.2   Some of Galileo's Arguments for the Heliocentric Theory

Discovery	Argument
Many faint stars	Difficult to reconcile with idea of stars as points attached to crystalline sphere
Craters on the moon	Moon is not perfect, immutable heavenly body
Moons of Jupiter	A body other than Earth as center of motion
Phases of Venus	Explained only if Venus orbits the Sun and shines by reflected sunlight
Sunspots	Spots are on the solar surface, showing that the Sun is not perfect
Variable planetary sizes	Angular size variations explained by motion of planets around Sun

Mostly as a result of the reputation Galileo earned with the publication of *The Starry Messenger,* he was able to negotiate successfully for the position of court mathematician to the grand duke of Tuscany, and in 1610 he moved to Florence, where he was to spend the remainder of his long career. Once established there, Galileo continued his observations and soon added new discoveries to his list. He found that Venus changes its appearance much as the Moon does, and showed that this proves that Venus orbits the Sun, undergoing phases as varying portions of the sunlit side are visible from Earth (Fig. 3.11). He analyzed sunspots, dark blemishes seen crossing the face of the Sun, and showed that they really were *on* the

Sun and were not small planets orbiting close to it as some had suggested. Both of these discoveries refuted ancient teachings.

Galileo began to draw increasingly heavy criticism from the Church, and he made efforts to develop good relations with high-ranking officials in Rome. Nevertheless in 1616 he was pressured into refuting the Copernican doctrine and for several years thereafter was relatively quiet on the subject. Except for a well-publicized debate on the nature of comets, Galileo spent most of his time preparing his greatest astronomical treatise, which was finally published, after some difficulties with Church censors, in 1632. To avoid direct violation of his oath not to support the Copernican heliocentric view, Galileo wrote his book in the form of a dialogue among three characters: one, named Simplicio, represented the official positions of the Church; another, Salviati, represented Galileo (although this was, of course, not stated explicitly); and a third, Sagredo, was always quick to see and agree with Salviati's arguments. In this treatise, called *Dialogue on the Two Chief World Systems,* Galileo, through the character Salviati and at the expense of Simplicio, systematically destroyed many of the traditional astronomical teachings of the Church. The book was published in Italian, rather than the scholarly Latin, and its contents were therefore accessible to the general populace.

Despite a lengthy preface in which Galileo disavowed any personal belief in the heliocentric doctrine, the Church reacted strongly, and within a few months Galileo was summoned before the Roman Inquisition, while further publication of the book was banned. It joined the works of Copernicus and Kepler on the *Index of Prohibited Books* (where it was to re-

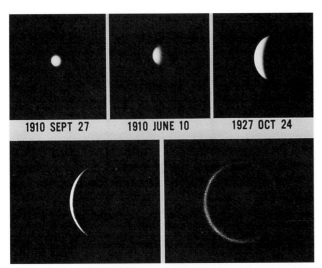

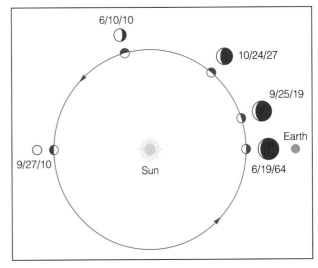

Figure 3.11   The phases of Venus.   As Venus orbits the Sun and its position changes relative to the Sun-Earth line, its phase varies as we see differing portions of its sunlit side. In addition, its apparent size varies because of its varying distance from Earth. (Photos from NASA)

ASTRONOMICAL INSIGHT
An Excerpt from the *Dialogue on the Two Chief World Systems*

Galileo got into trouble with the Roman Catholic church when he portrayed the official view of the nature of the universe as not only incorrect, but illogical and even silly. Reading a small sample from the *Dialogue on the Two Chief World Systems* might help us see how inflammatory it was.

During the course of three days of discussion, the three characters in Galileo's *Dialogue* thoroughly air all the available evidence and arguments bearing on the understanding of mechanics and the structure and nature of the universe. About midway through, the characters engage in a discussion that is central to Galileo's principal point: the Sun, not the Earth, is at the center of the universe. In the following excerpt,* we see an example of Salviati's persuasive style, Simplicio's dogged reluctance to give up the ideas of Aristotle, and Sagredo's ready comprehension of Salviati's arguments.

Salviati: *Now if it is true that the center of the world is the same about which the circles of the mundane bodies, that is to say, of the Planets, move, it is most certain that it is not the Earth but the Sun, rather, that is fixed in the center of the World. So that as to this first simple and general apprehension, the middle place belongs to the Sun, and the Earth is as far remote from the center as it is from that same Sun.*

Simplicio: *But from whence do you argue that not the Earth but the Sun is in the center of Planetary revolutions?*

Salviati: *I infer the same from the most evident and therefore necessarily conclusive observations, of which the most potent to exclude the Earth from the said center, and to place the Sun*

therein, are that we see all the planets sometimes nearer and sometimes farther off from the Earth, with so great differences, that, for example, Venus when it is at the farthest is six times more remote than when it is nearest, and Mars rises almost eight times as high at one time as at another.

See therefore whether Aristotle was somewhat mistaken in thinking that it was at all times equidistant from us.

Simplicio: *What in the next place are the tokens that their motions are about the Sun?*

Salviati: *It is shown in the three superior planets, Mars, Jupiter, and Saturn, in that we find them always nearest to the Earth when they are in opposition to the Sun and farthest off when they are towards the conjunction, and this approximation and recession imports thus much, that Mars near at hand appears sixty times greater than when it is remote. As to Venus, in the next place, and to Mercury, we are certain that they revolve about the Sun in that they never move far from it, and in that we see them sometimes above and sometimes below it, as the mutations in the figure of Venus necessarily prove. Touching the Moon, it is certain that it cannot in any way separate itself from the Earth, for the reasons that shall be more distinctly alleged hereafter.*

Sagredo: *I expect that I shall hear more admirable things that depend upon this annual motion of the Earth than were those dependent upon the diurnal revolution.*

In this exchange, Galileo, in the guise of Salviati, advances several arguments based on observations, including the well-known phases of Venus, and the less widely quoted arguments involving the varying distances of the planets from the Earth. In this and other passages, Galileo went out of his way to mock the followers of Aristotle, who, according

to Salviati in an earlier paragraph, "would deny all the experiences and all the observations in the world, nay, would refuse to see them, that they might not be forced to acknowledge them, and would say that the world stands as Aristotle writes and not as Nature will have it."

The Church did not take the drastic action of excommunicating Galileo for expressing his views in the *Dialogue*, but it did take severe measures, nevertheless. There the matter stood, as scientific and religious attitudes evolved over the subsequent centuries. In a spirit of reconciliation with science and in the interest of clearing up the clouded past, the Roman Catholic church recently reopened the case of its treatment of Galileo. After a study of the old records, the Church announced in 1983 that Galileo's name had been cleared and that the Church had erred in its treatment of him more than three hundred years ago.

This recent action may seem pointless, since Galileo himself is no longer alive. It was, however, an important statement by a major church that science and religion should tolerate each other—that there is philosophical room for both. In an age when religious attacks on science are occurring anew, this position by an important worldwide religion may have some helpful effect.

*Reprinted from the translation of Galileo's *Dialogue* by G. D. Santillana, by permission of the University of Chicago Press. Copyright 1953 by the University of Chicago. All rights reserved.

main until the 1830s), and Galileo himself was sentenced to house arrest, eventually serving this punishment at his country home near Florence. Here he spent the remainder of his days. Galileo died in 1642, having suffered blindness in the last four years of his life.

ISAAC NEWTON

In the year 1643, a few months after the death of Galileo, Isaac Newton (Fig. 3.12) was born in Woolsthorpe, England. Newton's childhood was unremarkable, except that he showed a growing interest in mathematics and science. At age 18 he entered college at Cambridge and received a bachelor's degree in early 1665. He then spent two years at his home in Woolsthorpe, largely because the plague made city living rather dangerous. During this time he made a remarkable series of discoveries in the fields of physics, astronomy, optics, and mathematics, in what surely must have been one of the most intense and productive periods of individual intellectual effort in human history.

Newton had a tendency to exhaust a subject, get bored with it, and go on to new fields, and none of his work was published until after some of his discoveries were repeated independently by others. Finally, after persistent urging by his friend and fellow astronomer Edmond Halley, in 1687 Newton published a massive work called *Philosophiae Naturalis Principia Mathematica* (Fig. 3.13), now usually referred to simply as the *Principia*. In this three-volume book, Newton established the science of mechanics (which he viewed as merely background material and relegated to an introductory section) and applied it to the motions of the Moon and the planets, developing the law of gravitation as well. His work in optics was published separately in 1704, although it was probably written much earlier than that.

The *Principia* received great notice, particularly in England. As a result, Newton's later life was a public one, with various government positions and less and less time for scientific discovery. With the help of younger associates, he did revise the *Principia* on two occasions, in 1713 and in 1726, making some improvements each time. Newton died in 1727 at the age of 84. His influence lives on in our modern understanding of physics and mathematics. Newton's

Figure 3.12   Isaac Newton.   Newton's experiments and brilliant mathematical intuition led him to profound new understandings of physics and mathematics. (The Granger Collection)

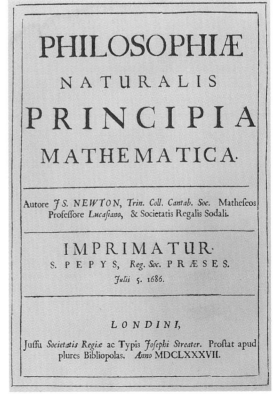

Figure 3.13   The title page from an early edition of the *Principia*. This massive volume is still considered one of the greatest and most influential books ever written. (The Granger Collection)

conclusions on the nature of motions and of gravity are still viewed as correct, although scientists now realize that in certain circumstances more complex theories (such as Einstein's theory of relativity) must be used.

The Laws of Motion

Newton put forth three laws of motion, principles he considered so self-evident that he relegated them to an introductory section of the *Principia*. The first of Newton's laws states the principle of **inertia,** a concept first recognized by Galileo, who realized that an object in motion tends to stay in motion unless something acts to slow or stop it. This was completely contrary to the teachings of Aristotle, who held that the natural tendency of any moving object was to stop, and that it would continue moving only if a force were applied. Aristotle was misled by his failure to recognize friction, a force that tends to stop motion in most everyday situations. Newton expanded the concept of inertia, recognizing it as just one in a series of physical principles that govern the motions of objects, and adding the all-important notion of **mass.**

The mass of an object reflects the amount of matter it contains, which in turn determines other properties such as weight and momentum. The other properties are easily observed, but the mass is not, so mass can be a difficult concept. We can make useful illustrations by considering situations in which we alter these other properties but not the mass. For example, the weight of an object varies according to the magnitude of the gravitational acceleration to which it is subjected, but its mass does not change. Astronauts in space may be weightless, but they contain just as much mass as they do on the ground. Mass is usually measured in units of **grams** or **kilograms.** One gram is the mass of a cubic centimeter of water, and a kilogram, which weighs about 2.2 pounds at sea level, is the mass of 1,000 cubic centimeters (or one **liter**) of water.

Inertia is very closely related to mass. The more massive an object is, the more inertia it has, and the more difficult it is to start it moving or to stop or alter its motion once it is moving (Fig. 3.14). Newton summarized the concept of inertia in what has become known as his first law of motion:

> A body at rest or in a state of uniform motion tends to stay at rest or in uniform motion unless an outside force acts upon it.

Having stated that a force is required to change an object's state of rest or of uniform motion, Newton went on to determine the relationship between force and the change in motion that it produces. To understand this, we must discuss the idea of acceleration.

Figure 3.14   Inertia.   An object in motion tends to remain in motion. (H. Roger Viollet)

Acceleration is a general word for *any* change in the motion of an object. Acceleration occurs when a moving object is speeded up or slowed down or when its direction of motion is altered. Acceleration also occurs when an object at rest is put into motion. A planet orbiting the Sun is undergoing constant acceleration; otherwise it would fly off in a straight line.

Newton's first law could also have been stated by saying that in order to accelerate an object, a force must be applied to it. Note that this force must be an unbalanced one; that is, there will be acceleration only when a force is applied to an object with no other force to counteract it. Thus, a crate sitting on the floor is subject to the downward force of gravity, but this is balanced by the upward force due to the floor, and there is no acceleration. Newton's second law spells out the relationship among an unbalanced force, the resultant acceleration, and the mass of the object:

> The acceleration of an object is equal to the force applied to it divided by its mass.

This may be written mathematically as $a = F/m$, where a is the acceleration, F is the unbalanced force, and m is the mass. More commonly, it is written in the equivalent form $F = ma$.

We can visualize simple examples to help illustrate the second law. If one object has twice the mass of another, for example, and equal forces are applied to the two objects, the more massive one will only be accelerated half as much. Conversely, if unequal forces are applied to objects of equal mass, the one to which the greater force is applied will be accelerated to a greater speed.

Newton's third law of motion is probably more subtle than the first two, although in some circum-

ASTRONOMICAL INSIGHT
The Forces of Nature

One of the most difficult concepts to grasp is that of a force, particularly when there is no concrete, visible object to exert it. We think of a force as a push or a pull, and we can visualize it easily in certain circumstances, such as when a gardener pushes a wheelbarrow and it moves. It is perhaps less obvious in the case where no movement results, and even less so when no tangible agent exerts the force. In the case of gravity, the force is exerted invisibly and over great distances.

In general terms, a force exerted by a concrete object is referred to as a **mechanical force,** whereas those exerted without any such agent are **field forces.** Gravity is the most familiar force that is created by a field.

There are four basic kinds of field forces in nature, and in fact these forces form the basis for *all* forces, mechanical or field. Gravity was the first of the four to be discovered, although in "discovering" gravity and mathematically describing its behavior, Newton did not develop a real fundamental understanding of how or why it works. The reason that gravity was the first to be discovered is simple; no special conditions are required for gravitational forces to be exerted (*all* masses attract each other), and it operates over very great distances. After all, Newton deduced its properties by noting how it controls the motions of planets that are separated from the Sun by as much as a billion kilometers.

The second most obvious type of field force, and the second to be discovered and described mathematically, is the **electromagnetic force**. There is an intimate relationship between electric and magnetic fields. The interaction of these fields with charged particles, first described mathematically by the Scottish physicist James Clerk Maxwell, creates forces. The electromagnetic force is actually much stronger than the force of gravity; the electromagnetic force binding an electron to a proton in the nucleus of an atom is 10^{39} times stronger than the gravitational force between the two particles. The electromagnetic force, like gravity, is inversely proportional to the square of the distance between charged particles. Therefore we might expect this force always to dominate over gravity, as it does on a subatomic scale. But it does not. The reason is that most objects in the universe, composed of vast numbers of atoms containing both electrons and protons, have little or no net electrical charge, whereas they always have mass and are therefore subject to gravitational forces. If the planets and the Sun had electrical charges in the same proportion to their masses as the electron and proton do, then electromagnetic forces, rather than gravity, would control their motions.

Besides being immensely stronger, the electromagnetic force also differs from gravity in that it can be either repulsive or attractive, depending on whether the electrical charges are the same or opposite.

The remaining two forces were not discovered until the 1930s. (Maxwell's work on electromagnetics was done in the late 1800s.) Both of these new forces involve interactions at the subatomic level, and they were not discovered until the development of the science of quantum mechanics. One of these is the **strong nuclear force,** which is responsible for holding together the protons and neutrons in the nucleus of an atom, and the other is the **weak nuclear force,** some 10^5 times weaker than the strong nuclear force. The strong nuclear force, in turn, is about 100 times stronger than the electromagnetic force, making it the strongest of all, but it operates only over very small distances. Within an atomic nucleus, where protons with their like electrical charges are held together despite their electromagnetic repulsion for each other, it is the strong nuclear force that acts as the glue that keeps the nucleus from flying apart. The weak nuclear force plays a more subtle role, showing its effects primarily in certain modifications of atomic nuclei during radioactive decay.

From the smallest scale to the largest, these four forces appear to be responsible for all interactions of matter. It is ironic that at the most fundamental level, the mechanism that makes the forces work is not understood. To understand the fundamental nature of forces is one of the principal goals of modern physics. One hope is to develop a mathematical framework that encompasses all of the forces, a framework first sought over 60 years ago by Albert Einstein, and referred to as a **unified field theory.** Progress has been made since Einstein's time, particularly toward unifying the electromagnetic and nuclear forces, but the ultimate goal still eludes us.

stances it is quite obvious. It states:

> For every action there is an equal and opposite reaction.

Put in other words, when a force is applied to an object, it pushes back with an equal force (Fig. 3.15).

This may sound confusing because the acceleration is not necessarily equal, and because in most common situations other forces such as friction complicate the picture. Furthermore, there are many static situations, in which forces are balanced and no acceleration occurs.

The third law can be most easily visualized by considering situations in which friction is not important. As an example, imagine standing in a small boat and throwing overboard a heavy object, such as an anchor. The boat will move in the opposite direction to the anchor, because the anchor exerts a force on you as you throw it. The "kick" of a gun when it is fired is another example of action and reaction. Technically speaking, when a person jumps off the ground by pushing against the Earth, both he and the Earth are accelerated by the mutual force, but, of course, the immensely greater mass of the Earth prevents it from being accelerated noticeably.

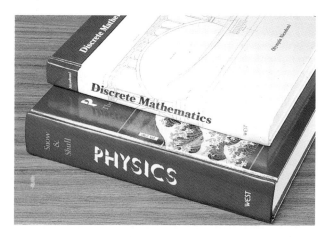

The third law of motion states the principle on which a rocket works. In this case hot, expanding gas is allowed to escape through a nozzle, creating a force on the rocket. The gas is accelerated in one direction, and the rocket is accelerated in the opposite direction (Fig. 3.16). Anyone who has inflated a balloon and then let go of it, allowing it to zoom through the air, is familiar with the operating principle of a rocket.

GRAVITATION AND WEIGHT

We return now to a point raised earlier—namely, that the planets would fly off along straight lines if no force were acting upon them. Newton's first law says that this should happen, yet obviously it does not. Newton realized that the planets must be undergoing constant acceleration toward the center of their orbits, that is, toward the Sun (Fig. 3.17). He set out to understand the nature of the force that creates this acceleration.

If you feel confused about the direction of this force, remember what you have just learned about acceleration and inertia: A planet needs no force to keep it moving, but it does require a force to keep its path curving as it travels around the Sun. What is needed is something to pull the planet inward, toward the Sun. This force can be compared to the tension in the string tied to a rock that you whirl about

Figure 3.15   Action and reaction.   Some manifestations of Newton's third law are rather subtle, while others are not. Here a book exerts a force on a table, and the table exerts an equal and opposite force on the book. This situation is static; there is no motion. When a cannon is fired, however, the shell is accelerated one way and the cannon the other. The force applied to each is the same, but the cannon has more mass than the shell and is therefore accelerated less, as Newton's second law states. (bottom: U.S. Army)

Figure 3.16   Newton's third law applied to the launch of a rocket.   Hot gases are forced out of a nozzle (or several, as in this case), and in return they exert a force on the rocket that accelerates it. (NASA)

your head; if you suddenly cut the string, the rock will fly off in whatever direction it happens to be going at the time.

What, then, is the string that keeps the planets whirling about the Sun? Newton realized that the Sun itself must be the source of this force, and he made use of Kepler's third law, as well as observations of the Moon's orbit and falling objects at the Earth's surface, to discover its properties. He was led to formulate his law of universal gravitation:

> Any two bodies in the universe are attracted to each other with a force that is proportional to the product of the masses of the two bodies and inversely proportional to the square of the distance between them.

The law of gravitation is one of the fundamental rules by which the universe operates. As we shall see in later chapters, it explains the motions of stars about each other or about the center of the galaxy, the movements of the Moon and planets, and the motions of galaxies about one another. Gravity, in fact, appears to be the dominant factor that will determine the ultimate fate of the universe. Table 3.3 shows the relative gravitational forces on a person on Earth due to various bodies in the universe.

TABLE 3.3 Gravitational Forces Acting on a Human Standing on Earth

Source of Force	Relative Strength of Force
Earth	1.0
Moon	3.4×10^{-6}
Sun	8.6×10^{-4}
Venus (at closest approach)	1.9×10^{-8}
Jupiter (at closest approach)	3.3×10^{-8}
Nearest star	1.4×10^{-14}
Milky Way galaxy	2.1×10^{-11}
Virgo cluster of galaxies	10^{-15}

It is useful to consider a few examples illustrating how the law of gravitation is applied. The weight of an object is simply the gravitational force between it and the Earth. If, for example, the diameter of the Earth were suddenly doubled (while its mass remained constant), our weight would decrease by a factor of $2^2 = 4$. If the Earth were three times smaller, our weight would be $3^2 = 9$ times greater. If we climb to the top of a high mountain, our weight decreases, but not very much, because even the highest mountains are small compared with the radius of the Earth. The Earth exerts a gravitational force on an astronaut in orbit, but there is no acceleration relative to the spacecraft, so the astronaut is weightless in the local environment (Fig. 3.18).

The law of gravitation states that the force also depends on the masses of the two objects that are attracting one another. It is easy to imagine that a person's weight would double if his or her mass doubled; similarly, the force between the Earth and the Moon depends on the masses of these two bodies. If we triple the mass of the Moon, the force is tripled; if we triple the mass of the Moon but decrease the mass of the Earth by a factor of 3 at the same time, the force remains unchanged.

Although it is useful for the purpose of illustration to work out simple thought examples as we have just done, a more quantitative approach is needed to deal with more realistic situations. As we might expect, the law of universal gravitation can be written in the form of an equation:

$$F = \frac{Gm_1 m_2}{r^2},$$

where F is the strength of the force between two objects whose masses are m_1 and m_2 and whose separation is r. The symbol G represents a constant that is required for the value of F to be expressed in normal units of force. In the system of units most commonly used by physicists (called the *Système Internationale*, or *SI* system), force is measured in **newtons**

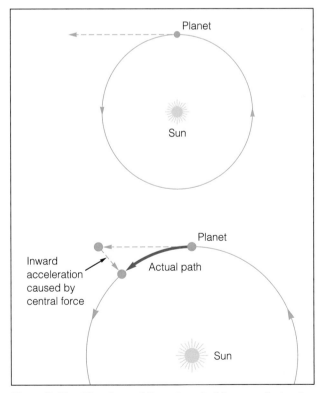

Figure 3.17   Planetary orbits and central force.   A planet would fly off in a straight line if no force attracted it toward the center of its orbit. From diagrams like the lower one shown here, Newton was able to determine the amount of acceleration required by an orbiting body to keep it in its orbit.

Figure 3.18   Weightlessness.   Although astronauts orbiting the Earth are subject to the Earth's gravitational force and are "falling around" the Earth along with their spacecraft, they are weightless because they experience no gravitational acceleration relative to their surroundings. This photo shows two astronauts aboard *Skylab,* an orbiting scientific research station. (NASA)

TABLE 3.4   Surface Gravities on Solar System Bodies

Body	Surface Gravity $(g)^a$
Earth	1.0
Sun	27.9
Moon	0.17
Mercury	0.38
Venus	0.90
Mars	0.38
Ceres (largest asteroid)	0.000167
Jupiter	2.64
Saturn	1.13
Uranus	0.89
Neptune	1.13
Pluto	0.08?

$^a g = 9.80$ m/sec^2.

Now recall that the acceleration on a body is equal to the force applied to it, divided by its mass; hence, the acceleration of the falling object is F/m_2. If we call this acceleration g, because it is due to gravity, then we have

$$g = \frac{F}{m_2} = \frac{Gm_1}{r^2} = \frac{GM}{R^2},$$

since the symbols M and R are generally used for the mass and radius of a planetary body.

At the surface of the Earth, the value of g is 9.80 meters per second per second, meaning that the speed of a falling object increases by 9.80 meters per second for every second of fall. In more familiar units, this corresponds to 32 feet per second per second. To find the surface gravity on some other planet or satellite, the same expression for g can be used, along with the appropriate values for the mass of the planet, M, and its radius, R (Table 3.4). For example, the Moon has 0.273 times the radius of the Earth, and 0.0123 times the mass. Thus, the acceleration of gravity at the Moon's surface is $0.0123/0.273^2 = 0.165$ that at the surface of the Earth. Therefore, astronauts on the Moon weigh approximately one-sixth as much as they do on the Earth.

(where 1 newton is the force exerted by a mass of 1 kilogram under an acceleration of 1 meter per second per second). A person who weighs 150 pounds (or a mass of 67 kilograms) has a weight in this system of 656.6 newtons, so we see that a newton is a rather small unit of force. The value of the constant G is 6.67×10^{-11} when the masses are expressed in kilograms, the separation between them in meters, and the force in newtons.

The law of universal gravitation can be used to show something that Galileo had postulated long before Newton's time: namely, that the acceleration due to gravity is independent of the mass of an object. Suppose, in the previous equation, that m_1 represents the mass of the Earth, and m_2 the mass of an object falling at the Earth's surface. In that case, r is the Earth's radius (it can be shown that the Earth acts gravitationally as though all its mass were in a single point at its center, so we say that the separation between the object and the Earth is equal to the Earth's radius).

ENERGY, ANGULAR MOMENTUM, AND ORBITS: KEPLER'S LAWS REVISITED

Even though Newton made use of Kepler's third law in deriving the law of gravitation, the latter is in fact more fundamental. It was soon possible for Newton to show that all three of Kepler's laws follow directly from Newton's laws of motion and gravitation. Kepler's studies of planetary motions had revealed the *result* of the laws of motion and gravitation, while Newton found the *cause* of the motions. To appreciate how he accomplished this, we must further discuss some basic physical ideas.

An important concept in understanding not only orbital motions but also many other aspects of the universe is **energy**. In an intuitive sense, energy may be defined as the ability to do work. Energy can take many possible forms, such as electrical energy, chemical energy, heat, and others. All forms of energy can be classified as either **kinetic energy,** the energy of motion, or **potential energy,** stored energy that must be released (i.e., converted to kinetic energy) if it is to do work. A speeding car has kinetic energy because of its motion; a tank of gasoline has potential energy in the form of its chemical reactivity, a tendency to produce large amounts of kinetic energy if ignited. Thus a car operates when this potential energy is converted to kinetic energy in its cylinders.

The units used for measuring energy can be expressed in terms of the kinetic energy of specified masses moving at specified speeds. The kinetic energy of a moving object is $\frac{1}{2}mv^2$, where m is its mass and v its speed. In the SI system the unit of energy is the **joule,** a small amount of energy that is equal to the kinetic energy of a mass that is moved a distance of 1 meter by a force of 1 newton. A 67-kg person walking at a brisk pace (2m/sec) has a kinetic energy of 135 joules.

We often speak in terms of **power,** which is simply energy expended per second. The SI system unit for power is the watt, which is 1 joule per second. Thus we will speak of the power (or, equivalently, the **luminosity**) of a star in terms of its energy output in watts, or joules, per second. In this system the Sun has a luminosity of 4×10^{26} watts.

Using our understanding of energy, we can now discuss orbital motions in a much more general way than in our previous discussions. Two objects subject to each other's gravitational attraction have kinetic energy due to their motions and potential energy due to the fact that each feels a gravitational force. Just as a book on a table has potential energy that can be converted to kinetic energy if it is allowed to fall, an orbiting body also has potential energy by virtue of the gravitational force acting on it.

Newton's laws and the concepts of kinetic and potential energy can be used to show that many types of orbits are possible when two bodies interact gravitationally (Fig. 3.19). Not all are ellipses, because if one of the objects has too much kinetic energy (exceeding the potential energy due to the gravity of the other), it will not stay in a closed orbit, but will instead follow an arcing path known as a hyperbola, and will escape after one brief encounter. Some comets have so much kinetic energy that after one trip close to the Sun, they escape forever into space, following hyperbolic paths.

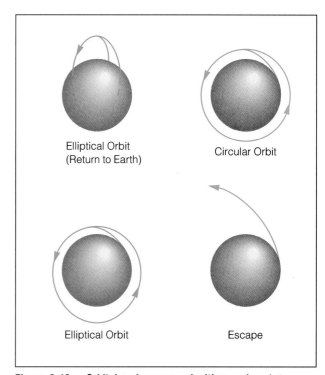

Elliptical Orbit (Return to Earth)

Circular Orbit

Elliptical Orbit

Escape

Figure 3.19   Orbital and escape velocities.   A rocket launched with insufficient speed for a circular orbit would tend to orbit the Earth's center in an ellipse, but would intersect the Earth's surface. Given the correct speed, it will follow a circular orbit. A somewhat larger speed will place it in an elliptical orbit that does not intersect the Earth. Given enough speed so that its kinetic energy is greater than its gravitational potential energy, however, it will escape entirely.

If the kinetic energy is less than the potential energy, as it is for all of the planets, then the orbit is an ellipse, as Kepler found. It is technically correct to say that a planet and the Sun orbit a common **center of mass** (Fig. 3.20), rather than saying that the planet orbits the Sun. The center of mass is a point in space between the two bodies where their masses are essentially balanced; more specifically, it is the point where the product of mass times distance from this point for the two objects is equal. Since the Sun is so much more massive than any of the planets, the center of mass for any Sun-planet pair is always very close to the center of the Sun, so the Sun moves very little, and we do not easily see its orbital motion (only Jupiter is sufficiently massive that the center of mass between it and the Sun lies barely above the Sun's surface). It is true, however, that the Sun's position wiggles a little as it orbits the centers of mass established by its interaction with the planets, especially the most massive ones. In a double star system, where the two masses are more nearly equal, it is easier to see that both stars orbit a point in space between them.

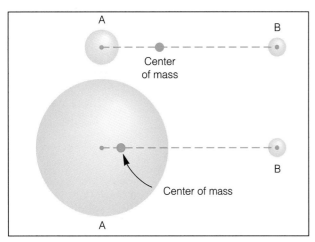

Figure 3.20   Center of mass.   The upper sketch depicts a double star where star A has twice the mass of star B, so the center of mass, about which the two stars orbit, is one-third of the way between the centers of the two stars. In the lower sketch, star A has 10 times the mass of star B, so the center of mass is very close to star A. The Sun is so much more massive than any of the planets that the center of mass for any Sun-planet pair is near the center of the Sun, so the Sun's orbital motion is very slight.

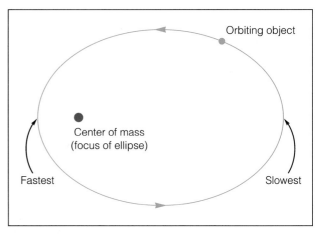

Figure 3.21   Conservation of angular momentum.   An object moves in an elliptical orbit with varying speed, and the product of its mass times its velocity times its distance from the center of mass (that is, its angular momentum) is constant (this statement is precisely correct only for a circular orbit; for an ellipse, the velocity in question is the component that is at right angles to the Sun-planet line). Kepler's second law of planetary motion is a statement of this fact.

Thus Kepler's first law as he stated it requires a slight modification: Each planet has an elliptical orbit about the *center of mass* between it and the Sun (the center of mass is at one focus).

The second law can also be restated in terms of Newton's mechanics. Any object that rotates or moves around some center has **angular momentum.** This depends on its mass, its velocity, and its distance from the center of motion. In the simple case of an object in circular orbit, the angular momentum is the product *mvr*, where *m* is its mass, *v* its speed, and *r* its distance from the center of mass.

The total amount of angular momentum in a system is always constant. Because of this, a planet in an elliptical orbit must move faster when it is close to the center of mass than when it is farther away, so that its velocity compensates for the changes in distance (Fig. 3.21). Thus a planet moves faster in its orbit near **perihelion** (its point of closest approach to the Sun) than at **aphelion** (its point of greatest distance from the Sun). Kepler's second law is really a statement that angular momentum is constant for a pair of orbiting objects.

Kepler's third law was also revised by Newton, and in this case the revision is especially important for our studies. Newton discovered that the relationship between period and semimajor axis depends on the masses of the two objects. Kepler had not realized this, primarily because the Sun is so much more mas-

sive than any of the planets that the differences between the masses of the planets has only a very small effect. Kepler's form of the third law can be written $P^2 = a^3$, where P is the planet's period in years and a is the semimajor axis in astronomical units. Newton revised this to

$$(m_1 + m_2)P^2 = a^3$$

where m_1 and m_2 represent the masses of the two bodies, for example, the Sun and one of the planets. The masses must be expressed in terms of the Sun's mass in this equation. If we use other units, such as kilograms for the masses, seconds for the period, and meters for the semimajor axis, the equation is complicated by the addition of a numerical factor and is written

$$(m_1 + m_2)P^2 = \frac{4\pi^2}{G}a^3$$

where G is the gravitational constant. Either of these forms of Kepler's third law can be solved for the sum of the masses, and we will see that this is a very important tool for astronomers in deducing the masses of distant objects.

The consideration of orbital motions in terms of kinetic and potential energy leads to the concept of an **escape speed.** If an object in a gravitational field has greater kinetic than potential energy, it can escape the gravitational field entirely. To launch a rocket into space (that is, completely free of the Earth) therefore requires giving it enough upward speed at launch to make its kinetic energy greater

ASTRONOMICAL INSIGHT
Relativity

In the text, Newton's work was characterized as being nearly modern in the sense that his laws of motion are still regarded today as adequate representations of the manner in which forces and bodies interact. Although it is true that Newtonian mechanics is sufficient for most practical applications, there are circumstances in which this is not so. By the late 1800s, scientists had begun to recognize some difficulties with Newton's laws, and in the first decades of this century these problems led Albert Einstein to develop his theories of special and general relativity.

The circumstances in which Newton's laws fail are those in which we deal with extremely high speeds (that is, speeds that are a significant fraction of the speed of light), in which case special relativity theory must be used; or when very strong gravitational fields are involved, when general relativity theory is required. Actually, relativity is always a more accurate representation than Newton's laws, but the two theories give virtually identical results except under the circumstances just cited.

Mathematically, both special and general relativity are too complex to be presented here in any detail. It is possible, however, to provide some intuitive description of what is involved. The dilemma that led Einstein to develop special relativity had to do with the speed of light. To see this, first let us consider a case involving lower speeds. Suppose a jet aircraft moving at 1,000 miles per hour fires a missile in the forward direction at a speed relative to the plane of 2,000 miles per hour. The speed of the missile with respect to the ground would then be the sum of the plane's speed and the speed with which the missile left the plane, that is, 3,000 miles per hour. This is the result expected from Newtonian mechanics.

Now imagine a train moving along a track, and suppose that it turns on its headlight. The light leaves the train at the normal speed of light relative to the train, and by analogy with the previous examples, we might expect that the speed of the light with respect to a bystander would be equal to the normal speed of light plus the speed of the train. For firm theoretical and experimental reasons, however, it is thought to be impossible for the normal speed of light to be exceeded. In other words, an observer on the train would actually measure the same speed for the light as would an observer standing by the track, in contradiction to the example of the plane and its missile. In Einstein's time, an actual measurement of this sort had not been made, but he had another reason for postulating that the speed of light should be the same for all observers: If this were not so, it would imply that there was some fundamental difference between one frame of reference and others. There would be a "fundamental" frame in which light would have the speed required by theory, and all other frames would be somehow less fundamental. This would imply that there was a preferred reference frame in the universe, an idea contrary to the lessons learned long before, when it was realized that the Earth was not the center of the universe.

Einstein's solution to the problem involved a surprising postulate: The rate of passage of time itself depends on the relative speed between observers. Let us think about the train example again. If we believe that the speed of light must be the same for someone on the train as for someone standing by the track, even though the train is moving, then we are forced to accept Einstein's postulate if both observers are to measure the same speed of light. For the two observers to measure the same speed for the light despite their motion with respect to each other, it is necessary for the time interval to be different for the two observers.

If the rate of time depends on the motion of the observers with respect to each other, there are a number of interesting consequences, some of which can be (and have been) tested experimentally. For example, people who travel at high speeds through space will return to Earth younger than they would have been if they had stayed home, because time will not pass as rapidly for them. Even the sizes and masses of objects are affected: The length of an object that is moving at high speed will appear smaller to an observer at rest than to someone moving with the object, and its mass will be greater as measured by the person at rest.

Whereas special relativity states that the laws of physics are the same for observers moving at different *speeds,* general relativity addresses the situation for observers undergoing different *accelerations.* We have seen that a gravitational field such as the Earth's creates acceleration: A falling object increases its speed with every second as it falls. Even without falling, an object is subject to a force because of the acceleration of gravity; we are all familiar with the fact that an object at the surface of the Earth has weight. Now suppose we are somewhere out in space, inside an enclosed spacecraft, and that the spacecraft is accelerating at a rate equivalent to the gravitational acceleration at the Earth's surface. Inside the spacecraft we would "feel" this acceleration, which would create a force (in the direction opposite the spacecraft's motion) exactly identical to the weight we would have on the

continued on next page

surface of the Earth. There is no experimental way to tell the difference between the two situations.

Now suppose that the person in the accelerating spaceship fires a gun. The bullet, being "left behind" as the ship continues to increase its speed, will follow a curving path with respect to the spacecraft, falling toward the rear. This is identical to the curving path of a bullet fired at the surface of the Earth; as the bullet travels toward its target, it drops toward the ground because of gravity. Now suppose that the "bullet" is a beam of light. In the spaceship, the beam would curve away from the direction of the ship's acceleration, just as a material bullet would. For the two situations to be fully equivalent the implication is that a beam of light on Earth also would curve downward because of gravity. Thus, Einstein predicted that a gravitational field should bend light, and this was later confirmed experimentally. Since light always follows the shortest possible path between two points, Einstein characterized the effect of a gravitational field as a "bending" of space itself. Hence, the light could travel the shortest path in the gravitational field, but this path would look like a curve to an observer not subjected to the same gravitational field.

Today we speak of **space-time,** indicating that time itself is regarded as a dimension because of the role it plays in governing the measured properties of an object; we also speak of **curved space-time** because of the curvature that occurs in the presence of a gravitational field. Throughout this text, but especially in the later chapters, where we discuss situations involving very high speeds or extreme gravitational fields, these concepts of special and general relativity will come into play.

than its potential energy due to the Earth's gravitational attraction (see Fig. 3.19). It so happens that the speed required to accomplish this is the same for any mass of object (as long as it is much smaller than the Earth's mass), and in equation form it is $v_e = \sqrt{2GM/R}$, where v_e is the escape velocity, G is the gravitational constant, and M and R are the Earth's mass and radius, respectively. For the Earth, this speed is 11.2 kilometers per second, or just over 40,000 kilometers per hour (or 24,000 miles per hour). As we will see in later chapters, a planet's escape speed plays a major role in determining the nature of its atmosphere.

TIDAL FORCES

We have seen that the gravitational force due to a distant body decreases with distance. This means that an object subjected to the gravitational pull of such a body feels a stronger pull on the side nearest that body, and a weaker pull elsewhere. For example, the part of the Earth on the side facing the Moon feels a stronger attraction toward the Moon than do other points on the Earth, and the point on the opposite side from the Moon feels a weaker force than a point on the near side. The Earth is therefore subjected to a **differential gravitational force,** which tends to stretch it along the line toward the Moon (Fig. 3.22). Of course, the Sun exerts a similar stretching force on the Earth, but it is too distant to have as strong an effect as the Moon (but its *total* gravitational force on the Earth is much greater than that of the Moon). A **tidal force,** as differential gravitational forces are called, depends on how close one body is to the other, because the key is how rapidly the gravitational force drops off over the diameter of the body subject to the tidal force, and it drops off most rapidly at small distances.

The Earth is a more or less rigid body, so it does not stretch very much due to the differential gravitational force of the Moon. Nevertheless, the tidal forces exerted on the Earth by the Moon create net forces that tend to make the liquid oceans flow toward the points facing directly toward the Moon and directly away from it. These forces are due to the fact that the near side of the Earth feels a stronger attraction toward the Moon than do other points on the Earth's surface, and the far side of the Earth feels a weaker attractive force toward the Moon. As the Earth rotates, the water in the oceans tends to follow the tidal forces created by the Moon, so that in effect the oceans have two bulges of water that remain fixed in position relative to the Earth-Moon direction, while the Earth rotates. The effect that we see, moving with the Earth's rotation, is two huge ridges of water that appear to flow around the Earth as it rotates. Since the two ridges of water are on opposite sides, and the Earth rotates in 24 hours, one of these ridges passes any given point on the Earth's surface every 12 hours. Thus we have the ocean's tides, with high tides at any given location separated by about 12 hours.

It is interesting to consider what is happening to the Moon at the same time. It is subjected to a more intense differential gravitational force than the Earth, because the Earth is more massive than the Moon. Even though the Moon is a solid body, its shape is deformed by this force, and it has tidal bulges. The

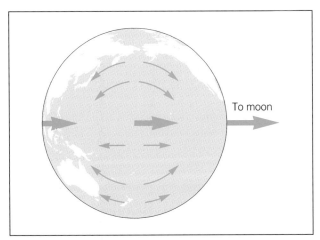

Figure 3.22   The Earth's tides.   The differential gravitational force caused by the Moon tends to stretch the Earth (large arrows). Seawater at any given point on the Earth is subjected to a combination of vertical and horizontal forces, causing it to flow toward either the side of the Earth facing the Moon or the side opposite it (curved arrows).

Moon is slightly elongated along the line toward the Earth. As we shall see, this has had drastic effects on the Moon's rotation.

There are many other examples of tidal forces, both in our solar system and outside it. The satellites of the massive outer planets are subjected to severe tidal forces, and in some double star systems the two stars are so close together that they are stretched out of round. We will discuss these in more detail later, along with star clusters and galaxies that are affected by tidal forces, sometimes even tearing each other apart.

PERSPECTIVE

We have now traced the development of astronomy from the introduction of the heliocentric concept to the point where the motions of the bodies in the solar system can be fully described in terms of a few simple laws of physics. The framework of these laws does more than that for us, however; it also provides a basis for understanding the motions of more distant astronomical objects. The law of universal gravitation will be invoked again and again in our study of the solar system and the rest of the universe because it explains so many important phenomena.

Newton's laws of motion and gravitation have stood the test of time rather well. Einstein's theory of general relativity may be viewed as a more complete description of gravity and its interaction with matter

than Newton's laws. For most situations, however, Newton's laws are perfectly adequate.

We are ready now to discuss other laws of physics, particularly those that govern the emission and absorption of light.

SUMMARY

1. Copernicus developed the heliocentric theory because it provided a unifying picture of the universe, not because it was more accurate or simpler than the Earth-centered theory. Copernicus's new system enabled him to calculate the relative distances of the planets from the Sun and arrive at the correct explanations of the seasons and precession.

2. Tycho Brahe made vast improvements in the quantity and quality of astronomical observations of stars and of the planets, accomplishing this by making more precise and complete measurements than his predecessors. Tycho was unable to detect stellar parallax and therefore rejected the heliocentric hypothesis.

3. Kepler sought the underlying harmony among the planets and in the process discovered that each planet orbits the Sun in an ellipse; that a line connecting a planet with the Sun sweeps out equal areas in space in equal intervals of time; and that the square of the period of a planetary orbit is proportional to the cube of its semimajor axis.

4. Galileo used telescopic observations and deductive reasoning to argue for the heliocentric concept of the universe. Despite Church opposition, Galileo brought his ideas before the public in the form of a fictional dialogue between characters of opposing points of view.

5. Isaac Newton, in the late 1600s, developed the laws of motion, the law of gravitation, and calculus, and he made many important contributions to our knowledge of the nature of light and telescopes.

6. Newton's three laws of motion describe the concept of inertia; state that acceleration is proportional to the force exerted on a body and inversely proportional to its mass; and state that for every action there is an equal and opposite reaction.

7. The law of universal gravitation states that any two objects in the universe attract each other with a force that is proportional to the product of their masses and inversely proportional to the square of the distance between them.

8. Newton's laws, along with the concepts of kinetic and potential energy and angular momentum, can be used to explain orbital motions.

9. Newton modified Kepler's third law to show that the relationship between the period and semimajor axis of an orbit is dependent on the sum of the masses of the two objects; this is an important tool for measuring the masses of distant objects.

10. Every body such as a planet or a star has an escape velocity, the upward speed at which a moving object has more kinetic energy than potential energy and will therefore escape into space.

11. Differential gravitational forces are responsible for tides on the Earth and in the interiors of other planets and satellites.

REVIEW QUESTIONS

1. Summarize the reasons Copernicus adopted the heliocentric view.

2. Describe how the beliefs of Copernicus conformed to the teachings of Aristotle and how they violated those teachings.

3. Suppose the true altitude of a certain star above the horizon was 27°14′. Suppose further that Tycho's measurements of the altitude of the star were 27°12′, 27°13′, 27°15′, 27°13′, and 27°16′. How close is the average of these values to the true value? Discuss the comparison between the average of several measurements and the accuracy of individual measurements.

4. There is a principle adopted by most scientists that the best explanation for something is the simplest one, the one requiring the fewest unprovable assumptions. Copernicus and Hipparchus had different explanations of retrograde motion. Which explanation better satisfies the principle of simplicity? Explain.

5. Discuss the philosophical outlook of Johannes Kepler. In what ways was he like the followers of Plato and Aristotle, and in what ways was he like a modern scientist?

6. What would be the period of a planet located 3 AU from the Sun? What would be the semimajor axis of a planet whose period was 11.18 years?

7. Contrast the explanation Newton would have given for the fact that a moving object stops unless you keep pushing it, with the explanation that Aristotle would have given.

8. Why does the law of inertia (Newton's first law) imply that there must be a force of attraction (rather than repulsion) between each planet and the Sun?

9. Jupiter is about 5 times farther from the Sun than the Earth is, and has about 300 times the mass of the Earth. Compare the gravitational force between Jupiter and the Sun with that between the Earth and the Sun. Perform the same calculation for Saturn, which is about 10 times farther from the Sun than the Earth is, and has about 100 times the mass of the Earth.

10. Explain why an astronaut in an orbiting spacecraft is "weightless." Is there really no force of gravity?

11. Why did Kepler not discover that his third law depends on the sum of the masses of the Sun and each planet?

ADDITIONAL READINGS

The references listed here are primarily sources of biographical and historical data on the people discussed in this chapter; many of them contain additional lists of references. Readings on the principles of physics described in this chapter can be found most easily in elementary physics texts, which are available at all levels ranging from completely non-mathematical to any degree of mathematical sophistication desired.

Beer, A., and P. Beer, eds. 1975. *Kepler. Vistas in astronomy.* Vol. 18. New York: Pergamon Press.

Christianson, G. E. 1987. Newton's *Principia:* A Retrospective. *Sky & Telescope* 74(1):8.

Christianson, G. E., and K. A. Strand, eds. 1975. *Copernicus. Vistas in astronomy.* Vol. 17 New York: Pergamon Press.

Cohen, I. B. 1960. Newton in light of recent scholarship. *Isis* 51:489.

———. 1974. Newton. In *Dictionary of scientific biography.* Vol. 10. C. G. Gillispie, ed. New York: Scribner's.

Drake, S. 1980. Newton's apple and Galileo's dialogue. *Scientific American* 243(2):150.

Draper, J. L. E. 1963. *Tycho Brahe: A picture of scientific inquiry in the sixteenth century.* New York: Dover.

Galileo, G. 1632. *Dialogue on the two chief world systems.* G. de Santillana, trans. Chicago: University of Chicago Press.

Gingerich, O. 1973. Copernicus and Tycho. *Scientific American* 229(6):86.

———. 1973. Kepler. In *Dictionary of scientific biography.* Vol. 7. C. G. Gillispie, ed. New York: Scribner's.

———. 1983. The Galileo affair. *Scientific American* 247(2):132.

Gingerich, O., ed. 1975. *The nature of scientific discovery.* Washington, D.C.: Smithsonian Institution Press.

Kuhn, T. 1957. *The Copernican revolution*. Cambridge, Mass.: Harvard University Press.

McCrea, A. W. 1991. Arthur Stanley Eddington. *Scientific American* 264(6):92.

Whiteside, D. T. 1962. The expanding world of Newtonian research. *History of Science* 1:15.

Will, C. M. 1990. Twilight time for the fifth force? *Sky and Telescope* 80(5):472.

ASTRONOMICAL ACTIVITY
Drawing an Ellipse

The properties of an ellipse can be described in many ways, including mathematical formulas that are not used in this text. But one definition of an ellipse that is used here lends itself to an easy method for drawing the figure. Recall that the total distance from any point on an ellipse to the two foci is always the same. In order to use this fact to draw an ellipse, you will need a board (a cork board will do nicely), two pins or nails, and some string and a pencil. Stick the two pins into the board to represent the two foci of the ellipse. Tie the ends of the string to the two pins, and then use the pencil to trace out an ellipse by keeping the string taut. (See the figure; note that it is easier to do this if you make a loop of string by tying the ends together and hanging it over the two pins. Then you can trace the full ellipse without getting the string tangled. The extra length of string connecting the two foci is fixed; thus the total distance from the foci to the ellipse also is constant.) The fixed length of string ensures that the total distance from any point on the ellipse to the two foci remains constant.

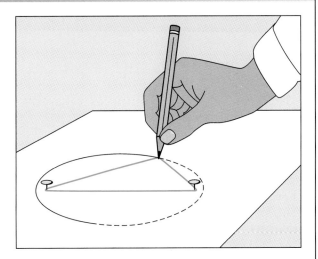

Now you can experiment by changing the length of the string relative to the distance between the foci (you can do this by changing the length of the string, by moving the pins closer together or farther apart, or by a combination of both). If the string is only a little longer than the distance between the foci, you get a very elongated, thin ellipse; if the string is very much longer than the distance between the foci, the ellipse is nearly a circle (think about what would happen if the two foci were moved so close together that they occupied the same spot; in that case, you would be drawing a precise circle).

The distance between the foci relative to the length of the major axis defines the *eccentricity* of an ellipse. In an intuitive sense, we say that a very elongated ellipse (with a major axis only a little longer than the separation between the foci) has a high eccentricity;

on the other hand, an ellipse whose major axis is much greater than the separation between its foci (i.e., an ellipse that is nearly circular) has a small eccentricity. Thus, the nearly circular orbits of the planets have small eccentricities (see Appendix 7), while the elongated orbits of comets (discussed in Chapter 9) have large eccentricities.

Mathematically, the eccentricity is defined as the ratio of the distance between the foci to the major axis; this may be written

$$e = F/A,$$

where e is the eccentricity, F is the distance between the foci, and A is the major axis. Note that eccentricities for closed, elliptical orbits always fall between 0 (for the circular case) and 1 (for the very elongated case).

Light and Telescopes

Stars over the Keck Telescope dome atop Mauna Kea (© 1990 Roger Ressmeyer—Starlight)

PREVIEW

Virtually everything we know about the universe beyond the Earth is based on our interpretation of the light that we receive from celestial objects. In this chapter we explore the nature of light and the processes by which it is emitted and absorbed. Like the laws of motion we presented previously, the principles discussed here will underlie the whole of our study of astronomy, and you should keep them in mind throughout the remainder of the course. The latter portion of the chapter describes the technology astronomers use to gather and analyze light from distant sources. The telescopes described here, both on the ground and in space, are the material monuments to our science, and they will help you to see how modern science depends on technology in order to advance.

Some of the tools for unlocking the secrets of the universe became available with the publication of Newton's *Principia* in the 1680s, but others had to wait two hundred years or more to be discovered. The laws of motion allowed astronomers to understand how the heavenly bodies move, and they were of fundamental importance in unraveling the clockwork mechanism of the solar system. To comprehend the essential nature of a distant object, however—to learn what it is made of and what its physical state is—one must understand what light is and how it is emitted and absorbed and must have tools to capture the light and analyze it. The only information we can obtain on the nature of a distant object is conveyed by the light that reaches us from it. Fortunately, an enormous amount of information is there, and astronomers have learned much about how to dig this information out.

THE ELECTROMAGNETIC SPECTRUM

One characteristic of light is that it acts like a wave (Fig. 4.1). It is possible to think of light as passing through space like ripples on a pond (although, as we will discuss shortly, the picture is actually somewhat more complicated than that). The distance from one wavecrest to the next, called the **wavelength,** distinguishes one color from another. Red light, for example, has a longer wavelength than blue light. It is possible to spread out the colors in order of wavelength, using a prism to obtain the traditional rainbow. Newton was the first to discover that sunlight contains all the colors, and he did so by carrying out experiments with a prism (Fig. 4.2). Whenever light is spread out by wavelength, the result is called a **spectrum**; that is, a spectrum is the arrangement of light from an object according to wavelength. The science of analyzing spectra is called **spectroscopy** and will be discussed later in this chapter.

The concept of **frequency** is often used as an alternative to wavelength in characterizing light waves. The frequency is the number of waves per second that pass a fixed point. It is determined by the wavelength and the speed with which the waves move. The speed of light, usually designated c, is constant, and the frequency of light with wavelength λ is $f = c/\lambda$. The standard unit for measuring frequency is the **hertz (Hz),** one hertz being equal to one wave per second. The frequency of visible light is typically about 10^{14} Hz.

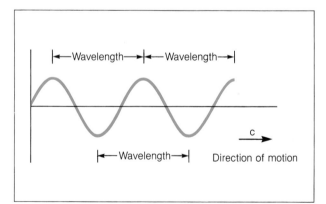

Figure 4.1   Properties of a wave.   Light can be envisioned as a wave moving through space at a constant speed, usually designated c. The distance from one wavecrest to the next is the wavelength, often denoted by the Greek letter lambda, λ. The frequency f is the number of wavecrests that pass a fixed point per second and is related to the wavelength and the speed of light by f = c/λ.

Figure 4.2   Prism dispersing light.   The rainbow of colors seen here is a spectrum. (Runk/Schoenberger, Grant Heilman Photography)

ASTRONOMICAL INSIGHT
The Perception of Light

Newton's experiments in optics are commonly thought to have marked the beginning of scientific inquiry into the nature of light, but in fact many scientists and philosophers preceded Newton in this area. Much of the early research, and even, to some extent, Newton's work, did not deal with the physical properties of light, but rather was aimed at understanding how the human eye senses light. The ancient Greeks, the earliest philosophers who studied the nature of light, did not recognize the distinction between light and the act of seeing it.

The Greeks studied the principles of vision in much the same way they studied astronomy: They performed very few experiments or even systematic observations, choosing instead to unravel the secrets of the universe by reason and logic. The prevalent concept, described by such notable figures as Hippocrates and Aristotle, was that the eye somehow emitted rays or beams with which it sensed things. Thus, the sense of sight was analogous to the sense of touch, where a hand is extended to feel objects. The Greeks knew that objects appeared distorted or bent when viewed under water, but they attributed this effect to bending of the rays that came from the eye, not to distortion of something coming from the object being viewed. It would have been possible for the Greeks to experiment with optics by using natural crystals or eyes removed from deal animals, but no such experimentation was done.

In a definitive summary of all that was known about vision, the Greek scientist Galen, in the second century A.D., wrote of an "animal spirit" that flowed from the brain along the optic nerve to the eye, where it was converted into a "visual spirit" in the retina, a membrane covering the rear interior portion of the eyeball. The lens, in the front of the eyeball, was thought to be responsible for sending out the beam with which the external world was perceived.

Like the astronomical work of Ptolemy, the concepts of the nature of light and vision that were summarized by Galen were accepted as doctrine for some fifteen centuries, until the time of the Renaissance. Some of the ancient Greek ideas, such as the notion that the optic nerve was hollow so that the animal spirit could flow along it from the brain, were especially persistent. Even Leonardo da Vinci, who carried out pioneering work in anatomy in the late fifteenth and early sixteenth centuries, accepted much of the old thinking. Da Vinci took a more modern view in that he believed that light rays passing from an object to the eye play an important role, but he was unwilling to entirely discard the idea that other rays emanate outward from the eye.

In the early seventeenth century, none other than Johannes Kepler and the great French philosopher René Descartes developed an understanding of refraction and the formation of an image by a lens. Both correctly viewed vision as the process of image formation on the retina by the lens in the front of the eyeball. Thus, when Christiaan Huygens and then Newton performed their experiments later in the seventeenth century, it was already well established that seeing was accomplished by sensing something that passes from the object to the eye.

Both men were more concerned with the nature of this radiation than with the properties of the eye, although Newton did discuss the perception of color, which he realized to be entirely a function of the eye. This is an important concept: The rays of light are not colored, nor, technically, is the object being viewed. It is the eye that senses wavelength differences and the brain that interprets them as colors. Light rays consist simply of alternating electric and magnetic fields.

Other advances in understanding the physical properties of light, made by scientists who followed Newton, are described in the text. Research has also been aimed at learning how the human eye perceives light. We now know that the eye responds logarithmically, meaning that we see differences in brightness in a way that is not directly proportional to the actual intensity ratios of the observed objects. This point is explained more fully in Chapter 12, in the discussion of the stellar magnitude system.

Entire books have been written on the complex interplay between what the brain perceives and physical reality, and, of course, this interplay is important in astronomy. To a large extent, modern observational astronomy is the science of recognizing and bypassing the limitations or distortions created by our natural light-gathering system.

In some situations, light is more like a stream of particles than a wave. Newton developed a "corpuscular" theory of radiation in which he assumed that light consisted of particles, but at the same time others, including most notably Christiaan Huygens, carried out experiments showing that light has definite wave characteristics. There are good arguments for both points of view; for example, the manner in which light waves seem to bend as they pass obstacles and the way in which they interfere with each other are wave characteristics. On the other hand, light waves can carry only discrete, fixed quantities of energy and can travel in a complete vacuum rather than requiring a medium, which are properties of particles.

Out of a variety of seemingly contradictory evidence has developed the concept of the **photon.** A photon is thought of as a particle of light that has a wavelength associated with it. The wavelength and the amount of energy contained in the photon are intimately linked; in general terms, the longer the wavelength, the lower the energy. Thus a red photon carries less energy than a blue one. Mathematically the energy can be expressed as $E = hc/\lambda$, where h is the Planck constant, c is the speed of light, and λ is the wavelength. It is important to understand that a photon carries a precise amount of energy, not some arbitrary or random quantity, and that when light strikes a surface, this energy arrives in discrete bundles like bullets, rather than as a steady stream. When a photon is absorbed, this energy can be converted into other forms, such as heat.

Let us consider for a moment what lies beyond red at one end of the spectrum or violet at the other end. By the mid-1800s, experiments had demonstrated that there is invisible radiation from the Sun at both ends of the spectrum. At long wavelengths, beyond red, is **infrared** radiation, and at short wavelengths is **ultraviolet** radiation. Later it was shown that the spectrum continues in both directions, without limit. Going toward long wavelengths, after infrared light come **microwaves** and **radio** waves; going toward short wavelengths, after ultraviolet come **X rays** and then **gamma (γ) rays**. All of these different kinds of radiation are just different forms of light, distinguished only by their wavelengths, and together they form the **electromagnetic spectrum** (Fig. 4.3). Electromagnetic radiation is a general term for all forms of light, whether it is visible light, X rays, radio waves, or anything else.

The reason for the name *electromagnetic radiation* is that the wave motion associated with this radiation consists of alternating electric and magnetic fields (Fig. 4.4). These fields propagate through space (or a medium) as they alternate, with the electric and magnetic fields always lying in planes that are perpendicular to each other. The speed of propagation in a vacuum is always the same (almost exactly 300,000 kilometers per second, or about 186,000 miles per second), but it is slightly slower in a medium such as air or glass. The electric and magnetic fields are said to be *in phase* with each other, meaning that both reach their maximum and minimum values

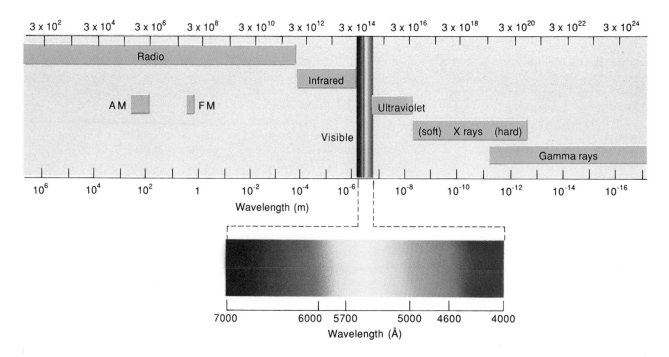

Figure 4.3   The electromagnetic spectrum.   All of the indicated forms of radiation are identical except for wavelength and frequency.

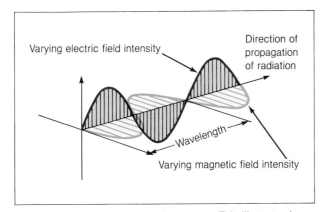

Varying electric field intensity

Direction of propagation of radiation

Wavelength

Varying magnetic field intensity

Figure 4.4   An electromagnetic wave.   This illustrates how the electric and magnetic fields oscillate in an electromagnetic wave. The directions of the fields and the direction of wave travel always have the relative orientations shown here.

at the same time as the wave propagates, with the relative orientation shown in Figure 4.4.

The light from a typical source, such as a light bulb or a distant star, consists of vast numbers of photons, each consisting of traveling electric and magnetic fields. Each photon has a characteristic orientation as it travels through space; that is, the planes in which its electric and magnetic fields oscillate remain fixed as the photon travels. Normally the orientations in a collection of photons from a source are random, but in some circumstances, they are not. When light is **polarized**, all the photons tend to have electric and magnetic planes with the same alignment. It is possible to determine from observations whether the light from a source is polarized. As we will learn in later chapters, certain astronomical objects produce polarized light, a fact that provides useful information on the nature of these objects.

Astronomers tend to identify different portions of the electromagnetic spectrum by specialized names

such as *infrared* or *ultraviolet,* but we must keep in mind that all are the same, consisting of alternating electric and magnetic fields and differing only in wavelength (or frequency).

The range of wavelengths from one end of the electromagnetic spectrum to the other is immense (Table 4.1). Visible light has wavelengths ranging from 0.0000004 to 0.0000007 m. In the case of light, a special unit called the **angstrom** is used, defined such that 1 m = 10,000,000,000 Å, or 1 Å = 0.0000000001 m = 10^{-10} m. Thus, visible light lies between 4000 Å and 7000 Å in wavelength. Recently physicists and astronomers have begun to use the *nanometer* (nm), which is equal to 10^{-9} m. In terms of this unit, visible light lies between 400 and 700 nm.

Infrared light has wavelengths between 7000 Å and a few million Å; that is, between 7×10^{-7} m and 2 or 3×10^{-4} m. Microwave radiation (which includes radar wavelengths) lies roughly between .001 and 0.5 m; no well-defined boundary separates this region from radio waves, which simply include all longer wavelengths, up to many meters or even kilometers. At the other end of the spectrum, ultraviolet light is usually considered to lie between 100 Å and 4000 Å, while X rays are in the range of 1–100 Å, anything shorter than that being considered gamma rays.

CONTINUOUS RADIATION

Researchers who followed Newton discovered that the Sun's spectrum contains a number of dark lines, each one corresponding to a particular wavelength. These **spectral lines** provide a great deal of information about a source of light such as a star, as does the **continuous radiation** (Figs. 4.5 and 4.6), the smooth distribution of light as a function of wavelength. Here we discuss continuous radiation, and in the following sections the spectral lines.

TABLE 4.1   Wavelengths and Frequencies of Electromagnetic Radiation

Type of Radiation	Wavelength Range[a]	Frequency Range
Gamma rays	$< 10^{-10}$	$>3 \times 10^{18}$ Hz
X rays	$1–200 \times 10^{-10}$	$1.5 \times 10^{16}–3 \times 10^{18}$
Extreme ultraviolet	$200–900 \times 10^{-10}$	$3.3 \times 10^{15}–1.5 \times 10^{16}$
Ultraviolet	$900–4000 \times 10^{-10}$	$7.5 \times 10^{14}–3.3 \times 10^{15}$
Visible	$4000–7000 \times 10^{-10}$	$4.3 \times 10^{14}–7.5 \times 10^{14}$
Near infrared	$0.7–20 \times 10^{-6}$	$1.5 \times 10^{13}–4.3 \times 10^{14}$
Far infrared	$20–100 \times 10^{-6}$	$3.0 \times 10^{12}–1.5 \times 10^{13}$
Radio	> 0.0001	$< 3 \times 10^{12}$
(Radar)	$(.02–0.2)$	$(1.5–15 \times 10^{9})$
(FM radio)	$2.5–3.5)$	$(85–120 \times 10^{6})$
(AM radio)	$(180–380)$	$(540–1600 \times 10^{3})$

[a]All wavelengths are given in meters; recall that 1×10^{-10} m = 1 angstrom; and 1×10^{-6} m = 1 micron.

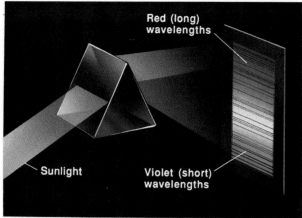

Figure 4.5   Continuous spectrum and spectral lines.
When sunlight is dispersed by a prism, the light forms a smooth rainbow of continuous radiation; one color gradually merges into the next, with a maximum intensity in the yellow portion of the spectrum. Superimposed on this continuous spectrum are numerous spectral lines, wavelengths at which little or no light escapes the Sun.

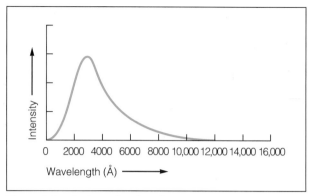

Figure 4.6   An intensity plot of a continuous spectrum.
This kind of diagram shows graphically how the brightness of a glowing object varies with wavelength. The curve shown roughly represents a star of surface temperature 10,000 K, whose continuous radiation peaks near 3000 Å in the ultraviolet.

Any object with a temperature above absolute zero emits radiation over a broad range of wavelengths, simply by virtue of the fact that it has a temperature. Thus, not only stars, but also such commonplace objects as the walls of a room or a human body, emit radiation. The continuous radiation produced because of an object's temperature is called **thermal radiation,** and in this chapter we will restrict ourselves to this type of continuous radiation. In other chapters we will discuss nonthermal sources of radiation.

Glowing objects such as stars emit thermal radiation over a broad range of wavelengths (technically, in fact, at least some radiation is emitted at *all* wavelengths), with a peak in intensity at some particular wavelength. A simple relationship between the wavelength of maximum emission and the temperature of an object was discovered in 1893 by W. Wien and is now known as Wien's law:

> The wavelength of maximum emission is inversely proportional to the absolute temperature.

The hotter an object is, the shorter the wavelength of peak emission, and the cooler it is, the longer its wavelength of maximum emission (Figs. 4.7 and 4.8). Thus, the variety of stellar colors is due to a range in stellar temperatures (Fig. 4.9). A hot star emits most of its radiation at relatively short wavelengths and thus appears bluish in color, whereas a cool star emits most strongly at longer wavelengths and appears red in color. The Sun is intermediate in temperature and in color.

When speaking of the colors of stars, we must keep in mind that a star emits light over a broad range of wavelengths, so we do not have pure red or pure blue stars. Our eyes receive light of all colors, and stars therefore are all essentially white. Our impression of color arises from the fact that there is a wavelength (given by Wien's law) at which a star emits more strongly than at other wavelengths. The star does not emit *only* at that wavelength.

A second property of glowing objects, known as either Stefan's law or the Stefan-Boltzmann law, has to do with the total amount of energy emitted over all wavelengths, and how this total energy is related to the temperature of an object:

> The total energy radiated per square meter of surface area is proportional to the fourth power of the temperature.

This shows that the total energy emitted is very sensitive to the temperature; if you change the temperature by a little, you change the energy by a lot. If, for example, you double the temperature of an object (such as the electric burner on your stove), you increase the total energy it radiates by $2^4 = 2 \times 2 \times 2 \times 2 = 16$. If one star is three times hotter than another, it emits $3^4 = 3 \times 3 \times 3 \times 3 = 81$ times more energy per square meter of surface area.

Notice that we have been careful to express this law in terms of energy emitted per square meter of surface area. The *total* energy emitted by an object therefore also depends on how much surface area it has. If we talk of stars or other spherical objects, which have a surface area of $4\pi R^2$, where R is the radius, then we can say that the total energy emitted is proportional to the fourth power of the temperature and to the square of the radius. This can be illustrated by considering two stars, one of which is

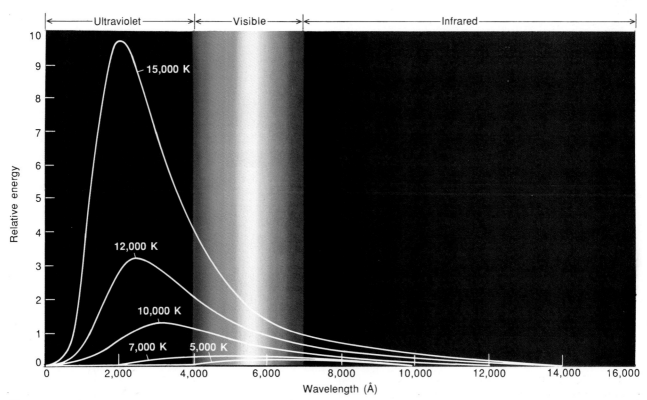

Figure 4.7   Continuous spectra for objects of different temperatures.   This diagram illustrates Wien's law, which says that the wavelength of maximum emission is inversely proportional to the temperature (on the absolute scale). It also illustrates another property of thermal emission: The hotter of the two objects is brighter at *all* wavelengths.

twice as hot but has only half the radius of the other. The hotter star emits $2^4 = 16$ times more energy per square meter of surface but has only $\frac{1}{2}^2 = \frac{1}{4}$ as much surface area; hence it is $16 \times \frac{1}{4}$ or 4 times brighter overall. If, on the other hand, this star were twice as hot and three times as large in radius as the other, it would be $2^4 \times 3^2 = 144$ times brighter.

Both Wien's law and the Stefan-Boltzmann law were first derived experimentally, in much the same manner as Kepler discovered the laws of planetary motion. In the case of planetary motions, it remained for Newton to find the underlying reasons for the laws, and he was able to derive them strictly on a theoretical basis. Analogously, Max Planck, the great German physicist who was active early in this century, found a theoretical understanding of thermal emission and was able to derive Wien's law and the Stefan-Boltzmann law from purely theoretical considerations. The basis of Planck's new understanding was the **quantum** nature of light; that is, light has a particle nature and carries only discrete, fixed amounts of energy. When characterizing the energy emitted by a star, astronomers often speak of the **luminosity,** which is the energy emitted per second. Luminosity is equivalent to **power** (discussed in Chapter 3).

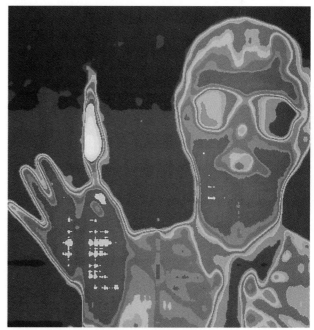

Figure 4.8   Giving Wien's Law a human touch.   This is an infrared image of a person holding a match. Colors are used here to represent different temperatures: white is the hottest (the flame); deep red, the warmest portions of the person (the palm of the hand and the chin area); and blue the coolest regions (eyeglass and necktie). (NASA/JPL)

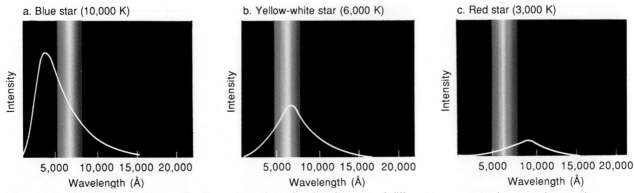

Figure 4.9   The colors of stars.   The figure shows intensity plots for stars of different temperatures in comparison to the colors perceived by the human eye. The variation in peak emission illustrates why stars have a range of colors as seen by the eye.

Finally, in addition to taking into account the temperature and size of a glowing object, we must consider the effect of its distance. So far we have discussed the energy as it is emitted at the surface, but not how bright it looks from afar. What we actually observe, of course, is affected by our distance from the object. For a spherical object that emits in all directions, the **inverse square law** tells us that the brightness decreases as the square of the distance (Fig. 4.10). Thus, if we double our distance from a source of radiation, it will appear $1/2^2 = 1/4$ as bright. If we approach the object, reducing the distance by a factor of two, it will appear $2^2 = 4$ times brighter. As we will see in later discussions of stellar properties, we must know something about the distances to stars before we can compare other properties connected with their brightness.

The Atom and Spectral Lines

The first detailed cataloging of the Sun's spectral lines was done in the early 1800s by Joseph Fraunhofer,

and the lines are known to this day as **Fraunhofer lines** (Fig. 4.11). In the late 1850s the German scientists R. Bunsen and G. Kirchhoff performed experiments and developed theories that made clear the importance of the Fraunhofer lines. Bunsen observed the spectra of flames created by burning various substances and found that each chemical element produces light only at specific places in the spectrum (Fig. 4.12). The spectrum of such a flame in this case is dark everywhere except at these specific places, as though the flame were emitting light only at certain wavelengths. The bright lines seen in this situation are called **emission lines** for that reason.

It was soon noticed that some of the dark lines in the Sun's spectrum coincide exactly in position with some of the bright lines seen by Bunsen in his laboratory experiments. Kirchhoff studied these lines in detail and was able to show that a number of common elements such as hydrogen, iron, sodium, and magnesium must be present in the Sun because of the coincidence in wavelengths of the lines. This was the first hint that the chemical composition of a distant

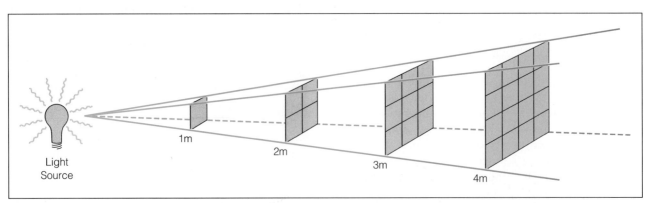

Figure 4.10   The inverse square law of light propagation.   This diagram shows how the same total amount of radiation energy must illuminate an ever-increasing area with increasing distance from the light source. The area to be covered increases as the square of the distance; hence, the intensity of light per unit of area decreases as the square of the distance.

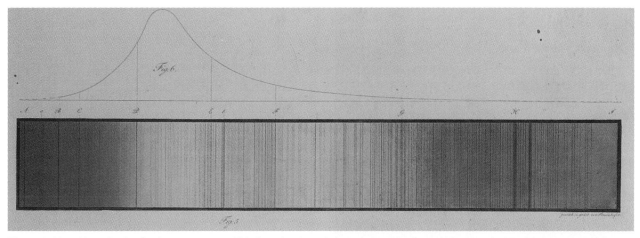

Figure 4.11   Lines in the Sun's spectrum.   This is a drawing of the solar spectrum made by Fraunhofer, the first to systematically study the dark lines in the spectrum (now called Fraunhofer lines). The colors illustrate roughly the human-eye response to the wavelengths in the spectrum. The graph at the top shows the relative intensity of sunlight as a function of wavelength. (Deutsches Museum, Munich)

object could be determined.

Further studies of the spectral lines revealed various regularities in the arrangement of the lines from a given element. Although scientists suspected that these regularities must reflect some aspect of the structure of atoms, it was not until 1913 that the true relationship was discovered by the Danish physicist Niels Bohr. By that time it had been established that an atom consists of a small, dense nucleus surrounded by a cloud of negatively charged particles called electrons. The nucleus contains positively charged particles called protons and neutral particles called neutrons. In a normal atom, the number of protons in the nucleus is equal to the number of electrons orbiting it, and the overall electrical charge is zero. The electrons are held in orbit by electromagnetic forces, with laws of the science called **quantum**

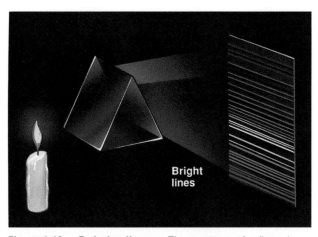

Figure 4.12   Emission lines.   The spectrum of a flame is devoid of light except at specific wavelengths where bright emission lines appear.

mechanics governing their motions. Bohr found that the electrons were responsible for the absorption and emission of light, and that they did so by gaining energy (absorption) or by losing it (emission) in the form of photons.

The key to the fixed pattern of spectral lines for each element lay in the fixed pattern of energy levels the electrons could be in (Fig. 4.13). We can visualize an atom as a miniature solar system, with electrons orbiting the nucleus. Each kind of atom (each element) has its own characteristic number of electrons, and in each case the electrons have a certain set of orbits. It may be helpful to visualize a ladder, with each rung representing an orbit, or energy level. The ladder for one element—hydrogen, for instance—has different spacings between the rungs than the ladder for some other element (Table 4.2). The energy associated with each level increases, the higher up the ladder, or the farther from the nucleus, the electron goes.

An electron can absorb a photon of light if the photon carries precisely the amount of energy needed to move the electron to some higher level than the one it is in. Returning to our ladder analogy, the electron can jump up only if it will land precisely on a higher rung; that is, it can only absorb a photon whose energy will boost it to another fixed energy level. Since the wavelength of a photon is determined by its energy, this means that an electron can absorb photons only with certain wavelengths. Because each kind of atom has its own unique set of energy levels, each has a unique set of wavelengths at which it can absorb photons.

Thus, each kind of atom has its own pattern of spectral lines. A spectral line is an absorption line when the electron receives energy from a photon and

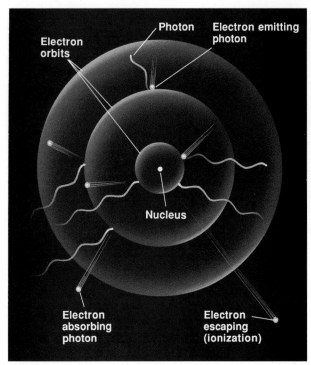

Figure 4.13   The formation of spectral lines.   An electron must be in one of several possible orbits, each representing a different energy level. If an electron absorbs a photon carrying precisely the amount of energy needed to jump to a higher (more outlying) orbit, it may do so. This is how absorption lines are formed: the wavelength of the photon absorbed corresponds to the energy difference between the two electron orbits, and that difference is fixed for any given kind of atom. Conversely, when an electron drops to a lower orbit, it emits a photon whose wavelength corresponds to the energy difference between the orbits. Emission lines are formed in this way. Note that it is possible for an electron to gain sufficient energy to escape from the atom altogether. This process is called ionization and is discussed in the next section.

TABLE 4.2   Principal Lines in the Spectrum of Hydrogen

Series	Lower Level	Line	Wavelength (Å)
Lyman (ultraviolet)	1	Alpha	1,216
		Beta	1,025
		Gamma	972
		Delta	949
		Epsilon	937
		(Limit)	(912)
Balmer (visible)	2	Alpha	6,563
		Beta	4,861
		Gamma	4,340
		Delta	4,101
		Epsilon	3,970
		(Limit)	(3,646)
Paschen (infrared)	3	Alpha	18,751
		Beta	12,818
		Gamma	10,938
		Delta	10,049
		Epsilon	9,545
		(Limit)	(8,204)
Brackett (infrared)	4	Alpha	40,512
		Beta	26,252
		Gamma	21,656
		Delta	19,445
		Epsilon	18,175
		(Limit)	(14,585)

and their lines are all blended together, and we see a continuous spectrum.[1]

2. In a hot, rarefied gas, the electrons tend to be in high energy states and create emission lines as they drop to lower levels.

3. In a relatively cool gas in front of a hot continuous source of light, the electrons tend to be in low energy levels and absorb radiation from the background continuous source.

Deriving Information from Spectra

A great wealth of information is stored in the spectrum of an object. From the analysis of the spectral lines, we can learn such things as the temperature and density of a gas and the velocity of a glowing object with respect to the Earth.

One important property of a gas is its degree of **ionization**. Ionization refers to the loss of one or more electrons from an atom; in this case, the electrons receive so much energy that they jump entirely free of the atom (Fig. 4.15; see also Fig. 4.13). The

jumps to a higher level and an emission line when an electron drops from a high level to a lower one, releasing a photon. For a given atom, the absorption and emission lines occur at the same wavelengths because the spacing of the energy levels determines the wavelengths. Whether the lines are emission or absorption depends simply on whether the electrons are dropping in energy level, giving off photons, or climbing in energy level, absorbing photons.

We can now state three rules of spectroscopy, first discovered experimentally by Kirchhoff (Fig. 4.14) and later put in the context of atomic structure:

1. In a hot, dense gas or a hot solid, the atoms are crowded together so that their energy levels overlap

[1]The production of a continuous spectrum is actually a bit more complex than this. When free electrons combine with ions, they emit at any wavelength, not just in spectral lines. See the discussion of ionization in the following section.

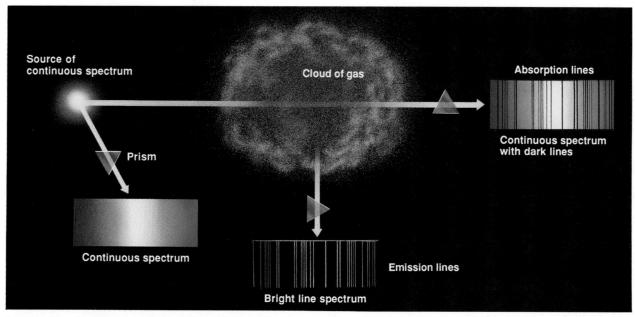

Figure 4.14   Continuous, emission-line, and absorption-line spectra.   The positions of the emission and absorption lines match because the same element emits or absorbs at the same wavelengths. Whether it emits or absorbs depends on the physical conditions, as described by Kirchhoff's rules.

gain in energy that frees an electron can come either from the absorption of a photon with energy exceeding that of the highest electron energy level associated with the atom or from a collision between atoms. In either case, the remaining atom, having lost one or more negative charges, has a net positive electrical charge and is called an **ion**. The likelihood of collisional ionization depends on the temperature of the gas, since the speed of collision depends on temperature. The hotter the gas, the more violent the collisions between atoms, and the greater the degree of ionization.

The spectrum of an atom changes drastically when it has been ionized because the arrangement of energy levels is altered and different electrons are now available to absorb and emit photons. The spectrum of atomic helium, for example, is quite different from that of ionized helium, so the astronomer not only can see that helium is present in the spectrum of a star, but also can tell whether it is ionized or not. This provides information on the temperature in the outer layers of the star where the absorption lines are formed. By analyzing the degree of ionization of all the elements seen in the spectrum of a star, it is possible to determine the gas temperature quite precisely.

When an electron moves to an energy state above the lowest possible one, it is said to be in an excited state (Fig. 4.15). This can happen as the result of the absorption of a photon of light with appropriate energy or as the result of a collision that is not energetic

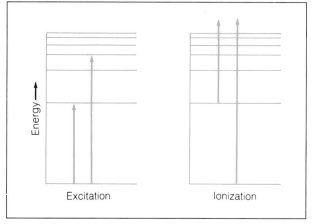

Figure 4.15   Excitation and ionization.   In the energy-level diagram at the left, electrons are jumping up from the lowest level to higher ones. This process is called excitation, and it can be caused either by the absorption of photons, or by collisions between atoms or between atoms and the free electrons. On the right, electrons are gaining sufficient energy to escape altogether, a process called ionization.

enough to cause ionization. The degree of excitation of a gas affects its spectrum because the spectral lines created in the gas depend on which energy levels the electrons are in. A hydrogen atom with its electron in the first excited level can produce absorption lines corresponding to the energy separation between this excited level and higher levels, whereas a hydrogen atom with its electron in the lowest state can only

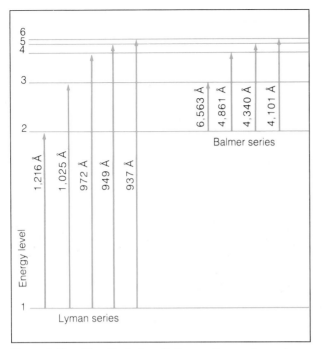

Figure 4.16   The effect of excitation on the spectrum of hydrogen.   As shown here, the absorption lines that can be formed by an atom depend on its degree of excitation. At the left are the wavelengths of absorption lines originating in the lowest energy level of hydrogen, while at the right are the wavelengths of absorption lines arising from an electron in the first excited level. Note that the Lyman lines are in ultraviolet wavelengths, while the Balmer lines are in the visible portion of the spectrum. This sketch shows only a few of the many energy levels of hydrogen.

produce absorption lines corresponding to transitions of the electron out of that state (Fig. 4.16). Therefore, analysis of the spectrum of a gas can tell us how highly excited the gas is, which in turn provides information on other properties such as density (the density affects the frequency of collisions between atoms, which in turn determines how many electrons are in excited levels).

From careful analysis of a star's spectrum, then, it is possible to determine both the temperature of the outer layers of the star (from the ionization) and the density (from the excitation). While most of the examples used in this discussion have referred to the Sun or the stars, the same techniques can be applied to the spectra of the planets. Because the visible light from a planet is reflected sunlight, however, its spectrum is the same as the Sun's spectrum, with some additional features contributed by the gas in the planet's atmosphere. These extra spectral lines must be identified and analyzed to derive properties of this gas.

Spectral lines can tell us something else about a distant object in addition to all the physical data that

we have been discussing. We can also learn how rapidly an object such as a star or a planet is moving toward or away from us.

When a source of light such as a star is approaching, the lines in its spectrum are all shifted toward shorter wavelengths than if the light source were at rest, and if the source is moving away from the observer, the lines are all shifted toward longer wavelengths (Fig. 4.17). These two cases are called **blueshift** (approach) and **redshift** (recession), respectively, because the spectral lines are shifted toward either the blue or the red end of the spectrum.

A general term for any wavelength shift due to relative motion between source and observer is **Doppler shift,** in honor of the Austrian physicist who first explored the properties of such shifts. The effect applies to other kinds of waves than light. Most of us, for example, have noticed the Doppler shift in sound waves when a source of sound passes by. The whistle on an approaching train suddenly changes to a lower pitch at the moment the train passes us, because the wavelength we receive suddenly shifts to a longer one (and the frequency drops).

In the case of the Doppler shift of light, it is possible to determine the speed with which the source of light is approaching or receding from the simple formula:

$$v = \left(\frac{\Delta\lambda}{\lambda}\right)c,$$

where v is the relative velocity between source and observer, $\Delta\lambda$ is the shift in wavelength (the observed wavelength minus the rest, or laboratory, wavelength of the same line), λ is the laboratory wavelength of the line, and c is the speed of light. If the observed wavelength is greater than the rest wavelength, we find a positive velocity, corresponding to a redshift; if the observed wavelength is less than the rest wavelength, the result is a negative velocity, indicating a blueshift.

It is important to notice that the Doppler shift tells us only about relative motion between the source and the observer: It is not possible to distinguish whether it is the star or the Earth that is moving or a combination of the two (which is most likely the case). It is also important to keep in mind that the Doppler shift tells us only about motion directly toward or away from the Earth, called the **radial velocity.** There is no Doppler shift due to motion perpendicular to our line of sight. If a star is moving with respect to the Earth at some intermediate angle, as is usually the case, then we can determine the part of its velocity that is directed straight toward or away from us (i.e., the radial velocity), but we cannot determine its true direction of motion or its speed transverse to our line of sight (Fig. 4.18).

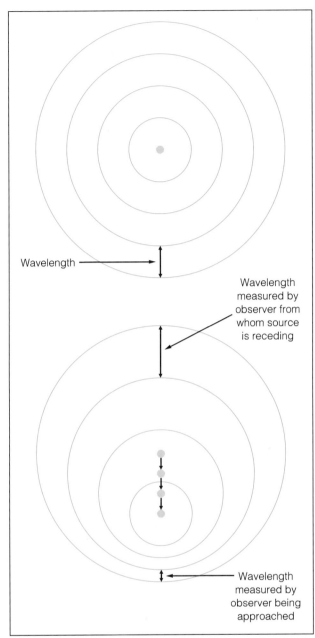

Figure 4.17 The Doppler effect. The light waves from a stationary source (upper) remain at constant separation (that is, constant wavelength) in all directions, whereas those from a moving source get "bunched up" in the forward direction and "stretched out" in the trailing direction (lower). This causes a blueshift or a redshift for an observer who approaches or recedes, respectively, from a source of light. Note that it does not matter which is moving, the source or the observer.

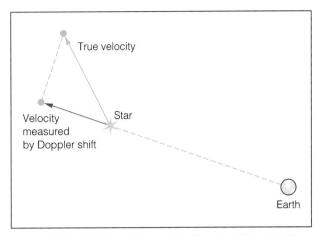

Figure 4.18 Radial velocity. The Doppler shift only indicates motion along the line of sight, called the radial velocity. Here a star is moving at an angle to the line of sight, so the Doppler shift measures only a fraction of its true velocity.

collecting light. The universe is filled with radiation of all wavelengths, so our discussion will necessarily include telescopes designed for all forms of radiation, not just visible light.

The Need for Telescopes

There are several basic benefits of using telescopes. For one thing, telescopes collect light from a large area and bring it to a focus, so that fainter objects than the eye could see unaided can be observed. The human eye has a collecting area only a tiny fraction of a meter in diameter, whereas the largest telescopes are 4 to 6 meters in diameter. The light-gathering power of a telescope depends on the area of its collecting surface; since the area of a circle of radius r is equal to πr^2, a 2-meter diameter telescope collects 4 times as much light as a 1-meter telescope, for example.

Another basic advantage of telescopes over the unaided eye is that the telescope can be equipped to record light over a long period of time by using photographic film or an electronic detector, while the eye has no capability to store light. A long-exposure photograph taken through a telescope reveals objects too faint to be seen with the eye, even by looking through the same telescope.

By combining both large size and the capability of making long exposures, the largest telescopes can detect objects more than a billion times fainter than the unaided eye can see, even under the best conditions.

A third major advantage of large telescopes is that they have superior angular **resolution,** the ability to discern fine detail. For visible-wavelength telescopes, the Earth's atmosphere creates practical limitations

TELESCOPES: TOOLS FOR COLLECTING LIGHT

Now that we know how to analyze light from distant objects, we can turn our attention to the methods for

ASTRONOMICAL INSIGHT
A Night at the Observatory

An astronomer's typical night at the observatory will vary quite a bit from one individual to another and especially from one observatory to another. There is a big difference between sitting in a heated room viewing an array of electronic panels and standing in a cold dome peering into an eyepiece while recording data on a photographic plate. Yet both extremes, and various possibilities in between, are part of modern observational astronomy.

Despite apparent differences in practice, however, the principles are much the same regardless of where the observations take place, or with what equipment. Here is a description of a typical night at a large observatory, where a modern electronic detector is used to record spectra of stars.

By midafternoon the astronomer and his or her assistants (an engineer who knows the detector and its workings and a graduate student) are in the dome, making sure the detector is working properly. They run various tests of the electronic instruments and record test spectra from special calibration lamps. Everything is in order, and they eat supper in the observatory dining hall a couple of hours before sunset, returning to the dome to complete preparations for the night's work. A night assistant (an observatory employee) arrives shortly before dark and is given a summary of the plans for the night. It is the night assistant's job to control the telescope and the dome, and he or she is responsible for the safety of the observatory equipment. At large observatories the telescope is so complex

and valuable that a specially trained person is always on hand, and the astronomer (who may use a particular telescope only a few nights a year) does not handle the controls except for fine motions required to maintain accurate pointing during the observations.

Before it is completely dark, a number of calibration exposures are taken, using lamps and even the Moon, if it is up. These measurements will help the astronomer analyze the data later, by providing information on the characteristics of the spectrograph and the detector.

Finally everything is in order, and if things are going smoothly, the telescope is pointed at the first target star just as the sky gets dark enough to begin work. Observing time on large telescopes is difficult to obtain and very valuable; not a moment is wasted.

A pattern is quickly established and repeated throughout the night. After the night assistant has pointed the telescope at the requested coordinates, the astronomer looks at the field of view to confirm which star is the correct one (often by consulting a chart prepared ahead of time). When the star is properly positioned so that its light enters the spectrograph, the observation begins. Depending on the brightness of the star and the efficiency of the spectrograph and the detector, the exposure time may be anywhere from seconds to hours. During this time, the astronomer or the graduate student assistant continually checks to see that the star is still properly positioned, making corrections as needed by pushing buttons on a remote control device.

If the observations are made with an instrument at the coudé focus,

the astronomer and assistants spend the entire night in an interior room in the dome building, emerging only occasionally to view the skies, give instructions to the night assistant, or go to the dining hall for lunch (usually served around midnight).

As dawn approaches and the sky begins to brighten, a last-minute strategy is devised to get the most out of the remaining time. Finally the last exposure is completed, and the night assistant is given the go-ahead to close the dome and park the telescope in its rest position (usually pointed straight overhead to minimize stress on the gears and support system). The astronomer and the assistants make final calibration exposures before shutting down the equipment and trudging off to get some sleep. The new day for them begins at noon or shortly thereafter.

The scenario just described depicts a typical night at a major modern observatory, such as Kitt Peak, which has extensive resources and large telescopes. At smaller facilities, such as a typical university observatory, it is more common to find the astronomer working alone, most often using a cassegrain-focus instrument. In such circumstances, the observer must spend the night out in the dome with the telescope (where it can be very cold in the winter) and be content with a bag lunch. The work pattern is similar, however, as is the desire not to waste any telescope time.

on how fine the resolution can be, but for radio telescopes and optical telescopes in space, the atmosphere is not a hindrance. The resolution of a telescope is normally expressed in terms of the smallest angular separation between a pair of objects that can be discerned by the telescope. Thus, small resolution is good. The resolution expressed in this way is proportional to the wavelength being observed and inversely proportional to the diameter of the telescope. For visible light (wavelength 5500 Å, for example), a telescope 0.1 meter in diameter (about 4 inches) has a resolution of about 1 arcsecond. (This is really quite good; an arcsecond is about the angular diameter of a dime seen at a distance of 2 miles!). The human eye, with a pupil diameter of about 0.001 m (1 millimeter), has a resolution of roughly 100 arcseconds, or a little over 1.5 arcminutes. In principle, a large telescope such as the 5-meter Palomar telescope has a resolution far smaller than 1 arcsecond, but, in practice, the Earth's atmosphere limits the resolution that is actually achieved. Normally it is impossible to separate objects closer than about 0.5 arcsecond when looking up through the turbulent atmosphere, and often this blurring effect, called "seeing," makes resolution much worse.

We can vastly improve resolution by using telescopes in pairs or groups. This technique, called **interferometry,** makes it possible to measure the precise direction toward a source of radiation by analyzing the **interference** of the waves arriving at separate telescopes. Interference refers to the fact that wavecrests and troughs can either add or cancel each other when waves overlap, depending on how the waves are shifted with respect to one another. The shift depends on the separation of the antennae and the direction of the source. Sources can be mapped with very high precision by using telescopes arranged in series along different alignments, as is done with the *Very Large Array* (Fig. 4.19), which can achieve resolutions of better than 0.1 arcsecond. Interferometry is obviously a very complex undertaking, but it has been successfully applied to radio astronomy, and there are now preliminary plans to do the same for infrared and perhaps visible light. (The shorter the wavelength being observed, the more instrumental precision is required in interferometry. The more widely separated the telescopes are, the better the resolution, and radio astronomers are today engaged in *Very Long Baseline Interferometry* (or simply *VLBI*), using telescopes separated by intercontinental distances.

One technical problem must be overcome in telescope construction: the Earth rotates, and if nothing is done to compensate for this, a star quickly moves out of the field of view. To avoid this problem, telescopes are mounted so that they can be moved by a motor in the direction opposite the Earth's rotation, keeping a target object centered in the field of view. Of course, these adjustments are not required for telescopes that are not attached to the Earth, such as the various observatories in space.

Principles of Telescope Design

To illustrate the general principles of telescope design, we begin by describing telescopes built for visible light. We will then see how those designed for other wavelengths have essentially the same features

Figure 4.19 The *Very Large Array.*   This arrangement of 27 radio dishes spread over a 25 km-wide section of New Mexico desert provides high-resolution radio maps through the use of interferometry. The individual telescopes can move along railroad tracks, allowing the separations to be adjusted, depending on the type of data needed. (National Radio Astronomy Observatory)

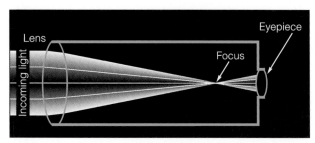

Figure 4.20   The refracting telescope.   The path of light is bent when it passes through a surface such as a glass lens. A properly shaped lens can thus bring parallel rays of light to a focus, where the image of a distant object may be examined. A second lens is used to magnify the image.

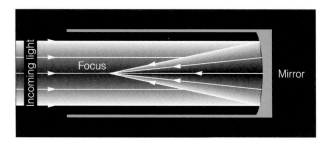

Figure 4.21   The reflecting telescope.   A properly shaped concave mirror can be used instead of a lens to bring light to a focus. Usually an additional mirror is used to reflect the image outside the telescope tube.

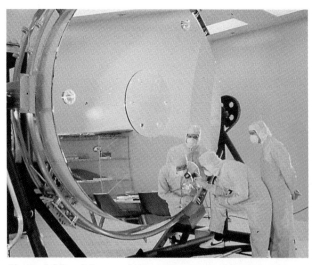

Figure 4.22   A large primary mirror.   As this photo shows, the proper shape for a telescope mirror is not very highly concave. In this case only the distorted reflections reveal that the mirror is not flat. This is the 2.4-meter primary mirror for the *Hubble Space Telescope*. (Perkin Elmer Corporation)

but differ in some details.

It is possible to construct a telescope, called a **refractor,** which focuses light with lenses (Fig. 4.20), and many of the earliest telescopes were of this type. Today nearly all telescopes are **reflectors,** which use mirrors to focus light. Reflectors can be built much larger and are not hampered by certain technical difficulties associated with lenses, such as the fact that a lens refracts different wavelengths of light at different angles.

In a reflector, light is focused by a concave mirror, called the **primary mirror** (Figs. 4.21 and 4.22). Because it is usually inconvenient to look at or record an image formed inside the telescope tube, most often a telescope has a secondary mirror that deflects the image outside the tube (Fig. 4.23). This can be a flat mirror, sending the image out the side (the so-called **Newtonian focus** arrangement); a convex mirror that reflects the image straight back out of a hole in the bottom of the telescope (the **cassegrain focus**); or series of flat mirrors that send the image to a remote

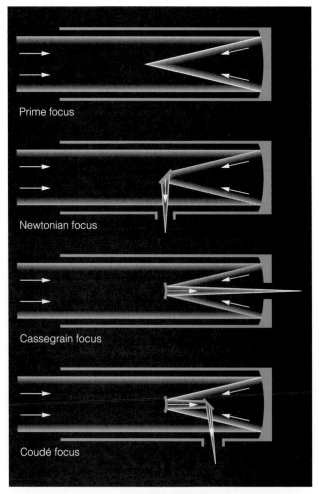

Figure 4.23   Various focal arrangements for reflecting telescopes.

location where heavy or large equipment can be used to analyze the light (the **coudé focus**). In very large telescopes, it is possible to observe at the **prime focus,** the point where the light is focused inside the telescope structure by the primary mirror. This has the advantage of reducing the number of reflections, so that light loss is minimized, but the disadvantage that the instrument used to collect and analyze the light cannot be very large or heavy. All of these focus arrangements have either a mirror or an instrument mounted inside the telescope, blocking some of the incoming light, but the percentage of the light that is lost this way is usually very small.

For many years the biggest telescope was the 5-meter reflector at Mt. Palomar in southern California (Fig. 4.24), but in the 1970s, a 6-meter telescope was completed in the Soviet Union (problems with the shape of the mirror, however, have prevented this telescope from taking full advantage of its size). Today there are several 3- to 5-meter telescopes (Fig. 4.23), and larger ones are under construction (discussed later in the chapter).

A very promising new kind of telescope was completed in 1979 at Mt. Hopkins, Arizona. This is the *Multiple-Mirror Telescope* (Fig. 4.25), which consists of six separate 1.8-meter primary mirrors with secondary mirrors arranged so that they each focus at the same point. The total collecting area of these six mirrors is equivalent to a single mirror 4.5 meters in diameter. This system is very complicated, and the difficulties of perfectly aligning the six mirrors are enormous, but the savings in cost and in the relative ease with which the smaller mirrors can be constructed made the effort worthwhile. (Interestingly, plans are now under way to replace the six small mirrors in the *Multiple-Mirror Telescope* with a single large one—6.5 meters in diameter!)

At the focus of any telescope (except for small ones used only for simple viewing of the sky) is an instrument that records light. This could be a simple camera, for making images of the sky, or a **photometer,** an instrument that measures intensities of light using a photocell (a device that converts light into an electrical current). A commonly used instrument is the **spectrograph,** which disperses light according to wavelength so that the spectrum can be recorded and analyzed. Film has traditionally been, and still is, used widely to record the light, but today many kinds

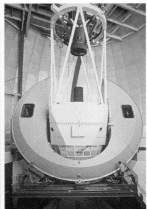

Figure 4.24 Large reflecting telescopes. Clockwise from upper left: the dome of the *Canada-France-Hawaii Telescope* on Mauna Kea; the 4-meter *Mayall Telescope* at Kitt Peak National Observatory; the 3.05-meter *Shane Telescope* at the Lick Observatory; and the 5-meter *Hale Telescope* on Palomar Mountain. (upper left: J. G. Timothy; upper right: National Optical Astronomy Observatories; lower right: Lick Observatory photograph; lower left: Palomar Observatory, California Institute of Technology)

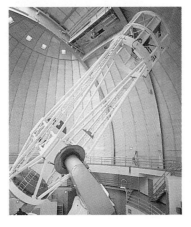

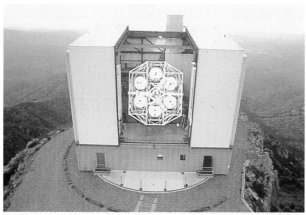

Figure 4.25   The *Multiple-Mirror Telescope*,   Mt. Hopkins, Arizona.   This observatory is operated by the University of Arizona and the Smithsonian Institute. Six 1.8-meter mirrors bring light to a common focus, giving the instrument the light-collecting power of a single 4.5-meter telescope. (Multiple-Mirror Telescope Observatory, University of Arizona and the Harvard-Smithsonian Center for Astrophysics)

of electronic **detectors** are being developed. These offer many advantages over film, both in accuracy and in such practical matters as storing the data electronically, so that they can be transmitted directly into computers for analysis. Auxiliary instruments used in astronomy range from rather compact, lightweight devices that can easily be mounted directly on the telescope for use at the prime or cassegrain focus to large, heavy instruments (usually spectrographs) that must be used at the coudé focus.

Telescopes for wavelengths other than visible light have the same basic components as the telescopes just described. There is an element such as a mirror for collecting and focusing light and an instrument at the focus for analyzing and recording the radiation. Telescopes for ultraviolet and infrared wavelengths are virtually identical in design to those for visible light, with a couple of exceptions: Ordinary mirrors do not reflect well in the ultraviolet, so special surfaces and chemical coatings have to be used; and infrared telescopes glow in the wavelengths they are built to observe, so the instrument must be cooled to reduce the glow (usually by circulating liquid nitrogen or liquid helium through the part of the instrument that contains the detector). Of course, ultraviolet light and large portions of the infrared spectrum do not penetrate the Earth's atmosphere (Fig. 4.26), so these telescopes have to be launched into space; therefore, the

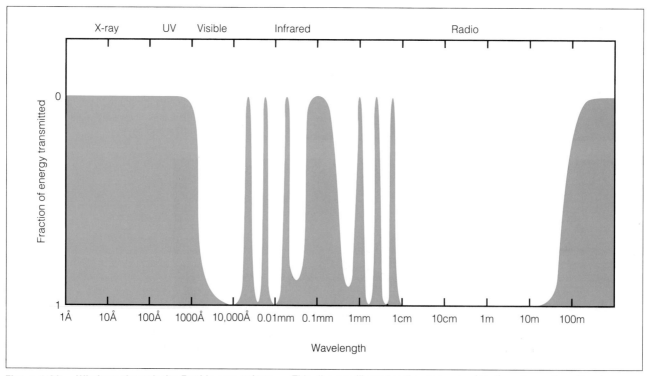

Figure 4.26   Windows through the Earth's atmosphere.   This diagram illustrates the many wavelength regions where the atmosphere blocks incoming radiation from space (shaded regions). The atmospheric "windows" allow ground-based observations only in visible and radio wavelengths, and some portions of the infrared spectrum. All other wavelengths must be observed from space.

TABLE 4.3   Major Telescopes in Space

Wavelength Region	Telescope[a]	Agency[b]	Diameter	Dates
Gamma ray	*Compton Observatory*	NASA		1991–
X-ray	*Einstein Observatory*	NASA	0.56 m	1978–1981
	Exosat	ESA		1982–1986
	Roentgen Satellite: ROSAT	German-NASA		1990–
	(Advanced X-Ray Astrophysics Telescope Facility: AXAF)	NASA		(1998)
	(X-Ray Multiple Mirror Mission)	ESA		(1998)
Extreme ultraviolet	*Extreme Ultraviolet Explorer: EUVE*	NASA		1992–
	(Far-Ultraviolet Spectroscopic Explorer: FUSE)	NASA-Canada	0.6	(2000)
Ultraviolet	*Copernicus*	NASA	0.9	1972–1980
	International Ultraviolet Explorer (IUE)	NASA-ESA-UK	0.41	1978–
	Hubble Space Telescope: HST	NASA-ESA	2.4	1990–
	(Far-Ultraviolet Spectroscopic Explorer: FUSE)	NASA-Canada	0.6	(2000)
Visible	*Hubble Space Telescope: HST*	NASA-ESA	2.4	1990–
	Hipparcos	ESA	0.29	1989–
Infrared	*Infrared Astronomical Satellite (IRAS)*	Dutch-UK-NASA	0.57	1983–1984
	(Shuttle Infrared Telescope Facility: SIRTF)	NASA		(2000 +)
	(Infrared Space Observatory: ISO)	ESA		(1994)
	(Large Deployable Reflector: LDR)	NASA	10	(?)
Radio	*Cosmic Background Explorer: COBE*	NASA		1989–
	(Orbital Very Long Baseline Interferometry: Space VLBI)	NASA		(1995)

[a]Instruments in parentheses are planned or under construction, but not yet launched.
[b]NASA is the U.S. National Aeronautics and Space Administration; ESA is the European Space Agency, a consortium of several countries.

fields of ultraviolet and infrared astronomy are relatively new, with most research occurring only in the past 30 years or so. Some existing and planned space observatories are listed in Table 4.3.

For wavelengths beyond infrared—that is, the radio portion of the spectrum—the radiation reaches the ground, and large Earth-based telescopes are used. Radio telescopes usually consist of very large dishes made of metal plating or a wire mesh, and they bring radiation to a focus at a point above the center of the dish (Fig. 4.27). A receiver, an electronic device that records the radio waves and turns them into electrical impulses, is suspended at the focus. Because the resolution of a telescope is proportional to the wavelength being observed, radio telescopes tend to have very poor resolution. Partly to compensate for this, they are built with gigantic dimensions, since resolution is also inversely proportional to telescope diameter. Even the largest radio dish, however, has relatively poor resolution compared with visible-light telescopes; the giant 300-meter Arecibo telescope (Fig. 4.28) can separate objects in the sky no closer together than about 3 arcminutes, whereas a 0.1-meter (4-inch) visible telescope has a resolution of about 1 arcsecond. As we discussed earlier, however, interferometry has been used very successfully with radio telescopes in pairs or groups, with the result

Figure 4.27   Radio telescopes at Green Bank, West Virginia.   This is one of the major facilities of the U.S. National Radio Astronomy Observatory. (National Radio Astronomy Observatory)

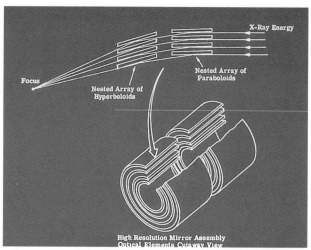

Figure 4.29   Design of an X-ray telescope.   This is a view of the nested x-ray reflections which formed the primary mirror assembly for the *Einstein Observatory,* a NASA spacecraft that operated in the early 1980s. (NASA/Harvard-Smithsonian Center for Astrophysics)

Figure 4.28   The giant Arecibo radio dish.   This is the largest single radio antenna in the world. The 300-meter dish is built into a natural bowl in the mountains of Puerto Rico. Since the telescope cannot be pointed in arbitrary directions, it must rely on the Earth's rotation to scan the sky. (Arecibo Observatory, part of the National Astronomy and Ionosphere Center operated by Cornell University under contract with the National Science Foundation)

that radio observatories have achieved much better resolution than visible-light telescopes.

For the shortest wavelengths, below ultraviolet, it is no longer possible to build a simple concave reflector to focus light. Extreme ultraviolet and X-ray radiation simply do not reflect well from any kind of mirror. These wavelengths can be reflected, however, if they strike a surface at a very oblique angle; this is called **grazing incidence,** and an X-ray telescope can be constructed using grazing-incidence reflections. Such a telescope has one or more ringlike primary mirrors at the front and can even have a secondary mirror to bring the radiation to a focus where a detector is located (Fig. 4.29). Like ultraviolet light, X rays do not penetrate the Earth's atmosphere, and observations must be made from space.

The shortest wavelength radiation, the gamma rays, are very difficult to focus. Even grazing-incidence reflections do not work for these extremely high energy photons, and the few crude gamma-ray telescopes built so far have consisted simply of detectors, with no primary light-gathering element to play the role of bringing the radiation to a focus from a large collecting area. A device called a **collimator** is placed in

front of the detector, so that only gamma rays from a certain direction are recorded; without this, there would be no way of knowing where the observed gamma rays came from. Again, gamma rays must be observed from space because they cannot pass through the Earth's atmosphere.

Major Observatories for All Wavelengths

Observatories for visible light, some portions of the infrared spectrum, and radio observations can be built on the Earth's surface because all these wavelengths reach the ground (see Fig. 4.26). Even so, most major observatories are located in remote regions, for a number of reasons. To minimize absorption of light by the atmosphere, it is desirable to locate an observatory on a high mountain, above as much of the atmosphere as possible. (This is especially important for infrared observations, since atmospheric water vapor absorbs much of the incoming infrared radiation.) Of course, a site with generally clear weather is usually imperative, for only radio telescopes can peer through clouds. It is also best to locate observatories well away from large cities, because the diffuse light from a densely populated area can drown out faint stars. Another consideration is the latitude of the site, which should not be too far north or south of the equator, since that would exclude a large portion of the sky.

Other factors that affect the quality of an observing site are the cleanliness of the air, for pollution can reduce the transparency of the atmosphere; and the amount of turbulence in the air above the site, because turbulence creates the twinkling effect known

ASTRONOMICAL INSIGHT
The Magic of Modern Detectors

At the heart of every astronomical instrument is a device that records the light that is brought to a focus by the telescope. Virtually no research in astronomy is done by the human eye alone. The light that reaches the focus of the telescope and its instrument, whether this is a camera, a spectrograph, a photometer, or some other device, must somehow be recorded so that detailed analysis can be done.

Until recently, the only technique known for recording light was to use photographic film. Film is remarkable for its ability to react to light and record the result of this reaction, thus creating images that can be examined and analyzed away from the telescope. Film is also capable of high resolution, meaning that it can record fine details in an image, and it can cover large areas as well (some photographic plates used in astronomy are as large as 25 centimeters square). But film also has some disadvantages: it is not very efficient, recording at best only 2 or 3 percent of the photons that strike it; it does not respond proportionally to varying light levels; and the range of brightness that can be recorded on a single piece of film is limited. Another important difficulty with film is that, in order to analyze images, it is usually necessary to store the information in a computer, and the process of converting photographic images into arrays of numbers suitable for computer analysis is very laborious and time-consuming.

The rise of modern electronic technology has enabled astronomers to develop alternatives to film. The simplest of these modern detectors are photoelectric cells, devices that produce an electrical current that is proportional to the intensity of the light that strikes them. Photocells, as they are commonly called, can record only the overall brightness of an object; they have no ability to record the variations of light from place to

place, creating an image. For this a detector must have many picture elements, or *pixels,* each capable of recording the intensity of light at its specific location.

Some early electronic detectors consisted of cameras similar to television cameras. While these offered some improvements over film, they retained some of the disadvantages as well. Their responses to different light levels tended to be non-proportional to the intensity of light, and they had very limited capability for simultaneously observing bright and faint objects in the same field of view.

The leading detectors in use today are *charge-coupled devices,* or *CCDs.* These combine virtually all of the desirable characteristics for astronomical observations. They are very efficient, recording up to 70 or 80 percent of the photons that strike them (even 100 percent is possible, for high-energy photons such as X rays). They respond precisely to widely varying light levels, so that the electrical signal generated is proportional to the intensity of light over a very large range, and they can simultaneously record the intensities of objects with very different brightnesses. Very small pixels can be made, so that the resolution, or ability to record fine details, rivals that of fine-grained film. Furthermore, CCDs can be used over an enormous range of wavelengths, all the way from the X-ray portion of the spectrum to the infrared. The one shortcoming of CCDs, as compared with film, is that they are not very large (usually less than 1 inch square), so that large fields of view cannot be recorded in a single exposure. Current development efforts are aimed at producing larger CCDs.

A CCD consists of layers of silicon with impurities that cause electrons to be freed when photons of light strike them. A CCD detector, or "chip," consists of a thin wafer of

silicon in which a pattern of electric fields (voltages) is embedded so that when a photon frees an electron, the electron is trapped at the location where the photon struck. Thus, as light reaches the CCD, a pattern of electrical charges builds up, corresponding to the intensity of light at each pixel. Then to record an image, it is necessary to "read out" the amount of electrical charge (i.e., the number of electrons) stored in each pixel.

The read-out procedure is accomplished by altering the voltages in adjacent rows of pixels, so that the stored electrons are transferred systematically from row to row. It is as though the image were cut into tiny strips, and then the strips were moved, one at a time, off to the edge of the picture. At the edge of the detector are electronic devices that measure the pattern of charges in each row and store this information numerically in a computer. To read out an entire image requires systematically marching all the rows of pixels off to the edge where their charges can be recorded. This process may seem very laborious, but it is possible to read out the entire detector very rapidly, even though the typical CCD in use today has up to a million individual pixels.

The use of CCDs has virtually revolutionized astronomy. Telescopes are now able to observe objects many times fainter than they could a few years ago, and to do so with greater accuracy as well. Because they are electronic devices that produce data in numerical form, CCDs are very well suited for space-based telescopes, which must be able to convert data into digital form to be radioed to the ground (the two cameras on the *Hubble Space Telescope* use CCDs, for example). While some other forms of electronic detectors are promising, CCDs appear likely to

continued on next page

dominate astronomical observations for some years to come.

CCDs are becoming readily accessible for casual or amateur astronomers, as the cost of small chips is going down (most modern video cameras use CCDs, but not of sufficient sensitivity or quality for astronomical purposes). Apart from cost, CCDs also require substantial computer equipment, which is needed both to operate the CCD and to record its output. Nevertheless ongoing advances in electronic technology, along with decreasing prices, may eventually bring about the day when every household, including that of the amateur astronomer, has several CCD-based devices.

to astronomers as "seeing," previously discussed. (Interestingly, it has recently been found that air currents inside the telescope or the dome building can contribute to poor seeing, so modern telescope design seeks to minimize these currents.

The largest observatories in the world tend to be located along the western portions of North and South America, in Hawaii, and in Australia; a fine site has also been developed in the Canary Islands by the British government (see Appendix 6). Through the National Science Foundation, the U.S. government supports large observatories in North America (Kitt Peak National Observatory, about 80 km west of Tucson, Arizona) and in South America (the Cerro Tololo Inter-American Observatory, on the crest of the Andes, some 300 km north of Santiago, Chile). The summit of Mauna Kea on the island of Hawaii is recognized as one of the finest sites in the world. Many of the most recently built major telescopes are there (Fig. 4.30), and several more are planned for the future. The telescopes on Mauna Kea represent several different international collaborations, as well as U.S. institutions such as the California Institute of Technology (Cal Tech), which has built a 10-meter radio telescope and (with the University of California) is building a pair of 10-meter visible-light telescopes; and NASA, which has a fine infrared telescope there.

The major radio observatories in the United States are located in Green Bank, West Virginia; near Socorro, New Mexico (these two sites are part of the *National Radio Astronomy Observatory*; the *Very Large Array* is near Socorro); and in the mountains of Arecibo, Puerto Rico (operated by Cornell University and the National Science Foundation, which also supplies major funding for the U.S. national observatories and several of the others that have been mentioned).

The majority of the ultraviolet, X-ray, gamma-ray, and far-infrared observatories built so far have been developed and launched into space by NASA (Table 4.3). Recently the European Space Agency (ESA), a consortium of several nations, has also been active in space astronomy. In ultraviolet astronomy the *International Ultraviolet Explorer* (a collaborative effort by NASA, ESA, and the United Kingdom), has been in successful operation for 14 years, and the *Hubble Space Telescope* has been in operation since 1990 (the

Figure 4.30   The Mauna Kea observatory.   This panorama shows the telescope domes on the summit of Mauna Kea. The large dome in the foreground houses the 10-meter *Keck Telescope.* The low white dome just beyond the Keck is the NASA-sponsored *Infrared Telescope Facility.* On the ridge in the background, from left to right, are the 3.6-meter *Canada-France-Hawaii* telescope, the University of Hawaii 2.2-meter telescope, and the 3.6-meter United Kingdom infrared telescope. (California Association for Research in Astronomy)

HST also operates in visible wavelengths where the space environment allows better resolution without the blurring effects of the atmosphere and better sensitivity without atmospheric absorption). Currently no major infrared telescope is in orbit, but scientists are still quite busy analyzing data from the *Infrared Astronomical Satellite*, a joint U.S.-British-Dutch venture that mapped the entire sky at far-infrared wavelengths. X-ray astronomers are studying data from the current *ROSAT* mission, a joint U.S.-German-British project, and high-energy observations are being carried out by the *Compton Observatory* (a gamma-ray mission; Fig. 4.31). The most recent addition to the family of active space observatories is the *Extreme Ultraviolet Explorer*, a U.S. telescope which began surveying the sky in the shortest ultraviolet wavelengths in mid-1992.

Future plans in all wavelength regions are described in the next section.

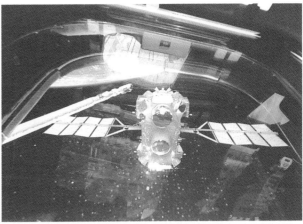

Figure 4.31   The *Compton Observatory* in the space shuttle. This view of the new gamma ray observatory was taken through a window from the flight deck of the shuttle, just before the satellite was released to follow its own orbit. (NASA)

THE NEXT GENERATION

What kinds of telescopes will supersede those already described in this chapter? We can speculate about this; astronomers have a flair for dreaming up new and better ways to overcome technical difficulties, so in some cases plans for new and wonderful instruments are already well developed.

Plans are now being made for a number of visible-wavelength telescopes in the 8- to 16-meter range

(imagine a telescope 16 meters, or about 50 feet, in diameter!). The most advanced of these plans, being developed by Cal Tech and the University of California, calls for the construction of two 10-meter telescopes on Mauna Kea. Rather than having a single 10-meter mirror, each *Keck Telescope* is being built using a **segmented mirror** design, in which the primary mirror consists of several segments (Fig. 4.32). Each segment can be separately controlled, so that all of them have a common focus. This is somewhat similar to the multiple-mirror concept, except that in this case the multiple mirrors are joined together. The

Figure 4.32   The 10-meter *Keck Telescope*.   The only ways to obtain a full view of this enormous instrument are either to hover above the dome in a helicopter, as was done in the left image, or to use a time-exposure photograph while the dome is rotated, as in the right image. The telescope was complete at the time of the photos, with all 36 hexagonal mirrors in place. (© 1992 Roger H. Ressmeyer/Starlight)

first *Keck Telescope* was completed in 1992; the second, which will be several years behind, will be adjacent to the first so that interferometry can be carried out using both instruments together.

The U.S. National Observatories, whose largest telescopes now are 4 meters in diameter, have been developing plans for larger national facilities. The *Gemini* project, now in final design stages, will build two 8-meter telescopes, one on Mauna Kea, and the other on Cerro Pachon, a second peak located within the bounds of the Cerro Tololo observatory in Chili. *Gemini* is a joint effort by the U.S., Canada, and the U.K.

The *European Southern Observatory,* operated by a consortium of several European nations (and already having several telescopes in Chile), is making plans for a 16-meter telescope using a multiple-mirror design, to be located in the Southern Hemisphere. This instrument, called the *Very Large Telescope,* will consist of four 8-meter telescopes operated together so that the total light-gathering power will be equivalent to that of a single mirror with a 16-meter diameter. It is hoped that the four will be operable with sufficient precision to allow for infrared and visible-wavelength interferometry, in the same manner as a radio observatory such as the *Very Large Array.* Meanwhile, several groups of U.S. universities are planning 8- to 10-meter telescopes, each using a single mirror. Mirrors this large may soon be made using a novel technique being developed at the University of Arizona. The molten glass for the mirror is poured into a rotating mold (Fig. 4.33), so that the rotation forces the glass into a concave shape as it hardens. This produces a mirror that is lightweight (because its internal structure is a honeycomb, rather than

being solid glass) and that is nearly in the correct shape to begin with, so that many months of grinding to the correct shape are eliminated. Recently a 6.5-meter mirror was successfully cast using this technique; this mirror will be installed in the *Multiple Mirror Telescope,* replacing its 6 smaller mirrors. Other novel techniques employing very thin glass mirrors, which must be actively controlled in order to maintain the correct shape, are also being developed.

In radio astronomy, there have been recent developments, particularly in observing the shortest radio wavelengths, around 1 millimeter and below. This is a rich region of the spectrum because many molecules in space emit at these wavelengths, but working in it has been difficult. The Earth's atmosphere provides a great deal of obscuration, and the short wavelengths require a more precisely shaped reflector than is normal for radio telescopes. Despite these difficulties, however, two submillimeter telescopes have recently been built on Mauna Kea and are now in full operation. One, having a diameter of 10 meters, is operated by Cal Tech; the other, a 15-meter dish, is a joint Dutch-British venture (Fig. 4.34). Another submillimeter telescope is under construction on a new site in Arizona, called Mt. Graham; this instrument will be operated by the University of Arizona.

In other developments, the U.S. National Radio Astronomy Observatory is developing plans to replace the large dish at Green Bank, West Virginia, which collapsed spontaneously in 1988 (this is the large dish shown at the left in Figure 4.27), and the *Australia Telescope,* a radio array comparable to the *VLA,* is now in operation in southeastern Australia.

In the more distant future, radio astronomers are

Figure 4.33   Casting a giant mirror in a spinning oven.   These two photos show steps in the creation of the 6.5-m mirror that will be installed in the *Multiple Mirror Telescope,* in place of the existing six 1.8-m mirrors. At left, chunks of glass are being placed in the mold before heating. Note the hexagonal pattern in the bottom of the mold, which creates a hollow honey-comb structure in back of the mirror. At right is a photo of the mirror after casting, just after removal from the oven. Note that the surface already has a concave shape, which is very close to the final curvature that is needed. (University of Arizona)

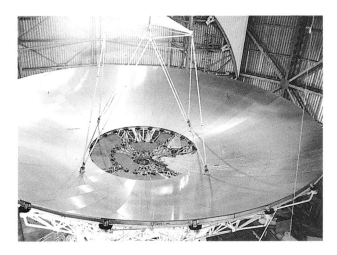

Figure 4.34   A new radio telescope.   This is the reflecting dish for the Dutch-British 15-meter radio telescope, now in operation on Mauna Kea. Telescopes of this size are usually designed for observing very short radio waves, with wavelengths of a few millimeters to a few centimeters, and the reflecting surface must be far more precisely shaped than in radio telescopes designed for much longer wavelengths. (T. P. Snow)

hoping to have telescopes orbiting the Earth, so that interferometry can be done with telescope separations even greater than the intercontinental distances now available.

In ultraviolet and visible astronomy, the newest instrument is the *Hubble Space Telescope (HST)* (Fig. 4.35), finally launched in April 1990, after years of delays due to slow funding, technical problems, and, above all, the *Challenger* disaster. The *HST* proved to be very complicated to operate, so there was a prolonged check-out phase before routine scientific studies got under way. In addition, a flaw in the shape of the primary mirror was discovered, with the result that some of the scientific goals will not be achieved until corrective optics are installed by astronauts in

1993 or 1994. Despite these difficulties, however, the *HST* has made many exciting discoveries and is living up to the expectation that it would revolutionize many areas of astronomy. Results from the *HST* are described in several sections of this book. The *HST* is being operated as a national observatory, accessible (through competitive proposal reviews) to astronomers from the United States. There is European participation as well: the European Space Agency (ESA) provided one of the principal cameras (the Faint Object Camera), and in return European astronomers are invited to compete for observing time on the *HST*.

Looking a bit farther into the future, another ultraviolet orbiting observatory is now under development, for launch in the late 1990s. This instrument, called the *Far Ultraviolet Spectroscopic Explorer (FUSE),* will observe wavelengths that are not reached by the *HST.* There are many important spectral lines of atoms, ions, and molecules at the shortest ultraviolet wavelengths, and it is the prime goal of *FUSE* to allow these features to be studied. *FUSE* is primarily a NASA venture with Canada, and possibly Germany, also participating.

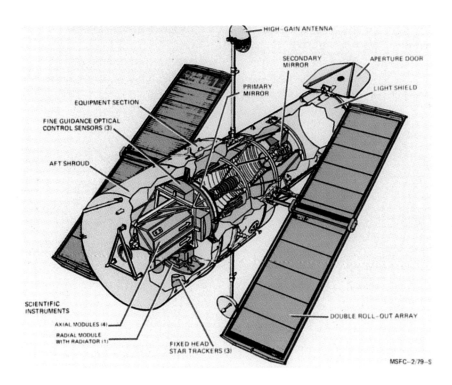

Figure 4.35   The *Hubble Space Telescope*.   This is a schematic drawing of the 2.4-meter telescope launched into Earth orbit in April 1990. The open end of the telescope is pointed away from us in this view; the near end houses control systems, cameras, and other scientific instruments. The large "wings" are solar panels that provide electrical power for the spacecraft. (NASA)

X-ray astronomers are awaiting two major instruments. One, called the *X-Ray Multiple Mirror Mission* (or *XMM*), is being planned for a late-1990s launch by ESA, and will also include some U.S. participation. The second, called the *Advanced X-Ray Astrophysics Facility,* or simply *AXAF*, will be similar to the *HST* in that it will be a large telescope with several interchangeable scientific instruments. Its launch date is expected early in the next decade.

Finally, in infrared astronomy, plans are under way for two major orbiting facilities. One, an ESA project called *Infrared Space Observatory* or *ISO*, is now being built and will be launched in 1994; the other, a major facility akin to the *HST* and called the *Space Infrared Telescope Facility* or *SIRTF*, is a NASA project now in the development stage. *SIRTF* will have a large telescope with several scientific instruments for observing infrared wavelengths that do not reach the Earth's surface. Even farther in the future is the *Large Deployable Reflector,* a 10- to 15-meter lightweight dish that will be unfolded in orbit, becoming the largest space telescope yet. This project may not get off the ground until around the turn of the century.

We have purposely discussed telescopes for all wavelengths together, in the preceding section on telescope design and in this one on future instruments, because the traditional segmentation of astronomers into one wavelength region or another is breaking down. Formerly an astronomer was usually considered an optical (visible-wavelength) astronomer or an infrared astronomer, a radio astronomer, and so on. Today, however, and certainly in the future, astronomers take a broader approach to scientific problems, using data from any wavelength region or instrument that is available. If astronomy is experiencing a revolution today, this is it.

PERSPECTIVE

We have now learned how light is absorbed and emitted by natural processes and how it is gathered and analyzed by various devices. We have seen how the continuous spectrum and overall intensity of light from a star depend on its temperature and size, and we have seen how the spectral lines provide information on the temperature, density, motion, and chemical composition of a star. Telescope technology allows astronomers to derive a remarkable amount of information about distant objects, something we will come to appreciate more and more in the chapters that follow. We are now ready to explore the universe, to see what astronomers have learned with the techniques we have been discussing.

SUMMARY

1. Visible light is just one part of the electromagnetic spectrum, which extends from gamma rays to radio wavelengths.

2. Light has properties associated with both waves and particles, and these aspects are combined in the concept of the photon.

3. The continuous radiation from a star provides information on its temperature, luminosity, and radius through the use of Wien's law, the Stefan-Boltzmann law, and the more general Planck's law.

4. The observed brightness of a glowing object is inversely proportional to the square of the distance between the object and the observer.

5. Spectral lines are produced by transitions of electrons between energy levels in atoms and ions. Absorption occurs when an electron gains energy, and emission occurs when it loses energy; in both cases, the wavelength corresponds to the energy gained or lost as the electron changes levels.

6. Each chemical element has its own distinct set of spectral line wavelengths, and therefore the composition of a distant object can be determined from its spectral lines.

7. The ionization and excitation of a gas can be inferred from its spectrum, the former yielding information on the temperature of the gas, and the latter providing data on its density.

8. Any motion along the line of sight between an observer and a source of light produces shifts in the wavelengths of the observed spectral lines, and measurement of this Doppler effect, as it is called, can be used to determine the relative speed of source and observer.

9. The principal reasons for using telescopes are to collect radiation from a large area; to allow radiation from a source to be collected over longer periods of time than is possible with the unaided eye; and to provide angular resolution, that is, the ability to discern fine detail.

10. The basic telescope design consists of a large mirror or reflecting surface to bring radiation to a focus, where an instrument containing film or an electronic detector records it. This general concept works at all wavelengths, from X ray to radio, but the details vary from one wavelength region to the next.

11. The instrument that receives the radiation at the telescope focus may be used to record images, measure the brightness of objects, or analyze the spectrum of the radiation.

12. Visible-light and radio telescopes can be located on the Earth's surface, but those for the gamma-ray, X-ray, ultraviolet, and much of the infrared spectrum must observe from above the Earth's atmosphere. On the ground, observatory sites must be chosen for clear weather, high atmospheric transmissivity, low turbulence in the air overhead, minimal pollution, remoteness from city lights, and the proper latitude for viewing the desired part of the sky.

13. Many new telescopes for all wavelength regions have recently been put into operation or are under construction or in the planning stage. The *Hubble Space Telescope,* a 2.4-meter telescope for visible and ultraviolet observations, was launched in April 1990. The first of several planned very large visible-light telescopes will be the 10-meter *Keck Telescope,* recently completed on Mauna Kea.

14. At all wavelengths, but most commonly in the radio portion of the spectrum, several telescopes can be used together to produce high-resolution images by interferometry.

15. Planned future telescopes include several giant reflectors that will be built on the ground and a variety of radio, X-ray, ultraviolet, infrared, and gamma-ray telescopes, many of them to be launched into orbit by the space shuttle.

REVIEW QUESTIONS

1. Summarize the various forms of electromagnetic radiation, explaining their differences and similarities. Think of everyday examples of each form of radiation.

2. Explain why the human eye is sensitive to the particular wavelengths of electromagnetic radiation that we can see. What wavelengths do you think beings on a planet circling a much hotter star than the Sun would be able to see?

3. Summarize the ways in which light acts as a stream of particles and the ways in which it acts as though it consists of waves.

4. Explain how the color of a star is related to its surface temperature.

5. Why are infrared-sensitive cameras useful devices for spotting people hiding in the dark?

6. Calculate the wavelengths of maximum emission for a star of surface temperature 25,000 K; a star of surface temperature 2,500 K; the Sun's corona, whose temperature is 2,000,000 K; and a human body, whose temperature is 310 K.

7. Pluto is approximately 25 times farther from the Sun than is Mars. Compare the intensity of sunlight reaching the two planets.

8. Explain how the spectral lines formed by a given type of atom are related to the structure of that atom.

9. Explain why an atom can form emission lines under some circumstances and absorption lines under other circumstances.

10. Summarize the information that can be gained about a distant object such as a star or a planet from the analysis of its spectral lines.

11. Summarize the benefits of using a telescope for astronomical observations, compared with using only the naked eye.

12. List the reasons why the summit of Mauna Kea, in Hawaii, is an excellent observatory site.

13. Explain why the *Hubble Space Telescope* offers advantages over ground-based telescopes, even for visible-wavelength observations.

ADDITIONAL READINGS

Bahcall, J. N. 1991. U.S. Astronomy's next decade. *Sky and Telescope* 81(6)584.

Bartusiak, M. 1984. Very Large Array (VLA). *Science 84* 5(6):64.

Beatty, J. K. 1990. ROSAT and the x-ray universe. *Sky and Telescope* 80:128.

Burns, J. O., N. Duric, G. J. Taylor, and S. W. Johnson. 1990. Observatories on the Moon. *Scientific American* 262(3):42.

Chaisson, E. J. 1992. Early results from the *Hubble Space Telescope. Scientific American* 266(6):44.

Chaisson, E., and R. Villard. 1990. Hubble Space Telescope: The mission. *Sky and Telescope* 79:378.

Chieu, P. 1992. *EUVE* probes the local bubble. *Sky and Telescope* 83(2):161.

Davis, J. 1992. The quest for high resolution. *Sky and Telescope* 83(1):29.

Dyer, A. 1991. *ROSAT*'s penetrating x-ray visions. *Astronomy* 19(6):42.

Eicher, D. J. 1991. New visions from CCDs. *Astronomy* 19(2):70.

Field, G., and D. Goldsmith. 1990. The space telescope: Eyes above the atmosphere. *Mercury* 19:34.

Fischer, D. 1989. A telescope for tomorrow. *Sky and Telescope* 78:248

Friedman, H. 1991. Discovering the invisible universe. *Mercury* 20(1):2.

Gillett, F. C., Gutley, I., and Hollenbach, D. 1991. Infrared astronomy takes center stage. *Sky and Telescope* 82(2):148.

Gordon, M. A. 1985. VLBA-A continent-size radio telescope. *Sky and Telescope* 69(6):487.

Gulkis, S., P. M. Lubin., S. S. Meyer, and R. F. Silverberg. 1990. The Cosmic Background Explorer. *Scientific American* 262(1):132.

Gustafson, J. R., and W. Argent. 1988. The Keck Telescope: 36 mirrors are better than one. *Mercury* 17:43.

Harrington, S. 1982. Selecting your first telescope. *Mercury* 11(4):106.

Harris, J. K. 1992. Seeing a brave new world (*Keck Telescope*). *Astronomy* 20(8):22.

Janesick, J., and M. Blouke. 1987. Sky on a chip: The fabulous CCD. *Sky and Telescope* 74(3):238.

Keel, W. C. 1992. Galaxies through a red giant (Soviet 6-m telescope). *Sky and Telescope* 83(6):626.

Kellerman, K. I. 1991. Radio astronomy: The next decade. *Sky and Telescope* 82(3):247.

Kellerman, K. I., and R. A. Thompson. 1988. The Very-Long-Baseline-Array. *Scientific American* 258(1):54.

King, H. C. 1979. *The history of the telescope*. New York: Dover.

Kniffen, D. A. 1991. The *Gamma Ray Observatory*. *Sky and Telescope* 81(5):488.

Krisciunas, K. 1989. Two astronomical centers of the world: Mauna Kea and LaPalma. *Mercury* 18(2):34.

Labeyrie, A. 1982. Stellar interferometry: A widening frontier. *Sky and Telescope* 63(4):334.

Maran, S. P. 1992. *Hubble* illuminates the universe. *Sky and Telescope* 83(6):619.

Margon, B. 1991. Exploring the high-energy universe. *Sky and Telescope* 82(6):607.

McAlister, H. A. 1988. Seeing stars with speckle interferometry. *American Scientist* 76:166.

McCaughrean, M. 1991. Infrared astronomy: pixels to spare. *Sky and Telescope* 82(1):31.

McLean, I. S. 1988. Infrared astronomy's new image. *Sky and Telescope* 75:254.

Mims, S. S. 1980. Chasing rainbows: The early development of astronomical spectroscopy. *Griffith Observer* 44(8):2

Nelson, J. 1989. The Keck Telescope. *American Scientist* 77:170.

Nichols, R. G. 1991. An eye on the violent universe (*Compton Observatory*). *Astronomy* 19(7):44.

Powell, C. S. 1991. Mirroring the cosmos. *Scientific American* 265(5):112.

Ridpath, I. 1990. The William Herschel Telescope. *Sky and Telescope* 80:136.

Robinson, L. J. 1992. Spinning a giant success. *Sky and Telescope* 84(1):26.

Shore, L. A. 1987. IUE: Nine years of astronomy. *Astronomy* 15(4):14.

Silk, J. 1990. Probing the primeval fireball. *Sky and Telescope* 79:600.

Sinnott, R. W. 1990. The Keck Telescope's giant eye. *Sky and Telescope* 80:15.

Spradley, J. L. 1988. The first true radio telescope. *Sky and Telescope* 76:28.

Strom, S. E. 1991. New frontiers in ground-based optical astronomy. *Sky and Telescope* 82(1):18.

Svec, M. T. 1992. The birth of electronic astronomy. *Sky and Telescope* 83(5):496.

Talcott, R. 1991. *Hubble* opens new vistas. *Astronomy* 19(2):30.

Teske, R. 1991. Starry, starry night: observing on Kitt Peak. *Mercury* 20(4):115.

Tucker, W., and R. Giacconi. The birth of x-ray astronomy. *Mercury* 14(6):178 (Part 1); 15(1):13 (Part 2).

Tucker, W., and K. Tucker. 1986. *The cosmic inquirers: Modern telescopes and their makers*. Cambridge, Mass.: Harvard University Press.

Verschuur, G. L. 1987. *The invisible universe revealed: The story of radio astronomy*. New York: Springer-Verlag.

Wearner, R. 1992. The birth of radio astronomy. *Astronomy* 20(6):46.

ASTRONOMICAL ACTIVITY
The Spectroscope

It is easy, with a modest investment in supplies, to build your own spectroscope. A spectroscope is a device that spreads out light according to wavelength so that when you look through it, you see the spectra of light sources (a spectrograph is a similar device except that it records the spectra, usually on film). The element of a spectroscope that spreads the light out according to wavelength (i.e., the part that *disperses* the light) can be a prism, as described in the text, or it may be a surface with closely spaced grooves called a *reflection grating* (this accounts for the colors you see when looking at light reflected from a grooved surface such as a phonograph record or a laser disk). Another type of grooved surface that disperses light is called a *transmission grating*; in this case the grooves are embedded in a transparent material so that light can shine through. Both types of gratings disperse light due to its wave nature; as light is reflected from (or passes through) the grooved surface, the waves create a pattern of interference in which they alternately cancel and reinforce each other. The transmitted light is brightest where the waves reinforce each other, and the direction in which this occurs depends on the wavelength; as a result, different wavelengths of light pass through the grating in different directions. Thus the light passing through the grating is dispersed according to wavelength.

Small transmission gratings are inexpensive and usually easy to obtain; a science supply store or scientific catalog will have them, or perhaps your instructor can get some. All that you need to build a spectroscope is a small piece of transmission grating (2-inch squares are often available), a cardboard tube (an inch or two in diameter and about a foot long), some opaque tape (such as electrical tape or duct tape), some flat cardboard (such as the back of a tablet of paper), and a pair of razor blades. Tape the grating across one end of the tube, being sure to seal it around the edges so that no light can seep in except by passing through the grating (see the drawing). Tape a piece of cardboard with a tiny slit formed by the razor blades across the other end of the tube so that light can pass only through the slit. The slit must be aligned in the same direction as the grooves in the grating. To make the slit, first cut a slit about an inch long and a quarter inch wide in the cardboard, then tape the razor blades over it so that the blades form new edges that are close together (it is helpful to insert a piece of cardboard between the blades while you are taping them to ensure that the spacing is both small and uniform).

Once the spectroscope is all sealed with tape, you are ready to try it out. To use it, point the slitted end toward a light source, and look through the grating at the other end, toward the slit. Try the spectroscope on a regular (incandescent) light bulb first; you should see a continuous spectrum, or rainbow (you will find that you have to look a little off to the side to see this). Then, at night, go outside and look at street lights, the Moon, stars, and any other light sources. What do you see? How does it relate to the discussions in this chapter about emission lines, continuous spectra, and absorption lines?

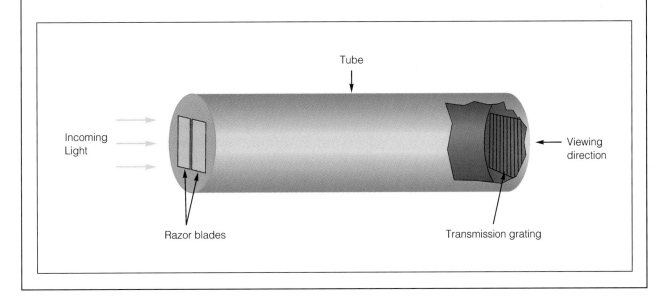

New Tools for an Old Science

Despite the many advantages of placing telescopes in space, instruments on the ground have remained essential to our science. Recent developments in ground-based telescope technology have shown that in some respects these instruments can approach the performance of orbiting observatories. If so, building a ground-based telescope is far more cost-effective than spending hundreds of millions of dollars to place instruments in orbit.

One of the important components of the new advances in telescope design is a growing capability for building very large mirrors. Shortly before this book went to press, two mirrors larger than any ever built before were successfully cast. One, mentioned in Chapter 4, is the 6.5-m spin-cast mirror that will be installed soon in the *Multiple-Mirror Telescope.* The other, which represents a different way of making big mirrors, is 8.2 m in diameter and was recently cast in Germany for the *Very Large Telescope.* This European project will consist of four such mirrors, acting together to match the light-gathering power of a single 16-m mirror.

Two distinct strategies are being used in manufacturing very large mirrors. One, used at the University of Arizona's Mirror Laboratory, involves making very lightweight, but rigid mirrors that will hold their shape even when the telescope is pointed in different directions. This technique was used to make the 6.5-m mirror. The casting is done in a rotating oven, which creates a curvature of the glass surface that approximates the final shape of the mirror, thus vastly simplifying the process of figuring the mirror to its final image-producing shape. The combination of light weight and rigidity is achieved by using a honey-combed mold for the back of the mirror; the resulting mirror structure has more empty space than glass, but the hexagonal shape of the cell walls makes it very rigid.

The second strategy for producing large mirrors is to make them relatively thin and compensate for the resulting flexibility by using active optics (see below). This method is being used in Germany for the 8.2-m *VLT* mirrors and in the United States for at least two projects. The German approach is to simply pour the molten glass in a single casting. Potentially, this method may lead to problems with stress and bubble formation, but at this writing three of the *VLT* mirrors have been cast, and one has nearly completed the subsequent annealing and crystallization steps that precede the polishing for final shaping. The alternative method for making large, thin mirrors is to build them from blocks of glass, which are melted and fused together in a large oven. The blocks are hexagonal in shape to help ensure a rigid bond to their neighbors, and they are arranged in a concave shape approximating the final figure of the mirror before being bonded together. This minimizes the amount of grinding and polishing needed. Soon the 7-m primary mirror for the Japanese National Telescope, to be located on Mauna Kea, will be built using this method.

An additional, very novel method for making large mirrors has also been proposed and was demonstrated recently for a 1.5-m mirror. This is the liquid mirror, which makes use of the fact that rotation will deform the surface of a circular body of liquid into a parabola, the desired shape for a telescope primary mirror. Experiments are under way in Canada using mercury, a liquid metal that has very high reflectivity from its surface. A shallow, circular tank of mercury is rotated at the correct speed to produce the desired degree of curvature, and this forms the primary mirror. This method is similar to the spin-casting technique, except that here the mirror remains liquid and is kept rotating while observations take place. A disadvantage is that the telescope can only observe in the vertical direction, so many kinds of astronomy cannot be done this way, but sky surveys and statistical studies of the strip of sky that passes overhead can be carried out. This method of making a large telescope mirror is very inexpensive and may find its place in the astronomy of the future.

The principal area in which ground-based telescopes may compete favorably with those in space is in achieving high-resolution images. Recall that resolution refers to the smallest angular separation between two objects that can be distinguished from each other and is normally measured in units of arcseconds. The theoretical resolution of a telescope, called the **diffraction limit,** is proportional to the observed wavelength and inversely proportional to the diameter of the telescope. For visible light (using a wavelength of 5500 Å, for example), a diffraction limit of 1 arcsecond is achieved with a telescope only 0.14 m (about 5.5 inches) in diameter. Thus large telescopes, in principle, are capable of much finer resolution than 1 arcsecond. But the Earth's atmosphere creates blurring (called "seeing") that normally limits the resolution to little better than 1 arcsecond.

Thus one of the goals of the *Hubble Space Telescope* is to achieve diffraction-limited image quality with its 2.4-m mirror by observing from above the atmosphere's blurring effect. At visible wavelengths, the diffraction limit for a 2.4-m mirror is about 0.06 arcseconds, and it is hoped that this level of image quality will be achieved when correcting mirrors are installed (in 1994) to compensate for the error in the shape of the primary mirror that was discovered after launch. But in the meantime recent developments have shown that, despite the effects of the atmosphere, this level of resolution may be approached using ground-based telescopes. You may see newspaper or magazine articles about some of these developments in the coming months.

Considerable improvement in image quality has been achieved at several observatories by reducing turbulent air flows inside the dome building and inside the telescope itself. Researchers have only recently recognized that a substantial portion of the seeing that blurs images is caused by localized air motions, rather than by motions in the column of air above the dome. Measures such as creating controlled, smooth air flows across the primary mirror, establishing pumping systems to draw air out of the building around the outer walls, and removing heat-producing equipment from the room where the telescope is mounted can substantially improve the resolution that can be attained. At fine, high-altitude sites such as Mauna Kea, Cerro Tololo, and Las Campanas, it is now possible to achieve resolutions better than 1 arcsecond, although not consistently.

The next step in improving resolution is to use "active optics" to compensate for the blurring caused by the atmosphere. As the name implies, with this technology the shape of the primary mirror is continually modified during observations. This can be done with a single, or "mon-

olithic," mirror, which is flexible enough to tolerate being bent a little, or with a segmented mirror. The flexible mirror idea is working effectively at the 3.5-m *New Technology Telescope (NTT)*, operated by the European Southern Observatory at Las Campanas, Chile, whereas the 10-m *Keck Telescope* on Mauna Kea has a segmented mirror design. The *NTT* routinely achieves subarcsecond seeing, while the *Keck* has reached a resolution of about 2 arcseconds in early tests. The *Keck* is expected to attain a resolution of 0.25 arcseconds when the activation system for the 36 mirror segments is fully operational and fine-tuned.

The flexible mirror concept can be carried a step further, by monitoring the mirror shape during observations and modifying it at a rapid rate, so that the fluctuations in the atmosphere can be counteracted as an observation proceeds. Here astronomers are borrowing technology originally developed for military and espionage applications in which very finely detailed photographs of the Earth's surface taken from great altitudes (or looking up, at Earth-orbiting satellites) are needed. The principal blurring effect caused by the atmosphere is created by cells of turbulent air whose diameters are typically 0.1 m or so, and whose time scale for changing is about a second. Thus, if you have a telescope whose shape can be controlled on scales smaller than 0.1 m and in times less than a second, it is possible to continually alter the shape so that the original image as it entered the top of the atmosphere is restored. This is a bit like placing a complex lens in front of the glass of a shower stall so that you get a clear view of the person in the shower.

But how does the telescope know what shape is needed each instant to

compensate for the atmospheric blurring? The answer is that you need an object whose intrinsic image shape is perfectly known. By tuning the telescope to keep the image of this "artificial star" in the correct shape, you ensure that you are also obtaining the correct image shape for astronomical objects. The artificial star must be a point-like light source that is observed through the atmosphere, and it must always be nearly in the same direction as the target object that is being observed. These are difficult requirements. At the present time, scientists are planning to send a laser beam upward, offset just a bit from the pointing direction of the telescope, so that the laser creates a bright spot in the upper atmosphere where it excites certain molecules to emit light. The results of this development will not be known for some time, but astronomers hope that resolutions as fine as 0.05 arcsecond can be achieved. One experiment, in which adaptive optics (as opposed to active optics) were installed on the *New Technology Telescope*, showed that diffraction-limited images can be obtained in the infrared, and nearly diffraction-limited images in visible wavelengths appear achievable as well. By combining this technology with the large sizes that are possible for ground-based telescopes, telescope power superior to that of the *Hubble Space Telescope* may eventually be achieved routinely.

Perhaps the ultimate technique for achieving high resolution is to use interferometry, in which two or more telescopes, widely separated from each other, are used together. In Chapter 4 we discuss interferometry using radio wavelengths, where

continued on next page

New Tools for an Old Science, *continued*

very high resolutions (0.1 arcsecond) can be attained. Using this technique in shorter wavelengths requires a much greater degree of precision in coordinating the telescopes and timing the arrival of photons of light, but this appears possible, at least at infrared wavelengths. Current experiments using small telescopes in France and Australia have achieved resolutions as fine as 0.05 arcseconds at visible wavelengths, and future plans call for using the separate telescopes of the *VLT* and the twin 10-m *Keck* instruments as interferometers.

Certainly, space-based observatories will always be needed to observe wavelengths that do not penetrate the atmosphere, but equally certainly, ground-based telescopes will always have a place as well. The combination of new mirror-making technologies and innovative methods of counteracting the effects of the atmosphere will ensure the future of "traditional" astronomy.

THE SOLAR SYSTEM

In this portion of the book, we will examine the properties of our system of planets, with a view toward understanding how it came to be, and we will study the fascinating assortment of phenomena that have been revealed through its exploration.

The solar system is a complex combination of a variety of individual parts. We will find, however, that a number of underlying, general processes are at work. As we go through the system examining individual objects, we will be able to make comparisons to see how these general processes have influenced different situations.

Among the effects that we will encounter again and again are tidal forces and other gravitational effects that are responsible for an astonishing assortment of phenomena, such as the spin rates of many of the planets and satellites, the locations of their orbits, and the failure of the asteroids and the ring particles of the outer planets to form large bodies.

We will often allude to the process by which the solar system formed from the collapse of a rotating interstellar cloud. Along the way on our tour we will encounter many recognizable artifacts of the process even before we undertake a detailed discussion of the formation and evolution of the system. Another phenomenon that we will often mention is the solar wind, a steady stream of subatomic particles from the Sun that pervades the solar system and creates important effects in the outer atmospheres of many of the planets.

As we study the planets, we will make several comparisons among them. To establish some of the standards for these comparisons, we will begin with an overview (Chapter 5) and then will discuss the best-studied and understood of all the planets, the Earth (Chapter 6). We will then tour the other three inner planets, whose overall properties are similar to those of the Earth (Chapter 7). Following this, we will turn our attention to the outer, giant planets (Chapter 8). Finally we will discuss other bodies in the solar system (Chapter 9), which are important leftovers from its formation, and will wrap up the section with a description of the formation and evolution of the solar system (Chapter 10).

Overview of the Solar System and Planetary Science

The eight planets that have been observed from close-up by space probes
(U.S. Geological Survey; data from NASA)

PREVIEW

Having gained a solid background in the methods of astronomy, we now embark on our exploration of the universe as it is understood today. In this chapter we take an overall look at our celestial neighborhood, the solar system. From these discussions, you will learn about the systematic trends among the planets and about the mechanisms that are common from one planet to the next. From this, you will gain a broad understanding of why the planets are the way they are, and why some are very similar to each other while others are very different. These discussions will prepare you for the detailed descriptions of the planets to come in the following chapters and will enable you to view each body from the perspective of the solar system as a whole.

As we prepare to explore the system of planets and interplanetary bodies that comprise the solar system, it is useful for us to begin with an overall description of the system and the general processes that shape the planets. This overview is particularly appropriate now, when a major era of planetary exploration has ended with the completion of the *Voyager* missions to the outer planets. With the exception of the currently operating *Magellan* radar mapper, which is orbiting Venus, it will be some time until the next major probe visits any of the planets and revises our current views of the Earth's neighbors. The *Mars Observer,* now on its way to Mars, will begin its survey of the red planet in late 1993, and the *Galileo* probe to Jupiter will not arrive at the giant planet until 1995.

In this chapter, we start with a general description of the solar system and follow this with a summary of the methods by which we learn about the planets and interplanetary bodies. The rest of the chapter is then devoted to an outline of the basic processes that mold the planets. We refer repeatedly to these general processes in the chapters that follow, as we discuss the individual objects that make up the solar system.

OVERVIEW OF THE SOLAR SYSTEM

The solar system is dominated by the Sun, which contains more than 99 percent of the total mass and provides the major source of energy that heats the surfaces of most of the planets. There are nine known planets (Table 5.1), which orbit the Sun at average distances ranging from about 58 million kilometers to almost 6 billion kilometers. In addition to these major bodies, there are innumerable smaller objects orbiting the Sun; these include the **asteroids** or **minor planets,** most of which orbit between Mars and Jupiter; the **comets,** which originate in a vast cloud far outside the orbit of Pluto and follow highly elongated paths through the solar system; interplanetary dust, a tenuous medium that is confined to the plane of the planetary orbits; and assorted other bodies, such as the icy asteroid-like objects occasionally found among the outer planets. In addition to this family of Sun-orbiting bodies, the solar system contains many satellites, moonlets, and rings, which orbit planets rather than the Sun directly.

Most of the major bodies in the solar system conform to a few general trends in location and motion. The system as a whole has a disklike structure, much like a record on a turntable, except that in this case the record is not rigid but instead acts like a fluid. Each planet and interplanetary object follows its own distinct orbit about the center, rather than being locked into a rigid overall rotation as a record is. The disk is very thin compared with its overall diameter (Fig. 5.1). Only the orbits of Mercury and Pluto deviate from the plane established by the Earth's orbit (i.e., the ecliptic) by more than about 3°, and even these two renegades have orbital planes lying within

TABLE 5.1 The Planets

Planet	Semimajor Axis	Sidereal Period	Mass[a]	Diameter[a]	Density	Temperature
Mercury	0.387 AU	0.241 yr	0.0558	0.381	5.50 g/cm^3	100–700 K
Venus	0.723	0.615	0.815	0.951	5.3	730
Earth	1.000	1.000	1.000	1.000	5.518	200–300
Mars	1.524	1.881	0.107	0.531	3.96	145–300
Jupiter	5.203	11.86	317.9	10.86	1.33	165
Saturn	9.555	29.46	95.2	8.99	0.69	134
Uranus	19.22	84.01	14.54	3.97	1.27	76
Neptune	30.11	164.79	17.15	3.86	1.64	74
Pluto	39.44	248.5	.0022	0.18	2.03	40

[a]The masses and diameters of the planets are expressed relative to the mass and diameter of the Earth, which are 5.974 $\times$ 10^{24} kg and 12,756 km, respectively.

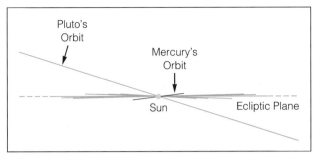

Figure 5.1   Alignment of planetary orbits.   As this edge-on view of the planes of the planetary orbits illustrates, seven of the nine planets have orbital planes closely aligned (within about 3° or less). Two planets, Mercury (7°) and Pluto (17°) deviate significantly. The relative sizes of the orbits are not drawn to scale here.

a few degrees of the ecliptic (the tilt of Mercury's orbit is about 7° and that of Pluto is just over 17°).

Despite the individual idiosyncrasies of the planets, they conform in several general ways. All of the planets orbit the Sun in the same direction, for example, and most of them spin in the same direction as well. We call this direction, which is counterclockwise as seen from above the North Pole, the **prograde** direction (Fig. 5.2). The rare exceptions in which planetary or satellite motions go against this trend are referred to as **retrograde** motions. As we shall see in Chapter 10, this general uniformity of orbital orientations and motions provides a powerful clue to the origin of the solar system.

Most of the planetary orbits are nearly circular; that is, they are ellipses with small **eccentricities,** or elongations (the formal definition of eccentricity is given in Appendix 7 and in the Astronomical Activity box in Chapter 2). The paths of Pluto and Mercury deviate significantly from circular, but even these two planets have orbits that are far closer to circular than

is the path of a typical comet. The comets, especially the long-period ones, usually follow very elongated elliptical paths around the Sun, indicating that the formation of these interplanetary bodies was significantly different from the origin of the planets.

The spin axes of most of the planets and their moons are oriented approximately perpendicular to their orbital planes, although there is quite a large range of **obliquities,** as the tilts of the planetary rotation axes are called. Except for Uranus and Pluto, all of the planets have obliquities of less than 30° (the Earth's tilt is 23°5). The unusual tilts of Pluto and Uranus (both over 90°) will invoke some singular explanation, some manner in which the formations of these two planets differed from the formation of the others.

Most of the interplanetary bodies, such as asteroids, the periodic comets, and interplanetary dust grains, obey the same general rules as the planets, following prograde orbits that lie near the plane of the ecliptic. There are some exceptions, however: the long-period comets (discussed in Chapter 9) trace out paths that are randomly oriented, and some of the interplanetary dust has apparently been disturbed enough to deviate from strict alignment with the ecliptic plane.

Classifying the Planets

Of the nine planets, eight fall cleanly into two general categories, based on their general characteristics. The innermost four are called the **terrestrial planets** because they are Earth-like in their overall nature. Like the Earth, these planets are relatively small (compared with the giant planets), have high densities, and have hard, rocky surfaces (Fig. 5.3). The Moon (Fig. 5.4) is often classified as a terrestrial planet due to its large size, its similar overall characteristics, and its similar geological history.

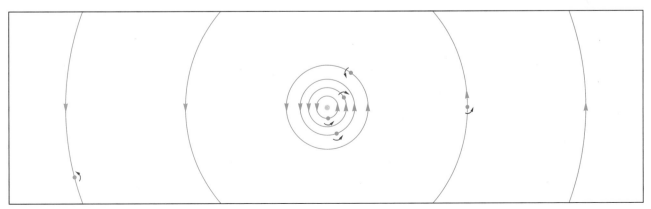

Figure 5.2   Prograde motions.   In this scale drawing of the inner six planets, arrows show the direction of orbital motion (blue arrows) and the direction of spin (red arrows). All of these motions are in the same direction, except that Venus (the second planet from the Sun) spins backwards (i.e., in the retrograde direction).

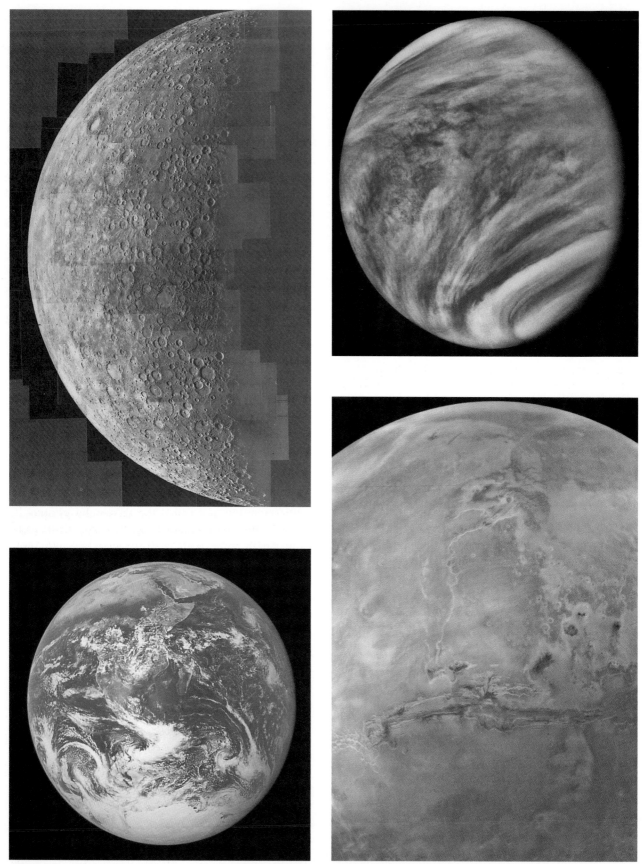

Figure 5.3　The terrestrial planets.　All of these images were obtained from space: Mercury (upper left) and Venus (upper right) were both photographed by *Mariner* 10; Earth (lower left) by one of the *Apollo* missions; and Mars (lower right) by one of the *Viking* orbiters. The terrestrial planets are all relatively small and dense, with rocky surfaces. (NASA)

Figure 5.4   The full Moon.   This is a high-quality photograph taken through a telescope on the Earth. Note that in this telescopic view, the Moon is upside down and reversed left-to-right, compared to the view with binoculars or the unaided eye. (Lick Observatory, University of California)

All of the giant planets have atmospheres; in fact, as noted already, they do not have solid surfaces, but consist instead of a fluid structure nearly all the way to their centers (it is thought that each of the giants has a small, rocky core). At the highest levels, these atmospheres are very cold and contain complex molecules consisting primarily of the simplest elements such as hydrogen, carbon, nitrogen, and oxygen (there is also a good deal of helium, in atomic form). By contrast, the terrestrial planets contain very little hydrogen and helium in their atmospheres because these lightweight gases can easily escape the low gravitational fields of the small, warm planets.

The ninth planet, Pluto, does not fit readily into either category, but instead more closely resembles some of the giant moons of the outer planets (Fig. 5.6). These bodies consist of ice and rock in comparable proportions (giving them overall densities ranging from 1 to 2 grams/cm^3). They have hard surfaces and are generally differentiated, and some of them (Pluto included, at least some of the time) have gaseous atmospheres. Pluto and the giant satellites are believed to have had rather different geological histories from either the gaseous giants or the terrestrial planets.

The terrestrial planets are all **differentiated,** meaning that the denser materials of which they are made have sunk to their centers. Differentiation can occur only if a planet is fully or partially molten inside, so we conclude that each of the terrestrial planets underwent some melting at some time in its past. In some the interiors are still quite hot today, and the Earth for one is still partially molten.

The terrestrial planets can have atmospheres if their surface gravities are sufficiently strong to trap gas particles. (Their ability to retain atmospheres depends also on the surface temperature, as discussed later in this chapter.) Only the Moon and Mercury fail to retain any significant atmosphere.

The next four planets beyond Mars are the **gaseous giant planets** (Fig. 5.5), also called the **Jovian** planets after the largest and closest one, Jupiter. These planets are radically different from the terrestrial planets, being very much larger, having no solid surfaces and having much lower densities. Where the terrestrial planets have densities ranging from around 3 grams/cm^3 to over 5 grams/cm^3, the densities of the gaseous giants range from 0.7 to 1.6 grams/cm^3 (for comparison, the density of water is 1 gram/cm^3, and that of ordinary rocks is in the range of 2–3 grams/cm^3).

Observational Techniques

In Chapter 4, we discussed general principles of observational astronomy along with the properties of light. In those discussions, we assumed that our only means of studying objects in the universe is to collect light that reaches the Earth and analyze it. This passive approach is indeed our only choice for most of the universe, but for the planets we can apply other, more direct techniques.

Clearly the best and most direct method of studying a planet is to go there and examine it at close hand. This direct sampling method, with human interaction, has now been applied to just four bodies in the solar system, the Earth, the Moon, Venus, and Mars (the latter two have been sampled by robot probes) (Fig. 5.7). Naturally, we find that these are the best-studied objects, though in the case of the Moon at least, perhaps not necessarily the best understood. Direct analysis of surface samples has been achieved for Mars through the U.S. *Viking* missions in the mid-1970s (Fig. 5.8) and for Venus through the Soviet *Venera* landers (see Chapter 7). Plans are being discussed, in both the United States and Russia, for future manned exploration of Mars, but this is certainly at least one or more decades away from reality at the present time (the United States has recently embarked on a course that will lead us back to the Moon before we go on to Mars).

Even if it is impractical as yet for us to visit the rest of the planets personally, we can send unmanned probes to visit them and make observations from close range. This has now been accomplished for all of the planets except Pluto; the August 1989 *Voyager 2* flyby of Neptune marked the completion of a remarkable era of planetary exploration. Both the United States and the Soviet space programs made stunning discoveries through their robot missions to the planets. Several advantages can be gained by making close-up observations: vastly more detail can be seen; objects can be observed from different angles (this is particularly useful in the interpretation of light-scattering properties); charged particles trapped in planetary magnetic fields can be captured and measured directly; direct measurements of the planetary gravitational field can be made, and from that the internal structure of the planet may be determined; and it may be possible to measure weak emissions (such as radio noise from particle belts) that cannot be detected from the Earth.

Even though the "grand tour" of the outer planets by *Voyager 2* is over, several other sophisticated planetary probes are in action or planned. Two spacecraft

Figure 5.5   The giant planets.   The images of Jupiter (upper left), Saturn (upper right), Neptune (lower left), and Uranus (lower right) were all obtained at close range by the Voyager probes. The giant planets are very different from the terrestrial planets, being much larger and far less dense and having no solid surfaces. (NASA)

have been orbiting Venus: one, the *Pioneer-Venus* probe, has just stopped after 14 years, and the *Magellan* spacecraft arrived in August 1990 and is mapping the cloud-covered planet using radar. Meanwhile the *Galileo* probe is on its lengthy and circuitous path to Jupiter, where it will arrive in late 1995. It will drop a probe into the atmosphere while the main spacecraft goes into orbit, allowing for long-term exploration of Jupiter and its major moons. The United States has just launched a new mission to Mars called *Mars Observer*, and the Russians are planning a series of probes (some expected to land and collect surface samples) to follow their recent *Phobos* missions to the red planet.

Farther in the future, there are plans for a joint U.S.–European Space Agency mission to Saturn, called *Cassini*. And another spacecraft, called the *Comet Rendezvous and Asteroid Flyby (CRAF)* mission, has been designed to make a close encounter with an asteroid and then accompany a comet in its orbit about the Sun. *Cassini*, expected to be launched in 1996, will put a probe into the atmosphere of the giant, cloud-shrouded moon Titan; it will also place an advanced spacecraft into orbit around Saturn to

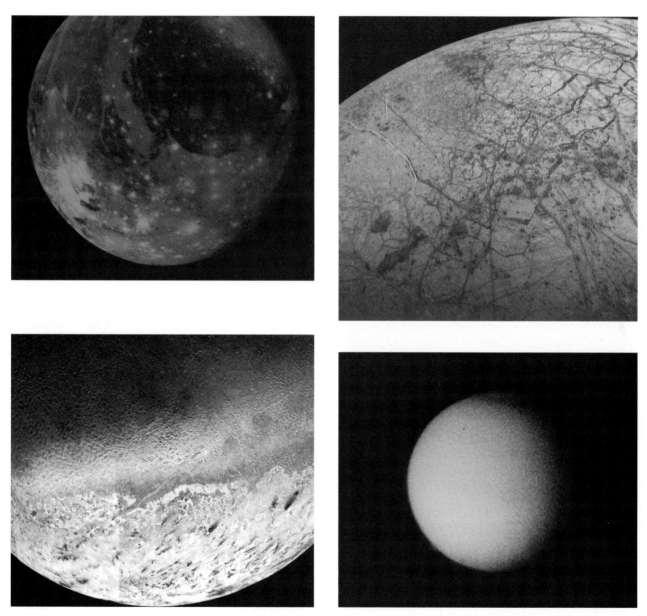

Figure 5.6   Giant moons of the outer planets.   These four satellites are all larger than the Earth's Moon and different from it in many respects. Where the Moon is rocky throughout, these giant satellites are partially composed of ice. While the Moon is inactive geologically, some of the giant satellites are quite active. Here we see Ganymede of Jupiter (upper left), Europa of Jupiter (upper right), Triton of Neptune (lower left), and Titan of Saturn (lower right). Ganymede and especially Europa show fault lines due to crustal shifting; Triton has active ice geysers and a dimpled crust due to slumping of the surface; and Titan has a thick atmosphere, possibly created by outgassing that is still active. (NASA)

explore the planet, its rings and moons and map Titan's surface with radar. *CRAF,* whose funding is now in doubt, is being designed to fly a spacecraft past an asteroid and then to match orbits with a comet. An instrumented penetrator will be shot into the comet's nucleus to make direct measurements of its composition. Perhaps the most ambitious project now being planned is a probe to Pluto by way of Neptune. This mission has been formally adopted as the next in line for the U.S. planetary program, following *Cassini* (and possibly *CRAF*).

Despite the enormous successes of the manned and unmanned probes to the Moon and the planets, there is still a lot to be learned from Earth-based observations. Of course, for a long time, telescopic data were all we had on any of the planets, so our first "explorations" of each body were of the traditional astronomical kind. From these observations, scientists were able to learn a great deal: surface temperatures could be derived from Wien's law (Chapter 4); surface compositions could be roughly analyzed from spectroscopy of reflected light (particularly at infrared wavelengths); and atmospheric composition could be deduced, again from spectroscopy. One of the great advantages of Earth-based observations, still important even in the age of space exploration, is the ability to make observations repeatedly over long time periods. This allows us to gather detailed information on seasonal variations, for example, and also allows us to use orbiting satellites as probes of the atmospheres of the planets (Fig. 5.9). Earth-based radio observations complement the visible and infrared data by providing information on charged particle belts (and indirectly on planetary magnetic fields, which create the belts).

Even space-based observations from the vicinity of the Earth are important. Orbiting spacecraft such as the *International Ultraviolet Explorer* (*IUE*) have obtained extensive ultraviolet spectroscopic observations of several of the planets, and sounding rocket observations have done ultraviolet work as well. The *Hubble Space Telescope,* orbiting the Earth since May 1990, is making enormous contributions to our

Figure 5.7   Human exploration of the Moon.   This is astronaut Harrison Schmidt on the last *Apollo* mission to the Moon (*Apollo 17*), collecting samples of small lunar rocks. (NASA)

Figure 5.8   Robots on Mars.   This is a U.S. *Viking* lander in a simulated Martian environment. Two of these robotic vehicles landed successfully on Mars in 1976, and carried out numerous experiments and observations. A mechanical arm was able to pick up soil samples and deposit them in test chambers where experiments were conducted in search of microscopic life forms. *Viking's* cameras took many panoramic photographs of the surface, and kept operating long enough to show how the dust and soil deposits vary with the seasonal winds. (NASA)

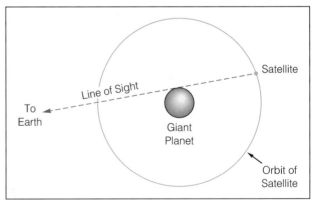

Figure 5.9   Using a satellite to probe a planetary atmosphere.   As the satellite passes behind the disk of its parent planet, the satellite is blocked by the planet. As the satellite passes behind the edge of the planet's disk as seen from the Earth, its light passes through the planetary atmosphere on its way to us. Scientists on the Earth can analyze the effects of the atmosphere on the light from the satellite to gain information on the gases in the atmosphere of the planet.

knowledge of the planets; it is not only providing both extended wavelength coverage (ultraviolet as well as visible) and unprecedented clarity of images, but is also revealing surface details and allowing them to be monitored over time.

Even though the completion of the *Voyager* mission signaled the end of a particular phase of planetary exploration, the wide assortment of ongoing and planned studies of the planets will continue to produce a steady stream of new information. There is no doubt that future astronomy textbooks will require extensive revisions as the exploration of the solar system continues.

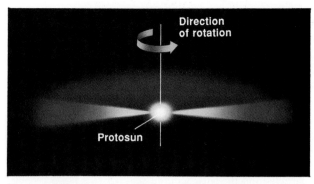

Figure 5.10   The solar nebula.   This is an edge-on view of the disk of gas and dust from which the Sun and planets formed. Rotation of a collapsing interstellar cloud caused the flattening. Gravitational compression created a hot concentration of gas at the center that was to become the Sun. The disk of surrounding material became clumpy as gas began to condense into solid form, and the clumps then merged, eventually creating the planets.

PLANETARY SCIENCE: THE PROCESSES THAT SHAPE THE PLANETS

The nine planets offer an amazing variety of appearances and phenomena, yet there are a few underlying physical mechanisms that explain most of the phenomena. In an important sense, the planets have more similarities than differences. As we prepare for our more detailed tour of the solar system, it is useful now to discuss the general processes that affect all the planets in varying degrees. This will not only save us from having to repeat some of the discussions from chapter to chapter, but more importantly, it will equip us to understand the underlying simplicity and the many connections and similarities among the planets as we go.

Formation of the Planets

A more complete discussion of the formation of the solar system will be found in Chapter 10, but here it is useful to gain an overview of the basic processes by which the planets formed. Certain general themes apply to all of the planets, but there are some differences as well, particularly between the terrestrial and giant groups.

The solar system apparently formed from a disk of gas and dust that orbited the young Sun at the time of its formation (Fig. 5.10). This disk, composed of material gathered together by gravitational forces, was initially dominated by hydrogen, the most abundant element in the universe. The composition (by mass fraction) of the original preplanetary material was approximately 73 percent hydrogen and 25 percent helium, with all the other elements comprising the remaining 2 percent of the mass. This is the composition of the Sun and most of the stars as well as of the interstellar matter in the galaxy today, and astronomers think it was about the same 5 billion years ago, when the solar system formed.

The disk of atoms and small, solid particles that orbited the young Sun had a natural tendency to become clumpy, which caused the condensation of larger solid objects than the original tiny dust grains. With time, these larger objects built up so that first snowball-sized, then basketball-sized, then even larger bodies were formed. The rate of growth of these objects increased until the disk became a collection of **planetesimals,** or preplanetary bodies, with diameters up to several thousand kilometers. During the buildup, or accretion, process, some elements were driven out by heat, particularly in the inner solar system. The lightweight, **volatile** (i.e., easily vaporized) elements were the least likely to become trapped in the solid state. Hence, the planetesimals that formed in the inner solar system never contained large quantities of lightweight elements such as hydrogen and helium.

The planetesimals themselves collided occasionally, often with small relative speeds (because they orbited the Sun in the same direction and at nearly the same speed at a given distance from the Sun). These collisions caused the planetesimals to coalesce, or stick together, eventually producing the planets of today. The entire accretion process is thought to have taken place fairly rapidly by astronomical standards; astronomers believe that the planets formed within the first few hundred million years after the Sun was born.

Note that the planets formed in the inner solar system started out with small quantities of the volatile elements, since these elements had been unable to condense into solid form in the first place, due to the high temperatures in the inner part of the system. In the outer solar system, however, the temperature was

ASTRONOMICAL INSIGHT
Voyager's Final Look Back

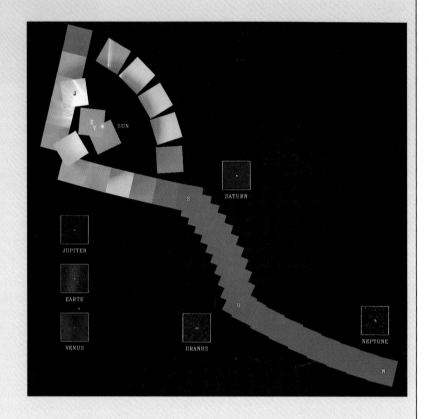

After *Voyager 2* encountered Neptune in August 1989, its mission was completed. Both *Voyagers* will continue to operate, measuring charged particles and the solar wind in the outermost reaches of the solar system, and making some observations of stars as well. But from the point of view of planetary research, the job was done. The cameras that obtained all the magnificent images of the planets and their satellites and ring systems would not be used further.

Mission scientists and engineers could not resist one final imaging project, however. The opportunity for an overview of the solar system, as seen from the outside, was unprecedented (and another may be a very long time in coming). Hence as *Voyager 2* raced beyond the orbit of Neptune, its scan platform was swiveled back for one last look at the Sun and planets. Final images of six of the nine planets were obtained, along with an image of the Sun (drastically enlarged due to multiple reflections of its brilliant light inside the camera optical system; even from 40 AU the Sun is about 8 million times brighter than Sirius, the brightest star in our sky other than the Sun).

The first illustration (above) presented here shows the positions of the six planets (indicated by the first letters of their names) when the images were obtained (Mercury, Mars, and Pluto were either in poor position relative to the Sun or were too faint). It also shows the sequence of 39 individual wide-angle images that were obtained in order to create a mosaic that encompasses the Sun and all six planets. This mosaic is shown as a jagged strip in the illustration. Next to the position of each planet is an enlarged version of that planet's color image, obtained with *Voyager's* narrow-angle camera.

The second illustration (on the facing page) shows, in greater detail, the six narrow-angle color images of the planets. Jupiter and Saturn were both resolved; that is, their images are larger than the size of the individual picture elements (pixels; see the discussion in the Astronomical Insight on page 87) of the camera. For Saturn it is possible to tell that the image is elongated due to the ring system. The images of the Earth and Venus are not resolved; their angular diameters from such a great distance were far smaller than the size of the picture elements of the camera, so these two planets appear as single points. The elongated shapes of Uranus and Neptune are not real, but are due to spacecraft motion during these exposures, which were relatively long.

While the individual images obtained from so far away are not very detailed or informative, the perspective this set of pictures gives us is invaluable. It is comparable in a way to the impact of the first photographs of the whole Earth that were obtained in the 1960s by the *Apollo* missions. Then we saw for the first time the outsider's view of our tiny planet, and for the first time many people gained a real appreciation of the limited extent and fragile resources of "spaceship Earth." Now we have a similar perspective on our entire solar system; it gives us a better understanding of the vastness of space and the very minor role played by planets. It is quite remarkable to think that our technology has made this view possible only 87 years after the first powered flight off of the Earth's surface.

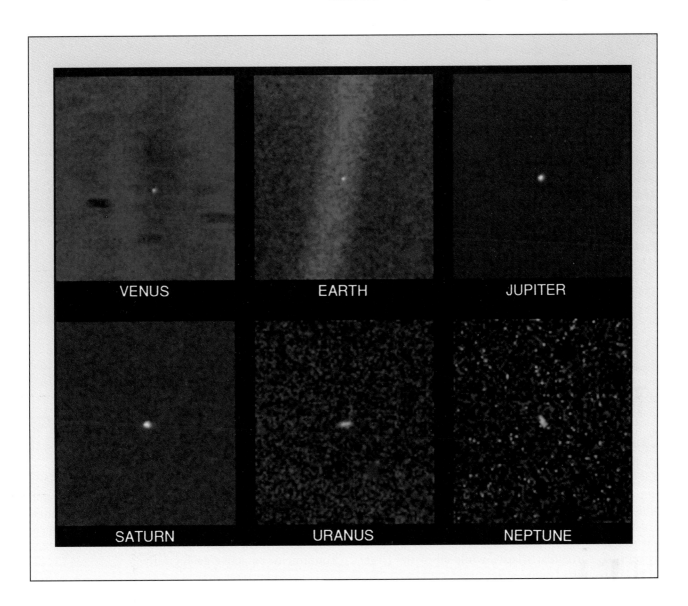

so low that even hydrogen and helium could become trapped; thus, the giant planets started out with a radically different composition from the terrestrial planets. The ability of the cold outer planets to trap these abundant gases gave the outer planets much larger initial masses than the inner planets. These larger masses helped the outer planets to gravitationally capture additional quantities of gas from their surroundings. Thus, the outer planets grew into much larger, more massive bodies than the inner planets and had rather different compositions as well.

In the inner solar system, proximity to the Sun created strong tidal forces (discussed later in this chapter) that prevented the planets from capturing surrounding gas and dust, and the higher temperature of this material made it more difficult to trap anyway. The result was a set of inner planets that contained mostly heavy elements and turned out to be dense and rocky, and a set of outer planets that

were far more massive, were cold, and contained ample quantities of lightweight gases. The trapped material surrounding the outer planets formed into extensive systems of rings and moons, while the inner planets tended not to have significant satellites (except for the Earth, whose large moon is thought to have formed largely by accident; see Chapter 6).

The asteroids, which orbit the Sun between the paths of Mars and Jupiter, along with a few outlying bodies (including the planet Pluto), are thought to be planetesimals that survived without merging into the larger planets. These bodies, along with the comets, represent primitive material from the early days of solar system formation.

All of the planets, and many of their satellites as well as the largest of the asteroids, are round. This is because the inward force of gravity controls the shapes of these bodies. Gravity presses inward from all sides, resulting in compression in the interior. As

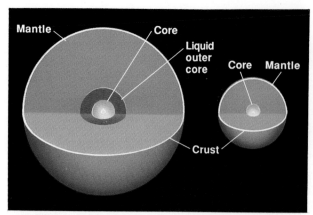

Figure 5.11   Internal structure of the terrestrial planets.
At the left is a cross-sectional view of a planet like the Earth, which has a liquid zone in its outer core. Other terrestrials, particularly the small ones such as Mars and the Moon, have internal structures more like the sketch at the right, with no liquid zone.

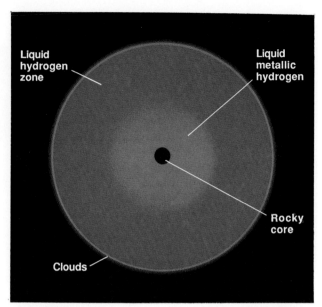

Figure 5.12   Internal structure of a gaseous giant planet.
Beneath the visible cloud tops, the density and temperature increase inward. As the density increases, hydrogen is forced into unusual forms, which are discussed in Chapter 8. Because the planet is fluid throughout, differentiation occurs readily, and virtually all of the planet's heavy elements sink to the center, forming a small, rocky core.

the material inside a developing planet is compressed, pressure builds up that counteracts gravity, eventually stopping the contraction. The result is a spherical body, denser in the center than in the outer region, in which gravity and pressure are balanced at every point. Only if there is another force to counteract gravity will a body end up in a nonspherical shape. For very small objects (i.e., the smaller asteroids or the tiniest moons), the strength of the material of which it is composed may exceed the inward force of gravity, allowing the body to retain a nonspherical shape. Thus, some of the smallest objects in the solar system have irregular shapes, while all of the larger ones are round.

While a planet formed, gravitational forces squeezed it together tighter and tighter, causing compression and heating in its interior. Radioactive elements in the interior of a planet released additional heat as they decayed, helping to keep the interior of the planet hot to this day. In the terrestrial planets, this heating caused the interior to be partially molten over long periods of time, which in turn allowed the bulk of the heaviest elements (such as iron, nickel, cobalt, and the superheavy species such as lead, uranium, and so on) to sink to the center of the planet. In this process, called **differentiation,** each terrestrial planet gained a dense core made primarily of iron and nickel. In some of the terrestrial planets (including the Earth and probably Mercury and Venus), the core remains at least partially molten, whereas others (probably Mars and the Moon) have cooled enough to be essentially solid throughout (Fig. 5.11).

The outer planets, composed as they are of lightweight gases, remained fluid throughout most of their interiors. Thus, these planets have no outer solid surfaces, and differentiation was able to segregate out the heavy elements very thoroughly, forming small, dense solid cores deep within (Fig. 5.12). The outer planets are thought to have been largely unchanged since their earliest days, while the terrestrial planets continue to be active geologically.

GEOLOGICAL EVOLUTION

Tectonic Activity

The terrestrial planets are still alive geologically in varying degrees. The Earth appears to be the most active, while the Moon (sometimes classified as a terrestrial body, even though it is a satellite) is probably the most inert. The current state of activity in Mercury, Venus, and Mars is still largely uncertain; perhaps these questions will be answered soon, through space probes such as *Magellan* to Venus and *Mars Observer* to Mars.

Let us consider the Earth as the prototype of a geologically active terrestrial planet. Recall that the interior is still quite hot, largely because of radioactive decay, and that a portion of the core is molten. The surface consists of a brittle crust overlying a thick layer that is in a semisolid state (Fig. 5.13). This underlying region, called the **mantle,** is the source of

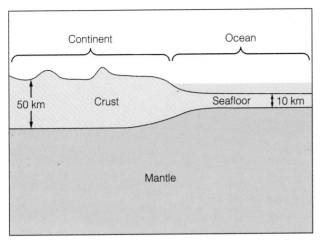

Figure 5.13   The Earth's crust. The outermost layer of the Earth is very thin (10–50 km) relative to the radius of the planet. The crust, unlike the mantle below, is quite rigid and brittle, and it has fragmented into several segments (plates), which move about in response to motions in the underlying mantle.

Figure 5.14   A recent volcanic eruption. This is a photograph of the explosive eruption of Mt. St. Helens, in Washington State, during its outburst in 1980. Events such as this provide graphic evidence that the Earth is very active geologically. (John H. Meehan/Science Source, Photo Researchers)

most geological activity such as earthquakes and volcanoes.

The crust is made of rocky material that is less dense than the mantle and therefore floats atop it. The crust has fractured into about a dozen segments, called **plates,** which are somewhat free to move about over the mantle. As they do so, the continents move (a phenomenon known as **continental drift**), and the plates rub against each other or collide, creating surface activity such as earthquakes and volcanic eruptions (Fig. 5.14). Where plates move alongside each other, such as along the famous San Andreas fault in California, they alternately stick together as stress builds up, then slip suddenly, causing earthquakes (Fig. 5.15). Volcanic eruptions also tend to occur near plate boundaries, where molten rock, or **magma,** can find its way to the surface (Fig. 5.16). Where a seafloor plate, which is denser than continental material, collides with a continental plate, it tends to slip underneath it, creating deep sea trenches (Fig. 5.17) and often volcanic activity. In places where two continental plates collide, the edges are uplifted, creating the loftiest mountain ranges on the Earth (Fig. 5.18).

The driving force behind these motions of the Earth's plates has not been completely established, but is thought most likely to be slow **convection** currents in the mantle. Convection is the overturning motion that can occur naturally when a fluid is heated from the bottom (and is subjected to a gravitational field at the same time). Under the right conditions, bubbles of heated fluid become more buoyant than their surroundings and rise. As they rise, they

Figure 5.15   Crustal shifting during an earthquake. This abrupt tear in the Earth's crust was the result of an earthquake that struck the Andes Mountains of Peru. (Carl Frank, Photo Researchers)

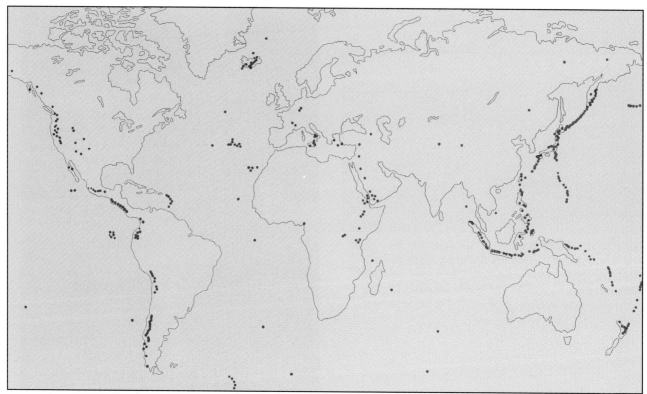

Figure 5.16   The distribution of active volcanoes.   This map shows the locations of volcanoes that have been active in recent years. Note that the volcanoes are concentrated along the boundaries of the Earth's crustal plates, particularly around the edges of the huge Pacific plate.

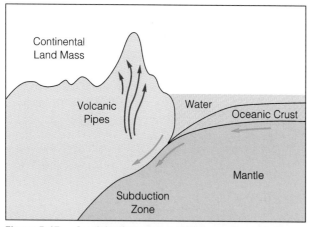

Figure 5.17   A subduction zone.   Here oceanic crust (sea-floor) is shown sinking below continental crust as two adjacent crustal plates move toward each other. Along the line of intersection of the two plates (i.e., along the subduction zone), the oceanic crust slips below the continental crust, creating a deep, undersea trench.

Figure 5.18   The result of a collision between crustal plates.   This is a portion of the Himalaya Mountains, the Earth's tallest range, including Mt. Everest (at right). These mountains are being uplifted by pressure created in a slow collision between the Indian plate and the Eurasian plate and are growing taller each year. (G. J. James/BPS)

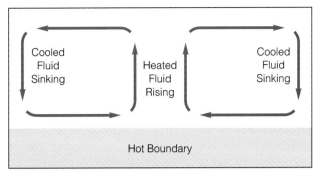

Figure 5.19   Convective overturn.   A fluid (i.e., a gas or a liquid) subjected to heating from below (in the presence of a gravitational field) may experience convective overturn, as shown here. Heated fluid, less dense than its surroundings, rises to higher levels, where it cools, becomes denser, and sinks. The result is a steady, overturning motion. In planetary atmospheres, the source of heat from below may be absorbed sunlight or internal heat from within the planet.

cool, eventually losing their buoyancy and then sinking again after moving a small distance sideways to make room for rising bubbles from below (Fig. 5.19). In a very viscous (i.e., thick) fluid such as the Earth's semisolid mantle, the process is slow, resulting in motions of the crustal plates that are measured in units of a few centimeters per year.

The **tectonic activity,** as the action of moving crustal plates is called, takes place in the mantle and the crust. Underneath the mantle is the core, part of which is molten (see Fig. 5.11). Here, too, fluid motions occur. Because some of the material in the inner Earth carries electrical charges, the motions in the outer core result in the creation of electrical currents. These in turn create a **magnetic field** that permeates the entire Earth and extends for some distance into space. Some of the other terrestrial planets also have magnetic fields, indicative that they, too, have partially molten cores. (A molten region is not the only requirement, however; the planet must also rotate rapidly enough to create sufficient electrical current.)

As we have already said, the other terrestrial planets are apparently not as active internally as the Earth. Nevertheless, both Mars and Venus show some signs of tectonic activity, and Mercury has a magnetic field, indicating that it must have a molten zone inside. Mars has surface structures that resulted from uplift and crustal fracture, but these processes may have become inactive early in the planet's history. In contrast, there are some indications that Venus may still be active today (the *Magellan* radar maps of Venus are helping to answer this question by providing detailed information on surface structures). The Moon is certainly inert geologically at present and

probably never underwent extensive tectonic activity, although lava flows blanketed much of the surface early in its history.

Because the giant planets have no real surfaces and no solid interiors (except for their cores), we do not speak of geological or tectonic activity in the same manner as we do for the terrestrial planets. On the other hand, some similar processes are taking place. For example, all of the giant planets have strong magnetic fields, indicating that electrical currents are flowing in their interiors. The presence of these currents is not surprising, since all of these planets are fluid throughout most of their interiors, and all of them rotate very rapidly.

Rocks and Surface Processes

In addition to tectonic activity, which can modify the surface of a terrestrial planet through quakes and volcanic eruptions, there are other forces that can act to alter the hard surfaces of the terrestrial planets. The surfaces of these planets are largely composed of rock, which may be covered by water over large regions (on today's Earth, for example, and perhaps on the early Mars and the young Venus as well), and which may be decomposed into soil, dust, or sand in other regions.

On the Earth, we recognize three general types of rock, based on origin: (1) **igneous rocks,** which formed from molten material that has hardened; (2) **sedimentary rocks,** which formed from the deposition of mud and clay deposits on the floors of oceans and lakes; and (3) **metamorphic rocks,** which have been altered by heat and pressure resulting from large-scale crustal movements. Continual geological activity can gradually change rocks from one of these forms to another. For example, a rock may first form as an igneous rock as the result of an eruption, but later may be broken down by wind and water into sand that is deposited underwater and forms sedimentary rock, and yet later may be uplifted by geological forces and squeezed and compressed so that it becomes a metamorphic rock.

Apart from its origin, we may also classify a rock according to its chemical composition. Thus, we define **minerals** based on composition. By far the most common minerals on the surface of the Earth and the other terrestrial planets are the **silicates,** which contain compounds of silicon and oxygen. The common basalts and granites on the Earth are silicates of igneous origin. Other common mineral types include the **carbonates,** which are dominated by carbon-bearing compounds, and **oxides,** consisting largely of oxygen compounds.

We have seen that tectonic activity can alter the face of a terrestrial planet significantly, as pieces of the crust move about, and at the same time that local processes may alter the nature of surface rocks. Additional mechanisms are at work as well. For example, the solar system still contains a large number of small bodies orbiting the Sun between the planets, and occasionally these objects strike the surface of a planet. We call this a **meteoroid impact,** and the surfaces of all the terrestrial planets, as well as the satellites of the giant planets, are scarred by the craters that result (Fig. 5.20). On the Earth, weathering processes such as wind and water erosion, as well as tectonic activity, have obliterated most of these craters, except those that have formed most recently.

On the Earth and Mars, erosion by water has been an important force affecting the structure of the surface. The flowing water that once carved channels and floodplains into the Martian surface (Fig. 5.21) has not been present for a very long time (perhaps 3 billion years), but on Earth, water is still a very important factor in surface modification. In addition, the atmospheres of the Earth and Mars alter the surface through winds and chemical processes, and the atmosphere of Venus affects its surface as well.

It is believed that the planets are all about the same age, having formed at about the same time some 4.5 billion years ago. On the other hand, the surfaces of the planets have been altered to varying degrees. Some surfaces (such as that of the Earth) are rela-

Figure 5.20   Impact craters throughout the solar system.   No planet or satellite is immune from impacts due to objects from space. Here we see a recent crater on the Earth, near Winslow, Arizona (upper left); a very ancient crater on the Moon (upper right); a crater on Venus (lower left), discovered to have an impact origin by recent *Magellan* radar images; and ancient craters on Saturn's icy moon Dione (lower right). Impact craters are characterized by high rim walls and central peaks, which can be seen in the examples shown here; volcanic craters may not have high walls and never have central peaks. (upper left: Meteor Crater Enterprises, Inc.; upper right: NASA; lower left: NASA/JPL; lower right: NASA/JPL)

were begun, and scientists found that a deficiency may also exist in that region. It now appears that thinning of the ozone layer is occurring globally and at a faster rate than was expected. This has prompted the U.S. government to hasten its plans to phase out all use of chlorofluoro- carbons in the hope that the ozone layer will stop eroding away and may perhaps even begin to recover.

The changes in the Earth's atmosphere caused by natural processes related to life-forms took millenia to occur, and they took place in concert with the development and evolution of those life-forms. Now it seems we must take measures to prevent much more rapid changes due to unnatural processes from altering our atmosphere in ways that would make the Earth inhospitable to life.

deposited in the stratosphere where ozone (O_3) absorbs ultraviolet light from the Sun. This absorption is responsible for the increase in temperature with height in the stratosphere.

The atmosphere exhibits many scales of motion, ranging from small gusts a few centimeters in size to continent-spanning flows thousands of kilometers in scale. Here we will discuss only the large-scale motions, since generally we are only able to study motions of similar scale in the atmospheres of other planets.

The primary influences on the global motions in the Earth's atmosphere are heating from the Sun, as already noted, and the rotation of the Earth. The heating creates regions where the air rises. The air must later cool and fall somewhere else, and a pattern of overturning motion called **convection** is created (Fig. 6.3). Generally, air rises in the tropics and descends at more northerly latitudes, although the situation is more complicated than that. For example, the continents tend to be warmer than the oceans in the summer, so that high-pressure regions (characterized by descending air) preferentially lie over water, whereas low-pressure regions (where air rises and cools) tend to be over land. In winter this pattern is reversed.

These tendencies, combined with the rotation of the Earth, create horizontal flows at several levels in the atmosphere. Air that is rising or falling is forced into a rotary pattern by the Earth's rotation, as described in Chapter 5. Where air rises in a low-pressure zone, the resultant swirling motion is called a **cyclone** (Fig. 6.4). In the Northern Hemisphere, the motion of a cyclone is counterclockwise; south of the equator, it is clockwise. Descending air flows in high-pressure regions are forced into oppositely directed circular patterns, called **anticyclones.** Cyclonic flows sometimes intensify into storms of great strength; hurricanes and typhoons are examples.

The net result of the vertical motions caused by the Sun's heating and the spiral motions caused by the Earth's rotation is the creation of a complex pattern of flows (Fig. 6.5) with vertical and horizontal components.

Seasonal shifts in the distribution of the Sun's energy input cause changes in the flow patterns, creating our well-known seasonal weather variations. Apparently other factors can influence the location of the high- and low-pressure regions, because the climate undergoes longer-term fluctuations whose cause

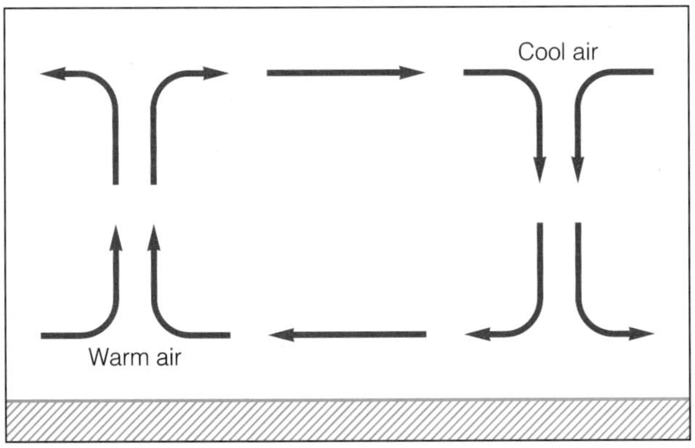

Figure 6.3 Convection in the Earth's atmosphere. This is a simplified illustration of the principle of convection, in which warm air rises, cools, and descends.

Figure 6.4 A storm system on Earth. This low-pressure region is in the Northern Hemisphere, so the circulation is counterclockwise. (NASA)

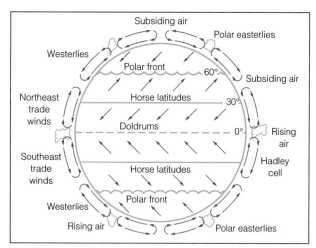

Figure 6.5   The general circulation of the Earth's atmosphere.   The circulation is made even more complex than shown here by temperature contrasts between seas and land and by surface features such as mountains. (Based on Fig. 7.3, p. 150, F. K. Lutgens and E. J. Tarbuck 1979, *The Atmosphere: An Introduction to Meteorology* [Englewood Cliffs: N.J.: Prentice Hall])

is not well understood. Some of these may be related to cyclic behavior in the Sun (see Chapter 11).

In later chapters, when we examine the flow patterns in the atmospheres of the other planets, we will see that in many ways they behave like the Earth, except for differences in the amount of solar-energy input and the rate of rotation.

THE EARTH'S INTERIOR AND MAGNETIC FIELD

At first it may seem surprising that we know much about the deep interior of the Earth. Substantial quantities of data have been collected, however, mostly by indirect means. The same indirect techniques could be applied (through the use of unmanned space probes) to other planets, and in principle we could know as much about their interiors as that of our own planet (this would not be true of the outer, gaseous planets, but it does apply to the other terrestrial planets and the Moon).

The primary probes of the Earth's interior are **seismic waves** created in the Earth as a result of major shocks, most commonly earthquakes. These waves take three possible forms, called P, S, and L waves. Studies of wave transmission in solids and liquids have shown that P waves, which are the first to arrive at a site remote from the earthquake location, are **compressional** waves, which means that the oscillating motions occur parallel to the direction of motion

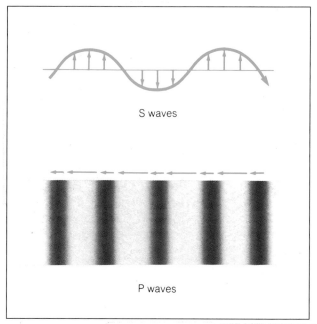

Figure 6.6   S and P waves.   An S, or shear, wave (top) consists of alternating motions transverse (perpendicular) to the direction of propagation. Waves in a tight string or on the surface of water are S waves. Compressional, or P, waves (bottom) have no transverse motion but consist of alternating dense and rarefied regions created by motions along the direction of propagation. The arrows at the top of the lower panel represent the relative speeds in and between the dense zones. Sound waves are P waves.

of the wave, creating alternating regions of high and low density without any sideways motions (Fig. 6.6). Sound waves are examples of compressional waves. The S waves are **transverse** or **shear**, waves, in which the vibrations occur at right angles to the direction of motion (see Fig. 6.6). These waves require that the material they pass through have some rigidity and, unlike the P waves, they cannot be transmitted through a liquid. The L waves travel only along the surface of the Earth and thus do not provide much information on the deep interior.

By measuring both the timing and the intensity of these seismic waves at various locations away from the site of an earthquake, scientists can determine what the Earth's interior is like. The speed of the P waves depends on the density of the material they pass through, and the distribution of the P and S waves reaching remote sites provides data on the location of liquid zones in the interior (Fig. 6.7).

The general picture that has emerged from these studies is of a layered Earth (Fig. 6.8 and Table 6.3). At the surface is a crust whose thickness varies from a few kilometers beneath the oceans to perhaps 60 km under the continents. There is a sharp break between the crust and the underlying material that is

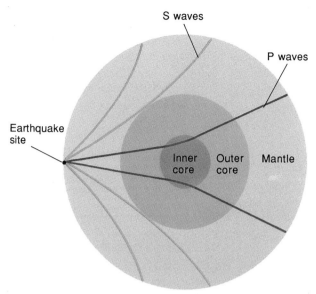

Figure 6.7   Seismic waves in the Earth.   This simplified sketch shows that P waves can pass through the core regions (although their paths may be bent), whereas S waves cannot. This led to the deduction that the outer core is liquid, because S waves, which require an elastic or solid medium, cannot penetrate liquids.

called the **mantle.** The mantle transmits S waves, so it must be solid; on the other hand, it undergoes slow, steady, flowing motions in its uppermost regions. Perhaps it is best viewed as a plastic material, one that has some rigidity, but can be deformed, given sufficient time (this is discussed more fully in the next section).

The uppermost part of the mantle and the crust together form a rigid zone called the **lithosphere.** The part of the mantle itself where the fluid motions occur, just below the lithosphere, is called the **asthenosphere.** Below the asthenosphere is a more rigid portion of the mantle that extends nearly halfway to the center of the Earth. The lower mantle is called the **mesosphere** (not to be confused with the level in the Earth's atmosphere bearing the same name).

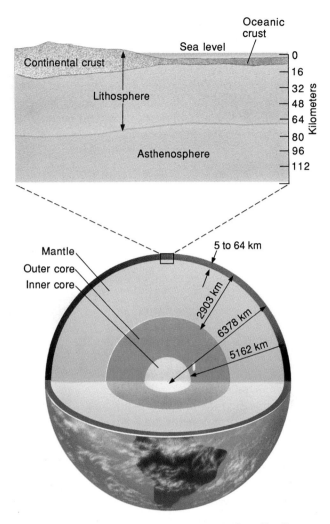

Figure 6.8   The internal structure of the Earth.   The illustration at the bottom shows the layers that are defined by contrasts in density or physical state, as deduced from observations of seismic waves. The expanded view at the top illustrates the relative thickness of the crust in continental areas as compared with the seafloor.

TABLE 6.3   Earth's Interior

Layer	Depth (km)	Density (g/cm³)	Temperature (K)	Composition
Crust	0–30	2.6–2.9	300–700	Silicates and oxides
Mantle				
Lithosphere	30–70	2.9–3.3	700–1,200	Basalt, silicates, oxides
Asthenosphere	70–1,000	3.9–4.6	1,200–3,000	Basalt, silicates, oxides
Mesosphere	1,000–2,900	4.6–9.7	3,000–4,500	Basalt, silicates, oxides
Core				
Outer core	2,900–5,100	9.7–12.7	4,500–6,000	Molten iron, nickel, cobalt
Inner core	5,100–6,378	12.7–13.0	6,000–6,400	Solid iron, nickel, cobalt

Beneath the mantle is the **core,** consisting of an **outer core** (between 2,900 and 5,100 km in depth) and an **inner core**. The S waves are not transmitted through the outer core, which is therefore thought to be liquid. As a result, the inner core can be probed only with the P waves. Since these waves travel more rapidly through the inner core, this innermost region is thought to be solid and to have a density greater than that of the outer core.

The overall density of the Earth is 5.5 times that of water (that is, 5.5 grams per cubic centimeter, or 5.5 g/cm^3). The rocks that make up the crust have typical densities of less than 3 g/cm^3, and the mantle material is thought to have relatively low density also, perhaps 3.5 g/cm^3. From these facts, scientists deduce that the core must have a density of roughly 13 g/cm^3. The most likely substance that could give the core this high density is iron, probably combined with some nickel and perhaps cobalt.

The high density and probable metallic composition of the core are quite significant, for they show that the Earth has undergone differentiation (Fig. 6.9). As explained in Chapter 5, differentiation can occur only when the planet is in a molten state; thus, the Earth must once have been largely liquid, probably early in its history, soon after it formed. The principal cause of the Earth's molten state most likely is radioactive heating in the interior. Several naturally occurring elements, such as uranium, thorium, and potassium, are radioactive, which means that their nuclei spontaneously emit subatomic particles and, over a long period of time, produce substantial quan-

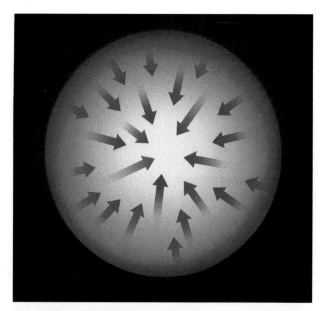

Figure 6.9   Differentiation.   Heavy elements sank toward the Earth's center, creating the dense core. This process, which took place very early in the Earth's history, required that the Earth be partially or wholly liquid.

tities of heat. Another source of heat early in the Earth's history was created by the energy of impact as the smaller bodies that merged to form the Earth came together. Additional heating was caused by the process of differentiation itself, as the heavy elements sank to the center of the Earth, releasing gravitational potential energy. We shall see that differentiation has not occurred in all the bodies in the solar system, which tells us something significant about their histories.

The fluid interior portions of the Earth give rise to its magnetic field. It is convenient to visualize the structure of the field by imagining magnetic lines of force connecting the two poles; these correspond to the lines along which iron filings lie when placed near a small bar magnet. In cross-sectional view, the Earth's magnetic field is reminiscent of a cut apple, but one that is lopsided because of the flow of charged particles from the Sun that constantly sweep past the Earth (Fig. 6.10; also see the discussion of the **solar wind** in Chapter 11). The region enclosed by the field lines is called the **magnetosphere,** and it acts as a shield, preventing the charged particles from reaching the Earth's surface.

The axis of the Earth's magnetic field is aligned closely (within $11\frac{1}{2}°$) with the rotation axis of the planet, but this has not always been the case. The past alignment of the magnetic field can be ascertained from studies of certain rocks that contain iron-bearing minerals whose crystalline structure is aligned with the direction of the magnetic field at the time the rocks solidified from a molten state. Thus, traces of the Earth's ancient magnetic field, called **paleomagnetism,** can be detected from the analysis of magnetic alignments of rocks. Such studies reveal that the magnetic poles have moved about during the Earth's history, and that the north and south magnetic poles have completely and rather suddenly reversed from time to time; that is, the north magnetic pole has moved to the south and vice versa. This flip-flop of the magnetic poles seems to have happened at irregular intervals, typically thousands to hundreds of thousands of years apart.

The source of the Earth's magnetic field is not fully understood, although it is nearly certain that it is related to the iron-nickel core. A magnetic field is produced by flowing electrical charges, as in the current flowing through a wire wound around a metal rod, which is the basis of an electromagnet. Convection and the Earth's rotation may combine to create systematic flows in the liquid outer core, giving rise to the magnetic field, if the core material carries an electrical charge. This general type of mechanism is called a **magnetic dynamo.** The cause of the reversals of the poles is not well understood but presumably could be related to occasional changes in the direction of the flow of material in the core.

Figure 6.10   The Earth's magnetic field structure.   This cross-sectional view shows the Earth's field and how its shape is affected by the stream of charged particles from the Sun known as the solar wind.

We have referred to the magnetosphere and its ability to control the motions of charged particles. When the first U.S. satellite was launched in 1958, zones high above the Earth's surface were discovered to contain intense concentrations of charged particles, primarily protons and electrons. There are several distinct zones containing these particles, and they are now called the **Van Allen belts** (Fig. 6.11) after the physicist who first recognized their existence and deduced their properties.

The charged particles, or ions, in the Van Allen belts were captured from space (primarily from the solar wind) and are forced by the magnetic field to spiral around the lines of force. The same phenomenon occurs in the uppermost portion of the outer atmosphere, called the **ionosphere,** which extends upward from a height of about 60 km. When a reversal of the Earth's magnetic poles is taking place, there is, for a short period of time, a much-weakened magnetic field and consequently a major disruption of the Van Allen belts and the ionosphere. The magnetosphere is greatly diminished, and charged particles from space are more likely to penetrate to the ground. These particles, particularly the very rapidly moving ones called **cosmic rays,** can cause important effects on life-forms, including genetic mutations. The sporadic reversals of the Earth's magnetic field may have played a major role in shaping the evolution of life on the surface of our planet.

The ionosphere has important effects for us on the

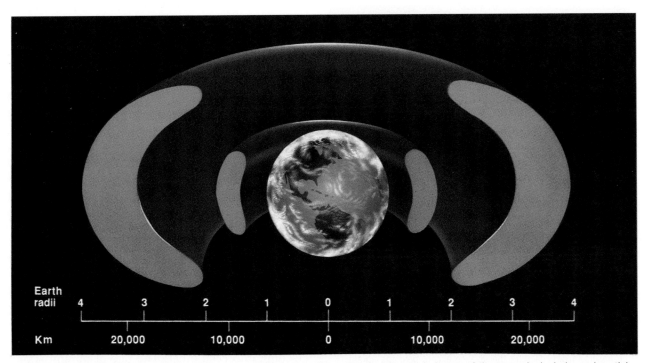

Figure 6.11   The Van Allen belts.   This cross-sectional view illustrates the sizes and shapes of the two principal charged-particle belts that girdle the Earth.

surface, including enhanced radio communications and the beautiful light displays (Fig. 6.12) known as **aurora borealis** (northern lights) and **aurora australis** (southern lights). The aurorae are caused by charged particles that enter the atmosphere and collide with atoms and molecules, exciting them so that they emit light. The aurorae occur most commonly near the poles, where the magnetic field lines allow particles to penetrate closest to the ground. The ionosphere contributes to radio communications because it reflects signals in the short-wave band, enabling them to travel around the Earth. When there are fluctuations in the solar wind, particularly following solar flares, enhanced fluxes of charged particles entering the ionosphere from space can disrupt radio communications.

Figure 6.12   The aurorae.   The upper photograph shows the aurora borealis, or northern lights, as seen from the ground at middle latitudes in the Northern Hemisphere. The lower image shows the aurora australis, or southern lights, as photographed from space by astronauts aboard the space shuttle *Discovery*. The Earth's magnetic field prevents the ions that create the aurorae from reaching the surface except near the poles, where the magnetic field reaches low altitudes. (Upper: © Ned Haines, Photo Researchers; lower: NASA)

A CRUST IN ACTION

Nearly three-fourths of the Earth's surface is covered by water, the rest taking the form of several major continents. As we explained in Chapter 5, the basic substance of the Earth's crust is rock, and rocks are classified into three basic groups according to their origin: **igneous, sedimentary,** and **metamorphic.** Igneous rocks, formed from volcanic activity, consist of cooled and solidified **magma,** the molten material that flows to the surface during volcanic eruptions. Sedimentary rocks are formed from deposits of gravel and soil that have hardened, usually in layers where old seabeds or coastlines lay. Metamorphic rocks have been altered in structure by heat and pressure created by movements in the Earth's crust. All three forms can be changed from one to another, in a continuous recycling process. Rocks are also classified according to their chemical compositions, as **minerals** of various types. The most common minerals are the silicates, which comprise some 90 percent of all rock on the Earth's surface.

Because of various evolutionary processes, the surface of the Earth is continuously being renewed. While the age of the Earth is thought to be some 4.5 billion years, the ages of most surface rocks can be measured in the millions or hundreds of millions of years. As we pointed out in Chapter 5, the distinction between the age of a planet and the age of its surface is an important one, and it will reappear as we discuss the surfaces of other bodies.

The crust of the Earth is not static, but is in constant motion. The continents themselves move about, and the world map is variable on geological time scales (Fig. 6.13). The idea that continental drift occurs is supported by a variety of evidence, ranging from the obvious fit between landmasses on opposite sides of the Atlantic to similarities of mineral types and fossils and the alignment of vestigial magnetic fields in rocks that once were together in the same place. In modern times more direct evidence, such as the detection of seafloor spreading away from undersea ridges and the actual measurement of continental motions using sophisticated laser-ranging techniques, has removed any doubt that pieces of the Earth's crust are in motion.

From all of this evidence has arisen a theory of **plate tectonics,** which postulates that the Earth's crust (the lithosphere) is made of a few large, thin pieces that float on top of the asthenosphere (Fig. 6.14). Due to flowing motions in the asthenosphere, these plates constantly move about, occasionally crashing into each other. The rate of motion is only a few centimeters per year at the most, and the major rearrangements of the continents have taken many millions of years to occur. (The Americas and the European-African system became separated from each other between 150 and 200 million years ago.)

The driving force in the shifting of the plates is not well understood. The most widely suspected cause is convection currents in the mantle (Fig. 6.15). This is the same process referred to earlier (in the discussion of the Earth's atmosphere), in which temperature differences between levels cause an overturning motion. If a fluid is hot at the bottom and much cooler at the top, warm material will rise, cool, and descend again, creating a constant churning. The speed of the overturning motions depends largely on the viscosity of the fluid; that is, the degree to which it resists flowing freely. The Earth's mantle, as we have already seen, is sufficiently rigid to transmit S waves and must therefore have a high viscosity. Hence if convection is occurring in the mantle, it is reasonable to expect that the motion is very slow.

Whether convection is the cause or not, the effects of tectonic activity are becoming well known. Where plates collide, one may submerge below the other in a process called **subduction** (see Fig. 6.15), creating an undersea trench and volcanoes along the adjacent shoreline; alternatively, particularly if both plates carry continents, the collision may force the uplifting

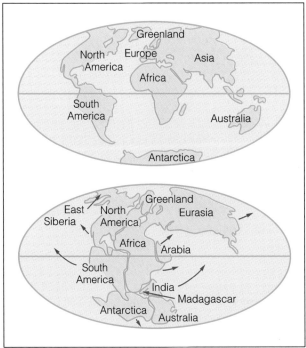

Figure 6.13   Continental drift.   These maps show the distribution of the continents today and as it was some 200 million years ago. The red arrows show the direction of continental motion.

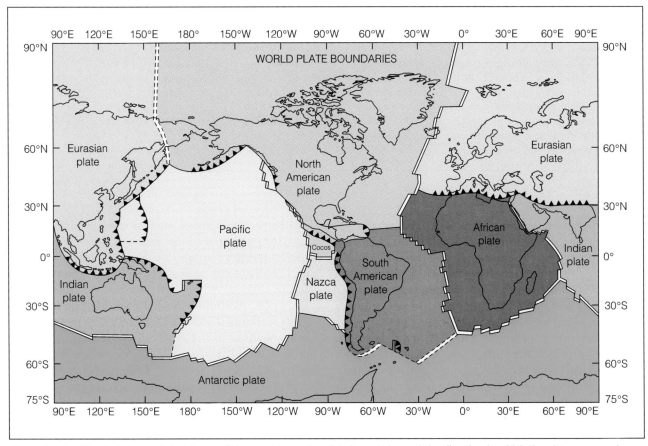

Figure 6.14   The Earth's crustal plates.   This map shows the plate boundaries and the direction in which the plates are moving.

ASTRONOMICAL INSIGHT
The Ages of Rock

We have spoken of the ages of rocks and of the Earth itself, but we have said little about how these ages are measured. The most direct technique, **radioactive dating,** involves the measurement of relative abundances of closely related atomic elements in rocks.

Recall from our earlier discussion of atomic structure (Chapter 4) that the nucleus of an atom consists of protons and neutrons, and that the number of protons (that is, the **atomic number**) determines the identity of the element. Different **isotopes** of an element have the same number of protons but differing numbers of neutrons. In most elements, the number of protons is not very different from the number of neutrons, but there are exceptions. If the imbalance between protons and neutrons is large enough, the element is unstable, meaning that it has a natural tendency to correct the imbalance by undergoing a spontaneous nuclear reaction. The reactions that occur in this way usually involve the emission of a subatomic particle, and the element is said to be **radioactive.** The emitted particle may be an **alpha particle** (that is, a helium nucleus consisting of two protons and two neutrons) or, as is more often the case, an

electron or a **positron,** a tiny particle with the mass of an electron but with a positive electrical charge. When a positron or an electron is emitted, the reaction is called **beta decay;** at the same time, a proton in the nucleus is converted into a neutron (if a positron is emitted) or a neutron is changed into a proton (if an electron is released). If the number of protons in the nucleus is altered, the identity of the element is changed. A typical example is the conversion of potassium 40 (^{40}K, with 19 protons and 21 neutrons in its nucleus) into argon 40 (^{40}Ar, with 18 protons and 22 neutrons).

Several elements are known to be radioactive, naturally changing their identity by emitting particles. The rate of change, expressed in terms of **half-life**, is known in most cases. The half-life is the time it takes for half of the original element to be converted. It may be as short as fractions of a second or as long as billions of years. The very slow reactions are the ones that are useful in measuring the ages of rocks.

If we know the relative abundances of the elements that were present in a rock when it formed, then measurement of the ratio of the elements in that rock at the present time can tell us how long the radio-

active decay has been at work; that is, the age of the rock. In a very old rock, for example, there might be almost no ^{40}K, but a lot of ^{40}Ar. The decay of ^{40}K to ^{40}Ar has a half-life of 1.3 billion years, so from the exact ratio of these two species in a rock, we can infer how many periods of 1.3 billion years have passed since the rock formed.

Other decay processes that are useful in dating rocks include the decay of rubidium 87 (^{87}Rb) into strontium 87 (^{87}Sr), which has a half-life of 47 billion years; and the decays of two different isotopes of uranium to isotopes of lead, ^{235}U to ^{207}Pb and ^{238}U to ^{206}Pb, with half-lives of 700 million years and 4.5 billion years, respectively. (Note that these decays each involve several intermediate steps and that the half-lives given represent the total time for all of those steps.)

Use of these dating techniques has shown that the oldest rocks on the Earth's surface are about 3.5 billion years old and that ages of a few hundred million years are more common. The age of the solar system, and therefore of the Earth itself, is estimated, from isotope ratios in meteorites, to be about 4.5 billion years.

---

of high mountain ranges The lofty Himalayas are thought to have been created when the Indian plate collided with the Eurasion plate some 50 million years ago. The boundaries where plates either collide or separate are marked by a wide variety of geological activity. Earthquakes are attributed to the sporadic shifting of adjacent plates along fault lines, and volcanic activity is common where material from the mantle can reach the surface, most often along plate boundaries. It has been long recognized, for example, that a great deal of earthquake and volcanic activity is concentrated around the shores of the Pacific, defining the so-called "Ring of Fire." Now it is understood that this region represents the boundaries of the Pacific plate.

Chains of volcanoes, such as the Hawaiian Islands, are also attributed to the action of plate tectonics. In several locations around the world, volcanic hot spots that lie deep and are fixed in place bring molten rock to the surface. As the crust passes over one of these hot spots, volcanoes are formed and then carried away, creating a chain of mountains as new material keeps coming to the surface over the hot spot. This process is still active, and the Hawaiian Islands, for example, are still growing. There is volcanic activity on the southeastern part of the largest (and youngest) island, Hawaii. Furthermore, a new island, already given the name Loihi, is rising from the seafloor about 20 km south of Hawaii. Its peak has already risen 80 percent of the way to the surface and only

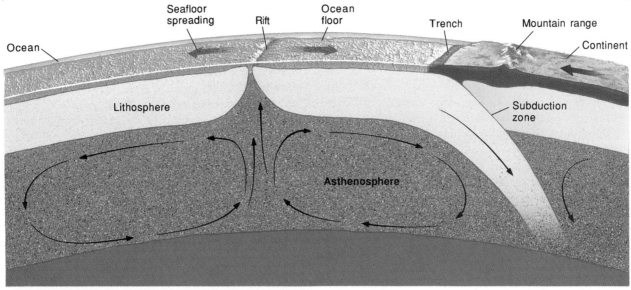

Figure 6.15   Schematic of the mechanisms of continental drift.   This sketch shows how continental drift is responsible for uplifted mountain ranges and parallel undersea trenches, where one crustal plate sinks below another (subduction zone), and how a mid-ocean ridge is built up where two plates move away from each other. The red arrows indicate the convection currents in the upper mantle that are thought to be responsible for continental drift.

has about 1 km to go, but it will take an estimated 50,000 years to break through.

Plate tectonics, then, accounts for many of the most prominent features of the Earth's surface. Of course, other processes, such as running water, wind erosion, and glaciation, are also important in modifying the face of the Earth.

tures, the most prominent of which are the light and dark areas. The latter, thought by Renaissance astronomers to be bodies of water, are called **maria** (singular: **mare**) after the Latin word for *seas*. Closer examination of the Moon with even primitive telescopes also revealed numerous circular features called **craters** after similar structures found on volcanoes. The

EXPLORING THE MOON

Seen from afar, the Moon (Table 6.4) is a very impressive sight, especially when full (Fig. 6.16). Its ½° diameter is pockmarked with a variety of surface fea-

TABLE 6.4   Moon

Mean distance from Earth: 384,401 km (60.4 $R_\oplus$)
  Closest approach: 363,297 km
  Greatest distance: 405,505 km
Orbital sidereal period: $27^d7^h43^m12^s$
Synodic period (lunar month): $29^d12^h44^m3^s$
Orbital inclination: 5°8'43"

Rotation period: 27.32 days
Tilt of axis: 6°41' (with respect to orbital plane)

Diameter: 3,476 km (0.273 $D_\oplus$)
Mass: 7.35×10^{25} grams (0.0123 $M_\oplus$)
Density: 3.34 grams/cm³
Surface gravity: 0.165 Earth gravity
Escape speed: 2.4 km/sec

Surface temperature: 400 K (day side); 100 K (dark side)
Albedo: 0.07

Figure 6.16   The full Moon.   This high-quality photograph was taken through a telescope on the Earth. (Lick Observatory, © Regents University of California)

lunar craters are not volcanic, however, but were formed by the impacts of bodies that crashed onto the surface from space.

The pattern of surface markings on the face of the Moon is quite distinctive, and because it never changes, observers recognized long ago that the Moon always keeps the same face toward the Earth. The far side cannot be observed from the Earth, but has now been thoroughly mapped by spacecraft (Fig. 6.17).

The lunar surface in general can be divided into lowlands, primarily the maria, and highlands, which are vast mountainous regions, not as well organized into chains or ranges as mountains on the Earth. With no trace of an atmosphere (because the escape velocity is so low that all the volatile gases were able to escape long ago), erosion due to weathering does not occur on the Moon, so large features such as craters and mountain ranges retain a jagged appearance when seen from afar. The surface everywhere consists of loosely piled rocks ejected and redistributed by impacts. Other types of features seen on the Moon include **rays,** light-colored streaks emanating from some of the large craters, and **rilles,** winding valleys that resemble earthly canyons.

The Moon's mass is only about 1.2 percent of the mass of the Earth, and its radius is a little more than one-fourth the Earth's radius. The Moon's density, 3.34 grams/cm^3, is below that of the Earth as a whole,

more closely resembling the density of ordinary surface rock. This indicates that the Moon probably does not have a large, dense core, and that it therefore has a low iron content. Even before the space program allowed direct measurements, infrared measurements showed that the lunar surface is subject to hostile temperatures, ranging from 100 K ($-279°$F) during the two-week night to 400 K (261°F) during the lunar day.

Of course, all of the long-range studies of the Moon were almost instantly made obsolete by the space program, which has featured extensive exploration of the Moon, first by unmanned probes, and then by manned landings (Figs. 6.18 and 6.19). The

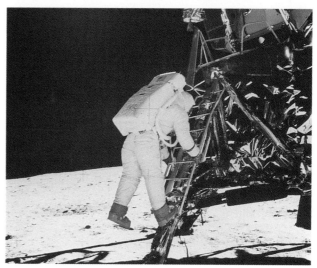

Figure 6.18   Man on the Moon.   The *Apollo* missions, six of which included successful manned landings on the Moon, are humankind's only attempt so far to visit another world. (NASA)

Figure 6.17   The Moon as seen from space.   This view, obtained by one of the *Apollo* missions, shows portions of the near (left) and far (right) sides. On the left horizon is Mare Crisium, and at left center are Mare Marginis (upper) and Mare Smythii (lower). No maria are visible in the right-hand half of this image; there are almost none on the entire lunar far side. (NASA)

Figure 6.19   The lunar rover.   The later *Apollo* missions used these vehicles to travel over the Moon's surface, allowing the astronauts to explore widely in the vicinity of the landing sites. (NASA)

historic first manned landing occurred on July 20, 1969. This was the *Apollo 11* mission; it was followed by five more manned landings, the last being *Apollo 17*, which took place in late 1972. Each mission incorporated a number of scientific experiments, some involving observations of the Sun and other celestial bodies from the airless Moon, but most devoted to the study of the Moon itself.

A Scarred Surface and a Dormant Interior

Viewed on any scale, from the largest to the smallest, the Moon's surface is irregular, marked throughout by a variety of features. We have already mentioned the maria, the large, relatively smooth, dark areas (Fig. 6.20). The maria appear darker than their surroundings because they have a relatively low albedo (that is, they reflect less sunlight). Despite their smooth appearance relative to the more chaotic terrain seen elsewhere on the Moon, the maria are marked here and there by craters.

Outside the maria, much of the lunar surface is covered by rough, mountainous terrain. Even though the maria dominate the near side of the Moon, there are almost none on the far side, and the highland regions actually cover most of the lunar surface.

Craters are everywhere (Figs. 6.21 and 6.22). They range in diameter from hundreds of kilometers to microscopic pits that can be seen only under intense magnification. In some regions, the craters are so

densely packed together that they overlap. The lack of erosion on the Moon allows craters to survive for billions of years, providing plenty of time for younger craters to form within the older ones. The fact that relatively few craters are seen in the maria indicates that the surface in these regions has been transformed after most of the cratering had already occurred.

All of the craters on the Moon are **impact craters,** formed by collisions of interplanetary rocks and debris with the lunar surface, rather than by volcanic eruptions. This conclusion is based on the shapes of

Figure 6.21   Craters on the Moon.   This photograph, taken by *Apollo 16* astronauts, shows cratered terrain on the lunar far side. A portion of the gamma-ray spectrometer carried by *Apollo 16* is visible at the right. (NASA)

Figure 6.20   The lunar "seas."   Here is a broad vista encompassing portions of three maria. Mare Crisium (foreground); Mare Tranquilitatis (beyond Mare Crisium); and Mare Serenitatis (on the horizon at the upper right). These relatively smooth areas are younger than most of the lunar surface, having been formed by lava flows after much of the cratering had already occurred. (NASA)

Figure 6.22   *Apollo 12* descending near the crater Herschel.   (NASA)

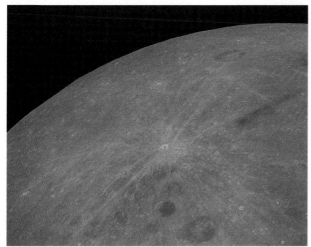

Figure 6.23   A crater with ejecta.   This crater on the lunar far side is a good example of a case in which material ejected by the impact has created rays of light-colored ejecta. Close examination of such features often reveals secondary craters, formed by the impacts of debris blasted out of the lunar surface by the primary impact. (NASA)

Figure 6.24   Rilles.   This photograph shows Hadley rille meandering through the Hadley-Appenine area. One of the *Apollo* landings was close enough to Hadley rille to allow it to be explored by astronauts. (NASA)

the craters, the central peaks in some of them, and the trails left by **ejecta,** or cast-off material created by the impacts (Fig. 6.23). The rays stretching away from craters are strings of smaller craters formed by the ejecta from the large, central crater.

Some of the lunar mountain ranges reach heights greater than any on Earth. They are more jagged than earthly mountains because there is no erosion, and they lack the prominent drainage features usually found in terrestrial ranges.

The **rilles,** which resemble dry riverbeds, are rather interesting features (Fig. 6.24). Apparently they were formed by flowing lava rather than water. In some cases there are even lava tubes that have partially collapsed, leaving trails of sinkholes. The rilles and the maria indicate that the Moon has undergone stages when large portions of its surface were molten.

The *Apollo* astronauts found a surface strewn in many areas with loose rock, ranging in size from pebbles to boulders as big as a house (Figs. 6.25 and 6.26). The rocks generally have sharp edges, due to the lack of erosion, and occasional cracks and fractures. In most cases the large boulders appear to have been ejected from nearly craters and are therefore thought to represent material from beneath the surface.

The *Apollo* astronauts brought almost 800 pounds of small lunar rocks back to Earth, giving scientists an opportunity to study the lunar surface characteristics in as much detail as is possible for Earth rocks and soils (Fig. 6.27). Thus, extensive chemical anal-

ysis and close-up observation of rocks in place on the Moon were both possible. The rock samples from the Moon are currently housed in numerous scientific laboratories around the world, where analysis continues. Specimens have also found their way into museums, and visitors to the Smithsonian's Air and Space Museum in Washington, D.C., can touch a lunar rock.

The lunar soil, called the **regolith,** consists of loosely packed rock fragments and small glassy mineral deposits probably created by the heat of meteor impacts. In addition to the loose soil, a few distinct types of surface specimens have been recognized on the basis of morphology. The most common of these are the **breccias** (Fig. 6.27), which consist of small rock fragments cemented together and resembling chunks of concrete. Similar kinds of rock are found on Earth, except that the Earth breccias are formed in streambeds where water plays a role in shaping them. The lunar breccias contain jagged, sharp rock fragments and were probably fused together by heat and pressure created in meteor impacts.

All lunar rocks are pitted, on the side that is exposed to space, with tiny craters called **micrometeorite craters.** These are formed by the impact of tiny bits of interplanetary material no bigger than grains of dust.

Figure 6.25   A field of boulders.   Rocks in a wide variety of sizes are strewn over much of the Moon's surface. Most have been blasted out of the surface by impacts. (NASA)

Figure 6.26   A large boulder.   Rocks on the lunar surface range in size from tiny pebbles to massive objects like this. (NASA)

Radioactive dating techniques showed the lunar rocks to be very old by earthly standards—as old as 3.5 to 4.5 billion years. Rocks from the maria are not quite so old, but still date back to 3 billion years ago or earlier.

Some of the experiments done on the Moon by the *Apollo* astronauts were aimed at revealing the interior conditions by monitoring seismic waves caused by earthquakes on the Moon. Therefore, the astronauts carried with them devices for sensing vibrations in the lunar crust, and, because it was not known whether natural quakes occurred frequently, they also brought along devices for thumping the surface to make it vibrate. It turned out that natural moonquakes do occur, although not with great energy. The seismic measurements continued after the *Apollo* landings, with data radioed to Earth by instruments left in place on the lunar surface.

The measurements showed that the regolith is typically about 10 meters thick and is supported by a thicker layer of loose rubble. The crust is 50 to 100 km thick at the *Apollo* sites and may be somewhat

Figure 6.27   Moon rocks.   One of the thousands of lunar samples brought back to Earth by the *Apollo* astronauts, the rock at the left is an example of a breccia. At the right is a thin slice of a Moon rock, illuminated by polarized light, which causes different crystal structures to appear as different colors. (NASA)

thicker than this on the far side where there are no maria (Fig. 6.28). Beneath the crust is a mantle, consisting of a well-defined lithosphere, which is rigid, and beneath that an asthenosphere, which is semiliquid. The innermost 500 km consists of a relatively dense core, but not as dense as that of the Earth.

The seismic measurements indicate no truly molten zones in the Moon at the present time, although temperature sensors on the surface discovered a substantial heat flow from the interior. This is probably caused by radioactive minerals below the surface.

The Moon has no detectable overall magnetic field, which is further evidence against a molten core. The chief distinction among the internal zones in the lunar interior is density, with the densest material closest to the center. Thus moderate differentiation has occurred in the Moon, implying that it was once at least partially molten.

There is no trace of present-day lunar tectonic activity, perhaps the greatest single departure from the geology of the Earth. The force that has had the greatest influence in shaping the face of the Earth has no role today on the Moon. Instead, the lunar surface is entirely the result of the way the Moon was formed (which may have involved some tectonic activity in the early stages) and the manner in which it has been altered by lava flows and by the incessant bombardment of debris from space.

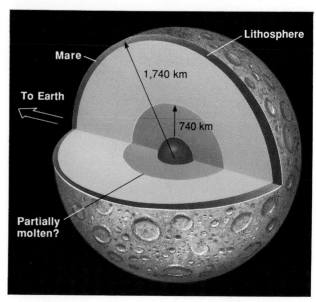

Figure 6.28   The internal structure of the Moon.   This cross-sectional view shows the interior zones inferred from seismological studies. The existence of a dense core is not certain. Note that the maria lie almost exclusively on the side facing the Earth, where the lunar crust is relatively thin.

THE DEVELOPMENT OF THE EARTH-MOON SYSTEM

While the formation of the Earth (and of planets in general) is reasonably well understood, there is a great deal of uncertainty about the Moon's formation. We will discuss the Earth's formation first, then delve into the complexities of the Moon's origin.

Formation of the Earth and Its Atmosphere

The age of the Earth-Moon system, estimated from geological evidence, is about 4.5 billion years. The Earth is thought to have formed from the coalescence of a number of **planetesimals,** solid bodies that were the first to condense in the earliest days of the solar system. Most or all of the Earth was molten at some point during the first billion years. The Earth became highly differentiated during its molten period, as the heavy elements tended to sink toward the planetary core. At the same time, volatile gases were emitted by the newly formed rocks at the surface. These gases, including hydrogen (H_2), ammonia (NH_3), methane (CH_4), and water vapor (H_2O), formed the earliest atmosphere of the Earth. According to some theories

of solar system formation, the solid material from which the Earth condensed did not contain much water vapor, so the Earth's oceans require another explanation. It is possible that most of the water originated in comets, which were much more numerous in the inner solar system during the first billion years than they are today. Hence, some scientists believe that most of the water in the Earth's oceans was accreted through the impacts of comets early in the Earth's history, rather than being present from the very beginning.

Whatever its origin, water collected on the surface, and from its earliest days our planet had seas. Carbon dioxide (CO_2) was probably abundant for a time but was eventually removed from the atmosphere by being absorbed into rocks. This absorption process, of critical importance to the further evolution of the Earth's atmosphere, depended on the presence of liquid water; if the early Earth had not had oceans, the carbon dioxide might never have left the atmosphere. This would have had fateful consequences for the Earth, which will become clear in our discussion of the evolution of Venus (Chapter 7).

Nearly all of the hydrogen escaped into space by the time the Earth was about a billion years old (about when the first simple life-forms apparently appeared). As discussed in Chapter 5, the individual molecules in a gas move around randomly at speeds that depend on both the temperature of the gas and the mass of the particles. At a given temperature the lightweight molecules move fastest and are therefore most likely to escape the planet's gravitational attraction. This is what happened to the hydrogen in the early atmosphere of the Earth, while the heavier gases that now dominate the atmosphere were too massive to escape. Another gas that escaped early in the Earth's history is helium, the second most abundant element in the universe, but one so rare on Earth that it was not discovered until spectroscopic measurements revealed its presence in the Sun.

The critical reactions that led to the development of life on Earth must have taken place before all the hydrogen escaped, because the types of reactions that were probably responsible involve this element. The earliest fossil evidence for primitive life dates back at least 3 billion years, when some hydrogen was still left.

Essentially no free oxygen was present in the atmosphere until the development of life-forms that released this element as a by-product of their metabolic activities. Most plants release oxygen into the atmosphere during photosynthesis; as a supply of this element built up, the opportunity arose for complex animal forms to evolve. Besides providing the oxygen necessary for the metabolisms of living animals, the buildup of oxygen created a reservoir of ozone (O_3) in the upper atmosphere. The ozone in turn began to screen out the harmful ultraviolet rays from the Sun, allowing life-forms to move onto the exposed land. Once life-forms had gained a toehold on the continental landmasses, the process of converting the atmosphere to its present state began to accelerate. Soon nitrogen from the decomposition of organic matter began to be released into the air in large quantities (some nitrogen was already present, due to volcanic activity), and by the time the Earth was perhaps 2 billion years old, the atmosphere had reached approximately the composition it has today.

By this time also, the mantle had solidified and the crust had hardened. (The oldest known surface rocks are nearly 3.5 billion years old.) The interior has remained warmer than the surface, because heat escapes slowly through the crust and because radioactive heating of the interior took place over a long period of time and is probably still effective today.

Origin of the Moon

The Moon's beginnings are much more obscure than the Earth's, despite the close-up examination of the Moon afforded by the *Apollo* missions. The reasons for the difficulty in explaining how the Moon formed lie in the very unusual chemical composition of the Moon, and the large amount of angular momentum of the Earth-Moon system. Recall that angular momentum is related to the mass, orbital speed, and separation of two bodies in mutual orbit. Compared with other planet-satellite pairs in the solar system, the Moon is very large relative to its parent planet, and the Earth-Moon system has a high angular momentum. This has been difficult to explain in any theory postulating that the Moon formed by splitting off from the Earth, for example, or that the Earth and Moon simply formed together as a double planet. In other words, scientists have had difficulty explaining why the material that formed the Moon remained separate from the Earth, instead of coalescing with it.

The chemical composition of the Moon is characterized by a low overall abundance of heavy elements, such as iron, and a higher relative abundance of the **refractory** elements, those that do not vaporize easily. The readily vaporized **volatile** elements are underabundant in comparison with the Earth, and water is almost completely absent. These contrasts might suggest that the Moon formed somewhere else in the solar system rather than near the Earth, yet there are some strong similarities, such as the relative proportions of different isotopes of oxygen, that argue for formation in the same general vicinity. It seems that most of the chemical properties of the Moon could be explained if the Earth and Moon formed out of very similar material originally, but then somehow the Moon lost much of its iron and was subjected to substantial heating early in its history.

Traditionally three classes of models for lunar formation have been advanced: (1) the fission hypothesis, in which the Moon somehow split off from the Earth; (2) the capture theory, in which the Moon formed elsewhere and was then captured by the Earth's gravitational field; and (3) the coeval formation, or double planet model, in which the two bodies formed together. Each of these has always had serious difficulties. The fission theory lacks a mechanism for causing the split of the young Earth, lacks an explanation for the high angular momentum of the Earth-Moon system, and cannot easily explain the low iron content of the Moon or the other chemical dissimilarities. The capture hypothesis can easily account for the high angular momentum, since the Moon would have approached the Earth with a substantial velocity that would have been transformed into orbital angular momentum upon capture, but that same velocity is the downfall of the theory, for it is difficult or impossible for an initially free Moon to be slowed enough to be caught by the Earth's gravity. This model also has difficulty explaining the chemical similarities between the Earth and the Moon, as well as the low iron content of the Moon. The coeval formation hypothesis has grave problems with the chemical contrasts between Earth and Moon, since this view holds that the two formed together from the same material, and this theory also founders on the angular momentum problem.

A recent suggestion appears to resolve all of the problems related to angular momentum and chemical composition and is therefore becoming the favorite among lunar scientists. This idea includes some elements of the capture model, in that it invokes a large body approaching the Earth, but there the similarity ends. In this model, the body that approaches is a large planetesimal, at least 10 percent the mass of the Earth (hence at least the size of Mars). It is proposed that this large body, already differentiated and having an iron core, collided with the young Earth at a grazing angle (Fig. 6.29) and was partially vaporized by the energy of the impact. Material from the body's iron-poor mantle formed a hot disk around the Earth, which subsequently coalesced to form the Moon. The result is a Moon having a high angular momentum,

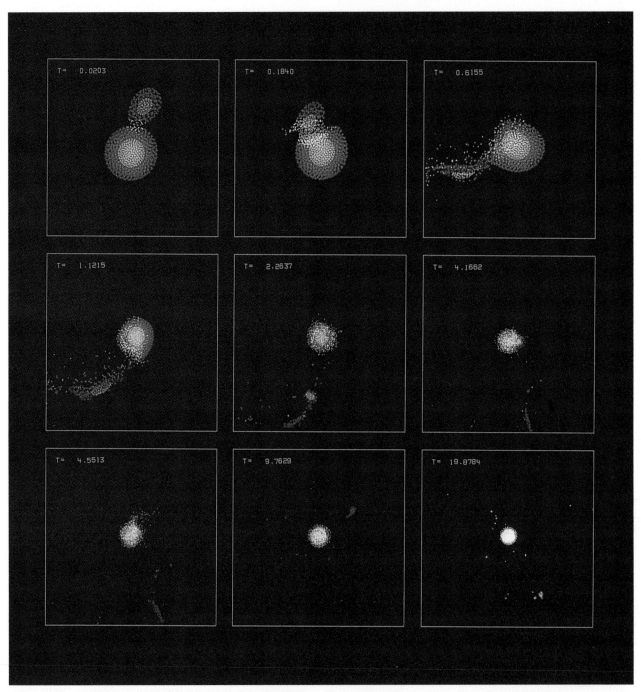

Figure 6.29   A computer simulation of the Moon's formation.   The newest theory of lunar formation envisions that a planetesimal as massive as Mars or larger collided with the young Earth and was partially vaporized on impact. This sequence of computerized illustrations shows how the resulting debris could then have coalesced to form the Moon. (W. Benz and W. Slattery, Los Alamos National Laboratories)

derived from the speed (about 10 km/sec) of impact; a low overall iron abundance, due to the fact that the incoming planetesimal had already differentiated and only its mantle material formed the Moon; and a very low content of volatile elements, since these were vaporized by the heating that occurred on impact, and lost to space.

Invoking a singular event such as a collision with a Mars-size planetesimal may seem contrived, but studies of the process by which early solar system material condensed into solid form and then built up planetesimals shows that this hypothesis is quite reasonable. There should have been a period of time when most of the preplanetary material was concentrated in a few rather large planetesimals, all orbiting the Sun in the same direction. The final formation of the planets came about as these large planetesimals merged, so the early history of each planet would have involved collisions between large bodies. Evidence for this is found in some other anomalous planetary characteristics: Venus has a backwards, slow rotation, for example, and Uranus is tipped over on its side.

History of the Moon

The Moon's evolution after its formation has not been as difficult to deduce. Detailed study of the surface structure of the Moon, along with analyses of surface rocks and information on the interior structure gleaned from seismic data, have revealed a comprehensive picture.

The Moon underwent a period when it was molten to a depth of about 400 km (this was another factor that was more easily explained by the planetesimal-collision formation model than by the others). As the crust cooled and solidified, impacts began to sprinkle it with craters. Within the first 1 or 2 billion years, extensive lava flows occurred on the surface, welling up in regions where the crust was relatively thin. Much of one side of the Moon was covered by the flows, which we know today as maria. The rate of impacts from space began to taper off at about the time the maria formed; hence, they have fewer craters than the highlands, where the high concentration of craters records the high frequency of collisions suffered by the young Moon. Impacts have continued throughout the history of the Moon, but following that early, intense period, the rate has been much slower and quite steady.

The Moon was much closer to the Earth when it formed than it is today, but tidal stresses exerted on it by the Earth have gradually slowed the Moon's spin, causing it to retreat from the Earth as angular momentum due to its spin has been converted into orbital angular momentum. The Moon's rotation slowed to the point where it keeps one face toward the Earth as it orbits; this is the face containing the maria, whose high density causes the Moon to be slightly nonsymmetric in its mass distribution. The relatively higher density of the side with the maria caused this side to "lock in" facing the Earth; thus it is no coincidence that the near side is the face with the maria.

Geologically, the Moon now is quite inert, having no molten core and insufficient internal heat to drive such processes as tectonic activity. The only changes that are expected to occur over the next few billion years will be the addition of new craters.

PERSPECTIVE

We have itemized the overall properties of the Earth and the Moon, the two best-studied objects in the solar system. In doing so we have discussed many principles of planetary structure and evolution that will be applied to the other planets in the coming chapters. We have seen that the Earth and Moon have had radically different evolutions, the Earth remaining a vital, dynamic body while the Moon has been nearly dormant for billions of years. In the next chapter we discuss the other terrestrial planets: Mercury, Venus, and Mars.

SUMMARY

1. The Earth's atmosphere is 80 percent nitrogen and about 20 percent oxygen, with only traces of water vapor and carbon dioxide.

2. The atmosphere is divided vertically into four temperature zones: the troposphere, the stratosphere, the mesosphere, and the thermosphere.

3. The Sun's heating and the Earth's rotation create the global wind patterns.

4. The Earth's interior, explored with seismic waves, consists of a solid inner core, a liquid outer core, the mantle, and the crust.

5. The Earth has a magnetic field, probably created by currents in the molten core; the magnetic field traps charged particles in zones above the atmosphere called radiation belts.

6. The crust is broken into tectonic plates that shift around, which accounts for continental drift and for most of the major surface features of the Earth.

7. The lunar surface consists of relatively smooth areas called maria, and mountainous regions, and is marked everywhere by impact craters.

8. The lunar soil is called the regolith, and rocks on the surface are all igneous, mostly silicates, with low abundances of volatile gases.

9. Seismic data show that the Moon has a crust 50 to 100 km thick, a mantle, and a core extending about 500 km from the center.

10. The Moon has no present-day tectonic activity and no magnetic field, which indicate that it probably does not have a liquid core.

11. The Earth's evolution from a largely molten planet with a hydrogen-dominated atmosphere to its present state was caused by the presence of liquid water on its surface; the loss of lightweight gases into space; and the development of life-forms on its surface, which helped convert the atmospheric composition to nitrogen and oxygen.

12. The Moon has significant chemical contrasts with the Earth, which argues against a common origin for the two. The true mechanism for the moon's formation is unknown, but the best current explanation appears to be the hypothesis that the Moon formed as the result of a collision between the young Earth and a large planetesimal.

13. The Moon's evolution consisted of a molten state, followed by hardening of the crust and subsequent large-scale lava flows that created the maria. Since that time (about 1 billion years after the Moon's formation), the Moon has been geologically quiet.

REVIEW QUESTIONS

1. Explain how heat from the Sun and rotation of the Earth create the global circulation pattern of the atmosphere. How do the oceans influence the circulation?

2. How do we know anything about the Earth's interior? Do you think the same methods can be applied to other planets?

3. Explain the differences between transverse and compressional waves. Can you think of everyday examples of each kind?

4. If the density of typical surface rocks is 3.5 grams/cm^3, and the Earth's average (overall) density is 5.5 grams/cm^3, what does this tell us about the density in the deep interior? How did this situation arise?

5. How are the motions in the Earth's atmosphere and those in its interior similar?

6. Explain the distinction between the age of the Earth's surface and the age of the Earth itself.

7. If North America is approaching Japan at a rate of 3 cm per year, and the present distance between North America and Japan is 5,000 km, how long will it take for the two to collide?

8. If the half-life of a radioactive element is 5 million years, and a rock sample contains one-half of its original quantity of that element, how old is the rock? How old is it if only ¼ of the original quantity of the element is left? How old is it if ¹⁄₆₄ is left?

9. Summarize the effects of life-forms on the evolution of the Earth's atmosphere.

10. What is the evidence that the craters on the Moon are formed by impacts, rather than by volcanic eruptions?

11. If impacts by objects from space occurred at a uniform rate over the entire history of the Moon, and the maria have only one-fourth as many craters (per square kilometer) as the rest of the Moon, how much younger than the rest of the lunar surface would the maria be? Is this consistent with what the text says about the ages of the maria and of the Moon itself? What does this imply about the rate of impacts over the Moon's history?

12. Summarize the similarities and contrasts between lunar rocks and those found on the Earth's surface.

13. Why does the Earth not have as many craters as the Moon?

ADDITIONAL READINGS

Anderson, D. L. 1974. The interior of the Moon. *Physics Today* 27(3):44.

Battan, L. J. 1979. *Fundamentals of meteorology.* Englewood Cliffs, N.J.: Prentice-Hall.

Beatty, J. Kelly, 1986. The making of a better moon. *Sky and Telescope* 72(6):558.

Ben-Avraham, Z. 1981. The movement of continents. *American Scientist* 69:291.

Benningfield, D. 1991. Mysteries of the Moon. *Astronomy* 19(12):50.

Bonatt, E. 1987. The rifting of continents. *Scientific American* 256(3):96.

Broadhurst, L. 1992. Earth's atmosphere: terrestrial or extraterrestrial? *Astronomy* 20(1):38.

Brownlee, S. 1985. The wacky theory of the Moon's birth. *Discovery* 6(3):65.

Burchfiel, B. C. 1983. The continental crust. *Scientific American* 249(3):86.

Burt, D. M. 1989. Mining the Moon. *American Scientist* 77:574.

Cadogan, P. 1983. The Moon's origin. *Mercury* 12(2):34.

Friedman, H. 1986. *Sun and Earth.* New York: W. H. Freeman.

Goldreich, P. 1972. Tides and the Earth-Moon system. *Scientific American* 226(4):42.

Hamill, P. and Toon, O. B. 1991. Polar stratospheric clouds and the ozone hole. *Physics Today* 44(12):34.

Hoffman, K. A. 1988. Ancient magnetic reversals: Clues to the geodynamo. *Scientific American* 258(5):76.

Ingersoll, A. P. 1983. The atmosphere. *Scientific American* 249(3):114.

Jeanloz, R. 1983. The Earth's core. *Scientific American* 249(3):40.

Jordan, T. H., and J. B. Minster. 1988. Measuring crustal deformation in the American west. *Scientific American* 259(2):48.

Kasting, J. F., B. Toon, and J. B. Pollack. 1988. How climate evolved on the terrestrial planets. *Scientific American* 258(2):90.

McKenzie, D. P. 1983. The Earth's mantle. *Scientific American* 249(3):50.

Morrison, D., and T. Owen. 1988. Our ancient neighbor, the Moon. *Mercury* 17:66 (Part 1); 17:98 (Part 2).

Murphy, J. B. and Nance, R. D. 1992. Mountain belts and the supercontinent cycle. *Scientific American* 266(4):84.

Nance, R. D., T. R. Worsley, and J. B. Moody. 1988. *Scientific American* 259(1):72.

Register, B. M. 1985. The fate of the Moon rocks. *Astronomy* 13(2):14.

Ruddiman, W. F. and Kutzbach, J. E. 1991. Plateau uplift and climatic change. *Scientific American* 264(3):66.

Runcorn, S. K. 1987. The Moon's ancient magnetism. *Scientific American* 257(6):60.

Siever, R. 1983. The dynamic Earth. *Scientific American* 249(3):30.

Taylor, S. R. 1987. The origin of the Moon. *American Scientist* 75:468.

Toon, O. B., and S. Olson. 1985. The warm Earth. *Science* 6(8):50.

Toon, O. B. and Turco, R. P. 1991. Polar stratospheric clouds and ozone depletion. *Scientific American* 264(6):68.

ASTRONOMICAL ACTIVITY
Observing the Moon

Even with the unaided eye, it is easy to see that the surface of the Moon has bright and dark regions. The bright regions were called *terrae*, or land, by medieval astronomers who thought the dark regions were *maria*, or seas. Today we know that the terrae are rugged, mountainous regions while the maria are low-lying lava plains that reflect light less efficiently. Without a telescope or binoculars, it is difficult to make out much detail, but with even a modest pair of binoculars, many of the more prominent features are easy to see.

The photographs on the following page will help you identify features on the Moon. Note that these are inverted with respect to the lunar photos shown in the chapter (Fig. 6.16), which show the view as seen through an astronomical telescope whose optics invert images (top-to-bottom and left-to-right). When binoculars (or a so-called terrestrial telescope) are used, however, the image is not inverted. Thus the images and drawings shown here are oriented the same way as you will see the Moon with the unaided eye or through binoculars.

It is preferable to view the Moon at first quarter or third quarter rather than when it is full. The reason is that when the Moon is full, sunlight strikes its surface from approximately the vertical direction, and no shadows are cast. This makes it difficult to see surface relief. At first and third quarter, sunlight strikes the Moon from an oblique angle, and surface features then create long shadows. Mountains and crater walls are much more prominent at these times.

Use the photos provided here to see how many of the major features on the Moon you can identify. To view the western hemisphere, you will need to observe during first quarter; the eastern hemisphere is visible at third quarter. Since the Moon does not rise until midnight when it is at third quarter, you may have to sacrifice some sleep to chart this portion of its surface.

continued on next page

ASTRONOMICAL ACTIVITY
Observing the Moon, *continued*

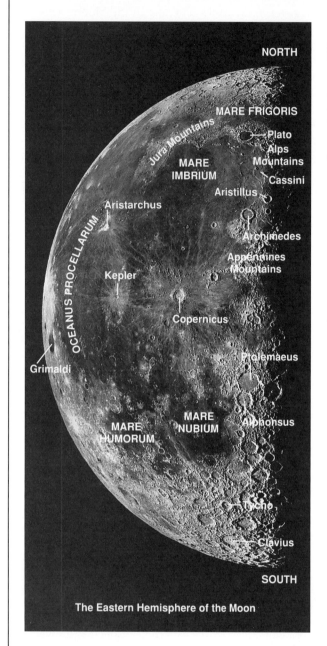

NORTH

MARE FRIGORIS

Plato

Jura Mountains

Alps Mountains

MARE IMBRIUM

Cassini

Aristillus

Aristarchus

Archimedes

OCEANUS PROCELLARUM

Appennines Mountains

Kepler

Copernicus

Grimaldi

Ptolemaeus

MARE NUBIUM

Alphonsus

MARE HUMORUM

Tycho

Clavius

SOUTH

The Eastern Hemisphere of the Moon

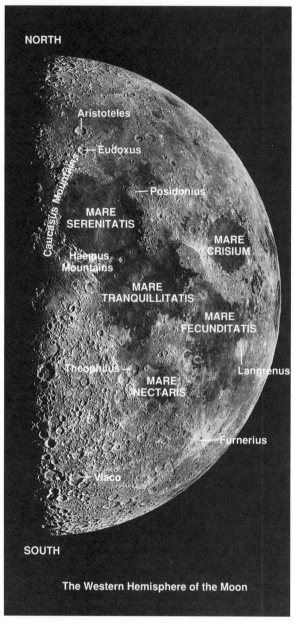

NORTH

Aristoteles

Eudoxus

Posidonius

MARE SERENITATIS

MARE CRISIUM

Caucasus Mountains

Haemus Mountains

MARE TRANQUILLITATIS

MARE FECUNDITATIS

Theophilus

Langrenus

MARE NECTARIS

Furnerius

Vlaco

SOUTH

The Western Hemisphere of the Moon

The Other Terrestrials:
Venus, Mars, and Mercury

Landslides in a Martian canyon (U.S. Geological Survey; data from NASA)

PREVIEW

In this chapter we explore the remaining terrestrial bodies: Venus, Mars, and Mercury. You will find that much of what we know about these planets can be understood by applying the same principles of planetary science that we invoked in the previous chapter, where we learned how the Earth and Moon came to be the way they are. Here you will see that Venus is a near-twin of the Earth, but that its closer proximity to the Sun has resulted in major contrasts. Mars, our nearest neighbor beyond the Earth's orbit, presents some intriguing hints of a once-watery world, a place where life might have arisen. Because of its smaller mass, Mars has not followed the same geological evolution as the Earth and is today a cold and barren world. Mercury, the final terrestrial planet to be discussed, may represent the remaining core of an Earth-like planet whose outer layers were blasted away by a celestial collision with another planetary object. Thus this innermost, Sun-scorched planet is a bit of a misfit among the terrestrial bodies.

In our discussions of the Earth and the Moon in the previous chapter, we encountered two extremes in the solar system's family of terrestrial bodies. The Earth is the largest and most active geologically, and the Moon is the smallest and least active. Now as we go on to discuss the other three terrestrial planets, we will find that in most properties they are intermediate between the Earth and the Moon.

The basic properties of Venus, Mars, and Mercury are given in Table 7.1. Here we see that Venus (Fig. 7.1) is closest to the Earth in its gross properties (and is closest in distance as well), while Mercury (Fig. 7.2) is much more like the Moon than the Earth. Mars (Fig. 7.3) represents the middle ground; its overall characteristics such as size and density are closer to halfway between the Earth and the Moon than to either of those bodies.

EXPLORATION OF THE TERRESTRIAL PLANETS

Venus, Mars, and Mercury have all been the subject of intensive observations from Earth since antiquity, and all three have been visited by unmanned probes from Earth in modern times. The earliest observers thought Venus and Mercury were each two separate planets, because each can appear in the morning sky, just before sunrise, or in the evening sky, just after sunset. Today we understand that those appearances occur because Mercury and Venus are closer to the Sun than the Earth is and are therefore always seen near the Sun, on one side of it or the other. Venus, which is the most prominent object in the sky after the Sun and the Moon, is quite easy to see much of the time. Mercury is more difficult to observe, because it is so close to the Sun that it is only visible

 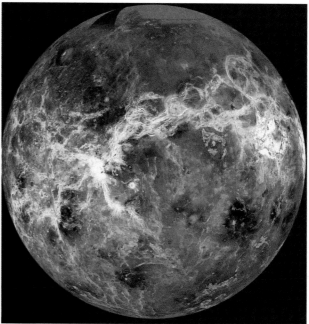

Figure 7.1   Venus.   At left is a visible-light image of the cloudtops of Venus, obtained by the *Pioneer-Venus* orbiter; at right is a full-disk image of Venus based on radar data from the *Magellan* mapper. The color of the *Magellan* image is synthetic, with areas of relative brightness and darkness indicating radar reflectivities and variations in elevation. The *Magellan* mapper, orbiting Venus since 1990, has now nearly completed its coverage of the entire surface of the planet. (NASA/JPL)

TABLE 7.1   Basic Properties of Mercury, Venus, and Mars

	Mercury	Venus	Mars
Orbital semimajor axis (AU)	0.387	0.723	1.524
Perihelion distance (AU)	0.308	0.718	1.381
Aphelion distance (AU)	0.467	0.728	1.667
Orbital period (yr)	0.241	0.615	1.881
Orbital inclination	7°0'15"	3°23'40"	1°51'0"
Rotation period (days)	58.65	−243*	1.026
Tilt of axis (obliquity)	28°	3°	23°59'
Diameter (relative to Earth)	0.382	0.951	0.531
Mass (relative to Earth)	0.0558	0.815	0.107
Density (g/cm³)	5.50	5.3	3.96
Surface gravity (g)	0.38	0.90	0.38
Escape speed (km/sec)	4.3	10.3	5.0
Surface temperature (K)	700/100	750	145–300
Albedo (average)	0.106	0.65	0.15
Satellites	None	None	2

*The minus sign indicates that the rotation of Venus is retrograde; i.e., in the direction opposite the spins of most of the other planets.

from Earth for a week or so when it is at its angle of greatest elongation (about 28° from the Sun).

An impressive array of telescopes, on the ground and in space, have been trained on the terrestrial planets. In addition, a number of probes have been sent to these bodies, providing information from close range or even from their surfaces (Table 7.2). Mars has been the subject of the greatest amount of attention from Earth-bound astronomers, largely because of a myth that reached its peak about a hundred years ago, concerning imagined evidence of life on the red planet. Some observers interpreted sketches of surface features made during the opposition of 1877 as showing a series of canals on the surface of Mars (Fig. 7.4), and this gave rise to broad speculation about a civilization that prospered there. The chief proponent of this idea was the wealthy American amateur astronomer Percival Lowell (Fig. 7.5), who wrote extensively on the subject—he thought the canals carried water from the ice caps to agricultural regions near the Martian equator. Lowell even established an observatory (in Flagstaff, Arizona)

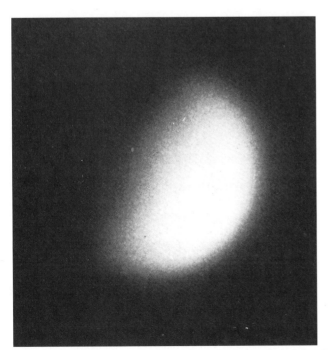

Figure 7.2   Mercury. On the left is an Earth-based photograph of Mercury, in which little detail can be seen. On the right is a mosaic of *Mariner 10* images covering a large portion of the sunlit portion of the planet at the time of the *Mariner* encounters. (NASA)

Figure 7.3   Mars.   At left is an image of Mars obtained from Earth orbit by the *Hubble Space telescope;* at right is a mosaic of *Viking* orbiter images showing the full disk of Mars in great detail. (left: NASA/Space Telescope Science Institute; right: U.S. Geological Survey; data from NASA)

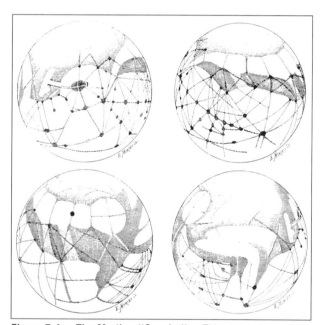

Figure 7.4   The Martian "Canals."   This sketch from 1896 shows the linear features eventually interpreted as artificial channels in which water supposedly flowed. (Historical Pictures Service, Chicago)

Figure 7.5   Percival Lowell.   A wealthy enthusiast for astronomy, Lowell became fascinated with the notion of a Martian civilization. (Lowell Observatory photograph)

dedicated to the study of Mars; later this observatory, which is still active today, was responsible for the discovery of the ninth planet, Pluto. Since Lowell's time, Earth-based observations of Mars have revealed seasonal changes in its surface markings (due to blowing dust), the composition of its atmosphere (largely carbon dioxide), and a lack of water vapor. Space probes, notably *Mariner 9* and the two *Viking*

landers, have allowed the surface to be examined and its rocks and soil to be chemically analyzed. Some of the *Viking* tests were also designed to search for evidence of microscopic life-forms in the soil.

Venus has been more difficult to study using Earth-based telescopes, because it is shrouded in dense clouds, which hide its surface. These clouds are no impediment to radio and radar observations, how-

TABLE 7.2   Probes to the Terrestrial Planets*

Spacecraft	Launch	Arrival	Remarks
Mercury			
Mariner 10	11/03/73	3/29/74	Trajectory allowed three working flybys
		9/21/74	
		3/16/75	
Venus			
Mariner 2	8/26/62	12/14/62	Flyby
*Venera 4***	6/12/67	10/18/67	Atmosphere probe
Mariner 5	6/14/67	10/19/67	Flyby
*Venera 5***	1/05/69	5/16/69	Atmosphere probe
*Venera 6***	1/10/69	5/17/69	Atmosphere probe
*Venera 7***	8/17/70	12/15/70	Lander, 23 minutes of data returned from surface
*Venera 8***	3/27/72	7/22/72	Lander (50 minutes)
Mariner 10	11/03/73	2/05/74	Flyby
*Venera 9***	6/08/75	10/22/75	Orbiter and lander
*Venera 10***	6/14/75	10/25/75	Orbiter and lander
Pioneer/Venus:			
Orbiter	5/20/78	12/04/78	Operated until late 1992
Multiprobe	8/08/78	12/09/78	Five atmosphere probes
*Venera 11***	9/09/78	12/21/78	Flyby and lander
*Venera 12***	9/14/78	12/25/78	Flyby and lander
*Venera 13***	10/30/81	3/01/82	Lander
*Venera 14***	11/04/81	3/05/82	Lander
*Venera 15***	6/02/83	10/10/83	Orbiter with imaging radar
*Venera 16***	6/07/83	10/14/83	Orbiter with imaging radar
*VEGA 1***	12/15/84	6/11/85	Lander and balloon atmosphere probe
*VEGA 2***	12/21/84	6/15/85	Lander and balloon atmosphere probe
Magellan	5/4/89	8/90	High-resolution radar
Galileo	10/18/89	2/90	Flyby on way to Jupiter
Mars			
Mariner 4	11/28/64	7/14/65	Flyby
Mariner 6	2/25/69	7/31/69	Flyby
Mariner 7	3/27/69	8/05/69	Flyby
Mariner 9	5/30/71	11/13/71	Orbiter (ceased functioning on October 27, 1972)
*Mars 2***	5/19/71	11/27/71	Orbiter and lander (lander returned no data)
*Mars 3***	5/28/71	12/02/71	Orbiter and lander (lander returned no data)
*Mars 5***	7/25/73	2/12/74	Orbiter and lander (lander failed within seconds of touchdown)
*Mars 7***	8/09/73	3/09/74	Orbiter (lander missed planet)
*Mars 6***	8/05/73	3/12/74	Orbiter and lander (lander crashed)
Viking 1	8/20/75	6/19/76	Orbiter (ceased functioning on August 17, 1980)
		7/20/76	Lander (ceased operating November 1982)
Viking 2	9/09/75	8/07/76	Orbiter (ceased functioning on July 24, 1978)
		9/03/76	Lander (ceased functioning on April 12, 1980)
*Phobos 2***	7/88	6/89	Lost contact with Earth and failed March 27, 1989, after two months in orbit gathering data

*Based on a compilation published by the Astronomical Society of the Pacific.
**Soviet spacecraft. All others are U.S. probes.

ever, and radio data have revealed some fundamental information, including two major surprises: the planet is rotating very slowly in the retrograde (backward) direction; and its surface is very hot, around 750 K (roughly 900°F, which is hot enough to melt lead!). The strange rotation of Venus is not fully understood, but is thought to have been caused by a late collision between Venus and a large planetesimal. The high surface temperature is well understood, however, as discussed later in this chapter.

Over 20 space probes have visited Venus, providing enormous amounts of data on the planet's atmosphere and surface. Both the U.S. and Soviet space programs have focused on Venus, as shown in Table 7.2, resulting in several probes deep into the atmosphere (including a few that have managed to survive on the surface long enough to send back pictures and data) and a couple that have been in long-term orbits about Venus, studying it at length from above the cloud layers. One of these, the *Pioneer-Venus* mission,

ASTRONOMICAL INSIGHT
Visiting the Inferior Planets

In Chapter 3 there was some discussion of the mechanics of placing a satellite in the Earth's orbit, but nothing on how to aim a spacecraft that is designed to visit another planet. The principles involved are particularly simple when an inferior planet is to be the target.

The concept of energy was discussed in Chapter 3, where we pointed out that in a system of two orbiting bodies, there is kinetic energy due to the motions of the bodies, and potential energy due to their gravitational attraction for each other. For a planet orbiting the Sun, the greater the total energy (the sum of kinetic plus potential), the larger the orbit. Thus for an object to go from the orbit of the Earth to that of an inner planet, it must lose energy.

A spacecraft sitting on the launchpad is moving with the Earth in its orbit, at a speed of 29 km/sec. To make the rocket fall into a path that intercepts the orbit of an inner planet, we must diminish this speed. Therefore, we launch the rocket backward with respect to the Earth's orbital motion, thereby lowering its velocity with respect to the Sun, and decreasing its orbital energy. If the launch speed is properly chosen, the spacecraft will fall into an elliptical orbit that just meets the orbit of the target planet (see figure). In order to reach Mercury from the Earth, the rocket must be launched backward with a speed relative to Earth of 7.3 km/sec.

There are, of course, a few additional considerations. For one thing, the rocket has to be launched with enough speed to escape the Earth's gravity. The escape speed for the Earth is 11.2 km/sec, so in fact we have to launch our Mercury probe with a speed in excess of this value, which is more than the speed

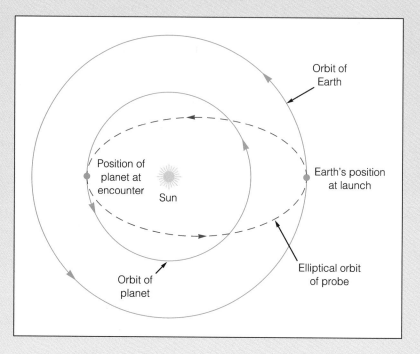

needed to attain our trajectory to Mercury. The launch speed must therefore be calculated to take the Earth's gravity into account, so that the rocket escapes the Earth, but in the process is slowed just the right amount to give it the proper course. The gravitational pull of the target planet must also be taken into account, for this speeds the spacecraft up as it approaches.

Another important consideration is timing; the target planet must be in the right spot in its orbit at just the moment when the spacecraft arrives. It is for this reason that the term **launch window** is used. The Earth and the target planet must be in specific relative positions at the time of the launch. The interval between launch windows for a given planet is simply its synodic period. For a probe to Mercury, the travel

time is about 211 days, so Mercury actually makes a little more than one complete orbit while the probe is on its way.

The gravitational pull of the target planet can be used to good advantage in modifying the orbit of a spacecraft that flies by. If the probe is aimed properly, its trajectory may be altered as it goes by in just the right way to send it on to some other target. This technique was used with *Mariner 10*, first to send it from Venus to Mercury, and then to modify its orbit again so that it returned to Mercury repeatedly thereafter.

Sending spacecraft to the outer planets is a bit more difficult, because the spacecraft has to have more energy than it gets from the Earth's motion. The launch is therefore made in the forward direction, so that the rocket has the speed of the Earth in its orbit plus its own launch speed with respect to the Earth.

Ironically, the seemingly simple tasks of launching a probe to the Sun is one of the most complex. The most straightforward solution would be to launch the spacecraft backward from the Earth with a speed of 29 km/sec, entirely canceling out the Earth's orbital speed, so that the probe would then fall straight in toward the Sun. This is a prohibitively high launch speed, however, so in practice we will probe the Sun by first sending the spacecraft out around Jupiter. The spacecraft will fly by the massive planet in such a way (backward with respect to Jupiter's orbital motion) that its own orbital energy is reduced; the craft can then fall into the center of the solar system. A mission called *Ulysses* to explore the outer layers of the Sun in this manner was launched in late 1990.

---

operated in orbit around Venus from 1978 until late 1992; the other, the *Magellan* radar mapper, has been at work since 1990 and is still operating, as of late 1992. *Pioneer-Venus* carried out infrared and ultraviolet observations of the clouds and upper atmosphere and studied the charged-particle environment near Venus; *Magellan* is making very detailed maps of the entire surface of the planet, peering through the clouds with its radar transmitters and receivers (interestingly, quite detailed maps of portions of the surface of Venus have been made from Earth as well, by sending radar beams to the planet and analyzing the return echo).

Mercury has been studied in the least detail of the three terrestrial planets; it is difficult to observe from Earth, and only one space probe has visited it. Earth-based observations showed some hints of surface markings, but led to confusion about the planet's rotation period until radar observations cleared up the picture. It is possible to use the Doppler effect to determine the rotation speed of a planet (Fig. 7.6), because the return radar echo from a moving body is Doppler-shifted due to the relative motion between the object and the observer (this is how police use radar units to determine the speed of cars). Astronomers used this technique to determine the very slow rotation speed (and long day) of Venus and then applied it to Mercury, where it again revealed a very slow rotation (this is discussed further in a later section). Only one space probe has been sent to Mercury—the U.S. *Mariner 10* mission, which first flew close to Venus (obtaining some very useful ultraviolet images in the process and also determining that Venus has no magnetic field). When *Mariner 10* arrived at Mercury, Earth-based controllers modified its orbit, adjusting the probe's direction and speed so that it went into a solar orbit with exactly twice the orbital period of Mercury. This orbit caused the spacecraft to have repeated close encounters with Mercury; the probe returned to the planet's vicinity once every two Mercurian years, or 176 days.

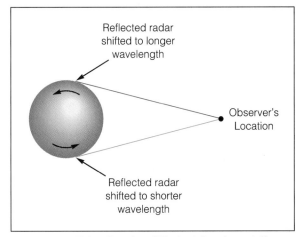

Figure 7.6   Using Doppler shifts to determine rotation speed.   The reflected radar waves are Doppler-shifted if they bounce off a portion of the planet that is moving away from us or toward us. By measuring the amount of the shift, the speed of approach or recession can be determined. Then the rotation period can be found if the planet's diameter is known.

Reflected radar shifted to longer wavelength

Observer's Location

Reflected radar shifted to shorter wavelength

COMPARATIVE PROPERTIES OF EARTH, VENUS, MARS, AND MERCURY

Before we discuss the individual planets in detail, it is useful to consider them as a group. This section highlights their similarities and differences, keeping in mind what we have learned about the Earth and the Moon in the previous chapter, as well as the general principles of planetary science gained from Chapter 5.

Atmospheres

In general we might expect that the more massive a planet is, the more extensive the atmosphere it can maintain (recall our discussion of escape speeds and the loss of atmospheric particles in Chapter 5). This expectation must be modified by the temperature of the planet, however, because we know that the hotter it is, the higher the speeds of individual atmospheric particles, and the greater the likelihood that they will escape into space. From these principles, we might expect that the Earth and Venus would have more extensive atmospheres than either Mars or Mercury. Indeed, Earth and Venus both have atmospheres, Mars has a far less dense atmosphere, and Mercury has virtually none at all. This makes sense, because the Earth and Venus are the most massive of the terrestrial planets, Mars is next, and Mercury is least massive (and also very hot).

The picture we have just painted is a bit too simple, however. For one thing, the atmosphere on Venus is far more massive and dense than that of the Earth, although Venus is much hotter than the Earth and has a slightly lower surface gravity and a lower escape speed. Furthermore, the atmospheres of Venus and Mars both have a composition very different from that of the Earth. Whereas the Earth's atmosphere is dominated by nitrogen and oxygen, those of Venus and Mars are almost pure carbon dioxide (CO_2).

The explanation for Venus's apparent deviation from the general principles we have learned lies in the very different composition of its atmospheric gases compared with those of the Earth (Table 7.3), which in turn is related to the differences in the formation and evolution of the two planets. The Earth is actually the misfit among the terrestrial planets, differing from them in two major ways: (1) the Earth has had liquid water on its surface for billions of years; and (2) life-forms have evolved on the Earth and modified its atmosphere. Mars may have had oceans (or at least extensive lakes) of water, but these have long since evaporated, and Venus apparently has long been too hot to allow liquid water to survive on its surface. The lack of oceans on Mars and Venus has allowed carbon dioxide to persist in the atmospheres of these planets, whereas on the Earth carbon

tends to be dissolved in seawater and then deposited in carbonate rocks. Thus, on the cooler Earth, much of the carbon dioxide is in the ground or in the oceans, but Venus is so hot that much of its carbon has been baked out of its surface rocks and released into the atmosphere as carbon dioxide. Furthermore, the Earth's atmosphere has been modified over time because life-forms have contributed vast quantities of oxygen (largely through photosynthesis by plants, especially the many forms of algae that live in the oceans) and much of the nitrogen as well (through the decay of organic matter).

Thus, mass and temperature are not the only factors that determine the nature of a planet's atmosphere: the quantity and type of gas injected into its atmosphere are also very important. Venus has a more massive and more extensive atmosphere than the Earth because more gas has been emitted into its atmosphere from the surface and interior and because it has no oceans to help rid the atmosphere of some gases.

The carbon dioxide in the atmosphere has also helped to heat the surface of Venus through the **greenhouse effect** (Fig. 7.7). The greenhouse effect refers to the trapping of heat near the surface of a planet, with the planetary atmosphere playing much

TABLE 7.3   The Atmospheres of Venus and Mars

Gas	Symbol	Fraction (by Number) Venus	Mars
Carbon dioxide	CO_2	0.96	0.95
Nitrogen	N_2	0.035	0.027
Argon	Ar	0.00007	0.016
Oxygen	O_2	—	0.0013
Carbon monoxide	CO	0.0004	0.0007
Water vapor	H_2O	0.0001	0.0003
Neon	Ne	5×10^{-6}	2.5×10^{-6}
Sulfur dioxide	SO_2	0.00015	—
Krypton	Kr	—	3×10^{-7}
Ozone	O_3	—	1×10^{-7}
Xenon	Xe	—	8×10^{-8}
Hydrogen chloride	HCl	4×10^{-7}	—
Hydrogen fluoride	HF	1×10^{-8}	—

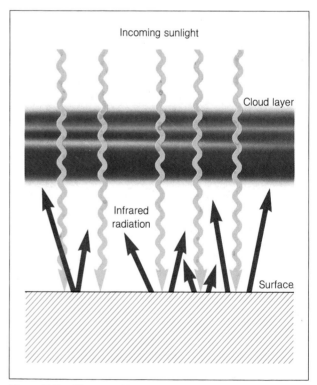

Figure 7.7   The greenhouse effect.   Visible light from the Sun reaches the surface of Venus and heats it, which causes the ground to emit infrared radiation. The carbon dioxide in the atmosphere efficiently absorbs infrared radiation and only re-radiates it very slowly, so heat is trapped near the surface.

the same role as the glass in a greenhouse. Solar radiation is absorbed at the surface, causing it to heat up. In effect, heat energy is "bottled up" in the atmosphere. This heat, in turn, causes the surface to emit infrared radiation (recall Wien's law, from Chapter 4). The infrared radiation is then absorbed by atmospheric gases, causing them to heat up. Carbon dioxide is a very effective absorber of infrared radiation, and it has caused Venus to become very hot, far hotter than the Earth. Water vapor and other trace gases in the Earth's atmosphere create some greenhouse heating, as does the thin carbon dioxide atmosphere of Mars.

In Chapter 5 we learned that the primary forces driving the circulation pattern in a planet's atmosphere are heating, which causes convection to occur, and rotation of the planet, which causes flows to curve due to the coriolis force. These mechanisms are at work on the terrestrial planets. We have already discussed the Earth's atmospheric circulation in some detail in the previous chapter, where we found that heating occurs at the surface and in the mesosphere (where ozone absorbs ultraviolet energy from the Sun). This heating, which is greatest near the equator, causes air flows from low latitudes toward the poles, and these flows are turned into rotary patterns by the Earth's rotation. Hence the general pattern of atmospheric motion on the Earth consists of rotary flows, with a strong dependence on latitude.

In general, the pattern of atmospheric motion on Mars is similar to that of the Earth. Currents flow from warm regions to cooler ones and are forced into rotary patterns by the coriolis force due to the planet's rotation (the Martian day is just a half hour longer than an Earth day). The major difference occurs at certain times of the Martian year when very strong temperature variations from place to place on the surface cause intense horizontal winds to develop. The temperature variations that drive these winds are caused by seasonal effects (discussed later in this chapter).

Venus, as we might expect, has a very different pattern of atmospheric motions than do the Earth and Mars. On the one hand, the very hot surface might be expected to create strong convection, causing an overturning motion, but on the other hand, the very dense atmosphere tends to suppress flows. The probes that have landed on the surface of Venus have discovered very little wind there; as yet, not much is known about vertical motions at higher elevations. The rotation of Venus is very slow, so we might not expect to find strong rotary patterns in its atmosphere, and indeed we do not. At high elevations (above the clouds, where it is possible to observe the motions), there are rapid horizontal motions, which are driven by the planet's rotation (the entire upper atmosphere flows uniformly in the same direction as the planet's rotation; Fig. 7.8).

Mercury has only a trace of an atmosphere, so thin that it defied detection until recently. The elements hydrogen, oxygen, sodium, and potassium have been measured spectroscopically. Astronomers believe these originated from the planet's crust and were released through meteorite impacts or gradual erosion due to the solar wind.

Interior Structure

We see immediately from the densities of Venus, Mars, and Mercury that all three are rocky planets that have undergone at least some differentiation. Remember that ordinary rock typically has a density of about 3 grams/cm^3, so a planet with an average density higher than this must have a concentration of mass in its core. Venus has an average density very similar to that of the Earth, whereas the density of Mars is considerably less, being closer to that of the Moon. From this we conclude that differentiation has been less extensive in Mars than in the Earth. If we consider the Earth, Venus, Mars, and the Moon, we

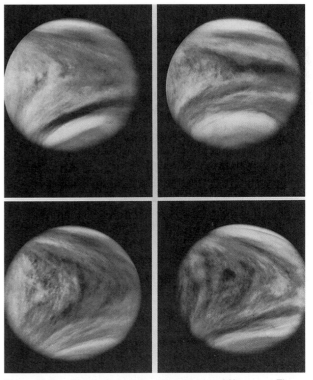

Figure 7.8   Circulation of the atmosphere of Venus.   These four ultraviolet views show two rotations of the planetary cloud cover. The upper two are separated by one day; the lower two were obtained about a week after the first pair and are also separated by a day. The atmospheric motion is from right to left in these images. (NASA)

see a clear trend toward lower density for smaller overall mass. This pattern makes some sense because, as we learned earlier, differentiation requires that a body be molten, so that the heavier elements can sift downward relative to the lighter ones. The degree of internal heating available to cause melting is related to the total mass of a planet, because the amount of gravitational compression and heating as the planet forms depends on its mass (interestingly, we will find the same pattern in stars; the more massive a star is, the hotter its core becomes due to gravitational compression). Following this reasoning, we might expect Mercury to have a lower overall density than Mars, since Mercury has less mass than Mars, but instead Mercury is very nearly as dense as the Earth. This suggests that something very different must have affected the internal structure and evolution of Mercury, something that did not happen to the Earth or the other terrestrials. We will return to this point later in the chapter.

Although we know less about the details of the internal structure of the other planets than about the Earth or the Moon, quite a bit of data is available due to probes that have encountered, orbited, or landed on each of the three (Table 7.2). Important information about the interior conditions can be inferred from the overall density and degree of differentiation; the presence and character of a magnetic field, if one is detected; and detailed studies of the surface, which provide clues to internal processes. For example, evidence of both volcanic activity and tectonic processes has been found on Venus, an indication that this planet is much like the Earth internally and probably has a molten zone in or around the core. Unlike the Earth, however, Venus has no detectable magnetic field, but this may be due to its very slow rotation (recall that the Earth's field is thought to be due to rotation-driven flows in its molten core).

Mars rotates nearly as rapidly as the Earth, but also has no detectable magnetic field. In this case, it is thought that the core does not have a molten zone; thus there are no regions where planetary rotation can cause the formation of the flowing electrical currents needed to produce a field. Recall that Mars has apparently undergone less heating in its interior because it is less massive than the Earth and Venus; perhaps it is not surprising then that Mars has no molten core.

Once again Mercury offers a surprise: it has a detectable magnetic field (about one-tenth the strength of the Earth's field) even though it is low in mass and also rotates very slowly. The presence of a magnetic field, along with Mercury's high density, has led astronomers to conclude that this small planet must have a very large core containing extensive molten regions.

Surface Characteristics

Like the Earth, the other terrestrial planets have highland and lowland regions on their surfaces. Most of Venus consists of rolling plains, but lowlands cover about a quarter of the surface, and there are three isolated highland regions (covering less than 10 percent of the surface). These highland regions may be likened to the continental areas of the Earth, and the lowlands in some ways resemble the seafloors of the Earth. There are craters on the surface of Venus, some due to impacts but mostly of volcanic origin. In addition, large regions of lava flows are found, where eruptions or noneruptive flows have covered the surface. Thus much of the surface of Venus is relatively young.

Impact craters are much more numerous on Mars, because its surface is older (Fig. 7.9). There are mountainous regions on Mars as well, and a major portion of the surface consists of a huge uplifted region called the Tharsis plateau. This area is quite old (around 3 billion years) and appears to be the result of an early period when convection in the mantle of Mars created uplift. Near the Tharsis plateau are several huge volcanic mountains, again dating from the early period of intense activity (there is some uncertainty about whether volcanic eruptions have occurred more recently as well; the controversy is based

Figure 7.9   Craters on Mars.   This mosaic of *Viking* images shows a portion of a cratered highland region on Mars (foreground). See also Fig. 7.3. (U.S. Geological Survey; data from NASA)

Figure 7.10   Olympus Mons.   This is the largest mountain known to exist in the entire solar system. Its base is comparable in size to the state of Colorado, and its height is about three times that of Mt. Everest. (NASA)

on interpretations of detailed shapes of mountain slopes, and some have argued for eruptions in rather recent times). These giant volcanoes on Mars include Olympus Mons (Fig. 7.10), with a height above its base of some 27 km (90,000 feet, about three times that of Mt. Everest) and a base about 700 km across (about the size of the state of Colorado!). Olympus Mons and the other giant volcanoes are thought to have formed from the same kind of upward lava flow that created the Hawaiian Islands on the Earth, except that on Mars there was no continental drift to carry the newly formed mountains away as lava continued

to be forced up from below. Thus, the effect is as though all the molten material that formed the Hawaiian Islands were piled together into one gigantic mountain instead of a chain of smaller ones.

Another major feature of the Martian surface is a huge canyon called Valles Marineris (Fig. 7.11; this canyon was named after the *Mariner 9* probe whose close-up photos first revealed its size). This valley is 5 to 10 times larger than the Grand Canyon in all dimensions, stretching across an expanse of some 4,500 km. Valles Marineris appears to have formed due to crustal fracturing that may have occurred at the time the Tharsis plateau was uplifted.

Mercury's surface is dominated by impact craters, giving the planet an uncanny resemblance to the Moon (Fig. 7.12). There are some subtle differences, though, which will be discussed later in the chapter. The high density of craters indicates that the surface of Mercury is quite old, much older than the surfaces of the Earth, Venus, or Mars.

The differences among the surfaces of the terrestrial planets can be explained largely in terms of tectonic activity; that is, the extent to which the surfaces have been modified by processes such as continental drift and volcanic eruptions. We have already seen that these processes have been very important for the Earth, with the result that most of the Earth's surface is relatively young (ages measured in the hundreds of millions of years, whereas the planet itself is several billion years old). The outlines of the continents and much of the surface topography can be understood

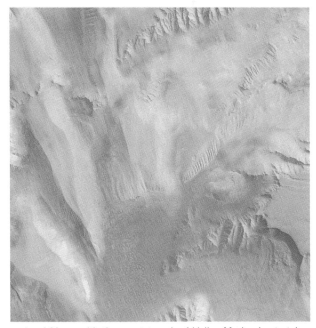

Figure 7.11   Valles Marineris.   At left we see a full-disk photomosaic of Mars, with the great trench of Valles Marineris stretching across the lower center. At right is a detailed view of a portion of the valley called the Candor Chasm. In the right-hand view, note the sections of canyon wall (center right) that have subsided or slumped; you may also be able to see a section of layered terrain (just left of center, above a dark region) that is thought to be an ancient lakeshore. (U.S. Geological Survey; data from NASA)

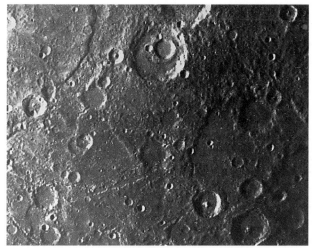

Figure 7.12   A moon-like surface.   This *Mariner 10* image of a portion of Mercury's surface shows its strong resemblance to the lunar surface. There are some differences, as explained in the text, but the general similarity is striking. (NASA)

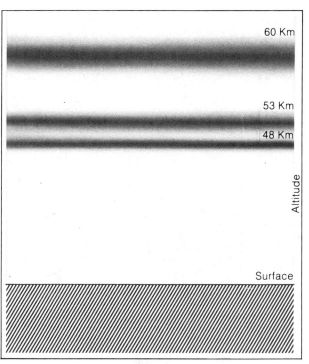

Figure 7.13   The clouds of Venus.   The sulfuric acid clouds are separated into three distinct layers. These layers occur at altitudes where the combination of temperature and pressure causes sulfuric acid to condense.

as the result of tectonic activity.

Mars and Venus, by contrast, show less evidence of these processes, particularly continental drift. Mars certainly has undergone periods of surface uplift and extensive volcanic eruptions, but apparently never formed crustal plates that were able to drift about. Venus shows some evidence of surface motions, in the form of linear trenches and ridges, and also appears to undergo volcanic eruptions, but like Mars it lacks an extensive system of tectonic plates that move about. The absence of crustal plates on Mars is probably due to the differences between its interior and that of the Earth: whereas the interior of the Earth is hot and pliable that it can undergo slow flowing motions, the inside of Mars is cooler and more rigid. In the case of Venus, where the internal structure might be expected to be very similar to that of the Earth, the explanation is not so simple. Current information, based on radar maps obtained by the *Magellan* mission, indicates that the mantle of Venus is undergoing the same kind of convective motions that are thought to be responsible for continental drift on the Earth; yet for some reason, on Venus these internal flows do not produce the same kind of surface motion. This point is discussed further in the next section.

VENUS: A CLOSER LOOK

Now that we have described the terrestrial planets as a class and examined their similarities and differences in terms of large-scale planetary processes, it is interesting to look a bit closer to see how some of the details vary from one body to another. We begin with Venus, the closest to the Earth, both literally and in its overall properties.

As we have already seen, the atmosphere of Venus consists mostly of carbon dioxide (Table 7.3), with traces of other species, primarily nitrogen. Some of the trace species play important roles, however: sulfur dioxide (SO_2), for example, acts as an absorber of solar ultraviolet radiation at high elevations in the atmosphere and therefore plays a role very similar to that of ozone in the Earth's atmosphere; and sulfuric acid (H_2SO_4) forms droplets at certain levels, creating the clouds of Venus and sulfuric acid rainfall, thus duplicating the role of water vapor in the Earth's atmosphere. The clouds lie at three distinct levels (Fig. 7.13) and appear featureless except when observed at ultraviolet wavelengths, which reveal dark streaks due to sulfur dioxide absorption (see Fig. 7.8). As mentioned previously, the clouds flow around the planet in the same (retrograde) direction as the planetary rotation.

At the surface of Venus, the atmospheric pressure is about 90 times that at sea level on the Earth (i.e., it is equivalent to the pressure some 3,000 feet deep in the Earth's oceans!), and the temperature is about 750 K due to greenhouse heating. Photos obtained at the surface by the Soviet *Venera* probes show that sunlight penetrates there, and that the surface rocks

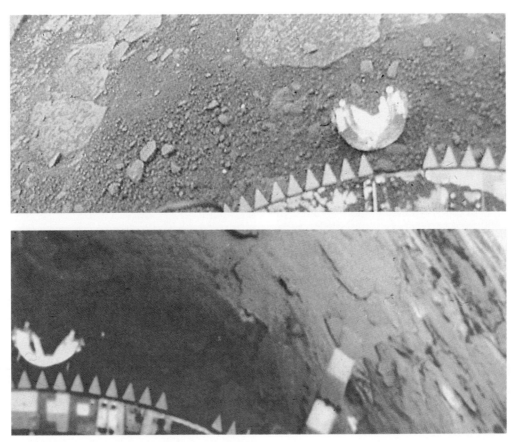

Figure 7.14   Color photos of surface rocks.   These photographs were obtained by the Soviet *Venera 13* and *14* landers. The yellowish color seen here is an artifact of the camera response; reprocessing of these images has shown that the surface rocks are gray, like similar rocks on Earth. (TASS from SOVFOTO)

appear sharp-edged (Fig. 7.14), indicating a lack of wind erosion. The rocks have been found to be silicates, probably volcanic basalts.

The Surface as Seen by *Magellan*

The *Magellan* spacecraft has been orbiting Venus since 1990, gradually creating a detailed map of the entire surface of the planet. *Magellan* uses radar to make this map, sending a beam to the ground, which is reflected back to the spacecraft and then analyzed to reveal details of the surface structure. Features as small as 100 meters in size can be detected, enabling astronomers to distinguish between impact and volcanic craters, for example, and to identify features due to lava flows and continental drift. In short, *Magellan* has offered scientists the first chance to carry out true geological studies of Venus.

One of the key questions before *Magellan*'s advent was the amount of volcanic activity on Venus. Some of the mountains detected by earlier probes (including a lower-resolution radar mapper aboard the *Pioneer-Venus* orbiter) were clearly volcanic in origin, but it was not known how recently the activity had taken place. Indirect evidence suggested that Venus

might be active currently, in that the *Pioneer-Venus* ultraviolet spectrometer found varying levels of atmospheric sulfur dioxide (SO_2), which is an abundant product of volcanic eruptions. The *Magellan* maps indicate clear-cut evidence of volcanic lava flows, craters, and mountains (Figs. 7.15 and 7.16), but do not definitively answer the question of whether eruptions are taking place at this time. Some of the volcanic structures are apparently quite young, however, so it is reasonable to think that eruptions are taking place sporadically. Thus Venus is probably a member of the small club of solar system bodies on which volcanic activity persists in modern times (the Earth is very active and, as we will see, so are one of the large satellites of Jupiter and the giant moon of Neptune).

The *Magellan* maps have revealed some very unusual surface features. The **tesserae** are large expanses of rock that is fractured by regular rectangular cracks (Fig. 7.17), probably as a result of uplifting after lava flows had hardened. Circular structures (Fig. 7.18), called **coronae,** probably were caused by upward flows of lava beneath the surface. They appear to be sites of recent or current volcanic eruptions.

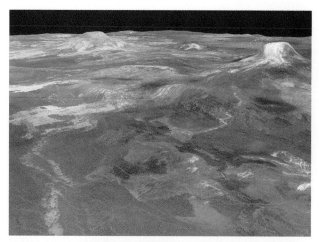

Figure 7.15   Volcanoes on Venus.   This is a *Magellan* reconstruction of a region in Eistla Regio. At left is the volcano Gula Mons; at right is Sif Mons. Lava flows extend from Gula Mons across toward the lower right. (NASA/JPL)

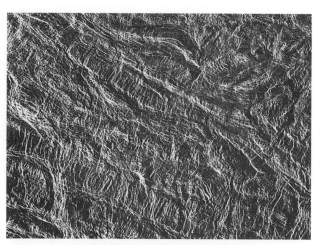

Figure 7.17   Tesserae.   This complex pattern of intersecting ridges and cracks is thought to be the result of repeated episodes of horizontal motion. (NASA/JPL)

Figure 7.16   A large volcanic caldera.   This is a depression called Sacajawea, in the Lakshmi Plenum region of Venus. The enormous caldera is 1 to 2 kilometers deep, 120 km wide, and 215 km in length. It is thought to have formed as the result of drainage and collapse of a large underground magma chamber. (NASA/JPL)

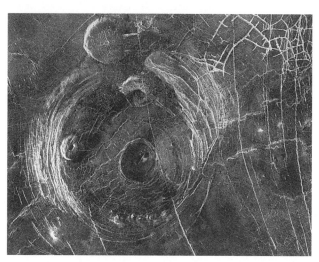

Figure 7.18   A corona.   The large, circular feature seen here is a corona. It is a raised structure, approximately 200 km in diameter, thought to be the result of uplift due to upwelling magma from below. The smaller circular feature is a "pancake" dome, about 35 km in diameter, formed by the eruption of very viscous lava. Another pancake is seen to the left; these features are seen nowhere else except on Venus. (NASA/JPL)

On the whole, the *Magellan* maps indicate that the entire surface of Venus has been modified by volcanic activity; that is, nearly all of the surface has been formed by volcanic eruptions or lava flows. The age of the surface, estimated from counts of impact craters, varies from place to place. Some regions have substantial crater densities and are therefore thought to be fairly old (up to about 1 billion years), while other areas have no impact craters at all, suggesting that these are sites of very recent volcanic activity. The overall density of impact craters is similar to the density of craters on the Earth. It is noteworthy that there are no small craters on Venus (Fig. 7.19); the

reason is that most meteors are broken apart or burned up by the thick atmosphere of Venus and do not reach the surface intact. Only large ones survive the fall.

On the question of tectonic activity, again it is apparent that Venus is active: its features include linear trench-and-ridge systems, mountain ranges that probably resulted from the upwelling of subsurface material and spreading of the surface (analogous to the formation of the mid-ocean ridges on the Earth), and regions of uplifted terrain that appear to have resulted from compression as adjacent portions of the surface were pushed together (Fig. 7.20). But instead of hav-

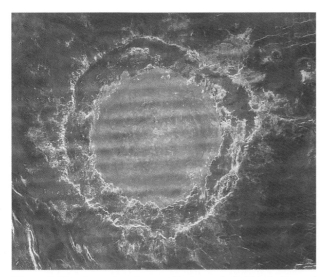 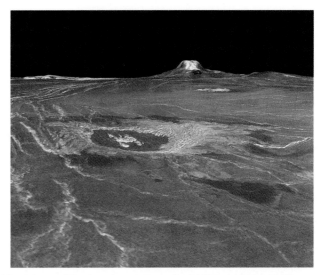

Figure 7.19   Impact craters on Venus.   Left: This is the crater Mead (named for American anthropologist Margaret Mead), the largest impact crater on Venus, having a diameter of 275 km. Right: In the foreground is the impact crater Cunitz; in the background is the volcano Gula Mons. Cunitz crater is roughly 48.5 km in diameter; it was named after the astronomer and mathematician Maria Cunitz. (NASA/JPL)

ing rigid crustal plates that slide underneath one another as on Earth, the crust of Venus simply stretches or is compressed while staying in place. One of the keys to understanding this behavior is the temperature and thickness of the crust, and these parameters are not well known as yet. Further analysis of *Magellan* data should help provide the needed information.

The Earth and Venus: So Near and Yet So Far

In this chapter we have been stressing the similarities between Venus and the Earth, and indeed many geological similarities can be found. But what of the enormous contrasts in surface conditions and atmospheric properties? We can begin to understand these differences by taking a look at the environments in which the two planets formed.

Like the Earth, Venus is thought to have formed from rocky debris orbiting the infant Sun in a great disk. The early evolution of Venus must have been similar to that of the Earth, with Venus undergoing a molten period during which its dense elements sank to the center, leaving a lighter crust composed largely of silicates and carbonates. Before the crust cooled, volatile gases escaped from the surface, forming a primitive atmosphere of hydrogen compounds and carbon dioxide, much like the earliest atmosphere on the Earth. At this point the evolution of the Earth and Venus began to diverge in a major way. On Earth, as oxygen escaped from the rocks and combined with hydrogen to form water, much of it could persist in the liquid state, and our planet had oceans from this time onward. Like the Earth, Venus may have had

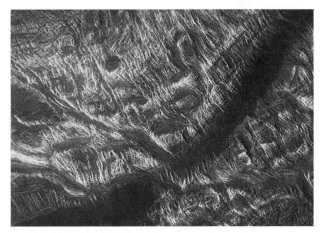

Figure 7.20   The effect of compression.   This is the highland region Ovda Regio, near the equator of Venus. The complex structure seen here is thought to be the result of compression (in the upper left-lower right direction) followed by stretching (upper right to lower left), caused by flowing motions in the mantle of Venus which are similar to the motions that cause continental drift on the Earth. (NASA/JPL)

large quantities of liquid water early in its history, either due to outgassing or to the impact of many water-bearing comets (there is currently some controversy about the role of comets in supplying the initial water for the oceans of the Earth as well; some scientists suggest that comets may have been a more important source than outgassing).

At this point a significant environmental difference between Venus and the Earth came into play. Venus is closer to the Sun, and the extra solar heating that resulted was apparently enough to prevent liquid water from remaining on the surface of the planet.

Venus is 0.72 AU from the Sun, so the intensity of sunlight at its surface is $1/.72^2 = 1/.52 = 1.92$ times greater than at the surface of the Earth. (Recall that the intensity of light from a source like the Sun drops off as the square of the distance; see the discussion in Chapter 4.) Thus Venus receives almost twice as much solar energy per square centimeter as the Earth.

As water on Venus evaporated due to the intense sunlight, water vapor was added to the atmosphere. Water vapor causes a mild greenhouse effect, so the temperature of the surface began to rise further. In addition, the lack of liquid oceans meant that there was no water in which atmospheric carbon dioxide could be dissolved, so the concentration of carbon dioxide continued to rise as this gas escaped from the planet's interior. Meanwhile, on the Earth the carbon dioxide was being removed from the atmosphere by reactions with ocean water and ended up deposited in carbonate rocks, where it still remains. The development of life on the Earth (at a much later time) also had a major effect that was absent on Venus; on the Earth plant life gradually converted carbon dioxide to oxygen through photosynthesis. The present-day atmosphere of the Earth, dominated by nitrogen and oxygen, does not have a strong greenhouse effect, and the atmosphere has stabilized with a moderate temperature (although human activity may be affecting this stability; see the Astronomical Insight on p. 126).

The large quantity of carbon dioxide in the atmosphere of Venus, on the other hand, caused intense greenhouse heating, which helped make the atmosphere very hot at a time when a lot of water vapor was present. The high temperature began to destroy the water vapor, separating (dissociating) the oxygen and hydrogen atoms from each other. The hydrogen then escaped into space (see the discussion of atmospheric escape in Chapter 5), while the oxygen became trapped in surface rocks through a process called **oxidation.** At the same time the surface of Venus probably continued to follow a geological evolution much like that of the Earth, with the exception of the differences noted in the previous section.

In contrast to this picture of a Venus that never had liquid oceans, some scientists have argued that the planet may once have had extensive, very hot seas. In this scenario, the oceans were able to persist because of the high atmospheric pressure; the boiling point of water rises with increasing pressure, so it is speculated that the high atmospheric pressure on Venus might have compensated for the high temperature and allowed oceans to remain. If so, these oceans would have eventually caused their own demise, however, because they would have gradually reduced the atmospheric pressure by gobbling up carbon dioxide. In due course the boiling point would have been reduced enough so that the oceans would have boiled away, allowing the atmosphere to evolve as described above. If this scenario is correct, then it is possible for a planet to lose oceans if the greenhouse effect becomes severe enough; therein lies a warning to us as we try to understand the effects of our own alterations of the Earth's atmosphere.

MARS: ANOTHER WATERY WORLD?

The question of water comes up again as we turn our attention to Mars, the next planet beyond the Earth as we go outward from the Sun. In its overall characteristics, Mars is quite distinct from the Earth, but at the same time its surface conditions are closer to those of the Earth than any other member of the solar system. Unlike the Earth, Mars has a carbon dioxide atmosphere (Table 7.3), but its daily cycles and seasonal variations closely resemble those of the Earth— at the height of the Martian summer, the temperature can reach comfortable values. Other Earth-like features include the polar ice caps, which exhibit seasonal variations, and the thin, wispy clouds that can be seen in the atmosphere.

The surface of Mars is distinctly reddish in color (this color is one reason for the planet's name; Mars was the god of war in ancient mythologies). The color is apparent in naked-eye and telescopic observations from the Earth (Fig. 7.21) and in close-up photos taken by the *Viking* landers and *Mariner* orbiters (Fig. 7.22). The color can be explained by the low level of

Figure 7.21   Mars as seen from the Earth.   This is thought to be the finest photograph ever obtained of the red planet as seen from the Earth's surface. The image was obtained with a CCD camera during the time of the 1988 opposition of Mars, when the planet was unusually close to the Earth. (J. Lecacheux, Meudon Observatory; CCD image obtained with the 1.05-m telescope, Pic-du-Midi Observatory)

Figure 7.22   A ground-level panorama on Mars.   This is the first photograph made by the *Viking 1* lander. It shows a rock-strewn plain extending in all directions. (NASA)

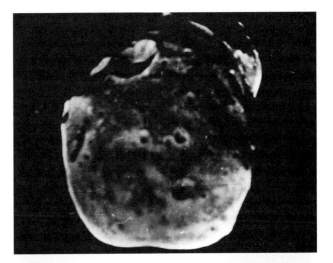

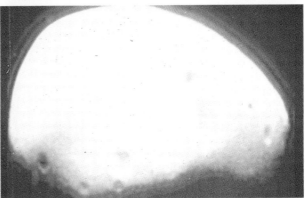

Figure 7.23   Phobos and Deimos.   At the top is an image of Phobos obtained by the Soviet mission *Phobos*. The irregular shape of the satellite is apparent, as are several impact craters. In contrast, Deimos (below), the more distant of the two moons of Mars, has a relatively featureless surface. (top: Soviet Academy of Sciences; bottom: NASA)

current and past geological activity; Mars never underwent as much differention as the Earth or Venus, and therefore large amounts of heavy elements such as iron have remained in the crust. When combined with oxygen, iron yields a reddish material known as iron oxide, or rust. Thus Mars is reddish in color because it is rusty.

Spectroscopic observations from Earth have shown that the atmosphere of Mars contains carbon dioxide and nitrogen (Table 7.3). Once scientists thought they had observed spectral lines due to chlorophyll, which would have confirmed the long-standing suspicion that plant life was present, but the spectral features turned out to have been misidentified. Speculation about life on Mars was also fueled by the linear features, once thought by some to be canals, and the seasonal variation in the surface markings, interpreted by some as annual growth of plant cover. All of these speculations were frustrated in 1965 when the U.S. spacecraft *Mariner 4* made the first close-up observations; they revealed Mars to be a desolate and barren world covered by craters. Nevertheless, the question of possible life forms on Mars, either past or present, still arouses substantial interest, as we will see in a later section.

In 1877, during its opposition, Mars made a particularly close approach to the Earth, allowing as-

tronomers to make out more detail than had previously been possible. The alleged canals were first sketched in detail at this time, setting off a wave of speculation about a Martian civilization, and the two tiny moons of Mars were discovered. These two satellites, called Phobos and Deimos (Table 7.4 and Fig. 7.23), are irregular chunks of rock that bear no resemblance to the Earth's Moon. They lie very close to Mars (hence the difficulty in detecting them from the Earth) and are very small and irregular in shape. It is thought that most likely they are captured asteroids.

TABLE 7.4   Satellites of Mars

No.	Name	Distance (R_M*)	Period (days)	Diameter (km)	Mass (g)	Albedo
1	Phobos	2.77	0.3189	27 × 21.6 × 18.8	9.6 × 10^{18}	0.06
2	Deimos	6.94	1.2624	15 × 12.2 × 11	1.9 × 10^{18}	0.07

*The distances given are orbital semimajor axes, in units of the radius of Mars, which is R_M = 3,384 km.

Seasons and Dust Storms

The rotation axis of Mars is tilted with respect to perpendicular, causing Mars to have seasonal variations much like those of the Earth. The obliquity of Mars (see Chapter 5) is 23°59′, very similar to the Earth's 23°27′. This might suggest a similar degree of variation from winter to summer, but the Martian seasons are a bit more complicated. The orbit of Mars is sufficiently noncircular that the variation in the intensity of sunlight reaching the planet is significant (in contrast, the Earth's orbit is so nearly circular that very little effect is felt; the Earth's closest approach to the Sun occurs in early January and certainly does not noticeably moderate winter in the Northern Hemisphere).

At its closest approach to the Sun, Mars is about 17 percent nearer than when it is farthest away. In other words, at this time the distance between Mars and the Sun is 83 percent as great as it is when the two are most widely separated. Using the inverse square law (Chapter 4), we can determine how much variation this causes in the intensity of sunlight reaching the surface of Mars: at closest approach the intensity is $1/.83^2 = 1.45$ times greater than at maximum separation—a large enough difference to have a pronounced effect on the climate.

Mars is closest to the Sun at the time of summer in the southern hemisphere and winter in the north (Fig. 7.24); as a result, the northern winter is mild while the southern summer is quite warm. Half a Martian year later, when it is summer in the north and winter in the south, Mars is at its farthest from the Sun, and the northern summer is mild while the southern hemisphere has a severe winter. Consequently, the south experiences extreme seasonal variations, while in the north they are only moderate.

The change in temperature is therefore very large and very rapid in the south, as one season turns into the next. In the southern spring, the enormous temperature contrasts create immense winds as air flows from the warm region toward the cooler surroundings. Since these winds are strong enough to pick up fine dust particles from the surface, Mars undergoes dust storms on a seasonal basis, every southern spring. Sometimes these storms become so severe and widespread that they cover the entire planet (Fig. 7.25). Such a storm was brewing when the first U.S. orbiter, *Mariner 9,* approached Mars (in 1971); the spacecraft was not able to photograph the surface of the planet for several weeks, until the storm finally subsided.

The seasons on Mars also affect the polar ice caps, a phenomenon that was observed through Earth-based telescopes long before any probes reached the red planet. The northern cap changes very little in size from winter to summer, but the southern cap grows very large in winter and almost disappears in summer. Spectroscopic measurements show that the caps are made of water ice and frozen carbon dioxide (known to us as dry ice). The outer surface of each cap is thought to be carbon dioxide ice, which overlies a layer of water ice.

Figure 7.25   A global dust storm.   This is the view of Mars that confronted the *Mariner 9* spacecraft for the first several weeks after it went into orbit around the planet.

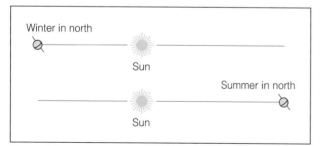

Figure 7.24   The Martian seasons.   As this exaggerated view shows, in the northern hemisphere both summer and winter are moderated by the varying distance of the planet from the Sun, whereas in the southern hemisphere both seasons are enhanced. The extreme temperature fluctuations in the southern hemisphere give rise to the winds that cause the seasonal dust storms.

The Story of Water on Mars

We have already mentioned that Mars may once have had large bodies of liquid water on its surface. Although scientists are still uncertain whether the early Mars had lakes or oceans, there is no doubt that water once flowed on the planet's surface. Ancient riverbeds have been found (Fig. 7.26), where large volumes of water obviously flowed, eroding the surface and carrying debris downstream. All of these channels are very old, lying in areas that are thought to date back some 3 billion years to the time of volcanic and tectonic activity on Mars. Since the atmosphere of Mars today could not sustain any large quantities of liquid water, we must ask how water was once able to exist there and what has happened to it.

Scientists have reached some agreement on the probable origin of the flow channels, if not on the eventual fate of the water (and its history). Most of the channels emanate from curious, jumbled regions known as **chaotic terrain** (Fig. 7.27). The ground in these regions apparently has collapsed, leading to the suggestion that there once were underground ice deposits, which suddenly melted. When the ice melted, the water flowed away in something like a flash flood, and the ground left behind collapsed, creating the

chaotic terrain that is seen today. The sudden melting is thought to have been caused by heating due to volcanic activity.

Today there is very little water on Mars, except for some water ice in the polar caps and traces of water vapor in the atmosphere (the concentration of the vapor varies; sometimes it forms thin clouds, and on cold days it creates early morning frost). But what happened to the much larger quantities of water that once were present? One possibility is that substantial quantities are still present in the form of underground

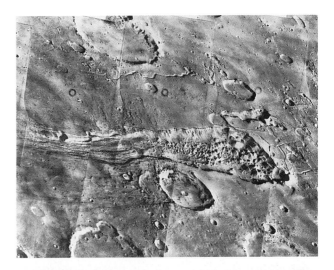

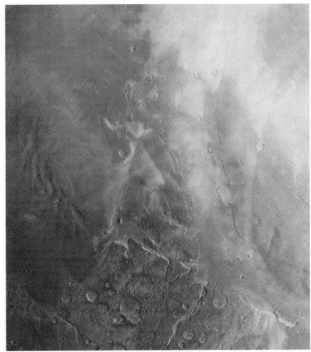

Figure 7.26   Ancient riverbeds on Mars.   At the lower center are several river channels showing northward flow (upward in the figure) from the edge of a highland scarp to a lowland plains region called Amazonis Planitia. (U.S. Geological Survey; data from NASA)

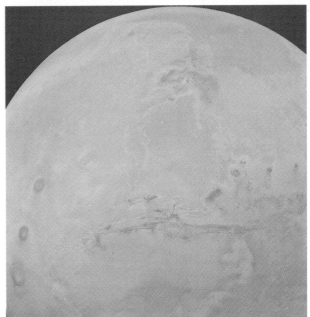

Figure 7.27   Chaotic terrain.   Top: This image shows a valley floor with chaotic terrain and an ancient river channel flowing to the left. Bottom: This full-disk mosaic shows Valles Marineris at lower center, and a region of chaotic terrain, just above Valles Marineris, at the center. River channels are seen flowing northward (up) from this chaotic region. (U.S. Geological Survey; data from NASA)

ice known as **permafrost.** According to another view, Mars formerly had a much thicker atmosphere and was able to sustain oceans or lakes, but changes in the climate reduced the atmospheric pressure and allowed the water to vaporize and escape. The causes of such a change in climate are unclear but one possibility is that the oceans themselves, if they existed, could have absorbed enough of the atmospheric carbon dioxide to reduce the pressure to the point where the water evaporated (this is similar to one scenario for the disappearance of water from Venus, as discussed in the previous section). In this view, not much water is left on Mars today. Thus one of the most interesting questions for future exploration of Mars will be to see whether substantial quantities of water are hidden beneath the surface. This will be a major goal of the *Mars Observer* mission, a U.S. probe due to arrive at Mars in 1993 that will be placed in a polar orbit around the red planet; the orbit will allow detailed observations of surface and atmospheric water vapor and seasonal changes in the water content of the ice caps.

Life on Mars, Past or Present

As we have seen, the possibility of life on Mars has been a recurrent theme of speculations about the red planet. Some of the speculation was fanciful, such as the myth of a civilization that built enormous canals that girdled the planet, but some was more logical, based on the knowledge that conditions under which life could form may once have existed on Mars. After all, life on the Earth is believed to have begun in the oceans; if Mars once had seas or large lakes, perhaps the same processes could have occurred there.

The most widely publicized aspect of the dual 1976 *Viking* missions was the attempt by the robot landers to detect evidence of life-forms on Mars. The robots conducted several tests, the first and most straightforward being simply to look with the television cameras for any large plants or animals. None were seen, so much more sophisticated tests were tried.

The searches for life were carried out by three different experiments on board the landers. The purpose of each experiment was to look for signs of metabolic activity in a soil sample scooped up by a mechanical arm and deposited in containers for analysis (Fig. 7.28). All living organisms on Earth, even microscopic ones, alter their environment in some way just by existing. Usually the effects involve chemical changes as the organism derives sustenance from its surroundings and ejects waste material.

Two of the three experiments, the *pyrolytic release experiment* and the *gas-exchange experiment*, showed no evidence of possible life-forms. The third, however, the *labeled release experiment*, aroused much excitement among scientists because its initial results duplicated the expected effects of active life-forms. In this experiment, a nutrient solution was added to a sample of Martian soil in a closed container, in the expectation that organisms in the soil would metabolize the nutrient and release waste gases in the chamber. To make the gases detectable, a small quan-

Figure 7.28   Sampling the Martian soil.   Here the scoop on *Viking 2* is shown as it digs up a sample of Martian soil for one of its life-detection experiments. (NASA)

tity of radioactive carbon (^{14}C, in which each atom consists of the usual six protons but has eight neutrons instead of the normal six) was added to the nutrient. If metabolic activity took place, then some ^{14}C should appear in gases emitted by the sample, and that is just what happened. As a check, the same experiment was performed on other samples that had been heated so that any life-forms present would have been killed, and indeed no tracer gases were emitted from those samples. The combination of activity in the normal samples and the lack of activity in the sterilized samples was consistent with the presence of life-forms in the Martian soil.

Still the evidence was not conclusive. The other experiments failed to show positive results, and the reactions in the labeled release experiment occurred much faster than seemed likely for metabolic activity. Gases containing ^{14}C were released more quickly and in greater quantity than would ever have been possible for any earthly microorganisms. Furthermore, and perhaps more telling, a given sample would react positively only *once*, even though nutrient was added to some samples several times. Real life-forms should be capable of eating more than one meal. Eventually scientists concluded that the observed activity must have been due to an unexpected type of chemical reaction between the soil and the nutrient. The reaction probably involved oxides in the soil whose chemical properties were altered by heating, so that the activity did not occur in the sterilized samples.

A further test for evidence of life was carried out by a device called a **mass spectrometer,** which is capable of analyzing a sample to determine the types of atoms and molecules it contains. The mass spectrometers aboard the *Viking* landers found no evidence of **organic molecules,** those containing certain combinations of carbon atoms that are always found in plant or animal matter on the Earth. Hence, no doubt with some reluctance, most *Viking* scientists concluded that no evidence for life on Mars had been found.

The lack of evidence for life is not the same as evidence for the lack of life, however. After all, these experiments could test samples at only two localities on an entire planet, and furthermore, the tests were predicated on the assumption that Martian life-forms, if they exist, would be in some way similar to those on Earth. Now that more is known about the chemistry of the Martian soil, new experiments could probably be devised that would be relatively free of the confusion created by the nonbiological activity detected in the *Viking* tests.

Even if no life exists on Mars today, the possibility remains that the planet once was the home of life-forms. We believe that life on Earth began through complex chemical reactions that took place in the early oceans (see Chapter 22). The oldest fossil evidence for life on Earth is found in rocks in western Australia that are about 3.5 billion years old. Thus life began on Earth within the first billion years after the formation of the planet. There is no obvious reason why it could not have done so on Mars as well, if the red planet underwent an early period when it had higher atmospheric pressure and liquid oceans or lakes, as is now suspected. The surface would have been warmer than it is now, due to the greenhouse effect, and we know that the surface of Mars contains the same elements from which life formed on Earth. Perhaps there were primitive life-forms on Mars that have since died out, as the planet lost most of its atmosphere and nearly all of its water. If so, fossil remains of life-forms might be found in the Martian soil. The discovery of such remains will require detailed scrutiny of soil samples, and experiments to do this are being proposed for future missions aimed at placing landers on the Martian surface. Both the U.S. and Russian space programs have placed high priority on future attempts to land spacecraft on Mars. Perhaps within a few years we will have found evidence that the Earth is not the only place where life has formed in the universe.

MERCURY: CLOSE ENCOUNTERS WITH THE SUN

The innermost of the planets, Mercury is dominated in many ways by its proximity to the Sun. As we will see, the Sun has played a major role in governing the orbital and rotational properties of Mercury, and the hot environment at the center of the solar system has largely determined the properties of the planet.

As already mentioned, Mercury is hard to observe because, from the vantage point of the Earth, the planet is never far from the Sun on the sky. Nevertheless, Earth-based observations revealed some hints of surface markings, which led to the conclusion that the planet was in synchronous rotation, always keeping the same face toward the Sun. This rotation was not unexpected, considering the enormous tidal forces that the Sun must be exerting on Mercury (see the discussion in Chapter 3).

Much more detailed observations are possible using radio wavelengths, however, and when radio telescopes were first turned toward Mercury, astronomers found some surprises. The dark side of the planet is not as cold as would be expected if the other side always pointed toward the Sun, and the planet's rotation is not synchronous. Measurements established that Mercury rotates once every 58.65 days and has an orbital period of almost 88 days. This rotation is very slow, but not as slow as would be required to keep one side facing the Sun. As a result, all portions of the surface of Mercury are exposed to direct sunlight at times, and this exposure is the reason the dark side is not as cold as had been expected.

Details of the surface structure of Mercury finally began to be obtained in 1974 and 1975 from the *Mariner 10* probe. As explained earlier in this chapter, the orbit of *Mariner 10* was adjusted so that it made three encounters with Mercury (Fig. 7.29). Much of the surface of the planet has now been imaged, revealing that it closely resembles the Moon (see Fig. 7.12). *Mariner 10* also made magnetic field measurements, which established that Mercury has a field despite its slow rotation rate. The presence of a magnetic field provides important clues to the internal structure of the planet.

Orbit and Rotation: Spin-Orbit Coupling

Precise measurements of the rotation period of Mercury, carried out through the analysis of radar echoes bounced off the planet, showed that the rotation period of 58.65 days is precisely equal to two-thirds of the orbital period of 87.97 days. Therefore Mercury spins one-and-a-half times during each orbit around the Sun. Thus, at the same point in its orbit, Mercury has opposite sides facing the Sun on consecutive trips around (Fig. 7.30).

This pattern is no coincidence; clearly the tidal force of the Sun is at work. The key is what happens at the point where Mercury is closest to the Sun,

called **perihelion,** for the tidal force is strongest here. If one side of Mercury is a bit denser than the other, as on the Moon, then the tidal force exerted by the Sun will act to ensure that this side is aligned toward the Sun when Mercury is at perihelion. Of course, with synchronous rotation, this side would always point toward the Sun, but because Mercury's orbit is quite elongated, the urge to keep its heavy side facing the Sun is much stronger at perihelion than at other points in its orbit.

Because Mercury spins one-and-a-half times each orbit, the heavy side of the planet faces either directly toward the Sun or directly away from it each time Mercury passes through perihelion. In either case the tidal force acting on Mercury is balanced, so that there is no tendency to alter the rotation further. Apparently the planet once rotated much faster, but whenever it passed through perihelion with the heavy side pointed in some random direction, tidal forces exerted a tug that tended to change its spin. This process went on until the rotation slowed to its present rate, so that the tidal force is always balanced when Mercury is closest to the Sun. If the orbit had not been so elongated, or if Mercury had been symmetric instead of having a relatively dense side, the tidal forces would have been more uniform throughout the orbit, and Mercury might be in synchronous rotation today.

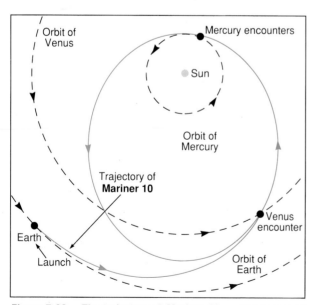

Figure 7.29   The trajectory of *Mariner 10*.   The spacecraft flew by Venus and then encountered Mercury repeatedly because its orbital period was adjusted to be equal to twice the orbital period of Mercury.

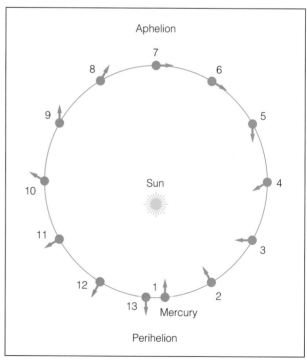

Figure 7.30   Spin-orbit coupling.   This sketch shows how Mercury spins one-and-a-half times while completing one orbit around the Sun. At perihelion it always has the dense side facing directly toward or away from the Sun.

In whole numbers, Mercury spins three times for every two orbits. Such a numerical relationship between the rotation period and the orbital period is called **spin-orbit coupling,** a general term applied to any situation where the rotation of a body has been modified by tidal forces so that a special relationship between the orbital and spin periods is maintained. The most common example is synchronous rotation (where the ratio is 1:1 instead of 3:2), and we will find several cases of synchronous rotation in the rest of the solar system and in the universe beyond.

The combination of orbital and rotational speeds on Mercury produces a very unusual effect for anyone who might visit the planet. At its closest approach to the Sun, when Mercury is moving most rapidly in its orbit, the orbital speed is actually greater than the rotational speed at its surface. For a short time (lasting a few hours), the Sun would appear to be moving backward in the sky, from west to east. Imagine the difficult time our ancestors would have had explaining this retrograde motion of the Sun if a similar phenomenon had occurred on the Earth!

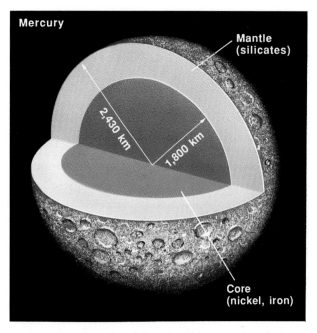

Figure 7.31   Mercury's internal structure.   The presence of a magnetic field, along with the planet's relatively high density, implies the presence of a large core.

Surface and Interior:
The Effects of Heat and Collisions

We have already mentioned that Mercury has a very high density and a detectable magnetic field, both perhaps a bit surprising in view of the low mass of the planet and its slow rotation. A partial explanation lies in the environmental conditions that prevailed in the central part of the solar system at the time the planets formed. The central region of the solar nebula was very hot, which made it impossible for the more volatile elements to condense into solid form. Thus the planetesimals that formed in the innermost part of the young solar system probably contained very small quantities of volatile elements from the beginning. A more important mechanism seems necessary to explain the extreme depletion of volatile elements, however. One suggestion is that soon after its formation, Mercury was struck by a planetesimal large enough to strip away most of its mantle. This collision would have occurred after differentiation—and the formation of a metallic core—had already taken place.

The end result of the collision is that Mercury has a very low abundance of the lightweight, volatile elements and thus a very high density. In addition, the planet's high overall density means that it must have a very large core (Fig. 7.31), and this core must be at least partially molten, producing the observed magnetic field. In effect, Mercury may be viewed as a terrestrial planet that has lost most of the outer layers that the other terrestrials have retained.

We commented at the beginning of this chapter that Mercury closely resembles the Moon. Now we see that this resemblance is only skin-deep; internally, the two bodies are very different. The Moon is geologically inert, has undergone little differentiation, and has no magnetic field, whereas Mercury has a large, partially molten core and internal currents that produce a magnetic field. In composition Mercury is much more iron-rich and lower in the volatile elements than the Moon (which, in turn, has lower abundances of volatile species than the Earth).

Despite the internal contrasts, on the surface Mercury can be mistaken for the Moon to the untrained eye (Fig. 7.32). There are craters everywhere and some smooth, dark areas that resemble maria. These features reflect some similarities in the evolutions of the surfaces of the two bodies: each has been bombarded by meteorites since its formation, with no erosion or surface geological processes to remove the craters, and each has had extensive lava flows in its past. Like the Moon, Mercury has a very old surface, which reflects the lack of crustal plate shifting and overturning, despite the hot, partially molten interior.

As already mentioned, the major differences between the surfaces of Mercury and the Moon are the lower vertical relief of the craters on Mercury (Fig. 7.33) and the global system of scarps (Fig. 7.34). The lower crater walls are due to the higher surface gravity of Mercury, and the scarps are thought to be the result of crustal shrinkage as volatile gases continued to escape from the interior after the crust had hardened.

Figure 7.32   Craters on Mercury.   Like the Moon, Mercury has a heavily cratered surface. Because Mercury has a greater surface gravity than the Moon, however, impact craters have lower rims and are shallower, and ejecta do not travel as far. (NASA/JPL)

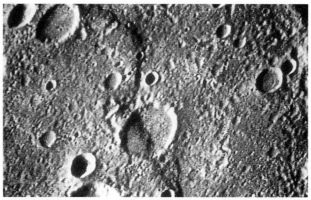

Figure 7.34   A scarp system on Mercury.   An enormous scarp (cliff) runs vertically through this *Mariner 10* image, cutting across a large crater at lower center (this illustrates that the scarp formed after the period of major cratering). (NASA)

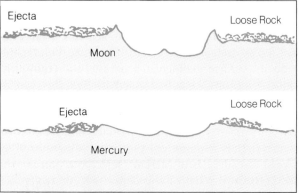

Figure 7.33   Craters on Mercury.   Top: This is a *Mariner 10* image showing a cratered region on the surface of Mercury. Bottom: This drawing illustrates the contrasts between craters on Mercury and those on the Moon: on Mercury, the crater walls are lower and the ejecta to not travel as far, due to Mercury's higher surface gravity. (Left: NASA)

The *Mariner 10* probe was only able to map half of the surface of Mercury, because the spacecraft's orbital period was equal to twice the period of Mercury. Due to Mercury's 3:2 spin-orbit coupling, the planet made exactly three rotations during the two orbital periods between its encounters with *Mariner 10*. Therefore every time the spacecraft came close to Mercury, the same side of the planet faced the Sun, and the other side, which was in darkness, could never be photographed.

As a result, the *Mariner 10* survey nearly missed a major surface feature. Just at the boundary between the daylight and dark sides of the planet at the time of the encounters, an enormous crater was observed (Fig. 7.35). This huge circular basin must be the result of a collision with a large body some time in the distant past (the floor of the crater is covered by many, more recent, craters, an indication of its great age). Interestingly, this basin lies at a position on Mercury that faces either directly toward or directly away from the Sun at perihelion, when the planet passes closest to the Sun. Thus, on every other orbit of the Sun, this crater is the hottest spot on the planet; accordingly, it was named Caloris Planitia (Latin for "Plain of Heat").

The location of Caloris Planitia is probably not a coincidence. Recall that Mercury's 3:2 spin-orbit coupling is thought to have come about because one side of the planet is slightly denser than the other. If the remains of the large meteorite that created Caloris Planitia are embedded in the surface of Mercury, they may well explain why that side of the planet has a higher density than the other. We may speculate that Mercury might never have fallen into its 3:2 spin-orbit coupling had it not been for the huge impact that created Caloris Planitia.

On the opposite side of Mercury from Caloris Planitia lies a region of strange, wavy surface structures,

Figure 7.35   Caloris Planitia.   This photo shows half of the immense impact basin known as Caloris Planitia. This region is directly facing the Sun at perihelion on every other orbit. (NASA/JPL)

quite unlike anything else found on the terrestrial planets. This region is so unusual that geologists call it the **weird terrain**. These structures are thought to be the result of the collision of seismic waves that traveled around Mercury when the impact that created Caloris Planitia occurred; as these waves came together on the far side of the planet, they created stresses in the crust that forced some regions upward and others downward, leaving a record of the waves permanently frozen into the surface. Interestingly, detailed examination of the Moon's surface has revealed similar features opposite major craters, but on the Moon these areas are much less pronounced.

PERSPECTIVE

We have completed our detailed look at the terrestrial planets, finding a wealth of comparisons among them. In gross properties they seem quite similar, being rocky bodies with comparable average densities, but in detail they provide many contrasts, including their atmospheres, the varied age of their surfaces, and their rather different internal structures.

Now we move on to discuss another class of planets, the gaseous giants, which resemble the terrestrials in very few ways. We will find that the environment of the outer solar system was so different from that of the inner regions that the basic nature of the planets that formed is strongly in contrast. But at the same time, we will find that the gaseous planets were molded by the same physical processes that governed the formation and evolution of the terrestrial bodies; consequently, in our discussion of the outer planets, we can apply much of what we have learned from our study of their terrestrial cousins.

SUMMARY

1. Venus, Mars, and Mercury have all been studied extensively from the Earth (radio observations being the most useful for Venus and Mercury), and all three have been visited by space probes from Earth.

2. Venus and Mars both have carbon dioxide atmospheres created by outgassing from the surface and a lack of liquid oceans to dissolve the gas; Mercury has only a very thin atmosphere.

3. The greenhouse effect is important in heating the surfaces of all the terrestrial planets, but has been extreme in the case of Venus; there the thick carbon dioxide atmosphere has raised the surface temperature to 750 K.

4. Venus is tectonically active (though without continental drift) and is thought to be volcanically active, and to have an internal structure much like that of the Earth. Mars underwent some volcanism and tectonic activity early in its history, but is probably not active today. Mercury has a relatively large core that must be partially molten and has had volcanic activity in the past, but the great age of the surface shows that this planet is not undergoing extensive volcanic or tectonic activity today.

5. The *Magellan* radar mapper has shown that the surface of Venus is dominated by volcanic structures and that surface stresses due to flows in the interior of the planet have created rift systems and regions of uplift; the planet has not experienced large-scale plate motions as on the Earth, however.

6. Scientists believe Venus has a dense carbon dioxide atmosphere and a high surface temperature largely because the planet's proximity to the Sun prevented it from retaining liquid oceans early in its history. The lack of seas allowed carbon dioxide to remain in the atmosphere, where it created the strong greenhouse heating that keeps the planet so hot.

7. Seasons on Mars are exaggerated by its elongated orbit; one result is annual dust storms (during spring in the southern hemisphere), which sometimes involve the entire planet and shift the surface dust cover, creating seasonal changes in the planet's coloration.

8. Mars almost certainly once had abundant liquid water on its surface, as evidenced by dry channels seen today, but the amount of the water and what happened to it are not known. It may have been lost to space by evaporation and escape, or much of it may remain on Mars in the form of subsurface permafrost.

9. Although the *Viking* landers failed to find evidence for life-forms in the Martian soil, it is possible that life could have existed there if the planet had a thicker atmosphere and liquid oceans early in its history.

10. Mercury is locked into 3:2 spin-orbit coupling, spinning three times for every two orbits around the Sun. This rotation pattern was caused by Mercury's noncircular orbit and the strong tidal force due to the Sun, which acted on Mercury's nonsymmetric mass distribution to slow its rotation so that the heavy side faces either toward or away from the Sun when the planet passes through perihelion.

11. Mercury's high average density and magnetic field indicate that the planet has a very large core and a relatively low abundance of lightweight, volatile elements. The sparsity of volatile elements is due to its formation in the inner solar system, where the volatile species were not able to condense into solid form, and, more importantly, to a suspected impact with a large planetesimal, which stripped off the planet's outer layers after differentiation had occurred.

REVIEW QUESTIONS

1. Using your knowledge of conditions on Venus and Mars, explain what the atmosphere of the Earth might be like today if life had never evolved.

2. Suppose that Mars and Mercury both are made of just two distinct kinds of material: crust, with a uniform density of 2.0 grams/cm^3; and core, with a constant density of 10.0 grams/cm^3. How large does the core of each planet have to be, relative to the total radius, to provide the average densities of 4.0 grams/cm^3 and 6.0 grams/cm^3 for Mars and Mercury, respectively? (The values for the average densities have been rounded to simplify the calculation.)

3. Discuss the general structure of the surfaces of Mercury, Venus, and Mars, explaining the similarities and differences among them. In your explanation, mention the role played by atmospheres, tectonic activity, volcanism, and impacts.

4. Using the inverse square law, calculate how much more intense sunlight is at Mercury's distance from the Sun than it is at Venus. Despite the contrast, the surface of Venus is hotter than the daylit side of Mercury. Explain why this is so.

5. Explain why the masses of Mercury and Venus could not be measured using the same technique that was used for the other planets.

6. Would you expect Mars, Venus, and Mercury to have charged-particle belts surrounding them similar to the Earth's Van Allen belts? Explain.

7. Why is it thought that life might once have existed on Mars? Would you expect that the same thing might have happened on Venus? Explain.

8. To illustrate why the Earth's seasonal variations are not affected by the varying distance of the Earth from the Sun, calculate the relative intensity of sunlight on the Earth when it is closest to the Sun and when it is farthest away. At closest approach, the Earth is 0.983 A.U. from the Sun; at its farthest, it is 1.017 A.U. away. Compare this variation in intensity with that on Mars during its year (figures are given in the text).

9. Summarize the similarities and differences between Mercury and the Moon.

10. Suppose another terrestrial planet with a mass equal to 80 percent of the mass of the Earth was in a circular orbit 0.9 A.U. from the Sun. Describe its likely properties; i.e., what its atmosphere, average density, degree of tectonic activity, and internal structure would be like. Support your answer by making comparisons with the properties of the actual terrestrial planets.

ADDITIONAL READINGS

Ainsworth, D. 1992. *Mars Observer:* return to the red planet. *Astronomy* 20(9):28.

Allen, D. A. 1987. Laying bare Venus' dark secrets. *Sky and Telescope* 74:350.

Bazilevsky, A. J. 1989. The planet next door. *Sky and Telescope* 77:360.

Beatty, J. K., B. O'Leary, and A. Chaikin, eds. 1990. *The new solar system.* 2nd. Ed. Cambridge, England: Cambridge University Press.

Burnham, R. 1991. Venus, planet of fire. *Astronomy* 19(9):32.

Carr, M. S. 1983. The surface of Mars: a post-*Viking* view. *Mercury* 12(1):2.

Cordell, B. M. 1986. Mars, Earth, and ice. *Sky and Telescope* 7(1):17.

Dick, S. J. 1988. Discovering the moons of Mars. *Sky and Telescope* 76:242.

Eicher, D. J. 1991. *Magellan* scores at Venus. *Astronomy* 19(1):54.

Goldman, S. J. 1987. In search of Martian seas. *Sky and Telescope* 73:6.

_____. 1990. The legacy of *Phobos. Sky and Telescope* 79:156.

_____. 1992. Venus unveiled. *Sky and Telescope* 83(3):258.

Hartmann, W. K. 1989. What's new on Mars. *Sky and Telescope* 77:471.

Horowitz, N. H. 1977. The search for life on Mars. *Scientific American* 237(5):52.

Kasting, J. F., O. B. Toon, and J. B. Pollack 1988. How climate evolved on the terrestrial planets. *Scientific American* 258(2):90.

Leovy, C. B. 1977. The atmosphere of Mars. *Scientific American* 227(1):34.

Saunders, S. 1991. The exploration of Venus: a *Magellan* progress report. *Mercury* 20(5):130.

Strom, R. G. 1990. Mercury: the forgotten planet. *Sky and Telescope* 80(3):256.

ASTRONOMICAL ACTIVITY
Looking for Mercury

Even though Mercury is a very bright object, easily bright enough to be seen with the naked eye, many people never see it. The reason is that Mercury is always close to the Sun, never straying more than 28° from it (this is Mercury's angle of greatest elongation). Thus, to see Mercury, you have to look for it low in the sky near its time of greatest elongation, either just before sunrise or just after sunset, depending on where the planet is in its orbit. When Mercury is near its greatest western elongation, we see it as a morning object, rising a little before the Sun. When the planet is near its greatest eastern elongation, it sets just after the Sun, and we see it as an evening object.

The accompanying table lists the times of Mercury's greatest elongations for the coming few years. Because Mercury's synodic period is short (116 days), there are frequent opportunities to see it near one of its maximum elongations (there are two per synodic period, or one every 58 days). Use this table to plan your observations; all you need to do is go outside at the appropriate time (before sunrise for a western elongation; after sunset for an eastern one) and look for Mercury low in the sky toward the diffuse glow caused by the Sun's presence just below the horizon. You must be prompt, because the conditions for best viewing do not last for long.

Greatest Elongations of Mercury

Eastern Elongations	Western Elongations
February 21, 1993	April 5, 1993
June 17, 1993	August 4, 1993
October 14, 1993	November 22, 1993
February 4, 1994	March 19, 1994
May 30, 1994	July 17, 1994
September 26, 1994	November 6, 1994
January 19, 1995	March 1, 1995
May 12, 1995	June 29, 1995
September 9, 1995	October 20, 1995
January 2, 1996	February 11, 1996
April 23, 1996	June 10, 1996
August 21, 1996	October 3, 1996
December 15, 1996	January 24, 1997
April 6, 1997	May 22, 1997
August 4, 1997	September 16, 1997
November 28, 1997	January 6, 1998
March 20, 1998	May 4, 1998
July 17, 1998	August 31, 1998
November 11, 1998	December 20, 1998

The Gas Giants and Pluto

A crescent Uranus as seen by *Voyager 2* (NASA)

PREVIEW

The gaseous outer planets present many contrasts with the terrestrial bodies, yet we can understand their nature by invoking the same principles of planetary science. Their atmospheric motions, internal structures, and even the very features that set the giant planets apart from the terrestrials are all derived from the same mechanisms discussed in Chapters 3 and 5. By now you should be so well versed in these physical processes that you could almost predict the properties of a planet simply by knowing its location in the solar system. The giant planets, having formed under very cold conditions, have retained even the lightweight elements such as hydrogen and helium; they are therefore far more massive, yet at the same time much less dense, than the terrestrials. Each of the giant planets was surrounded by a cloud of debris as it formed, with the result that today each has an extensive system of moons and rings. The outermost planet, Pluto, is like no other among the nine planets, but we will find that it may be just the most prominent member of an enormous class of outer solar system bodies that are left over from the time of planet formation.

In moving to a discussion of the outer five planets in the solar system, we are making a major transition not only in distance, but also in style. The outer planets bear little resemblance to the terrestrial bodies, either superficially or on close examination (Fig. 8.1). Furthermore, in origin and evolution, the outer planets differ markedly from the inner four. Yet we will find that many of the same physical processes at work in the inner regions occur in the outer solar

system. We need only understand the different starting conditions to see how these processes led to such different final results.

Table 8.1 summarizes the basic properties of Jupiter, Saturn, Uranus, and Neptune; Pluto is omitted for now because it has very little in common with the other four. A quick glance at the table reveals how the bulk properties of the four giant planets differ from those of the terrestrials: the masses of the outer planets are far greater, their average densities are much lower, and they are much colder. In addition, these planets tend to have many satellites, and they all rotate very rapidly. In the discussions that follow, we will find many more contrasts as well.

EXPLORING THE OUTER SOLAR SYSTEM

Only two of the outermost five planets—Jupiter and Saturn— are easily seen by the unaided eye, so the ancient astronomers were not aware of Uranus, Neptune, or Pluto. Uranus was noticed by some observers and included in a few star charts, but was not recognized as a planet until its discovery by William Herschel in 1781. Using a fine telescope for the time, Herschel was able to discern the disk of the new body and, by tracking its motion for some time, established that it is a Sun-orbiting body located far from the center of the solar system.

The subsequent discovery of Neptune in 1840 came about because the motion of Uranus appeared to violate Newton's laws of motion and gravitation. From its discovery in 1781 until 1840, Uranus wan-

TABLE 8.1   Basic Properties of the Gas Giants

	Jupiter	Saturn	Uranus	Neptune
Orbital semimajor axis (AU)	5.203	9.555	19.22	30.11
Perihelion distance (AU)	4.951	9.023	18.31	29.85
Aphelion distance (AU)	5.455	10.086	20.13	30.37
Orbital period (yr)	11.86	29.46	84.01	164.79
Orbital inclination	1°18'17"	2°29'33"	0°46'23"	1°46'22"
Rotation period (hr)	9.925	10.657	−17.240*	16.05
Tilt of axis (obliquity)	3°5'	26°44'	97°52'	29°34'
Mean diameter (relative to Earth)	10.86	9.00	3.97	3.86
Polar diameter	10.50	8.54	3.92	3.82
Equatorial diameter	11.23	9.47	4.01	3.89
Mass (relative to Earth)	318.1	95.12	14.54	17.15
Density (g/cm³)	1.327	0.69	1.27	1.64
Surface gravity (g)	2.64	1.13	0.89	1.13
Escape speed (km/sec)	60	36	21.2	23.5
Surface temperature (K)	165	134	76	74
Albedo	0.52	0.47	0.50	≈0.5
Satellites	16	18	15	8

*The minus sign indicates that the rotation of Uranus is technically in the retrograde direction (i.e., opposite of the spins of most of the other planets), due to the fact that its north pole is tipped over so far that it points below the plane of the planet's orbit.

dered some 2′ away from its predicted path around the Sun. Astronomers in England (John C. Adams) and in France (Urbain Leverrier) independently reached the conclusion that another planet could be causing the discrepancy due to its gravitational pull on Uranus, and in 1840 the German astronomer Johann Galle found Neptune within 1° of the predicted position. Thus the discovery of the eighth planet turned a seeming failure of Newton's laws into a triumph and added another complex world to the known family of the Sun.

The list of planets was not extended again until 1930, when Lowell Observatory astronomer Clyde Tombaugh found Pluto after an extended search (Fig. 8.2). Ironically, the motivation for the search came from the late Percival Lowell himself (the amateur astronomer who believed in a Martian civilization; see Chapter 7), who thought he had found discrepancies in the motions of both Uranus and Neptune that could only be explained by a ninth planet. Astronomers later determined that Pluto is far too small to have caused any such effect, so its discovery was fortuitous.

Telescopic observations of Jupiter and Saturn are capable of revealing some of the details of their surfaces (actually, since these planets do not have solid

Figure 8.1 The giant planets. The images of Jupiter (upper left), Saturn (upper right), Neptune (lower left), and Uranus (lower right) were all obtained at close range by the Voyager probes. The giant planets are very different from the terrestrial planets, being much larger and far less dense and having no solid surfaces. (NASA)

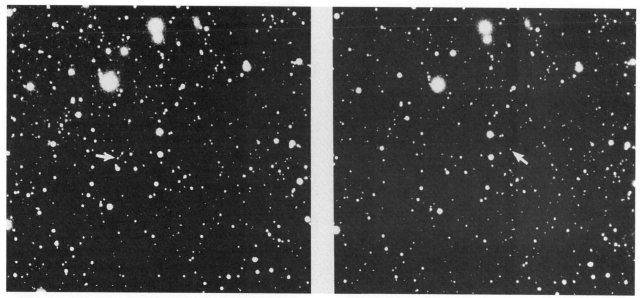

Figure 8.2   The discovery of Pluto.   These are portions of the original photographs on which the ninth planet was discovered in 1930. The position of the planet in each is indicated by an arrow. (Lowell Observatory photograph)

surfaces, it is better to speak of their outer atmospheres). But Uranus, Neptune, and Pluto are all too distant for much of their surface markings to be seen, and rather little was known about them until very recently. Jupiter displays a vivid banded appearance as well as a large reddish oval region known as the Great Red Spot (Fig. 8.3), whereas Saturn has much more muted bands and a generally less colorful appearance (Fig. 8.4). As if to compensate for this drab appearance, it is distinguished by a spectacular sys-

tem of rings that girdle the planet. The rings show intricate structure, even when viewed from the remote vantage point of the Earth. In recent decades, both Jupiter and Saturn have been found to have many satellites. Recall (from Chapter 2) that the four large moons of Jupiter were first observed by Galileo, who used the fact that they clearly do not orbit the Earth as an argument against the Earth-centered solar system.

Little about Uranus and Neptune could be determined from Earth, except for the discovery of some satellites (five for Uranus and two for Neptune) and some evidence of atmospheric clouds and banding, primarily from infrared images (Fig. 8.5). From spec-

Figure 8.3   A *Voyager* image of Jupiter.   The *Voyager* spacecraft obtained this photo when it was still millions of kilometers from Jupiter. Already a vast amount of detail is evident in the atmospheric structure. (NASA)

Figure 8.4   A *Space Telescope* portrait.   This image of Saturn, which was obtained by the *Hubble Space Telescope* in Earth orbit, shows much more detail than can be seen in the best ground-based images. (NASA/Space Telescope Science Institute)

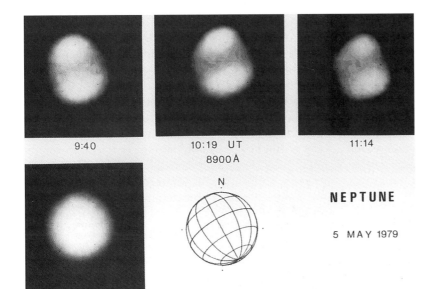

9:40   10:19 UT   11:14
8900A

10:07

N

NEPTUNE

5 MAY 1979

Figure 8.5   Earth-based views of Neptune.   These images, obtained with the use of special filters that isolate the wavelengths at which methane molecules absorb light, represent the greatest amount of detail observed from the Earth. Despite the great distance to Neptune and the tiny size of its image, data such as these demonstrated the presence of atmospheric structure, even before the *Voyager 2* encounter. (NASA)

troscopic data, scientists learned that lightweight gases such as hydrogen and its compounds ammonia (NH_3) and methane (CH_4) dominate the outer layers of four of the giant plants (they presumed that helium, the second most common element after hydrogen, was also abundant, but helium has few spectral lines in visible wavelengths and hence was difficult to measure directly).

Radio and infrared observations of the outer planets revealed some major phenomena that are not found on the terrestrial planets. First Jupiter and then Saturn were found to be strong emitters of radio waves. The emissions include bursts of static that are apparently caused by lightning discharges in the clouds, the normal thermal emission of any body whose temperature is greater than absolute zero (see Chapter 4), and a strange kind of continuous radio emission called **synchrotron emission.** This radiation, which is caused by free electrons moving very rapidly in a magnetic field, was the first clue that the giant planets are surrounded by extensive belts of charged particles. These radiation zones dwarf the Earth's Van Allen belts in both size and intensity. It was expected (and later confirmed by space probes) that Uranus and Neptune would exhibit similar phenomena, but their greater distances from the Earth made the radiation zones difficult to detect.

Infrared observations of the outer planets revealed surface features, primarily those due to methane in the cloud tops. The infrared data also provided accurate measurements of the energy emitted by the cold outer planets, whose peak radiation occurs in the infrared, according to Wien's law (Chapter 4). One major surprise was the discovery that three of the four planets (Uranus being the exception) emit considerably more energy than they receive from the

Sun. This implies that the planets have some sort of internal heat source, which is either generating energy or releasing excess energy stored in the planetary interiors. The source of the extra heat is now thought to be different for each planet, as discussed later in the chapter.

Starting in the 1970s, several successful space probes added immensely to our knowledge of the outer planets (Table 8.2). The two *Pioneer* missions (*Pioneer 10* and *Pioneer 11*), and especially the *Voyagers* (*Voyager 1* and *Voyager 2*), revealed that the four giant planets and their rings and moons were fantastically complex and beautiful worlds when seen close up. We may expect further revelations with the

TABLE 8.2   Probes to the Gas Giants*

Spacecraft	Launch	Arrival	Remarks
Jupiter			
Pioneer 10	3/03/72	12/03/73	Flyby
Pioneer 11	4/05/73	12/02/74	Flyby
Voyager 1	9/05/77	3/05/79	Flyby
Voyager 2	8/20/77	7/09/79	Flyby
Galileo	10/18/89	12/95	Orbiter and atmosphere probe
Saturn			
Pioneer 11	4/05/73	9/01/79	Flyby
Voyager 1	9/05/77	11/12/80	Flyby
Voyager 2	8/20/77	8/25/81	Flyby
Uranus			
Voyager 2	8/20/77	1/24/86	Flyby
Neptune			
Voyager 2	8/20/77	8/24/89	Flyby

*Based on a compilation published by the Astronomical Society of the Pacific.
NOTE: All are U.S. missions.

advent of the *Galileo* mission (now on its way to Jupiter, but suffering from communication problems that may limit its success), the *Cassini* mission to Saturn (now under construction), and a planned mission to Pluto (now in its very early developmental stages). The space probes revealed that each of the gas planets was unique and intrinsically very complex. Each has more moons than previously detected and a ring system, though none is as brilliant as Saturn's rings (the rings of Uranus and Neptune were discovered through Earth-based observations, but little was known about them before *Voyager*).

In the following sections we will discuss and compare the important properties of the four gas giants as a group, before moving on to more detailed descriptions of them individually. Pluto will be discussed later in the chapter, and most of the comments on rings and moons will be deferred until Chapter 10, where we will see how these phenomena share a common origin that is linked to the birth of the solar system itself.

THE INTERNAL STRUCTURE OF THE GAS GIANTS

As already noted, the four giant planets do not have solid surfaces, but instead are gaseous. Theoretical models of their interiors show that they probably have no surface below the visible cloud layers and, in fact, probably have little solid material anywhere inside.

Before we describe the internal structure further, it is useful to consider how astronomers can know anything at all about what goes on inside a planet that is far away and very different from the Earth and its closer neighbors. It is possible to construct mathematical models of the planets (or, for that matter, of stars or galaxies or anything else where we think the laws of physics are at work). To do this requires equations that describe the forces acting on the matter contained in the planet, as well as information on conditions such as rotation rate, surface temperature, overall size and density, and chemical composition. For example, one equation would state that the inward force of gravity at any point inside the planet is balanced by an equal and opposite (outward) pressure force (these forces must be equal, or the planet would be collapsing or swelling up, which is not observed to be happening). Another equation would describe the flow of energy throughout the interior of the planet. Others would involve additional physical processes. The goal is to write a set of equations that include all the important phenomena; then solving these equations will yield values for conditions at dif-

ferent positions throughout the interiors of the planets. It is possible to show that such a set of equations can have only one solution. The challenge is to use the correct equations; i.e., equations that include all the important processes with sufficient accuracy. The validity of a computed planetary model can be checked by comparing the results of the calculation with aspects that can be observed. For example, a model might predict the surface temperature or the total mass of a planet, and the computed values can be compared with data obtained from observations. If the model succeeds in matching the observed surface values, then we can have some confidence that its values for the interior are also realistic.

The results of model calculations for Jupiter, Saturn, Uranus, and Neptune show that the four share some common internal properties (Fig. 8.6). Each has an outer layer of clouds above a warmer, denser region where the immense gravitational compression has forced hydrogen into its liquid state. In the two larger planets, Jupiter and Saturn, the pressure is so great in the deeper interior that hydrogen enters a new phase, known as **liquid metallic hydrogen**, which requires such extreme conditions that it is very difficult or impossible to create in earthly laboratories. In this state hydrogen is predicted to have a semirigid crystalline structure and to act like a liquid in some ways and like a metal in others. Below these layers of exotic forms of hydrogen, each of the giant planets is thought to have a solid core, composed chiefly of rock and metallic elements. These cores are the remainders of the original solid planetesimals from which the planets formed, with high-density material added by differentiation. This process, by which heavier elements sink to the center of a planet (discussed in Chapters 5 and 7), is much more efficient in a body that is gaseous throughout than in a rocky body such as a terrestrial planet. Hence virtually all of the heavy, dense material in the giant planets now resides in their cores, whereas in the terrestrial worlds substantial quantities of heavy elements are still found in the crust.

ATMOSPHERIC CIRCULATION

Heat rises from the interiors of the gaseous planets, and additional heating occurs within their atmospheric layers due to sunlight. The presence of heating sets the stage for convection to occur, creating overturning motions as warm gas rises and cooler gas falls. At the same time, each of the giant planets rotates very rapidly, with days ranging from just under 10 hours (for Jupiter) to a little more than 17 hours (for Uranus). Thus any horizontal motions that take

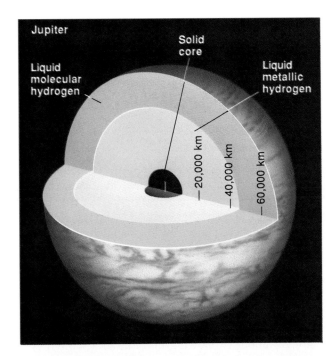

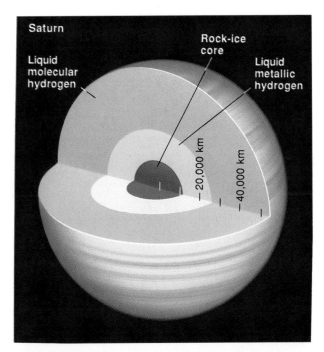

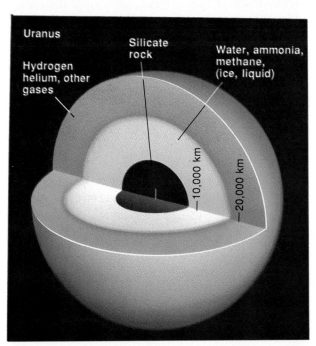

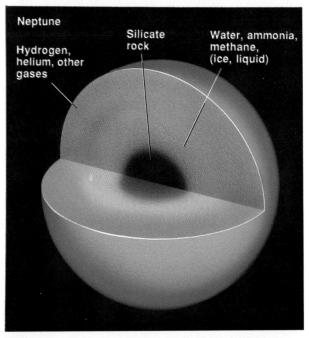

Figure 8.6   The internal structures of Jupiter, Saturn, Uranus, and Neptune.   The two larger planets, Jupiter (top left) and Saturn (top right), have small, solid cores surrounded by a layer of liquid metallic hydrogen. Uranus (bottom left) and Neptune (bottom right) do not have liquid metallic hydrogen zones because the pressure is not sufficient, but as on Jupiter and Saturn, differentiation has caused virtually all their heavier elements to sink into the core. This model of Uranus, based on *Voyager 2* data, shows a gaseous outer zone underlain by a thick ice and liquid zone extending from about 7,800 km out to 18,000 km from the center. This zone contains about 65 percent of the planet's mass, while the rocky core contains roughly 24 percent of the mass.

place will be converted into curved or rotary patterns by a very strong coriolis force (see Chapter 5). As a result, the giant gas planets have very active weather systems.

In effect the rapid rotation of these planets causes

what would otherwise be circular flows to stretch into planet-girdling bands, with adjacent bands moving in opposite directions relative to each other (Fig. 8.7). The speed of motion increases toward the equatorial zones, as do the widths of the bands. Astronomers

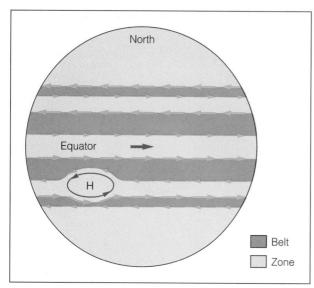

Figure 8.7   Circulation of the Jovian atmosphere.   The global circulation pattern shown here indicates the location and designations of the belts and zones in Jupiter's cloud layer.

and will allow continuous observations over a period as long as 20 years.

MAGNETIC FIELDS AND PARTICLE BELTS

As we have noted, the four gaseous planets are surrounded by belts of charged particles. The existence of these belts indicates that each of the planets has a magnetic field, since it is the field that traps the particles. Those regions are analogous to the Earth's magnetic field and the Van Allen belts, except for some differences caused by the rapid rotation of the giant planets.

Understanding why the gaseous planets, with their fluid interiors, have magnetic fields is not difficult. Recall from our discussions of the terrestrial planets that magnetic fields are thought to arise from electrical currents in planetary interiors, which in turn are brought about by flowing motions in a molten zone near the core. The presence of a field depends on whether a planet has an electrically conducting molten zone and on the rotation rate of the planet. The fluid nature and rapid rotation of the gaseous planets easily satisfy both conditions. Hence all four of these planets have strong magnetic fields.

The rapid rotation not only helps create the magnetic fields, but it also strongly affects their shapes (Fig. 8.8). The rapid rotation of the giant planets forces their magnetic fields into flattened, doughnut-like shapes, rather than the more nearly spherical shape of the Earth's field. The charged particles are therefore confined to a flattened torus (doughnut) as well, where the magnetic field is most intense. Within these disks the intensity of the particle belt is far stronger than that of the Van Allen belts; in fact, it is

suspect that the widths and speeds of the moving regions vary seasonally, but so far monitoring such changes has proved difficult because the *Voyager* probes made only very brief observations as they sped by the planets, and the seasonal cycles on the outer planets are very long (equal to the orbital periods of the planets, ranging from almost 12 years for Jupiter to about 165 years for Neptune). Both the *Galileo* and *Cassini* missions include orbiters that will be able to observe conditions on Jupiter and Saturn for extended periods of time. The *Hubble Space Telescope* will also provide a better opportunity to observe long-term seasonal changes on these planets, since it can resolve enough detail to detect the expected effects

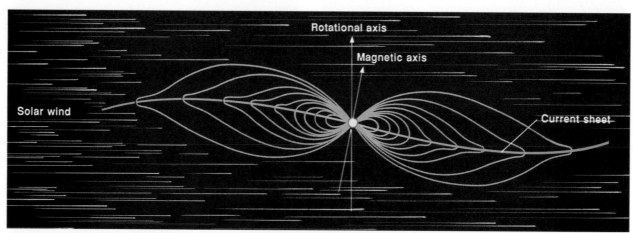

Figure 8.8   The Jovian magnetosphere and radiation belts.   The shape of Jupiter's magnetosphere is influenced by the solar wind and by the planet's rapid rotation. The result is a sheet of ionized gas that is closely confined to the equatorial plane but wobbles as the planet's off-axis magnetic field rotates. (NASA/IPL)

strong enough to represent a serious threat to space-craft that pass through the belts during encounters with the outer planets (charged particles can damage electronic components, for example, and some minor failures in on-board computers in the *Viking* probes were thought to be due to charged particles).

Another factor that influences the shapes of the particle belts around the outer planets is the degree of alignment between the rotation axis and the magnetic axis of each planet. On the Earth there is a small offset between the rotational poles (the true north and south poles) and the magnetic poles (the positions where the poles of the Earth's magnetic field lie on the surface). For the Earth the misalignment between the rotational axis and the magnetic axis is about 11°. Jupiter has a similar 11° misalignment, while Saturn's is nearly zero. Uranus and Neptune, however, have much larger misalignments, about 60° and 55°, respectively.

If the magnetic field and the rotation axis are misaligned, then as a planet spins, the magnetic field and the particle belts oscillate; consequently, the torus of charged particles oscillates as first one side and then the other is tipped up and down by the magnetic field as it rotates. Adding further complexity to the motion is the solar wind, which, as it streams past, pulls the field out into a long tail in the downwind direction (away from the Sun). The result is that the magnetic fields and particle belts of the giant planets have torus-like shapes, with overall oscillation and extended tails in the downwind direction; for Uranus and Neptune these tails gyrate about due to the misalignment between the magnetic equator and the direction in which the solar wind is flowing.

The principal source of the electrons and protons that make up the particle belts is the solar wind, which provides a steady stream of charged particles from the Sun. Some particles, however, have a local source. For example, volcanic activity on Jupiter's moon Io produces free atoms that become ionized and trapped in the particle belts.

MOONS AND RINGS

All four of the gaseous giants have numerous satellites, ranging from 8 for Neptune to 18 for Saturn (with more suspected). This is in stark contrast to the terrestrial planets, which seem to acquire moons only by accident, if at all (recall the Moon is now thought to have formed as the result of a grazing collision between a huge planetesimal and the Earth, and that the two tiny moons of Mars are thought to be captured asteroids).

In addition to having many satellites, the outer planets have ring systems, which are disks of solid particles (mostly ice and dust) that orbit the planets in the same plane (the equatorial plane) as the satellites. We know the rings are made of countless tiny particles because observations (first made long ago from the Earth) show that light passes through the rings. Furthermore, the speed of rotation of the rings is not constant, as would be expected from a rigid ring, but instead varies with distance from the parent planet in accordance with Kepler's third law. In reality, the ring particles are nothing more than a horde of very tiny moons orbiting each of the giant planets. The size of the particles, inferred from the manner in which they reflect light, ranges from microscopic dust to many meters in diameter. Thus each of the giant planets has a large family of orbiting bodies, with tiny particles (the rings) closest in and the larger moons farther out. Clearly, these orbiting systems must have formed as a result of some process that occurred in common for all four of the planets. Astronomers now think that several processes are at work, and these will be discussed in some detail later (primarily in Chapter 10). For now, we simply note that the origins of the rings themselves, as well as most of their structural details, are attributed to the immense gravitational and tidal force of the parent planets.

As a by-product of the strong tidal forces exerted by the giant planets, all the satellites of these planets, even those orbiting farthest out, are in synchronous rotation, always keeping the same side facing inward. Additional phenomena attributed to tidal forces acting on the satellites will be described later in this chapter.

In the following sections, we will discuss the planets individually, emphasizing their unique qualities but also noting the features they have in common. We will discuss Jupiter and Saturn together, and then Uranus and Neptune, because these pairs have many commonalties.

JUPITER AND SATURN

The fifth and sixth planets from the Sun, Jupiter and Saturn are the largest as well. In gross properties they are in a class by themselves; Jupiter contains over 300 times the mass of the Earth, and Saturn is almost 100 times more massive than the Earth (see Table 8.1). They are also very unusual in another important way, having densities far lower than that of any terrestrial planet. Jupiter's average density is about 1.4 grams/cm^3 (as opposed to roughly 5.5 grams/cm^3 for Venus and the Earth), and Saturn's density, at about 0.7 grams/cm^3, is actually less than that of water. These low densities are a clear indication that the gaseous planets must have a very different composition from the terrestrials, as we have already seen. Recall that

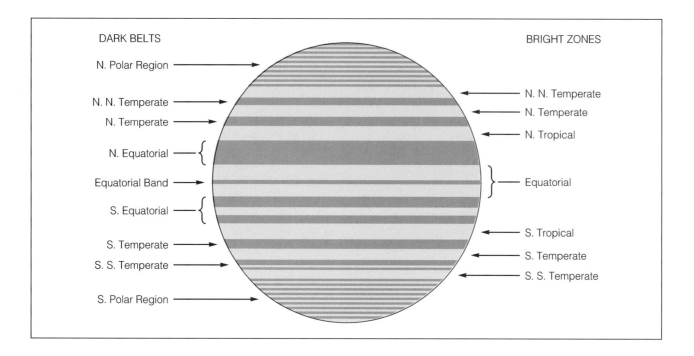

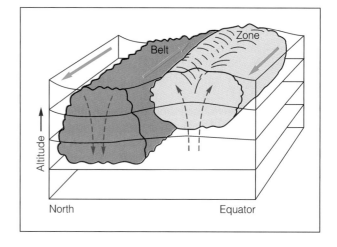

Figure 8.9   Motions in the Jovian atmosphere.   These drawings indicate both the horizontal (above) and vertical (right) circulation in the clouds of Jupiter. (NASA)

the gas giants are composed chiefly of hydrogen and its compounds, along with helium, with only very small amounts of the elements that dominate the makeup of the rocky planets.

Despite the similarities between the two largest planets, they also exhibit some notable contrasts. These will be explored in the following sections.

Atmospheres and Interiors

Both Jupiter and Saturn have light-colored bands called **zones** and darker, reddish brown bands known as **belts,** but Jupiter's bands are more vivid and colorful than Saturn's (see Fig. 8.7). The zones are regions where gas is rising and is converted into a rotary flow by the planet's rotation. On Jupiter the flow moves clockwise in the northern hemisphere, analogous to that of a high-pressure region in the Earth's atmosphere (Fig. 8.9). The belts, in contrast, are regions where gas is descending and is forced into a counterclockwise flow in the north (the directions of the belts and zones are reversed in the southern hemisphere). The color of the belts is thought to be caused by molecular species that persist under the temperature and pressure conditions of the belts, but not under the conditions that prevail in the zones. The identity of these molecules is not known.

The belts and zones on Saturn are less prominent than those on Jupiter because the clouds on Saturn, subjected to a lower surface gravity than that of Ju-

piter, extend to a greater height and limit our view of the atmospheric bands. The speed of the flow patterns is much greater on Saturn than on Jupiter. The equatorial flows on Jupiter, for example, move at about 100 meters/sec (360 km/hr), while on Saturn the corresponding winds move at 400 to 500 meters/sec (up to 1,800 km/hr). In addition, the belts and zones on Saturn exhibit contrasts between the northern and southern hemispheres that may be seasonal effects, since Saturn has an obliquity of almost 27° compared with Jupiter's 3°. On Saturn the belts and zones are broader in the north, and a concentration of spots and eddies occurs at about 40° north latitude, where there is an abrupt change in flow speed at the northern edge of the equatorial belt.

Jupiter has a large, reddish brown oval-shaped area in its southern hemisphere, known as the **Great Red Spot** (Fig. 8.10). This feature has been present for hundreds of years and was seen by Galileo and

Huygens in the seventeenth century. Within the Great Red Spot, the gas is rising and rotating in a counterclockwise direction (i.e., the same as the zones in the southern hemisphere of Jupiter, except that the zones are light colored). Modern studies of the Great Red Spot indicate that it is basically a storm system, much like the great rotary storms we have on Earth (such as hurricanes and cyclones) but much longer-lasting. Theoretical models have succeeded in reproducing the Great Red Spot quite accurately, starting with a pair of vortices that are modified by the global wind system on Jupiter (Fig. 8.11).

Both Jupiter and Saturn display a large number of smaller, more ephemeral spots and rotary patterns (see Figs. 8.3 and 8.4). These generally tend to be white, rather than dark like the Great Red Spot, but dark examples are seen on both planets as well. Occasionally these spots can grow to large sizes; starting in late 1990, for example, Saturn, developed a "Great White Spot" that was observed by the *Hubble Space Telescope* (this spot was sufficiently prominent to be seen from Earth through modest telescopes, but

Figure 8.10   The Great Red Spot.   This is a *Voyager* image of the rotating column of rising gas that has been present on Jupiter for at least 300 years. (NASA)

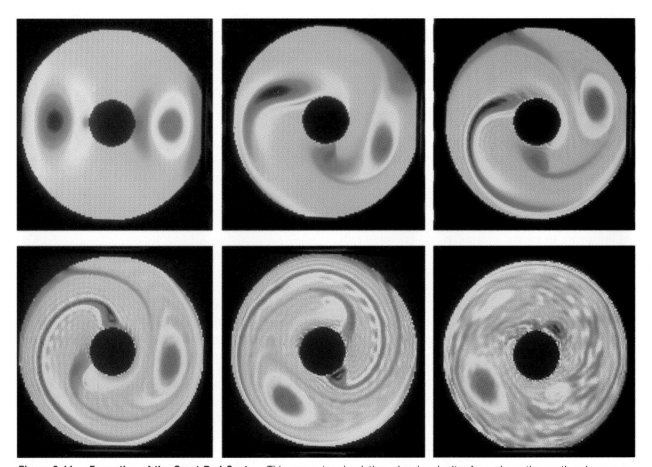

Figure 8.11   Formation of the Great Red Spot.   This computer simulation, showing Jupiter from above the south pole, successfully recreates the Great Red Spot. A pair of vortices (upper left), rotating in opposite directions are modified by the strong global winds so that a clockwise blue vortex is destroyed while a counterclockwise (red) vortex grows and persists. (P. Marcus and N. Socci, University of California, Berkeley and Harvard-Smithsonian Center for Astrophysics; computations done at the National Center for Atmospheric Research.)

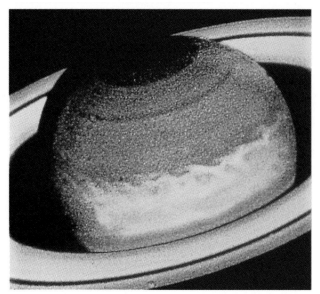

Figure 8.12   A storm on Saturn.   This *Hubble Space Telescope* image shows an elongated whitish region just above Saturn's equator. This turbulent storm system appeared quickly and then persisted for weeks. (NASA/Space Telescope Science Institute)

Figure 8.13   The Galilean satellites of Jupiter.   These are the four moons discovered by Galileo, shown in correct relative size. They are Io (upper left), Europa (upper right), Ganymede (lower left), and Callisto (lower right). Ganymede is the largest satellite in the solar system. Moving from Callisto, the farthest from Jupiter, to Io, the innermost of the four, we find a wide range of geological and physical properties, as discussed in the text. (NASA)

much more detail is visible in the *HST* images; Fig. 8.12).

Models of the interiors of Jupiter and Saturn show a similar overall structure, but with Jupiter having a much more extended liquid metallic hydrogen zone due to its greater internal pressure. As mentioned earlier, both planets emit excess infrared radiation, indicating that both are producing more heat in their interiors than can be explained by the amount of sunlight that they are receiving. Interior models suggest that this excess heat has a different source on each planet. For Jupiter, it appears likely that this is simply heat left over from the formation of the planet, which has been able to escape only very slowly and is still trickling out today. For Saturn, however, with its lower mass and lower density, the calculations show that less heating would have occurred initially and that any primordial heat should have dissipated long ago. Hence Saturn must have an internal source that is producing heat today. One suggestion is that Saturn is still undergoing differentiation and releasing gravitational potential energy in the form of heat. Internal turbulence on Jupiter, caused by its higher interior temperatures, is thought to prevent the same kind of differentiation from occurring there.

Tidal Forces and the Galilean Satellites of Jupiter

As the compilation of the properties of the known satellites throughout the solar system in Appendix 7 indicates, Jupiter is currently known to have 16

moons, while 18 have been confirmed for Saturn. Each planet probably has additional, small undiscovered moons as well.

Among the moons of the giant planets are several that are very large (some rival or even exceed the diameter of Mercury). Jupiter has four such satellites, first observed by Galileo in the early 1600s, while Saturn and Neptune have one each. Many astronomers consider Pluto to be the same type of body, a point that we will reinforce later in the chapter. All of these moons consist of mixtures of rock and ice and are therefore quite unlike the terrestrial planets, despite their hard surfaces. The four Galilean satellites of Jupiter present a fascinating and complex series of worlds (Fig. 8.13) that are modified in many ways by the gravitational influence of their parent planet.

The four Galilean moons, from outermost to innermost, are Callisto, Ganymede, Europa, and Io. The detailed observations of these bodies by the *Voyager* probes revealed that they vary in spectacular fashion. The colorful images and fine detail disclosed several clear trends among these satellites (Table 8.2). For example, the surface density of impact craters declines progressively from a very high density for Callisto, the outermost satellite, to a much lower density for Io, the innermost. This decline shows that the

Figure 8.14   An eruption on Io.   This image of an eruption from a volcanic vent with a plume of ejected gas dramatically illustrates Io's present state of dynamic activity. (NASA)

ages of the surfaces decrease from the outer part of the system toward the center, with Callisto having a very old surface and Io a very young one. The observations also revealed that the densities of the moons increase with proximity to Jupiter. In addition, the albedos (the reflectivity of the surfaces) of the Galilean satellites decrease smoothly from Callisto, the shiniest of the four, to Io, the least reflective.

Combining these observations leads to the conclusions that the outer Galilean moons contain more ice than the inner satellites (this explains the lower densities and higher reflectivities of the outer moons) and that the inner moons undergo more geological activity (this is apparent from the youth of their surfaces). Even before the first visit by a *Voyager* spacecraft in 1979, it had been suggested that this trend might be found. The reason: tidal forces acting on the inner moons were expected to cause internal heating that, in turn, would cause the geological differences that were discovered. The first images showing volcanic eruptions on Io provided dramatic confirmation of this prediction (Fig. 8.14).

As we have already mentioned, the giant planets exert very strong tidal forces on their satellites. These tides are sufficient to keep all the moons of the gas planets locked into synchronous rotation and to force them into slightly elongated shapes due to the immense stretching forces to which they are subjected. Pressing and squeezing a body internally can cause heating and, in turn, internal melting and volcanic activity. Such activity would not occur in a satellite in stable synchronous rotation, however, because the stresses acting on it would be constant. But the stresses are not constant for the innermost two Gal-

ilean satellites; instead these two bodies are subjected to additional, varying forces. The two satellites exert these forces on each other. As it happens, the orbital period of Io is almost exactly half that of Europa, meaning that the two moons are regularly aligned in the same configuration relative to Jupiter. At these times their gravitational attraction for each other adds to the tidal stresses they already feel due to Jupiter, and the two are subjected to additional internal stress. A situation like this, in which two orbiting bodies have periods that are simple multiples of each other, in called **orbital resonance.**

The extra tidal stress exerted on Io each time it is aligned with Europa and Jupiter has created sufficiently high temperatures in its interior to melt it. The satellite consequently is undergoing continual volcanic eruptions and has lost most of its volatile materials, which explains its high average density (and low ice content). The volcanic outpourings have continually resurfaced Io, obliterating any impact craters and giving the moon a very young surface. In addition, the gas being released by the volcanoes can escape Io's gravitational field and go into orbit around Jupiter, where it has created a doughnut-shaped ring of gas called the **Io torus** (Fig. 8.15). Particles released from the volcanoes on Io become ionized as they interact with the planetary particle belts; once ionized, these particles are forced to orbit Jupiter at the speed of rotation of the planet's magnetic field, thereby creating the torus. The 11° tilt of the magnetic axis of Jupiter relative to its spin axis causes the Io torus to wobble up and down by 11° with a period of just under 10 hours, the rotational period of Jupiter.

Figure 8.15   The Io torus.   This remarkable series of photographs, taken through a ground-based telescope on Earth, shows the cloud of sodium atoms surrounding Io as it orbits Jupiter. The yellow color of the cloud is due to the fact that sodium atoms emit most strongly in a pair of emission lines in the yellow part of the spectrum (this is why certain types of street lights appear yellow). Once this gas, which also includes other atoms such as sulfur, escapes from Io, it becomes ionized and then is spread all the way around Io's orbit by magnetic forces. This ionized gas is not visible here, so we do not see the full extent of the torus. The sizes of Jupiter, Io (dot inside the crosshair), and Io's orbit are all to correct relative scale. (B. A. Goldberg, G. W. Garneau, and S. K. LaVoie, JPL)

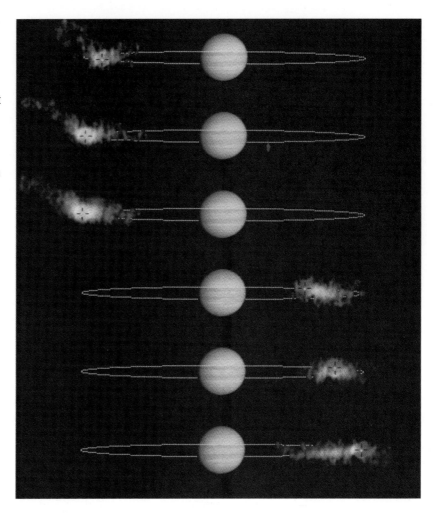

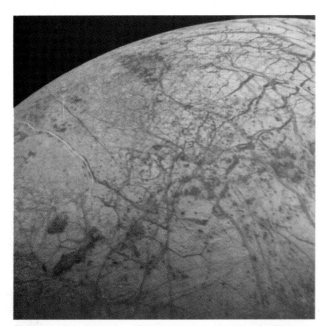

Figure 8.16   Europa.   This is the second of the Galilean satellites in progression outward from Jupiter. Its linear features and relative lack of craters indicate that it is a tectonically active moon, with a young surface. (NASA)

Europa is also subjected to extra internal stress and heating due to its orbital resonance with Io, but the effects are not as severe. The surface of Europa is quite young, but no active volcanoes have been seen. It is known that Europa still contains a substantial quantity of volatiles; interestingly, models of its interior and observations of its surface features suggest that the interior of this satellite may contain large quantities of liquid water. The surface is marked by linear features that appear to be cracks between crustal plates that are floating on the liquid water; the cracks themselves appear to be filled with ice, formed as water flowed to the surface and froze (Fig. 8.16).

The small moons of Jupiter were not observed closely by the *Voyager* spacecraft, whose trajectories were optimized for making close passages by the Galilean satellites. Hence few details are known about the small moons. The *Galileo* mission is to make close visits to several of the smaller satellites, so if the current problems with its communications systems can be overcome, we should know more about these moons after 1995.

The *Voyagers* did obtain images revealing that Jupiter has a single thin ring (Fig. 8.17). This news

Figure 8.17   The geometry of the Jovian ring.   Here the ring has been drawn in, showing its size relative to Jupiter. The ring radius is about 1.8 times the radius of the planet; the ring is estimated to be no more than 30 kilometers thick. (NASA)

Figure 8.18   Titan.   The largest of Saturn's moons, Titan is almost unique among the satellites in the solar system in having an atmosphere. (NASA)

came shortly after the discovery (from Earth-based observations) that Uranus has rings as well, so at that point astronomers knew that at least three of the four gas giants had rings. At the same time, rings of Neptune were suspected from Earth-based observations and were finally confirmed when *Voyager 2* reached the eighth planet in 1989. Thus the detection of the ring around Jupiter contributed to a growing realization that rings are the norm, rather than the exception, in the outer solar system.

Titan and the Moons of Saturn

Saturn has 18 known satellites, most of them very small and close in (several orbit within the ring system; these are discussed in the next section). There are seven intermediate-sized moons (diameters of a few hundred kilometers) and one giant satellite, Titan (Fig. 8.18).

The intermediate moons (see Appendix 7) are composed primarily of ice, indicated by the fact that their average densities are about 1 gram/cm^3, the density of water ice. Most are heavily cratered, indicating that their surfaces are quite old (Fig. 8.19), but there are exceptions. Enceladus (Fig. 8.20), for example, has a very shiny surface (albedo near 1, which means that it reflects light about as well as a mirror), which probably means that some process is continually re-coating its surface with fresh ice. Some suspect that Enceladus may be volcanically active, spewing out water vapor that falls to the surface and freezes; this satellite is in orbital resonance with another of the intermediate moons, Dione, which might account for the volcanic activity. Yet another satellite, Iapetus, has a large dark region on one side (Fig. 8.21), evidently the result of the deposition of some kind of

Figure 8.19   Dione.   This intermediate-sized moon of Saturn has wispy, light-colored features on its surface that may be icy deposits of material that escaped from the interior and crystallized. This side of Dione, which is in synchronous rotation, always trails as the satellite orbits Saturn; the leading side is more heavily cratered. (NASA)

sooty substance of unknown origin. Large fissures and wispy, frosted regions on other moons indicate past tectonic activity (Fig. 8.22), as crustal cracking and outgassing appear to have occurred. The satellite Hyperion is very asymmetric in shape, appearing as a flattened, almost disk-like object (Fig. 8.23).

Some of the intermediate moons have smaller bodies (sometimes called "moonlets") sharing their orbit (see Appendix 7). These co-orbital satellites are found either 60° ahead or 60° behind the larger satellite. The locations of these satellites coincide with a prediction

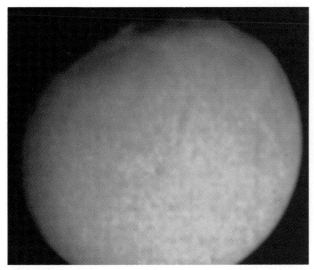

Figure 8.20   Enceladus.   This satellite is the shiniest object in the solar system, with an albedo of approximately 1, just like a mirror. The surface may have been coated with deposits from internal melting and outgassing. (NASA)

Figure 8.21   Iapetus.   This satellite has very unusual bright and dark areas. It appears that the dark material may be some sort of deposit. (NASA)

Figure 8.22   Rhea.   One of the seven intermediate-sized moons, Rhea shows evidence of cratering and light-colored areas that are probably composed of ice. (NASA)

Figure 8.23   Hyperion.   This satellite has a very unusual shape and is marked by several impact craters. (NASA)

made centuries ago by the French mathematician J. L. Lagrange. Working with Newton's laws of motion, Lagrange found that in a circular orbit there are stable positions that lead or trail the orbiting body by 60°. At these positions the combined effect of the central object (Saturn in this case) and the orbiting body (one of the intermediate moons, such as Dione or Tethys) acts to keep smaller bodies trapped, so that they remain in orbit at those positions. In the next chapter we will see that groups of asteroids are orbiting the Sun at the corresponding positions in the orbit of Jupiter.

Saturn's giant satellite, Titan, is nearly as big as Ganymede, which is the largest moon in the solar system. Titan is distinguished by its thick atmosphere (see Fig. 8.18), found by the *Voyager* probes to be composed chiefly of nitrogen. The atmospheric pressure is about 50 percent greater than sea level pressure on the Earth, and the atmosphere of Titan is deeper as well (Fig. 8.24); cloud layers made it impossible for the *Voyager* cameras to obtain images of the surface. Nevertheless, some information was obtained by radio instruments, and surface conditions were measured. At the surface the combination of pressure and temperature corresponds to the conditions under which methane (CH_4), which is observed in the atmosphere, can be in vapor, liquid, or solid form. Hence astronomers speculate that there may be methane lakes on Titan, perhaps with methane icebergs floating in them.

We do not know why Titan has an atmosphere, while other large satellites do not (although Triton, a moon of Neptune, does have a thin atmosphere, none of the Galilean satellites of Jupiter does). That the atmosphere is composed primarily of nitrogen is also unusual; in the rest of the solar system, only the Earth and Triton have nitrogen-dominated atmospheres. In both those cases, volcanic eruptions are a leading

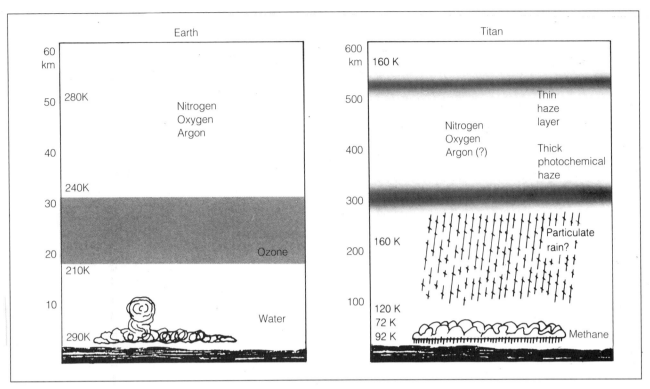

Figure 8.24   Titan's atmosphere.   This diagram compares conditions in the atmospheres of Titan and the Earth. Note that the vertical scales are not the same; Titan's atmosphere is much deeper. (Data from J. K. Beatty, B. O'Leary, and A. Chaikin, eds. 1990. *The new solar system.* 2nd ed. Cambridge, England: Cambridge University Press.)

source of nitrogen, so it is possible that Titan may be volcanically active. Perhaps the *Cassini* mission, which will place a radar-equipped spacecraft in orbit around Saturn, will find evidence of eruptions.

The Rings of Saturn

As we have seen, the rings of Saturn are thought to consist of ice and dust particles that formed a disk around the planet. Such a disk probably formed early in solar system history, as Saturn accreted material from the surrounding solar nebula. In the outer disk, the particles coalesced to form satellites, but in the inner portion, tidal forces due to Saturn would have prevented this. Inside a point called the *Roche limit,* the tidal forces are strong enough to prevent coalescence. While it is tempting to think that today's rings are simply leftover debris from the original disk, astronomers now believe this is unlikely because a disk of particles cannot survive for times as long as the age of the solar system. Instead, they now suspect that the ring material is occasionally replenished, probably through the collisional destruction of small moons.

The rings of Saturn exhibit a highly complex structure (Fig. 8.25). They include prominent gaps, most notably the **Cassini division,** very thin "ringlets," and sections that vary widely in brightness. Most of these

Figure 8.25   The rings of Saturn.   This *Voyager* image reveals some of the fantastically complex structure of the ring system. (NASA)

features are attributed to gravitational effects of one sort or another. For example, the Cassini division, as well as some other gaps, is thought to be due to orbital resonance (recall that we invoked orbital resonance in the case of Io and the relationship of its orbital period to that of Europa). Ring particles orbiting Saturn at the location of the Cassini division would have exactly one-half the orbital period of the satellite Mimas; thus, a particle at this position would be periodically subjected to enhanced gravitational forces, which eventually would nudge it out of that orbit. Thus, the orbital resonance between Mimas and the particles at that position has created and maintained a gap. Another gravitational mechanism that creates structure in the rings involves some of the small moons. In cases where two of these bodies orbit at nearly the same distance from Saturn, their combined gravitational effects can act to keep ring particles trapped between them. They are called **shepherd satellites** for this reason. Finally, the detailed ring images provided by the *Voyager* spacecraft revealed that in some sections of the ring system, the thin ringlets actually consist of spirals rather than discrete circles of particles. The rings in these regions are part of a **spiral density wave,** a stable oscillation pattern in which particles are concentrated in a spiral pattern. Another example of a spiral density wave is the structure of our galaxy and others like it, which are characterized by spiral arms in which much of the interstellar gas and dust, along with bright young stars, are concentrated. The formation of a spiral density wave requires a rotating, fluid disk that is subjected to an outside gravitational force. In the case of Saturn's rings, the outside disturbance is supplied by the larger moons that orbit outside the ring system.

In addition to these gravitational influences on the rings of Saturn, electromagnetic forces also appear to be at work. The *Voyager* images showed several curious "dark spokes," zones extending outward from the planet where the rings appeared darker. These are now thought to be formed by very tiny ring particles that carry electrical charges and are therefore influenced by Saturn's magnetic field. These particles appear to be suspended above and below the plane of the main rings, and their appearance changes with time as the planetary magnetic field rotates.

The mechanisms thought to be responsible for the formation and maintenance of rings systems are discussed more generally in Chapter 10, where we will consider the origin of the solar system.

URANUS AND NEPTUNE

The seventh and eighth planets from the Sun appear at first glance to be very similar to each other (see Fig. 8.1). Both are bluish in color as seen from afar, and they are comparable in size, mass, and rotation period. Close examination, however, has revealed that they are quite distinct from each other in many ways.

One difference is very obvious: the tilt (obliquity) of Uranus's rotational axis is so large that the pole of

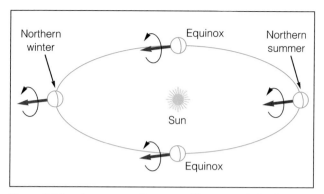

Figure 8.26   Seasons on Uranus. The unusual tilt of Uranus creates bizarre seasonal effects. The solid arrow in each case indicates the direction of the planet's north pole, and the curved arrow, the direction of its rotation.

the planet actually lies below the plane of its orbit (Fig. 8.26). The obliquity of Uranus is 98°, meaning that the north pole points 8° below the orbital plane. This strange tilt probably came about as the result of a collision with a large planetesimal during the time when Uranus was forming. The rings and moons of Uranus lie in its equatorial plane, meaning that their orbital plane is almost perpendicular to the orbital plane of the planet as it circles the Sun. Seasons on Uranus are quite unusual; the northern hemisphere has full-time sunlight during the height of its summer, when the north pole is pointed almost directly at the sun, and permanent darkness during winter (42 years later). On the other hand, when Uranus is at intermediate positions in its orbit, so that the Sun lies over its equator, days and nights are very short, equaling half the rotation period of just over 17 hours. Neptune has a much smaller obliquity, and therefore the seasonal variations in the amount of sunlight reaching its surface are much less pronounced.

Atmospheres and Interiors

Like Jupiter and Saturn, Uranus and Neptune have atmospheres in which hydrogen and its compounds are dominant, and helium is also abundant. Heavy elements have sunk into the core through differentiation. The blue color of the planets is created by methane, which absorbs red light, permitting blue light from the Sun to be reflected much more efficiently (Fig. 8.27). The color of Neptune is much deeper, and belts, zones, and spots are more easily seen because the atmosphere is not as deep and hazy (Fig. 8.28). Neptune's atmosphere is also more dynamic; its more active flows are probably due to a greater temperature contrast between the poles and the equator (recall that temperature differences are one of the driving forces behind atmospheric circulation). The *Voyager 2* data on Uranus showed a surprisingly small temperature contrast from pole to equator.

The general circulation patterns in the atmospheres of both Uranus and Neptune are similar to those of Jupiter and Saturn, with alternating belts and zones representing rising and descending currents that are forced into elongated circulatory patterns by rapid planetary rotation. Astronomers were not surprised by Neptune's similarity to the circulation patterns of Jupiter and Saturn, but the circulation on Uranus was unexpected because it has little or no excess internal heat to help drive convection. Furthermore, astronomers had expected that any flows on Uranus would go the other way. On Jupiter, Saturn, and Neptune, the main equatorial stream goes around the planet in the same direction as the rotation. But on Uranus, the north pole is currently pointed toward the Sun, and the pole was therefore expected to be warmer than the equator; models of atmospheric circulation show that in that case the equatorial flow should move opposite to the planet's rotation. It turns out that the models may well be correct, but that the equator on Uranus is not cooler than the pole (the temperature is nearly constant over the entire surface of the planet, with the coldest temperatures at intermediate latitudes). Hence the question is not so much why the equatorial flow is in the normal direction, but why the equator is not colder than the pole. The answer is not known.

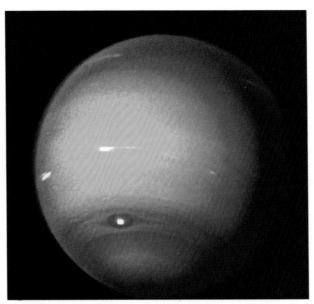

Figure 8.27   The effect of methane on the appearance of Neptune.   This is a false-color image of Neptune, designed to illustrate the effect of methane in the atmosphere. The blue color of most of the disk is natural, but the red around the edges has been enhanced to show light scattered by methane molecules. This gas absorbs red photons, causing the remaining color to be blue in regions where we see light that has passed through a region of methane absorption, such as the central portions of the disk. In regions where we see the reemitted (scattered) light from the methane, we see red. (JPL/ NASA)

Figure 8.28   Uranus and Neptune.   These full-disk *Voyager* images show the intense blue color of Uranus (left) and Neptune (right) that is caused by methane in the atmospheres of these two planets (methane absorbs red light more efficiently than blue). Note also that atmospheric circulation and clouds are more readily seen on Neptune. (NASA)

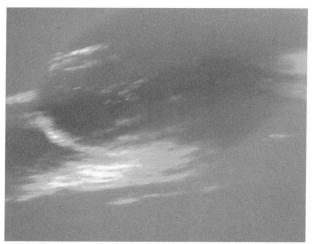

Figure 8.29   The Great Dark Spot. This *Voyager* image shows the large, dark oval feature that is very reminiscent of Jupiter's Great Red Spot, as well as a couple of smaller features. South of (below) the Great Dark Spot are a white feature, known as the "scooter" because of its rapid motion around the planet, and the "dark spot 2," a second counterclockwise storm system. Each of these features moves around the planet in the eastward direction (to the right in the view), but at different speeds, so they are not often grouped together as in this image. (JPL/NASA)

Figure 8.30   Vertical structure in the atmosphere of Neptune. This close-up image of Neptune shows that the white "cirrus" clouds lie above other levels of the atmosphere that are observed. Here we see that the sides of the clouds that face the Sun are brighter, and that the clouds cast shadows on the lower atmosphere in the direction away from the Sun. (JPL/NASA)

The many spots and storm systems on Neptune include a large, dark one called, naturally enough, the Great Dark Spot (Fig. 8.29), which may be related to the phenomenon responsible for Jupiter's Great Red Spot. The upper atmosphere of Neptune is quite transparent, allowing features to be seen at great depth. Suspended at considerable heights in the clear atmosphere are wispy, white clouds, probably composed of crystalline methane (Fig. 8.30).

Uranus and Neptune differ internally as well as externally. For example, Uranus has little or no excess heat from its interior (i.e., the amount of energy it radiates is essentially equal to the amount it receives from the Sun), whereas Neptune has a substantial amount of excess radiation. Neptune emits more excess energy relative to its mass than either Jupiter or Saturn.

Both planets have magnetic fields, in each case strongly misaligned with the rotation axis (Fig. 8.31). The magnetic axis of Uranus is tilted by some 60° with respect to its rotation axis and is offset as well, so that the magnetic axis does not pass through the center of the planet. The tail of the planet's magnetosphere, which points away from the Sun due to the solar wind, oscillates around as the planet rotates. The magnetic field of Neptune is weaker than those of the other gas giants, and its existence was not confirmed until the encounter of *Voyager 2,* which was able to measure its strength directly. The magnetic axis of Neptune, like that of Uranus, is highly tilted relative to the planet's rotation axis; in this case the

tilt is about 55°. Once again, the magnetic axis is offset so that it does not pass through the center of the planet. The fact that this offset occurs in two planets means that it is not accidental or temporary, as had been suggested for Uranus. Instead some aspect in the formation of the magnetic fields of these two planets must cause the strange tilts and offsets.

Moons and Rings

Before the *Voyager 2* flybys, Uranus was known to have five satellites and Neptune two. The spacecraft found several additional moons; some of them are very small moonlets, but a few are of substantial size. Now 15 moons are known for Uranus and 8 for Neptune, and evidence for additional small ones has been found as well.

Each planet also has a ring system. Uranus has 9 thin, dark rings which are very difficult to see from Earth (they were originally discovered from the Earth when Uranus passed in front of a distant star, and the star was seen to alternately dim and brighten as the rings passed in front of it). The rings of Neptune, suspected from Earth-based observations but not confirmed until the arrival of *Voyager 2,* are not symmetric circles; instead they are uneven and lumpy, resembling arcs and segments of rings rather than complete circles.

The five relatively large satellites of Uranus (Fig. 8.32) have varied geological properties and present some interesting puzzles. Their diameters range

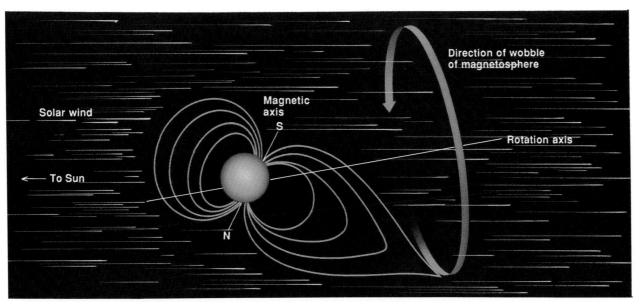

Figure 8.31   The magnetosphere of Uranus.   Due to the combination of the high tilt of the rotation axis of Uranus and the misalignment between its rotation and magnetic axes, the magnetosphere of Uranus resembles that of a planet whose rotation axis is more normal. The tail of the magnetosphere rotates as the planet spins around its highly tilted axis.

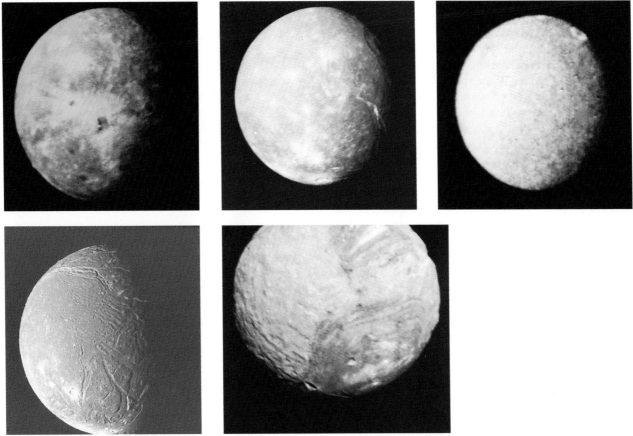

Figure 8.32   The satellites of Uranus.   Starting with the outermost, the satellites are Oberon (upper left), Titania (upper middle), Umbriel (upper right), Ariel (bottom left), and Miranda (bottom right). Titania shows signs of more recent cratering and thus a younger surface than Oberon and Umbriel, the moons on either side of it. Umbriel is the darkest of the five, an indication that its surface has probably been coated by some sort of low-reflectivity deposit. Ariel's bright surface and evidence of recent ice flows indicate that it is probably undergoing internal motions and outgassing. Miranda also shows signs of recent geological activity. Tidal stresses are insufficient to explain this satellite's youthful surface, and astronomers are looking for more exotic explanations. (NASA/ JPL)

from just under 500 km (Miranda) to over 1,600 km (Titania), and their densities lie between 1.3 and 1.7 grams/cm³, indicating that they have substantial quantities of rocky material embedded within an icy structure. The outermost of the five is Oberon, which has a heavily cratered (hence very old) surface and appears to be inactive geologically. Moving inward, the next is Titania, which has surface fissures that may be the result of freezing and expansion in the interior; oddly enough, Titania appears to have relatively young impact craters on its surface, as though it had recently experienced a period of heavy bombardment by debris from space. By contrast, Umbriel has a very old, dark surface that may be covered by some sort of dirty deposit, whereas Ariel has a very bright, young surface that appears to have been recently coated with icy deposits, perhaps due to water vapor escaping from its interior. Finally, Miranda, the innermost of the five, has a very strange surface, covered in places by craters and in others by very unusual uplifted regions indicative of recent tectonic activity. This satellite appears to have a very unsettled geology, with very old regions coexisting with young ones on its surface.

The strange properties of the satellites of Uranus may be related to some unusual aspects of its rings (Fig. 8.33). The inner portion of the ring system consists of a broad, dark ring of dust particles (as opposed to ice). Calculations show that this ring cannot persist for a very long time without being replenished; astronomers therefore speculate that this ring either is very young or has a source of new particles to replace the ones that spiral into Uranus. Other evidence of instability among the rings of Uranus includes the fact that some of them are not perfectly circular; over the long term, gravitational forces should produce round rings. Thus the evidence suggests that the moons and rings of Uranus are changing over time. One possibility is that the satellites occasionally collide with large bodies; the collisions not only disrupt the geology of the moons but also cause them to fragment, thereby providing new debris for the rings. Specifically, it has been proposed that an earlier satellite at Miranda's location was shattered by a collision some time ago. The fragments from this collision then created a dense ring that soon reformed into a satellite, but one with an unsettled geology as Miranda has today. As we will see, similar episodes of collision and ring renewal are now thought to have occurred at Neptune and possibly Saturn as well.

The two moons of Neptune that were known before the *Voyager 2* encounter both have very unusual properties. Nereid, the smaller moon (340 km diameter), has a very elongated orbit with a period equal to about one Earth year. Triton, which is larger (2,705 km diameter), has a circular orbit but goes

Figure 8.33   The rings of Uranus. Both views show the outermost or epsilon ring at the top and include all the other major rings inward toward the planet. The color-enhanced image brings out greater detail. (NASA/JPL)

around Neptune in the retrograde direction. This satellite, like Saturn's Titan, has an atmosphere consisting of nitrogen and methane.

Triton provided a wealth of surprises when the *Voyager* images came in (Fig. 8.34). The surface could be seen clearly through the atmosphere and revealed signs of both tectonic and volcanic activity. The surface albedo is high, especially in the southern polar region, which is consistent with a young surface. Few impact craters are visible. Linear fractures provide evidence of crustal plate shifting, and curious, depressed circular basins may be sites where the crust has slumped in the wake of outgassing episodes. Confirmation of current geological activity came with the discovery of several dark streaks (see Fig. 8.34), which turned out to be deposits from recent eruptions. Detailed analysis of the images showed that at least two of these regions are sites of vents that are

Figure 8.34   The surface of Triton. This close-up view shows a variety of surface features that indicate that some reprocessing has taken place recently. The surface is bright, unlike old surfaces, and there are relatively few impact craters. There are also linear, grooved features suggestive of crustal plate motion. Most intriguing of all are the dark streaks near the south pole (bottom), which resemble volcanic plumes or ejecta. It has been suggested that these are either current or recently active geysers, which emit frozen nitrogen. If so, Triton becomes the third body in the solar system known to be volcanically active. (The Earth and Io are the other two, but recall that Venus may also have active volcanoes; see Chapter 7.) (JPL/NASA)

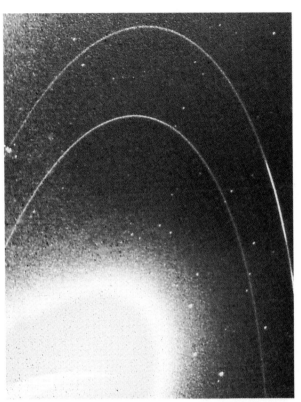

Figure 8.35   The rings of Neptune. This *Voyager* image shows the thin, dim rings of Neptune in some detail. Here we see two distinct rings, one of which has three bright segments, which may have led to earlier indications that the rings are not complete circles. Other views show that an indistinct, broad region, known as the "plateau," encompasses some of the thin rings and that an inner belt of ring particles extends close to the planet. (JPL/NASA)

currently active, spewing gases out of the interior and creating the dark deposits. Thus Triton was added to the short list of solar system bodies undergoing current volcanic activity (the Earth and Io are the only others where eruptions have been directly observed, but as we have seen, indirect evidence suggests that Venus and at least two satellites of Saturn are also active). The eruptions on Triton are more like geysers than volcanoes; geysers are sites where subsurface gases are heated and forced out by expansion (as opposed to the upwelling of molten material from deep within a planet or moon). No crustal melting is involved in the outbursts of water vapor from Triton, which scientists call "ice volcanoes."

Triton has an average density of about 2 grams/cm^3, indicating that it is composed of about equal parts rock and ice. As we will see in the next section, this density is very similar to Pluto's, and it is now thought that several bodies of this type may have formed in the outer solar system. Triton's similarity to Pluto and its unusual motion have led to the suggestion that Triton may have been captured by Neptune rather than being formed in place as its satellite.

As already mentioned, the rings of Neptune consist of asymmetric circles and arcs (Fig. 8.35), rather than complete, symmetric circles. These arcs are difficult to explain as stable features, although recent model calculations do show that they can last for a long time as a result of orbital resonance with satellites. Once again, however, evidence indicates that the rings are not permanent features, but instead change with time. Like Uranus, Neptune also has a dusty ring that cannot last without being replenished, leading scientists again to suggest that collisions between larger bodies occasionally yield new supplies of small ring particles. This new picture of ring systems as dynamic, changing entities is discussed further in Chapter 10.

PLUTO: PLANETARY MISFIT

The little that we know about Pluto, the ninth planet, indicates that it is a nonconformist. Pluto (Table 8.3) simply does not fit into the systematic trends that we have found for the other planets: it has a solid surface, but is certainly not rocky throughout; therefore it does not qualify as a typical terrestrial planet. At the same time, it is not gaseous like the other cold,

ASTRONOMICAL INSIGHT
The Search for Planet X

The discovery of Pluto was by no means the end of the search for new planets. Quite to the contrary, this event sharpened the interest in continuing the search, and it was carried on for an extended period.

The technique by which Pluto was discovered was tedious but efficient, in that it was thorough. Tombaugh accomplished the task by comparing photographs taken at different times (usually a few days apart), to see whether any of the objects in the field of view moved. This would ordinarily be very difficult to do, particularly in areas of the sky that are crowded with stars, but was made easier with the aid of a special instrument called a **blink comparator.** This is a microscope for viewing two photographic plates simultaneously, with a small mirror that flips back and forth, providing, in rapid succession, alternating views of the two plates. If the plates are mounted so that the images of fixed stars are perfectly aligned, then an object that moved between the times of the two exposures will appear to blink on and off at two separate locations, while all the fixed objects are steady. Examining a region of the sky using this technique, therefore, consists of mounting in the blink comparator two plates taken at different times, then systematically scanning them with the movable eyepiece so that each object could be checked, to see if it is blinking. The entire job could take days or weeks for each pair of plates.

The blink comparator technique, while time-consuming, has the capability of allowing thorough searches of large regions of the sky. The search for a tenth planet, often referred to as Planet X, continued at Lowell Observatory for 13 years after the discovery of Pluto. Clyde Tombaugh, who had discovered Pluto, was the principal worker in the extended search. Nothing was found.

During and since that time, occasional reports of evidence for a tenth planet have surfaced, but always the evidence has been indirect. Reanalysis of the motion of Neptune has convinced some astronomers that there really may be discrepancies attributable to gravitational perturbations caused by another planet. Other researchers have pointed to the motions of certain comets as indicative of gravitational influence exerted by Planet X. Most of these predictions have suggested a distance from the Sun of 50 to 100 AU. The motion of Halley's comet, a bright and very famous object that visits the inner solar system every 76 years, has been analyzed by several scientists, who find that its arrival near the Sun is usually several days later than it should be, according to the laws of motion. This has been attributed by some to the gravitational influence of a tenth planet, but no such object could be found at the predicted position. One of the predictions based on the motion of Halley's comet was rather spectacular: The inferred Planet X had a mass 3 times that of Saturn, and an orbit inclined by 120° to the ecliptic!

Today the discrepancy in the motion of Halley's comet has been explained as due to the eruption of gases from the comet's interior as it heats up on approach to the Sun. These gases stream outward, toward the Sun, and slow the comet's motion.

Comets have played a prominent role in the modern search for new objects in the outer solar system. In 1977 a Sun-orbiting body was discovered in an elliptical orbit whose distance from the Sun ranges between the orbits of Saturn and Uranus. This object, named Chiron, subsequently flared up in brightness for a brief period, probably due to a sudden eruption of gases from its interior. It is thought that Chiron is a large cometary nucleus.

New theories about the origin of comets have stimulated renewed interest in the search for objects beyond Pluto. As will be explained in the next chapter, it is now thought that some comets originate in a disk of icy objects that orbit between 50 and 100 A.U. from the Sun. This disk, called the Kuiper belt, may in some sense be viewed as a second asteroid belt, consisting of small bodies formed in the early days of the solar system, and then ejected into distant orbits by gravitational encounters with the giant planets. Today several astronomers are searching for objects in the Kuiper belt, but until recently with no success.

In September of 1992, however, a very faint, reddish object was found in the constellation Pisces. Careful observations over the next several weeks confirmed that it is orbiting the Sun, and that its distance lies between 37 and 59 A.U., making it the most faraway solar system object yet discovered, and the first one lying beyond Pluto. In order to be seen at such a great distance, the object must be moderately large (at least 200 km in diameter), but not large enough to be considered a true planet. Thus the new object is thought to be either an asteroid or a large cometary body.

There is little evidence today for the existence of a tenth planet; indeed some astronomers would prefer to think of Pluto as a large asteroidal body or remnant planetesimal, rather than a full-fledged planet in its own right. But there are good reasons to expect that a large number of Sun-orbiting bodies exist beyond Pluto, and that the recent discovery of object 1992 QB1 may signal the beginning of a new era of solar system exploration, as additional members of the Sun's family are found.

TABLE 8.3 Pluto

Orbital semimajor axis: 39.44 AU (5,900,000,000 km)
  Perihelion distance: 29.58 AU
  Aphelion distance: 49.30 AU
Orbital period: 248.5 years (90,700 days)
Orbital inclination: 17°10'12"

Rotation period: 6.39 days
Tilt of axis: 122°

Diameter: 2302 km (0.18 $D_\oplus$)
Mass: 1.3×10^{25} gm (0.0022 $M_\oplus$)
Density: 2.03 gm/cm³
Surface gravity: 0.06 Earth gravity
Escape speed: 1.1 km/sec

Surface temperature: 40 K
Albedo: 0.6:

Satellites: 1

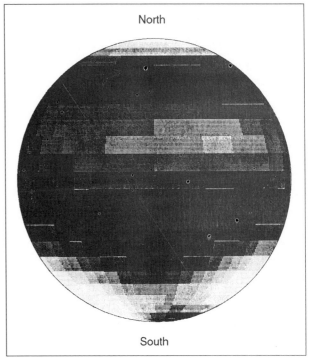

North

South

Figure 8.36 Surface map of Pluto. Analysis of brightness variations as Pluto was repeatedly eclipsed by Charon allowed this map of Pluto's surface to be constructed. The light and dark areas are regions of higher and lower albedo; the bright areas are thought to be ice deposits. The presence of a polar cap is interpreted as evidence that Pluto has seasons. (R. P. Binzel and E. F. Young, Massachusetts Institute of Technology)

outer planets and therefore does not qualify for membership in that group, either. As we will see, these unique aspects of Pluto have led to some rather exotic theories about its origin, but lately scientists have begun to suspect that Pluto is not so strange after all. It may represent a class of bodies that are actually fairly common in the outermost reaches of the solar system.

Earth-based observations have historically discovered very little about Pluto's nature, revealing only a dim point of reflected sunlight that is difficult to see without a substantial telescope. But even these limited images were enough to reveal some of the planet's major peculiarities. For example, the orbit of Pluto is very unusual, being highly tilted (by 17°) relative to the plane of the ecliptic and also highly noncircular. At its greatest distance from the Sun, Pluto is nearly 70 percent farther away than when it is at perihelion (when it actually moves closer to the Sun than Neptune; in fact, Pluto now is temporarily the eighth planet from the Sun, having moved inside Neptune's orbit in 1979, not to reemerge until 1999).

Initially establishing the size or the mass of Pluto was very difficult, but even rough estimates showed it to be a very small planet, the smallest of the nine by a wide margin. The mass was not measured until Pluto's satellite was discovered (discussed in the next section), but the small size made it clear that Pluto could not have enough mass to affect the orbital motions of Uranus or Neptune significantly, as some had suspected (recall the comments earlier in this chapter about Percival Lowell's original motivation to search for a ninth planet).

Spectroscopic measurements showed that Pluto has solid methane on its surface, which indicated that it has a methane atmosphere as well. In 1988, when Pluto passed in front of a background star, the star's light dimmed slowly, demonstrating that the atmosphere of the planet is quite extended. Scientists de-

duced that a second, heavier gas must also be present, but this gas has not been identified (it is thought to be either carbon monoxide or nitrogen).

Because Pluto has just passed through its perihelion (closest approach to the Sun), it is speculated that the atmosphere has been enhanced by the evaporation of surface ices (Fig. 8.36). The intensity of sunlight on Pluto is 2.8 times greater at perihelion than at aphelion (the point where a planet is farthest from the Sun). It is suspected that the atmosphere of Pluto may largely freeze out during the planet's passage through the remote parts of its orbit, remaining entirely in frozen form except around the time of perihelion. If this is true, it is quite a fortunate coincidence that the planet happened to be approaching perihelion when it was discovered; otherwise, its atmosphere and strange seasons might not have been discovered for another 250 years!

Variations in Pluto's surface coloration indicated that its rotation period is 6.39 days, and its obliquity was estimated (with great difficulty) to be very unusual, probably over 90° (later the discovery of Pluto's satellite would enable the obliquity to be established more precisely).

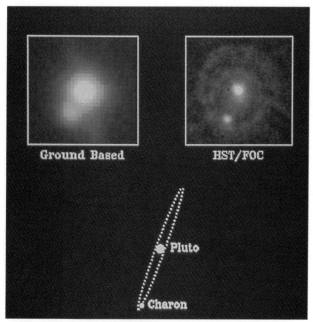

Figure 8.37   Pluto and Charon.   The image at the upper right, obtained by the *Hubble Space Telescope,* shows Pluto and Charon clearly separated, whereas the best ground-based photo (upper left) barely resolves the two bodies. (NASA/ European Space Agency: Space Telescope Science Institute)

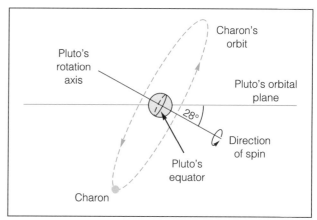

Figure 8.38   The orientation of Pluto's spin axis and Charon's orbit.   Like Venus and Uranus, Pluto has its rotation axis tipped over so far that it points below the plane of the planet's orbit (which is itself tipped by an unusually large angle, 17°, relative to the ecliptic). Thus the spin of the planet is technically retrograde (i.e., backwards). Charon orbits in the equatorial plane of the planet. During the late 1980s, the plane of Charon's orbit was aligned with the Pluto-Earth direction so that Pluto and Charon alternately passed in front of each other. Observations of these repeated transit events provided a wealth of information on the nature of both bodies and the transient atmosphere of Pluto.

Charon: The Key to Understanding Pluto

In 1978 Pluto was discovered to have a satellite (Fig. 8.37), which was named Charon. This satellite is rather large (about half the size of Pluto), making Pluto and Charon a double planet in effect, but was difficult to detect due to its proximity to Pluto and its great distance from the Earth. Both Pluto and Charon are locked into synchronous rotation, so that each keeps the same side always facing the other. This would create an interesting effect for anyone visiting the surface of Pluto: Charon would be seen to hang in the same spot in the sky, day and night, as the background stars streamed past due to Pluto's spin. Thus Charon is analogous to the communications satellites that have been launched into synchronous orbits about the Earth, so that they always remain over the same spot above the equator. Charon's orbital and rotation periods are both 6.39 days, the same as Pluto's rotation period.

Charon's orbital plane is highly tilted with respect to the orbital plane of Pluto (Fig. 8.38). By assuming that Charon orbits in the equatorial plane of Pluto (which it must, according to the laws of motion, if the two-body system is stable), astronomers were able to deduce that the pole of Pluto points about 28° below its orbital plane, meaning that the planet's obliquity is approximately 118°.

The discovery of Charon gave astronomers an opportunity to learn a great deal more about Pluto than had been possible from previous observations. First, it became possible to apply Kepler's third law to the orbit of Charon and to determine the combined masses of the two bodies. Recall (from Chapter 3) that Kepler's third law, which relates the orbital period to the semimajor axis of the orbit, makes it possible to solve for the sum of the masses of the two bodies. Normally the mass of the satellite can be ignored, and the sum can be assumed to be approximately equal to the mass of the planet alone. This is not true for the Pluto-Charon system, however, because Charon is so large compared with Pluto. Kepler's third law provides only the sum of the masses, but we can estimate the individual masses by assuming that the two bodies have similar densities; then the masses are proportional to their volumes. This calculation yields a mass for Pluto of about 0.0022 times the mass of the Earth, or about half the mass of Neptune's great moon, Triton.

Charon provides other information about Pluto as well. Recently, for a period of five years, the orbital plane of Charon was aligned with our sightline from the Earth, so that the two bodies eclipsed each other repeatedly for a period of time, as seen from the Earth. As Pluto moves around the Sun, the angle between it, the Earth, and the Sun changes. Every half-year on Pluto (i.e., about every 125 years) the orbital plane of Charon is aligned with the direction of the Sun, which means that, as seen from the inner solar system, Pluto and Charon undergo mutual eclipses.

Once again it was fortuitous that this rare event occurred not long (only 60 years or so) after the discovery of Pluto and very soon (about a dozen years) after the discovery of Charon.

The eclipses have provided astronomers with a series of opportunities to learn more about the characteristics of both bodies and their surfaces or atmospheres. When the duration of an eclipse and the orbital speeds of the two bodies are known, their sizes can be determined. For example, when Charon passes in front of Pluto as seen from the Earth, the duration of the eclipse tells us how long Charon takes to travel a distance equal to Pluto's diameter; combining this figure with the known orbital speed of Charon tells us how far the satellite travels during an eclipse, which is, in fact, the diameter of Pluto. Conversely, the diameter of Charon can be measured from the duration of its eclipses by Pluto (this technique is also used to measure diameters of double stars; see Chapter 12). Once the diameter of Pluto is known, it can be combined with the planet's mass to derive its density, which turned out to be approximately 2.0 grams/cm^3, meaning that Pluto is more than half rock, the remainder being ice. The numerical values for Charon are less certain, but apparently its composition is similar. Spectroscopic data show that its surface is covered with water ice.

The mutual eclipses also meant that the atmosphere of Pluto could be probed, using Charon as a source of (reflected) background light. As Charon moves behind Pluto and is seen through its atmosphere, the manner in which Charon dims in brightness reveals the density and extent of Pluto's atmosphere. By the same token, the lack of dimming of Pluto as it is about to be eclipsed by Charon shows that Charon has no significant atmosphere.

The Origin of Pluto

Pluto's sharp differences from all of the other planets suggest that its formation and evolution must be different as well. Some theories for the origin of Pluto have been quite exotic; for example, several have suggested that Pluto is an escaped satellite of one of the giant planets, most likely Neptune. Recall the strange orbits of the two major moons of Neptune, Triton and Nereid. According to this view, Pluto once also orbited Neptune and then had a near-collision with Triton and Nereid, which altered their orbits and ejected Pluto from Neptune's gravitational clutches, allowing Pluto to enter into a solar orbit of its own. Although a set of circumstances could be devised in which the present-day motions of all three bodies could have resulted from such a collision, this theory received a major setback when Charon was discovered. The problem was that a former moon had a moon of its own, a phenomenon never observed and probably not

possible from the viewpoint of the laws of motion.

The similarity between Pluto and some of the large satellites of the gaseous planets has already been mentioned. In particular, Triton seems to be very much like Pluto in many respects; both have comparable average densities, and hence probably similar compositions, and thin methane atmospheres. We cannot compare their surface structures and geological status until data from the planned Pluto encounter mission are available (early in the next century at the soonest), but the overall similarity between the two is quite striking. Furthermore, they share many characteristics with some of the other giant moons, such as the Galilean satellites of Jupiter.

Apparently the outer solar system is the home of a class of orbiting bodies that are composed of mixtures of rock and ice and have comparable sizes and masses (actually, there may well be many more that are smaller and therefore not detectable from Earth as yet). Some of these are satellites of the giant planets, while others, such as Pluto, are in their own solar orbits. These bodies appear to be planetesimals left over from the time of solar system formation. They are unlike the planetesimals from which the terrestrial planets formed in that they contain high abundances of volatile elements (hence their icy compositions), but they may be similar to the planetesimals that formed the cores of the gaseous planets. It has been proposed that these planetesimals were ejected into the outer solar system through gravitational encounters with the giant planets, causing most of them to remain very far out, well beyond the orbit of Pluto. In this scenario, Pluto, and perhaps Triton, are close enough to Earth to be readily observed only because random gravitational events caused them to be trapped in their present orbits instead of being expelled from the solar system or pushed out to much greater distances.

PERSPECTIVE

With this chapter, we have finished our survey of the major bodies in the solar system. We have discussed all nine of the planets, always keeping in mind the overall scheme of the solar system and how the varied properties of the planets fit into the general trends. Now we are ready to move on to a discussion of the minor bodies that orbit the Sun: the asteroids, the comets, and the interplanetary dust and gas. In doing so, we will see that even these bits of space flotsam and jetsam fit into the picture of the solar system that we have been developing—a solar system in which all the bodies are related to each other, and in which each can be understood in terms of a grand design that was established by the formation of the Sun. The minor bodies are discussed in the next chapter; the grand design is the subject of Chapter 10.

SUMMARY

1. The five outermost planets are all very different from the terrestrials: the four giants are large, gaseous bodies containing high abundances of lightweight elements and having many rings and moons; Pluto is more like one of the major moons of the giant planets, being composed of ice and rock.

2. Uranus was discovered accidentally, while Neptune and Pluto were found as the result of deliberate searches, based on observed irregularities in the motion of Uranus (and of Neptune, in Pluto's case; Pluto, however, turned out to be too small and distant to have caused any such irregularities, so its discovery is now regarded as fortuitous).

3. Detailed information on Jupiter, Saturn, Uranus, and Neptune was obtained through close encounters with spacecraft sent from Earth, especially the *Voyager 1* and *Voyager 2* probes. Only Pluto has yet to be observed in this fashion, and plans are now being developed to do so early in the next century.

4. The four giant planets have no solid surfaces, but instead are gaseous at high levels, with zones of liquid hydrogen below the clouds and then (in the cases of Jupiter and Saturn) liquid metallic hydrogen; all have small, solid cores of rock and metal.

5. Atmospheric circulation on the four giants is governed by convection, due to solar heating and heat rising from the interior, and rotation, which is so rapid that rotary circulation patterns are stretched into planet-girdling belts and zones.

6. All four of the giant planets have magnetic fields and charged particle belts, created by electrons and protons trapped in the magnetic fields. The particle belts are forced into flat disk-like shapes by the rapid rotation of the planets. Radio emission due to the synchrotron process occurs in these belts and is especially strong from Jupiter.

7. Three of the four giant planets emit excess radiation, in the sense that the amount of energy radiated away (in the infrared, due to thermal emission) is greater than the amount received from the Sun. Only Uranus does not show this effect strongly. For Jupiter and possibly Neptune, the extra energy may come from heat that was created and trapped internally at the time the planets formed; for Saturn, the excess heat is thought to be created by an ongoing differentiation process.

8. The ring systems of the four giant planets lie inside the Roche limit for large satellites, implying that tidal forces prevent the formation of moons and thus maintain the disks of debris that we see as rings.

9. In general, Jupiter and Saturn are very similar, but they do offer some contrasts: Saturn has a lower average density, resulting in a deeper atmosphere that obscures its atmospheric bands; probably creates excess internal heat through differentiation (see item 7, above); and probably undergoes seasonal changes in its atmospheric circulation pattern, whereas the obliquity of Jupiter is so small that seasonal effects are minimal.

10. The four Galilean satellites of Jupiter are strongly influenced by tidal forces and by orbital resonance, with the result that geological activity increases from the outermost of the four to the innermost. Io, which is closest to Jupiter, undergoes so much internal stress that it is in a constant state of volcanic eruption, re-coating its surface and contributing particles to the Io torus.

11. Saturn has one giant moon (Titan), seven intermediate satellites, and a number of very small ones. Titan has a thick atmosphere of nitrogen and methane and may have liquid or solid methane on its surface. The intermediate satellites are composed largely of ice, and some show signs of ongoing outgassing and surface re-coating. The tiny moons, many of them located within the rings, are responsible for creating some of the intricate ring structure through their gravitational effects.

12. The rings of Saturn (and of the other giant planets) are shaped by gravitational forces: the tidal force due to the parent planet; spiral density waves; and shepherding forces created by small moons. In the case of Saturn's rings, electrical forces appear to play a role in creating the dark "spokes"; these are thought to be due to tiny ice particles carrying electrical charges that are acted upon by the planetary magnetic field.

13. Uranus and Neptune are very similar in gross properties, but differ in detail: Uranus has a highly tilted rotational axis, resulting in strange seasonal effects; Neptune has a more active atmosphere, with more vivid belts and zones and more spots; and the orbital and geological properties of the two planets' satellites are quiet different. The rings and moons of both show indications of current or very recent dynamic processes that can renew and reshape the rings.

14. Pluto appears to have formed as a planetesimal in the region of the giant planets and was then moved to its present orbit as a result of a near-collision with one of the giant planets. Pluto has a moderate density, indicating a composition of ice and rock. Charon, Pluto's moon, has provided reliable information on Pluto's properties. Recently Charon's orbital plane was aligned with our line of sight from the Earth, causing alternate eclipses of Pluto and Charon.

REVIEW QUESTIONS

1. How is the circulation pattern in the atmospheres of the giant planets different from the atmospheric circulation of the terrestrial planets? What would you change about the terrestrial planets if you wanted to create wind patterns in their atmospheres similar to those of the giant planets?

2. Explain why more can be learned about some properties of a giant planet from Earth-based telescopes than from probes that fly by the planet at close range (*note:* "Earth-based" here includes telescopes in orbit about the Earth as well as those on the ground). For what types of information are Earth-based observations more useful?

3. If the radius of Jupiter's solid core is 12,000 km and the mass of the core is 15 Earth masses, what is the average density of the core? Compare your answer with the typical densities of the terrestrial planets and their cores.

4. Explain why the giant planets all have very intense belts of charged particles as compared with the Earth's Van Allen belts.

5. In Chapter 5 (and again in Chapter 10) the text stresses the many regularities in the motions of the planets and satellites in the solar system: that is, the orbital motions are mostly in the same direction, the orbital planes are mostly co-aligned, and the planets and satellites mostly spin in the same direction. The outer planets, however, exhibit a few irregularities in which planetary or satellite motions do not fit the overall pattern. List these, and give possible causes for them.

6. Io has the highest density among the Gaililean satellites of Jupiter, largely because it has little or no ice in its interior, whereas the others have retained their original ice in varying proportions. What would the density of Io be if its volume were doubled by adding ice to it? (Ice has a density of approximately 1 gram/cm³.) (*Note:* it might also be interesting to compute Io's radius in this case and compare it with the radii of the largest satellites, such as Ganymede and Titan.)

7. How is methane on Titan comparable with water on the Earth?

8. Summarize the geologies of the five major moons of Uranus, and compare their geologies with the geologies of the Galilean satellites of Jupiter.

9. Compare the overall properties (density, composition, size, and so on) of Pluto with those of Triton, the large satellite of Neptune. What does this comparison suggest about the origin of Pluto?

10. Explain how the discovery of Charon has helped astronomers learn about the properties of Pluto.

ADDITIONAL READINGS

Allen, D. A. 1983. Infrared views of the giant planets. *Sky and Telescope* 65(2):110.

Beatty, J. K. 1985. Pluto and Charon: the dance begins. *Sky and Telescope* 69(6):501.

———. 1986. A place called Uranus. *Sky and Telescope* 71(4):333.

———. 1987. Pluto and Charon: the dance gones on. *Sky and Telescope* 74(3):248.

———. 1990. Getting to know Neptune. *Sky and Telescope* 79:146.

Beatty, J. K., and A. Killian. 1988. Discovery of Pluto's atmosphere. *Sky and Telescope* 76:624.

Beatty, J. K., B. O'Leary, and A. Chaikin, eds. 1990. *The new solar system.* 2nd. Ed. Cambridge, England: Cambridge University Press.

Beebe, R. F. 1990. Queen of the giant storms. *Sky and Telescope* 80(4):359.

Binzel, R. 1990. Pluto. *Scientific American* 262(6):50.

Carroll, M. W. 1987. Project *Galileo:* the phoenix rises. *Sky and Telescope* 73:359.

Chaikin, A. 1986. *Voyager* among the ice worlds. *Sky and Telescope* 71(4):338.

Croswell, K. 1986. Pluto: enigma at the edge of the solar system. *Astronomy* 14(7):6.

Cuzzi, J. N., and L. W. Esposito. 1987. *Scientific American* 257(1):52.

Dyer, A. 1991. Tracking the Great White. *Astronomy* 19(3):36.

Elliott, J. L., E. Dunham, and R. L. Mills. 1977. The discovery of the rings of Uranus. *Sky and Telescope* 53(6):412.

Elliott, J., and R. Kerr. 1985. How Jupiter's ring was discovered. *Mercury* 14(6):162.

Esposito, L. 1987. The changing shape of planetary rings. *Astronomy* 15(9):6.

Ingersoll, A. P. 1981. The meterology of Jupiter. *Scientific American* 245(6):90.

———. 1981. Jupiter and Saturn. *Scientific American* 145(6):90.

———. 1987. Uranus. *Scientific American* 256(1):38.

Johnson, T. V., R. H. Brown, and L. A. Soderblom. 1987. The moons of Uranus. *Scientific American* 256(4):48.

Johnson, T. V., and L. A. Soderblom. 1983. Io. *Scientific American* 249(6):60.

Kinashita, J. 1989. Neptune. *Scientific American* 261(5):82.

Lanzerotti, L. C., and C. Uberoi. 1989. The planets' magnetic environments. *Sky and Telescope* 77:149.

Limaye, S. S. 1991. Neptune's weather forecast: cloudy, windy, and cold. *Astronomy* 19(8):38.

Littman, M. 1989. Where is planet X? *Sky and Telescope* 78:596.

Miner, E. D. 1990. *Voyager 2's* encounters with the gas giants. *Physics Today* 43(7):40.

Moore, P. 1989. The discovery of Neptune. *Mercury* 18:98.

Morrison, D. 1985. The enigma called Io. *Sky and Telescope* 69(3):198.

Pollack, J. B., and J. N. Cuzzi. 1981. Rings in the solar system. *Scientific American* 145(5):104.

Sanchez-Larega, A. 1989. Saturn's great white spot. *Sky and Telescope* 78:141.

Soderblom, L. A. 1980. The Galilean moons of Jupiter. *Scientific American* 242(1):68.

Soderblom, L. A., and T. V. Johnson. 1982. The moons of Saturn. *Scientific American* 246(1):100.

Squyres, S. W. 1983. Ganymede and Callisto. *American Scientist* 71(1):56.

Tombaugh, C. W. 1979. The search for the ninth planet. *Mercury* 8(1):4.

————. 1986. The discovery of Pluto. *Mercury* 15(3):66.

————. 1991. Plates, Pluto, and Planet X. *Sky and Telescope* 81(4):360.

ASTRONOMICAL ACTIVITY
Observing Jupiter and Saturn

The two largest of the giant planets are readily observable without sophisticated equipment, and each is situated favorably for nighttime observations at some time during each year. As the faster-moving Earth "catches up" with one of the giants, at first the planet will be up in the nighttime sky only after midnight, but as the Earth pulls even with the giant and then passes it by, the planet will become visible in the evening after sunset. You can predict when this situation will occur by noting the times of opposition for Jupiter and Saturn in the accompanying table. The best period for evening observations will be during the three months or so following opposition; but if you are an early-morning person, you can observe the planets before sunrise two or three months before opposition.

Dates of Oppositions

Jupiter	Saturn
March 30, 1993	August 19, 1993
April 30, 1994	September 1, 1994
June 1, 1995	September 14, 1995
July 4, 1996	September 26, 1996
August 9, 1997	October 10, 1997
September 16, 1998	October 23, 1998
October 23, 1999	November 6, 1999
November 28, 2000	November 19, 2000

You can easily see Jupiter and Saturn without binoculars or a small telescope, but will be unable to make out any detail. In this case you can observe the motions of the planets by noting their positions relative to the fixed stars and then seeing how the positions change with time. Since Jupiter and Saturn move slowly, you will need weeks or months to notice significant changes. To do this, make a sketch each time you observe Jupiter and Saturn, showing the relative positions of a few nearby stars. In due course your sketches will begin to reveal the motions of the planets with respect to the stars. Which way do the planets appear to move? What do you find if you observe the motion through the time of opposition?

If you have access to a pair of binoculars or a small telescope, you can see some of the surface details, particularly on Jupiter, and you can see satellites as well. You should be able to see Saturn's rings, unless they happen to be aligned directly with the line of sight to the Earth; in that case their plane is so thin that they become invisible to us as we view them edge-on. For Jupiter, make a sketch of the relative positions of the four Galilean satellites, and see if you can identify them by noting their order going outward from the planet. If you take care to make these sketches as accurately as possible, and if you make a few during the course of each night you observe, you may be able to make your own estimates of the satellites' orbital periods. This will work best if you observe for several hours during each of several consecutive nights, perhaps an arduous task (but something that astronomers do routinely!). Such a strategy works best because the orbital periods of these satellites are as short as a few hours, and it is necessary to measure the position several times per orbit to see how long it takes for the satellite to go around Jupiter once.

Space Debris

Comet Halley, with star trails due to telescope motion
(Anglo-Australian Telescope Board; photo by D. Malin)

PREVIEW

The planets dominate the family of Sun-orbiting objects in size and mass, but not in numbers. There are countless interplanetary bodies, ranging from asteroids hundreds of kilometers in diameter to microscopic dust grains, and they are the subject of this chapter. Despite the many contrasts in their properties, there are also some unifying themes: all of these bodies represent primordial material left over from the time of solar system formation; and all of them are influenced not only by the gravity of the Sun, which dominates the motions of the planets, but also by the gravitational influences of the planets themselves. Jupiter especially plays an important role; the origin and motions of asteroids, comets, and even the meteorites that reach the Earth's surface all show the signature of Jupiter's enormous gravitational force. When you finish this chapter, you should have a solid understanding of orbits, tidal forces, and gravitation that will help you to understand not only the solar system, but also many phenomena related to stars and galaxies, to be discussed later in the book.

The planets are the dominant objects among the inhabitants of the solar system (except, of course, for the Sun), but they are not entirely alone as they follow their regular paths through space. The abundant craters on planetary and satellite surfaces have shown us that there must have been a time when interplanetary rocks and gravel were more abundant than they are today, covering any exposed surface with impact craters. Today the rate of cratering is much lower than it once was, but some vestiges of the space debris that caused it still remain, orbiting the Sun and occasionally becoming obvious to us as they pass near the Earth or enter its atmosphere.

There are at least four distinct forms of interplanetary matter, some of which are closely related. In this chapter we will discuss the **asteroids** (more commonly, the **minor planets**), myriad rocky chunks up to several hundred kilometers in diameter that orbit the Sun between Mars and Jupiter; the **comets,** whose ephemeral and striking appearances have caused us to pause and marvel since ancient times; **meteors,** the bright streaks often visible in our nighttime skies, along with the objects that create them; and the **interplanetary dust,** a collection of very fine particles that occupy the void between the planets.

THE MINOR PLANETS

As we have already seen, some of the early students of planetary motions were concerned with the distances of the planets from the Sun. Kepler spent enor-

mous amounts of time and energy seeking a mathematical relationship that would describe the distances of the planets in terms of geometrical solids. What he finally did discover was something different, a relationship between the distances and orbital periods. Kepler's third law can be used to predict the period of a planet, given its distance from the Sun; or its distance, if its period is known. But it did not provide any underlying basis for explaining why planets are found only at certain distances from the Sun.

In 1766, a German astronomer named J. D. Titius found a simple mathematical relationship that seemed to accomplish what Kepler had set out to do. Titius discovered that if we start with the sequence of numbers 0, 3, 6, 12, 24, 48, and 96 (obtained by doubling each one in order), then add 4 to each and divide by 10, we end up with the numbers 0.4, 0.7, 1.0, 1.6, 2.8, 5.2, and 10.0, which correspond closely to the observed planetary distances in astronomical units. A few years after Titius found this numerological device, it was popularized by another German astronomer, Johann Bode, and eventually became known as Bode's law, or the Titius-Bode relation.

The sequence of numbers dictated by Bode's law included one, 2.8 AU, where no planet was known to exist. The discovery of Uranus in 1781 and the recognition that its distance fits the sequence (the next number is $(192 + 4)/10 = 19.6$, and the semimajor axis of Uranus's orbit is 19.2 AU) aroused a great deal of interest in the "missing" planet at 2.8 AU, since Bode's law was doing so well in predicting the positions of the others.

A deliberate search for such a planet began in 1800, but the sought-after object was discovered accidentally, when an Italian astronomer named G. Piazzi noticed a new object on the night of January 1, 1801, and within weeks found from its motion that it was probably a solar system body.

When the orbit of the new object was calculated, its semimajor axis turned out to be 2.77 AU. Thus the object was in solar orbit between Mars and Jupiter, where Bode's law had predicted that a new planet might be found. The new planet was named Ceres.

A little over a year after the discovery of Ceres, a second object was found orbiting the Sun at approximately the same distance and was named Pallas. It was clear from their faintness that both Ceres and Pallas were very small bodies and not respectable planets. By 1807, two more of these **asteroids,** as they were then being called, had been found and were designated Juno and Vesta. A fifth, Astrea, was discovered in 1845, and in the next decades, vast numbers of these objects began to turn up. The process of finding them became much more efficient when photographic techniques began to be used. The orbital motion of a minor planet or asteroid causes it

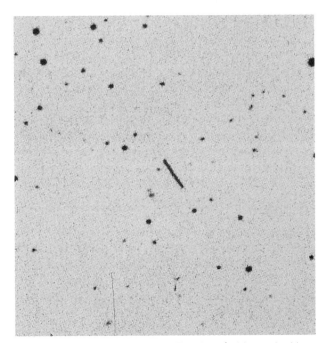

Figure 9.1 Asteroid motion. The elongated image in this photo is the trail made by an asteroid that moved relative to the fixed stars during the 20-minute exposure. Most new asteroids are discovered on photos of this type, although the first few were spotted visually. (Eleanor F. Helin, Palomar Observatory)

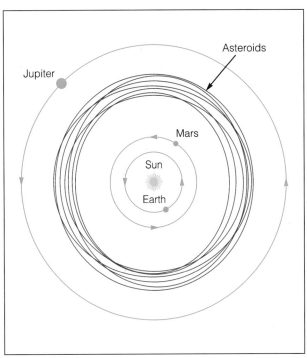

Figure 9.2 Asteroid orbits. This sketch shows typical orbital paths for the majority of asteroids, which orbit the Sun between Mars and Jupiter.

to leave a trail on a long-exposure photograph (Fig. 9.1).

Today thousands of asteroids are known, with almost 5,000 of them sufficiently well observed to have had their orbits calculated and logged in catalogs. The total number is probably much higher, perhaps 100,000. The vast majority have orbits lying between the orbits of Mars and Jupiter (Fig. 9.2).

During the past 20 years, a number of techniques have been developed for determining the properties of asteroids. Most of these techniques involve the measurement of sunlight that is reflected from the surfaces of the asteroids. The brightness of an asteroid can reveal information on its surface reflectivity (albedo, or the fraction of light that is reflected) and on the size of the reflecting body. Most asteroids are irregular in shape, and as they rotate, their reflected brightness varies. Analysis of the variations in brightness can yield information on the shape of an asteroid and the orientation of its rotation axis.

Another technique for establishing the sizes and shapes of asteroids is the observation of stellar occultations, times when asteroids happen to pass in front of background stars. When this happens, the star is temporarily obscured; measurement of the length of time the star is invisible, coupled with the known orbital speed of the asteroid, provides a measure of the size of the asteroid. Today astronomers use extensive computer calculations to predict when

known asteroids will pass in front of stars.

Asteroid sizes and shapes can also be determined from radar measurements, a technique now being applied to many asteroids by using the giant Arecibo radio telescope (see Chapter 4); by performing interferometric observations (again, see Chapter 4); and by making direct observations of asteroids that pass sufficiently close to the Earth or are visited by spacecraft (Fig. 9.3).

The launch of the *IRAS* satellite in 1983 signaled a new era in the study of asteroid sizes. *IRAS* mapped nearly the entire sky in far-infrared wavelengths, where cold bodies such as asteroids glow due to their own surface temperature (see the discussion of thermal emission in Chapter 4). The Stefan-Boltzmann law relates the luminosity of a glowing object to its surface temperature and its total surface area (recall that the luminosity, or total power emitted, is proportional to the surface area and to the fourth power of the surface temperature). The infrared data from *IRAS* provided the luminosity and the surface temperature (through the application of Wien's law), making it possible to solve for the surface area. Then the diameter can be computed if the shape of the asteroid is known; for simplicity, it is usually assumed to be spherical, with the result that the diameter is only a representative dimension, not an exact figure. A by-product of these studies is an estimate of the surface albedo in visible wavelengths of light.

Figure 9.3   Portrait of an asteroid.   This fine image of the asteroid Gaspra was obtained by the *Galileo* spacecraft on its way to Jupiter. The irregular shape and cratered surface are thought to be typical of most asteroids. (NASA/JPL)

Asteroid compositions can be determined by analyzing visible light that is reflected from their surfaces. Atoms and molecules that are bound into solid surfaces retain their ability to absorb light only at specific wavelengths, just as free atoms or molecules do (see the discussion of spectral lines in Chapter 4). In solids, however, the mixture of different types of particles, along with their close proximity in the material, leads to much broader spectral features than the narrow lines that characterize free particles. Thus the spectrum of light reflected from an asteroid's surface has several broad regions where less light is reflected due to absorption by surface compounds. If these broad absorption features can be identified with known mineral compounds, based on laboratory studies, the composition of the surface can be determined. Many of the features seen in asteroid spectra have been identified, at least as belonging to a broad class of minerals, although some remain unidentified.

When spectra of large numbers of asteroids began to be observed, researchers noticed that they tended to fall into distinct classes, based on the wavelengths and strengths of the broad absorption bands. These classes were assigned letters of the alphabet, based (where known) on the composition they indicated. Thus, the M asteroids are metallic, the C class is carbonaceous, and so on. More refined measurements have revealed many more classes than can easily be lettered according to composition, however, so the modern classification system includes many letters that have no obvious connection with the overall chemical makeup.

The C-class (carbonaceous) asteroids are the most common. About three-fourths belong to this group, with most of the rest falling into an assortment of classes containing metals and silicates. As we will see in a later section of this chapter, it is thought that the meteorites are fragments of asteroids that have been fractured in collisions. Thus the compositions of meteorites provide detailed information on the composition of the asteroids. Overall the results of

Once the total surface area is known, the albedo can be derived by comparing the amount of energy absorbed (as indicated by the infrared luminosity) with the amount reflected (as indicated by direct brightness measurements).

From these assorted techniques astronomers have established that asteroids have a wide range of diameters (Table 9.1), including a few as large as several hundred kilometers. Ceres, the largest, is nearly 1,000 km in diameter, but the majority are rather small, having diameters of 100 km or less. The largest are apparently spherical, while the smaller ones are often jagged, irregular chunks of material; their irregularity is indicated by their variations in brightness as they spin. A few asteroids are thought to be binary, consisting of two objects that orbit each other as they circle the Sun.

TABLE 9.1   Selected Asteroids

No.	Name	Year of Discovery	Diameter (km)	Mass (g)	Period (years)[a]	Distance (AU)[a]
1	Ceres	1801	933	1×10^{24}	4.60	2.766
2	Pallas	1802	538	3×10^{23}	4.61	2.768
3	Juno	1804	200	2×10^{22}	4.36	2.668
4	Vesta	1807	561	2×10^{23}	3.63	2.362
6	Hebe	1847	220	2×10^{22}	3.78	2.426
7	Iris	1847	200	2×10^{22}	3.68	2.386
10	Hygiea	1849	320	6×10^{22}	5.59	3.151
15	Eunomia	1851	280	4×10^{22}	4.30	2.643
16	Psyche	1852	280	4×10^{22}	5.00	2.923
51	Nemausa	1858	80	9×10^{20}	3.64	2.366
511	Davida	1903	260	3×10^{22}	5.67	3.190

[a]Periods are orbital periods; distances are orbital semimajor axes.

laboratory studies of meteorites and the inferred compositions of asteroids agree, with one major exception: very few asteroids are found with the same composition as the most common meteorites. These meteorites are chondrites, composed of rocky material with inclusions indicating that they formed under conditions of low temperature. Asteroid researchers today are trying to solve this mystery by finding the asteroids from which the chondritic meteorites originated. One suggestion is that the parent bodies are simply too small to have been detected in asteroid searches, possibly because only very small bodies could have avoided being heated during their formation during the era of planetesimal accumulation in the early solar system.

KIRKWOOD'S GAPS: ORBITAL RESONANCES REVISITED

As increasing numbers of asteroids were discovered and cataloged throughout the nineteenth century, calculations of their orbits showed remarkable gaps at certain distances from the Sun. One such gap is at 3.28 AU, and another is at 2.50 AU.

In 1866 Daniel Kirkwood observed that these distances correspond to orbital periods that are simple fractions of the period of Jupiter (Fig. 9.4). The giant planet, at its distance of 5.2 AU from the Sun, takes 11.86 years to make a trip around it, whereas an asteroid at 3.28 AU, if one existed there, would have a

period exactly half as long, 5.93 years. Thus, every time Jupiter made an orbit, this asteroid and the planet would be lined up in the same way, and Jupiter's gravity would subject the asteroid to regular tugs. Apparently this effect has prevented asteroids from staying in orbits at this and other distances where the orbital periods would result in regular alignments. The gap at 2.50 AU corresponds to orbits with a period one-third that of Jupiter. These orbital resonances are exactly analogous to the ones created by the moons of Saturn, which are responsible for some of the gaps in the ring system of that planet (see Chapter 8). Since Jupiter is the most massive of the outer planets and the nearest to the asteroid belt, it has by far the greatest effect in producing gaps, but in principle the other outer planets could do the same thing.

THE ORIGIN OF THE ASTEROIDS

For a long time the most natural explanation of the origin of the asteroids seemed to be the breakup of a former planet, creating a swarm of fragments that continued to orbit the Sun. This argument was weakened when the total mass of the asteroids was estimated and found to be much less than that of any ordinary planet (the total mass of the asteroids is estimated to be only about 0.04 percent of the mass of the Earth). Another rather strong argument against the planetary remnant hypothesis is that scientists

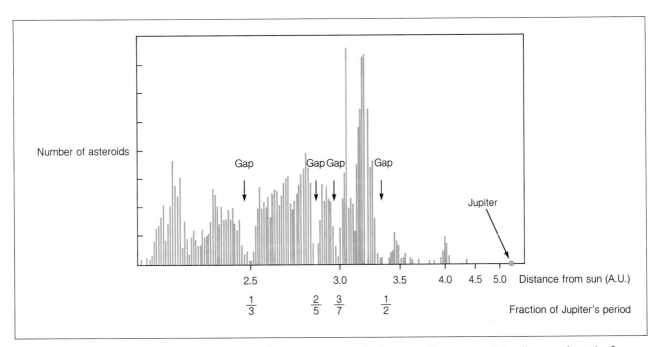

Figure 9.4   Kirkwood's gaps.   This graph shows the distribution of asteroid orbits. Gaps appear at the distances from the Sun where asteroids would have exactly one-half, one-third, or other simple fractions of the orbital period of Jupiter.

know of no reasonable way for a planet to break apart once it has formed, whereas it is easy to understand why material orbiting the Sun at the position of the asteroid belt could never have combined into a planet in the first place. It became simpler to accept the idea that the debris never was part of a planet than to explain how it first formed a planet and then broke apart.

Thus it is now considered likely that the asteroids represent material from the early solar system that never coalesced to form planets. According to modern theories of solar system formation (mentioned in Chapter 5 and discussed more fully in Chapter 10), there was a time when the system consisted of a disk of objects called planetesimals, which orbited the Sun in more-or-less circular orbits. These bodies would occasionally collide and, if the relative velocity was low enough, would stick together. Rather quickly (relative to the age of the solar system), the planetesimals accreted into a small number of larger bodies, today's planets. Planetary bodies built up more quickly in the outer portion of the solar system than in the inner portion, so Jupiter and Saturn grew to nearly their present sizes before the planetesimals closer to the Sun had grown very much. Once the two giants had formed, their gravitational forces (particularly that of Jupiter) strongly influenced the further evolution of the planetesimals near them.

Jupiter's gravity accelerated the motion of planetesimals that came near, causing these bodies to have relative speeds far too high for them to collide gently and stick together. Thus the planetesimals just inside the orbit of Jupiter, extending all the way inward to the orbit of Mars, were unable to coalesce to form a planet. Today's asteroids, therefore, are yesterday's planetesimals, and their study affords us an opportunity to learn a great deal about the early solar system.

Typical relative speeds measured today within the asteroid belt are around 5 km/sec. Not only do these speeds prevent the asteroids from forming a planet, but they also cause them to collide and occasionally to break apart. These collisions are the source of meteorites that reach the Earth, and over the billions of years since the solar system formed, they have reduced the total mass of the asteroid belt by a factor between 3 and 5 (many of the fragments escape entirely due to gravitational perturbations by Jupiter).

It is interesting that some of the asteroids are almost purely metallic, as are the meteorites that come from them. This composition implies that some of the original planetesimals were large enough to have undergone differentiation, forming metallic cores and rocky mantles and crusts. These large bodies were sufficiently heated during their formation to have melted internally, allowing the heavier metallic elements to separate out and sink to the core.

The asteroids are not uniformly mixed throughout the asteroid belt, but instead tend to fall into zones where different classes of asteroids are dominant. This distribution must reflect conditions that existed in the early solar system. For example, the inner asteroid belt (around 2 AU) is dominated by E- and S-class asteroids, which are made of silicate and metallic materials; these materials, which are thought to resemble igneous rocks on the Earth, require substantial heating for their formation. In the central asteroid belt, which peaks around 3 AU, the C asteroids, consisting of carbonaceous materials, are most common, and they persist all the way out to 5 AU, near the orbit of Jupiter. In these outer regions the D and P asteroids, consisting of assorted carbonaceous and organic silicate compounds, become the most common types. These are more primitive materials, which would not have survived substantial heating.

This distribution of compositions within the asteroid belt can be understood in the context of modern theories of solar system formation. It is thought that the compounds that condensed into solid form at any given distance from the Sun were determined by the temperature at that distance. At high temperatures, for example, metallic elements can condense while the lighter elements cannot. At intermediate temperatures some carbonaceous materials can condense, while the organic silicate mixtures can condense only at low temperatures. Thus the observed range of compositions within the asteroid belt tells us something about the distribution of temperatures as a function of distance from the Sun in the early solar system. Not surprisingly, the distribution indicates that it was hotter in the inner regions and colder in the outer portions. The rather rapid decrease in heating from the inner portion of the asteroid belt to the outer region may have been the result of heating due to the impact of the outflowing wind from the Sun in its early formation (this is discussed in the next chapter).

COMETS: FATEFUL MESSENGERS

Among the most spectacular of all the celestial sights are the comets. With their brightly glowing heads and long, streaming tails, along with their infrequent and often unpredictable appearances, these objects have sparked the imagination (and often the fears) of people through the ages.

In antiquity, when astrological omens were taken very seriously, great import was attached to a cometary appearance (Fig. 9.5). Ancient descriptions of comets are numerous, and in many cases observers associated these objects with catastrophe and suffering.

Edmond Halley. Aware of the power of Newton's laws of motion and gravitation, Halley reviewed the records of cometary appearances and noted one outstanding regularity. Particularly bright comets seen in 1531, 1607, and 1682 seemed to have similar properties, and Halley suggested that in fact all three were appearances of the same comet, orbiting the Sun with a 76-year period.

Calculations using Kepler's third law showed that for a period of 76 years, this object must have a semimajor axis of nearly 18 AU. Halley realized, therefore, that in order to appear as dominant in our skies as the comet does, it must have a highly elongated orbit (Fig. 9.6) that brings it close to the Sun at times, even though its average distance is well beyond the orbit of Saturn, nearly as far out as Uranus. Such an eccentric orbit, as a very elongated ellipse is called, had not previously been observed, even though Newton's laws clearly allowed the possibility.

Since Halley's time, searches of ancient reports of comets have revealed that Halley's comet has been making regular appearances for many centuries. The earliest records are provided by the ancient Chinese astronomers, who apparently observed its every appearance for well over 1,000 years, possibly beginning as early as the fifth century B.C.

The most recent visit of Halley's comet was in 1985 and 1986, when it did not pass very close to the Earth, and the best views were from the Southern Hemisphere (Fig. 9.7). A much more spectacular appearance occurred on the comet's previous visit in 1910 (Fig. 9.8), when the Earth actually passed through the rarefied gases of its tail. This will happen again during the 2062 appearance of Halley.

Other spectacular comets have been seen (Fig. 9.9), and a number have rivaled Halley's comet in brightness. Traditionally, a comet is named after its discoverer, and there are astronomers around the world who spend long hours peering at the nighttime sky through telescopes, looking for a piece of immortality.

When a new comet is discovered, a few observations of its position are sufficient to allow computation of its orbit. The results of many years of comet watching have shown that many comets have orbits so incredibly stretched out that their periods are measured in thousands or even millions of years. These comets, for all practical purposes, are seen only once. They return thereafter to the void of space beyond the orbit of Pluto, there to spend millenia before visiting the inner solar system again.

The orbits of comets, particularly these so-called long-period ones, are randomly oriented. These comets do not show any preference for orbits lying in the plane of the ecliptic, in strong contrast with the planets, and about half go around the Sun in the retrograde direction.

Figure 9.5   Calamity on Earth associated with the passage of comets.   This drawing is from a seventeenth-century book describing the universe. (The Granger Collection)

Included among the teachings of Aristotle was the notion that comets were phenomena in the Earth's atmosphere. Aristotle had no good evidence for this idea, and apparently adopted it because he believed that the heavens were perfect and immutable; therefore any changeable phenomena were associated with the Earth. In any case, this idea was accepted for centuries to come. Tycho Brahe in 1577 was able to prove that comets were too distant to be associated with the Earth's atmosphere, because they do not exhibit any parallax when viewed from different positions on the Earth. If a comet were really located only a few kilometers or even a few hundred kilometers above the surface of the Earth, its position as seen from the Earth would change from one location to another. Tycho was able to show that this was not the case, and that therefore comets belonged to the realm of space.

Halley, Oort, and Cometary Orbits

A major advance in the understanding of comets was made by a contemporary and friend of Newton,

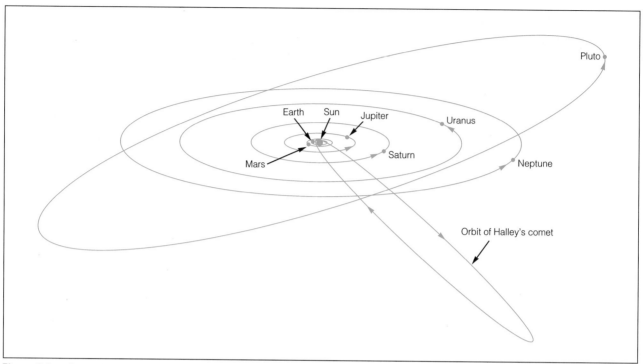

Figure 9.6   A cometary orbit.   This is a rough scale drawing of the orbit of Halley's comet. Most comets actually have much more highly elongated orbits than this, and correspondingly longer periods.

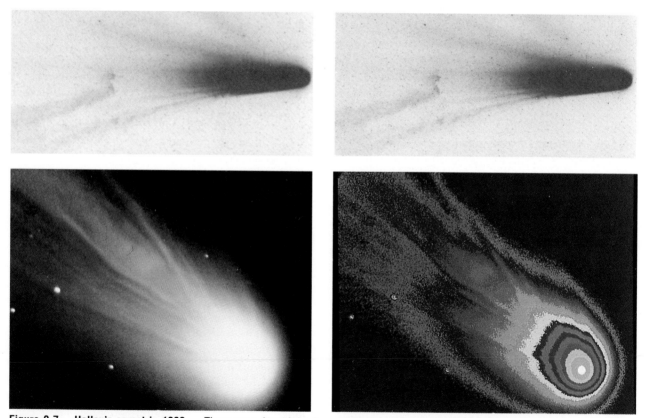

Figure 9.7   Halley's comet in 1986.   These are a few of the fine photographs made from Earth during the apparition of Halley's comet in 1985 and 1986. The two negative images on top, obtained only days apart, show changes in the detailed structure of the tail, including a disconnection event, as part of the ion tail broke free of the comet (top right). The two images at the bottom were both made from the same photograph, the color-coded version being made to enhance slight brightness variations. (top: © 1986 Royal Observatory, Edinburgh; bottom: U. Fink, University of Arizona)

Figure 9.8   Halley's comet as it appeared in 1910. (Palomar Observatory, California Institute of Technology)

Figure 9.9   Another bright comet. This is Comet IRAS-Araki-Alcock, which appeared in 1983 and was first discovered in infrared images obtained with the *IRAS* satellite. The colored streaks are star trails created by the telescope's motion as it tracked the comet, using color filters to make multiple exposures. (National Optical Astronomy Observatories)

Consideration of these orbital characteristics, especially the large orbital sizes, led Dutch astronomer Jan Oort to suggest that all comets originate in a cloud of objects that surrounds the solar system. He envisioned the **Oort cloud,** as it is now called, as a spherical shell with a radius of about 100,000 AU, extending, therefore, a significant fraction of the distance to the nearest star, which is almost 300,000 AU from the Sun. Modern calculations show that the Oort cloud lies primarily between 1,000 and 30,000 AU from the Sun, although its tenuous outer portions may extend as far out as Oort thought.

Occasionally, a piece of debris from the Oort cloud is disturbed from its normal path, either by a collision with another object or perhaps by the gravitational tug of a nearby star, and it begins to fall inward toward the Sun. If undisturbed by other forces, a comet falling in from the Oort cloud would follow a highly elongated orbit with a period of millions of years, appearing to us as one of the long-period comets

when it made its brief, incandescent passage near the Sun. In many cases, a comet is not left undisturbed, however. Instead it runs afoul of the gravitational pull of one of the giant planets—most often Jupiter. When this happens, the comet may be speeded up, so that it will escape the solar system entirely after it loops around the Sun. Or it may be slowed down, dropping into a smaller orbit with a shorter period, so that it becomes one of the numerous comets that reappear frequently (Table 9.2).

TABLE 9.2   Selected Periodic Comets

Comet[a]	Period (years)	Semimajor Axis (AU)	Year of Next Appearance
Temple (2)	5.26	3.0	1994.2
Schwassmann-Wachmann (2)	6.52	3.50	1994.3
Encke	3.30	2.21	1994.3
Wirtanen	6.65	3.55	1994.6
Reinmuth (2)	6.72	3.6	1994.5
Finlay	6.88	3.6	1995.1
Borrelly	7.00	3.67	1995.5
Whipple	7.44	3.80	1993.1
Oterma	7.89	3.96	1990.0
Schaumasse	8.18	4.05	1993.0
Wolf	8.42	4.15	1992.9
Comas Sola	8.58	4.19	1995.6
Vaisala	10.5	4.79	1991.9
Schwassmann-Wachmann (1)	16.1	6.4	1998.6
Neujmin (1)	17.9	6.8	2002.7
Crommelin	27.9	9.2	2012.6
Olbers	69	16.8	2025
Pons-Brooks	71	17.2	2025
Halley	76.1	18.0	2062.5

[a]A number in parentheses following the name of a comet indicates cases where a single observer has discovered more than one comet.

ASTRONOMICAL INSIGHT
A Fateful Comet and Women in Science

The first prominent woman astronomer in the United States was Maria Mitchell, who became a leader of women in all fields of science. In 1848 she became the first woman member (by 95 years!) of the American Academy of Arts and Sciences, and in 1850 she was the first woman elected to the American Association for the Advancement of Science. In 1873 she helped found the American Association for the Advancement of Women, and later she served as its president for two years.

Who was Maria Mitchell and how did she reach such prominence? The daughter of an amateur astronomer, she was born in 1818 and grew up on Nantucket, off the coast of Cape Cod, Massachusetts. From her father Maria learned about the fascination of the nighttime sky. Her formal schooling stopped when she was 16, but she continued to live at home, working days as a librarian. At night she had time to scan the skies using her father's small telescope, and in 1847 she discovered a new comet. For this she achieved considerable recognition, most notably a gold medal awarded by the king of Denmark. Her discovery instantly made her the most widely recognized and respected woman in American science. Election to prestigious scientific societies followed, and in 1849 Mitchell took a position as a "computer" at the U.S. Naval Observatory, assisting in the calculation of positions and timings for a publication called the *American Ephemeris and Nautical Almanac* (this book was needed at the time for navigation and for keeping track of positions of the Moon and planets; it is still published today and remains a fundamental source of basic information on motions in the sky, although it is no longer in quite so much demand as an aid to celestial navigation).

When Vassar College was founded in 1865, Mitchell was asked to become its first professor of astronomy and director of its observatory (despite her own lack of advanced training), and she accepted. At Vassar she dedicated her efforts to the education of young women and used unorthodox teaching methods with outstanding results. She emphasized observation, deductive reasoning, and mathematical analysis and discouraged loose and inaccurate thinking as she taught her students to strive for objectivity and accuracy. She and her students carried out many significant observations, including the analysis of several solar eclipses, observations of sunspots, and deductions on the nature of the atmosphere and satellites of Jupiter.

Maria Mitchell spoke for the role of women in science and did much to advance this cause. She believed that women were as capable as men of scientific insight and endeavor, but characterized herself as being of ordinary intellect while having extraordinary persistence. Several of her students went on to achievements in science; one of them succeeded her as professor of astronomy at Vassar.

Following Mitchell's death in 1889, an observatory was built in her memory on Nantucket and is still active today. The Maria Mitchell Observatory follows the traditions of education for women that were begun by its namesake; all the directors have been women, and the observatory sponsors an annual program for visiting undergraduate students that for many years was for women only (a recent expansion of the living quarters has allowed young men to participate in this program as well). The observatory operates under a modest endowment and serves as a highly visible reminder of the pioneering role of one of the earliest and greatest American women of science.

Detailed simulations of the process by which comets fall into the inner solar system from the Oort cloud show that most of the short-period comets, whose orbits tend to lie close to the plane of the planetary orbits, cannot originate in a spherical Oort cloud. Instead these comets must arise in a region that is disk-shaped, so that they start out with orbits that are aligned with the planetary orbits. This has led astronomers to deduce the existence of an inner cloud of comets, called the **Kuiper belt** after the American astronomer Gerard Kuiper, lying just outside the orbits of the outer planets.

Thus, the modern picture of cometary origins distinguishes between the long-period comets, those with very elongated, randomly oriented orbits and periods of millions of years, and short-period comets, which have smaller orbits lying close to the plane of the planets. The long-period comets arise in the inner Oort cloud, a spherical shell concentrated between 1,000 and 30,000 AU from the Sun, while the short-period comets originate in the Kuiper belt, a disk some 30 to 50 AU in radius. The total mass in the Kuiper belt is estimated to be about equal to the mass of the Earth, while the Oort cloud is thought to contain about 10^{11} objects totaling 100 or more Earth masses.

Both the Oort cloud and the Kuiper belt consist of bodies left over from the time of solar system formation, as will be discussed in the next chapter. The Kuiper belt objects are thought to have formed at roughly their present positions, while the Oort cloud bodies were ejected from the region of the outer planets by gravitational encounters and by tidal forces due to the Milky Way galaxy.

THE ANATOMY OF A COMET

A remarkably accurate picture of the nature of a comet was developed many years ago by American astronomer Fred Whipple. Whipple's model, sometimes called the "dirty snowball," envisions that for most of its life, a comet is just a frozen chunk of icy material, probably consisting of small particles like gravel or larger boulders, embedded in frozen gases. This picture was developed on the basis of observations of many comets over the years, and its basic features were verified by the close-up examination of Halley's comet in early 1986, when several space probes were able to make observations from close range (Fig. 9.10). Of course, a great many new details were discovered as well.

As a comet passes through the outer reaches of its orbit, far from the Sun, it does not glow, has no tail, and is not visible from Earth. As it approaches the Sun, however, it begins to warm up as it absorbs sunlight, and the added heat causes volatile gases to escape. A spherical cloud of glowing gas called the

coma develops around the solid **nucleus** (Fig. 9.11). The principal gases found in the coma of Halley's comet (Table 9.3) are H_2O (about 80 percent by number), carbon dioxide (CO_2; about 3.5 percent), and carbon monoxide (CO; most of the remaining 16.5 percent). Other slightly more complex species, such as ammonia (NH_3) and methane (CH_4) are probably present also, along with hydrogen molecules (H_2). The observed species glow by a process called **fluorescence.** The molecules absorb ultraviolet light from the Sun, causing them to be excited to high energy levels; then they emit visible light as they return to low energy states. A cloud of hydrogen atoms, resulting from the breakup of the molecules in the coma, extends out to great distances from the nucleus. The visible coma may be as large as 100,000 km in diameter, while the halo of hydrogen atoms may extend as much as 10 times farther from the nucleus, as demonstrated by ultraviolet observations of Halley's comet in 1986 (Fig. 9.12). The solid nucleus is relatively tiny, having a diameter of a few kilometers. The nucleus of Halley was found to be irregularly shaped, with a long axis of about 15 km and a width ranging up to about 10 km (Fig. 9.13). The mass of the nucleus is estimated to be about 10^{14} kilograms, and its density is extremely low—0.5 to 1.0 grams/cm³—less than the density of ordinary ice. The nucleus was found to have a coating of very dark material (albedo only 0.04), probably a matrix of car-

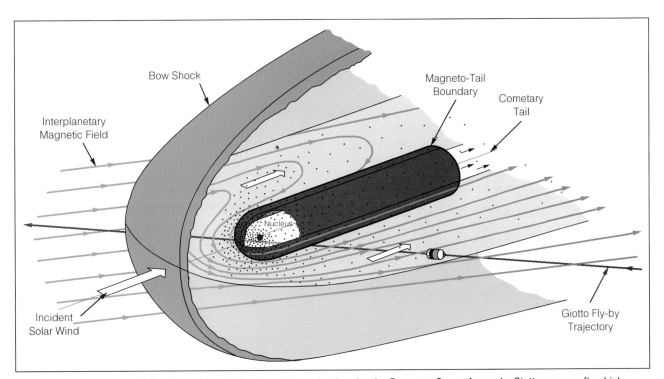

Figure 9.10 Probes to Halley's comet. This shows the path taken by the European Space Agency's *Giotto* spacecraft, which made the closest approach to the nucleus of the comet. Several other probes flew along similar paths, crossing ahead of the nucleus, but not as close to it. (European Space Agency)

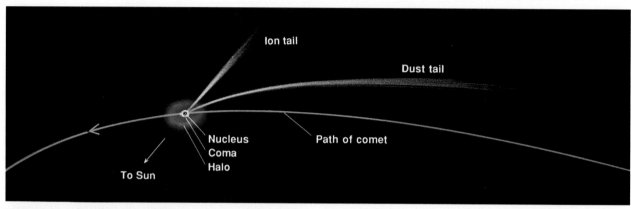

Figure 9.11   Anatomy of a comet.   This sketch illustrates the principal features of a comet, though not necessarily to correct scale.

Figure 9.12   The halo of Halley's comet.   This ultraviolet image, made by the *Pioneer Venus* spacecraft from Venus orbit, shows (in yellow, red, and orange) the extent of the hydrogen cloud surrounding the comet's nucleus. The hydrogen is a by-product of water vapor that is ejected from the nucleus and then dissociated by the Sun's ultraviolet radiation. The halo was about 500,000 km, or roughly ⅓ AU in diameter, when this image was obtained (the white circle represents the size of the Sun). (LASP, University of Colorado, sponsored by NASA)

TABLE 9.3   Detected Constituents of Comets

Coma		Ion Tail	
Atoms	Molecules	Ions	Molecular Ions
Hydrogen (H)	H_2O	C^+	CO^+
Oxygen (O)	CO_2	Ca^+	CO_2^+
Sulfur (S)	C_2	O^+	H_2O^+
Carbon (C)	C_3	S^+	OH^+
Sodium (Na)	CH	Fe^+	CH^+
Iron (Fe)	CN	Na^+	CN^+
Potassium (K)	CO		N_2^+
Calcium (Ca)	CS		H_3O^+
Vanadium (V)	NH		CS_2^+
Chromium (Cr)	OH		S_2^+
Manganese (Mn)	NH_2		CS^+
Cobalt (Co)	NCN		
Nickel (Ni)	CH_3CN		
Copper (Cu)	S_2		
	HCO		
	NH_3		

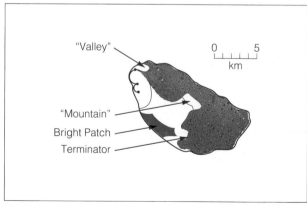

Figure 9.13   The nucleus of Halley's comet.   This sketch, based on images obtained by the *Giotto* spacecraft, illustrates the size and shape, as well as some surface features, of the nucleus. (Based on data from the European Space Agency).

bonaceous dust left behind by the evacuation of ices that evaporated previously. Hence, the dirty snowball idea has been modified to incorporate a complete outer coating of dirt, along with the traditional picture of mixed dirt and ice in the interior of the nucleus.

Another modification of the picture was that the surface of the nucleus is quite hot instead of cold and icy. The high temperature had been suspected from Earth-based infrared measurements and was confirmed by the spacecraft observations made at close range. The high temperature is the result of the blackness of the crust; low reflectivity means that sunlight is efficiently absorbed, causing heating.

Apparently the gas escaping from the nucleus of a comet is released in jets that can be so powerful that they alter the orbital motion of the comet. Astronomers had known for years that comets can be slowed in this way as they approach the Sun, but the close encounters with Halley's comet provided dramatic confirmation; the images revealed spectacular fountains of gas and dust erupting from beneath the crust of the nucleus (Fig. 9.14). Apparently these gas jets are created by heat that is absorbed by the dark crust and conducted into the icy interior; there the heat causes ices to evaporate in a process called **sublimation,** in which solid ice is converted directly into gas. The released gases expand and create outward pressure, which is relieved as the gases break through the crust, forming the jets that are observed. As the nucleus rotates, jets are active only on the sunlit side, so each individual jet turns on and off as it rotates toward and then away from the Sun. Surprisingly,

photos of Halley's comet in 1910 indicate that the jets occurred in the same locations then as they did in 1986. Apparently the jets occur at weak points in the crust of the comet, creating craters that persist over the years between encounters with the Sun's warming radiation.

As a comet nears the Sun, the solar radiation and the solar wind force some of the gas from the coma to flow away from the Sun, forming the tail, which in some cases is as long as 1 AU. Often there are two distinct tails (Fig. 9.15): one, formed of gas from the coma, usually contains molecules that have been ionized, such as CO^+, N_2^+, CO_2^+, and CH^+; the other is formed of tiny solid particles released from the ice of the nucleus. The **ion tail,** the one formed of ionized gases, is shaped by the solar wind and therefore points almost exactly straight away from the Sun at all times. As Halley and other comets have demonstrated on many occasions, all or part of the ion tail can detach itself and dissipate and then be replaced by a new tail (see Fig. 9.7). These "disconnection events" may be related to variations in the Sun's magnetic field, which exert forces on charged particles in the ion tail.

The other tail, called the **dust tail,** usually takes on a curving shape, as the dust particles are pushed away from the Sun by the force of the light they absorb. This **radiation pressure** is not strong enough to force the dust particles into perfectly straight paths away from the Sun, so they follow curved trajectories that are a combination of their orbital motion and the outward push caused by sunlight. The dust particles ejected from the nucleus are apparently quite fragile

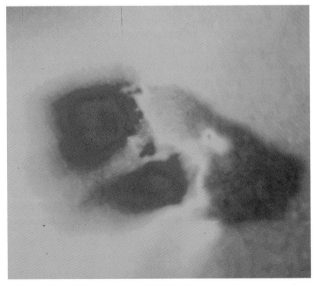

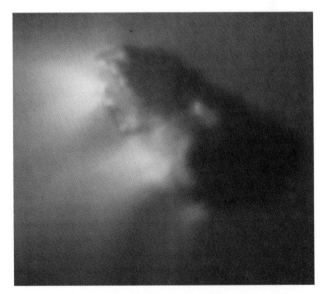

Figure 9.14   The nucleus of Halley's comet.   These are two versions of the same composite of some 60 images obtained by the *Giotto* spacecraft as it flew past the nucleus of the comet. Detailed surface markings in the crust of the nucleus can be seen, as well as the bright jets of warm dust that were being ejected toward the Sun (to the left). (Max Planck Institute for Aeronomie, courtesy of A. Delamere, Bell Aerospace Corporation)

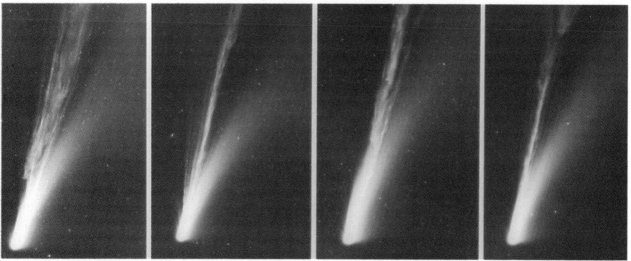

Figure 9.15   Two tails.   These photos of Comet Mrkos (1957) show distinctly the two types of tails that characterize many comets. The ion tail points straight up in these views, while the dust tail curves to the right. (Palomar Observatory, California Institute of Technology)

and readily break apart in collisions or when heated. Streams of gas resulting from the evaporation of icy dust particles were observed to extend as far as 50,000 km from the nucleus of Halley. Particles captured by the *Giotto* spacecraft appeared to be low-density, fluffy grains, and the farther from the nucleus they were captured, the smaller they tended to be. Infrared images obtained by the *IRAS* satellite (see Chapter 4) revealed extensive trails of dust particles racing out the orbits of several periodic comets. The quantity of dust in these trails indicates that a typical cometary nucleus is at least half rock and dirt. In an old comet like Halley, that has made many passages near the sun where its gases were evaporated, the relative proportion of rock and dirt can be even higher.

One of the major discoveries made by the probes that encountered Halley's comet and Comet Giacobini-Zinner was that the ionized gases surrounding a comet create a trapped magnetic field. A "bow shock," analogous to a boundary of a planet's magnetosphere, builds up on the sunward side of a comet, where the solar wind encounters the trapped magnetic field and flows around it. In the direction away from the Sun, the magnetic field that is wrapped around the comet forms a sheet of trapped charged particles called a **current sheet.**

The gases that escape from the nucleus of a comet as it approaches the Sun are highly volatile and would not be present in the nucleus if it had ever undergone any significant heating. This tells us that comets must have formed and lived their entire lives in a very cold environment, probably never even getting as warm as 100 K before falling into orbits that bring them close to the Sun. If a comet is so easily vaporized, then once it has begun to follow a path that regularly brings it close to the Sun, its days are numbered. It may take many round trips, but eventually it will dissipate all of its volatile gases, leaving behind nothing but rocky debris. It has been estimated that Halley's comet was losing some 50 tons per second of water ice when it passed nearest the Sun in early 1986! Despite this high mass-loss rate, the comet should last at least another 100,000 years before the nucleus is entirely dissipated. There is some speculation that a comet as large as Halley may eventually build up a crust thick enough to halt further sublimation of its interior ices. If this is so, Halley's comet may be immortal, but it will eventually stop developing a coma and tail on its visits to the inner solar system. Several times, a comet has failed to reappear on schedule, but has been replaced by a few pieces or perhaps a swarm of fragments (Fig. 9.16). In time, the remains of a dead comet will be dispersed all along the orbital path, so that each time the Earth passes through this region, it encounters a vast number of tiny bits of gravel and dust, and we experience a meteor shower.

METEORS AND METEORITES

Occasionally one of the countless pieces of debris floating through the solar system enters the Earth's atmosphere, creating a momentary light display as it evaporates in a flash of heat created by the friction of its passage through the air. The streak that is seen in the sky is called a **meteor** (Fig. 9.17). Most of us are familiar with this phenomenon, commonly called a shooting star, since it is often possible to see one in

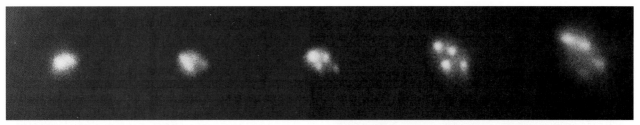

Figure 9.16   The breakup of Comet West.   This dramatic sequence shows the nucleus of Comet West fragmenting into four pieces. (New Mexico State University)

just a few minutes of sky-gazing on a clear night. On rare occasions an especially brilliant meteor is seen, possibly persisting for several seconds, and these spectacular events are called **fireballs** or **bolides.**

The piece of solid material that causes a meteor is called a **meteoroid.** Most are very small, amounting to nothing more than tiny grains of dust or perhaps fine gravel. A few, however, are larger, solid chunks, which are responsible for the bright fireballs.

Occasionally one of the larger meteoroids survives the arduous trip through the atmosphere and reaches the ground intact. Such an object is called a **meteorite** (Fig. 9.18), and examples can be found in museums around the world. Meteorites have been the subject of intense scrutiny, for until the past 20 years or so, they were the only samples of extraterrestrial material scientists could get their hands on.

Historically, scientists scoffed at the notion that rocks could fall from the sky, until a meteorite that was seen to fall near a French village in 1803 was found and examined just after it fell. Such falls are rare, but they are observed occasionally and have even been known to cause damage (but so far, few injuries).

Primordial Leftovers

Meteorites are old, much older than most surface rocks on the Earth. The age of the meteorites provides us with a glimpse into the history of the solar system and therefore makes them especially interesting to scientists.

Meteorites generally can be grouped into three classes: the stony meteorites, which comprise about 93 percent of all meteorite falls; the iron meteorites, accounting for about 6 percent; and the stony-iron meteorites, which are the rarest. These relative abundances were determined indirectly, because the different types of meteorites are not equally easy to find on the ground. The majority of all meteorites found are the iron ones, although they comprise only a small fraction of those that fall. The stony meteorites look so much like ordinary rocks that they are usually difficult to pick out, and some are burned up on their way through the atmosphere. A particularly productive place to search for meteorites is Antarctica, where a thick layer of ice conceals the native rock. Meteorites that fall there are relatively easy to find and are unlikely to be confused with Earth rocks.

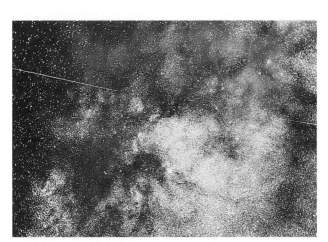

Figure 9.17   Two bright meteors.   The streaks of light in this photo are created by tiny particles entering the Earth's atmosphere from space. (Yerkes Observatory)

Figure 9.18   A meteorite.   This is a stony meteorite; the black coloring is due to heating as the object passed through the atmosphere. The light-colored spots are breaks in the fusion crust where interior material is exposed. (NASA)

Figure 9.19   Cross section of a chondrite.   This photo shows the many chondrules (light patches) embedded within the structure of this type of stony meteorite. (Griffith Observatory, Ronald A. Oriti Collection)

The stony meteorites are mostly of a type called **chondrites,** so named because they contain small spherical inclusions called **chondrules** (Fig. 9.19). These are mineral deposits formed by rapid cooling, most likely at an early time in the history of the solar system, when the first solid material was condensing. A few of the stony meteorites are **carbonaceous chondrites** (Fig. 9.20), thought to be almost completely unprocessed since the solar system was formed, and therefore representative of the original stuff of which the planets were made. The primordial nature of the carbonaceous chondrites, like the carbonaceous asteroids mentioned in an earlier section, is deduced from their highly volatile contents, which indicates that they were never exposed to much heat. One particularly fascinating aspect of these meteorites is that, in at least one case, complex organic molecules called **amino acids** were found inside a carbonaceous chondrite meteorite, showing that some of the ingredients for the development of life were apparently available even before the Earth formed.

The iron meteorites have varying nickel contents and sometimes show an internal crystalline structure (Fig. 9.21) that indicates a rather slow cooling process in their early histories. This has important implications for their origin, as we shall see.

Dead Comets and Fractured Asteroids

The origins of the meteoroids that enter the Earth's atmosphere can be inferred from what we know of the properties of meteorites and of the asteroids and comets.

As we have seen, most meteors are caused by relatively tiny particles that do not survive their flaming entry into the Earth's atmosphere. During **meteor showers,** when meteors can be seen as frequently as once per second, all seem to be of this type. As noted earlier, these showers are associated with the remains of comets that have disintegrated (Table 9.4), leaving behind a scattering of gravel and dust. Therefore, it is thought that the most common meteors, those created by small, fragile meteoroids, are a result of cometary debris.

The larger chunks that reach the ground as meteorites may have a different origin. It is likely that there are occasional collisions among the asteroids, sometimes sufficiently violent to destroy them and disperse the rubble that is left over throughout the solar sys-

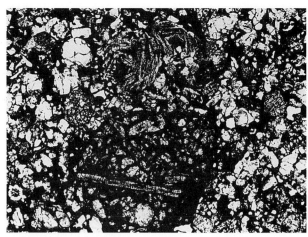

Figure 9.20   A carbonaceous chondrite.   This example, which is not the type in which amino acids have been found, shows a large chondrule (the light-colored spot, upper center), about 5 millimeters in diameter. (Griffith Observatory, Ronald A. Oriti Collection)

Figure 9.21   Cross section of a nickel-iron meteorite.   This example shows the characteristic crystalline structure indicative of a slow cooling process from a previous molten state. Such meteorites are thought to have once been parts of larger bodies that differentiated. (Griffith Observatory, Ronald A. Oriti Collection)

TABLE 9.4 Major Meteor Showers

Shower	Approximate Date	Associated Comet
Quandrantid	January 3	—
Lyrid	April 21	Comet 1861 I
Eta Aquarid	May 4	Halley's comet
Delta Aquarid	July 30	—
Perseid	August 11	Comet 1862 III
Draconid	October 9	Comet Giacobini-Zinner
Orionid	October 20	Halley's comet
Taurid	October 31	Comet Encke
Andromedid	November 14	Comet Biela
Leonid	November 16	Comet 1866 I
Geminid	December 13	—

tem. Most meteorites are probably fragments of asteroids. The iron meteorites apparently originated in asteroids that had undergone differentiation, while the stony ones either came from the outer portions of differentiated asteroids or from smaller bodies that never underwent differentiation at all. The chondrites probably fall into the latter category, since the chondrules reflect a rapid cooling that would be characterized by very small bodies. This is why the chondrites are thought to be the most primitive of the meteorites, having undergone no processing in the interiors of large bodies.

From our studies of the other planets and satellites, we know that there was a time long ago when frequent impacts occurred, forming most of the craters seen today. Certainly the Earth was not immune and no doubt was also subjected to heavy bombardment. The difference, of course, is that the Earth has an atmosphere, along with flowing water and glaciation, all of which combine to erase old craters in time. A few traces are still seen, however; a very large basin under the Antarctic ice is probably an ancient impact crater, and a portion of Hudson's Bay in Canada shows a circular shape thought to have a similar origin. A number of other suspected ancient impact craters have been found throughout the world (Fig. 9.22 and Table 9.5).

Although the frequency of impacts has decreased, major impacts still occur on rare occasions. The Barringer Crater (Fig. 9.23) near Winslow, Arizona, was formed only about 25,000 years ago, for example, and the possibility exists that other large bodies could still hit the Earth. Given a long enough time, it is almost inevitable.

MICROSCOPIC PARTICLES: INTERPLANETARY DUST AND THE INTERSTELLAR WIND

The space between the planets plays host to some very tiny particles in addition to the larger ones we have just described. There is a general population of small solid particles, perhaps a millionth of a meter

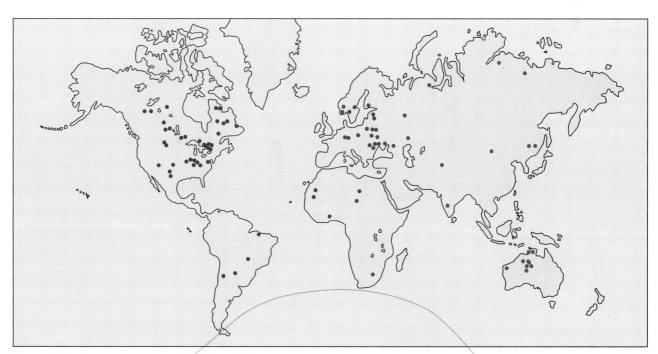

Figure 9.22 Impact craters on the Earth. This map shows the locations of major craters thought to have been created by impacts of massive objects. (Griffith Observatory)

ASTRONOMICAL INSIGHT
The Impact of Impacts

It is well established that impacts by objects from space have occurred throughout the solar system: indeed every exposed surface has craters. Recently scientists have found evidence that impacts may have played a far more important role in the shaping of solar system bodies than just creating craters. Impacts may have modified biological and geological processes on Earth, might have caused the formation of our Moon, may explain various anomalies in the motions of other planets, and might provide evidence for a companion star orbiting the Sun. All of these possibilities are unproven, but the amount of attention they are receiving is testimony to their appeal among scientists in many fields.

Mathematical analyses of the process by which planetesimals formed and built up by coalescence during the early history of the solar system show that there was a time when most of the preplanetary material was in the form of very large planetesimals several thousand kilometers in diameter. These orbited the Sun in a disk, so that planetesimals near each other traveled in the same direction with only relatively small velocity differences between near neighbors. Planets formed when the largest planetesimals collided and merged. Even with low relative speeds, vast amounts of heat were released upon impact, causing much of the material to melt and allowing the two bodies to merge into a single spherical one. In most cases, the planet that resulted had its rotation axis roughly perpendicular to the plane of the solar system disk, and it rotated in the prograde direction. It is easy to see, however, that the tilt of the axis and the direction of spin could have been modified by the direction of impact of the major planetesimals that merged to form the

planet. We see that the slow retrograde spin of Venus and the large tilt (98°) of Uranus could have resulted from the off-center impacts of large planetesimals. It has also been suggested (as discussed in Chapter 6) that the Earth's Moon might have formed as the result of the impact of a Mars-size planetesimal during the latter stages of formation of the Earth. In addition, as mentioned in Chapter 7, some researchers have suggested that Mercury lost most of its volatile elements due to a collision that stripped away its outer layers.

While the very massive impacts of planetesimals occurred only during the very early stages of solar system formation, we know that impacts by much smaller bodies still take place in modern times. The Barringer Crater in Arizona is only about 25,000 years old, a major impact occurred in Siberia in 1908, and we find traces of craters scattered over the Earth (these are described in the text). In the past few years, some have suggested that impacts of large meteorites or cometary nuclei on the Earth have been responsible for major modifications in the climate, which in turn may have been linked to mass extinctions of lifeforms on the Earth. Geological evidence for impacts (consisting of thin layers of debris deposited over large areas of the Earth) coincides in age with times of major extinctions. The most widely discussed event seems to have occurred at just about the time (65 million years ago) when the dinosaurs became extinct. This suggestion is highly controversial, because although the extinctions were apparently not as sudden as the impact scenario supposes, there is strong evidence that such an impact occurred. Very recently a large crater resulting from a major impact 65

million years ago has been identified on the coast of the Yucatan penninsula in Mexico, virtually proving that a catastrophic event did occur at roughly the time of the dinosaur extinctions. It is still controversial whether this impact actually was responsible for the extinctions, however.

Very recently it has hypothesized that impacts have been responsible for the sporadic reversals in the Earth's magnetic field (described in Chapter 6). There is a pattern of very striking coincidences between the times of major impacts in the past and the times when the Earth's magnetic field has reversed. The impacts are dated by the deposition of thin layers of debris, and the magnetic field reversals are dated by measuring the magnetic orientation of rocks, which shows what the field direction was when the rocks formed. If indeed impacts somehow triggered magnetic field reversals, the actual mechanism is not well understood. It has been suggested that changes in climate caused by dust in the atmosphere could have trapped more of the world's water in the form of polar ice, thus changing the mass distribution of the Earth enough to affect its rotation. The change in rotation of the Earth, it is argued, could disrupt the internal currents that give rise to the magnetic field, causing the field to vanish temporarily. When the field reestablished itself later, its orientation would have an even chance of being the opposite of what it had been before the impact.

Some scientists have suggested that there is a regular pattern of major impacts in the Earth's geological record, as though the bodies that caused the impacts had a higher chance of striking the Earth at specific time intervals. Some fascinating explanations for this periodic behavior have been offered. The most well known is the suggestion that the Sun has a companion star orbiting it, with a period of some 26 million years. This star, dubbed Nemesis, would regularly pass near the Oort cloud where it would disturb the cometary bodies and cause some of them to fall in toward the inner solar system; thus Nemesis would increase the likelihood that the Earth and the other planets would experience impacts. Many astronomers do not believe that the impacts of the past exhibit a regular pattern, however, and no evidence for a companion star has yet been found, despite intensive searches.

TABLE 9.5   Some Known or Suspected Impact Craters on the Earth

Location	Diameter (km)	Location	Diameter (km)
Amirante Basin, Indian Ocean	300	Deep Bay, Saskatchewan	13.7
Sudbury, Ontario	140	Deep Bay, Saskatchewan	12
Vredefort, Organge Free State, South Africa	100	Lake Bosumtwi, Ashanti, Ghana	10.5
Manicougan	100	Chassenon Structure, Haut-Vienne, France	10
Sierra Madera, Texas	100	Wolf Creek, Western Australia	8.5
Charlevoix Structure, Quebec	46	Brent, Ontario	3.8
Clearwater Lake West, Quebec	32	Chubb (New Quebec), Quebec	3.2
Mistastin Lake, Labrador	28	Steinem, Swabia, Germany	2.5
Gosses Bluff, Northern Territory, Australia	25	Henbury, Northern Territory, Australia	2.2
Clearwater Lake East, Quebec	22	Boxhole, Central Australia	1.8
Haughton, Northwest Territories	20	Barringer Crater, Winslow, Arizona	1.2
Wells Creek, Tennessee	14		

in diameter, called interplanetary dust grains. In addition, a very tenuous stream of gas particles flows through the solar system from interstellar space.

The presence of the dust has been known for some time from two celestial phenomena, both of which can be observed with the unaided eye, though only with difficulty. The dust particles scatter sunlight, so under the proper conditions a diffuse glow can be seen where the light from the Sun hits the dust. This phenomenon is analogous to seeing the beam of a searchlight stretching skyward; you see the beam only where there are small particles (either dust or water vapor) that scatter its light, so that some of it reaches your eye.

One of the phenomena created by the interplanetary dust is the **zodiacal light** (Fig. 9.24), a faintly illuminated belt of hazy light that can be seen stretching across the sky (along the ecliptic) on clear, dark nights, just after sunset or before sunrise. The second observable phenomenon created by the dust is a small bright spot seen on the ecliptic in the direction opposite the Sun. This diffuse spot, called the **gegenschein** (Fig. 9.25), is created by sunlight that is reflected straight back by the interplanetary dust, which is concentrated in the plane of the ecliptic. This is analogous to seeing a bright spot on a cloud bank or

Figure 9.23   Meteor crater near Winslow, Arizona.   The impact that created this crater occurred about 25,000 years ago. (Meteor Crater Enterprises)

Figure 9.24   The zodiacal light.   The diffuse glow of reflected sunlight from dust grains in the ecliptic plane is clearly visible in this photograph made from the summit of Mauna Kea. (W. Golisch)

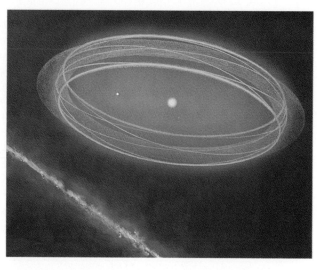

Figure 9.26   Dust bands in the outer solar system.   The *IRAS* satellite's infrared sensors detected emission from this spiral-shaped tail of dust, which may have been created by a collision between an asteroid and a comet. At the lower left is the Milky Way, as seen in the infrared by *IRAS*. (NASA)

Figure 9.25   The gegenschein.   This photo shows the Milky Way stretching across the upper portion and a diffuse concentration of light at lower center, which is the gegenschein. It is created by light reflected directly back to Earth from interplanetary dust in the direction opposite the Sun. (Photo by S. Suyama, courtesy of J. Weinberg)

low-lying mist when you look at it with the Sun directly behind you; the bright spot is just the reflected image of the Sun and is the counterpart of the gegenschein.

The *IRAS* satellite (see Chapter 4), which mapped the sky in infrared wavelengths, observed interplanetary dust. The grains of dust are cold and emit light at infrared wavelengths, so these observations reveal much more information about the dust than observations at other wavelengths. The *IRAS* sky maps show that the zodiac glows at infrared wavelengths due to the concentration of dust in the ecliptic plane.

In addition, the infrared maps revealed streaks of dust following cometary orbits and a huge spiraling ring of dust between the orbits of Mars and Jupiter (Fig. 9.26). This band of dust, which extends above and below the plane of the ecliptic, is thought to have been created by a collision between asteroids or between a comet and an asteroid.

Scientists have succeeded in collecting interplanetary dust particles for direct examination (Fig. 9.27). High-altitude balloons were once used to gather particles, but now new techniques have been developed. Interplanetary particles that fell into the oceans can be scooped off the seafloor, and very recently rich deposits of the particles have been found in Arctic lakes, particularly in Greenland, where they can be dredged up in vast numbers. The Earth is constantly being pelted with dust particles (which add about 8 tons per day to its mass!), and those that fall into the oceans or on Arctic ice sheets can lie undisturbed for long times. Studies of the grains show that they are probably of cometary origin, having been dispersed throughout space from the dead nuclei of old comets. These grains have a complex structure, with many segments stuck together (see Fig. 9.27). The grains are low in density and fragile, having the general characteristics attributed to grains ejected from Comet Halley (as discussed earlier in this chapter).

As we will learn in Chapter 16, the space between stars in our galaxy is permeated by a rarefied gas medium. In the Sun's vicinity, the average density of this gas is far below that of any artificially created vacuum; it amounts to only about 0.1 particle per cubic centimeter (i.e., there is one atom, on average, in

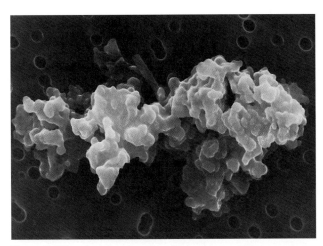

Figure 9.27   An interplanetary dust grain.   This is a microscopic view of a tiny particle from interplanetary space. The amorphous structure is highly variable form one grain to another. (NASA photograph, courtesy of D. E. Brownlee)

every volume of 10 cm³, corresponding to a cube about 1 inch on each side). Because of the motion of the Sun, the interstellar gas streams through the solar system with a velocity of about 20 km/sec. The presence of this ghostly breeze, consisting mostly of hydrogen and helium atoms and ions, was discovered in the early 1970s, when observations made from satellites revealed very faint ultraviolet emission from the hydrogen and helium atoms in the gas. The interstellar wind, tenuous as it is, has very little effect on the other components of the solar system but is nevertheless studied with some interest for what it may tell us about the interstellar medium.

PERSPECTIVE

The interplanetary wanderers discussed in this chapter have given us insight into the history of the solar system and have told us much about its present state as well. We have found two primary origins of the various objects: comets and asteroids. The former account for most of the meteors and for the interplanetary dust, and the latter are responsible for the meteorites, including the massive bodies that formed the major impact craters in the solar system.

In the next chapter we consider the overall properties of the solar system as we discuss theories of its formation and evolution.

SUMMARY

1. Thousands of minor planets have been discovered and cataloged; they display a variety of sizes (up

to 1,000 km in diameter) and compositions (ranging from metallic ones to rocky minerals).

2. Gaps in the asteroid belt are created by orbital resonances with Jupiter, whose gravitational influence was probably also responsible for preventing the asteroids from coalescing into a planet in the first place.

3. Comets are small, icy objects that develop their characteristic comae and tails only when in the inner part of the solar system.

4. Comets apparently originate in a cloud of debris very far from the Sun, occasionally falling in, either to bypass the Sun and return to the distant reaches of the solar system for millenia, or to be perturbed by the gravitational influence of one of the planets and become periodic comets.

5. When near the Sun, a comet ejects gases that glow by fluorescence. Periodic comets eventually lose all of their icy substance in this process and disintegrate into swarms of rocky debris.

6. A comet may have two tails, one created by ionized gas, and the other made of fine dust particles.

7. A meteor is a flash of light created by a meteoroid entering the Earth's atmosphere from space, and a meteorite is the solid remnant that reaches the ground in some cases.

8. Meteorites are either stony, stony-iron, or iron in composition; they are very old and thus provide information on the early solar system.

9. Most meteors are created by fine debris from comets, but most meteorites are fragments of asteroids.

10. Interplanetary space is permeated by fine dust particles and by an interstellar wind of hydrogen and helium atoms from the space between the stars.

REVIEW QUESTIONS

1. How is the asteroid belt similar to the rings of Saturn?

2. What is the orbital period of an asteroid whose semimajor axis is 2.8 AU?

3. Why is the study of asteroids, comets, and meteorites of particular interest to scientists concerned with the history of the solar system?

4. Compare Bode's law with scientific theories; that is, how is it like a theory and how is it not?

5. Summarize the role of Jupiter in influencing asteroids, meteorites, and comets.

6. Suppose a comet in the Oort cloud has a semimajor axis of 100,000 AU. The nearest star like the Sun is Alpha Centauri, 4 light-years (about 300,000 AU) from the Sun. Compare the gravitational force on the comet due to the Sun with that due to Alpha Centauri, when the comet lies in the same direction from the Sun as Alpha Centauri.

7. Discuss how the "dirty snowball" model of comets was supported or modified by the detailed closeup observations of Comet Halley in 1986.

8. Why are carbonaceous chondrites especially important clues to he early history of the solar system?

9. Would an observer on the surface of Mercury see meteors in the nighttime sky? Might he or she find meteorites on the ground there?

10. Summarize the evidence for the existence of interplanetary dust in the solar system.

ADDITIONAL READINGS

Beatty, J. K. 1986. An inside look at Halley's comet. *Sky and Telescope* 71(5):438.

Beatty, J. K. 1991. Killer crater in the Yucatan? *Sky and Telescope* 82(1):38.

Binzel, R. P., Barucci, M. A., and Fulchignoni, M. 1991. The origins of the asteroids. *Scientific American* 265(4):88.

Cassidy, W. A., and L. A. Rancitelli. 1982. Antarctic meteorites. *American Scientist* 70(2):156.

Chapman, C. R. 1975. The nature of asteroids. *Scientific American* 23(91):24.

Chapman, R. P., and J. C. Brandt. 1985. An introduction to comets and their origins. *Mercury* 14(1):2.

Cunningham, C. 1992. The captive asteroids. *Astronomy* 20(6):40.

Cunningham, C. J. 1992. Giuseppe Piazzi and the "missing planet." *Sky and Telescope* 84(3):274.

Dietz, R. S. 1991. Demise of the dinosaurs—a mystery solved? *Astronomy* 19(7):30.

Gehrels, T. 1985. Asteroids and comets. *Physics Today* 38(2):32.

Gingerich, O. 1986. Newton, Halley and the comet. *Sky and Telescope* 71(3):230.

Greenstein, G. 1985. Heavenly fire: Tunguska. *Science 85* 6(6):70.

Grieve, R. A. F. 1990. Impact cratering on the Earth. *Scientific American* 262(4):66.

Hartmann, W. K. 1975. The smaller bodies of the solar system. *Scientific American* 233(3):142.

Larson, S., and D. H. Levy. 1987. Observing Comet Halley's near nucleus features. *Astronomy* 15(5):90.

McFadden, L. A., and Chapman, C. 1992. Interplanetary fugitives. *Astronomy* 20(8):30.

McSween, H. Y., Jr. 1989. Chondritic meteorites and the formation of planets. *American Scientist* 77:146.

Morrison, D. 1976. Asteroids. *Astronomy* June 1976, p. 6.

———. 1992. The Spaceguard survey: Protecting the Earth from cosmic impacts. *Mercury* 21(3):103.

Morrison, D., and C. R. Chapman. 1990. Target Earth: It will happen. *Sky and Telescope* 79:261.

Opallco, J. 1992. Maria Mitchell's haunting legacy. *Sky and Telescope* 83(5):505.

Ronan, C., and M. Mohs. 1986. Scientist of the year: Edmund Halley. *Discover* 7(1):52.

Sagan, C., and A. Druyan. 1985. *Comet.* New York: Random House.

Smith, F. 1992. A collision over collisions: a tale of astronomy and politics. *Mercury* 21(3):97.

Spratt, C., and Stephens, S. 1992. Against all odds—meteorites that have struck home. *Mercury* 21(2):50.

Tatum, J. B. 1982. Halley's comet in 1986. *Mercury* 11(4):126.

Van Allen, J. A. 1975. Interplanetary particles and fields. *Scientific American* 233(3):160.

Verschuur, G. L. 1991. The end of civilization? *Astronomy* 19(9):50.

———. 1992. Mysterious sungazers. *Astronomy* 20(4):46.

Wagner, J. K. 1984. The sources of meteorites. *Astronomy* 12(2):6.

Whipple, F. L. 1974. The nature of comets. *Scientific American* 230(2):49.

———. 1986. Flying sandbanks or dirty snowballs: Discovering the nature of comets. *Mercury* 15(1):2.

ASTRONOMICAL ACTIVITY
A Scale Drawing of Halley's Orbit

Using what you learned in the Astronomical Activity in Chapter 3, you can make a scale drawing of the orbit of Halley's comet (or, for that matter, of any other Sun-orbiting body whose eccentricity and semimajor axis you know). To do this, you will need a cork board, two pins, some string, and a ruler.

You will need to decide on a scale for your drawing. A convenient one would be 1 cm = 1 AU; on this scale, the Earth's orbit would be a near-circle with a radius of 1 cm. Recall (from the Astronomical Activity in Chapter 3) that the eccentricity e is given by

$$e = F/A,$$

where F is the separation between the two foci of the ellipse and A is the major axis (twice the semimajor axis). The eccentricity for Halley's comet is $e = 0.97$, and the major axis is $A = 35.6$ AU. Therefore your string should be marked at two points 35.6 cm apart,

and the pins, which represent the foci, should be stuck in the board with a separation of 0.97 × 35.6 = 34.5 cm. Now tie the string to the pins at the two marked points, and trace out the ellipse by holding the string taut with a pencil. The result will be a very elongated ellipse with a major axis of 35.6 cm (semimajor axis of 17.8 cm) and an eccentricity of 0.97. If you now take out the pins and mark the position of one of them as being the location of the Sun, you can draw additional ellipses around that point to represent the orbits of the planets and see how they compare with the orbit of Halley's comet. This is a very effective way to see how the cometary and planetary orbits contrast, but it cannot easily show that the cometary orbits normally do not lie in the same plane as the planetary orbits. Can you think of a way to show Halley's orbit with both the correct size and shape and the correct orientation?

Formation of the Solar System: Disks, Rings, and Moons

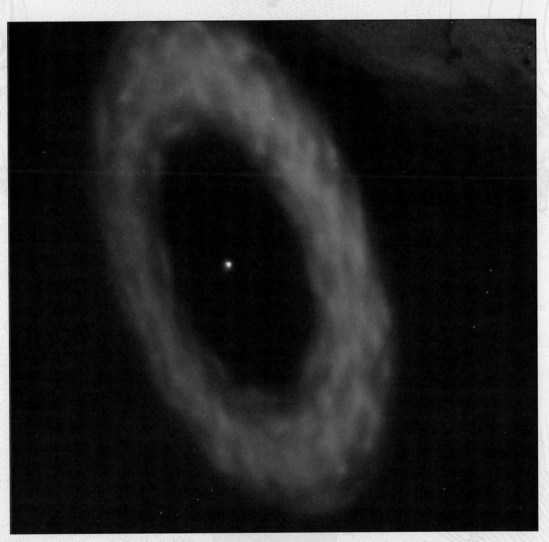

An artist's view of the dust around Vega. (NASA)

PREVIEW

In this chapter we tie together all the diverse solar system phenomena that were discussed in the previous five chapters, as we develop a scenario for the formation of the Sun and planets. Starting with the overall similarities and systematic properties of the planets and interplanetary bodies, we will bring into play the many gravitational mechanisms that have been invoked along the way. Not only will this discussion help to cement in your mind the modern understanding of how our solar system came to be, but it will also set the stage for the search for planetary systems that may be orbiting other stars, a major quest of today's astronomers. Following this chapter, you will be ready to undertake the next step on your odyssey through the universe: the study of stars and stellar systems, the subject of the next section of this book.

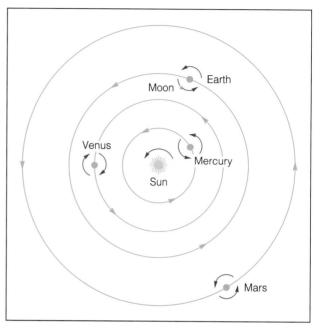

Figure 10.1   Prograde motions in the inner solar system. The rotations and orbital motions of the planets are generally in the same direction. This sketch of the orbits of the inner four planets shows that Venus has retrograde rotation but that the other three planets and their satellites obey the rule. The outer planets (except for Pluto) and nearly all their satellites do also.

We have collected a considerable quantity of information about the solar system and its contents. Along the way, we have uncovered various systematic trends, such as the similarities in internal structure and composition of the terrestrial planets and the comparisons among the outer planets, and in Chapter 5 we briefly outlined the modern picture of solar system formation. In discussing the histories of the individual planets, we have referred to some of the important processes thought to have been involved, such as the condensation of the first solid material from a cloud of gas surrounding the young Sun, but it remains for us to put together a coherent story of the formation and evolution of the entire system.

We have so far said little about the Sun, the central object in the solar system. For the purposes of this chapter, we need to know something about the Sun, although most of our discussion will be deferred until the next chapter, which opens the section on stars and their properties. For our present discussion, we need to keep in mind that the Sun is by far the most massive body in the solar system, having a mass over 300,000 times that of the Earth. It is composed primarily of hydrogen (about 70 percent, by mass) and helium (about 27 percent), with other elements in trace amounts, carbon, nitrogen, and oxgen being the most abundant. (Table 11.2 shows the composition of the Sun in detail.) The Sun is gaseous throughout, with a surface temperature of just under 6,000 K and a central temperature of roughly 10 million degrees. It has hotter layers above the surface; of particular interest for this chapter is the **corona,** an extended zone as hot as 1 to 2 million degrees, and the **solar wind,** a stream of charged particles (electrons and protons) flowing outward from the Sun throughout the solar system.

Before we describe the modern theory of solar system formation, we will review the overall properties of the system that must be accounted for, and then we will have a look at the historical developments and alternative origin theories that scientists have considered over the years.

A SUMMARY OF THE EVIDENCE

Any successful theory of the formation of the solar system must explain several facts. In a way, this is much like a mystery story, in which the detective (the scientist seeking the correct explanation) has certain clues that reveal isolated parts of the story and must reconstruct past events from them.

One category of clues in our mystery has to do with the orbital and spin motions of the planets and satellites in the solar system (Figs. 10.1 and 10.2, Table 10.1). First and most obvious, the planetary orbits lie in a common plane, all are nearly circular, and all go around the Sun in the same direction. Pluto violates these rules to an extent because its orbit is both elongated and tilted (by 17°) with respect to the ecliptic, but even this misfit goes around the Sun in the same direction as the other planets. The spins of nearly all the planets are in this same direction, as

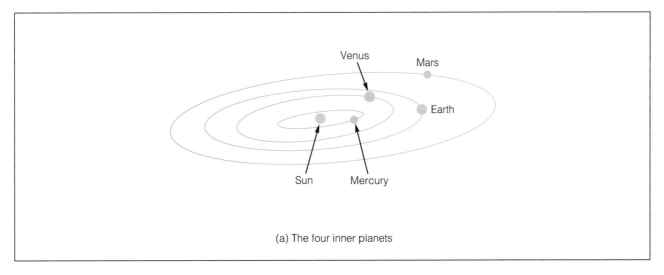

(a) The four inner planets

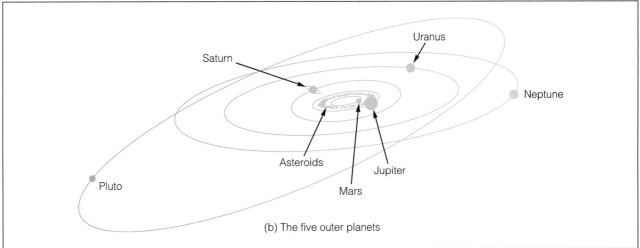

(b) The five outer planets

Figure 10.2   The coalignment of planetary orbits.   All but one of the nine planets have nearly circular orbits in a common parallel plane. The exception is Pluto, whose orbit is tilted 17° with respect to the ecliptic, and is sufficiently eccentric (elongated) that Pluto is actually closer to the Sun than Neptune at times.

are the orbital and spin motions of the vast majority of the satellites in the solar system. The direction of these motions, counterclockwise when viewed from above the North Pole, is said to be **direct,** or **prograde,** motion.

A successful theory must also explain why the planets formed where they did, and not elsewhere. A final clue involving the motions of the planets is that most of them have small **obliquities,** meaning that their equatorial planes are nearly aligned with their orbital planes. Even the Sun has its axis of rotation nearly perpendicular to the plane of the planetary orbits. Uranus, with its 98° tilt, does not fit the pattern, nor do Venus and Pluto.

A second general type of clue in the mystery concerns the natures of the individual planets and the systematic trends from planet to planet that we have discussed (Table 10.2). First, we must consider the planetary compositions, seeking an explanation for

TABLE 10.1   Properties of Planetary Orbits

Planetary distances from the Sun follow a regular progression.

Planetary and satellite orbits lie in a common plane.

Nearly all orbital and spin motions are in the same direction.

Spin axes of most planets and satellites are nearly perpendicular to the ecliptic.

Planetary and satellite orbits are nearly circular.

why the inner planets—the terrestrials—have high densities, indicating that they are made mostly of rock and metallic elements; in contrast, the outer planets have low densities and consist primarily of lightweight gases such as hydrogen and helium, as well as ices composed of these and similarly volatile gases. Within each of the two groups of planets, there

TABLE 10.2   Physical Properties of the Planets

There are two major types of planets: terrestrials and gaseous giants.

Giant planets are much larger and more massive than the terrestrial planets but have lower densities.

Terrestrial planets have low abundances of volatile elements; giant planets have plentiful volatiles.

Giant planets have more satellites than terrestrial planets.

Giant planets tend to have ring systems; terrestrial planets do not.

Giant planets rotate more rapidly than do terrestrial planets.

are relatively minor differences in chemical composition, such as the various isotope ratios we have discussed, which must also be accounted for.

The Sun has a rotation period of about 25 days, which posed a major mystery for a long time. According to many early theories, the Sun should be spinning much faster. The difficulty arises in theories that suggest that the solar system formed from the collapse of a cloud of gas and dust; these theories, as we shall soon see, are the most successful ones. The laws of physics tell us that the angular momentum of an object (see Chapter 3) must remain constant. This means that, if a spinning object shrinks in size, it must spin faster to compensate, thereby maintaining constant angular momentum. Thus the Sun, thought to have formed at the center of a collapsing cloud, should have a very rapid rotation rate, rather than the leisurely 25-day period it has. As we will see, the explanation for this slow rotation was a long time in coming, as astronomers were sidetracked for a while in their effort to understand the formation of the solar system.

A final category of information that bears on the origin of the solar system is the distribution and nature of the various kinds of interplanetary objects that were described in Chapter 9. We learned there that asteroids, comets, meteroids, and interplanetary dust inhabit the space between the planets, and that the motions and compositions of these objects reflect conditions very early in the history of the solar system. Somehow this information must fit into our overall picture.

Let us now take a look at some of the ideas that have been developed historically in the face of all this evidence.

CATASTROPHE OR EVOLUTION?

Since the time of Copernicus, many scientists and philosophers have tackled the problem of explaining the origin of the solar system. Although many hypotheses have been proposed, most of them have fallen into one or the other of two general types: **catastrophic theories,** which invoke some singular event; and **evolutionary theories,** which explain the formation of the solar system through gradual, natural processes.

The first speculations and hypotheses involved evolutionary theories that suggested that the Sun and the planets formed from a gigantic whirlpool or vortex that flattened into a disk from which the planets condensed. First the French philosopher René Descartes (in 1644) and later the German Immanuel Kant (in 1755) developed this picture (Fig. 10.3).

Figure 10.3   The hypothesis of Kant and Laplace.   Descarte's simple vision of a vortex was refined by Kant, who realized that rotation should cause a collapsing cloud to take a disk-like shape. A little later, at the end of the eighteenth century, Pierre Simon de Laplace, a French mathematician, hypothesized that a rotating disk would form detached rings that could then condense into planets.

Ironically, even though the evolutionary model was then pushed aside for some time by catastrophic theories, it was the evolutionary hypothesis that eventually won out. The chief reason it temporarily fell out of favor was the realization that if the Sun formed at the center of a collapsing, spinning cloud of material, conservation of angular momentum (discussed in Chapter 3) should have given it a much faster spin rate than it has. The catastrophic scenarios that were suggested (mostly during the nineteenth and early twentieth centuries) all invoked a near-collision between the Sun and some other body, which resulted in gaseous material being dragged out of the Sun and then left to condense into planets that continued to orbit it.

Soon several serious objections were raised. For example, gas from the Sun's interior should have too much internal energy to condense into solid material, and furthermore such material should be completely devoid of the hydrogen isotope **deuterium,** yet deuterium is found in the planets. Deuterium is a form of hydrogen having a neutron and a proton in the nucleus; it is much less abundant than normal hydrogen (with no neutrons and only a single proton), but is present on the Earth and in meteoritic material. Deuterium undergoes nuclear reactions very easily and consequently cannot survive at temperatures typical of the Sun's interior. Therefore, any gas pulled out of the Sun should have no deuterium in it. Because the planets contain deuterium, it has been argued that planetary material could not have formed from gas released from the solar interior.

The slow spin of the Sun remained a mystery until the 1960s, when the solar wind was discovered. This stream of charged particles from the Sun consists of mostly electrons and protons (see the discussion in Chapter 11). The Sun's magnetic field exerts forces on the charged particles in its vicinity, creating a form of drag that acts to slow the Sun's rotation (Fig. 10.4). As the Sun spins, its magnetic field spins with it, but the field is held back by its interaction with the charged particles. The result can be likened to trying to spin a pinwheel under water; drag causes the pinwheel to turn more and more slowly. The drag caused by the action of the solar magnetic field on solar wind particles is called **magnetic braking,** and the process can easily explain why the Sun spins today with a leisurely 25-day period.

Given the difficulties with the catastrophic theories for the formation of the solar system and the success of the evolutionary models in accounting for the observed nature of the Sun and planets, only the evolutionary theory is seriously considered today. It is interesting to note, however, that the modern picture of solar system formation and evolution includes several very important events that could only be called catastrophes.

A MODERN SCENARIO

Table 10.3 summarizes the current theory of solar system formation. Although aspects of this theory are uncertain, particularly the role of magnetic fields in initiating the process, the general picture appears to be correct and is supported by observations. The first step was the collapse of an interstellar cloud, one that must have been rotating before it began to fall in on itself. Our galaxy has a large quantity of interstellar material that contains both gas and small dust grains. The interstellar medium is not uniformly distributed

Figure 10.4   Magnetic braking.   The young, rapidly spinning Sun had the magnetic field structure shown here, with the field lines rotating with the Sun. The early solar system was permeated by large quantities of debris and gas, some of which was ionized and therefore subject to electromagnetic forces. The Sun's magnetic field exerted a force on the surrounding gas, which in turn created drag that slowed the Sun's rotation.

ASTRONOMICAL INSIGHT
The Explosive History of Solar System Elements

Even according to the evolutionary theory that is now widely accepted, the development of the solar system required explosive events. The evolution of the universe itself may be viewed as the result of a series of explosions, including the initial one from which the universe has been expanding ever since. A variety of astronomical evidence (discussed thoroughly in Chapter 21) shows that the universe began some 10 to 15 billion years ago in a single point and, at some time in the past, began to expand. The initial temperature and density were so extreme that the birth of the universe must have been an explosive event, commonly called the big bang.

For a brief time during the early stages of the universal expansion, conditions were suitable for nuclear fusion reactions to occur, and the primordial substance was converted from a soup of subatomic particles to one of recognizable elements, primarily hydrogen and helium, with almost no trace of heavier species. Eventually, as the universe continued to expand, galaxies and individual stars formed. Just as nuclear reactions in the Sun form a heavier element (helium) from a light one (hydrogen), similar reactions take place inside all stars, gradually enriching the universal content of heavy elements. Stars more massive than the Sun undergo more reaction stages, going from the production of helium to that of carbon, and from

carbon to other elements. When these stars die—often explosively—their newly formed heavy elements are returned to interstellar space and are available for inclusion in new stars that form later. Thus, during the 5 to 10 billion years that passed between the time of the big bang to the time the solar system was formed, the chemical makeup of our galaxy gradually changed, so that roughly 2 percent of the total mass was in the form of elements heavier than helium.

Since this roughly represents the composition of the Sun and the primordial material in the solar system, it might seem that there is no more to the story of the origin of the elements. We have compelling reasons to believe, however, that the formation of the solar system was influenced by at least two local explosive events. Certain atomic isotopes that are present in solar system material are the result of explosive nuclear reactions that must have occurred relatively recently, compared with the time that had passed since the big bang. These isotopes, one a form of aluminum (with 26 instead of the normal 27 neutrons and protons) and the other a form of plutonium (with 244 protons and neutrons instead of 242) are radioactive, with half-lives short enough that they could not have been present when the solar system formed unless they had been created relatively recently in explosive events. The plutonium

isotope in question probably formed in a stellar explosion called a supernova that occurred a few hundred million years before the solar system formed, whereas the aluminum must have been created in a nearby supernova only a million years or so before the solar system was born.

The second of these inferred explosions may have played a crucial role in triggering the formation of our Sun and planetary system. We have said that the solar system formed from an interstellar cloud that collapsed. What we have not discussed, however, is what caused this collapse to occur. Normally, an interstellar cloud is quite stable, and it will not fall in on itself unless something happens to compress it. Various mechanisms can bring this about, one of which is a shock wave from an explosive event such as a supernova. Thus, in a singular event sometimes called the "bing bang," our solar system may have been born as the direct consequence of the death of a nearby star. We derive from this event not only some minor anomalies in the abundance of nuclear isotopes, but also the very existence of the solar system. It is interesting to note that, even in an evolutionary origin as discussed in this chapter, the solar system has been strongly influenced by what can only be described as catastrophic events.

throughout space, but instead is concentrated here and there in large amorphous regions called interstellar clouds. Some of the clouds are sufficiently dense to block out starlight and appear as dark regions in photographs of the sky (Fig. 10.5). Even in these areas of relatively high concentration, the material is so rarefied that a quantity of mass equal to that of the Sun is spread over a volume up to 10 or more light-years across.

The composition of the interstellar material is ap-

parently quite uniform, consisting of about the same mixture of elements as the Sun. Hence the Sun and other stars are born with a standard composition of about 73 percent hydrogen (by mass), about 25 percent helium, and just traces of all the other elements. The planets must also have formed out of material with this composition, yet, as we have seen, their present makeup is quite different from this, especially in the case of the terrestrial planets.

Somehow the extended, tenuous gas in interstellar

TABLE 10.3   Timetable for Solar System Formation

Approximate Time	Event
0	Collapse of interstellar cloud; disk formation
400,000 years	Central condensation becomes hot enough for pressure to balance gravity; matter still falling in
1 million years	Second collapse stage is followed by new nearly stable phase, with heating by slow gravitational contraction; central object is now called the protosun; infalling gas and dust shroud protosun from exterior view
1–10 million years	Planetesimals form and coalesce to form planets
1–100 million years	T Tauri wind
100 million years	Start of nuclear reactions in core
0.1–1 billion years	Magnetic braking slows Sun's rotation; major cratering occurs throughout solar system

Figure 10.5   Regions of star formation.   At the left is the Orion nebula, in the "sword" of the constellation Orion, and at the right is the region near the M supergiant Antares (bright object at right center). From infrared observations, we know that newly formed stars lie in both regions within the dark, dusty clouds seen here. (ROE/AAT Board; photos by D. Malin)

space had to become concentrated in a very small volume on its way to becoming a star. Gravitational forces were responsible for pulling the cloud together, but there must have been an initial push or a chance condensation to get the process started. (One possibility is that a shock wave from a violent event such as a supernova explosion caused the presolar cloud to begin its collapse; see Chapter 14 for details.) Once the cloud began falling in on itself, the rest of the process leading to the formation of the Sun was in-evitable, dictated by physical laws.

A combination of theory and observation of the formation of other stars tells us how the collapse proceeded. The innermost portion of the cloud fell in on itself very quickly, leaving much of the outer material still suspended about the center. The rotation of the cloud sped up as its size diminished, and if the cloud had a magnetic field to begin with (as most do), the field was intensified in the central part as a result of the condensation.

The core of the cloud began to heat up from the energy of impact as the material fell in. Eventually it began to glow, first at infrared wavelengths, and finally, after a prolonged period of gradual shrinking, at visible wavelengths. During this period of slow contraction, the central object is called the **protosun;** other stars in this stage are known as **protostars.** Nuclear reactions began in the center when the temperature and pressure were sufficiently high, and the Sun began its long lifetime as a star, powered by these reactions.

The steps in the formation of the Sun described up to this point are fairly well understood and have been observed to be taking place today in many cloudy regions of the galaxy. In a number of stellar nurseries, infrared sources are found embedded inside dark interstellar clouds, indicating that newly formed stars are hidden there, still heating up. The details of planet formation are sketchier, however, because it is difficult to observe the process as it occurs. All that we know about it we have had to infer from observations of the end product: the present-day solar system.

As the central part of the interstellar cloud collapsed to form the Sun, the outer portions were forced into a disk shape by rotation, just as Kant had argued. At this stage there was an embryonic Sun surrounded by a flattened, rotating cloud called the **solar nebula** (Fig. 10.6). The inner portions of the nebula were hot, but the outer regions were quite cold.

Throughout the solar nebula, the first solid particles began to form, probably by the growth of the interstellar grains that were mixed in with the gas. For every element or compound, there is a combination of temperature and pressure at which it "freezes out" of the gaseous form, just as frost forms on a cold night on Earth. The **volatile elements** condense only at very low temperatures, so these materials tended to stay in gaseous form in the inner portions of the solar nebula but condensed to form ices in the outer portions. The elements that condense easily, even at high temperatures, are called **refractory elements,** and they formed the first solid material in the warm inner portions of the solar nebula. This material therefore consisted of rocky debris containing only low abundances of the volatile species.

In due course, rather substantial objects built up that resembled asteroids in size and composition; these are referred to as **planetesimals.** Mathematical analysis of the process by which particles orbiting the young Sun collided and built up their sizes shows that the large planetesimals grew rather quickly (in only a few thousand years), and that most of the material in the disk would have gone into large planetesimals. Thus a stage was reached in which collisions between large bodies were likely. Because the planetesimals were all orbiting the Sun in a disk, their rel-

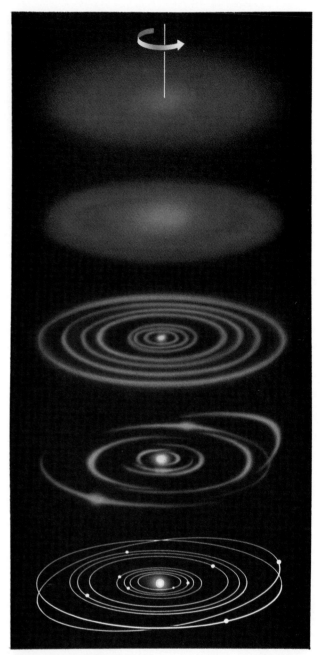

Figure 10.6   Steps in the Sun's formation.   An interstellar cloud, initially very extended and rotating very slowly, collapses under its own gravitation. This happens most quickly at the center. As the collapse occurs, the internal temperature rises and the rotation rate increases. Eventually the central condensation becomes hot and dense enough to be a star, with nuclear reactions in its core.

ative speeds would have been rather small; therefore, when collisions occurred, the colliding bodies would stick together. Even at low speeds (perhaps 10 km/sec, typically), enough energy would be released in a collision between large planetesimals to partially vaporize or melt both objects. This would help them merge into a single spherical body and would also aid the process of differentiation, which requires that a body be molten. In most cases, the planet that was formed in this way rotated in the same direction as the overall rotation of the disk, but it is possible that an off-center collision could alter the direction of spin. This may have happened during the formation of Venus and Uranus, accounting for the slow retrograde rotation of Venus and the large tilt of Uranus. As we learned in Chapter 6, many scientists now think that the Earth's Moon formed as the result of a grazing collision between the young Earth and a Mars-sized planetesimal. Recently some have even suggested that Mercury was involved in such a collision during its formation.

The scenario just described apparently applies to the terrestrial planets, which therefore seem to have formed in two stages: (1) the condensation of refractory elements, leading to the development of planetesimals; and (2) the accretion of the planetesimals to form planets. The low quantity of volatile elements that characterizes the terrestrial planets was already established when the planets formed, and then was exaggerated by the release of volatile gases that occurred during their early histories, when they underwent molten periods.

There is more uncertainty about the sequence of events that led to the formation of the outer planets. These planets contain a much higher proportion of volatile gases, which would naturally be expected because of the lower temperatures in the outer portions of the solar nebula. Planetesimals that formed there would have contained higher relative abundances of volatiles and would thus have been partly rock and partly ice, much like the large satellites of the giant planets or Pluto.

The most recent theory of the formation of the giant planets suggests that their cores formed from the coalescence of planetesimals, just as the terrestrial planets formed. In the outer solar system, however, where the temperatures were much lower than in the inner portions of the system, the solid objects that formed from the coalescence of planetesimals were able to gravitationally trap gas from large volumes in their vicinities. These planets therefore accreted extended, gaseous atmospheres, including the very light-weight and volatile gases hydrogen and helium. Tidal forces from the Sun played an important role in this scenario: for the inner planets, these forces helped prevent the accretion of gases from the surrounding nebula, but in the outer solar system, farther from the Sun, the tidal forces were weaker and did not prevent the giant planets from trapping gases.

It is thought that the outer planets were able to gain mass very rapidly due to a runaway effect created by their own gravitation. Once a massive body began to grow by accretion of planetesimals, its gravitational field would tend to pull in additional planetesimals, which would merge, adding mass to the original plan-

etary body and further enhancing its gravitational attraction for new material. Thus, once a major planet began to form, it would have grown very rapidly, quickly providing it enough mass to disrupt the remaining solar nebula. In other words, Jupiter and Saturn quickly reached sufficient mass to disrupt the asteroid belt and to eject cometary bodies to the Oort cloud. Current models suggest that Uranus and Neptune formed somewhat later (perhaps 100 million years) than Jupiter and Saturn. Because the two giant planets had already trapped much of the volatile gas in their vicinity, less was available for Uranus and Neptune, which consequently ended up with smaller masses and higher proportions of heavy elements.

This view is prompted in part by the extensive ring and satellite systems of the outer planets, which resemble miniature solar nebulae. The suggestion is that these planets, in accreting gas from their surroundings, formed disks much like the solar nebula itself. Gravitational and rotational effects would have caused the material falling in to form a disk in the equatorial plane of each planet, and then instabilities in the disk would have caused planetesimals to form, leading to the formation of satellites (Fig. 10.7). Ring systems formed wherever the disks extended closer to the parent planet than the Roche limit, where tidal forces prevented the formation of large satellites. The rapid rotation rates of the giant planets are explained in this view by the fact that angular momentum conservation would force them to spin more rapidly as the disks contracted (and by the fact that no magnetic braking process would have acted, in contrast with the early Sun).

It is interesting that Uranus, despite its large (98°) tilt, has its satellites and rings lying in its equatorial plane. This indicates that the planet accumulated its disk of icy debris after the event that caused its highly tilted orientation, and this scenario is consistent with

Figure 10.7   Formation of a giant outer planet.   Here a rotating disk has formed around a condensation in the outer solar system. Lumps in the disk grow to become satellites, except in the innermost portion where tidal forces prevent this, and a ring system forms instead. The entire system of planet, rings, and moons orbits the Sun.

the picture of outer planet formation just described. In this view, material trapped by Uranus would be forced into a disk in the equatorial plane of the planet, regardless of the orientation of that plane. Hence the tilt of Uranus must have been established before it trapped the material that was to form its satellites and rings.

The asteroids probably formed as planetesimals, similar to those that eventually created the terrestrial planets, but they were prevented from coalescing into a planet by the gravitational effects of Jupiter, as we noted in Chapter 9. Jupiter's gravity would also have altered the orbits of many of these planetesimals, causing them to be ejected from the solar system. Hence the total mass of the asteroids is small. The comets probably formed farther out, through not at the distance of the Oort cloud, where they now reside. At such a great distance (up to 30,000 AU) from the center of the solar nebula, it is doubtful that any condensation could have occurred, because the density would have been too low. Therefore the comets are believed to have condensed at intermediate distances, probably near the orbits of Uranus and Neptune, and then were forced out to their present great distances by the gravitational effects of the major planets. Both Jupiter and Saturn, lying closer to the Sun, would have forced the cometary bodies farther out by speeding them up a little each time they passed by. Eventually most of the comets retreated to a great distance where they formed the Oort cloud. The Kuiper belt, a disk of cometary bodies lying between 30 and 50 AU, probably represents leftover planetesimals that were not ejected to greater distances, but have remained in the region where they formed. The Kuiper belt is thought to be the source of periodic comets, which have orbits closely aligned with the ecliptic plane.

Once the planets and satellites were formed, the solar system was nearly in its present state, except that the solar nebula had not totally dissipated. A lot of gas and dust was still swirling around the sun, along with numerous planetesimals that had not yet accreted onto planets. During this time the Sun's magnetic field, pulling at ionized gas in the nebula, was very effective in slowing the solar rotation through the magnetic braking process that we have already described. Meanwhile, the remaining planetesimals were orbiting about, now and then crashing into the planets and their satellites. Most of the cratering in the solar system occurred during this period, in the first billion years.

The leftover gas in the solar nebula was dispersed rather early, when the infant Sun underwent a period of violent activity, developing a strong wind that swept the gas and tiny dust particles out into space (Fig. 10.8). Only the larger solid bodies, the newly formed planets and the planetesimals, were left be-

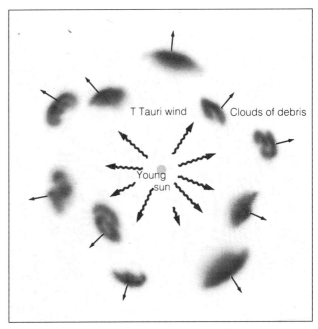

Figure 10.8   The T Tauri phase.   A star like the sun is thought to go through a phase very early in its lifetime during which it violently ejects material in a high-velocity wind. This wind sweeps away matter left over from the formation process.

hind. This wind phase has been observed in other newly formed objects called **T Tauri** stars and is apparently a natural stage in the development of a star like the Sun. Because of rotational forces, the wind escaped primarily in directions perpendicular to the equatorial plane of the young Sun, creating a bipolar flow similar to those seen today coming from newly formed stars.

Once the majority of the remaining interplanetary debris was eliminated, either by the T Tauri wind or by accretion onto planetary bodies, the solar system looked much as it does today. Each planet went its own way, either continuing to evolve geologically, as most of the terrestrial planets have, or remaining perpetually frozen in nearly its original condition, as the outer planets apparently have.

MINIATURE SOLAR SYSTEMS: RINGS AND MOONS

We have seen that although the giant planets and the terrestrial bodies began to form in a similar fashion, conditions in the outer solar system caused some very significant differences as well. Light-weight elements were able to condense in the outer solar system, and gravitational trapping allowed the giant planets to gain enormous quantities of mass, becoming the gaseous, cold worlds we see today. Along the way, the giant planets acquired disks of material that produced

ASTRONOMICAL INSIGHT
Catastrophe Revisited: Violent Origins of the Planets

In the text we reviewed some early theories that hypothesized that the planets formed from material pulled out of the Sun by some catastrophic event such as a near-collision with another star. These theories are now considered untenable, largely because scientists have demonstrated convincingly that solar matter could not and did not condense to form the planets. Furthermore, the need to invoke such theories was reduced when the solar wind was discovered, and it was realized that the Sun's slow spin could be explained by magnetic braking. With this discovery, the so-called evolutionary theory of solar system formation became widely accepted, because it appeared to explain most of the properties of the system quite well. Some nagging problems remained, however, and today scientists are modifying the evolutionary theory by adding some catastrophes back in.

The most noticeable departures from the expectations of the evolutionary theory are the unusual spins of Venus, Uranus, and Pluto. Venus spins very slowly in the backward, or retrograde, direction, while Uranus is tilted so far over that its axis lies almost in the plane of its orbit around the Sun. Both planets are quite anomalous because the evolutionary theory would predict that each planet's axis would be nearly perpendicular to its orbital plane, and that the spin of each planet would be in the prograde, or forward, direction, the same direction as its orbital motion about the Sun. Thus the strange properties of both Venus and Uranus have long been attributed to unusual events that might have occurred during their formation, such as a collision with a large planetesimal.

Now similar collisions are being invoked to explain much more subtle anomalies in other planets. The Moon is quite unusual among inner solar system bodies, being the only major satellite. Furthermore, it has many properties that have proven difficult to explain: it carries far more angular momentum than can be understood on the basis of standard models for the formation of the Earth, and it has chemical and isotopic anomalies that distinguish its material from that of the Earth's outer layers. Hence, theories hypothesizing that the Moon formed by splitting off from the Earth, or that the two bodies formed together, are faced with fatal difficulties. The large angular momentum of the Moon could be explained if the Moon were somehow captured by the Earth, but this theory has other problems in that such a capture is either difficult or impossible. (Models for the formation of the Moon are reviewed in Chapter 6).

The most recent suggestion for the formation of the Moon calls for a catastrophic origin. It is now thought that the Moon may be the remnant of a very large planetesimal, perhaps the size of Mars or larger, that underwent a grazing collision with the young Earth, vaporizing its own mantle and part of the Earth's outer layers. The debris from the collision would have formed a disk around the Earth, which subsequently condensed to form the Moon. This theory is still quite new and therefore subject to possible revision or rejection, but so far it appears to explain the Moon's formation better than any of the others that have been proposed.

The planet Mercury also presents anomalies that have been difficult to explain. Foremost among them are its high density and detectable magnetic field, which have led to the conclusion that Mercury's nickel-iron core may be much larger (relative to its overall diameter) than for any other terrestrial planet (see Fig. 7.31, where a model of Mercury's interior is shown). Although Mercury formed closer to the center of the solar system than the other planets and therefore should have had a lower abundance of lightweight volatile elements to begin with, scientists have had difficulty explaining how Mercury could be so deficient in these elements. Now a new theory suggests that a major impact that occurred just as Mercury was forming could explain the anomaly. This impact, involving the young Mercury and a large planetesimal, would have vaporized much of the material of Mercury's outer layers, leaving behind a very dense planet with an anomalously large core and thin outer layers.

There is little dispute that the early solar system underwent a period when impacts were important. The scarred surface of the Moon and other satellites alone is sufficient to establish that there must have been a time when bombardment from space was far more common than it is today. The evolutionary theory of solar system formation has always included an early period when planetesimals formed in the disk of solid material that orbited the young Sun; what is new today is the realization that during the age of planetesimals some very catastrophic collisions between large bodies must have occurred. These catastrophes have molded the planets into the form they have today.

rings and satellites. In this section, we discuss the formation and evolution of these disks and the rings and moons in further detail.

Satellites of the Giant Planets

The satellites of Jupiter, Saturn, Uranus, and Neptune seem to fall into two general classes: those that follow nearly circular orbits lying in the equatorial plane of the parent planet; and those that have orbital peculiarities such as high tilts to the equatorial plane or retrograde directions. All are composed largely of ice, although many have densities greater than 1 gram/cm^3, indicating that rocky material is mixed in to varying degrees. Pluto may be more like one of the giant satellites than a proper planet; note, for example, the many similarities between Pluto and Triton.

We have seen that each giant planet is thought to have been surrounded by a disk of residual material when it formed. This disk consisted of small bodies that could coalesce to form satellites. Thus the formation of satellites is not difficult to understand, given the prior formation of the disks. The satellites formed this way would have circular orbits lying in the equatorial plane of the parent planet, placing them in the first of the two categories discussed above.

The satellites with irregular orbits may have had a different origin. These may be planetesimals that have been captured gravitationally; such captures would lead to orbits that do not necessarily conform to the normal pattern. It was pointed out earlier (Chapter 6) that capture of a free body is not easy and in fact cannot happen unless there is some way for the incoming body to lose some of its kinetic energy. Energy can be lost through a near-collision with a third body or through friction if the incoming body encounters an atmosphere or a medium of fine particles, such as an equatorial disk. Most of the satellites that have irregular orbits are small moons such as Phoebe of Saturn or Nereid of Neptune, but at least one giant satellite also falls into this category. Triton, a very large satellite having a thin atmosphere and an assortment of geological scars on its surface, follows a retrograde orbit around Neptune and may have been captured.

If Triton is indeed a captured satellite, it may resemble the planetesimals that had built up in the outer solar system at late stages in the formation of the planets. Hence this giant moon may represent a sample of the intermediate stages in the creation of planets such as Jupiter, Saturn, Uranus, and Neptune. Some other examples of this stage have been found. In 1977, a Sun-orbiting body in the outer solar system was discovered, and for a while astronomers speculated that it might be a new planet. Further analysis showed that it was not large enough to be a planet,

and this body, named Chiron, came to be thought of as an asteroidlike or cometary body much farther from the Sun than the main asteroid belt. There may be other such bodies in the outer solar system, yet to be discovered. Further, as already noted, Pluto is far more similar to the giant satellites (Triton in particular) than to any of the planets and may also represent a planetesimal, a leftover from the era of planet formation.

The larger satellites are spherical, indicating that gravitational compression, rather than internal rigidity, dominates their structure. Their inner conditions are difficult to analyze, but probably most have a central core region, where most of the rocky material resides, and outer mantles and crusts consisting of mixtures of ice and rock (Fig. 10.9).

Geological activity has proceeded in varying degrees, for reasons that are not always well understood. Tidal stresses can account for many of the contrasts from one satellite to another, as we have seen in the case of the four Galilean moons of Jupiter. In that instance, we saw that the amount of heating and hence of geological activity increases from the outermost, Callisto, to the innermost, Io. Triton is active geologically, with evidence of slushy outflows on its surface as well as active geyser eruptions in which plumes of nitrogen or methane gas are released. Enceladus, the very bright satellite of Saturn, may also

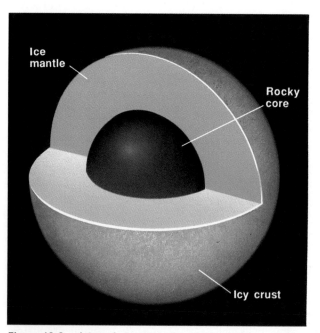

Figure 10.9   Internal structure of a giant satellite. Substantial differentiation has occurred, so the heavy elements have congregated to form a rocky core. The lightweight elements, in the form of ice, constitute the mantle and crust (if there is significant internal heating, because of tidal stress, for example, the mantle may be partially liquid water). In a typical case, the mass of the satellite is about half rock and half ice.

be undergoing volcanic activity in which water is released, re-coating the surface with ice.

Water, particularly in the form of ice, plays a major role in the geology of the outer satellites. The degree to which a moon undergoes internal convection, hence the movement of crustal plates at its surface, depends on whether the ice in the interior is heated enough to melt and flow. Among the Galilean satellites of Jupiter, this apparently happened inside Ganymede, which has old surface fractures due to plate motion (Fig. 10.10), but not in Callisto, where no crustal shifting has apparently ever taken place. The inner moons, Europa and Io, probably never had as much water in the first place (although Europa has a thin water or ice mantle), since the inner part of the disk that surrounded Jupiter when it formed is thought to have been too hot for water to condense. These two satellites are heated enough due to tidal stresses that their interiors are sufficiently elastic to undergo convective flows, even in the absence of large quantities of ice.

Other satellites undergoing ice-related geological activity include Enceladus of Saturn, as already mentioned, Triton of Neptune, possibly Titan of Saturn, and perhaps Pluto (which, as we have noted, closely resembles the giant, icy satellites of the outer planets). The source of internal heating for these objects is not well understood; Triton may undergo suf-

ficient tidal stress due to Neptune and may have some residual heating due to radioactive elements in its interior. It appears unlikely that sufficient tidal stress is taking place on Enceladus, yet it is clear that water is occasionally being ejected from the interior.

Ice may play another important role related to the formation of atmospheres on some of the large satellites, as well as on Pluto. Ice that is very cold (below 135 K) can trap large quantities of other gases within its crystal structure. If the ice is then heated to a temperature above 135 K, the trapped gas escapes. It has been suggested that this mechanism helped to form the atmospheres of Titan, Triton, and Pluto. If these bodies were cold enough at the time of their formation to trap gases in their icy material, then subsequent heating (by tidal stresses or internal radioactivity) could have released the gases that had been trapped, forming the atmospheres of these bodies.

One additional mechanism may be affecting the geological evolution of the moons of the giant planets, a mechanism that has only recently been suggested in the light of some of the bizarre findings of the *Voyager* probes. Some of the moons show perplexing varieties and degrees of geological activity, which is difficult to explain by the conventional processes such as tidal stress. Miranda, the innermost of the five large moons of Uranus, is the best example of this; its surface exhibits a variety of unusual and dramatic features that must represent recent and energetic activity (Fig. 10.11). At the same time these features were being discovered, a growing body of

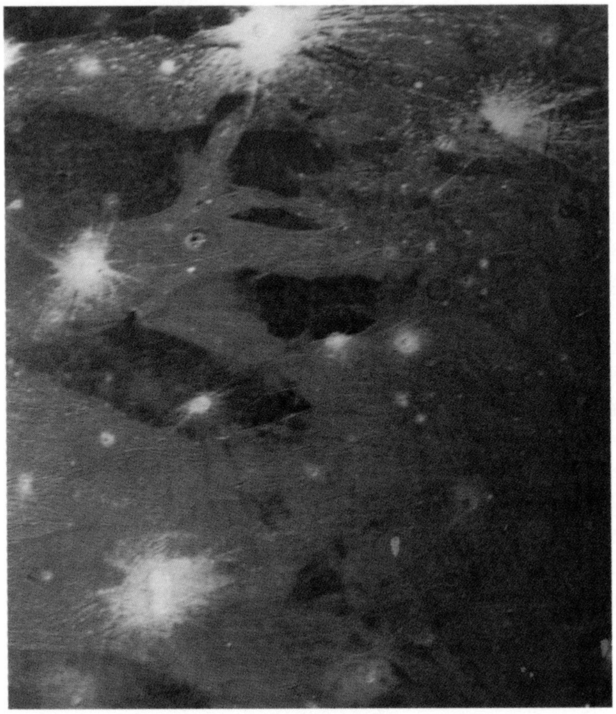

Figure 10.10 Features on Ganymede. This close-up of the surface shows a complex pattern of grooves and ridges, which were probably created by deformations of an icy crust. Some impact craters are also visible. (NASA)

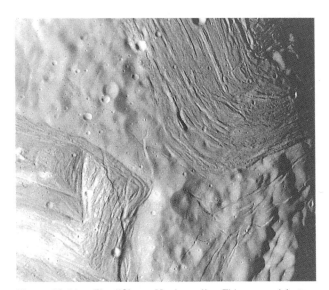

Figure 10.11 The "Circus Maximus." This unusual feature (right) on Miranda consists of an elevated inner portion surrounded by sharply sloping terraces; its name derives from its resemblance to a Roman racetrack. The cause of this feature is unknown, but it would require some source of uplift in the interior. At the lower left is a similar, chevron-shaped feature. (NASA/JPL)

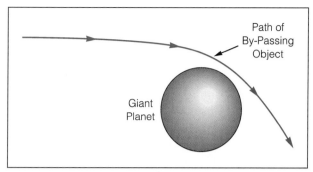

Figure 10.12   Gravitational focusing.   The immense gravitational field of a giant planet causes objects such as asteroids, comets, or large meteoroids (see Chapter 9) to veer in close to the planet. This focusing effect increases the likelihood of collisions between such bypassing bodies and satellites orbiting the giant planet.

Figure 10.13   Mimas.   This satellite shows many impact craters, including a very large one that is about one-fourth the diameter of the satellite itself. This moon is probably composed chiefly of ice, as are several of the other satellites of Saturn. (NASA)

evidence was mounting to indicate that planetary ring systems are transient, changing with time and somehow being replenished occasionally. Putting these observations together has led to the suggestion that the satellites of the giant planets, especially the inner satellites, are occasionally destroyed by major impacts. If the fractured moon is outside the Roche limit, it can soon re-form, as appears to have happened in the case of Miranda; if the debris left over from the collision is too close to the parent planet, then tidal forces prevent reaccumulation of the material, and a new ring results.

The innermost moons are most likely to undergo such major collisions because the immense gravity of the parent planet attracts incoming meteoroids, pulling them close in. Each giant planet can be viewed as a gravitational trap, altering the paths of bypassing objects so that they speed up and pass by at very close approaches (Fig. 10.12). This **gravitational focusing** process may account for the many gigantic craters found on inner satellites such as Mimas of Saturn (Fig. 10.13), for the strange geology of Miranda, and for the very existence of the ring systems.

Rings and Moonlets

We now know that all four of the giant planets have rings. The detailed characteristics of these rings vary from planet to planet, ranging from the broad, bright rings of Saturn to the single thin, dim ring of Jupiter. The differences in appearance are in part due to the varying sizes of the ring particles; the very small particles in the rings of Saturn are much more effective at reflecting light in the backward direction than are the forward-scattering large particles that dominate the rings of the other planets.

In Chapter 8, we suggested that rings originate in the disks of debris that surrounded each giant planet when it formed. Close to the parent planet (inside the Roche limit), tidal forces prevent free particles from merging together, and they remain forever as a thin sheet or series of belts confined by gravity and rotational forces to the equatorial plane of the parent planet. In this picture, the rings are primordial features, left over from the time of planetary formation some 4.5 billion years ago.

Recent evidence suggests, however, that this picture needs to be modified. It is likely that rings started in the way just described, but now it is apparent that they are dynamic, evolving entities, occasionally replenished with new injections of particles. The evidence for this first became apparent after the *Voyager* encounters with Saturn, when it was discovered that some of the structure in the ring system is unstable, unable to persist for very long times. The noncircular rings (Fig. 10.14) and one that is irregularly shaped (Fig. 10.15) cannot be maintained indefinitely; these features must be varying with time. At Uranus, *Voyager 2* found a broad inner belt or ring that is losing its material to friction and infall toward Uranus and must either be relatively new or receiving new material from some source. The rings of Neptune include some that are noncircular and some containing clumpy portions, again suggesting that variations with time are occurring. The very fine structure found in the rings of Saturn (Fig. 10.16) is by itself convincing evidence that the rings must change with time, because these structures are probably caused by variable phenomena, as discussed shortly.

Figure 10.14 Asymmetries in the rings. This composite shows that Saturn's rings are not perfectly circular; here the thin ring within the dark gap is thinner on one side of the planet than on the other and slightly displaced. (NASA)

Figure 10.15 The F ring. Here we see the unusual braided appearance of this thin, outlying ring. The cause of its asymmetric shape is not well understood. (NASA)

Figure 10.16 The rings of Saturn. This *Voyager* image reveals some of the fantastically complex structure of the ring system. (NASA)

The idea of fragmented satellites again comes into our discussion. If, as we have already mentioned, inner satellites of the giant planets are susceptible to destruction from impacts of large meteoroids, then such collisions can be a source of new ring particles. A satellite orbiting close to or within the Roche limit, if destroyed, would not easily be able to re-form because of tidal forces and would remain a cloud of debris orbiting the planet. In this way, a ring system can be replenished from time to time.

The *Voyager* images revealed an astounding array of detailed structure in the rings of Saturn (see Fig. 10.16). Major gaps had been seen from Earth, but the vast and intricate detail of the many tiny ringlets and small gaps came as a surprise. Many mechanisms have been proposed to explain the fine structure.

First, as already discussed, orbital resonance is probably responsible for some of the major features. Most notable among these is the Cassini division, the prominent dark gap between the two major outer rings (this "gap" is not truly empty, however; *Voyager* images reveal several dim, forward-scattering rings there). The Cassini division lies at the distance from Saturn where ring particles would have orbital periods exactly equal to half the period of Mimas. Hence, even though the mechanism is not fully understood, it appears that orbital resonance with Mimas is responsible for maintaining this gap.

In addition to orbital resonance, at least three other mechanisms have been invoked to explain the fine structure in the rings of Saturn. One of these is also caused by gravitational effects of Saturn's moons. The clue to this mechanism is the behavior of the two small satellites, discovered by *Voyager 1*, that orbit just inside and just outside the F ring (Fig. 10.17). The gravitational interaction between the ring particles and these satellites acts to keep the particles confined to orbits between them (Fig. 10.18). The moon just outside the ring, according to Kepler's laws, moves a little more slowly than the ring particles. Whenever a ring particle overtakes this satellite at a close distance, the particle loses energy due to the gravitational tug of the satellite and drops to a slightly lower orbit. Meanwhile the satellite on the inside of the ring slowly overtakes the ring particles, and when it passes close by one, the particle gains energy and moves out to a higher orbit. In this way the two satellites ensure that the ring particles stay in orbit between them, because any particle that strays too close to either one will have its orbit altered in such a way that it moves back into the ring. For this reason the two moons are called "shepherd" satellites, because they behave like sheepdogs, keeping their flock in line by nipping at their heels.

Voyager 2 data showed that another mechanism, suspected before *Voyager* missions, was also influencing the structure of the rings. Mathematical calcula-

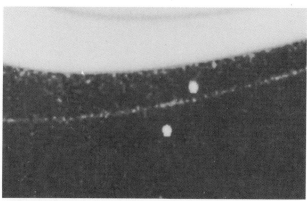

Figure 10.17   The shepherd satellites and the F ring.   The two small satellites (Pandora and Prometheus) that maintain the F ring through gravitational interaction are both visible in this image. (NASA)

Figure 10.18   The action of shepherd satellites.   This figure illustrates how a pair of small moons can keep particles trapped in orbit between them. The inner shepherd satellite overtakes the particles, accelerating them if they fall inward and thus moving them out. Similarly, the outer shepherd slows particles that move too far out, so that they fall back in. (NASA/JPL)

tions show that, in any flattened disk system such as Saturn's rings, small gravitational forces can create a spiral wave pattern, something like the grooves on a phonograph record. The waves take the form of alternating dense and rarefied regions, and the entire spiral pattern rotates about the center of the system. Similar spiral density waves have long been thought to be responsible for the spiral arm structure of galaxies like our own, and so the *Voyager* data have shown that small-scale structures in the rings of Saturn bear similarities to enormous structures in galaxies like our own.

Figure 10.19   Very fine structure in the rings.   This is a synthetic image of a small section of the ring system, reconstructed from stellar occultation data obtained by the photopolarimeter experiment on *Voyager 2*. The finest details visible here are only about 100 meters in size, whereas the best *Voyager* camera images have a resolution of about 10 kilometers. Data like these have established that spiral density waves play a role in shaping the rings. (NASA)

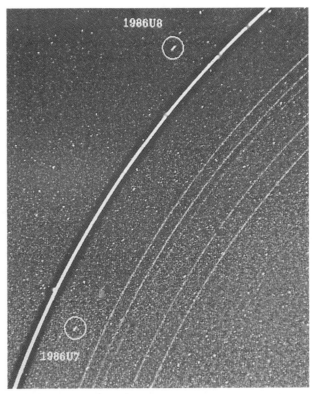

Figure 10.20   Shepherd satellites.   Here we see two small moons on either side of a thin ring around Uranus. The gravitational effects of these moons keep the ring confined. (NASA/JPL)

Scientists calculated that the spacing between spiral density waves should increase steadily with distance from the center. The photopolarimeter data from *Voyager 2* (Fig. 10.19) revealed that the expected increase in spacing between ringlets occurs in several portions of the ring system, confirming that at least some of the ring structure is created by spiral density waves. The ringlets that are formed by this mechanism are not separate, circular structures, but instead are part of a continuous, tightly wound spiral pattern that slowly rotates. Individual ring particles, by contrast, follow elliptical orbits and alternately pass through the ringlets and the spaces between. (It is important here to distinguish between the motion of the wave and the motions of the particles, which do not move with it; a reasonable analogy would be a cork bobbing on water waves, but not moving along with them.) The reason more particles are in the ringlets than between them at any given moment is that they move slowly as they pass through the density waves and tend to congregate there, just like a traffic jam that builds up at a point where the flow of cars is slowed down.

The gravitational disturbances that create the density waves in portions of Saturn's ring system are caused by satellites. The strongest disturbances occur at resonance points where the particles have orbital periods that are a simple fraction of the periods of satellites orbiting at greater distances. These are also the locations of gaps in the rings in some cases, as explained earlier, so we sometimes find that a gap lies just at the inner edge of a series of ringlets that are created and maintained by a spiral density wave.

Although the ring system of Saturn is the most extensive and complex, those of the other giant planets exhibit some of the same features and are probably governed by many of the same processes. For example, the rings of Uranus tend to be thin, which indicates that shepherd satellites are responsible for confining them. Indeed some of the shepherd satellites were found in the *Voyager 2* images (Fig. 10.20), and the presence of others is suspected. Asymmetric ring shapes were found for both Uranus and Neptune, indicating that perturbations due to moons are acting on the rings, as well as contributing to the hypothesis that the rings are changing with time.

In all the ring systems, structures were found that cannot be permanent features, indicating that all the systems are evolving. If a new *Voyager*-like mission is launched in the distant future, the images sent back might look very different from those we have today.

Figure 10.21   The dark spokes.   At the left the spokes appear bright because they scatter light in the forward direction, and they were photographed looking toward the Sun. At the right they are viewed looking away from the Sun, so they appear dark. (NASA)

Possibly even the *Cassini* probe, planned to orbit and observe the ringed planet for several years, will find some minor differences in the structure of Saturn's rings.

The *Voyager* probes found one additional new phenomenon, which is associated only with the rings of Saturn. Several dark, tenuous features were seen in the rings, pointing away from Saturn (Fig. 10.21). It appears that these so-called "dark spokes" lie above and below the plane of the rings, rather than in the plane. This suggests that the forces holding them there are not gravitational, since gravity (combined with the rotation of the system) acts to keep the rings tightly confined to the equatorial plane of Saturn. The spokes are highly variable; one may appear very quickly (in a matter of minutes), then gradually dissipate over several hours. It is suspected that the spokes consist of very tiny particles that are electrically charged; as a result, the magnetic field of Saturn can exert forces on them to suspend them above the ring plane. The major question is where these tiny particles come from; apparently, they are injected into the ring system every now and then. One suggestion is that each spoke's appearance is triggered by the impact of a small meteoroid (a few centimeters in diameter) on the ring particles, pulverizing some of them into a fine dust. The dust particles accumulate electrical charge from free electrons in the vicinity and are then spread out into the observed spoke-like pattern as the planetary magnetic field acts on them.

The many discoveries made by the *Voyager* probes about the moons and rings of the outer planets have initiated a revolution in our understanding of the outer solar system; in revealing their complexities and their continuing evolution, the *Voyagers* have taught us that our solar system is still changing. Next we consider the possibility that there may be other solar systems, just as complex, elsewhere in the galaxy and the universe.

ARE THERE OTHER SOLAR SYSTEMS?

As far as we know, the entire process by which our planetary system formed was a by-product of the Sun's formation, requiring no singular, catastrophic event. We might expect, therefore, that the same processes have occurred countless times as other stars were born, and therefore there may be many planetary systems in the galaxy. Even if we rule out double- and multiple-star systems, in which perhaps all the material was either ejected or included in the stars, a vast number of candidates remains.

Detecting planets orbiting distant stars is not easy. For a long time astronomers suspected one particular star of having a planet. The gravitational pull of a very massive planet would cause its parent star to wobble a little in position as the planet orbited it. A nearby star called Barnard's star was thought for a long time to be wobbling in such a way, but a recent reexamination of the data found no clear-cut indication of such a motion, and at the moment the case for the existence of a massive planet circling Barnard's star is not well established.

Another technique is to search carefully for variations in the velocities of stars suspected of having

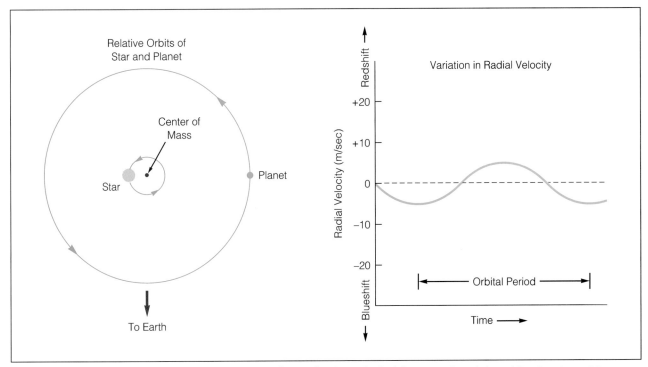

Figure 10.22   Using radial velocity measurements to detect planets.   At the left we see the relative orbits of a star and its planet about their common center of mass (here the relative size of the star's orbit has been greatly exaggerated). As the star orbits the center of mass, its spectrum undergoes small Doppler shifts (alternately blueshifts and redshifts) as the star approaches us and then recedes. At the right is a graph showing how the Doppler shift (radial velocity; see Chapter 4) changes with time. Note that the size of the velocity shifts is very small, only a few meters per second in this hypothetical case. Such small shifts are very difficult to detect and measure.

planets. Here the Doppler shift is used to measure the speed of the star relative to Earth (see Chapter 4). Small, regular changes in this velocity could indicate that the star is orbiting a small companion (Fig. 10.22). Very recently, a study using this technique revealed several nearby stars with dim companions, some of which may be massive planets. Planets as small as the Earth are much more difficult to detect in this way, because they would cause only minute variations in the velocity of their parent star.

Infrared observation shows some promise as a technique for detecting other planetary systems. This was illustrated by the *Infrared Astronomical Satellite* (*IRAS*), which mapped most of the sky in far-infrared wavelengths and found several stars that are surrounded by disk containing tiny solid particles of the sort thought to be the precursors of planetesimals (Fig. 10.23). Very recently, some of these disks have been shown to contain gaps, which may indicate the presence of planets (the gaps were detected by a very indirect technique, however, based on the infrared spectra of the disks, which showed that there was no dust emission corresponding to temperatures expected at certain distances from the center of the disk).

Figure 10.23   A star with a dust disk.   This image of the star β Pictoris, obtained with an infrared telescope on the ground, shows a disk of solid particles edge-on, much like the solar nebula. Infrared data have revealed a number of stars with surrounding material, supporting the suggestion that planetary systems may form around many stars. (NASA/JPL, courtesy of B. A. Smith)

Other kinds of new technology may also provide new information in this area. The *Hubble Space Telescope (HST)*, for example will be (when new cameras are installed) free of the distortion and fuzziness that plagues ground-based telescopes and may be able to detect bright planets that are widely separated from their parent stars. The *HST* is expected to allow astronomers to see objects comparable to Jupiter, if they orbit nearby stars. There is little hope of directly detecting smaller objects such as the terrestrial planets, however.

The fine resolution of the *HST* images will also allow much more precise measurements of stellar motions than have been achieved from the ground. Consequently tiny oscillations in the positions of stars will be more easily detected. Again, this will probably not allow the detection of small planets like the Earth, but could locate objects as large as Jupiter. The use of interferometry at ground-based observatories may also provide new hope of detecting small oscillations in star positions indicative of the presence of planets. Astronomers are now waiting eagerly for the chance to use these new instruments and techniques in a search for planets in other solar systems. It will be fascinating to learn the results in the coming years.

PERSPECTIVE

In this chapter, we have outlined a natural, evolutionary process that seems to account for all of the observed properties of the solar system. The orbital characteristics of the planets, their compositions, the Sun's slow rotation, and the nature and motions of the interplanetary material have all been explained. Many details are yet to be worked out, but the overall scenario is probably correct in essence.

We have tied together the many facets of the satellites and rings of the outer planets and have found that several important underlying mechanisms explain their complexity. One of the most important themes has been the growing realization that the ring systems are evolving, dynamic entities that change with time.

With this chapter, we have completed our examination of the solar system and are ready to move on. The next logical step is to explore the realm of the stars.

SUMMARY

1. A successful theory of the formation of the solar system must explain a number of facts including the systematics of the orbital and spin motions of the planets and satellites, the contrasts between the terrestrial and giant planets, the slow rotation of the Sun, and the existence and properties of the interplanetary bodies.

2. There are two general classes of theories: catastrophic theories and evolutionary theories.

3. Catastrophic theories require some singular event, such as a near-collision between stars, to distort the Sun by tidal forces, pulling out matter that can then condense to form the planets.

4. Evolutionary theories postulate that the planets formed as a natural by-product of the formation of the Sun. These theories, although they are simpler, were not accepted for a while because they failed to explain the slow spin of the sun.

5. When the solar wind and its creation of magnetic braking were discovered in the 1960s, the evolutionary theory became widely accepted.

6. In the modern theory, the planets formed from condensations in the solar nebula, the flattened disk of gas and dust that formed around the young Sun.

7. The terrestrial planets formed from the coalescence of planetesimals, solid, asteroidlike objects that condensed from the hot inner portions of the solar nebula.

8. The giant planets are thought to have formed as planetesimals merged to form their cores, which then trapped surrounding gas, forming their atmospheres. The extensive ring and satellite systems of these planets formed from disks of gas and solid particles in their equatorial planes.

9. Most of the moons of the giant planets probably formed by accretion in disks that were left around the planets when they formed; these moons have circular orbits lying in the equatorial plane of their parent planets. Other moons, having irregular orbits, may be captured objects left over from the era of planetesimals during planet formation.

10. Some of the large satellites are or have been geologically active, due to internal heating from tidal stresses or from radioactivity. The geological activity consists of crustal plate motions or volcanic eruptions in which ice or gas is released.

11. Some satellites show evidence that occasional large impacts shatter inner moons of the giant planets; this would account for the strange surface geology in some cases (especially Miranda of Uranus) and can also serve to replenish the rings.

12. Many kinds of evidence indicate that the ring systems of the giant planets are not stable over long periods of time, but instead must be evolving and changing. The evidence includes the existence of very fine structure in the rings, the presence of asymmetric rings, and the inference that particles are falling inward from some rings toward the parent planets.

13. It is thought likely that other planetary systems exist, orbiting other stars. Careful radial velocity measurements have begun to reveal evidence confirming this, in the form of possible detections of Jupiter-class planets circling a few nearby stars.

REVIEW QUESTIONS

1. Summarize the overall pattern of orbital and spin motions in the solar system that must be explained by any successful theory of solar system formation. Describe any exceptions to the normal pattern.

2. We already know (from Chapter 5) that Jupiter contains more mass than all the other planets combined. Now compare the orbital angular momentum of Jupiter with the total orbital angular momentum of the other planets. Assume that each planetary orbit is a circle having a radius equal to the semimajor axis given in Appendix 7, and recall that the orbital angular momentum for a circular orbit is mvr, where m is the body's mass, v is its orbital speed, and r is the radius of the orbit. Discuss your result.

3. The acceptance of evolutionary models of solar system formation was delayed considerably because of the difficulty in explaining why the Sun's rotation is so slow. Explain why this was difficult. Discuss the parallels between this slow acceptance, the difficulty early astronomers had in accepting the heliocentric model, and the long delay in the general acceptance that continental drift occurs.

4. If the solar system formed from an interstellar cloud of uniform chemical composition, explain how the terrestrial planets developed compositions that are so different from that of the Sun.

5. The total mass of the rings of Saturn is estimated to be about 3×10^{22} grams. If all the ring particles came from the destruction of a single moon made of pure ice (thus having a density of 1 gram/cm^3), how large would this satellite have been (assume it was spherical)? Does this size correspond to that of any of Saturn's moons? Which one?

6. Discuss the many ways in which the moons and rings of the outer planets interact with each other.

7. How much dimmer would Jupiter appear if it orbited the nearest star, 4.3 light-years, or 270,000 AU away? (Compare its brightness as seen from 270,000 AU with its brightness when it is closest to Earth, 4.2 AU away.) In view of your answer, discuss the feasibility of directly observing a planet like Jupiter orbiting another star.

8. Explain why it is sometimes argued that Pluto may have more in common with the large satellites of the giant planets than with other planets, either terrestrial or giant.

9. Discuss the methods that can be used to search for planets orbiting other stars. Which method do you think is most sensitive, that is, most likely to succeed?

10. Based on the information in this chapter regarding the formation of the Sun, how do you think astronomers can observe other stars in the process of formation?

ADDITIONAL READINGS

Beatty, J. K., B. O'Leary, and A. Chaikin, eds. 1990. *The new solar system.* 2d ed. Cambridge, England: Cambridge University Press.

Black, D. C. 1991. Worlds around other stars. *Scientific American* 264(1):76.

Bruning, D. 1992. Desperately seeking Jupiters. *Astronomy* 20(7):36.

Cameron, A. G. W. 1975. The origin and evolution of the solar system. *Scientific American* 233(3):32.

Crosswell, K. 1991. Does Alpha Centuuri have intelligent life? *Astronomy* 19(4):28.

Falk, S. W., and D. N. Schramm. 1979. Did the solar system start with a bang? *Sky and Telescope* 58(1):18.

Goldsmith, D., and N. Cohen. 1991. The great molecule factory in Orion. *Mercury* 20(5):148.

Head, J. W., C. A. Wood, and T. A. Mutch. 1977. Geologic evolution of the terrestrial planets. *American Scientist* 65:21.

Killian, A. M. 1989. Playing dice with the solar system. *Sky and Telescope* 78:136.

Knacke, R. 1987. Sampling the stuff of a comet. *Sky and Telescope* 73:246.

McSween, H. Y., Jr. 1989. Chondritic meteorites and the formation of the planets. *American Scientist* 77:146.

Reeves, H. 1977. The origin of the solar system. *Mercury* 7(2):7.

Sagan, C. 1975. The solar system. *Scientific American* 233(3):23.

Schild, R. 1990. A star is born. *Sky and Telescope* 80(6):592.

Schramm, D. N., and R. N. Clayton. 1978. Did a supernova trigger the formation of the solar system? *Scientific American* 237(4):98.

Stom, A. 1992. Where has Pluto's family gone? *Astronomy* 20(9):40.

Stahler, S. W. 1991. The early life of stars. *Scientific American* 265(1):48.

Tombaugh, C. W. 1991. Plates, Pluto, and Planet X. *Sky and Telescope* 81(4):360.

Waldrop, M. 1985. First sightings: Solar systems elsewhere. *Science 85* 6(5):26.

Wetherill, G. W. 1981. The formation of the earth from planetesimals. *Scientific American* 244(6):162.

To help illustrate the challenge of detecting planets orbiting other stars, you can perform some simple experiments involving the solar system. The calculations you carry out will show how difficult it would be to detect even the giant Jupiter from the distance of a nearby star.

Recall that two of the methods mentioned in the text for seeking other planetary systems are looking for the slight wobbling motion of the parent star as it orbits the center of mass of its system of planets, or looking for the Doppler shifts in the spectrum of the parent star that are caused by its orbital motion. The first method requires measurements of very slight shifts in the positions of stars; the second calls for very accurate measurements of the radial velocities of stars.

To simplify the job, assume that the solar system consists only of the Sun and Jupiter, and calculate the amount the Sun is displaced from side to side as it orbits the center of mass between it and Jupiter. First, calculate the position of the center of mass by computing the ratio of the masses of the two bodies (use the values given in appendixes) and noting that the ratio of distances from the center of mass is the inverse of the ratio of masses. In other words, if the Sun had twice the mass of Jupiter, then Jupiter would be twice as far from the center of mass; in this case the center of mass would be one-third of the distance from the Sun to Jupiter. Now do the calculation using the correct values for the two masses. How far is the center of mass from the center of the Sun? Is it inside or outside the Sun (i.e., is your value less than or greater than the Sun's radius)?

The Sun orbits the center of mass with a period of 11.86 years (Jupiter's orbital period). The full distance across the Sun's orbit is twice the distance D from the Sun to the center of mass that you just calculated. Now the question is how large an angular motion this would represent, as seen from a nearby star. Let's take the closest Sun-like star, Alpha Centauri, which has a distance of $d = 4.3$ light-years $= 4.1 \times 10^{13}$ km, and determine how much motion would be observed from there. The accompanying drawing should help you visualize this situation (the drawing is not to scale).

The problem is to determine angle A, the full angle over which the Sun would appear to move, as seen from Alpha Centauri (this angle is greatly exaggerated in the drawing). In a long, skinny triangle such as this, the angle A can be expressed in terms of the lengths of the two sides $2D$ and d:

$$A = (206,265)(2D/d),$$

where A is in units of arcseconds.

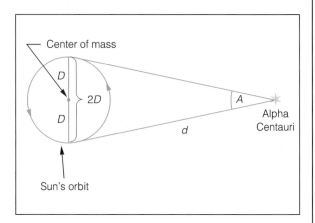

Now insert the values for D and d. How does the angle A compare with the accuracy of stellar position observations? (This is discussed early in Chapter 12; for now, assume that the best measurements of small position shifts are accurate to about 0.01 arcsecond.) Would it be feasible to detect a Jupiter-like planet orbiting Alpha Centauri by measuring small shifts in position? What if Alpha Centauri were 10 times farther away? One hundred times farther?

The other method for detecting a planet is to look for small shifts in velocity due to the Doppler effect, as the parent star orbits the center of mass. You can calculate the size of the expected velocity shift, again using the Sun-Jupiter pair as an example. To do this, note that the Sun has to travel a distance equal to the circumference of its orbit (i.e., equal to $2\pi D$) in a time of 11.86 years. To find the orbital speed of the Sun, calculate the circumference and divide by the number of seconds in 11.86 years. The most accurate Doppler shift measurements today can measure speeds as low as 100 m/sec, or 0.1 km/sec, in a single measurement. Would the Sun's orbital motion be detectable in this way? What if the Sun were 10 times or 100 times farther away?

The accuracy of either kind of measurement can be improved by averaging many individual measurements together. If the errors in individual measurements are random, then the accuracy improves in proportion to the square root of the number of measurements. How many individual measurements must be made using the positional and Doppler shift techniques to attain the degree of accuracy needed to detect a Jupiter-like planet orbiting Alpha Centauri?

ASTRONOMICAL UPDATE

Mission to Mars

Mars is not the closest planetary neighbor to Earth in distance, but the red planet is the Earth's nearest companion in other ways. Mars has the nearest approximation to an Earth-like climate and has historically stimulated the greatest speculation and expectations concerning possible extraterrestrial life in the solar system. From the time of Schiaparelli's "canali" to Lowell's fantasies of a Martian canal-building civilization, from Orson Welles's "War of the Worlds" to the superb photographic surveys by the *Mariner* and *Viking* spacecraft, Mars has held a special place in the minds of humans. Now we are embarking on a new phase of exploration of the red planet, starting with the *Mars Observer* mission, already well along on its interplanetary passage at this writing. Since the *Mars Observer* is expected to take nearly a year to arrive at the red planet and then is expected to keep working there for at least two additional Earth years, this earthly visitor to Mars may well be producing regular news bulletins during the time you are taking astronomy classes and using this book.

Mars Observer is a U.S. spacecraft, designed to study Mars for at least one full Martian year. It was launched on September 15, 1992, on a Titan III booster, which placed it into Earth orbit. From there the spacecraft was accelerated by a second rocket motor into a trajectory that will take it to Mars in just under a year. When it arrives at Mars in August 1993, it will be braked in several steps so that it drops into orbit around the red planet.

The initial orbit will be highly elliptical, but over a four-month period, it will be gently altered into a circular orbit passing over the Martian poles. A polar orbit offers several advantages for surveying the surface of Mars and monitoring the

yearly climate cycle. One advantage is that it brings the spacecraft over the polar caps, so that these repositories of water and carbon dioxide ice can be observed directly, something previous missions to Mars were unable to do. Another advantage of the polar orbit is that its rate of precession, or gradual shifting of orientation, can be adjusted so that the orbital plane remains aligned with the Mars-Sun direction as Mars follows its orbit around the Sun. This means that the spacecraft will always pass over the sunlit side of Mars at local noon and over the dark side at local midnight. This timing helps astronomers separate seasonal changes from daily effects. If, for example, we want to know how the daily temperature varies with the seasons, we must make sure that we are measuring the temperature at the same time each day and are not confused by the normal cycle of temperature changes from day to night.

The principal scientific aims of the *Mars Observer* mission are to survey the surface of the planet in great detail, seeking information on its mineral composition and on the role of water and ice; to determine the annual pattern of water and carbon dioxide migration between the polar caps and the equatorial zone; and to gain new insight into the physical structure and geology of Mars. All of these goals are a prelude to further exploration of Mars, first by unmanned probes (including landers) and, ultimately, by human expeditions. It is an official (though as-yet unfunded) goal of the U.S. space program to land people on Mars by the year 2019. Another long-term goal of great interest to scientists is to conduct an on-site search for evidence of fossil organisms—living cells that might have flourished at earlier times, when

Mars had a denser atmosphere and freestanding liquid water on its surface.

Aboard *Mars Observer* are seven scientific instruments, which are expected to go into operation once the final, circular orbit is achieved at the very beginning of 1994. Five of the instruments are mounted on the main body of the spacecraft, and two are on 6-m booms that will be extended once the probe is in its final orbit. The spacecraft will be rotated synchronously with its orbital period, so that the instruments will be pointed downward, toward the planet's surface, at all times.

One of the seven instruments is a versatile camera, designed to obtain the finest photographs yet of the Martian surface. The camera will have three configurations, ranging from wide-angle, relatively low-resolution capabilities to very narrow-angle, high-resolution capabilities. Even the low-resolution mode, which will be used to map the entire surface every Martian day, will provide quite detailed images; the resolution in this mode is 7.5 km. The highest-resolution mode, by contrast, will provide images with a resolution of 1.4 m (about 4.6 feet!). The high-resolution mode will be used sparingly, because of the enormous quantity of data that it produces (data from all seven instruments have to be combined and relayed via radio to Earth, so the rate of data transmission must be considered in planning the scientific observations).

Two instruments will observe and analyze infrared emission from the surface and atmosphere of Mars. The surface emission is ordinary thermal emission, as described by Wien's law (Chapter 4), but modified by absorp-

continued on next page

tion bands due to mineral constituents of the surface rocks. Thus this instrument, called the Thermal Emission Spectrometer, will be capable of determining the chemical makeup of surface materials. The second infrared instrument, called the Pressure Modulator Infrared Radiometer, will measure emission due to molecules in the Martian atmosphere. These emissions will be analyzed to determine the vertical structure of the atmosphere; that is, to measure the pressure, temperature, water vapor content, and dust abundance as a function of altitude above the Martian surface.

The Gamma Ray Spectrometer will detect very high-energy photons ejected from the surface of the planet due to natural radioactivity and the impacts of cosmic rays, which are energetic subatomic particles from space. The energy (or wavelength) of a gamma ray is determined by the structure of the atomic nucleus from which it was emitted, so spectral analysis of the gamma rays emitted from the Martian surface will reveal the identity and relative quantities of the elements that are present. This is analogous to using spectral lines in visible or ultra-

violet wavelengths to identify elements, except that the gamma rays are emitted and absorbed by the nuclei of atoms rather than by the electrons that orbit the nucleus.

Additional instruments include the Mars Observer Laser Altimeter, which will measure the surface topography of Mars by sending pulses of light to the surface and timing their return as they are reflected; the Magnetometer/Electron Reflectometer, which will make an intensive search for a Martian magnetic field, heretofore undetected, which would provide new information on the planet's interior structure; and, finally, a Radio Science instrument, which really is not a distinct instrument at all, but instead consists of the radio system placed on the spacecraft for communication with scientists on Earth. Detailed analyses of the modulation of the radio signals as they pass through the Martian atmosphere on their way to Earth will provide unique information on the vertical density structure of the Martian atmosphere and will also (through Doppler shifts) reveal subtle changes in the orbital motion of the spacecraft that can be analyzed to reveal the structure of the

planet's gravitational field.

One additional piece of instrumentation on *Mars Observer* was put there for the purpose of assisting planned Russian missions that are expected to reach Mars in 1994 or 1996. This is a radio system, supplied by the French national space agency, which will be used to relay data from the Russian probes. These probes will land on the surface and make detailed measurements of surface rocks and the Martian atmosphere (one of the Russian missions includes penetrators, spear-like probes that will plunge into the surface, providing the first opportunity to analyze underground rocks and minerals on Mars). If *Mars Observer* is still alive and well by the time the Russian probes arrive and go to work, this radio relay system will enhance the productivity of the Russian missions by allowing for much greater data transmission rates than the Russian instruments alone could handle. Thus the *Mars Observer* mission, in addition to its broad scientific contributions, may also help set the stage for cooperative international exploration of our neighbor planet when the time comes for human exploration.

THE STARS

The material in this section is devoted to the stars and their properties. In its five chapters we will learn first about our own star, the Sun, and then about how stars are observed, how their fundamental properties are deduced from observations, and how they work. We will see what processes control the structure of stars, and we will learn how they live, evolve, and die. We will see that stars are dynamic, changing entities and seem fixed and immutable only because the time scales on which they evolve are vastly longer than the human lifetime.

The first chapter of the section summarizes what is known of the Sun, a typical star in many respects, but very atypical in that we can observe it in much more detail than any other star. Information on the Sun will be used as we discuss stars in general in subsequent chapters.

The second chapter of the section discusses the three basic types of stellar observations and how the data obtained from these techniques are used to derive fundamental properties of

stars. We systematically consider several stellar properties, discuss how each is determined, and summarize the typical values that are found.

Chapter 13 initiates the discussion of how stars function and evolve by describing the physical processes that occur in stars and how these processes govern the stellar parameters previously discussed. The final two chapters of the section describe the evolution of stars, from their beginnings to their deaths. In Chapter 14 we base our discussion on observations of star formation regions and especially of clusters of stars, as well as theoretical calculations. In Chapter 15 we discuss the remnants of stars that have finished their lives, some of the most bizarre and exciting objects in the universe.

Having studied the stars, we will be ready to consider large groups of stars like our galaxy. A galaxy contains many billions of stars, and its evolution is governed by the evolution of the individual stars within it.

CHAPTER

11

The Sun

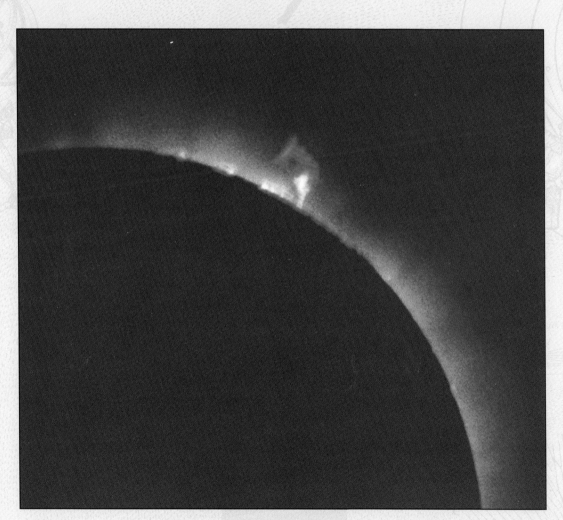

A prominence observed during the solar eclipse of 1991 (K. Krisciunas)

PREVIEW

With this chapter we begin our study of stars by investigating the nature of the Sun. We will find that its overall properties are governed by gravity, along with the laws of physics that describe the behavior of gases and the properties of atomic nuclei. Application of these principles allows us to determine the conditions and processes taking place inside the Sun, even though we cannot observe directly what is happening there. From this perspective we can understand and interpret the many surface phenomena: the overturning motions in the Sun's atmosphere; the very hot outer layers known as the chromosphere and corona; the steady stream of charged particles that form the solar wind; and the 22-year cycle of sunspots and solar activity. Our detailed treatment of the local star in this chapter will prepare us for discussions of other stars, beginning in the next chapter.

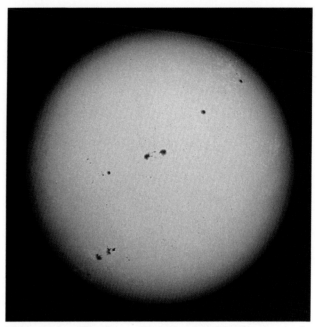

Figure 11.1   The Sun.   This is a visible-light photograph of the Sun at a time when only a few sunspots were present. The overall yellow color reflects the surface temperature of almost 6,000 K. The spots and the darkening of the disk near the edges are discussed in the text. (National Optical Astronomy Observatories)

Our star, the Sun, is rather ordinary by galactic standards. Its mass and size are modest—there are stars as much as a few hundred times larger in diameter and a million times more luminous. Its temperature is also moderate, as stars go. In many respects then, the Sun is a very run-of-the-mill entity.

Because the Sun is an ordinary star, and because we have far more detailed knowledge of it than of other stars, it is useful to begin our discussion of stars with a chapter on the Sun. In subsequent chapters, we will see how the information on the Sun helps us to understand the nature of other stars. Conversely, we will also find that studies of other stars help us to learn more about the nature of the Sun.

BASIC PROPERTIES AND INTERNAL STRUCTURE

Perhaps the most obvious attribute of the Sun is its brightness; it emits vast quantities of light. Its **luminosity,** the amount of energy emitted per second, is about 4×10^{26} watts. The intensity of sunlight reaching the Earth (above the atmosphere) is about 1400 watts/m^2. This quantity is known as the **solar constant,** although it varies slightly according to activity in the Sun's outer layers (discussed later in this chapter).

The Sun is a ball of hot gas (Fig. 11.1) It dwarfs all the planets in mass and radius (Table 11.1). Its density, on average, is 1.41 g/cm^3, not much more than that of water, but its center is so highly compressed that the density there is about 10 times greater than that of lead. The interior is gaseous rather than solid, because the temperature is very

high, around 15 million degrees at the center, diminishing to just under 6,000 K at the surface. At these temperatures, the gas is partially ionized in the outer layers of the Sun and completely ionized in the core, where all electrons have been stripped free from their parent atoms.

The Sun is held together by gravity. All of its constituent atoms and ions attract each other, with the net effect that the solar substance is held in a spherical shape. Gas that is hot exerts pressure on its surroundings, and this pressure, pushing outward, balances the force of gravity, which pulls the matter inward. This balance is called **hydrostatic equilibrium,** with gravity and pressure equaling each other everywhere. The deeper we go into the Sun, the greater the weight of the overlying layers, and the more the gas is compressed. The higher the pressure,

TABLE 11.1   The Sun

Diameter: 1,391,980 km (109.3 D$_\oplus$)
Mass: 1.99×10^{30} kg (332,943 M$_\oplus$)
Average density: 1.41 g/cm^3
Surface gravity: 27.9 Earth gravities
Escape speed: 618 km/sec
Luminosity: 3.83×10^{33} erg/sec
Surface temperature: 5,780 K
Rotation period: 25.04 days (at equator)

TABLE 11.2 The Composition of the Sun's Photosphere

Element	Symbol	Number of Atoms[a]	Fraction of Mass
Hydrogen	H	1.0000	7.35×10^{-1}
Helium	He	8.51×10^{-2}	2.48×10^{-1}
Lithium	Li	1.55×10^{-9}	7.85×10^{-9}
Beryllium	Be	1.41×10^{-11}	9.27×10^{-9}
Boron	B	2.00×10^{-10}	1.58×10^{-9}
Carbon	C	3.72×10^{-4}	3.26×10^{-3}
Nitrogen	N	1.15×10^{-4}	1.18×10^{-3}
Oxygen	O	6.76×10^{-4}	7.88×10^{-3}
Fluorine	F	3.63×10^{-8}	5.03×10^{-7}
Neon	Ne	3.72×10^{-5}	5.47×10^{-4}
Sodium	Na	1.74×10^{-6}	2.92×10^{-5}
Magnesium	Mg	3.47×10^{-5}	6.15×10^{-4}
Aluminum	Al	2.51×10^{-6}	4.94×10^{-5}
Silicon	Si	3.55×10^{-5}	7.27×10^{-4}
Phosphorus	P	3.16×10^{-7}	7.14×10^{-6}
Sulfur	S	1.62×10^{-5}	3.79×10^{-4}
Chlorine	Cl	2.00×10^{-7}	5.17×10^{-6}
Argon	Ar	4.47×10^{-6}	1.30×10^{-4}
Potassium	K	1.12×10^{-7}	3.19×10^{-6}
Calcium	Ca	2.14×10^{-6}	6.26×10^{-5}
Scandium	Sc	1.17×10^{-9}	3.84×10^{-8}
Titanium	Ti	5.50×10^{-8}	1.92×10^{-6}
Vanadium	V	1.26×10^{-8}	4.68×10^{-7}
Chromium	Cr	5.01×10^{-7}	1.90×10^{-5}
Manganese	Mn	2.51×10^{-7}	1.01×10^{-5}
Iron	Fe	3.98×10^{-5}	1.62×10^{-3}
Cobalt	Co	3.16×10^{-8}	1.36×10^{-6}
Nickel	Ni	1.91×10^{-6}	8.18×10^{-5}
Copper	Cu	2.82×10^{-8}	1.31×10^{-6}
Zinc	Zn	2.63×10^{-8}	1.25×10^{-6}
(all others combined)		(less than 10^{-8} of total)	

[a]The numbers given are relative to the number of hydrogen atoms.

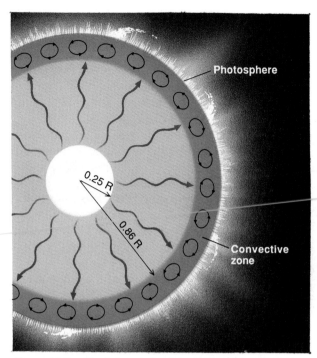

Figure 11.2   The internal structure of the Sun.   This drawing shows the relative extent of the major zones within the Sun, except that the depth of the photosphere is greatly exaggerated.

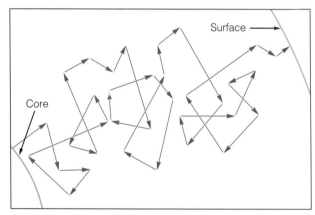

Figure 11.3   Random walk.   A photon is continually being absorbed and reemitted as it travels through the Sun's interior. Because each reemission is in a random direction, the photon's progress from the core, where it is created, to the surface is very slow.

the greater the temperature required to maintain the pressure, so we find that the pressure and temperature both increase as we approach the center.

The composition of the Sun (Table 11.2) is the same as that of the primordial solar system and of most other stars: about 73 percent of its mass is hydrogen, 25 percent is helium, and the rest is made up of traces of other elements. As we have already seen, this is similar to the chemical makeup of the outer planets and would represent the terrestrial planets as well, except that they have lost most of their volatile elements such as hydrogen and helium. It is apparent that all the components of the solar system formed together, from the same material.

The ultimate source of all the Sun's energy is located in its core, within the innermost 25 percent or so of its radius (Fig. 11.2). Here nuclear reactions create heat and photons at gamma-ray (γ-ray) wavelengths. This radiation eventually reaches the surface and escapes into space, but it is a laborious journey. Each photon is absorbed and reemitted many times

along the way (Fig. 11.3), gradually losing energy. In the process, most γ-ray photons become visible-light photons, and the energy they lose heats the surroundings. Because a photon travels only a short distance before it is absorbed and then reemitted in a random direction, progress toward the surface is very slow.

Throughout most of the solar interior, the gas is quiescent, without any major large-scale flows or currents. The energy from the core is transported by the radiation wending its slow way outward except in the layers near the surface, where convection occurs, and heat is transported by the overturning motions of the gas. As we shall see, the bubbling, boiling action in the outer portions of the Sun creates a wide variety of dynamic phenomena on the surface.

The Sun rotates, and it does so differentially. This is apparent from observations of its surface features, which reveal that, like Jupiter and Saturn, the Sun goes around faster at the equator than near the poles. The rotation period is about 25 days at the equator, 28 days at middle latitudes, and even longer near the poles. As we will see, the differential rotation probably plays an important role in governing variations in the solar magnetic field, which in turn have a lot to do with the behavior of the most prominent surface features, the sunspots.

There is evidence that the core of the Sun rotates a little more rapidly than the surface. The only clue for this is the presence of very subtle oscillations on the solar surface, which may be wave motions affected by the more rapid spin of the interior. The spin of the solar core is probably a direct result of the collapse and accelerated rotation of the interstellar cloud from which the Sun formed. (The origin of the Sun's rotation was discussed in Chapter 10.)

NUCLEAR REACTIONS

One of the biggest mysteries in astronomy in the early decades of this century was presented by the Sun. The problem was how to account for the tremendous amount of energy it radiates, which was particularly perplexing in view of geological evidence that the Sun has been able to produce this energy for at least 4 or 5 billion years. Two early ideas, that the Sun is simply still hot from its formation, or that it is gradually contracting, releasing stored gravitational energy, were both ruled out, because neither mechanism could possibly supply the energy needed to run the Sun for a long enough time.

The first hint at the solution came in the first decade of the 1900s, when Albert Einstein developed his theory of special relativity. He showed that matter and energy are equivalent, according to the famous formula $E = mc^2$, where E is the energy released in the conversion, m stands for the mass that is converted, and c for the speed of light. If mass could be converted into energy, enormous amounts of energy could be produced. Physicists began to contemplate the possibility that somehow this energy was being released inside the Sun and stars.

By the 1920s, following the pioneering work on atomic structure by Max Planck, Niels Bohr, and others, the concept of nuclear reactions began to emerge. These are transformations, like chemical reactions, except that in this case it is the subatomic particles in the nuclei of atoms that react with each other. Some reactions are **fusion** reactions, in which nuclei merge to create a larger nucleus, representing a new chemical element; and others are **fission** reactions, in which a single nucleus, usually of a heavy element with a large number of protons and neutrons, splits into two or more smaller nuclei. In either type of reaction, energy is released as some of the matter is converted according to Einstein's formula. In the 1920s, many physicists explored these possibilities, and by the 1930s Hans Bethe had suggested a specific reaction sequence that might operate in the Sun's core.

Bethe envisioned a fusion reaction in which four hydrogen nuclei (each consisting of only a single proton) combine to form a helium nucleus, made up of two protons and two neutrons. The reaction actually occurs in several steps (see Appendix 11): (1) two protons combine to form **deuterium,** a type of hydrogen that has a proton and a neutron in its nucleus (one of the two protons undergoing the reaction converts itself into a neutron by emitting a positively charged particle called a **positron**) and another particle called a **neutrino,** which has very unusual properties (see the Astronomical Insight on the facing page); (2) the deuterium combines with another proton to create an isotope of helium (^{3}He) consisting of two protons and one neutron; and (3) two of these ^{3}He nuclei combine, forming an ordinary helium nucleus (^{4}He, with two protons and two neutrons in the nucleus) and releasing two protons. At each step in this sequence, heat energy is imparted to the surroundings in the form of kinetic energy of the particles that are produced, and in step (2) a γ-ray photon is emitted as well.

The net result of this reaction, which is called the **proton-proton chain,** is that four hydrogen nuclei (protons) combine to create one helium nucleus. The end product has slightly less mass than the ingredients, 0.007 of the original amount having been converted into energy. We will learn in Chapters 13 and 14 that other kinds of nuclear reactions take place in some stars, but the vast majority produce their energy by the proton-proton chain.

It is easy to see that the proton-proton chain can produce enough energy to keep the Sun shining for billions of years. The total mass of the Sun is about 2×10^{30} kilograms. Only the innermost portion of this, perhaps 10 percent, undergoes nuclear reactions, so the Sun started out with about 2×10^{29} kg of mass available for nuclear reactions. If 0.007 of this, or 1.4×10^{27} kg, is converted into energy, then to find the total energy the Sun can produce in its lifetime, we multiply this mass times c^2, finding a total

ASTRONOMICAL INSIGHT
The Solar Neutrino Mystery

Nearly all that we know about the reactions in the Sun's core is based on theoretical calculations, and, of course, astronomers and physicists have been interested in verifying the results of these calculations. We think that the theory is basically correct, because it neatly accounts for the Sun's observed energy output, but nevertheless a more direct confirmation has been sought.

One way to do this was suggested by the early work of Enrico Fermi, who had postulated the existence of a tiny subatomic particle called the **neutrino,** which, his calculations showed, ought to be released at certain stages in nuclear reactions. The neutrino is a strange little particle, having no mass and no electrical charge; nevertheless it can carry away small amounts of energy, thus balancing the equations describing the reactions. Because of its ethereal properties, a neutrino hardly interacts with anything at all (it can interact with other matter only through the *weak nuclear force,* which was mentioned in the Astronomical Insight on p. 000). Thus, neutrinos produced in the Sun's core ought to escape directly into space at the speed of light, in sharp contrast to the photons from the core, which, as we have seen, take a very long time to get out.

A possible test of the reactions going on inside the Sun, therefore, would be to measure the rate at which neutrinos are coming out. Unfortunately, the same properties that allow them to escape the Sun so easily also make them very difficult to catch and count. There is an indirect technique, however, based on the fact that certain nuclear reactions can be triggered by the impact of a neutrino.

A chlorine atom, for example, can be converted into a radioactive form of argon when it encounters a neutrino. Accordingly, an experiment was set up a few years ago, and is still in operation, in which a huge underground tank containing a chlorine compound is monitored to see how many of its atoms are being converted into radioactive argon atoms, which in turn indicates how many neutrinos are passing through the tank.

Much to everyone's surprise, neutrinos have not been detected coming from the Sun in the expected numbers, and at present no satisfactory explanation has been found. Clearly something is not yet understood about nuclear reactions in the Sun, but it is not clear whether the problem is in our understanding of nuclear reactions or in our understanding of the internal properties of the Sun. Unfortunately, the neutrinos that can react with chlorine are produced by a minor offshoot of the main energy-producing reactions in the Sun, so it is possible that the main reaction sequence occurs as expected, but that the rate at which this other reaction occurs is slower than thought.

Ideally, we would like to detect the neutrinos produced directly in the proton-proton chain. These neutrinos are even harder to catch, however, and the first attempts to detect them began only recently (in 1990). The neutrinos produced by the principal reaction ($p + p \rightarrow {}_1^2H + e^+ + v$) in the Sun can trigger reactions in a rare element called gallium, and accumulating the needed gallium took time (and substantial funds). The first results from this experiment, announced in mid-1992, showed that neutrons produced in

the proton-proton reaction have been detected, but still not at the predicted level. About two-thirds of the expected number were detected during the first year of the experiment, which is continuing.

The shortage of observed neutrinos must indicate that either we don't fully understand the nuclear reactions going on in the Sun, or we don't understand the properties of neutrinos well enough. Recently there have been several suggestions that it is our understanding of neutrinos that is incomplete. Several scientists have suggested that neutrinos may have a property that would allow a significant fraction of them to escape detection by the methods involving chlorine or gallium detectors. There are three types of neutrinos (particle physicists often speak of them as three "flavors" of neutrinos), and the current detectors can detect only one of these types. It has been suggested that neutrinos traveling through space can change types, so that only a fraction of them would have the right "flavor" when they reached the Earth from the Sun. The others would escape detection, thus accounting for the low number being seen in the experiment. These theoretical ideas have been revitalized by the recent failure of the gallium detectors to find the predicted number of solar neutrinos. The interplay between solar physics and elementary particle physics that has been stimulated by the neutrino problem has been fascinating and will almost certainly lead to new physical insights.

ASTRONOMICAL INSIGHT
Measuring the Sun's Pulse

In the early 1960s, measurements of spectral lines formed in the solar photosphere revealed small Doppler shifts that alternated between blue- and redshifts. This seemed to indicate that the Sun's surface was rising and falling with a period of about 5 minutes. The reported Doppler shifts were quite small (less than one one-hundredth of an Ångstrom), near the limit of detectability, and the measurements were correspondingly difficult. Today the existence of this and other periodic oscillations of the Sun's surface is the basis for an entire research field, sometimes called solar seismology.

Oscillations of the solar surface can be likened to a musical instrument with a resonant cavity. When waves of any kind are created inside a cavity with reflecting walls, certain wavelengths or frequencies persist, while all others die out. The reason is that waves reflected from the walls interfere with incoming waves, the valleys and crests either adding together or canceling each other out. In an enclosed space, the size of the cavity determines which wavelengths

will add together constructively and therefore persist, and which wavelengths will interfere destructively and therefore will die out. The wavelengths that persist correspond to the so-called resonant frequency of the cavity. In a musical wind instrument, this frequency determines the pitch of the tone that is emitted. The musician can control the pitch by altering the dimensions of the internal cavity (by opening and closing valves in most cases).

Inside the Sun, there are no walls, but physical conditions that vary with depth can create barriers as real as walls for certain types of waves. It has been deduced that sound waves (i.e., compressional waves) can persist in the Sun's convective zone and are confined above and below by the manner in which the Sun's temperature varies with depth. Sound waves in a certain range of wavelength are reflected back as they travel farther in or farther out, and this creates a resonant cavity in the outermost portion of the Sun for sound waves whose period of pulsation is about 5 minutes.

These waves create a standing pattern of crests and valleys on the solar surface, and at any particular point on the surface, the gas rises and falls with this period.

This phenomenon is quite interesting for its own sake but is worthy of careful study for another reason as well: the properties of the waves or oscillations of the solar surface can reveal information on conditions in the solar interior, where the waves arise. For example, study of the frequency range of the waves can yield information on exactly how the temperature varies with depth inside the Sun. Since this variation determines the location of the lower boundary of the resonant cavity, this information can be helpful. Perhaps even more interestingly, analysis of the surface oscillations can provide information on the internal rotation of the Sun. This is possible because waves traveling over the solar surface in opposite directions are either spread out or squeezed closer together, depending on whether they move in the direction of the Sun's rotation or in the opposite direction.

---

energy of $E = (1.4 \times 10^{27}$ kg$) \times (3 \times 10^8$ m/s$)^2 = 1.3 \times 10^{44}$ joules. The rate at which the Sun is losing energy is its luminosity, which is about 4×10^{26} watts $= 4 \times 10^{26}$ joules/sec. At this rate, the Sun can last for

$$\frac{1.3 \times 10^{44} \text{ joules}}{4 \times 10^{26} \text{ joules/sec}} = 3.2 \times 10^{17} \text{ seconds,}$$

or just about 10 billion years. Geological evidence shows that the solar system is now about 4.5 billion years old, so we can expect the Sun to keep shining for another 5 billion years or more. (In Chapter 14, we will see what happens to stars like the Sun when the nuclear fuel, hydrogen, runs out.)

Nuclear fusion reactions can take place only under conditions of extreme pressure and temperature, because of the electrical forces that normally would keep atomic nuclei from ever getting close enough together to react. Nuclei, which have positive

charges, must collide at extremely high speeds to overcome the repulsion caused by their like electrical charges. The speed of particles in a gas is governed by the temperature, and only in the very center of the Sun and other stars is it hot enough (around 15 million degrees) to allow the nuclei to collide fast enough to fuse. This is why only the innermost portion of the Sun can ever undergo reactions. The high pressure in the Sun's core causes nuclei to be crowded together very densely, and this means that collisions will occur very frequently, another requirement if a high reaction rate is to occur.

MODELING THE SUN

Astronomers have learned a great deal about the interior of the Sun by combining observations of its

Consequently, waves going in opposite directions have different frequencies, and the difference in frequency depends on the speed of rotation of the Sun in the layer where the waves arise. In effect, it is possible to infer the Sun's internal rotational velocity at the lower boundary of the resonant cavity in which the waves oscillate.

The determination of internal rotational speeds using observations of surface oscillations is very complex and requires very sophisticated treatment of extensive quantities of data. At present, scientists have not been able to carry this effort very far, but they have determined from the 5-minute oscillations that the Sun is rotating more rapidly near the bottom of the convective zone than at the surface, leading to speculations that the rotation near the core may be even more rapid. If so, this could have important implications for the Sun's overall structure and evolution, and a great deal of importance is being attached to further study of this phenomenon.

While the 5-minute oscillation of the Sun is now well established and understood, other reported oscillations are not. There is a 3-minute oscillation that may be caused by a similar resonant cavity at the level of the chromosphere, and apparently oscillations lasting much longer are driven by waves in very deep resonant cavities, near the solar core. These controversial oscillations, having periods of 40 and 160 minutes, have the potential of revealing conditions such as the rotation rate very deep inside the Sun, and their study is therefore of great importance.

For a variety of reasons, the detection of long-period oscillations is very difficult. It is important, for example, to observe the Sun continuously over many cycles of the oscillations, which means it is necessary to observe without interruption for several days. This may sound impossible, because of the Sun's daily rising and setting, but in fact it can be done from one of the Earth's poles during local summer. Therefore, solar studies have been done from the scientific research station at the South Pole. During one of these studies, involving six days of continuous observations of the Sun, the 160-minute oscillation was convincingly detected.

Another approach to making continuous observations of the oscillations is to establish a worldwide network of observers, so that the Sun can be observed, with similar instruments, throughout the Earth's 24-hour rotation period. One such network, called the Global Oscillation Network Group (GONG), has been established by the U.S. National Solar Observatory. GONG, which will consist of several identical observing stations located at sites around the world, is expected to provide data on solar oscillations for years to come.

Beyond simply detecting the existence of the long-period oscillations, further refinements are needed if the oscillations are to be exploited as tools for probing the Sun's deep interior. To observe them in enough detail requires entirely new technology, and at present several research groups are at work developing the necessary instruments. The problem is to find a way to observe very small Doppler shifts in spectral lines at many positions on the Sun's surface, and to make these measurements repeatedly at very closely spaced intervals of time. The velocities involved are very small, only a meter per second or less, so the Doppler shift in a typical visible-wavelength line is only a few millionths of an Ångstrom. Measuring such small shifts accurately is a formidable task, but one that probably will be successfully completed within a few years. When this has been accomplished, a wealth of new information on the internal properties of our local star will have been mined.

---

surface with theoretical calculations of the internal structure. The theoretical models are constructed by writing equations to represent the known physical processes that take place in the interior, and then solving these equations for many different positions inside the Sun. For example, one equation expresses the relationship that the inward force of gravity is equal to the outward force of pressure. Another gives the rate of energy flow outward (this depends on the density and temperature at each point), and another provides the rate of energy production by nuclear reactions (which depends on temperature, density, and composition).

It can be shown that the set of equations has only one possible solution. The major uncertainty in the process lies in the fact that some of the physical parameters are either too complex to be represented accurately or are not well enough known, so that approximations have to be used. Furthermore, it is possible that some physical processes, such as internal rotation or the role of the magnetic field inside the Sun, are more important than allowed for in the equations at present.

Despite these uncertainties, astronomers believe that the results of model calculations for the Sun are at least approximately correct, because they reproduce the observed surface conditions quite well. One exception is the failure to detect the predicted number of neutrinos, and this has caused some consternation about the models (see the Astronomical Insight above).

The results of one such model calculation are summarized in Table 11.3, which lists the temperature and density at several points inside the Sun (and extending throughout its atmosphere, which is discussed in the next section). The fraction of the Sun's mass contained inside each distance from the center is listed also.

TABLE 11.3   Structure of the Sun

Zone	Radius[a] (R)	Radius[a] (km)	Temperature (K)	Density (g/cm³)	Mass Fraction (M)[a]
Interior	0.00	0	1.6×10^7	160	0.000
	0.04	28,000	1.5×10^7	141	0.008
	0.1	70,000	1.3×10^7	89	0.07
	0.2	139,000	9.5×10^6	41	0.35
	0.3	209,000	6.7×10^6	13	0.64
	0.4	278,000	4.8×10^6	3.6	0.85
	0.5	348,000	3.4×10^6	1.0	0.94
	0.6	418,000	2.2×10^6	0.35	0.982
	0.7	487,000	1.2×10^6	0.08	0.994
	0.8	557,000	7.0×10^5	0.018	0.999
	0.9	627,000	3.1×10^5	0.0020	1.000
	0.95	661,000	1.6×10^5	0.0004	1.000
	0.99	689,000	5.2×10^4	0.0005	1.000
	0.995	692,000	3.1×10^4	0.00002	1.000
	0.999	695,300	1.4×10^4	0.0000001	1.000
Photosphere	1.000	695,990	6.4×10^3	3.5×10^{-7}	1.000
	1.000	+ 280	4.6×10^3	4.5×10^{-8}	1.000
Chromosphere	1.000	+ 320	4.6×10^3	3.1×10^{-8}	1.000
	1.001	+ 560	4.1×10^3	3.6×10^{-9}	1.000
(Transition)	1.002	+ 1,900	8.0×10^3	3.4×10^{-13}	1.000
	1.003	+ 2,400	4.7×10^5	4.8×10^{-15}	1.000
Corona	1.003	+ 2,400	5.0×10^5	1.7×10^{-15}	1.000
	1.2	+140,000	1.2×10^6	8.5×10^{-17}	1.000
	1.5	+348,000	1.7×10^6	1.4×10^{-7}	1.000
	2.0	+696,000	1.8×10^6	3.4×10^{-18}	1.000

[a]The radii are expressed in fractions of the Sun's radius $R_\odot$ as well as in kilometers. Above 695,990 km, the kilometer values are heights above the Sun's "surface," or lower boundary of the photosphere. The mass fractions are expressed in units of the sun's mass $M_\odot$, which has the value 1.989×10^{30} kg. (Data from C. W. Allen, 1973, *Astrophysical Quantities,* London: Athlone Press.)

STRUCTURE OF THE SOLAR ATMOSPHERE

Observations of the Sun at different wavelengths of light make it clear that the outer layers are divided into several distinct zones (Fig. 11.4). The "surface" of the Sun that we see in visible wavelengths is the **photosphere,** with a temperature ranging between 4,000 and 6,500 K. Viewing the Sun at the wavelength of the strong line of hydrogen at 6563 Å, we see the **chromosphere,** a layer above the photosphere whose temperature is 6,000 to 10,000 K. Outside the chromosphere is the very hot, rarefied **corona,** best observed at X-ray wavelengths, whose temperature is 1 to 2 million degrees. Between the chromosphere and the corona is a thin region, called the **transition zone,** where the temperature rises rapidly. Overall, the tenuous gas within and above the photosphere is referred to as the solar atmosphere. Considering the solar atmosphere as a whole, we find that the temperature decreases as we move outward through the photosphere, reaching a minimum value of about 4,000 K. From there the trend reverses itself, and the temperature begins to rise as we go farther out. The chromosphere, immediately above the temperature

minimum, is perhaps 2,000 km thick. Above this point the temperature rises very steeply within a few hundred kilometers to the coronal value of over a million degrees. Clearly something is adding extra heat at these levels; shortly we will consider where this heat comes from.

First let us discuss the photosphere, the "surface" of the Sun as we look at it in visible light. We see this level because here the density becomes great enough for the gas to be opaque, preventing us from seeing farther into the interior. The photosphere is where the Sun's absorption lines (Fig. 11.5) are formed, as the atoms in this relatively cool layer absorb continuous radiation coming from the hot interior.

A photograph of the photosphere reveals a cellular appearance called **granulation** (Fig. 11.6). Bright regions, thought to represent areas where convection in the Sun's outer layers causes hot gas to rise, are bordered by dark zones where cooler gas is apparently descending back into the interior. Recent detailed studies of granules and their changes over time have shown their behavior is more complex than previously thought: granules can change appearance rapidly and can disintegrate quickly and disappear.

The temperature of the photosphere, roughly 6,000 K, is measured from the degree of ionization in the gas there. The density, roughly 10^{17} particles per cubic centimeter in the lower photosphere, was found from the degree of excitation, as described in Chapter 4. This is lower than the density of the Earth's atmosphere, which is about 10^{19} particles per cubic centimeter at sea level.

Most of what we know about the Sun's composition is based on the analysis of the solar absorption lines, so strictly speaking, the derived abundances represent only the photosphere. We have no reason to expect strong differences in composition at other levels, however, except for the core, where a significant amount of the original hydrogen has been converted into helium by the proton-proton reaction.

The photosphere near the edge of the Sun's disk looks darker than in the central portions. This effect, called **limb darkening** (see Fig. 11.1), occurs because we are looking obliquely at the photosphere when we look near the edge of the disk. We therefore do not see as deeply into the Sun as we do when looking near the center of the disk. Therefore, at the limb, the gas we are seeing is cooler than at disk center and radiates less.

The chromosphere lies immediately above the temperature minimum (Fig. 11.7). The fact that this region forms emission lines tells us, according to Kirchhoff's laws, that the chromosphere is made of hot, rarefied gas, hotter than the photosphere behind it. When viewed through a special filter that allows only light at the wavelength of the hydrogen emission line at 6563 Å to pass through (see Fig. 11.7), the

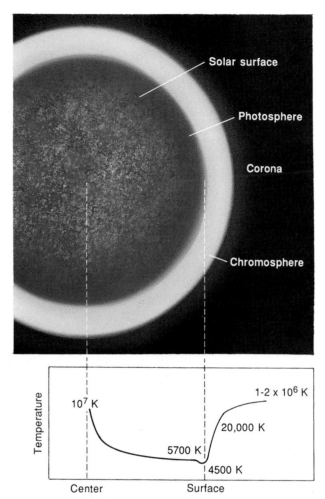

Figure 11.4 The structure of the Sun's outer layers. This diagram shows the temperatures of the convective zone, the photosphere, the chromosphere, and the corona.

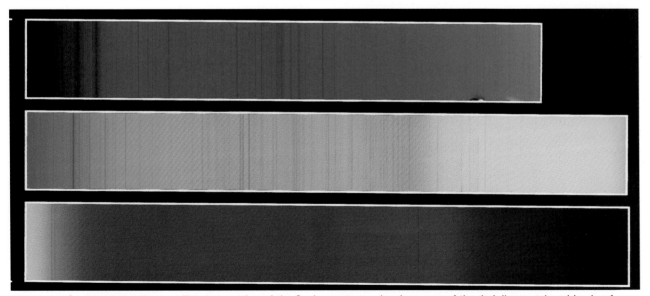

Figure 11.5 Fraunhofer lines. This is a portion of the Sun's spectrum, showing some of the dark lines cataloged by Josef Fraunhofer, a German physicist. The two lines near the far left in the upper segment are due to ionized calcium. The center segment shows several lines of the Balmer series of hydrogen, and the strongest Balmer line, H-alpha, is seen in the red portion of the lower segment. (K. Gleason, Sommers Bausch Observatory, University of Colorado)

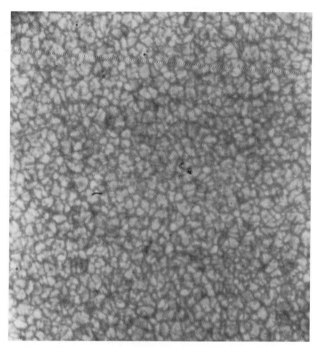

Figure 11.6   Solar granulation.   This photograph, obtained by a high-altitude balloon above much of the atmosphere's blurring effect, distinctly shows the granulation of the photosphere, which is due to convective motions in the Sun's outer layers. (Project Stratoscope, Princeton University, sponsored by the National Science Foundation)

chromosphere has a distinctive cellular appearance referred to as **super-granulation,** which is similar to the photospheric granulation, but with cells some 30,000 km across instead of about 1,000 km. The chromosphere also exhibits fine-scale structure in the form of spikes of glowing gas called **spicules** (Fig. 11.8). These come and go, probably at the whim of the magnetic forces that seem to control their motions.

The outermost layer of the Sun's atmosphere is the corona, which extends a considerable distance above the photosphere and chromosphere. The corona is irregular in form, patchy near the Sun's surface, but with radial streaks at great heights suggestive of outflow from the Sun (Fig. 11.9). The density of the coronal gas is very low, only about 10^9 particles per cubic centimeter.

As we have already seen, the corona is very hot and contains highly ionized gas. The source of the energy that heats the corona to such extreme temperatures is not well understood, although a general picture has emerged, as we will see.

X-ray observations reveal that the corona is not uniform, but instead has a patchy structure (Fig. 11.10). In large regions that appear dark in an X-ray photograph of the Sun, the gas density is even lower than in the rest of the corona. These regions

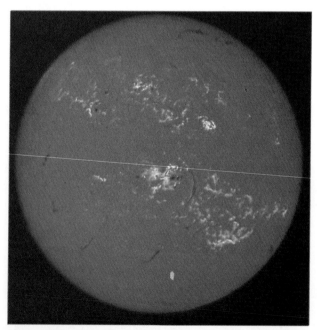

Figure 11.7   The solar chromosphere.   This is an image of the Sun, taken through a filter that allows only the wavelength of the red emission line of hydrogen to pass through. The Sun's chromosphere emits strongly at this wavelength whereas the photosphere does not; hence this image shows primarily the chromosphere and its structure. (National Optical Astronomy Observatories)

Figure 11.8   Spicules.   This photograph shows the transient features in the chromosphere known as spicules. These spikes of glowing gas, which are apparently shaped by the Sun's magnetic field, come and go irregularly. (Sacramento Peak Observatory, National Optical Astronomy Observatories)

are called **coronal holes** and, as we shall see, are probably created and maintained by the Sun's magnetic field. The coronal holes, as well as the overall shape of the corona, vary with time (Fig. 11.11), showing that the corona in general is a dynamic region. Another impressive sign of this dynamic nature is presented by the **prominences** (Figs. 11.12 and 11.13), great streamers of hot gas stretching upward from the surface of the Sun to take on an arc-shaped appearance. These are usually associated with sunspots, and both phenomena are linked to the solar activity cycle (to be discussed shortly).

The hot outer layers of the Sun have provided astronomers with a second major mystery regarding the solar energy budget. Unlike the mystery of the Sun's internal energy source, the question of the mechanism for heating the chromosphere and corona has not yet been entirely resolved. A great deal of energy is available in the form of gas motions in the convection zone just beneath the solar surface, and it is generally accepted that the heating of the corona somehow comes from this energy. It is not clear how the energy is transported to such high levels, however, although there are several ideas, each invoking some form of waves. Waves of any kind carry energy, and it has been suggested that either sound waves or magnetic waves of some kind are the agents that transfer

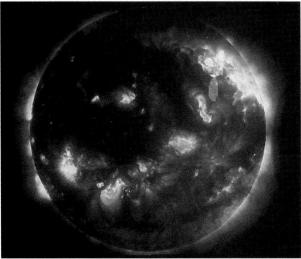

Figure 11.10   An X-ray portrait of the corona.   This image was obtained by a small, rocket-borne x-ray telescope on the day of the total solar eclipse in 1991 (July 11). The bright regions are relatively dense, whereas the dark regions, called coronal holes, are less dense. All of the emission seen here originates in the corona, where the gas temperature is between 1 and 2 million degrees. (L. Golub, Harvard-Smithsonian Center for Astrophysics)

Figure 11.9   The corona.   This photograph, obtained during a total solar eclipse, shows the type of structure commonly seen in the corona. There are giant loop-like features and an overall appearance of outward streaming. This image shows the total eclipse of July 11, 1991, as seen from the Baja penninsula, in Mexico. Reddish spots along the edge of the solar disk are prominences, colored by the red emission line of ionized hydrogen. (S. Albers, G. Emerson, D. Dicicco, and D. Sime)

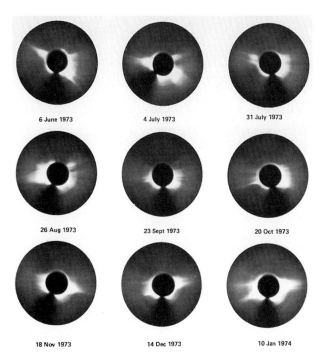

Figure 11.11   Changes in the corona.   This series of photographs was obtained using a special shutter device that blocks out the solar disk and allows the corona to be observed without waiting for a total solar eclipse. Here we see that the structure of the corona changes quite markedly over a period of a few months. (High Altitude Observatory, National Center for Atmospheric Research, sponsored by the National Science Foundation)

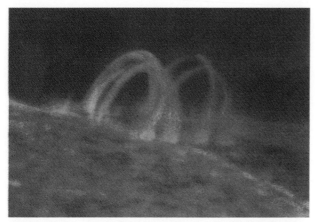

Figure 11.12   A prominence.   Here the looplike structures thought to be governed by the Sun's magnetic field can be readily seen. This photograph was obtained by *Skylab* astronauts using an ultraviolet filter. (NASA)

Figure 11.13   An eruptive prominence.   The striking sequence at the top shows an outburst of ionized gas from a prominence on the Sun's limb. The photographs were obtained by using a special device to block out the Sun's disk. At the bottom is an ultraviolet view of another eruptive prominence, showing the extent it reached in less than two hours. (top: High Altitude Observatory, National Center for Atmospheric Research, sponsored by the National Science Foundation; bottom: NASA)

energy from the convection zone to the corona. There is certainly enough energy in the convective motions in and below the photosphere; the problem is how this energy is transported into the higher levels.

Recent satellite observations at ultraviolet and X-ray wavelengths have shown that stars similar to the Sun in general type also have chromospheric and coronal zones, so if we can discover how the Sun operates, we will also gain a better understanding of how other stars work. Similarly, observations of other stars, particularly of the relationship of chromospheric and coronal activity to stellar properties such as age and rotation, can help us learn more about the Sun.

THE SOLAR WIND

The long, streaming tail of a comet always points away from the Sun, regardless of the comet's direction of motion. The significance of this was fully realized in the late 1950s, when the first U.S. satellites revealed the presence of the Earth's radiation belts and the fact that they are shaped in part by a steady flow of charged particles from the Sun. Near the Earth the solar wind reaches a speed of 300 to 400 km/sec. It is this flow of charged particles that forces cometary tails always to point away from the Sun.

The existence of the solar wind is evidently a natural by-product of the same heating mechanisms that produce the hot corona of the Sun. Astronomers originally thought that particles in the high-temperature region move about with such great velocities that a steady trickle escapes the Sun's gravity, flowing outward into space. Subsequently, however, X-ray observations of the Sun have shown that the situation is not that simple. The solar magnetic field governs the outward flow of charged particles. The coronal holes, mentioned earlier (see Fig. 11.10), are regions where the magnetic field lines open out into space. Charged particles such as electrons and protons, constrained by electromagnetic forces to follow the magnetic field lines, therefore escape into space only from the coronal holes. The speed of the solar wind is relatively low close to the Sun, but accelerates outward, quickly reaching the speed of 300 to 400 km/sec already mentioned, after which it is nearly constant. The wind has nearly reached its maximum speed by the time it passes the Earth's orbit, and beyond there it flows steadily outward, apparently persisting at least beyond the orbit of Neptune. (The *Pioneer 10* spacecraft, on its way out of the solar system, still detected solar wind particles as it crossed the orbit of Neptune in mid-1983, and it has not yet found a limit to the solar wind.) At some point in the outer solar system, the wind is thought to come to an abrupt halt where it runs into an invisible and tenuous wall of matter swept up from the interstellar medium that surrounds the Sun.

Most of the direct information we have on the solar wind comes from satellite and space probe observations (Fig. 11.14), because the Earth's magnetosphere shields us from the wind particles. Solar wind monitors are placed on board the majority of spacecraft sent to the planets. One striking discovery has been that the density of the wind is not uniform; instead the wind flows outward from the Sun in sectors, indicating that it originates only from certain areas on the Sun's surface. As already mentioned, X-ray data have shown that the wind emanates only from the coronal holes. Because the base of the wind is rotating with the Sun, the wind sweeps out through space in a great curve, similar to the trajectory of water from a rotating law sprinkler (Fig. 11.15).

Occasionally, explosive activity occurring on the Sun's surface releases unusual quantities of charged particles, and some three or four days later, when this burst of ions reaches the Earth's orbit, we experience disturbances in the ionosphere that can interrupt short wave radio communications and cause auroral displays. These **magnetic storms,** as they are often called, are outward manifestations of a much more complex overall interaction between the Sun and the Earth.

SUNSPOTS, SOLAR ACTIVITY, AND THE MAGNETIC FIELD

Centuries ago, Chinese astronomers observed dark spots on the Sun's disk. More recently, Galileo cited spots as evidence that the Sun is not a perfect, unchanging celestial object, but instead has occasional flaws. Over the centuries since Galileo, observations of the spots, which individually may last for months, have revealed some very systematic behavior. The number of spots varies, reaching a peak every 11 years, and during the interval between peak numbers of spots, their locations on the Sun change steadily from middle latitudes to a concentration near the equator. When the sunspot number is high, most of them appear in activity bands about 30° north or south of the solar equator. During the next 11 years, the spots tend to lie ever closer to the equator, and by the end of the cycle they are nearly on it (Fig. 11.16). By this time, the first spots of the next cycle are already forming at middle latitudes. A plot of sunspot locations from cycle to cycle clearly shows this progression toward the equator; the plot is called a "butterfly diagram" from the shape of the pattern formed by the spots (Fig. 11.17).

A hint at the origin of the spots was found when their magnetic properties were first measured. This is accomplished by applying spectroscopic analysis to the light from the spots, taking advantage of the fact that the energy levels in certain atoms are altered by the presence of a magnetic field. As a result, spectral lines formed by those levels are split into two or more

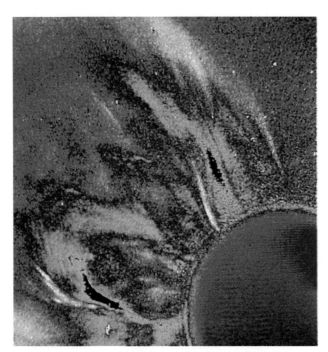

Figure 11.14   The solar wind.   This far-ultraviolet image was obtained by the Solar Maximum Mission in Earth orbit. A special color separation technique was used to show the outward flow of gas. (Data from the Solar Maximum Mission, NASA and the High Altitude Observatory National Center for Atmospheric Research, sponsored by the National Science Foundation)

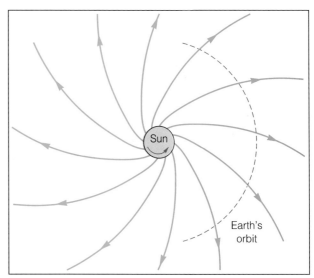

Figure 11.15   The solar wind.   This schematic diagram illustrates how ionized gas from the Sun spirals outward through the solar system in a steady stream. The solar magnetic field creates sectors of variable density in the wind.

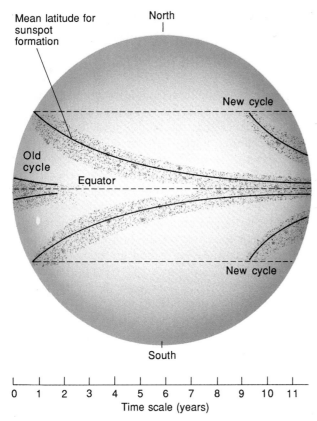

Figure 11.16   **Sunspot locations.**   This drawing shows where sunspots appear at different times in the sunspot cycle.

When the first measurements of the Sun's field were made in the first decade of this century, the field was found to be especially intense in the sunspots, where it was about 1,000 times stronger than in the surrounding gas. Scientists now think that the intense magnetic field associated with a sunspot inhibits convection and thus prevents heated gas from below from rising up to the surface. Hence the spot is cooler than its surroundings and is therefore not as bright (Fig. 11.18). The typical temperature in a spot is about 4,000 K, compared with the roughly 6,000 K of the photosphere. Using Stefan's law, we see that the intensity of light emitted in a spot compared with the surroundings is $(4,000/6,000)^4 = 0.2$; i.e., the brightness of the solar surface within a sunspot is only about one-fifth of the brightness in the photosphere outside.

When the sunspot magnetic fields were measured (Fig. 11.19), they were found to act like either north or south magnetic poles; that is, each spot has a specific magnetic direction associated with it. Furthermore, pairs of spots often appear together, with the two members of a pair usually having opposite magnetic polarities. During a given 11-year cycle, in every sunspot pair in the same hemisphere the magnetic polarities always have the same orientation. For example, during one 11-year cycle, in one hemisphere the spot to the east in each pair will generally have north magnetic polarity, while the one to the west has south magnetic polarity. During the next cycle, all the pairs will be reversed, with the south magnetic spot to the east, and the north polar spot to the west. In the other hemisphere, these relative orientations are exactly the opposite. Between cycles, when this arrangement is reversing itself, the Sun's overall magnetic field also reverses, with the solar magnetic poles exchanging places. The Sun's magnetic field and sunspot patterns actually take 22 years to repeat them-

distinct, closely spaced lines. The degree of line splitting, which is referred to as the **Zeeman effect**, depends on the strength of the magnetic field, so the field can be measured from afar simply by analyzing the spectral lines to see how widely split they are. This technique works for distant stars as well as for the Sun (except that for distant stars we cannot measure different portions of the disk separately).

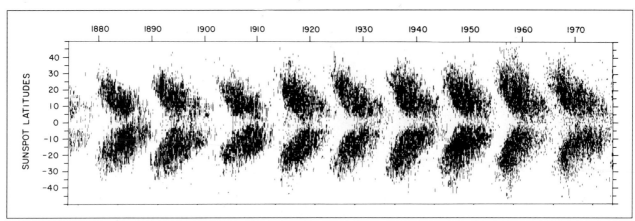

Figure 11.17   **The butterfly diagram.**   This is a plot of the latitudes of observed sunspots through several cycles of solar activity. At the beginning of each cycle, the spots tend to appear at mid-latitudes, and then as the cycle progresses, they arise at positions closer and closer to the equator. (Prepared by the Royal Greenwich Observatory and reproduced with the permission of the Science and Engineering Council, courtesy of J. A. Eddy)

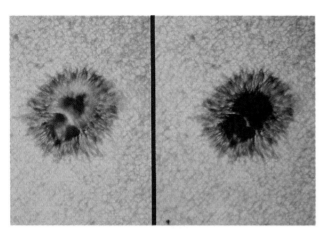

Figure 11.18   Sunspots.   This telescopic view of a group of spots shows their detailed structure. They appear dark only in comparison with their much hotter surroundings. (Mt. Wilson and Las Campanas Observatories, Carnegie Institution of Washington)

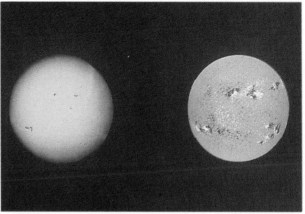

Figure 11.19   Magnetic fields on the Sun.   The magnetogram, or map of magnetic field strength (right), was made by measuring the splitting of spectral lines by magnetic fields (the Zeeman effect). Here colors (dark blue and yellow) are used to indicate regions of opposite magnetic polarity. Note that the polarities of east-west pairs in the southern hemisphere are reversed from those in the north. At the left is a visible-light photograph of the sun taken at the same time, so that the correspondence between the surface magnetic field and the visible sunspots can easily be seen. (National Optical Astronomy Observatories)

selves, so the solar cycle is really 22 sunspot years long (but note recent evidence that things are even more complicated, as discussed a little later in this chapter).

Sunspot groups, often called **solar active regions** (Fig. 11.20), are the scenes of the most violent forms of solar activity, the **solar flares.** These are gigantic outbursts of charged particles, as well as visible, ultraviolet, and X-ray emission, created when extremely hot gas spouts upward from the surface of the Sun (Fig. 11.21). Flares are most common during sunspot

maximum, when the greatest numbers of spots are seen on the solar surface. Close examination of flare events shows that the trajectory of the ejected gas is shaped by the magnetic lines of force emanating from the spot where the flare occurred. Charged particles flow outwards from a flare, some of them escaping

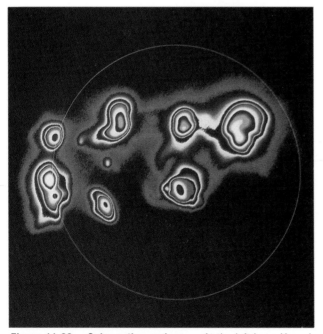

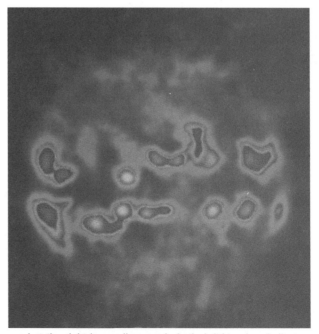

Figure 11.20   Solar active regions.   At the left is an X-ray image and at the right is a radio map. In both, bright colors indicate the solar active regions, which emit strongly at X-ray and radio wavelengths. (left: NASA; right: National Radio Astronomy Observatory)

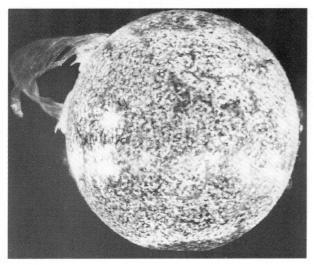

Figure 11.21   A major flare.   The gigantic looplike structure in this ultraviolet photograph obtained from *Skylab* is one of the most energetic flares ever observed. Supergranulation in the chromosphere is easily seen over most of the disk. (NASA)

into the solar wind. If the solar wind flowing from the site of the flare reaches the Earth, then the Earth is bathed by an extra dose of solar wind particles some three days later, with the effects on radio communications that have already been described. The extra quantity of charged particles entering the Earth's upper atmosphere can also be responsible for unusually widespread and brilliant displays of aurorae. Apparently flares occur when twisted magnetic field lines in the Sun suddenly reorganize themselves, releasing heat energy and allowing huge bursts of charged particles to escape into space.

Because of the connection between the solar activity cycle and reversals in the Sun's magnetic field, a model for the activity cycle was developed some years ago in which sunspots are envisioned as locations where tubes of magnetic field lines break through the solar surface. According to the model, these magnetic flux tubes connect the Sun's north and south poles and become twisted and kinked when the poles

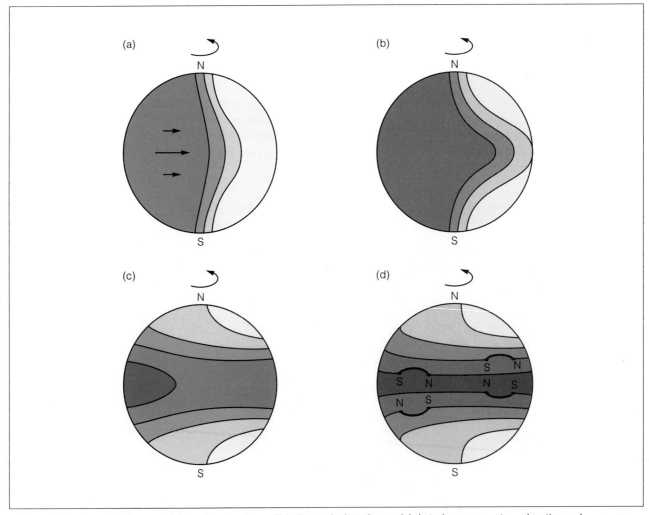

Figure 11.22   Flux tube model for solar activity.   This figure depicts the model that views sunspots as locations where magnetic flux tubes break through the surface. In this scenario, after a reversal of the magnetic poles, magnetic lines connecting the Sun's poles are bent by differential rotation (a), then stretched around the Sun's interior (b and c), until they become tangled and twisted so that they break through, forming pairs of oppositely polarized sunspots (d).

switch places. In the process they break through the surface in more and more places, ever closer to the equator, until the overall reversal of the field is completed and the flux tubes can become relatively smooth and straight again (Fig. 11.22). In this way, the flux tube model accounts for the main features of the solar activity cycle, particularly the changes in number, latitude, and magnetic field orientation of the spots during each cycle.

Unfortunately for this theory, however, recent observations have suggested that the solar activity cycle is more complicated than previously thought, and that it must be driven by some other mechanism. Astronomers have found that, 11 years before the appearance of the first spots of a new cycle, regions of intense magnetic fields occur at very high solar latitudes, close to the poles. The polarities of the magnetic fields are arranged in east-west pairs, with the same orientation of north and south magnetic polarities that will appear when the next spot cycle starts, 11 years later. Thus it appears that these regions of magnetic enhancement are early precursors of the next sunspot cycle. If so, then a typical cycle consists of an initial period where no spots are seen, but pairs

of magnetic-enhanced regions form near the poles. After 11 years, these regions have drifted down to middle latitudes, where they begin to form spots, which then increase in number as they continue moving toward the equator over the next 11 years. In this picture, it takes 22 years for the magnetic regions in one hemisphere to form and migrate toward the equator as they increase in number, and the full cycle time is 44 years (Fig. 11.23). Furthermore, two cycles are present on the Sun at the same time, because as the new magnetically enhanced regions form near the poles to begin one cycle, the similar regions from the previous cycle are just beginning to appear as spots at middle latitudes.

A possible mechanism for this large-scale solar activity cycle has been suggested, based on the discovery of surface flows on the Sun near the poles. These flows may indicate that convection in the Sun's outer regions begins near the poles with very large cells of hot gas from the interior that roll over as they spread, like gigantic doughnuts (Fig. 11.24). In this model, these rolls form every 22 years and have alternating flow directions, which causes the surface magnetic field to become compressed between them; these

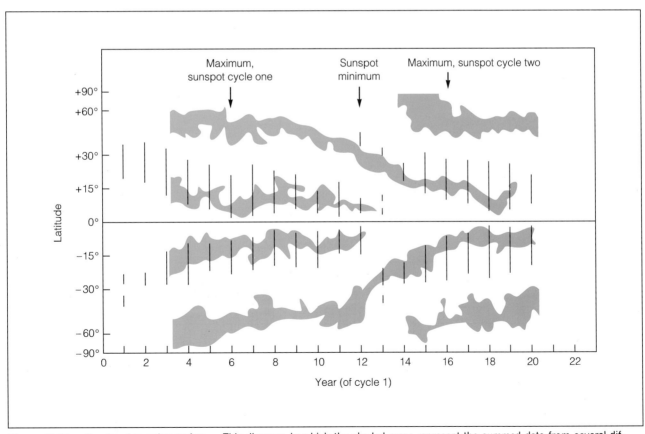

Figure 11.23 Overlapping solar cycles.   This diagram, in which the shaded areas represent the summed data from several different types of observations of solar activity, shows that two 22-year cycles may be occurring at all times. The two cycles are separated in time by one-half cycle, or 11 years, so that their patterns of active regions overlap as shown. (Adapted from a diagram in *Sky and Telescope*, 73:590, 1987)

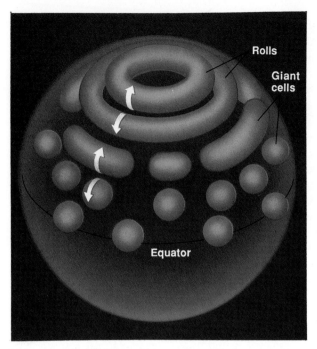

Figure 11.24   Structure within the convective zone.   These gigantic doughnut-shaped eddies, rotating in alternating directions, may be responsible for the sunspot patterns that are observed. These eddies form near the poles and then migrate toward the equator, breaking up into smaller cells in the process. The surface flows of gas that are created can compress the magnetic field in the regions where the rolls converge, causing concentrations of sunspots. (Adapted from a figure in *Sky and Telescope* 73:591, 1987)

zones where the magnetic field is compressed start at high latitudes and drift toward the equator, forming first the magnetically enhanced regions and then sunspots. At the same time, the Sun's overall magnetic field is reversing with a 22-year cycle time, explaining the behavior of the magnetic polarities of the spots.

Whether this new model of the solar activity cycle is correct or not, a fundamental question remains: What causes the 22-year reversals of the solar magnetic field? Whatever the cause, it probably originates very deep in the solar interior, so we must look there to find the underlying explanation of the activity cycle. Perhaps new information derived from studies of solar oscillations will eventually yield the missing link.

There is some evidence for long-term changes in solar behavior, at intervals longer than 22 (or 44) years, and these cycles will also require an explanation from deep inside the Sun, where the magnetic field is formed. The most striking known variation from the regular cycle occurred in the late 1600s, when sunspots seemed to stop altogether for over 50 years (Fig. 11.25). This period, called the **Maunder Minimum** after Edward W. Maunder, its discoverer (who also invented the butterfly diagram), was accompanied by extreme behavior of the Earth's climate, including a very cold period known in Europe as the "Little Ice Age." Only time will tell whether the Maunder Minimum was part of some very long-term cyclic behavior in the Sun. The implied link between sunspot activity and weather patterns on Earth is also quite uncertain; recent studies have suggested that a correlation may exist.

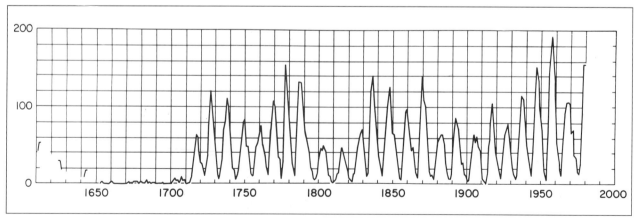

Figure 11.25   The Sun's long-term activity.   This diagram illustrates the relative level of activity (in terms of sunspot numbers) over three centuries. It is clear that the level varies, including a span of about 50 years (1650–1700), known as the Maunder Minimum, when there was little activity. There is evidence that the Sun has long-term cycles that modulate the well-known 22-year period (J. A. Eddy)

PERSPECTIVE

Our Sun, the source of nearly all our energy, is a very complex body. As stars go, it is apparently normal in all respects, so we imagine that other stars are just as complex, even though we cannot observe them in such detail.

We have explored the Sun, both in its deep interior and in the outer layers that can be observed directly. We know that nuclear fusion is the source of all the energy, and that the size and shape of the Sun are controlled by the balance between gravity and pressure throughout the interior. Somehow heat is transported above the surface, keeping the chromosphere and the corona hotter than the photosphere. Perhaps most intriguing of all are the solar activity cycle and its relationship to the Sun's complex magnetic field.

We are now prepared to study stars in general, using the Sun as an example to help us understand some of what we learn about other stars.

SUMMARY

1. The Sun is an ordinary star, one of billions in the galaxy.

2. The Sun is a spherical, gaseous object whose temperature and density increase toward the center. From its surface, the Sun emits light, whose ultimate source is nuclear reactions in the core.

3. Energy is produced in the core by nuclear fusion and is slowly transported outward by radiation, except near the surface, where energy is transported by convection.

4. The nuclear reaction that powers the Sun is the proton-proton chain, in which hydrogen is fused into helium.

5. The outer layers of the Sun govern the nature of the light it emits. These layers consist of the photosphere, with a temperature of about 6,000 K; the chromosphere, where the temperature ranges between 6,000 and 10,000 K; and the corona, where the temperature is over 1,000,000 K.

6. The photosphere, which is the visible surface, creates absorption lines, while the hotter chromosphere and corona create emission lines.

7. The excess heat in the outer layers is somehow transported there from the convective zone just below the surface, but the mechanism for transporting the heat is not fully understood.

8. The Sun emits a steady outward flow of ionized gas called the solar wind. The wind originates from coronal holes and is controlled by the solar magnetic field.

9. The sunspots occur in 11-year cycles that are part of the 22-year cycle of the Sun's magnetic field reversals. The spots are regions of intense magnetic fields where magnetic field lines from the solar interior break through the surface. Some preliminary evidence indicates that the 22-year cycle of solar activity is part of a longer-term cycle (possibly a 44-year cycle).

REVIEW QUESTIONS

1. Explain in your own words why hydrostatic equilibrium causes the Sun to have a spherical shape. Do you think the rotation of the Sun might affect its shape?

2. Why do nuclear reactions occur only in the central region of the Sun?

3. How would the lifetime of the Sun be altered if its luminosity were 10 times greater than it is and its mass were twice as large?

4. Explain why the use of special filters to isolate the wavelengths of certain spectral lines allows astronomers to examine different layers of the Sun separately.

5. Explain, in the context of Kirchhoff's rules (Chapter 4), why the chromosphere forms emission lines while the photosphere forms absorption lines.

6. Calculate the wavelengths of maximum emission for the solar photosphere (temperature 6,500 K), the chromosphere (20,000 K), and the corona (2 × 10^6 K). How can each of these layers be best observed?

7. If the average speed of the solar wind between the Sun and the Earth is 250 km/sec, how long does it take for a particle to reach the Earth once it is emitted by the Sun? How is your answer related to the time it takes for a solar flare to begin to affect radio communications on the Earth?

8. Suppose the temperature in a sunspot were one-half the temperature in the surrounding photosphere. How much less intense would the radiation be from the sunspot than from the photosphere?

9. Why do we say that the solar activity cycle is 22 years long, even though the sunspot maxima occur every 11 years? (For now, ignore the recent suggestion concerning a possible 44-year cycle.)

10. Describe a research program that you could undertake to determine the effects of solar variations on the Earth's climate.

ADDITIONAL READINGS

Bahcall, J. N. 1990. Neutrinos from the Sun: An astronomical puzzle. *Mercury* 19:53.

————. 1990. The solar neutrino problem. *Scientific American* 262(15):54.

Foukal, P. V. 1990. The variable Sun. *Scientific American* 262(2):34.

Friedman, H. 1986. *Sun and Earth* (New York: W. H. Freeman).

Giampapa, M. S. 1987. The solar-stellar connection. *Sky and Telescope* 74:142.

Gough, D. 1976. The shivering Sun opens its heart. *New Scientist* 70:590.

Harvey, J. W., J. R. Kennedy, and J. W. Leibacher. 1987. GONG: To see inside our Sun. *Sky and Telescope* 74:470.

Newkirk, G., and K. Frazier. 1982. The solar cycle. *Physics Today* 35(4):25.

Parker, E. N. 1975. The Sun. *Scientific American* 233(3):42.

Pusachoff, J. 1991. The Sun: A star close up. *Mercury* 20(3):66.

Walker, A. B. C., Jr. 1982. A golden age for solar physics. *Physics Today* 35(11):61.

Wallenhorst, S. G. 1982. Sunspot numbers and solar cycles. *Sky and Telescope* 64(3):234.

Wentzel, D. 1991. Solar chimes: Searching for oscillations inside the Sun. *Mercury* 20(3):77.

Williams, G. E. 1986. The solar cycle in precambrian times. *Scientific American* 255(2):88.

Wilson, O. C., A. H. Vaughan, and D. Mihalas. 1981. The activity cycle of stars. *Scientific American* 244(2):104.

Wolfson, R. 1983. The active solar corona. *Scientific American* 148(2):104.

Zirin, H. 1988. Heading toward solar maximum. *Sky and Telescope* 76:335.

ASTRONOMICAL ACTIVITY
Imaging the Sun

The brilliance of the Sun makes it difficult and dangerous to try to look at it directly. But with care, you can create an image of the Sun that can be viewed safely. There are at least two ways to do this: making a pinhole "camera" or using a pair of binoculars to project a solar image.

Use of binoculars is easier and has the further advantage of easily creating a large enough image of the Sun to allow features such as sunspots to be seen. But if binoculars are not available, it is not difficult to make a pinhole device for viewing a solar image. To do this, mount two pieces of thin, stiff cardboard onto a board so that they are parallel to each other (see the drawing). Make a small hole in one piece of cardboard; the other will be used as a screen onto which the Sun's image will be projected. The hole acts as a lens, forming an image of the Sun. The size of the image depends on the distance between the pinhole and the screen, so it is best to mount the two pieces of cardboard as far from each other as possible. Alternatively, you can use a cardboard box, with the pinhole in one end and the other end acting as the screen; the only difficulty is that unless the box is very large, the resulting image will be rather small. Whether you use two separate pieces of cardboard mounted on a board or a large box, you will need to add a shade over the screen, so that the solar image will contrast more strongly with the surrounding screen.

Using binoculars to create a solar image is very simple and requires nothing more than the binoculars and a piece of stiff paper or cardboard to act as a screen onto which the image can be projected. First, place the screen on the ground with one side propped up so that it is approximately perpendicular to the incoming sunlight (see the drawing). Then hold the binoculars about two feet above the screen, with the large end pointing toward the Sun. The binoculars will project an image of the Sun onto the screen (note: you will get two solar images if you take all the lens caps off, or you can leave the caps on one side of the binoculars and obtain a single image). You will probably have to move the binoculars around a bit until they are properly aligned so that the solar image falls onto the screen, but when you do see the image, you will find that it is large enough to allow some detail, such as major sunspots, to be seen.

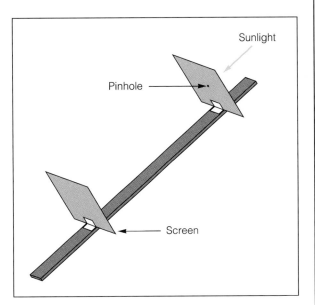

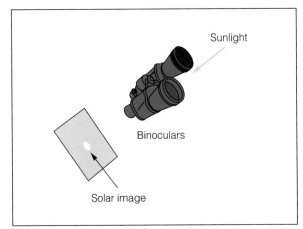

You will also find that the image is very bright, perhaps too bright to be viewed without discomfort. An easy way to reduce the intensity of the image is to block out some of the light entering the binoculars; this can be done by making a hole in a small piece of cardboard and taping it over the large end of the binoculars. The intensity of the image will be reduced by the ratio of the area of the hole in your cover piece to the area of the lens when it is unblocked.

Observations and Basic Properties of Stars

Orion setting, showing the colors of the stars (NOAO)

PREVIEW

In order to understand the stars, we begin in this chapter by discussing how they are observed. Here we confine ourselves to the information that comes directly from the analysis of the light that reaches us; in later chapters we will see how astronomers combine this information with the laws of physics to develop an understanding of the structure and evolution of the stars. In this chapter you will learn about three basic types of observations: positional measurements, brightness determinations, and the measurement of spectra. Then you will see how astronomers use these measurements to deduce the basic properties of stars; that is, how we determine the sizes and masses of stars, their composition, their temperatures, and other properties. In these discussions you will learn that double or binary stars are both very common and very useful to astronomers because they provide important data on the nature of stars.

Everything that we can learn about a star is contained in the light that we receive from it. Fortunately, a lot of information is there, and astronomers have learned how to extract much of it.

In this chapter we will first discuss the three basic types of astronomical observations and then learn how fundamental parameters of stars are measured using these techniques.

THREE WAYS OF LOOKING AT STARS: POSITIONS, MAGNITUDES, AND SPECTRA

Nearly all observations of stars fall into one of three categories, involving measurements of position, brightness, or spectra. The first two methods have rather long histories. Positional measurements date back to the first human observations of the skies, and brightness measurements (although rather crude ones) were made by Hipparchus in the second century B.C. In the following sections, each method is discussed separately.

Positional Astronomy

The science of measuring star positions is called **astrometry.** Ancient astronomers used simple devices such as quadrants and sextants (Fig. 12.1) to measure angular positions on the sky, but today astronomers usually photograph the sky and carefully measure the positions of the star images on the photographic plates. If several photographs of a given portion of the sky are taken and measured separately, the results can be averaged to produce a more accurate deter-

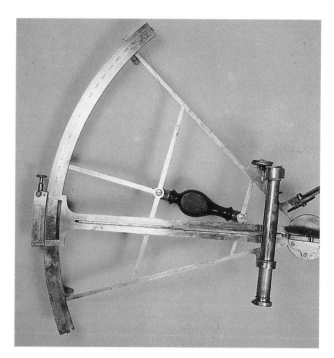

Figure 12.1   A sextant.   Instruments of this type were used for centuries to measure star positions relative to each other. In modern times, sextants are used as tools for navigation. (The Granger Collection)

mination than is possible from a single photograph. Today a stellar position can be measured to a precision of less than 0.01 arcsecond. When the European astrometric satellite *Hipparcos* went into operation (in 1990), even more accurate measurements became possible, because the fuzziness of star images caused by the Earth's atmosphere was eliminated.

Modern astronomers make astrometric measurements for a variety of reasons. For example, such measurements are used in analyzing stellar motions and in determining what these motions reveal about the structure of the galaxy (see Chapter 16). Astrometric data are also very important in cataloging stars with sufficient accuracy that they can be observed by other telescopes. (A major task was the preparation of star lists for *Hubble Space Telescope* observations, which will involve much fainter stars than were listed in previous catalogs.) Finally, and perhaps most importantly, positional measurements can be used to measure distances to stars by making use of their parallax motions.

Recall from Chapter 2 that stellar parallax is the apparent shifting in a nearby star's position due to the orbital motion of the Earth (Fig. 12.2). Ancient and medieval astronomers were unable to detect stellar parallax, leading most of them to reject the idea that the Earth orbits the Sun. Eventually the heliocentric theory won out, but stellar parallax remained undetected until 1838. The reason stellar parallaxes

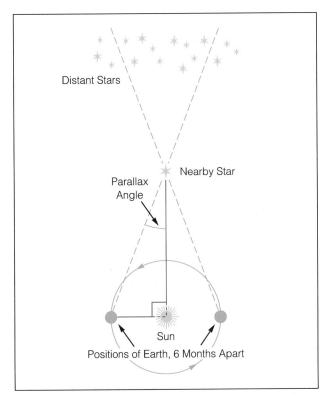

Figure 12.2 Stellar parallax. As the Earth orbits the Sun, the direction of our line of sight to a nearby star varies, so that the star appears to move back and forth against the more distant background stars. The parallax angle *p* is defined as one-half of the total angular motion. If *p* is one arcsecond, the distance to the star is 206,265 AU, or one parsec. This drawing is not to scale.

had defied earlier observers now became obvious: even for the closest stars, the maximum shift in position was less than 1 arcsecond. The stars are simply much farther away than the ancient astronomers had dreamed possible. The annual parallax motion of the star α Centauri, the nearest to the Sun, is comparable to the angular diameter of a dime as seen from a distance of about 2 miles, a very small angle indeed.

The successful detection of stellar parallax led to a direct means of determining distances to stars. The amount of shift in a star's apparent position due to our own motion about the Sun depends on how distant the star is; the closer it is, the bigger the shift. The parallax angle is defined as one-half the total angular shift of a star during the course of a year (see Fig. 12.2). Once it is measured, this angle tells us the distance to a star in a unit of measure called the **parsec (pc)**. A star whose parallax is 1 arcsecond is, by definition, 1 parsec away (the word *parsec,* in fact, is a contraction of *parallax-second,* meaning a star whose parallax is 1 arcsecond). In mathematical terms, the distance to a star whose parallax is *p* is *d* = 1/*p*. Thus, if a star has a parallax of 0.4 arcsecond,

it is 1/0.4 = 2.5 pc away. Recall that the closest star has a parallax of less than 1 arcsecond; we conclude therefore that even this star is more than 1 pc away.

In more familiar terms, a parsec is equal to 3.26 light-years, or 3.08×10^{16} meters, or 206,265 astronomical units (AU). The use of parallax measurements to determine distances is a very powerful technique; it is the only direct means astronomers have for measuring how far away stars are. Unfortunately, only stars rather close to us in the galaxy have parallaxes large enough to be measured. The smallest parallax that can be measured accurately is about 0.01 arcsecond, corresponding to a distance of 100 pc. Our galaxy, on the other hand, is more than 30,000 pc in diameter! Clearly other methods of determining distance are needed if we are to probe the entire galaxy. One very powerful method will be described later in this chapter.

Stellar Brightness

The second general type of stellar observation is the measurement of the brightness of stars. This was first attempted in a systematic way over 2,000 years ago when Hipparchus established a system of brightness rankings that is still used today. Hipparchus ranked the stars in categories called **magnitudes,** from first magnitude (the brightest stars) to sixth (the faintest visible to the unaided eye). In his catalog of stars, and in most since then, these magnitudes are listed along with the star positions.

The magnitude system has been modernized, and all astronomers use the same technique for measurement rather than relying on subjective impressions. In the mid-1800s scientists discovered that what the eye perceives as a fixed *difference* in intensity from one magnitude to the next actually corresponds to a fixed intensity *ratio*. Measurements showed that a first-magnitude star is about 2.5 times brighter than a second-magnitude star; a second-magnitude star is 2.5 times brighter than a third-magnitude star; and so on (Fig. 12.3). The brightness ratio between a first-magnitude star and a sixth-magnitude star was found to be nearly 100. In 1850 the system was formalized by the adoption of this ratio as exactly 100; thus, the ratio corresponding to a one-magnitude difference is the fifth root of 100, or $(100)^{1/5}$ = 2.512. Hence a first-magnitude star is 2.512 times brighter than a star of second magnitude, a sixth-magnitude star is 2.512 × 2.512 = 6.3 times fainter than a fourth-magnitude star, and so on.

Magnitudes are most commonly measured through the use of **photometers;** these devices produce an electric current when light strikes them and are used in many familiar applications, such as door openers in modern buildings. The amount of electrical current produced is determined by the intensity of light, so

ASTRONOMICAL INSIGHT
Star Names and Catalogs

From the earliest times, when astronomers systematically measured star positions, they began to compile lists of these positions in catalogs. Hipparchus developed an extensive catalog, as did early astronomers in China and other parts of the world, more than 2,000 years ago (see Chapter 2).

Listing stars in a catalog of positional measurements requires some kind of system for naming or numbering the stars, and a variety of systems have been employed. The ancient Greeks designated stars by the constellation and the brightness rank within the constellation; for example, the brightest star in Orion is α Orionis, the next brightest is β Orionis, the next is γ Orionis, and so on. (The constellation names are given in their Latin versions, since this system was perpetuated by the Romans and the Catholic church after the time of the Greeks.) Today many stars are still referred to by their constellation rankings, using the Greek alphabet to indicate the rank.

After the time of Ptolemy, when Western astronomy went into decline for some 1,300 years, Arab astronomers, occupying northern Africa and southern Europe, carried on astronomical traditions. The Arabs assigned proper names to many of the brightest stars, and these names, such as Betelgeuse (α Orionis), Rigel (β Orionis), Vega (α Lyrae), and Spica (α Virginis), are also still in use.

With the advent of the telescope, when many new stars were discovered that were too faint to be seen with the naked eye, catalogs rapidly outgrew the old naming systems. Generally each catalog now assigned numbers to the stars, usually in a sequence related to their positions. Thus, modern catalogs, such as the *Henry Draper Catalogue,* the *Boss General Catalogue,* the *Yale Bright Star Catalogue,* and the *Smithsonian Astrophysical Observatory Catalog,*

list stars by increasing right ascension (i.e., from west to east in the sky, in the order in which the stars pass overhead at night). In some cases, separate listings are made for different zones of declination (i.e., for different east-west strips of sky, separated in the north-south direction). Because each of these catalogs includes a different set of stars, each has its own numbering system, and often no cross-reference to other catalogs is provided. Hence a given star may have a constellation-ranked name, an Arabic name, and numbers in a variety of catalogs, but only by matching up the coordinates is it possible to be sure you are referring to the same star in different catalogs. Actually things are not quite so bad as that, since some modern catalogs do provide cross-references to others. Nevertheless, one of the principal tasks of a would-be astronomer is to become familiar with catalogs and their uses.

an astronomer need only measure the current to determine the brightness of a star (Fig. 12.4) and hence its magnitude.

Once magnitudes could be measured precisely, astronomers found that stars have a continuous range of brightnesses and do not fall neatly into the various magnitude rankings. Therefore fractional magnitudes must be used; Deneb, for example, has a magnitude of 1.26 in the modern system, although it was formerly classified simply as a first-magnitude star. Each of the former categories has been found to include a

range of stellar brightnesses. This is especially the case for the first-magnitude stars, some of which turned out to be as much as two magnitudes brighter than others. To measure these especially bright stars in the modern system, magnitudes smaller than 1 must be adopted. Sirius, for example, which is the brightest star in the sky, has a magnitude of −1.45. By using negative magnitudes for very bright objects, astronomers can extend the system to include such objects as the Moon, whose magnitude when full is about −13, and the Sun, whose magnitude is −26

Figure 12.3   Stellar magnitudes.
Magnitude differences here are indicated above the line; below the line are the corresponding brightness ratios.

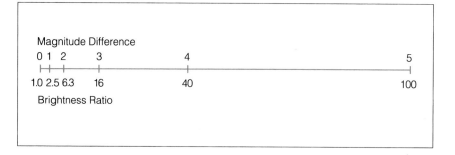

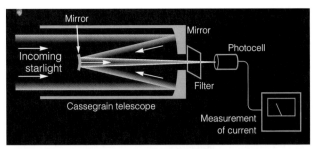

Figure 12.4   Measuring stellar magnitudes.   This is a schematic illustration of photometry, the measurement of stellar brightnesses. Light from a star is focused (often through a filter that screens out all but a specific range of wavelengths) onto a photocell, a device that produces an electric current in proportion to the intensity of light. The current is measured and, by comparison with standard stars, converted into a magnitude.

TABLE 12.1   Apparent Magnitudes of Familiar Objects

Object	Apparent Magnitude
Sun	−26.74
100-watt bulb at 100 ft	−13.70
Moon (full)	−12.73
Venus (greatest elongation)	−4.22
Jupiter (opposition)	−2.6
Mars (opposition)	−2.02
Sirius (brightest star)	−1.45
Mercury (greatest elongation)	−0.2
Alpha Centauri (nearest star)	−0.1
Large Magellanic Cloud	+0.1
Saturn (opposition)	+0.7
Small Magellanic Cloud	+2.4
Andromeda galaxy (farthest naked eye object)	+3.5
Brightest globular cluster (47 Tucanae)	+4.0
Orion nebula	+4
Uranus (opposition)	+5.5
Faintest object visible to naked eye	+6.0
Neptune (opposition)	+7.9
Crab nebula	+8.6
3C273 (brightest quasar)	+12.8
Pluto (opposition)	+14.9

(thus the Sun is about 25 magnitudes—or a factor of $2.512^{25} = 10^{10}$—brighter than Sirius). Table 12.1 lists the magnitudes of a variety of objects.

Of course, some stars are fainter than the human eye can see, so the magnitude scale must also extend beyond sixth magnitude. With moderately large telescopes, stars as faint as fifteenth magnitude can be measured, and images obtained with modern electronic detectors at large telescopes can detect objects as faint as magnitude 31. A star of magnitude 30 is 24 magnitudes—or a factor of $2.512^{24} = 4 \times 10^9$—fainter than the faintest star visible to the unaided eye.

The stellar magnitudes we have discussed so far all refer to visible light. As we learned in Chapter 4, however, stars emit light over a much broader wavelength band than the eye can see. We also learned that the wavelength at which a star emits most strongly depends on its temperature. By measuring a

star's brightness at two different wavelengths, it is therefore possible to learn something about its temperature. For this purpose astronomers use filters that allow only certain wavelengths of light to pass through. A star's brightness is typically measured through two such filters, one that passes yellow light and one that passes blue light, resulting in the measurement of V (for visual, or yellow) and B (for blue) magnitudes (Fig. 12.5). Since a hot star emits more light in blue wavelengths than in yellow, its B magnitude is *smaller* than its V magnitude. For a cool star, the situation is reversed, and the B magnitude is

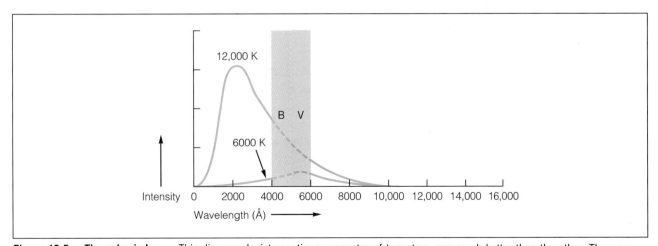

Figure 12.5   The color index.   This diagram depicts continuous spectra of two stars, one much hotter than the other. The wavelength ranges over which the blue (*B*) and visual (*V*) magnitudes are measured are indicated. We see that the hot star is brighter in the *B* region than in the *V* region of the spectrum; therefore, its *B* magnitude is *smaller* than its *V* magnitude, and it has a negative *B — V* color index. The opposite is true for the cool star.

larger than the *V* magnitude (Remember that the magnitude scale is backward in the sense that a lower magnitude corresponds to greater brightness.)

The difference between the *B* and *V* magnitudes is called the **color index.** The exact value of this index is a function of the temperature of a star, so that stellar temperatures may be estimated simply by measuring the *V* and *B* magnitudes. A very hot star might have a color index $B - V = -0.3$, while a very cool one might typically have $B - V = +1.2$.

One additional type of magnitude should be mentioned, although measuring it directly is very difficult. This is the **bolometric magnitude,** which includes all the light emitted by a star at all wavelengths (Fig. 12.6). To determine a bolometric magnitude requires ultraviolet and infrared, as well as visible, observations and can best be done by using telescopes in space. This is particularly true for hot stars, which emit a large fraction of their light at ultraviolet wavelengths. For cool stars, which emit little ultraviolet but a lot of infrared radiation, bolometric magnitudes can be measured from the Earth's surface with fair accuracy. The bolometric magnitude of a star is always smaller than the visual magnitude, because more light is included when all wavelengths are considered. Often bolometric magnitudes are estimated by comparison with "standard" stars, stars whose properties are well known and assumed to represent all stars of similar type. The normal means of comparison is to measure the visual magnitude of the target star and then add a **bolometric correction,** a negative correction factor based on standard stars. We will see later in this chapter how bolometric magnitudes are related to other properties of stars, particularly luminosity.

Measurements of Stellar Spectra

The first observations of stellar spectra, made in the mid-1800s (before scientists understood how spectral

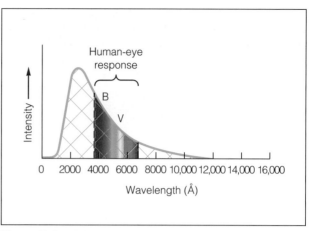

Figure 12.6   Bolometric magnitude.   The bolometric magnitude includes all the energy emitted by a star at all wavelengths, not just the visible spectrum.

lines are formed), were accomplished using **spectroscopes,** simple devices that allow the observer to see the light from a star spread out according to wavelength, but not to record it. In the late 1800s introduction of the technique of photographing spectra allowed a systematic study of stellar spectra to get under way. A photograph of a spectrum is called a **spectrogram** (Fig. 12.7). Spectra of many stars can be photographed at one time when a thin prism is placed in front of a telescope (Fig. 12.8).

Among the first astronomers to examine spectra of a large number of stars systematically was the Roman Catholic priest Angelo Secchi, who in the 1860s cataloged hundreds of spectra using a spectroscope. He found that the appearance of the spectra varied considerably from star to star, although they were consistent in one respect: they all showed continuous spectra with absorption lines. The work of Kirchhoff soon showed that this was due to the relatively cool outer layers of a star, which absorb light from the

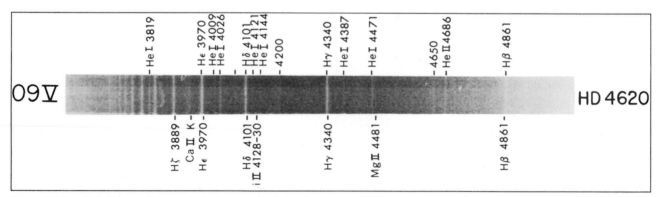

Figure 12.7   A stellar spectrum.   This photograph of a stellar spectrum shows the manner in which spectra are usually displayed, as negative prints. Hence light features are absorption lines, wavelengths at which little or no light is emitted by the star (From Abt. H., W. W. Morgan, and R. Tapscott. 1968. *An Atlas of Low-Dispersion Grating Stellar Spectra* Tuscon-Kitt Peak National Observatory)

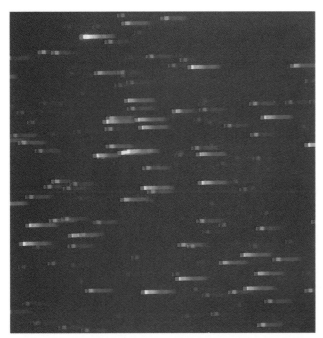

Figure 12.8   Stellar spectra.   This photograph was made through a telescope with a thin prism in the light beam, so that each stellar image was dispersed into a spectrum. It is possible to measure and classify the spectra of large numbers of stars using this technique. (University of Michigan Observatories)

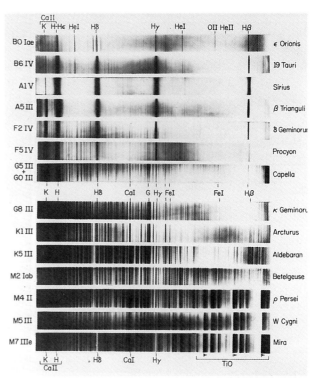

Figure 12.10   A comparison of spectra.   Here we see several spectra representing different spectral classes. A few major absorption lines are identified. The spectra are arranged in order of decreasing stellar temperature (top to bottom). (University of Michigan Observatories)

hotter interior. At first it was thought that the differing appearances of stellar spectra were caused by differences in the chemical composition of the stars, and in one of the early classification schemes, stars were assigned to categories based on their compositions. The basis of the modern classification system for stellar spectra was established by a group of astronomers at Harvard, most notably Annie J. Cannon (Fig. 12.9). Cannon found a smooth sequence in which the pattern of strong absorption lines changed gradually from one type of spectrum to the next. Having already assigned letters of the alphabet to the various

types, she placed them in the sequence O, B, A, F, G, K, M (Fig. 12.10).

It was later realized that the differing appearances of stellar spectra were due not to differences in chemical composition, but to differences in temperature. The hotter a star, the more highly ionized the gas in its outer layers is. The degree of ionization in turn governs the pattern of spectral lines that will form.

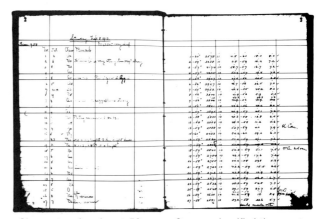

Figure 12.9   Annie J. Cannon.   A member of the Harvard College Observatory for almost 50 years, Cannon classified the spectra of several hundred thousand stars. At the right is a page from one of her notebooks. Today she is recognized as the founder of modern spectral classification. (Harvard College Observatory)

ASTRONOMICAL INSIGHT
Stellar Spectroscopy and the Harvard Women

There is some irony in the fact that Harvard University, long a stronghold of the all-male tradition in American colleges, was the institution that nurtured some of the nation's first leading women astronomers. Even today there are few women among professional astronomers, but they were even more underrepresented at the turn of the century when the foundations of modern stellar spectroscopy were developed at the Harvard College Observatory.

When Edward C. Pickering became director of the observatory in 1877, Secchi's work (described in the text), based on visual inspection of stellar spectra through a spectroscope, had been the only attempt to classify stars according to the appearance of their spectra. In 1872 Henry Draper, an American amateur astronomer, had become the first to photograph the spectrum of a star. Upon Draper's death in 1882, his widow endowed a new department of stellar spectroscopy at Harvard. As director, Pickering hired among his assistants a number of women, several of whom went to work on the problem of classifying spectra of stars.

An innovative technique was used to photograph spectra of large numbers of stars. A thin prism was placed in front of the telescope, so that each star image on the photo-

graphic plate at the focus was stretched, in one direction, into a spectrum. If color film had been used (it wasn't), each stellar image would have looked like a tiny rainbow (see Fig. 12.8). Pickering and his group refined this **objective prism** technique to the point where all stars in the field of view as faint as ninth or tenth magnitude would appear as spectra sufficiently exposed for classification. Thus it became possible to amass stellar spectra in vast quantities.

The problem of sorting out and classifying the spectra fell to Pickering's associates. One of them, Williamina Fleming, published the first *Draper Catalog of Stellar Spectra* in 1890. This catalog assigned some 10,351 stars in the Northern Hemisphere to spectral classes A through N, in a simple elaboration of a rudimentary classification scheme adopted earlier by Secchi. A number of Fleming's classes were later dropped, however.

While the first catalog was being prepared, a niece of Henry Draper, Antonia Maury, joined the staff and set to work on the analysis of spectra of bright stars. Spectra for these objects could be photographed using thicker prisms (actually, a series of two or three thin prisms), a technique that spread out the spectra more widely according to wavelength and allowed greater detail to

be seen. Maury concluded, on the basis of these high-quality spectra, that stars should be grouped into three distinct sequences rather than just one. She had found that some stars had unusually narrow spectral lines, others were rather broad, and a third group was in between. It was later found that her sequence *c*, the thin-lined stars, are giant stars. (The lower atmospheric pressure in these stars causes the line to be less broadened than in main-sequence stars.) Maury's discovery helped the Swedish astronomer Ejnar Hertzsprung (discussed later in the chapter) confirm the distinction he had recently found between giant and normal stars, and he thought her work to be of fundamental importance. Unfortunately Maury's separate sequences were not adopted in the further work on classification at Harvard, which was thereafter based on a single sequence.

The most important of the Harvard workers in spectral classification arrived on the scene in 1896. Her name was Annie J. Cannon, and she gradually modified the classification system to the present sequence, finding that the arrangement O, B, A, F, G, K, M was a logical ordering, with smooth transitions from one type to the next. (At this time, no one was yet aware that this was a temperature sequence.) Cannon was able to distinguish the gradations so

A large group of Harvard astronomy assistants, about 1917. Sixth from the left is Henrietta Leavitt, who discovered the period-luminosity relationship for variable stars (see Chapter 18); and fifth from the right is Annie J. Cannon, whose contributions to spectral classification are described in this chapter.

finely that she established the ten subclasses for each major division that are in use today. In 1901 she published a catalog of classifications for 1,122 stars and then embarked on her major task: the classification of over 200,000 stars whose spectra appeared on survey plates covering both hemispheres. By this time she was so skilled that she could reliably classify a star in a few minutes, and most of the job was done in a four-year period between 1911 and 1914. The resulting *Henry Draper Catalog* appeared in nine volumes of the *Annals of the Harvard College Observatory*, the final one being published in 1924. Pickering died in 1919, before the catalog was complete, and was succeeded as director by Harlow Shapley.

Cannon continued her work, later publishing a major *Extension* of the catalog, along with a number of other specialized catalogs. She died in 1941, and the American Astronomical Society subsequently established the Annie J. Cannon prize for outstanding research by women in astronomy.

The successes of the Harvard women were not confined to stellar spectroscopy. Another major area of interest to Pickering and later to Shapley was the study of variable stars, and Henrietta Leavitt, who joined the group as a volunteer in 1894, played a leading role in this area. Following some early work on the establishment of standard stars for magnitude determinations, by 1905 Leavitt was at work identifying variables from comparisons of photographic plates. (She was eventually to discover over 2,000 of them.) In the process she examined the Magellanic Clouds for variables and noticed that their periods of pulsation correlated with their average brightnesses. This was a discovery of profound importance; the period-luminosity relationship for variable stars was to become an essential tool for establishing both the galactic distance scale and the intergalactic scale (see the discussion in Chapter 16).

Another major figure, who came to Harvard later (1920s), was Cecilia Payne, who was a pioneer in the development of stellar astrophysics. Payne was the first to demonstrate, through detailed and complex analyses of stellar spectra, combined with theoretical considerations of ionization and radiative processes, that the chemical compositions of most stars are dominated by hydrogen, and that the makeup of most stars is the same.

The Harvard women, including those discussed here and several whose names are now obscure, were a remarkable group. They were responsible for a number of major advances in the science of astronomy at a time when its basis in physics was just becoming clear. In this day of awakening recognition of the proper role of women in all areas, it is fitting to consider and appreciate the pioneering work done by this group.

(If this is confusing, review the discussion of ionization in Chapter 4.) Therefore Cannon's sequence of spectral types was a *temperature* sequence: the hottest stars are the O stars, and the coolest are the M stars. She developed a fine enough eye for subtle differences between spectra to assign subclasses to each of the major classes. In this system, the Sun is a G2 star, being an intermediate between types G and K.

In the modern classification system, a few key spectral lines establish the type of a given star (Fig. 12.11). For the O stars, ionized helium, which requires a very high temperature for its formation, is the principal species that reveals the spectral type. For the slightly cooler B stars, it is atomic helium; for the A stars, atomic hydrogen. The F stars have strong hydrogen lines, along with lines due to certain metallic elements that are ionized once (that is, these atoms have lost just one electron). The G stars have a mixture of ionized and atomic metals, and in the K and M stars these elements are nearly all in atomic form. The cooler M stars also have strong molecular lines.

Having established this classification system around the turn of the century, Cannon cataloged nearly a quarter of a million stars, a monumental task that took about five years (although the publication of the results, called the *Henry Draper Catalog*, took

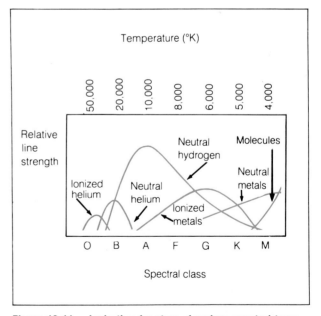

Figure 12.11   Ionization for stars of various spectral types. This diagram shows which ions appear prominently in the spectra of stars of different classes. Note that the degree of ionization evident for the hot stars (left) is much greater than for the cool ones (right); for example, helium is much more difficult to ionize than are the metallic elements. Hence an O star, which has ionized helium, is hotter than the F, G, and K stars, which have ionized metals but no ionized helium.

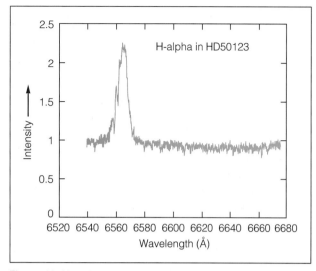

Figure 12.12   A stellar emission line.   This graph shows an emission line due to hydrogen in a star belonging to a peculiar class of stars known to have hot gas surrounding them. (T. P. Snow; data obtained at the Canada-France-Hawaii Telescope)

place over a much longer period). This catalog has been a fundamental reference for generations of astronomers, and the system of classification established by Cannon has, with some modification, been in use since its development.

Some stars do not fit neatly into the standard spectral classes, and these are often referred to as "peculiar" stars. In most cases these are stars whose chemical composition is unusual, at least at the surface where the spectral lines form. Most stars have the same basic composition. (See the list of relative abundance of the elements in the Sun in Table 11.3). Others are unusual because they have emission lines in their spectra (Fig. 12.12), which, according to Kirchhoff's laws, means that they must be surrounded by hot, rarefied gas (Fig. 12.13). The so-called peculiar stars are probably quite normal but in short-lived stages of evolution, so there are not many of them around at any one time.

The variable stars represent another kind of unusual star. The majority of these are stars whose brightness fluctuates regularly as they alternately expand and contract (Fig. 12.14). As in the case of the peculiar stars, the variables are normal stars in special stages of their lifetimes where particular combinations of atmospheric pressure and ionization conditions produce instabilities that cause the pulsations. The most widely known pulsating variable stars are **δ Cephei** stars, giants whose spectral type varies between F and G as they pulsate. As we will see in Chapter 18, these stars, as well as the less luminous **RR Lyrae** variables, are very useful tools in measuring distances, because their luminosities can be inferred from their periods of pulsation. Other pulsating stars include the **Mira** stars, or **long-period variables,** M supergiants that take a year or longer to go through a complete cycle; and a variety of shorter-period variables that are not quite as regular as the δ Cephei and RR Lyrae stars. Some stars vary erratically, even explosively, and these will be discussed in Chapter 15.

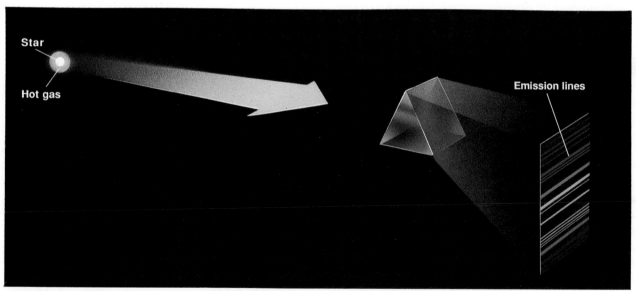

Figure 12.13   Stellar emission lines.   Very few stars have emission lines in their visible-wavelength spectra (but many do at ultraviolet wavelengths; see Chapter 13). When such lines are present, it usually signifies the existence of hot gas above the surface. According to Kirchhoff's second law, a rarefied, hot gas produces emission lines.

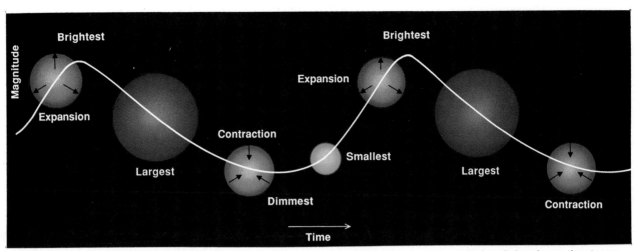

Figure 12.14   A pulsating variable star.   The curved line shows how the brightness (in magnitude units) varies as the star expands and contracts. The sequence of sketches illustrates how the expansion and contraction phases are related in sequence to the variations in brightness. The surface temperature also varies.

BINARY STARS

About half of the stars in the sky are members of double or **binary** systems, where two stars orbit each other regularly. All types of stars can be found in binaries, and their orbits also come in many sizes and shapes. In some systems the two stars are so close together that they are actually touching, and in others they are so far apart that it takes hundreds or thousands of years for them to complete one revolution.

Binary systems can be detected by each of the three different types of observations we have discussed (Table 12.2). Positional measurements, of course, tell us when two stars are very close to each other in the sky. Sometimes this proximity occurs by chance, when a nearby star happens to lie almost in front of one that is in the background (Fig. 12.15); this is

TABLE 12.2   Types of Binary Systems

Type of System	Observational Characteristic
Visual binary	Both stars visible through telescope (usually requires photograph); change of relative position confirms orbital motion.
Astrometric binary	Positional measurements reveal orbital motion about center of mass.
Spectrum binary	Spectrum shows lines from two distinct spectral classes, revealing presence of two stars.
Spectroscopic binary	Spectral lines undergo periodic Doppler shifts, revealing orbital motion. If two sets of spectral lines are seen shifting back and forth (which occurs only if the two stars are comparable in magnitude), it is a double-lined spectroscopic binary; if only one set of lines is seen (in the more common situation where one star is much brighter than the other), it is a single-lined spectroscopic binary.
Eclipsing binary	Apparent magnitude varies periodically because of eclipses, as stars alternately pass in front of each other.

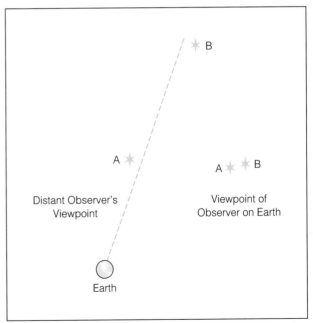

Figure 12.15   An optical double.   At the left we see the line of sight from the Earth passing close to two stars (A and B) that are far apart but in nearly the same direction from Earth. At the right we see that these two stars appear very close together on the sky. This is not a true binary system.

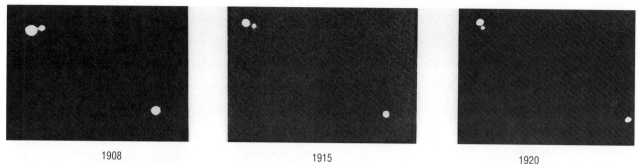

1908   1915   1920

Figure 12.16   Binary star Krueger 60.   Between 1908 and 1920, the visual binary, in the upper left corner of each photograph, completed about a quarter of a revolution. (Photographs by E. E. Barnard, Yerkes Observatory)

called an **optical double** and is not a true binary system, since the two stars do not orbit each other. A pair of stars that appear close together and are in motion about one another is called a **visual binary** (Fig. 12.16). Accurate positional measurements are needed to reveal the orbital motion, because the two stars are very close together in the sky and appear to move very slowly about each other. The two stars must be separated by many AUs to appear separately as seen from the Earth, and therefore the orbital period is many years.

It is possible for binary systems to be recognized even when only one of the two stars can be seen. If positional measurements over a long period of time reveal that a star exhibits a wobbling motion as it moves through space (Fig. 12.17), it may be inferred

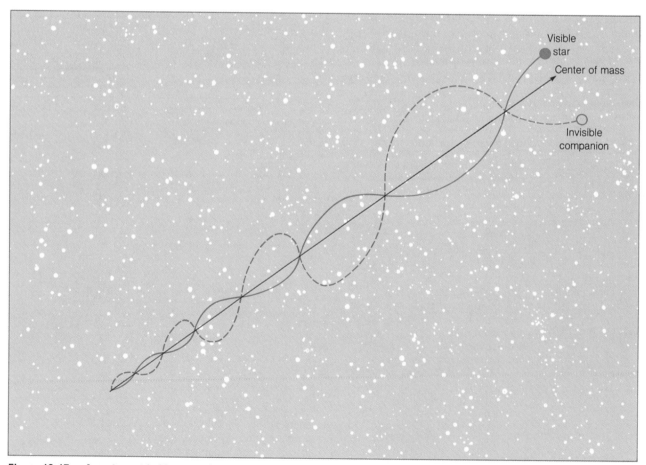

Visible star

Center of mass

Invisible companion

Figure 12.17   An astrometric binary.   Careful observation of the motion of a star across the sky in some cases reveals a curved path such as the one shown here. Such motion is caused by the presence of an unseen companion star; the visible star orbits the center of mass of the binary system as it moves across the sky.

that the star is orbiting an unseen companion. Such systems, detected because of variations in position, are called **astrometric binaries.**

Simple brightness measurements can also tell us when a star, which may appear to be single, is actually part of a binary system. If we happen to be aligned with the plane of the orbit, so that the two stars alternately pass in front of each other as we view the system, the observed brightness will decrease each time one star is in front of the other. This is called an **eclipsing binary** (Fig. 12.18).

Finally, spectroscopic measurements can be used to recognize binary systems even when a star appears single. Most often this is made possible by the Doppler effect, which causes the spectral lines to shift slightly in position as the stars orbit each other, alternately approaching and receding from the Earth (Figs. 12.19 and 12.20). Star pairs known by this technique to be binaries are called **spectroscopic binaries.** In many cases, one of the two stars is too dim for its light to be detected, but the system is still recognized as binary because of the shifts in the spectral lines of the brighter star.

A given double star system may fall into more than one category. For example, a relatively nearby system seen edge-on could be a visual binary if both stars can be seen in the telescope; it may also be an eclipsing binary if the two stars alternately pass in front of each other; and it almost surely will be a spectroscopic binary since the motion back and forth along our line of sight is maximized when we view the orbit edge-on.

Binary stars merit our attention partly because there are so many of them, but even more importantly because of what they can tell us about the properties of individual stars. As we will see in the next section, much of our basic information on the nature of stars comes from measurements of binary systems.

FUNDAMENTAL STELLAR PROPERTIES

In order to understand how stars work, astronomers need to determine their physical properties and how they are related to each other. A number of fundamental quantities characterize a star, and in these sections we will see how the observations described previously can be used to determine those quantities. They include the luminosities, the temperatures, the radii, and, above all, the masses of stars. At several steps along the way it is also important to know the distances to stars, and we will learn about a new, very powerful distance-determination method.

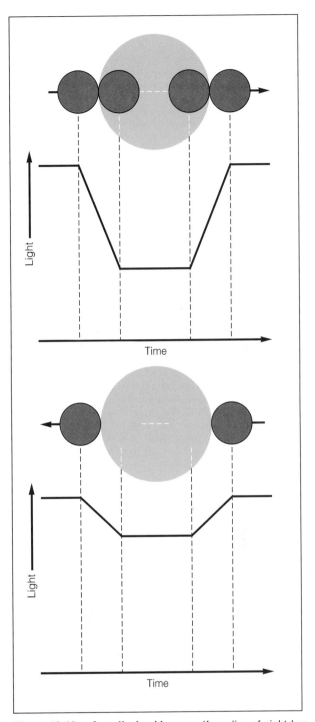

Figure 12.18   An eclipsing binary.   If our line of sight happens to be aligned with the orbital plane of a double star system, the two stars will alternately eclipse each other, causing brightness variations in the total light output from the system, as shown. In the upper figure, the smaller star is passing in front of the larger one; in the lower sketch, the smaller one is passing behind. The segmented graph in each case, called the light curve, illustrates how the total brightness of the system varies as these eclipses occur.

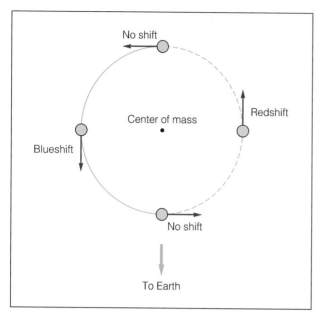

Figure 12.19   A spectroscopic binary.   As a star orbits the center of mass of a binary system, its velocity relative to the Earth varies. This produces alternating redshifts and blueshifts in its spectrum, as the star recedes from and approaches us. For simplicity, only one of the stars is moving in this diagram, although in reality both do.

Absolute Magnitudes and Stellar Luminosities

The luminosity of a star is the total amount of energy it emits from its surface, in all wavelengths, and is usually measured in units of joules per second, or watts (see Chapter 4). To measure luminosity, astronomers must take into account the distance to a star and the light received from it at *all* wavelengths, not just those to which the human eye responds. To eliminate the effect of distance, astronomers use the **absolute magnitude,** defined as the magnitude a star would have if it were seen at a standard distance of 10 pc (Fig 12.21; Table 12.3). This distance was chosen arbitrarily but is used by all astronomers. Thus a comparison of absolute magnitudes reveals differences in luminosities, because all the effects of distance have been canceled out.

It helps to consider a specific example. Suppose a certain star is 100 pc away and has an **apparent magnitude** (observed magnitude) of 7.3. To determine the absolute magnitude, we must find out what this star's magnitude would be if it were only 10 pc away. The inverse square law tells us that since the star would be a factor of 10 closer, it would appear a factor of $10^2 = 100$ brighter, or exactly 5 magnitudes

Figure 12.20   Spectroscopic binaries.
These spectrograms of spectroscopic binaries show periodic Doppler shifts due to orbital motion. The sequence of three spectra at the top shows a single-lined spectroscopic binary, in which lines of only one star are seen to shift back and forth. Below are spectra of a double-lined spectroscopic binary, in which lines of both stars are visible when the two are moving in opposite directions along the line of sight. The emission-line spectra at the top and bottom of each set of stellar spectra are used to provide a reference for the wavelength scale. (Yerkes Observatory)

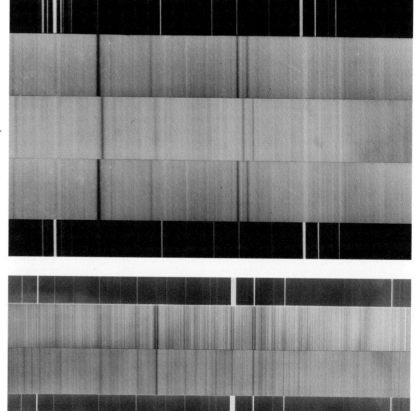

brighter. Therefore, its absolute magnitude is 7.3 − 5 = 2.3. The absolute magnitude can always be derived from the measured apparent magnitude and the distance in this way. Other examples are given in the questions at the end of this chapter and in Appendix 10.

Let us now return to the question of luminosities. By determining the absolute magnitudes of stars, we can compare their luminosities, since the distance effect has been removed. Since we must allow for light emitted by a star at all wavelengths, the bolometric magnitude is the quantity that is used. Thus, if one star's bolometric magnitude is 5 magnitudes smaller than another's, we know that its luminosity is a factor of 100 greater. The luminosity can be expressed in terms of watts by comparing it with luminosity of a standard star that has been measured directly by measuring the intensity of light at all wavelengths and allowing for distance.

When astronomers determine stellar luminosities, they find that the values from star to star vary over an incredible range. Stars are known with luminosities as small as 10^{-4} the luminosity of the Sun and as great as 10^6 that of the Sun, a range of 10 billion from the faintest to the most luminous! The luminosity is by far the most highly variable parameter for stars; the others that we shall discuss only cover ranges of a few hundred or less from one extreme to the other.

Stellar Surface Temperatures

Earlier in this chapter, we saw that the color of a star, like its spectral class, depends on its temperature, so that the temperature can be deduced by observing either. Recall that the color index, the difference between the blue (B) and visual (V) magnitudes, is a measured quantity that indicates temperature. A negative value of $B − V$ means that the star is brighter in blue than in visual light and therefore is a hot star. A large positive value indicates a cool star. A specific correlation of color index with temperature has been developed and is used to determine temperatures from observed values of $B − V$.

More refined estimates of temperature can be made from a detailed analysis of the degree of ionization, which is done by measuring the strengths of spectral lines formed by different ions. This is basically the same as estimating the temperature from the spectral class, since in either case the point is that the strengths of spectral lines of various ions depend on how abundant those ions are, which in turn depends on the temperature.

The temperature referred to here may be called the surface temperature, although stars do not have solid surfaces. We are really referring to the outermost layers of gas, where the absorption lines form. This

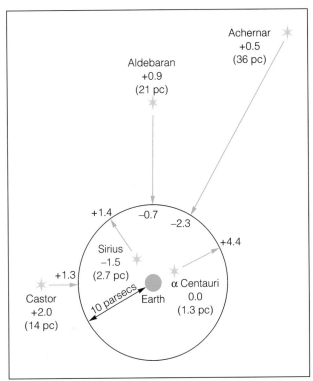

Figure 12.21 Absolute magnitudes. A few stars are shown at their correct relative distances (actual distances in parentheses). The arrows indicate the imaginary movement of these stars to a 10-pc distance from the Earth. The numbers indicate the apparent (actual distance) and absolute (10-pc distance) magnitudes.

TABLE 12.3 Visual Absolute Magnitudes of Selected Objects

Object	Absolute Magnitude
Typical bright quasar	−28
Brightest galaxies	−25
Milky Way galaxy	−20.5
Andromeda galaxy	−21.1
Type I supernova (maximum brightness)	−18.8
Large Magellanic Cloud	−18.7
Type II supernova (maximum brightness)	−17
Small Magellanic Cloud	−16.7
Typical globular cluster	−8
The most luminous stars	−8
Typical nova outburst	−7.7
Vega (bright star in summer sky)	+0.5
Sirius	+1.41
Alpha Centauri (nearest star)	+4.35
Sun	+4.83
Venus (greatest elongation)	+28.2
Full moon	+31.8
100-watt bulb	+66.3

region is called the atmosphere of a star, and it actually has some depth, although it is very thin compared with the radius of the star.

Stellar temperatures, as we have already seen, range from about 2,000 K for the coolest M stars to 50,000 K or more for the hottest O stars.

The Hertzsprung-Russell Diagram and a New Distance Technique

We have seen that temperature and spectral class are intimately related, and we have learned how astronomers deduce the luminosities of stars. In the first decade of this century, the Danish astronomer Ejnar Hertzsprung and, independently, the American astrophysicist Henry Norris Russell (Fig. 12.22) began to consider how luminosity and spectral class might be related to each other. Each gathered data on stars whose luminosities (or absolute magnitudes) were known and found a close link between spectral class (temperature) and absolute magnitude (luminosity). This relationship is best seen in the diagram constructed by Russell in 1913 (Fig. 12.23), now called the **Hertzsprung-Russell** or **H-R diagram.** In this

plot of absolute magnitude (on the vertical scale) versus spectral class (on the horizontal axis), stars fall into narrowly defined regions rather than being randomly distributed (Fig. 12.24). A star of a given spectral class cannot have just any absolute magnitude, and vice versa.

The great majority of stars fall into a diagonal strip running from the upper left (high temperature, high luminosity) to the lower right (low temperature, low luminosity) of the H-R diagram. This strip has been given the name the **main sequence.**

One group of a few stars fails to fall on the main sequence but appears in the upper right (low temperature, high luminosity) of the diagram. Since the spectra of these stars indicate that they are relatively cool, their high luminosities cannot be due to greater temperatures than the main-sequence stars of the same type. The only way one star can be much more luminous than another of the same temperature is if it has a much greater surface area. (Recall that the two stars will emit the same amount of energy per square meter of surface.) Hertzsprung and Russell realized that these extra-luminous stars sitting above the main sequence must be much larger than those

Figure 12.22   Henry Norris Russell.   One of the leading astrophysicists of the era when a physical understanding of stars was first emerging, Russell made many major contributions in a variety of areas. (Princeton University)

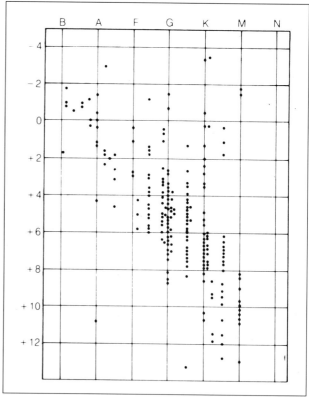

Figure 12.23   The first H-R diagram.   This plot showing absolute magnitude versus spectral class was constructed in 1913 by Henry Norris Russell. (Estate of Henry Norris Russell, reprinted with permission)

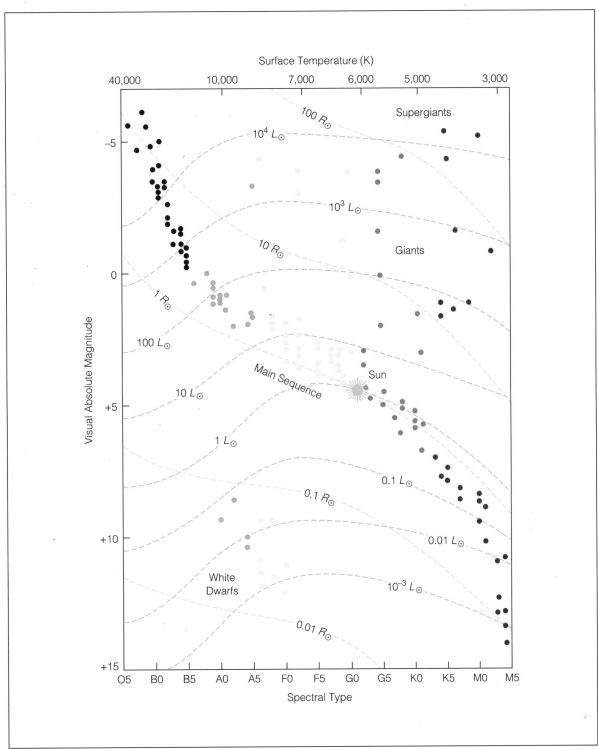

Figure 12.24   A modern H-R diagram.   This diagram shows the three commonly used temperature indicators on the horizontal axis. Blue dashed lines indicate stellar radii in solar units, and red dashed lines indicate luminosities, also in solar units (both have been derived using bolometric corrections appropriate for each spectral type; see the discussions earlier in this chapter).

on the main sequence, and they called them **giants** and **supergiants.**

The distinction among giants, supergiants, and main-sequence stars (commonly known as **dwarfs**) has been incorporated into the spectral classification system used by modern astronomers. A **luminosity class** (Fig. 12.25) has been added to the spectral type with which we are already familiar. The luminosity classes, designated by Roman numerals following the spectral type, are I for supergiants (this group is further subdivided into classes Ia and Ib), II for extreme giants, III for giants, IV for stars just a bit above the main sequence, and V for main-sequence stars, or dwarfs (not to be confused with white dwarfs, which are discussed later). Thus, a complete spectral classification for the bright summertime star Vega, for example, is A0V, meaning that it is an A0 main-sequence star. Betelgeuse, the red supergiant in the shoulder of Orion, has the full classification M2Iab (because it is intermediate between luminosity classes Ia and Ib). It is usually possible to assign a star to the proper luminosity class by examining subtle details of its spectrum. For example, giant and especially supergiant stars have relatively low atmospheric pressures, which affect the spectral line widths and the state of ionization of certain elements.

Another group of stars (which has become known mostly since Russell first plotted the H-R diagram) does not fall into any of the standard luminosity classes but appears in the lower left (high temperature, low luminosity) corner of the diagram. Since these stars are hot but not very luminous, they must be very small, and they have been given the name **white dwarfs.** These objects have some very bizarre properties, which will be discussed in Chapter 15.

Using the H-R Diagram to Find Distances

The H-R diagram can be used to find distances to stars, even stars that are very far away (Fig. 12.26). The idea is really very simple: If we know how bright a star is intrinsically (how much energy it is actually emitting from its surface), and we measure how bright it appears to be, we can determine how far away it is because we know that the difference between its intrinsic brightness and its observed brightness is due to its distance from us. The only problem lies in knowing the intrinsic brightness (luminosity) of the star, which is where the H-R diagram comes in.

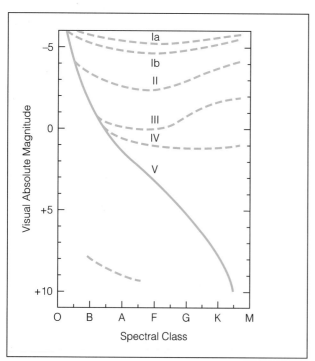

Figure 12.25   Luminosity classes.   This H-R diagram shows the locations of stars of the luminosity classes described in the text. A complete spectral classification for a star usually includes a luminosity class designation, if it has been determined.

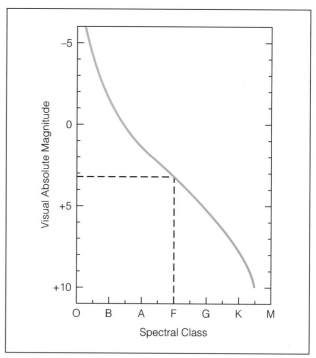

Figure 12.26   Spectroscopic parallax.   Knowledge of a star's spectral class (including the luminosity class) allows the absolute magnitude, and hence the distance, to be determined. This figure shows how the absolute magnitude of an F0 main-sequence star is read from the H-R diagram. The distance can then be found by comparing the absolute and apparent magnitudes of the star (as explained in the text and Appendix 10).

Once we determine the spectral class of a star, we can place it on the H-R diagram (as long as we know its luminosity class, which indicates whether it is on the main sequence or is a giant or supergiant). Once we have placed the star on the diagram, the vertical axis tells us its absolute magnitude, which is a measure of its luminosity. A comparison of the absolute magnitude with the observed apparent magnitude then amounts to the same thing as a comparison of the intrinsic and apparent brightnesses of the star, and from such a comparison the distance can be found.

The calculation of a star's distance from the difference between the apparent and absolute magnitudes can best be illustrated by considering a few simple cases. For example, if the difference $m - M$ (the apparent minus the absolute magnitude), which is called the **distance modulus,** is 5, the star appears 5 magnitudes, or a factor of 100, fainter than it would at the standard distance of 10 pc. A factor of 100 in brightness is created by a factor of 10 change in distance, so this star must be 10 times farther away than it would be if it were at 10 pc distance; therefore, it is $10 \times 10 = 100$ pc away. Similarly, a star whose distance modulus $m - M$ is 10 is 1,000 pc away. If $m - M = 15$, the distance is 10,000 pc. It should be obvious that if $m = M$ (that is, $m - M = 0$), the distance to the star must be 10 pc, because this is the distance that defines the absolute magnitude. As a rule of thumb, it helps to remember that for every 5 magnitudes of difference between the apparent and absolute magnitudes, the distance increases by a factor of 10.

This method is very powerful, because it can be used for very large distances. All that is needed is to be able to place a star on the H-R diagram so that its absolute magnitude can be determined, and to measure its apparent magnitude. Because this distance-determination technique depends on placing a star on the H-R diagram by classifying its spectrum, it is called the **spectroscopic parallax** method. (Astronomers use the word *parallax* as a general term for distances, even though, technically speaking, no parallax is measured in this case.)

Stellar Diameters

We have already seen that a star's position on the H-R diagram depends partly on its size, since luminosity is related to total surface area. If two stars have the same surface temperature (and therefore the same spectral class), but one is more luminous than the other, we know that it must also be larger. The Stefan-Boltzmann law (see Chapter 4) specifically relates the three quantities luminosity, temperature, and radius; use of this law allows the radius to be

determined if the other two quantities are known.

Eclipsing binaries provide another means of determining stellar radius that is independent of other properties (Fig. 12.27). Recall that these are double star systems in which the two stars alternately pass in front of each other as we view the orbit edge-on. The eclipsing binary is also very likely to be a spectroscopic binary, so that the speeds of the two stars in the orbits can be measured from the Doppler effect. We therefore know how fast the stars are moving, and from the duration of the eclipse we know how long it takes one to pass in front of the other; then the simple formula *distance = speed × time* gives us the diameter of the star that is being eclipsed. Even if no information on the orbital velocity is available, the relative diameters of the two stars can be deduced from the relative time they need to *enter* eclipse compared with the time *in* eclipse, as indicated by the sloped and flat segments of the light curve (see Fig. 12.18).

Eclipsing binaries provide the most direct means of measuring stellar sizes, but unfortunately they are not numerous. In most cases the radii are estimated from the luminosity and temperature as described above. In a few cases stellar radii have been measured directly by use of speckle interferometry, a sophisticated technique for clarifying the image of a star by removing the blurring effects of the Earth's atmosphere. Only relatively near, large stars can be measured this way, however. This technique measures the angular diameter; to convert this to true diameter requires knowledge of the distance to the star.

Stars on the main sequence do not vary greatly in radius, ranging from perhaps 0.1 times the Sun's radius for the M stars at the lower right-hand end to

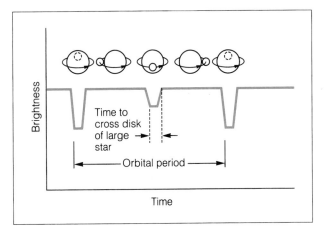

Figure 12.27   Measuring stellar diameters in an eclipsing binary.   The duration of each eclipse is combined with knowledge of the orbital speeds of the stars (from the Doppler effect) to yield the diameters of the stars. (In this figure, the smaller star is hotter, making the eclipses deeper when it is obscured.)

10 or 20 solar radii at the upper left. Of course, large variations in size occur as we go away from the main sequence, either toward the giants and supergiants, which may be 100 times the size of the Sun, or toward the white dwarfs, which are as small as 0.01 times the size of the Sun.

Binary Stars and Stellar Masses

The mass of a star is the most important of all its fundamental properties, for it is the mass that governs most of the others. This point will be discussed at some length in the next chapter.

The only way to measure a star's mass is by observing its gravitational effect on other objects, and this is possible only in binary star systems, where the two stars hold each other in orbit by their gravitational fields. Because binary systems are common, astronomers have many opportunities to determine masses by analyzing binary orbits.

The basic idea is rather simple, although the application may be quite complex, depending on the type of binary system. Kepler's third law, in the form derived by Newton, is used. Remember that if the period P is measured in years, the average separation of the two stars (the semimajor axis a) is measured in astronomical units, and the masses m_1 and m_2 are in units of solar masses, then Kepler's third law is

$$(m_1 + m_2)P^2 = a^3$$

We need only observe the period and the semimajor axis to solve for the *sum* of the two masses. Careful observations of the sizes of the individual orbits (actually, of the relative distances of the two stars from the center of mass; Fig. 12.28) also yield the ratio of the masses; when both the sum and the ratio are known, it is simple to solve for the individual masses.

Complications arise when some of the needed observational data are difficult to obtain. The period is almost always easy to measure with some precision, but not so the semimajor axis a. One problem is that the apparent size of the orbit is affected by our distance from the binary (Fig. 12.29), so the distance must be well known if a is to be accurately determined. Another problem is that the orbital plane is inclined at a random, unknown angle to our line of sight (so the apparent size of the orbit is foreshortened by an unknown amount). In some cases it is possible to unravel these confusing effects by carefully analyzing the observations. Even in cases where this is not possible, some information about the masses of the stars can still be gained, but usually only in terms of broad ranges of possible values rather than precise answers.

The masses of stars vary along the main sequence from the least massive stars in the lower right to the most massive at the upper left. The M stars on the

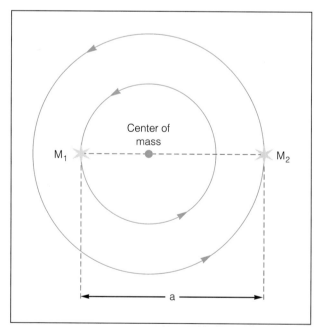

Figure 12.28   Binary star orbits.   This figure illustrates the terms used in Kepler's third law. Two stars of masses m_1 and m_2 (m_1 is larger than m_2 in this case) orbit a common center of mass, each making one full orbit in period P. The semimajor axis a that appears in Kepler's third law is actually the sum of the semimajor axes of the two individual orbits about the center of mass; this sum corresponds to the average distance between the two stars.

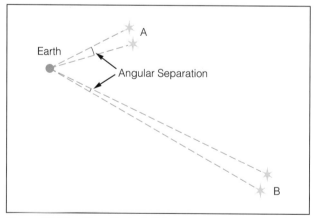

Figure 12.29   The effect of distance on measurements of binary star orbits.   Here two binary systems have the same actual separation, but from Earth the nearer system (A) appears more widely separated than the more distant system (B). Therefore the distance to a binary system must be known in order to analyze the stellar orbits.

main sequence have masses as low as 0.1 solar mass or a bit less, while the O stars reach values as great as 60 solar masses. It is likely that stars occasionally form with even greater masses (perhaps up to 100 solar masses), but as we will see in the next chapter, such massive stars have very short lifetimes, so they are rarely encountered.

The giants, supergiants, and white dwarfs have masses comparable to those of main-sequence stars. Hence their obvious differences from main-sequence stars in other properties such as luminosity and radius have to be caused by something other than extreme or unusual masses. This is discussed in the next chapter.

Main-sequence stars exhibit a smooth progression of all stellar properties from one end to the other (Table 12.4). The mass and the radius vary by similar factors, while the luminosity changes much more rapidly along the sequence. The mass of an O star is perhaps 100 times that of an M star, while the luminosity is greater by a factor of 10^8 or more. Main-sequence stars follow a so-called **mass-luminosity relation**, a numerical expression in which the luminosity of a star is proportional to an exponential power of the mass. In its simplest form, this relation states that the luminosity is proportional to the cube of the mass. In more precise versions of the mass-luminosity relation, the exponent varies somewhat along the main sequence.

Other Properties

Several other properties of stars can be determined primarily by analyzing their spectra. Perhaps the most fundamental of these is the composition. We have already seen that stars are generally made of the same material in the same proportions, but it required a sophisticated analysis to show this. First, techniques had to be developed that take the effects of temperature into account, because these effects, as we have already learned, play a dominant role in controlling the strengths of lines in a stellar spectrum. In modern work on stellar composition, complex computer programs are used to calculate simulated spectra with different assumed compositions until a match with the observed spectrum is found (Fig. 12.30).

Another stellar property that can be learned from the spectrum is the rotational velocity at the surface, because the Doppler effect causes broadening of the spectral lines (Fig. 12.31). A spectral line forms over the entire disk of a star, and in general one edge of the disk is approaching the Earth, and the other is receding. Hence some of the gas creates a blueshift, and some a redshift (the gas in the central portions of the disk has little or no shift), so the rate of rotation of the star determines the degree of broadening of the spectral lines.

A final property, whose importance is hard to assess because of the difficulty of measuring it in many cases, is the magnetic field of a star. In certain stars the Zeeman effect (described in Chapter 11) can be used; in this effect some spectral lines are split due to the presence of a magnetic field (Fig. 12.32). The Zeeman effect can be applied only to very slowly spinning stars whose spectral lines are very narrow, so that the splitting can be seen. In the many stars where the lines are too broad for this technique, we have no good method of measuring the magnetic field strength.

TABLE 12.4   Properties of Main-Sequence Stars

Spectral Type	Mass ($M_\odot$)	Temperature	$B - V$	Luminosity ($L_\odot$)	M_v	B.C.	Radius ($R_\odot$)
O5V	40	40,000	−0.35	5×10^5	−5.8	−4.0	18
B0V	18	28,000	−0.31	2×10^4	−4.1	−2.8	7.4
B5V	6.5	15,500	−0.16	800	−1.1	−1.5	3.8
A0V	3.2	9,900	0.00	80	+0.7	−0.4	2.5
A5V	2.1	8,500	+0.13	20	+2.0	−0.12	1.7
F0V	1.7	7,400	+0.27	6	+2.6	−0.06	1.4
F5V	1.3	6,580	+0.42	2.5	+3.4	0.00	1.2
G0V	1.1	6,030	+0.58	1.3	+4.4	−0.03	1.1
G5V	0.9	5,520	+0.70	0.8	+5.1	−0.07	0.9
K0V	0.8	4,900	+0.89	0.4	+5.9	−0.19	0.8
K5V	0.7	4,130	+1.18	0.2	+7.3	−0.60	0.7
M0V	0.5	3,480	+1.45	0.03	+9.0	−1.19	0.6
M5V	0.2	2,800	+1.63	0.008	+11.8	−2.3	0.3

Note: Masses are in units of $M_\odot$, the solar mass (whose value is 1.99×10^{30} kg); luminosities are in the units of $L_\odot$, the solar luminosity (having a value of 3.83×10^{26} watts), and radii are in units of $R_\odot$, the solar radius (whose value is 6.96×10^8 m). The temperatures are **effective temperatures,** which are a measure of surface temperature. The heading $B - V$ stands for the color index, M_v stands for visual absolute magnitude, and B.C. stands for bolometric correction.

Figure 12.30   Determination of stellar composition.   This diagram illustrates a modern technique for measuring the chemical composition of a star. A theoretical spectrum (dashed line) is compared with the observed spectrum (dotted and dashed line). The abundances of elements assumed present are varied in the computed spectrum until a good match is achieved. (D. L. Lambert)

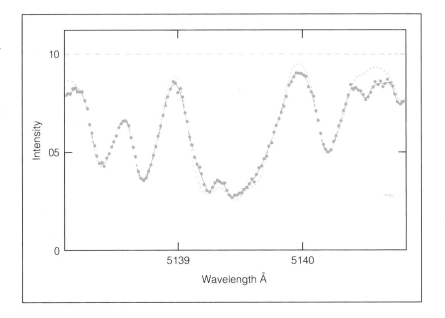

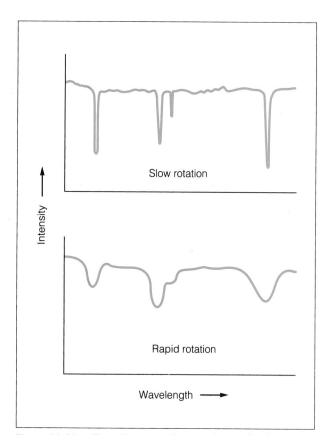

Figure 12.31   The effect of stellar rotation.   The Doppler effect causes spectral lines in a rapidly rotating star (bottom) to be broadened, because one edge of the star's disk is rapidly approaching us while the other edge is receding. Little broadening occurs in a slowly rotating star (top).

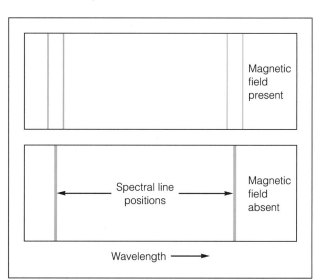

Figure 12.32   The measurement of a magnetic field.   The presence of a magnetic field splits certain absorption lines of some elements, in a process called Zeeman splitting. The amount of separation is a measure of the strength of the magnetic field.

ASTRONOMICAL INSIGHT
Cecilia Payne and the Composition of the Stars

One of the most brilliant astrophysicists of the twentieth century was Cecilia Payne (later Payne-Gaposchkin), who enrolled as a graduate student at Harvard College Observatory in 1923. There she worked under the direction of Harlow Shapley, who had succeeded Edward Pickering as director. Having arrived at Harvard just as the immense *Henry Draper Catalog* was being published, Payne had access to the largest collection of stellar spectra available anywhere (see the discussion of Annie Cannon and spectral classification in the Astronomical Insight in Chapter 12). For her doctoral dissertation, Payne combined this wealth of observational data with recent advances in atomic physics, particularly the work of the Indian scientist Meghnad Saha. Saha had investigated the effects of ionization and excitation on the spectra of atoms, showing how the lines that are formed depend on physical conditions such as temperature and pressure. Payne undertook to interpret the spectra of the stars in light of Saha's results.

One of her conclusions was that the stars all have the same chemical composition. Astronomers had already recognized that the spectral classes represent a sequence of varying temperatures, rather than a sequence of stars with different compositions, but until Payne's work no one had been able to determine the relative quantities of the various elements. She was able to do so by first using Saha's equation to make allowance for the effect of temperature and pressure on the strengths of the spectral lines; then she used the line strengths to determine the abundances of the elements.

Payne's conclusions were published in 1925 in her dissertation, *Stellar Atmospheres,* which is today widely regarded as one of the foundation pillars on which this field was built. Interestingly, she disclaimed one of her most important conclusions because it ran counter to the general belief of most astronomers of the day. Her analysis of stellar spectra had not only showed that all stars have the same composition, but that hydrogen was, by far, the most abundant element. But when Payne described this result in correspondence with Henry Norris Russell at Princeton, Russell persuaded her that her values for hydrogen (and helium, which she had found to be next most abundant after hydrogen and much more common than any other element) must be incorrect. This prejudice was based largely on the fact that on the Earth other elements such as oxygen, carbon, nitrogen, and iron are comparable in abundance to hydrogen (and far more abundant than helium). Payne, no doubt influenced by Russell's preeminence in the field of stellar astrophysics, became convinced that he must be right. Accordingly, she said little about the hydrogen and helium abundances in her treatise, except to comment that "although hydrogen and helium are manifestly very abundant in stellar atmospheres, the actual values derived . . . must be regarded as spurious."

It is most ironic that it was Russell himself, in 1929, who published the first study showing that hydrogen is the most common element in stars (his conclusion was initially based on the analysis of a solar spectrum). To his credit, Russell did recognize and quote Payne's earlier discovery, although he did not explain his own role in convincing her that it must be wrong.

Payne stayed on at Harvard following the completion of her degree (which was awarded by Radcliffe College, since Harvard did not officially award degrees to women until several decades later; her doctorate is now recognized as the first ever earned by a woman at Harvard). She continued to study stellar atmospheres, but when she married Sergei Gaposchkin in 1934, she began working with him on variable stars, particularly novae and other explosive variables. Together they performed many of the pioneering studies in this field.

In due course Cecilia Payne-Gaposchkin earned wide respect in the astronomical community, which bestowed on her many awards and other forms of recognition, and at Harvard, which made her full professor and chair of the Astronomy Department in 1956 (another first for the university!). Today the importance of her work is receiving new recognition, as the role of women in science is gaining greater exposure than was typical in earlier times.

PERSPECTIVE

In this chapter we have learned the basics of stellar observations by discussing the three fundamental types of observations and the characteristics of stars revealed by each. Considerable astronomical terminology has been introduced in the process.

We have gone on to learn how all the basic properties of stars are derived from observational data. We can categorize, classify, and describe stars in any way we wish. Now we are ready to see how they work—*why* the various quantities are related the way they are and no other way.

SUMMARY

1. There are three basic types of stellar observations: position, brightness, and spectroscopy.

2. Positional astronomy (astrometry) has developed techniques capable of measuring positions to an accuracy approaching 0.01 arcsecond, which is sufficient to detect binary motions in some cases, and stellar parallaxes for stars up to 100 pc away.

3. Stellar brightness measurements are commonly done using the stellar magnitude system, in which a difference of one magnitude corresponds to a brightness ratio of 2.512.

4. Magnitudes can be measured at different wavelengths, allowing the determination of color indices; or over all wavelengths, resulting in the determination of bolometric magnitudes.

5. Stellar spectra contain patterns of absorption lines that depend on the surface temperatures of stars and therefore can be used to assign stars to spectral classes that represent a sequence of temperatures.

6. Peculiar stars are those that do not conform to the usual spectral classes, most often because of unusual surface compositions but sometimes because they have emission lines.

7. Binary star systems can be detected by positional variations of one or both stars (astrometric binaries), brightness variations (eclipsing binaries), or periodically Doppler-shifted spectral lines (spectroscopic binaries).

8. Stellar luminosities are determined from knowledge of the distances to stars and their apparent magnitudes. The absolute magnitude, a measure of luminosity, is the magnitude a star would have if it were seen from a distance of 10 pc.

9. Stellar temperatures can be inferred from the $B - V$ color index, estimated from the spectral class, or determined from the degree of ionization in the star's outer layers.

10. The Hertzsprung-Russell diagram shows that the luminosities and temperatures of stars are intimately related and that stars that do not fall on the main sequence are either larger (red giants or supergiants) or smaller (white dwarfs) than those on the main sequence.

11. The distance to a star can be measured by determining its spectral class and then using the H-R diagram to infer its absolute magnitude, which is then compared with its apparent magnitude to yield the distance; this technique is called spectroscopic parallax.

12. Stellar diameters can be determined directly in eclipsing binaries from knowledge of the orbital speed and the duration of the eclipses.

13. Masses of stars are derived in binary systems by observing the period and the orbital semimajor axis and using Kepler's third law.

REVIEW QUESTIONS

1. Compare the techniques used today for positional measurements in astronomy with those used by the ancient Greeks. How much more accurate are today's techniques?

2. Briefly discuss the role of stellar parallax in the development of the heliocentric theory.

3. Convert the following distances from parsecs to light-years: 100 pc; 1.33 pc (the distance to the nearest star); 8,000 pc (the distance to the center of our galaxy).

4. What is the distance (in parsecs) to the following stars: (a) a star whose parallax is $p = 0''.25$; (b) a star whose parallax is $p = 0''.04$; (c) a star whose parallax is $p = 0''.005$?

5. How much fainter is a twelfth-magnitude star than an eleventh-magnitude star? How does a fourth-magnitude star compare with a fifth-magnitude star?

6. Compute the brightness ratio for (a) a fourth-magnitude star and a seventh-magnitude star; (b) a star with magnitude $m = 17$ and a star with $m = 22$; (c) a star of $m = 6$ and a star of $m = -1$.

7. How much fainter are the faintest stars detectable to modern telescopes ($m = 31$) than the brightest stars in the sky ($m = -1$)?

8. Explain, in terms of what you learned about ionization in Chapter 4, why stars of different temperature have different absorption lines in their spectra.

9. Why is it difficult to directly measure a star's brightness at all wavelengths?

10. Determine the absolute magnitude of the following stars: (1) a star 1,000 pc from the Sun with apparent magnitude $m = 12.3$; (b) a star 10,000 pc from the Sun with $m = 12.3$; (c) a star 10 pc away with $m = 6.8$; (d) a star 1 pc away with $m = 2.1$

11. If a star has absolute bolometric magnitude $M_{bol} = 1.5$, and another star has $M_{bol} = 4.5$, which is more luminous, and by how much?

12. Explain why a G2 star that lies above the main sequence must have a larger diameter than a G2 star that lies on the main sequence.

13. Explain the spectroscopic parallax method for finding the distances to stars. Why do you think the word *parallax* is used?

14. If one star is four times hotter than another and has twice as large a radius, which is more luminous, and by how much?

15. What is the sum of the masses in a binary system whose orbital period is 1 year and whose semimajor axis is 2 AU? What are the individual masses of the two stars if the center of mass between them is one-fourth of the way from the more massive star (star A) to the less massive one (star B)?

ADDITIONAL READINGS

Probably the best sources of additional information on stars and their observational properties are general astronomy textbooks, of which a number are readily available in nearly any library. A few more specific resources are listed here as well.

Beyer, S. L. 1986. *The star guide*. Boston, Mass.: Little, Brown.

De Vorkin, D. H. 1989. Henry Norris Russell. *Scientific American* 260(5):26.

_____. 1978. Steps towards the Hertzsprung-Russell diagram. *Physics Today* 31(3):32.

Dobson, A. K., and K. Bracher. 1992. Urania's heritage in a historical introduction to women in astronomy. *Mercury* 21(1):4.

Evans, D. S., T. G. Barnes, and C. H. Lacy. 1979. Measuring diameters of stars. *Sky and Telescope* 58(2):130.

Getts, J. 1983. Decoding the Hertzsprung-Russell diagram. *Astronomy* 11(10):16.

Giampapa, M. S. 1987. The solar-stellar connection. *Sky and Telescope* 74:42.

Kaler, J. B. 1991. The brightest stars in the galaxy. *Astronomy* 19(5):30.

_____. 1991. The faintest stars. *Astronomy* 19(8):26.

_____. 1987. The B stars: Beacons of the sky. *Sky and Telescope* 74(2):174.

_____. 1987. The spectacular O stars. *Sky and Telescope* 74:464.

_____. 1987. The temperate F stars. *Sky and Telescope* 73(2):131.

_____. 1987. White Sirian stars: Class A. *Sky and Telescope* 73(5):491.

_____. 1986. Cousins of our Sun: The G stars. *Sky and Telescope* 72(5):450.

_____. 1986. The K stars: Orange giants and dwarfs. *Sky and Telescope* 72(2):130.

_____. 1986. M stars: Supergiants to dwarfs. *Sky and Telescope* 71(5):450.

_____. 1986. The origins of the spectral sequence. *Sky and Telescope* 71(2):129.

McAlister, H. A. 1977. Binary star speckle interferometry. *Sky and Telescope* 53(5):346.

Michalas, D. 1973. Interpreting early type stellar spectra. *Sky and Telescope* 46(2):79.

Philip, A. G. D., and L. C. Green. 1978. The H-R diagram as an astronomical tool. *Sky and Telescope* 55(5):395.

Rubin, V. 1986. Women's work: Women in modern astronomy. *Science* 86 7(6):58.

Steffey, P. C. 1992. The truth about star colors. *Sky and Telescope* 84(3):266.

Welther, B. 1984. Annie Jump Cannon: Classifier of the stars. *Mercury* (1):28.

ASTRONOMICAL ACTIVITY
Observing Binary Stars

As explained in the text, a visual binary is a system in which both stars are visible. In most visual binaries, however, a telescope is required to see the two stars as separate objects. Nevertheless, several visual binaries can be seen with binoculars, and one can be made out by the naked eye under good conditions. This is the pair named Mizar and Alcor, and you might enjoy having a look.

Mizar and Alcor together form one of the stars in the handle of the Big Dipper (also known as Ursa Major). The position of Mizar/Alcor is shown in the accompanying sketch. Because the Big Dipper is very close to the north celestial pole, it is up and accessible for observation all year.

Both stars are bright enough for naked-eye detection (Mizar has an apparent magnitude of 2.3; Alcor's apparent magnitude is 4.0), and their separation on the sky is large enough (about 15 arcminutes) so that both stars can be seen as distinct objects by the human eye (recall the discussion of resolving power in Chapter 4). On a clear night, test your eyes by looking at Mizar and trying to spot Alcor (North American Indians once used this as an eye test, as did several European armies). Take another look using binoculars. You will see that both stars are white in color (the spectral types are A1V for Mizar and A5V for Alcor).

Another very interesting binary system, which can be seen with the naked eye (but is better viewed through binoculars or a small telescope), is β Cygni, also known as Albireo. The two stars that form Albi-

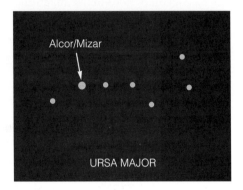

reo have very different spectral types (the brighter star, whose apparent magnitude is 3.1, is a K3II star;* the fainter, having apparent magnitude of 5.1, is a B8V star). Thus the brighter star has a red-orange color, while the fainter companion is blue white. The human eye tends to exaggerate color contrasts, so the colors look very different. This pair, separated by about 35 arcseconds, is a very pretty sight. The accompanying chart indicates where to look; Cygnus is up during the summer and early fall.

*The brighter member of the Albireo pair is itself a binary star; the K3II star has a B0V companion that is two magnitudes fainter.

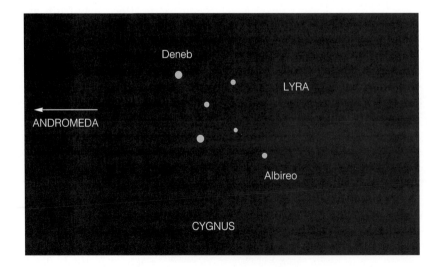

Stellar Structure:
What Makes a Star Run?

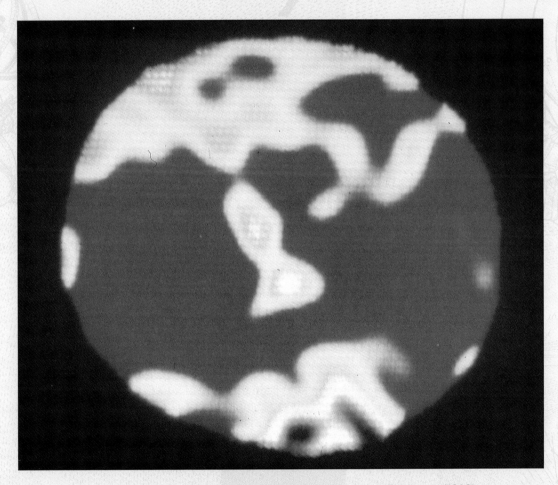

Convective flow patterns on the surface of the red supergiant Betelgeuse (NOAO)

PREVIEW

We have no direct means of probing the interiors of stars, but nevertheless we know a great deal about conditions and processes that occur there. In this chapter you will learn how astronomers obtain this information by applying simple physical principles. For example, the fact that a star is stable, neither contracting nor expanding, allows us to equate the inward force of gravity with the outward force of pressure, and this in turn leads to information on the pressure, temperature, and density inside the star. Other similar equations yield further information. A key element in the structure and evolution of a star is the production of energy through nuclear reactions in the core; these reactions not only power the star, but also force it to change with time, as its elemental composition is altered. In this chapter you will learn about these physical processes, the methods astronomers use to create models of the structure and evolution of stars, and the ways in which the models are tested by comparison with observations. This will set the stage for our discussions in the following chapters of the lifetimes of stars and the remnants they leave behind.

We have learned how astronomers determine the physical characteristics of stars from observations. These observations tell us a great deal about the surface properties of stars, but very little about what goes on inside them or how they evolve. To understand these secrets, we must apply the laws of physics and calculate the internal structure and evolution theoretically. If these calculations successfully reproduce the observable properties, then we can have some confidence that the calculations also accurately describe the internal conditions. We have already seen what can be learned, in our discussions of the Sun in Chapter 11. In this chapter, we will rely on our previous discussions, for the same physical principles that were useful for the Sun can be applied to other stars.

WHAT IS A STAR, ANYWAY?

As we found for the Sun, a star is a spherical ball of hot gas. The outer layers are partially ionized, and the interior, where the temperature and pressure are much higher, are fully ionized, so that the gas there consists of bare atomic nuclei and free electrons. Continuous radiation is generated at the core of a star, and from there photons of light make their way slowly out to the surface, where they are emitted into space from a layer called the **photosphere,** just as in

the Sun. Absorption lines are formed in the photosphere, so that the spectra of normal stars are absorption-line spectra, just as the Sun's is. The interior regions of a star are very dense because of gravitational compression, but they are still gaseous because of the high temperature. Temperatures in the cores of stars range from 10 to 100 million degrees absolute.

Like the Sun, most stars consist primarily of hydrogen, although, as we will see, the abundance of hydrogen changes gradually over a star's lifetime because of nuclear reactions. (Recall that in the Sun's core, hydrogen is gradually being converted into helium.)

The internal structure of a star, like that of the Sun, is governed by the balance between gravity and pressure that we call **hydrostatic equilibrium.** The gas particles all exert gravitational forces on each other, so that the entire star is held together by its own gravity. This force is always directed toward the center; thus, the star is forced into a symmetric, spherical shape. The fact that gas is compressible explains how gravity creates a state of high density inside.

A gas that is compressed heats up, causing it to exert greater pressure on its surroundings. Thus, a star's interior is very hot and the internal pressure is high. If it were not for this pressure, gravity would cause a star to keep shrinking. A balance is struck between gravity, which is always trying to squeeze a star inward, and pressure, which pushes outward (Fig. 13.1). This balance, which can be stable for long periods of time only if there is a source of energy inside the star, plays a dominant role in determining the internal structure. Whenever the balance is disturbed, as it is during some phases of a star's lifetime, the internal structure changes (this is discussed in the next chapter).

It is the mass of a star that causes it to reach a certain state of internal compression and no other. The amount of gravitational force is set by the total mass, and this force in turn determines how much pressure is needed to balance gravity. The star will be compressed until this pressure is reached, and then it will become stable. The pressure required to balance gravity dictates the temperature inside the star. Hence a star's mass determines the internal density and temperature as well as the overall size of the star, since size is a function of the mass and the density.

Luminosity is also governed largely by the mass. The luminosity of a star is simply the amount of energy per second generated inside the star that eventually reaches the surface and escapes into space, and it is determined by the temperature in the interior. As we have just seen, the temperature is set primarily by the mass. The dependence of luminosity on mass

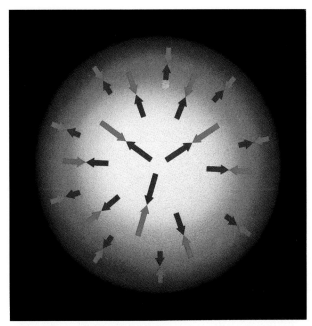

Figure 13.1   Hydrostatic equilibrium.   This cutaway sketch of a star shows its spherical shape. The arrows represent the balanced forces of gravity (inward) and pressure (outward), and the shading indicates that the density increases greatly toward the center. Equilibrium is reached when the core becomes sufficiently hot to attain the pressure necessary to counterbalance gravity.

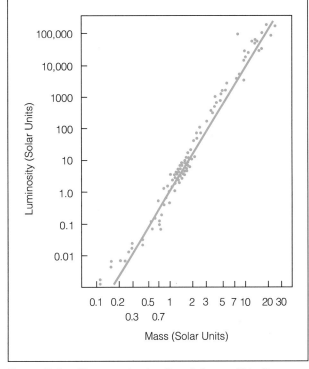

Figure 13.2   The mass-luminosity relation.   This diagram, based on observed luminosities and masses for stars on the main sequence, shows how well the two quantities correlate with each other.

for main-sequence stars can be demonstrated by making a graph of luminosity versus mass (Fig. 13.2). This is the **mass-luminosity relation,** and it was discovered observationally before it could be fully explained theoretically.

Thus, virtually all the observed properties of stars depend on their masses (Fig. 13.3). This explains why the main sequence represents a smooth run of masses increasing from the lower right to the upper left in the H-R diagram: The luminosity and temperature, which define a star's place in the diagram, both depend on mass.

Of course, some stars do not fall on the main sequence, yet their masses are not different from those of main-sequence stars. This indicates that something other than mass can influence a star's properties. This other factor is the chemical composition. As we shall see, this varies as a star ages. The supergiants, giants, and white dwarfs have different chemical makeups from main-sequence stars. In the next section we will see how these differences arise.

The amount of internal compression, and hence the temperature and luminosity of a star, depend on the average mass of the individual nuclei in the star's

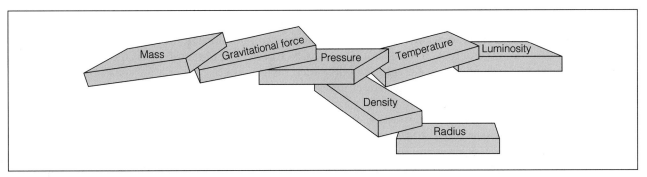

Figure 13.3   The importance of mass.   A star's mass is the single quantity that governs all its other properties for a given composition. The sequence shown here is general, but the details vary for different chemical compositions.

core. The heavier the particles that make up the gas, the more tightly they are compressed by gravity, the hotter the gas becomes, and the greater the star's luminosity. When a star is formed, it consists mostly of hydrogen, but as it ages its core material is converted to helium at first and later possibly to other, even heavier, elements. In the process, the core heats up and the star becomes more luminous.

The fact that all properties of a star depend on just its mass and composition was recognized several decades ago and is usually referred to as the **Russell-Vogt theorem,** after the astrophysicists who first stated it. Astronomers now know that other influences, such as magnetic fields and rotation, must play roles in governing a star's properties as well. Just what these roles are is not yet well understood, however.

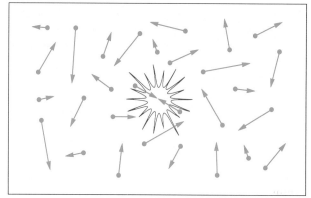

Figure 13.4   Nuclear fusion.   The core of a star is a sea of rapidly moving atomic nuclei. Occasionally a pair of these nuclei, most of which are simple protons (hydrogen nuclei), collide with sufficient velocity to merge and form a new kind of nucleus (deuterium in this case, composed of one proton and one neutron). Energy is released in the process.

NUCLEAR REACTIONS AND ENERGY TRANSPORT

We have seen that the core of a star must be very hot to maintain the pressure required to counterbalance gravity. Since energy flows outward from the core, there must be some source of heat in the interior. If there were not, the star would gradually shrink.

Nuclear fusion reactions provide the only source of heat capable of maintaining the required temperature over a sufficiently long period of time. A particular type of reaction was described in Chapter 11, where we discussed the Sun's interior. Recall that in atomic fusion reactions, nuclei of light elements such as hydrogen combine to form nuclei of heavier elements. In the process, a small fraction of the mass is converted into energy according to Einstein's famous equation $E = mc^2$. It is this energy, in the form of heat and radiation, that maintains the internal pressure in a star.

Fusion reactions can occur only under conditions of extremely high temperature and density. Protons all carry positive electrical charges and therefore repel each other. The force (called the **strong nuclear force**) that holds the protons and neutrons together in a nucleus and causes fusion to occur can overcome this repulsive electric force only over very short distances. Therefore, to combine, the nuclei must collide at very high velocities (Fig. 13.4), so that they can get close enough together despite their electromagnetic repulsion for each other. The high temperature of a stellar core imparts high speeds to the nuclei, so that they collide with great energy, and the high density means that collisions will be frequent. Even so, only occasionally do two nuclei combine in a fusion

reaction; a single particle may typically collide and bounce around inside a star for millions or even billions of years before it reacts with another. There are so many particles, however, that reactions are continually occurring despite the low probability of reaction for any individual particle.

The amount of energy released in a single reaction between two particles is small, about 10^{-12} joule. A joule (defined in Chapter 3) is itself a small quantity; a 100-watt light bulb radiates 100 joules per second. Hence, a light bulb would require 10^{14} reactions per second to keep glowing, if nuclear reactions were its energy source. A star like the Sun emits more than 10^{26} joules per second, so a tremendously large number of reactions (more than 10^{38} per second) must be occurring in its interior at all times.

In Chapter 11 we learned that the reaction that powers the Sun is the proton-proton chain, in which hydrogen is converted to helium, with 0.007 of the initial mass being converted to energy. The net result of this reaction is the combination of four hydrogen nuclei into one helium nucleus; two of the protons must be converted into neutrons in the process (see Chapter 11 or Appendix 11 for details of this reaction sequence). Another reaction sequence, called the **CNO cycle** (see Appendix 11), has the same net effect, but operates primarily in stars hotter and more massive than the Sun. *All* stars on the main sequence produce their energy by one of the two processes that convert hydrogen into helium in their cores; the specific reaction that is dominant depends on the internal temperature, which in turn depends on the star's mass. The point on the main sequence that divides proton-proton chain stars from CNO stars corresponds roughly to spectral class F0; stars above that point produce most of their energy by the CNO cycle, and those below that point do so by the proton-proton chain.

Once it is produced in the core, energy must somehow make its way outward to the stellar surface. For most stars, two different energy-transport mechanisms are at work at different levels (Fig. 13.5). One of these is **radiative transport,** meaning that the energy is carried outward by photons of light. In the core, where it is very hot, the photons are primarily gamma rays, but as they slowly move outward, being continually absorbed and reemitted on the way, they are gradually converted to longer wavelengths. When the light emerges from the stellar surface, it is primarily in the visible-wavelength region (or the ultraviolet or infrared, if the star is very hot or very cool). Radiative transport is a very slow process, as we learned in the case of the Sun, where it takes thousands of years for a single photon to make its way from the core to the surface.

The second means of energy transport inside stars is **convection,** a process already discussed briefly in Chapter 11. In convection the gas is overturned, as heated material rises and cooled material sinks. The same process causes warm air to rise toward the ceiling of a room and is responsible for the overturning of water in a pot that is being heated from the bottom. When conditions are right for convection to occur, it is much more efficient than radiative transport. Energy travels outward much faster in convective zones than in zones where radiation is the chief means of energy transport.

Whether radiative transport or convection will be the dominant energy-transport mechanism in a star depends on the temperature structure (specifically, on how rapidly the temperature decreases with distance from the center). Model calculations show that most stars have both a convective zone and a radiative region. In stars like the Sun (i.e., stars on the lower half of the main sequence, of spectral types F, G, K, and M), convection occurs in the outer layers, while radiative transport is the principal means of energy transport in the interior. For stars on the upper main sequence, the situation is reversed: Convection occurs in the central core but not in the rest of the star, where radiative transport is responsible for conveying the energy to the surface.

STELLAR LIFE EXPECTANCIES

The lifetime of a star is determined by the amount of energy it can produce in nuclear reactions and the

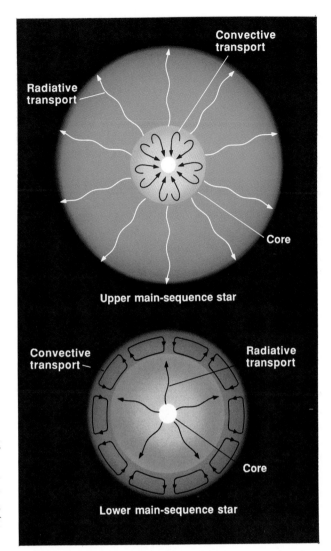

Figure 13.5   Energy transport inside stars.   In a star on the upper portion of the main sequence, energy is transported by convection in the inner zone and radiation in the outer regions. The opposite is true of lower main-sequence stars.

rate at which the energy is radiated away into space (Fig. 13.6). When all the available nuclear fuel has been used up, the star undergoes major changes in its structure and properties, and it leaves the main sequence (this will be discussed in detail in the next chapter). In Chapter 11 we found that the Sun has a life expectancy of about 10 billion years, assuming that 10 percent of its mass undergoes reactions, and that 0.007 of that mass is converted into energy. For stars in which the CNO cycle is the dominant reaction, the calculation is made the same way, and the fraction of the initial mass converted into energy is the same. The calculation involves determining the total amount of energy available to the star during its

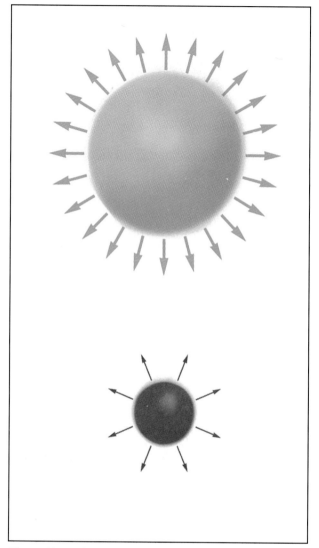

Figure 13.6 Stellar lifetimes. Even though the more massive star (top) has a greater supply of nuclear fuel, it loses energy so much more rapidly than the lower-mass star (bottom) that its lifetime is far shorter.

TABLE 13.1 Main-Sequence Lifetimes

Spectral Type	Hydrogen-Burning Lifetime (years)
O5V	1×10^6
B0V	5×10^6
B5V	8×10^7
A0V	4×10^8
A5V	1×10^9
F0V	3×10^9
F5V	5×10^9
G0V	9×10^9
G5V	1×10^{10}
K0V	2×10^{10}
K5V	4×10^{10}
M0V	2×10^{11}
M5V	3×10^{11}

estimate; it is a few million years.) By astronomical standards, such massive stars exist for only an instant before using up all their fuel and dying. This is one reason why stars of this type are very rare; only a few may be around at any given moment, regardless of how many may have formed in the past (as it happens, very massive stars also do not form very often, another reason for their rarity).

Now let us consider a lower main-sequence star, for example, an M star with mass 0.08 solar masses and luminosity 10^{-4} times that of the Sun. Its lifetime is $0.08/10^{-4} = 800$ times *greater* than the Sun's lifetime, or about 8×10^{12} years. This is a very long time—greater than the age of the universe. No star formed with such a low mass has had time to use up all of its fuel; all such stars that have ever been born are still on the main sequence. For this reason, the vast majority of all stars in existence today are low-mass stars. (It is also true that these stars form in greater numbers, adding further to their numerical dominance.)

lifetime (derived from the formula $E = mc^2$) and dividing this by the luminosity, which is the rate at which energy is expended.

We can apply similar calculations to other stars (Table 13.1). A star near the top of the main sequence, for example, may have 50 times the mass of the Sun, but at the same time it uses its energy as much as 10^6 times more rapidly. (Recall how rapidly the luminosity of a star varies with its mass, as discussed in the last chapter.) These numbers lead to an estimated lifetime that is $50/10^6 = 5 \times 10^{-5}$ times that of the Sun. This star will last only half a million years! (Actually, in such a massive star, more than 10 percent of the total mass can undergo reactions, and the lifetime is accordingly longer than this simple

HEAVY ELEMENT ENRICHMENT

Clearly nuclear reactions have an effect on the chemical composition of a star, because they change one element into another. We have stressed that all stars consist of about the same mixture of hydrogen and helium, with a trace amount of other elements, but now we find that, in the core of a star, hydrogen is gradually converted into helium. While this change may not immediately affect the surface composition, which is what astronomers can measure directly from spectral analysis, it does affect the internal composition. It is this change that causes a star to evolve, because the core density is altered as hydrogen nuclei combine into the heavier helium nuclei. Furthermore, when the hydrogen runs out, the star must make major structural adjustments as its primary source of internal pressure disappears.

These adjustment are discussed in Chapter 14; for now, let us consider only the change in the abundances of the elements. There are other nuclear reactions that can take place in stars after the hydrogen-burning stage has ended, if the core temperature reaches sufficiently high levels. One such reaction is the conversion of helium into carbon by a sequence called the **triple-alpha reaction** (see Appendix 11; helium nuclei are called alpha particles, and in this reaction, three of these combine to form one carbon nucleus). When the triple-alpha reaction takes place, the composition of the star's core changes from helium to carbon.

Many additional reactions can occur, in later stages, if the core temperature goes even higher (Fig. 13.7). These reactions produce ever-heavier products, so that as a star goes through successive stages of nuclear burning, elements as heavy as iron (which has twenty-six protons and thirty neutrons in its nucleus) replace the hydrogen that was originally predominant.

These reactions require ever-higher temperatures, because as the particles become more massive, more energy is needed to keep them moving fast enough to react. Furthermore, the heavier nuclei have greater electrical charges and therefore stronger repulsive forces tending to keep them apart. For both of these reasons, reactions involving the fusion of heavy nuclei require higher temperatures than those in which lightweight nuclei such as hydrogen undergo fusion.

The most massive stars go through the greatest number of reaction stages and, as we have just seen, these stars live only a short time. In later chapters we will discuss the role played by these massive stars in the enrichment of chemical abundances in the galaxy.

STELLAR CHROMOSPHERES AND CORONAE

Let us now consider other processes that may affect a star's structure and evolution. Some very important ones take place near the surface rather than deep inside.

As we have seen, in most stars convection occurs at some level inside. For stars on the upper portion of the main sequence, convection takes place near the center and has no directly observable consequences. For stars on the lower portion of the main sequence, however, where convection occurs in the outer layers, there are important effects. Stars of spectral types F,

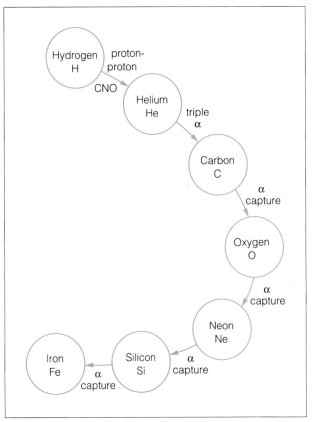

Figure 13.7   The enrichment of heavy elements.   As a star ages, each time one nuclear fuel is exhausted, another ignites if the temperature in the core rises sufficiently (which depends on the mass of the star). The net result is an enrichment of heavy elements.

G, K, and M, where surface convection is thought to occur, also are found to have chromospheres and coronae. (Recall from Chapter 11 that chromospheres and coronae produce ultraviolet emission lines and X-ray emission, and these signatures are generally found in *all* cool stars; Figs. 13.8 and 13.9).

The presence of chromospheres and coronae seems to be linked with the presence of convection in the outer layers (Fig. 13.10). Stars on the lower portion of the main sequence generally have both phenomena. Somehow, the kinetic energy of the turbulent motions in the convective layer is transported into higher zones, where it causes heating. As we learned in our discussion of the Sun, the precise mechanism for converting the energy of convection into heat is not understood but almost certainly involves magnetic fields. Observations of flare activity and ultraviolet chromospheric and coronal emission lines show that all cool stars have activity cycles similar to that of the Sun (see the discussion in Chapter 11). Detailed observations of brightness variations in some stars demonstrate the presence of "starspots," similar to sunspots. X-ray data show that virtually all cool

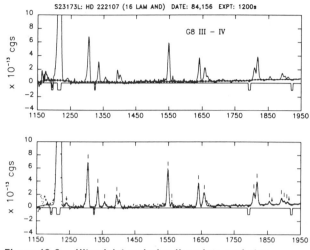

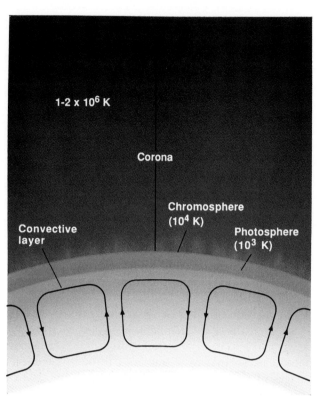

Figure 13.8   Ultraviolet emission lines in a cool star.
These graphs show the intensities of several emission lines observed with the *International Ultraviolet Explorer,* in two different stars. (J. Bennett)

Figure 13.10   A stellar chromosphere and corona.   Stars in the lower portion of the main sequence all have chromospheres and coronae, analogous to the Sun (see Chapter 11 for more details). The photosphere is the region in which the star's continuous spectrum and absorption lines are formed; the chromosphere is a thin, somewhat hotter region just above the photosphere; and the corona is a very hot, extended region outside the chromosphere. The source of heat for the chromosphere and corona is probably related to convective motions in the star's outer layers.

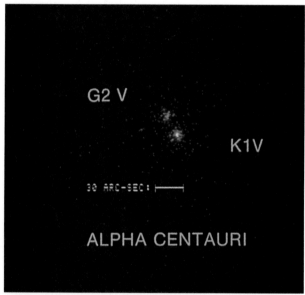

Figure 13.9   An X-ray image of cool stars.   This image, obtained by the *Einstein Observatory's* X-ray telescope, shows our nearest neighbor, Alpha Centauri. This is a multiple star system, and here we see that two of the stars, a G star and a K star, emit X rays from their coronae. (Harvard-Smithsonian Center for Astrophysics)

internal structure is rather different, since they are so much larger. These stars also have convection in their outer layers, however, and indeed they also have chromospheres, although it is not certain that the extremely high temperatures characteristic of coronae are present.

STELLAR WINDS AND MASS LOSS

So far we have said nothing about the possibility that the hot stars might have chromospheres or coronae. Some of the upper main-sequence stars are known to have emission lines in their spectra, indicating that they have some hot gas above their surfaces. Visible-wavelength emission lines are found in only a few extremely hot or luminous stars, however.

In the late 1960s, with the first observations of ultraviolet spectra, it was immediately found that

stars have chromospheres and coronae. Thus it seems likely that magnetic fields are important in all these stars, playing key roles in transporting energy into high levels of the atmospheres.

What of the red giants and supergiants? These stars can have surface temperatures comparable with the lower main-sequence stars, but obviously their

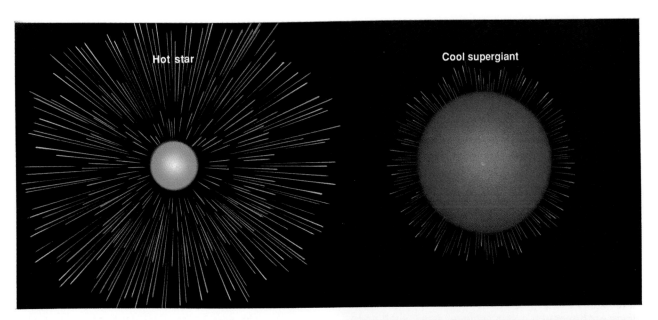

Figure 13.11   Stellar winds.   In the drawing at the top, a luminous hot star (left) ejects gas at a very high velocity; the lengths of the arrows indicate that the gas accelerates as it moves away from the star. A luminous cool star (especially a K or M supergiant) is so extended in size that the surface gravity is very low, and material drifts away at relatively low speeds (right). In both types of stars, radiation pressure helps accelerate the gas outward. At the right is a photograph of a hot star surrounded by an extensive bubble of gas created by a stellar wind. (© 1984 ROE/AAT Board)

many hot stars (nearly all the O stars and many of the hotter B stars) have ultraviolet emission lines. Furthermore, there are absorption features that show enormous Doppler shifts, indicating that the stars are ejecting material at speeds as great as 3,000 kilometers per second! The most extreme **stellar winds,** as these outflows are called (Fig. 13.11), occur in the O stars, but most cool giants and supergiants also have winds, usually with lower speeds.

The cause of the winds is not known, although there are indications of how the high velocities are reached. Once gas begins to move outward, light from the star can exert sufficient force to accelerate it to high speeds. This force, called **radiation pressure,** is very weak (you certainly can't feel the breeze from a light bulb, for example), but O and B stars are so luminous that strong acceleration of the wind is possible. For the most luminous of these stars, radiation pressure may be sufficiently strong to initiate the winds, but for the slightly less luminous O and B stars, calculations show that radiation pressure is not capable of creating the initial outflow but that it can accelerate the gas to high speed once the flow begins. Some other mechanism must start the outflow first. We do not know what this mechanism is, but recently

it has been suggested that subtle pulsations might be responsible.

Very high degrees of ionization are observed in the winds from O and B stars. Recently X-ray emission from stars with winds has been discovered, and it is possible that the X rays are responsible for ionizing the gas (X-ray photons have high energies and cause ionization when they are absorbed by atoms). The X rays must be produced in some very hot region either in gas in the wind that is heated by turbulence or in a hot coronal region in the star's outer layers. Theorists would be surprised if these hot stars have coronae, however, because they do not think the O and B stars have convection in the outer layers. Perhaps this view is incorrect, and the stars do have convection zones near their surfaces, or perhaps some other process is creating the coronae in these stars.

Whatever the cause of the winds, they have important consequences. Analysis of the ultraviolet emission lines shows that, in some cases, stars are losing matter at such a great rate that a large fraction of the initial mass may be lost during their lifetimes. An O star might lose as much as one solar mass every 100,000 years. If such a star begins life with 20 or 30 solar masses and lives a million years, it could lose 10 solar masses, a significant fraction of what it had to begin with. This loss of matter has important effects on how such a star evolves, as we shall see.

The supergiants in the upper right of the H-R diagram also lose mass through stellar winds (Fig. 13.12), but these winds have a distinctly different nature. The red supergiants are so large that their surface gravity is very low, and the gas in the outer layers is not tightly bound to the star. Radiation pressure can easily push the gas outward, at the relatively low speed of 10 or 20 kilometers per second. The amount of mass lost can be just as great as in the hot stars, however, because these low-velocity winds are much denser than the high-speed winds from the O and B stars. It is thought that slow pulsations in red giant and supergiant stars may play a role in triggering mass loss, just as rapid pulsations may help create the high-speed winds in hot stars.

Because the red supergiants are relatively cool objects, the gas in their outer layers is not ionized, and even molecular species form there. In addition, small solid particles called dust grains can condense, shrouding the star in a cloud. In extreme cases the dust cloud becomes so thick that little or no visible light escapes to the outside. The dust grains become heated, however, and emit infrared radiation, so the star can still be detected with an infrared telescope.

Some of the dust that forms in the outer layers of red supergiants escapes into the surrounding void and contaminates interstellar space with a kind of interstellar haze (the interstellar dust is discussed at greater length in Chapter 16).

STELLAR MODELS: TYING IT ALL TOGETHER

Once astrophysicists have understood all the phenomena that govern a star's internal structure and its outward appearance, they can combine their knowledge of all these processes into a theoretical calculation (Fig. 13.13) that will tell them what goes on inside the star and what will happen to it in time. When the calculations correctly reproduce the observable quantities, it is assumed that they also correctly illustrate the nonobservable aspects, such as the star's interior conditions. We can have reasonable confidence in such calculations, because it can be demonstrated that the resulting solutions are unique; that is, no other solution to the mathematical equations can exist. The most difficult question is whether we know all the correct equations and physical parameters to begin with.

The calculation of a model star involves simultaneously solving a set of basic mathematical equations describing the physical processes and then solving them repeatedly for different depths inside the star. In this way, the calculations may proceed from the observed surface conditions to the center, or from assumed central conditions to the surface, with adjustments made until the calculations match the observed properties of the star. To carry out such a calculation requires a substantial amount of time on a large computer.

To see how a star evolves, it is necessary to compute the model many times over, each time taking

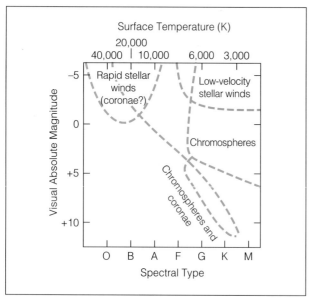

Figure 13.12   Chromospheres, coronae, and stellar winds in the H-R diagram.   Here we see that most stars of type F or cooler have chromospheres, although coronae may be confined to the main sequence. Very luminous stars, hot and cool alike, have sufficiently strong winds to result in significant loss of mass.

ASTRONOMICAL INSIGHT
Stellar Models and Reality

In this chapter we have discussed the physical processes that are believed to govern the internal conditions and external properties of stars. Despite the remoteness of stars and the fact that we must derive all of our observational information from the analysis of light emitted from their surfaces, we know quite a bit about their insides. It may seem surprising that we can have any confidence at all in our theories of stellar interiors and atmospheres, but the fact that we do stems from our faith that the same laws of physics that apply on Earth govern distant stars. Given that, stars are actually rather simple objects compared with, for example, a solid planet or a biological system.

Stars are made entirely of gas, which is the major reason that describing them theoretically is relatively simple. Gases obey certain well-defined laws, which are verified through experiment and theory. Add that stars are basically symmetric objects and stable (not changing with time), and we can invoke additional physical laws in the form of equations for the forces acting on gas particles in a star and for the production and transport of energy. Within the context of the adopted physical laws, the set of equations that describe a star's structure and properties has only one solution. This is a very powerful statement, for it implies that all we need to do is be sure to include the correct laws

of physics in our calculations, and we will get the correct solution.

Aside from the fact that our known laws of physics may be incomplete, the major difficulty in calculating theoretical models of stars lies in the fact that some of the processes occurring in stars are quite complex when examined in sufficient detail. For example, it is easy to treat the transport of energy from the core to the surface for a star when the energy is carried solely in the form of radiation, but very difficult when the energy is transported by flowing streams of heated gas, in the process we call convection. The calculations are simple when a star is spherically symmetric, but they become more complex when the star is rotating rapidly, so that its shape becomes somewhat flattened. Magnetic fields in stars are probably very complex, if our Sun is any example, and theory has not yet fully treated even the Sun's magnetic field. These are cases in which we know the laws of physics well enough, but their application is complicated by the vast amount of detail that must be incorporated into our calculations.

Despite the complications, stellar model calculations are quite good. We know this because it is possible to achieve very good agreement between observed quantities (such as stellar luminosity, radius, surface temperature, and others) and those calculated from theory. The first test of any theoretical model for a star is to see whether the model successfully reproduces the observed surface conditions. If it does, then we are confident that the model must also represent the interior fairly well.

Today's stellar structure theorists are busily refining their models, not so much because the models are incorrect, but because more and more details can be added to the calculations. Just as new telescope technology increases the ability of astronomers to observe astronomical objects, new advances in computer technology increase the capability of theorists to represent reality in ever-greater detail. The first stellar models were calculated with slide rules or slow, mechanical desk calculators, and it took weeks or months to calculate a single model. In the past two or three decades, model complexity has increased in proportion to computer capacity and speed, and it is now possible to calculate many models (for different stages in the evolution of a star, for example) in hours. Today we are using a new generation of "supercomputers," machines with vastly greater speed and capacity for massive calculations, and no one will benefit from this development more than the stellar structure theorist. As the capabilities of the computers increase, so will the ability of astrophysicists to represent reality in their calculations.

POINT	RADIUS	DENSITY	TEMP	PRESSURE	ENERGY GEN		LUMIN	OPACITY	DELAD	DELRAD	BETA
1	0.	2.8526E+00	3.5455E+07	1.7651E+16	1.9682E+05	1	0.	3.4602E-01	2.8868E-01	3.0013E+00	7.7421E-0
2	5.4734E+08	2.8526E+00	3.5455E+07	1.7651E+16	1.9680E+05	1	3.0509E+32	3.4602E-01	2.8868E-01	3.0013E+00	7.7421E-0
3	7.3124E+08	2.8525E+00	3.5455E+07	1.7650E+16	1.9679E+05	1	9.1397E+32	3.4602E-01	2.8868E-01	3.0011E+00	7.7421E-0
4	8.8730E+08	2.8525E+00	3.5455E+07	1.7650E+16	1.9677E+06	1	1.6376E+33	3.4602E-01	2.8868E-01	3.0009E+00	7.7421E-0
5	1.0365E+09	2.8525E+00	3.5455E+07	1.7650E+16	1.9675E+05	1	2.6200E+33	3.4602E-01	2.8868E-01	3.0008E+00	7.7421E-0
6	1.1927E+09	2.8524E+00	3.5455E+07	1.7649E+16	1.9673E+00	1	3.9844E+33	3.4602E-01	2.8868E-01	3.0006E+00	7.7421E-0
7	1.3548E+09	2.8524E+00	3.5455E+07	1.7649E+16	1.9671E+05	1	5.4460E+33	3.4602E-01	2.8869E-01	3.0001E+00	7.7421E-0
8	1.5283E+09	2.8523E+00	3.5456E+07	1.7648E+16	1.9668E+05	1	1.0303E+34	3.4602E-01	2.8869E-01	2.9990E+00	7.7421E-0
9	1.7159E+09	2.8523E+00	3.5457E+07	1.7648E+16	1.9664E+05	1	1.1801E+34	3.4602E-01	2.8869E-01	2.9990E+00	7.7421E-0
10	1.9200E+09	2.8522E+00	3.5457E+07	1.7647E+16	1.9659E+05	1	1.6659E+34	3.4602E-01	2.8869E-01	2.9994E+00	7.7421E-0
11	2.1449E+09	2.8521E+00	3.5456E+07	1.7646E+16	1.9654E+05	1	2.3202E+34	3.4602E-01	2.8869E-01	2.9989E+00	7.7422E-0
12	2.3910E+09	2.8519E+00	3.5455E+07	1.7644E+16	1.9647E+05	1	3.2162E+34	3.4602E-01	2.8869E-01	2.9988E+00	7.7422E-0
13	2.6639E+09	2.8518E+00	3.5454E+07	1.7643E+16	1.9639E+05	1	4.4432E+34	3.4602E-01	2.8869E-01	2.9976E+00	7.7422E-0
14	2.9647E+09	2.8515E+00	3.5453E+07	1.7641E+16	1.9628E+05	1	6.1231E+34	3.4602E-01	2.8869E-01	2.9967E+00	7.7423E-0
15	3.2977E+09	2.8513E+00	3.5452E+07	1.7639E+16	1.9615E+05	1	8.4230E+34	3.4602E-01	2.8869E-01	2.9956E+00	7.7423E-0
16	3.6666E+09	2.8510E+00	3.5450E+07	1.7636E+16	1.9600E+05	1	1.1571E+35	3.4602E-01	2.8869E-01	2.9942E+00	7.7426E-0
17	4.0755E+09	2.8506E+00	3.5448E+07	1.7632E+16	1.9580E+05	1	1.5800E+35	3.4602E-01	2.8870E-01	2.9925E+00	7.7426E-0
18	4.5291E+09	2.8501E+00	3.5445E+07	1.7628E+16	1.9557E+05	1	2.1477E+35	3.4602E-01	2.8870E-01	2.9903E+00	7.7427E-0
19	5.0324E+09	2.8496E+00	3.5443E+07	1.7623E+16	1.9527E+05	1	2.9050E+35	3.4602E-01	2.8870E-01	2.9878E+00	7.7428E-0
20	5.5916E+09	2.8488E+00	3.5439E+07	1.7616E+16	1.9491E+05	1	4.0871E+35	3.4602E-01	2.8870E-01	2.9847E+00	7.7430E-0
21	6.2112E+09	2.8480E+00	3.5434E+07	1.7608E+16	1.9446E+05	1	5.5949E+35	3.4602E-01	2.8870E-01	2.9808E+00	7.7432E-0
22	6.8990E+09	2.8469E+00	3.5428E+07	1.7599E+16	1.9391E+05	1	7.6590E+35	3.4602E-01	2.8870E-01	2.9760E+00	7.7433E-0
23	7.0640E+09	2.8455E+00	3.5421E+07	1.7585E+16	1.9324E+05	1	1.0408E+36	3.4602E-01	2.8871E-01	2.9701E+00	7.7434E-0

Figure 13.13   Building a model star.   Modern astrophysicists reconstruct what is happening inside stars with complex computer programs (above) that simultaneously solve large numbers of equations describing the physical conditions. Often the result is plotted (right), as a convenient means of showing the implications of these calculations. (C. Hansen, University of Colorado)

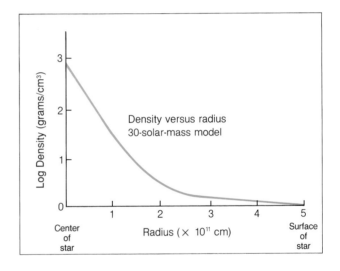

into account changes in the star that occurred in the previous step in the calculation. For example, a certain amount of hydrogen is converted into helium at each step, and this change in the chemical composition of the core must be taken into account in the next step. The conversion of hydrogen into helium uses up the hydrogen fuel, and the star will eventually exhaust its supply. Furthermore, the density of the core will increase, because helium nuclei are more massive than hydrogen nuclei. These changes alter the star's overall structure, and the calculations keep track of the changes, allowing the evolution to be traced. In the next chapter we will discuss the evolution of stars, and how it is determined from a combination of theory and observation.

PERSPECTIVE

We now are acquainted with all the physical processes that affect a star; we know what a star is and what makes it run. We have seen that mass plays a dominant role in dictating all the other properties, although the composition, which changes with time, is also important. We are now prepared to discuss the life story of a star—how it is born, how it lives, and how it ends its days.

SUMMARY

1. A star is gaseous throughout, with temperature and density increasing toward the center.

2. The balance between gravity and pressure within a star, called hydrostatic equilibrium, governs the internal structure.

3. The mass of a star, along with its chemical composition, governs its central temperature and pressure (through hydrostatic equilibrium), luminosity, radius, and internal structure.

4. Nuclear fusion reactions occur in the core of a star, where the temperature and density are high enough to allow nuclei to collide with sufficient velocity and frequency.

5. In most stars (those on the main sequence), hydrogen is converted into helium by reactions in the core.

6. The energy inside a star is transported by convection or by radiation. On the lower main sequence (including the Sun), radiative transport dominates in the inner parts of the star, while convection operates in the outer layers. On the upper main sequence, convection is dominant in the core, while radiative transport dominates in the outer layers.

7. The lifetime of a star depends on its mass and how rapidly it uses up its nuclear fuel; the lifetime decreases dramatically from the lower main sequence to the upper main sequence.

8. Nuclear reactions in stars, and the recycling of matter between stars and interstellar space, result in a gradual increase in the abundance of heavy elements in the universe.

9. Stars in the cool half of the H-R diagram have chromospheres and coronae analogous to those of the Sun.

10. Stars in the hot portion of the H-R diagram, and very luminous cool supergiants as well, have stellar winds that can cause the loss of large fractions of the initial stellar mass. The loss of mass can have significant effects on how the stars evolve.

11. To determine the interior conditions in a star, astronomers must compute them from known laws of physics and observations of the surface conditions.

REVIEW QUESTIONS

1. How do we know that a star must be in a state of balance between gravity and pressure? What would happen if this balance were disturbed?

2. Explain in your own words why mass is a star's most fundamental property; that is, explain how the mass of a star is the most important factor in determining its other properties.

3. Why do nuclear reactions in stars occur only in the innermost region—that is, in the core?

4. Estimate the hydrogen-burning lifetime of a star of 15 solar masses and luminosity 10,000 times that of the sun, and a star of 0.2 solar masses and luminosity 0.008 that of the Sun.

5. Summarize the ways in which energy produced in a star's core is transported outward to the surface. Point out ways in which the transport varies with the type of star.

6. Explain why the lifetime of a star is *shorter,* the more massive the star.

7. As a star ages and undergoes successive nuclear reaction stages, a higher temperature is required to initiate each reaction stage. Why?

8. Summarize how an upper main-sequence star (an O or B star) and one on the lower main sequence (K or M) differ in structure, energy generation and transport, and outer layers.

9. If a star has an initial mass of 20 solar masses and a lifetime of 5 million years, how much of its mass will it lose if the rate of mass loss is one solar mass every 500,000 years?

10. Suppose the Sun becomes a red supergiant star with a radius 200 times its present radius. By how much would its surface gravity and escape velocity decrease? Discuss the importance of your answer in view of the stellar winds from red supergiants.

ADDITIONAL READINGS

Aller, L. H. 1971. *Atoms, stars, and nebulae.* Cambridge, Mass.: Harvard University Press.

Baliunas, S., and S. Saar. 1992. Unfolding mysteries of stellar cycles. *Astronomy* 20(5):42.

Boss, A. P. 1985. Collapse and formation of stars. *Scientific American* 252(1):40.

Cox, A. N., and J. P. Cox. 1967. Cepheid pulsations. *Sky and Telescope* May, p. 278.

Giampapa, M. S. 1987. The solar-stellar connection. *Sky and Telescope* 74(2):142.

Graham, J. 1984. A new star becomes visible in Chile. *Mercury* 13(2):60.

Hoyle F. 1975. *Astronomy and cosmology: A modern course.* San Francisco: W. H. Freeman.

Jones, B., D. Lin, and R. Hanson. 1984. T Tauri and TIRC: A star system in formation. *Mercury* 13(3):71.

McCrea, A. W. 1991. Arthur Stanley Eddington. *Scientific American* 264(6):92.

Perry, J. R. 1975. Pulsating stars. *Scientific American* 232(6):66.

Shu, F. 1982. *The physical universe.* San Francisco: W. H. Freeman.

Smith, E. P., and K. C. Jacobs. 1973. *Introductory astronomy and astrophysics.* Philadelphia: Saunders.

Snow, T. P. 1981. Dieting for stars, or how to lose 10^{25} grams per day (and feel better too!). *Griffith Observer* 45(5):2.

Weymann, R. J. 1978. Stellar winds. *Scientific American* 239(2):34.

ASTRONOMICAL ACTIVITY
Building a Model Sun

Table 11.3 provides information on interior conditions in the Sun, which can be used to develop an intuitive understanding of the internal structure of a typical star. To do this, you will need three sheets of graph paper and a ruler. Preferably, the graph paper should have relatively heavy lines marking 1-cm squares and lighter grid lines marking 1-mm square as well.

Make three graphs, each with the relative radius of the Sun on the horizontal axis, from 0 (at the left) to 1.0 (at the right). On the vertical axis of the first graph, plot the values for temperature; on the second graph, plot the values for density; and on the third, plot the values for mass fraction. All of these values are given in Table 11.2 Note that you should choose your horizontal and vertical scales before beginning to plot the points for each graph. If your graph paper has 1-cm squares, a convenient horizontal scale would be 1-cm = 0.1 $R_{\odot}$, so that your horizontal axis will be 10 cm long (plot only the points out to R = 1.0 $R_{\odot}$; ignore the remaining values, which represent the atmosphere of the Sun rather than the interior). For your vertical scales, note the maximum value you will have to plot, and find a value per centimeter that will provide a convenient vertical height in the range of 15 to 20 cm. For example, for the temperature plot, you could choose a scale of 1 cm = 1.0×10^6 K, since the maximum value of 1.6×10^7 K is 16 times greater than this, providing a vertical height of 16 cm. Similarly, you could choose a scale of 1 cm = 10 g/cm^3 for density and a scale of 1 cm = 0.05 $M_{\odot}$ for mass fraction.

Once you have made points on the graphs for each of the values in the table, connect the points in each graph with a smooth line. This line shows how the value of each plotted quantity varies with radius, from the center of the Sun to its surface. Right away you will find some interesting results. For example, you will see that the temperature peaks sharply toward the center of the Sun, as does the density, and that the mass fraction grows quickly as you move outward from the center. On all three graphs you may have difficulty distinguishing the values in the outer portion of the Sun, because they are so low relative to the central value (for temperature and density) or are very close to 1.0 throughout the outer portion (for mass fraction). For this reason graphs like this are often plotted in logarithmic units, which can more easily show points covering a very wide range of values.

To help you visualize some of the important aspects of the internal structure of a star, use your graphs to answer some questions. First, as we already noted, the temperature drops off very quickly with distance from the Sun's center. To illustrate this, find the relative radius where the temperature has dropped to one-half its central value, to one-tenth of the central value, and to one-hundredth of the central value. In case you are unfamiliar with reading graphs, here is how to do this: For the first example, find the point on the vertical axis where the temperature equals one-half of the central value; then follow the horizontal grid lines on the graph paper over to the curve you have drawn, then go down to the horizontal axis, where you can read the value of R that corresponds to this point.

In similar fashion, use your other graphs to find where the density drops off to one-half, one-tenth, and one-hundredth of its central value; and where the mass fraction reaches one-half, three-fourths, and nine-tenths of the total mass of the Sun.

How would these results differ if the Sun were uniform throughout its interior with constant density and temperature? Indicate on the same graphs how the curves would look in that case. To do this, you will have to choose values for the temperature and the density; the average between the central and surface values will do. For mass fraction, note that the fraction will vary from 0 at the center to 1.0 at the surface, but the graph will not be a simple straight line, because the mass inside a spherical volume of constant density grows with the cube of the radius. Calculating the exact shape of this curve may be a bit complicated but you may be able to deduce its general nature with some simple experiments of plotting mass fraction versus R, where the mass fraction grows with the cube of R.

Now that you have done this exercise for the Sun, think about what a similar model for a red giant star might be like. Based on the discussions in this chapter, how do you think the graphs of temperature, density, and mass fraction for a red giant might differ from the graphs you have made for the Sun?

Life Stories of Stars

The planetary nebula NGC2346 (ROE/AAT Board; photo by D. Malin)

PREVIEW

As stars age, they must change because the nuclear reactions taking place in their cores gradually alter their composition and internal structure. In this chapter you will learn how stars evolve, at first by reviewing the observational evidence and then by following step-by-step descriptions of the stages that stars go through during their lives. The observational information on which the theory of stellar evolution is based comes largely from observations of stars in groups and clusters, so the first part of the chapter is devoted to these systems and the biographical information on stars that they provide. Following this you will learn how stars are thought to form, and then you will embark on the life story of a star like the Sun, following it in detail from birth through its red giant stage to its final state as a white dwarf. Next you will see how stars of greater mass than the Sun differ from our local star in their lives and deaths as bizarre supercompact objects. You will find that one of the most energetic and spectacular phenomena in the universe, the mighty supernova explosion, results from the evolution of a massive star. Finally, you will see how the evolutionary process can be affected by the presence of a companion star in a close binary system. Throughout all these discussions, the importance of mass loss and mass transfer will be emphasized, because alterations in the mass of a star can dramatically alter the course of its evolution.

How can we watch a star evolve? After all, even the shortest-lived ones are around for hundreds of thousands of years, while human studies of astronomy date back only four or five millennia. Occasionally we catch a star in the act of change, as it makes a transition from one stage in its evolution to another, but clearly we cannot hope to see a star through its whole lifetime, watching it form, live, and die.

We can, however, piece together a story of stellar evolution by examining many stars of different ages, deducing from this the sequence of events that occur in a single star, and then computing theoretical models that reproduce the observed stellar properties. These models take into account the physical processes that govern stellar structure and allow us to understand the developmental sequence that stars go through in their lives.

THE OBSERVATIONAL EVIDENCE

Most of the observational evidence we have on how stars form and evolve comes from deducing the relative ages of stars and piecing together their life stories. This process is analogous to determining how a tree grows by observing seeds, young saplings, mature trees, and dead logs in a forest. On rare occasions astronomers actually get to see stars changing from one phase of their lives to another, because some stages in the life of a star are quite sudden. There are violent deaths, known as **supernovae,** in which stars explode, and less violent but still dramatic occasions when newborn stars emerge from the cloak of gas and dust from which they formed. We will discuss some of these singular events, but will concentrate primarily on the less dramatic but more informative observational evidence based on the study of stars of different ages.

Stars are not uniformly distributed within our galaxy; many are found in locally concentrated regions

TABLE 14.1   Selected Clusters of Stars

Cluster	Type	m_v	Distance (pc)	Number of Stars	Diameter (pc)	Age (years)
47 Tuc	Globular	4.0	5,100	10^4	10	1.5×10^{10}
M103	Open	6.9	2,300	30	5	1.6×10^7
h Persei	Open	4.1	2,250	300	16	1.0×10^7
χ Persei	Open	4.3	2,400	240	14	1.0×10^7
M34	Open	5.6	440	60	30	1.3×10^8
Perseus	Open	2.2	167	80	12	1.0×10^7
Pleiades	Open	1.3	127	120	4	5.0×10^7
Hyades	Open	0.6	42	100	5	6.3×10^8
S Mon	OB	4.3	740	60	6	6.3×10^6
τ CMa	Open	3.9	1,500	30	3	5.0×10^6
Praesepe	Open	3.7	159	100	4	4.0×10^8
o Velorum	Open	2.6	157	15	2	2.5×10^7
M67	Open	6.5	830	80	4	4.0×10^9
θ Carinae	Open	1.7	155	25	3	1.3×10^7
Coma	Open	2.8	80	40	7	5.0×10^8
ω Cen	Globular	3.6	5,000	10^5	20	1.5×10^{10}
M13	Globular	5.9	7,700	10^5	11	1.5×10^{10}
M22	Globular	5.1	3,000	10^5	9	1.5×10^{10}
M15	Globular	6.4	14,000	10^6	11	1.5×10^{10}

Figure 14.1   The Pleiades.   This relatively young galactic cluster is a prominent object in the Northern Hemisphere during the late fall and winter. (© 1985 Anglo-Australian Telescope Board; photograph by David Malin)

called clusters. Comparisons among the stars in a cluster, which evolve at different rates because of their differing masses, provide a wealth of information on how stars change as they age.

A few clusters are sufficiently prominent to be visible to the unaided eye, while others can be viewed with a small telescope or a good pair of binoculars (Table 14.1). The Pleiades (Fig. 14.1), a bright cluster to the west of Orion, is perhaps the best-known cluster to observers in the Northern Hemisphere where it is easily visible in the evening sky throughout the late fall and early winter. Other clusters easy to find with a small telescope are the Hyades, the double cluster in Perseus, and M13, a fantastic, spherically shaped collection of hundreds of thousands of stars (Fig. 14.2). It is apparent that the stars in a cluster are gravitationally bound to it, with each star orbiting the common center of mass.

There are several distinct types of star clusters. Those that contain a modest number of stars (up to a few hundred) and are found in the disk of a galaxy are referred to as **galactic,** or **open,** clusters (Figs. 14.3 and 14.4). The Pleiades and the Hyades are examples. Loose groupings of hot, luminous stars, also found in the plane of the galaxy, are called **OB**

Figure 14.2   The Globular Cluster 47 Tucanae.   (© 1992 AAT Board; photo by D. Malin)

Figure 14.3   The Open Cluster NGC3293.   (© 1977 Anglo-Australian Telescope Board; photo by D. Malin)

Figure 14.4   The Open Cluster NGC4755.   (National Optical Astronomy Observatories)

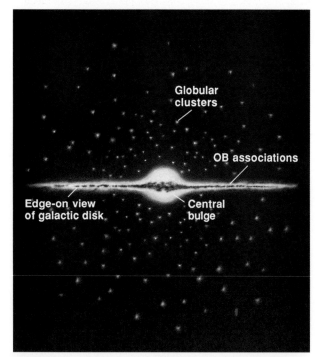

Figure 14.5   The locations of clusters in the galaxy.   The open or galactic clusters are found in the disk of the Milky Way, as are the OB associations, which are always located in spiral arms (see Chapter 16). In contrast, the globular clusters inhabit a large spherical volume and are not confined to the plane of the disk.

associations because they are dominated by stars of spectral types O and B. Unlike the other types of clusters mentioned here, these groups are probably not gravitationally bound together, but appear grouped simply because the stars have recently formed together. The giant **globular clusters** (see Fig. 14.2) are found in a spherical volume about the galactic center (Fig. 14.5) and are not confined to the plane of the galaxy.

Astronomers make two important assumptions about a cluster of stars:

1. The stars in a cluster are all at the same distance from us.
2. The stars formed together, so they are all the same age and have the same composition initially.

The first assumption means that any observed differences from star to star within a cluster must be real variations and not false effects created by differing distances from us. Brightness differences, for example, have to be due to differences in the luminosities of stars within the cluster.

Because distance effects are eliminated when we compare stars within a cluster, it is possible to plot an H-R diagram for a cluster without first determining the absolute magnitudes of the stars. We simply plot apparent magnitude against spectral type or, more commonly, against the color index $B - V$, which is easier to measure for a large number of stars. To do this requires only the determination of the visual (V) and blue (B) magnitudes of the cluster stars, a rather straightforward observational procedure, and then the construction of a plot of V versus $B - V$.

The result, called a **color-magnitude diagram,** looks just like an H-R diagram except that the vertical scale is an apparent magnitude scale. This is because the *differences* in magnitude of the cluster stars are the same, whether we use apparent or absolute magnitudes.

A nice spin-off of plotting cluster H-R diagrams is that they can be used to determine the distances to clusters. A comparison of a cluster H-R diagram with a standard diagram (one with an absolute magnitude scale) allows us to see what the absolute magnitude scale for the cluster diagram should be. This in turn tells us the difference between the absolute and apparent magnitudes for the cluster, and the difference (the distance modulus; see Chapter 12) tells us the distance. This procedure, known as **main-sequence fitting,** depends only on the assumption that the cluster has a main sequence that is identical to that of the standard H-R diagram, a safe assumption as long as the chemical composition of the cluster is not unusual.

The second major assumption made about clusters—that all the stars are the same age—is very important in studying stellar evolution. The basic premise is that the stars did not just happen to come together by chance, but formed together out of a common cloud of interstellar material. Binary systems as well as clusters must *form* as binaries or clusters. A corollary of this assumption is that all the stars in a cluster must begin with the same chemical composition, since they form from a common source of material.

The knowledge that all the stars in a cluster have the same age and initial composition gives astronomers tremendous leverage in deducing the evolutionary histories of stars. The only major feature that differs from one star to another within a cluster is mass, so the observed differences among stars in a cluster give us clues to how the properties of stars depend on mass. Comparison of H-R diagrams of clusters of different ages then gives us a picture of how these different masses evolve.

Striking differences are found between H-R diagrams of different clusters (Fig. 14.6). The Pleiades, for example, has an almost complete main sequence but no red giants. In contrast, M13, a globular cluster, has no stars on the upper portion of the main sequence, but it has a large number of red giants. A wide variety of intermediate cases can be found.

As we discussed in the last chapter, we expect massive stars to evolve most rapidly. These stars start out near the top of the main sequence and have such tremendous luminosities that they use up their nuclear fuel rapidly. Tying this together with the observed differences among cluster H-R diagrams leads to the conclusion that the upper main-sequence stars must evolve into red giants (Figs. 14.7 and 14.8)

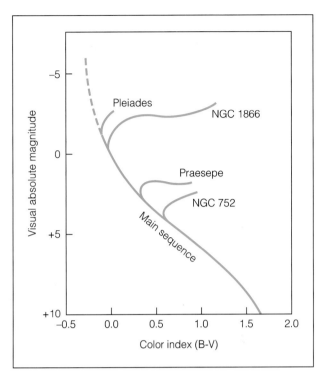

Figure 14.6   H-R diagrams for star clusters.   Here H-R diagrams (using color index rather than spectral type) of several clusters are plotted on the same axes. Different symbols are used in the right-hand portion to distinguish among the giant stars in different clusters. The solid lines indicate the main-sequence turnoff points, which are related to the ages of the clusters.

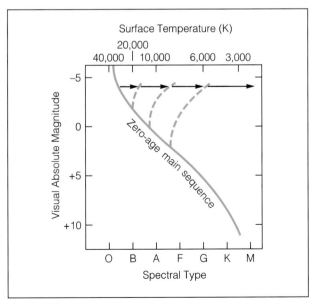

Figure 14.7   The evolution of the main sequence.   As a cluster ages, the stars on the upper main sequence move to the right. The farther down the main sequence this has happened, the older the cluster.

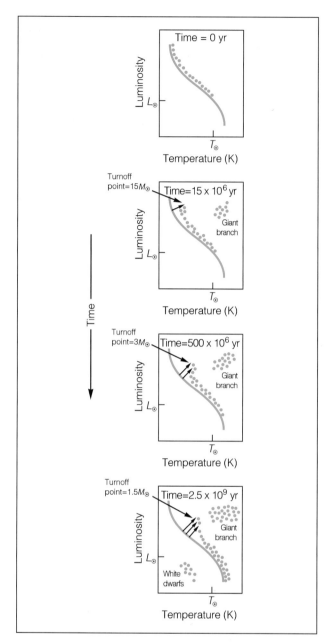

Figure 14.8   Evolution of the H-R diagram for a star cluster. This shows how the H-R diagram for a cluster of stars changes as the cluster ages and the most massive stars evolve to become red giants.

the red giant region. This process creates a definite cutoff point on the main sequence, above which there are no stars. The location of this cutoff depends almost entirely on the age of the cluster. The position of this **main-sequence turnoff,** as it is commonly called, indicates the age of a cluster. If, for example, we find a cluster whose main sequence stops at the position of stars like the Sun (that is, it has no stars above G2, or $B - V = 0.6$, on the main sequence), we know that all stars more massive than one solar mass have burned their hydrogen and evolved toward the red giant region. In the last chapter we learned that the Sun's lifetime for hydrogen burning is about 10 billion years; we conclude, therefore, that such a cluster is just about 10 billion years old. If it were younger, its main sequence would extend farther up, and if it were older, even the Sun-like stars would have had time to evolve into red giants, and the main-sequence turnoff would be farther down.

YOUNG ASSOCIATIONS AND STELLAR INFANCY

Earlier in this chapter we mentioned OB associations, loosely bound groups of O and B stars. We now know that these are very young clusters, probably the youngest in the galaxy, because they contain stars on the extreme upper portion of the main sequence. In many cases the stars are moving away from the center and escaping the association, so that within a few million years no concentration of stars will remain.

Stars form from interstellar gas and dust; young clusters may still be found embedded in the remains of the cloud from which they formed (Figs. 14.9–14.12). The spaces between stars in our galaxy are permeated with a very rarefied gas and fine solid particles known as interstellar dust. This material is discussed in Chapter 16; here we note that there are large concentrations known as interstellar clouds, where the density may build up to the point where stars can form. The best way to probe these dark and dusty regions where visible light cannot penetrate is to observe them at infrared wavelengths. Infrared radiation penetrates much farther through clouds of gas and dust than does visible light; in addition, the dust around a newborn star is hot and glows at infrared wavelengths. To see infant stars, therefore, we must look for infrared sources buried in dark clouds (Fig. 14.13).

Many regions where star formation is apparently taking place have been found in our galaxy. One of the best observed of these is a great, glowing cloud of gas in the sword of Orion within the Orion nebula.

when they have used up all their hydrogen fuel. In a very young cluster, even the most massive stars may still be on the main sequence, while in an older cluster these massive stars will have had time to use up their core hydrogen and become red giants. Hence the Pleiades cluster is relatively young, while M13 is very old.

As a cluster ages, more and more of its stars use up their hydrogen and leave the main sequence. The upper main sequence gradually erodes away as its stars move to the right in the H-R diagram, toward

Figure 14.9   Regions of young stars and nebulosity.   Here are two regions populated by recently formed stars and their associated interstellar gas and dust. At left is the emission and reflection nebula NGC6188; at right is a dark cloud showing bright rims illuminated by starlight. (Both © 1992 AAT Board; photos by D. Malin)

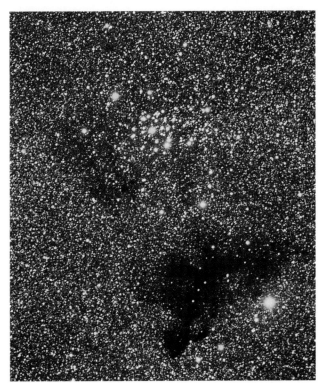

Figure 14.10   A very young cluster with a nearby dust cloud.   The cluster, NGC6520, contains several hot, young blue stars. The nearby dust cloud is so dense that it blocks out the light from the stars behind it. (© 1980 Anglo-Australian Telescope Board)

Figure 14.11   A dark cloud.   This is the easily recognizable Horsehead nebula. Many stars in the process of formation have been detected, by infrared techniques, within the dark and dusty regions seen here. (© 1984 Anglo-Australian Telescope Board)

Figure 14.12   Sites of star formation embedded in emission nebulosity.   The small, dark concentrations seen here are globules of interstellar material whose interiors are dense and well-shielded from radiation. It is thought that low-mass stars can form in these globules. (© 1992 AAT Board; photo by D. Malin)

Figure 14.14   The Orion nebula.   At the top is a normal photograph showing the familiar nebula. Below is an infrared image of the same region. Here little of the glowing gas can be seen, but the young, hot stars embedded within the outer portions are clearly visible. Note that the Trapezium, a close group of four brilliant young stars, is visible in both images. (top: U.S. Naval Observatory; bottom: © 1984 Anglo-Australian Telescope Board)

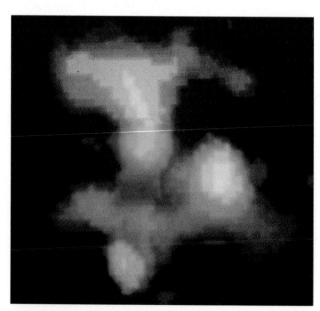

Figure 14.13   An embedded infrared source.   In this infrared photograph of a region within the Orion nebula, blue indicates the location of an intense, relatively warm source that is thought to be a newborn star. (R. D. Gehrz, J. Hackwell, and G. Grasdalen, University of Wyoming)

·A very young cluster of stars is located there, and infrared observations reveal a number of infrared sources embedded within the dark cloud associated with the cluster (Fig. 14.14).

In other, similar regions we find faint variable stars called **T Tauri** stars. These appear to be newborn stars in the process of shedding the excess gas and dust from which they formed; in fact, spectroscopic measurements indicate that some material may still

TABLE 14.2   Steps in Star Formation (one solar mass)

1. Gravitational collapse of interstellar cloud: Collapse proceeds most rapidly at the center, where the density increases most quickly. The cloud core remains cool because heat escapes in the form of infrared radiation.

2. Slow contraction of core: Once the core density is high enough so that the cloud becomes opaque to infrared radiation, heat is trapped, and the temperature and pressure in the core rise. The central portion of the cloud contracts very slowly for a while.

3. Infall: Material from the outer portions of the cloud continues to fall in, creating heating and shock waves. When the hydrogen molecules in the core begin to break up because of the additional heating, the breakup of molecules takes up heat from the gas, reducing the pressure, so that a new collapse phase occurs.

4. Final contraction: After the second collapse, the internal pressure again rises to balance gravity, and the central condensation enters a new phase of very slow contraction. At this point, the protostar is still surrounded by gas and dust and is usually observable only as an infrared source.

5. Nuclear ignition: The protostar continues to heat gradually as the contraction goes on. Eventually the core becomes hot enough to start nuclear fusion reactions, and the contraction stops because there is an internal source of energy to provide pressure that balances gravity fully. The star is now on the main sequence. Observationally it may still be associated with gas and dust and may still undergo a T Tauri phase.

be falling into these stars, as though the collapse of the cloud from which they are forming is not yet complete.

STAR FORMATION

From the assorted observations described here, a picture has emerged of how stars form (Table 14.2). This picture is based in part on theoretical calculations, as well as on observations.

The process begins with the gravitational collapse of an interstellar cloud (Fig. 14.15). It is not certain what causes a cloud to begin to fall in on itself, but once it starts, gravity takes care of the rest. The process of collapse may commonly begin in a large cloud, far more massive than an individual star, which then breaks into fragments that become single stars (or star pairs).

Calculations show that the innermost portion of the collapsing cloud falls in most quickly, while the outer parts of the cloud are still slowly picking up speed. The temperature in the core builds up as the density increases, but for a long time the heat escapes as the dust grains in the interior radiate at infrared

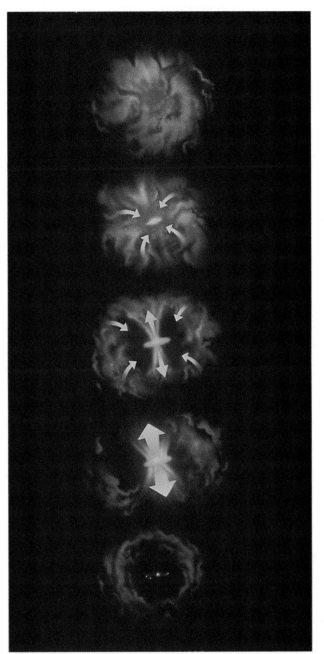

Figure 14.15   Steps in star formation. A rotating interstellar cloud collapses most rapidly at the center. At first infrared radiation escapes easily and carries away heat, but eventually the central condensation becomes sufficiently dense that this radiation is unable to escape, and the core becomes hot. The continued infall of gas, combined with rotation, produces a disk, and eventually a bipolar outflow. The outflow, in time, carries away enough material to negate the infall, leaving a cavity. The remaining disk either dissipates or forms planets as the outer envelope is gradually dispersed. (Painting by Rob Wood, in collaboration with Charles J. Lada.)

wavelengths. Eventually, however, the formation of hydrogen molecules causes the cloud to become opaque in the center, and radiation can no longer escape directly into space. After this the heat builds up much more rapidly, and the resulting pressure causes the collapse to slow to a very gradual shrinking. Material still falling in crashes into the dense core, creating a violent shock front at its surface. The luminosity of the core, which by this time may have a temperature of 1,000 degrees or more, depends on its mass. If the mass is large—that is, several times the mass of the Sun—the luminosity may be sufficient to blow away the remaining material by radiation pressure. For a lower-mass **protostar,** as the central dense object is now called, material continues to fall in. The infall creates heat, which can break up the hydrogen molecules, allowing radiation to escape and leading to a new collapse of the core. The pressure then builds again, allowing the protostar to nearly stabilize as it enters a new phase of very gradual contraction. Eventually the density of the infalling material becomes so low that the protostar can be seen through it. Regardless of mass, a point is reached where a starlike object becomes visible to an outside observer; this stage may be identified with the **T Tauri stars,** unstable young stars that eject material in strong winds (discussed in Chapter 10). If the obscuring matter is swept away violently, the object may appear to brighten dramatically in a very short time. In a few cases a previously rather faint star has brightened suddenly by several magnitudes.

In any case, the protostar continues to shrink slowly, growing hotter in its core. The shrinking stops when a sufficiently high temperature is reached that nuclear reactions begin in the center. This point is a major landmark in the process of forming a star; once the reactions have started, pressure can balance gravity, and the star is stable (no further shrinking). It lies on the main sequence, where it will spend most of its life converting hydrogen into helium in its core.

As we learned in Chapter 10, when we discussed the formation of the Sun, the material surrounding a young star takes on a disk-like shape due to the rotation of the system. This disk, which we called the solar nebula, was the progenitor of the planets, as the dust and gas in the disk condensed and coalesced into solid bodies.

Recently astronomers have found that young stars and prestellar objects are usually surrounded by disks. This discovery was not surprising, since we believe all stars form in much the same way—through the collapse of a spinning gas cloud—but observationally it had been very difficult to see the disks. Infrared observations have revealed the presence of disks around many protostars, however, because the disks contain cold dust, which glows at infrared

wavelengths. Radio observations have become very useful as well, since the gas in the disks is cold enough to be largely molecular, and molecules in space form spectral lines in the radio (specifically, the microwave) portion of the spectrum (this is discussed further in Chapter 16).

Not only do these observations reveal disks, but in many cases they also show streams of material flowing outward from the protostar along the rotation axes (Figs. 14.15 and 14.16). As gas and dust in the disk slowly spiral inward, the angular momentum of the disk is transferred to the protostar, causing it to spin faster and faster. This would eventually prove fatal to the protostar, because rotational forces would break it apart at some point, but it is rescued by the fact that it develops a strong outward-flowing wind. This wind is directed along the axes (i.e., perpendicular to the disk), forming a **bipolar outflow.** Such

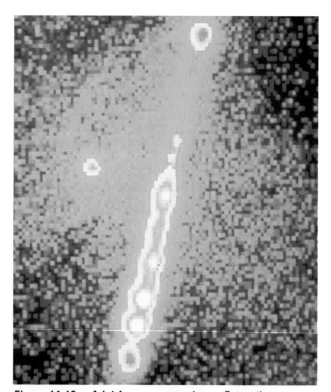

Figure 14.16   A jet from a young star.   Energetic mass outflows seem to be an intrinsic part of the star formation process. Part of the outflow occurs in narrow "jets" of ionized gas, which can extend to distances as great as 100,000 times the extent of the solar system. Here the faint, red star in the upper portion of the figure is the source of the jet, which is characterized by a series of knots of shocked, glowing gas moving outward at hundreds of km/sec. Jets such as this are believed to emerge perpendicular to the plane of the disk of gas and dust that forms as the new star condenses (the disk, about the size of the solar system, is too small to be seen here). (R. Mundt, Max Planck Institut für Astronomie; photo obtained at the Calar Alto 3.5-m telescope, in Spain)

flows are now recognized as integral steps in the process of star formation.

It is convenient to trace the star's path on the H-R diagram as it forms (Fig. 14.17). When it is a cold, dark cloud, its position is far to the right and down below the bottom of the diagram, well off the scales of temperature and luminosity appropriate for stars. As the cloud heats up in the core and begins to glow in infrared wavelengths, it moves up and to the left, eventually attaining sufficient luminosity to move onto the standard H-R diagram, but still to the right of the main sequence. When it reaches the phase of slow contraction, it moves gradually to the left and down, finally reaching the main sequence when the reactions begin.

The entire process, from the beginning of cloud collapse until the newly formed star is on the main sequence, takes many millions or even billions of years for the least massive stars but only a few hundred thousand years for the most massive stars. In a cluster with stars of various masses, the most massive stars may form, live, and die before the least massive ones even reach the main sequence. The H-R diagram for such a cluster shows stars off to the right of the main sequence at the lower end (Fig. 14.18).

THE EVOLUTION OF STARS LIKE THE SUN

When a star with the mass of the Sun arrives on the main sequence, it is similar to the Sun in all its properties. Thus it is a G2 star, with a surface temperature around 6,000 K and a luminosity of about 4×10^{26} watts. Its composition initially is about 73 percent hydrogen by mass, with roughly 25 percent helium and only about 2 percent other elements. It has convection in the outer layers, and it has a chromosphere and a corona. The star must change with time, because its core composition changes. Table 14.3 summarizes the stages in the life of a star like the Sun.

In this star's core, the proton-proton chain converts hydrogen into helium. As we saw previously (in Chapters 11 and 13), this process lasts some 10 billion years for a star of one solar mass. During this time, the core gradually shrinks as the hydrogen nuclei are replaced by a smaller number of the heavier helium nuclei, and the internal temperature rises. This causes an increase in luminosity, and the star gradually moves up in the H-R diagram. The main sequence, therefore, is not a narrow strip but has some breadth, as stars move slowly upward on it as

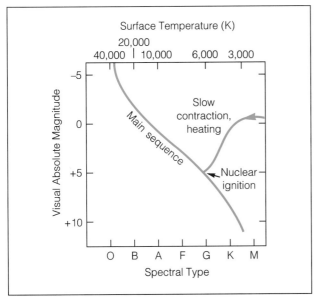

Figure 14.17   The path of a newly forming star on the H-R diagram.   As a protostar heats up, it eventually becomes hot and luminous enough to appear in the extreme lower right-hand corner of the H-R diagram. When the core becomes opaque, the rapid collapse slows to a gradual shrinking, and the protostar moves cross to the left as it heats. Eventually nuclear reactions begin, and the star is on the main sequence.

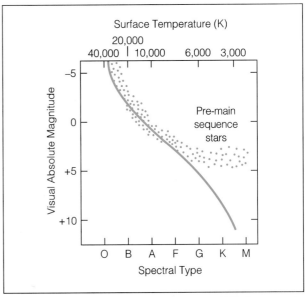

Figure 14.18   Pre–main-sequence stars.   H-R diagrams of very young clusters often show stars that have not yet reached the main sequence, but are still moving toward it. At the same time, some of the massive, short-lived stars at the top of the main sequence may have already evolved away from it, having used up their hydrogen fuel.

ASTRONOMICAL INSIGHT
Star Tracks

We have seen that as a star evolves, its position in the H-R diagram changes. The correct deduction of how a star's position changes was not easily made, however. During the years following the development of the H-R diagram, great intellectual effort was concentrated on discovering how stars evolve, but because the necessary information was not yet complete, many of the first ideas were wrong. As a result, an assortment of views on the direction of motion in the H-R diagram were offered, and evolutionary tracks thought to be made by stars during their lifetimes changed directions several times.

By the second decade of this century, it was well established that the red giants formed a distinct grouping of stars above the main sequence in the H-R diagram. Nothing was known of the true energy source that powers stars, and little was known of stars' chemical composition. Astronomers thought that stars glowed by virtue of energy released in a slow gravitational contraction, which caused the stars to shrink and heat, at least in their early stages. They also suspected that within the main sequence, the hot stars were the youngest and the cool stars the oldest. In 1913 Henry Norris Russell, in an address to the American Astronomical Society, put these ideas together to suggest a comprehensive theory of stellar evolution. He proposed that a star begins life as a red giant; for a while it contracts and heats, moving across the top of the H-R diagram from right to left. Eventually the temperature gets so hot inside the star that the gas there becomes highly ionized (now we know this is true from the start). Because ionized gas (Russell supposed

incorrectly) cannot compress and heat further, the star begins to cool while still contracting and moves down the main sequence from upper left to lower right. Thus Russell proposed that the evolutionary track of a star resembled a backward figure seven, and that the main sequence was just an age sequence.

In the decade following Russell's hypothesis, several astrophysicists, including most notably Sir Arthur Eddington and Sir James Jeans, developed theoretical models of stellar structure. These models, which of necessity incorporated many approximations, appeared to be consistent with the evolutionary scenario of Russell. But by the early 1920s, discrepancies began to arise. The chief problem was that there appeared to be no reason for the ionized gas in a star's interior to become incompressible, and therefore no reasonable justification for assuming that a contracting star should stop heating up and start cooling when its interior became ionized. The new theoretical work, along with developing evidence for a close relation between a star's mass and its luminosity, led to the first suggestion (again by Russell) that the main sequence represented a mass sequence, where stars are stable and spend a long time with no movement in the H-R diagram. Russell suggested that the difference between the main-sequence stars and the red giants was that the two groups had different internal energy sources, although his ideas of what those sources might be were rather vague. He still thought that the red giants were young stars, and that they moved down to the main sequence as their first energy source was exhausted and the second took over. Thus, in 1923 (when

this was hypothesized), the evolutionary tracks attributed to stars were much like those known today, but with the motion in the exact opposite direction.

The proper sequence of events in a star's lifetime finally began to be sorted out in the late 1920s and early 1930s, with Russell once again leading the way. Two major clues were developed: (1) Cluster H-R diagrams (studied by Hertzsprung, among others) showed that as a cluster ages, it loses stars from its upper main sequence and gains red giants; and (2) the Russell-Vogt theorem was developed, which demonstrated that a star's properties are all determined specifically by its mass and composition. It finally became clear that a star moves from the main sequence to the red giant region, and that this movement must be caused by a change in its internal composition. Some even suggested that the change of composition involves the conversion of hydrogen into helium, although it was not until the late 1930s that the exact nature of this process was understood.

The study of star tracks in the H-R diagram demonstrates how plausible, but incorrect, theories can be deduced from incomplete information, yet it also shows how patient and careful collection of data will eventually lead to the correct picture. The incorrect hypotheses developed along the way were important, for they helped inspire the continued efforts to solve the problem.

TABLE 14.3 Evolution of a Star (one solar mass)

1. Hydrogen burning: During the 10^{10} years the star spends on the main sequence, hydrogen is converted to helium through the proton-proton chain. The luminosity gradually increases as the core becomes denser and hotter.

2. Development of degenerate core and evolution to red giant: When hydrogen in the core is used up, the core contracts until it is degenerate. A hydrogen shell source outside the core still undergoes reactions, causing the outer layers to expand and cool. The star becomes a red giant.

3. Helium flash: The core eventually gets hot enough to ignite the triple-alpha reaction, in which helium is converted to carbon. Because the core is degenerate and cannot expand and cool to counteract the new source of heat, the reaction rapidly takes place throughout the core. So much heat energy is added that degeneracy is quickly destroyed, and the reactions become stable.

4. Stable helium burning: Once reactions are taking place in the core again, the major energy production occurs there, and the shell source becomes less important. The outer layers contract, and the surface heats up; the star moves back to the left on the H-R diagram.

5. Second red giant stage: When the core helium is used up, the energy again comes from a shell (an inner shell, where helium is converted to carbon, and an outer shell, where hydrogen is still converted to helium). The outer layers expand again, and the star becomes a red giant for the second time.

6. Mass loss: Through a stellar wind and pulsational instabilities, the star loses its outer layers, exposing the hot core. The core continues shrinking but is never again hot enough for nuclear reactions. The star becomes a planetary nebula as the outer layers are periodically shed.

7. White dwarf: The core again becomes degenerate as it shrinks. Eventually the outer layers are gone altogether, leaving only the hot, degenerate core, which is a white dwarf.

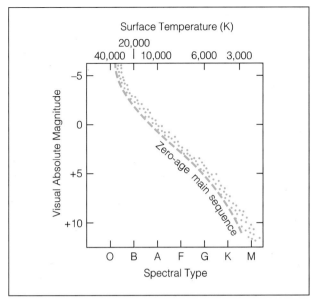

Figure 14.19   The width of the main sequence.   As stars on the main sequence gradually convert hydrogen into helium in their interiors, the cores shrink a little and get hotter. This increases the luminosity, and the stars gradually move upward on the H-R diagram. As a result, the main sequence, consisting of stars of a variety of ages, is not a narrow strip, but has some breadth.

their luminosities increase (Fig. 14.19). The starting point, the lower edge of the main sequence, is called the **zero-age main sequence (ZAMS)**, because this is where newly formed stars are found.

When the hydrogen in the core of the star is gone, reactions there cease. By this time, however, the temperature in the zone just outside the core has nearly reached the point at which reactions can take place, and with a little more shrinking and heating of the core, the proton-proton chain begins again in a spherical shell surrounding the core (Fig. 14.20). At this point, the star has an inert helium core and is producing all its energy in the hydrogen-burning shell, which steadily moves outward inside the star, eating its way through the available hydrogen.

The core continues to shrink and heat. This heating enhances the nuclear reactions in the shell, causing it to produce more and more energy. As the zone

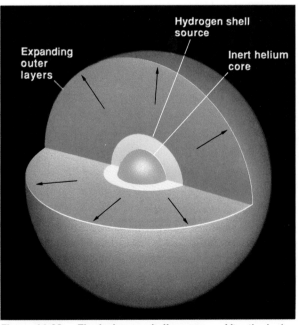

Figure 14.20   The hydrogen shell source.   After the hydrogen in the core of a star is completely used up, the core, now composed of helium, continues to shrink and heat. This causes a layer of gas outside the core to reach the temperature required for nuclear reactions. A shell source, in which hydrogen is converted into helium, is ignited. The star's outer layers expand and cool, and the star becomes a red giant.

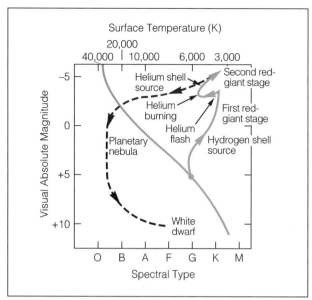

Figure 14.21   Evolutionary track of a star like the Sun.
This H-R diagram illustrates the path a star of one solar mass is thought to follow as it completes its evolution. The dashed portion, following the second red giant stage, is less certain than the earlier stages.

of energy production moves toward the surface, the outer layers of the star are forced to expand, and as this occurs, the surface cools. The luminosity of the star increases because of its increased surface area, so it moves upward on the H-R diagram (Fig. 14.21). It also moves to the right, because the surface temperature is decreasing. The star rapidly becomes a red giant, reaching a size of 10 to 100 times its main-sequence radius (Figs. 14.22 and 14.23). The outer layers are constantly overturning as convection occurs to a great depth.

The helium core becomes extremely dense, but at first the temperature is not high enough for any new nuclear reactions to start. (The triple-alpha reaction, in which helium nuclei combine to form carbon, requires a temperature of about 100 million degrees.) As the core continues to shrink, the matter there takes on a very strange form. Electrons have a property that prevents them from being squeezed too close together, and this resistance creates a new kind of pressure, which becomes the principal force support-

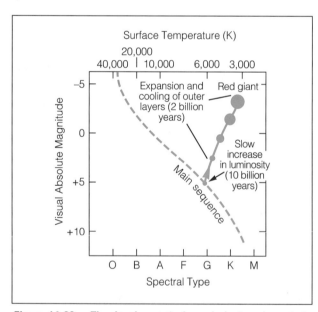

Figure 14.22   The development of a red giant.   As a star's outer layers expand because of the shell hydrogen source, they cool. At the same time the surface area increases, raising the star's luminosity. The star therefore moves up and to the right on the H-R diagram.

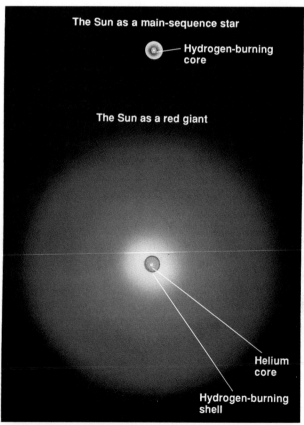

Figure 14.23   Structure of a red giant.   When a star like the Sun becomes a red giant, its diameter grows much larger, but most of its mass remains concentrated in a small inner region. Energy is supplied by nuclear reactions in the hydrogen shell source. This drawing is not to scale.

ing the star against the inward force of gravity. The gas in the stellar core contains many free electrons, and eventually their resistance to being compressed becomes the dominant pressure that supports the core against further collapse. When this happens, the gas is said to be **degenerate.**

A degenerate gas has many unusual properties. One of them is that the pressure no longer depends on the temperature, and vice versa. If the gas is heated further, it will not expand to compensate, as an ordinary gas would. As we will soon learn, there are important consequences if nuclear reactions start in the degenerate region of a star.

We now have a red giant star, with highly expanded outer layers. Near the center is a spherical shell in which hydrogen is burning in nuclear reactions, and inside this shell is a degenerate core containing helium nuclei and degenerate electrons. The core is small, but it may contain as much as a third of the star's total mass.

During the red giant phase, the helium core continues to be heated by the reactions going on around it. Eventually the temperature becomes sufficiently high for the triple-alpha reaction to begin. When it does, the consequences are spectacular.

Ordinarily, when a gas is heated, it expands, which limits how hot it can get, because an expanding gas tends to cool. In a degenerate gas, however, no such expansion occurs, because pressure does not depend on temperature in the usual way. Hence, when the reactions begin in the degenerate core of a red giant, the temperature rises quickly, but the core retains the same density and pressure. The increased temperature speeds up the reactions, producing more heat, which in turn accelerates the reactions even further. There is a rapid snowball effect, and in an instant (literally seconds), a large fraction of the core is involved in the reaction. This spontaneous runaway reaction is called the **helium flash.** Although it has dramatic consequences for the star's interior, calculations show that few if any effects would be immediately visible to an observer. The overall direction of the star's evolution is changed, however, as would be apparent after thousands of years.

The helium flash quickly disrupts the core, destroying its degeneracy (as temperature increases, more energy states become available to the electrons, and they can be squeezed closer together). The core then returns to a more normal state in which temperature and pressure are linked, and the triple-alpha reaction continues, now in a more stable, steady fashion. The outer layers of the star begin to retract as the star reverts to a more uniform internal structure, and as they do so, they become hotter. The star moves back to the left on the H-R diagram (see Fig.

14.22). Old clusters, particularly globular clusters, are found to have a number of stars in this stage of evolution, forming a sequence called the **horizontal branch** (Fig. 14.24). This is seen only in clusters old enough for stars as small as one solar mass to have evolved this far; as we learned earlier, this requires an age of 10 billion years or more.

In due course, the helium in the core becomes exhausted, leaving an inner core of carbon (Fig. 14.25). Helium still burns in a shell around the core; there could even be an active hydrogen-burning shell farther out in the star at the same time. The star expands and cools, becoming a red giant for the second time. This second red giant stage is short-lived, however, lasting perhaps 1 million years. A degenerate core again develops, but without another dramatic flare-up in its interior. In fact, no more nuclear reactions ever occur in the core of this low-mass star, because the temperature never reaches the extremely high levels needed to cause heavy elements like carbon to react.

The star in its second red giant stage may be viewed as two distinct zones: the relatively dense interior, consisting of a carbon core inside a helium-burning shell, which in turn is surrounded by a hydrogen-burning shell; and the very extended, diffuse outer layers. The inner portion may contain up

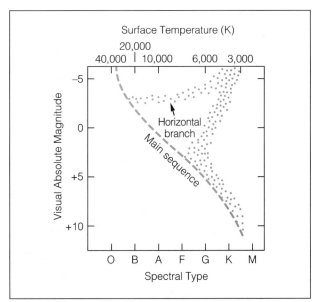

Figure 14.24   The H-R diagram for a globular cluster. These star clusters are very old, and all the upper main-sequence stars have evolved to later stages. Here we see a number of stars in or approaching the red giant stage, and a number on the horizontal branch, where they move following the helium flash. Only very old clusters have a horizontal branch.

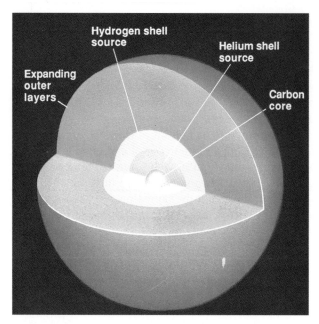

Figure 14.25   Internal structure late in a star's life.   The core of this star has undergone hydrogen burning and helium burning and is composed of carbon. Hydrogen and helium shell sources are still undergoing reactions and producing energy. This is not to scale; the stellar core is actually smaller than shown here.

to 70 percent of the mass, but occupies only a small fraction of the star's volume.

As the nuclear fuel in the interior runs out, the core shrinks. The details of what happens to the outer layers are not clear, but apparently the star ejects portions of this material in a series of minor outbursts that could be likened to blowing smoke rings. Thus the star gently sheds its outer layers.

As the hot core of the star is exposed, its high surface temperature and luminosity trigger a new high-velocity wind phase. The gas from this rapid wind sweeps up the slower-moving gas from the earlier ejection phase, creating a shell of hot gas surrounding the remnant of the star (Fig. 14.16). These shells often have a bluish green color because of the emission lines (primarily those of twice-ionized oxygen) by which they glow, and through a telescope or on a photograph they bear some resemblance to the planets Uranus and Neptune. Because of this resemblance, they are called **planetary nebulae.** Some, such as the Dumbbell nebula (see Fig. 14.26) or the Ring nebula (Fig. 14.27), are well-known objects visible with a small telescope.

The star that remains in the center of a planetary nebula lies far to the left in the H-R diagram (see Fig. 14.22); it has a surface temperature that may be as high as 100,000 K or more, far hotter than any main-sequence star. When the shell is being ejected, the star's surface temperature increases as hot inner gas becomes exposed on the surface. In essence, such a star is a naked core that is left behind when the outer layers are cast off.

Some nuclear reactions may still be going on in a shell inside the star, but they do not last much longer, since the fuel in the core is depleted. The density, already very high, increases as the stellar remnant condenses under the force of gravity. A larger and larger fraction of the star's interior becomes degenerate. As the star shrinks, it stays hot, but its luminosity decreases because of its diminishing surface

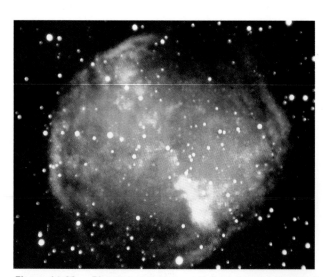
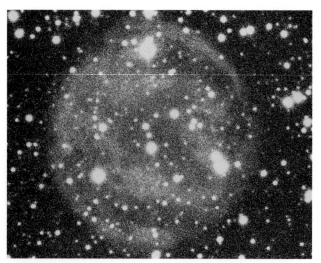

Figure 14.26   Planetary nebulae.   At the left is the Dumbbell nebula; at the right, the faint planetary nebula Ack 277-03. The blue-green color at left is produced by emission lines of twice-ionized oxygen. The central stars, which are responsible for ejecting the gas of the nebulae, are visible at the center of each. (Left: Lick Observatory photograph; right: © 1992 Anglo-Australian Telescope Board.)

area, and it moves downward on the H-R diagram (see Fig. 14.22). Eventually the pressure of the degenerate electron gas stops the collapse, and the star becomes stable again, but now it is a very small object, so dim that it lies well below the main sequence in the lower left-hand corner of the H-R diagram. It is called a **white dwarf.**

A white dwarf is a bizarre object in many ways. It is supported by degenerate electrons whose peculiar properties govern its internal structure. It has a mass as great as that of the Sun (some white dwarfs are even slightly more massive), yet it is approximately the size of the Earth. This means that its density is incredibly high, roughly a million times that of water. A cubic centimeter of white dwarf material would weigh a ton at the surface of the Earth!

A white dwarf does not do much. It undergoes no more nuclear reactions, so it does not evolve further. It just sits there, radiating away the heat energy still contained inside and slowly cooling off. A white dwarf can be stirred into action under certain circumstances (as described in the next chapter), but in most cases, this is the end of the line for a star like the Sun.

THE EVOLUTION OF MASSIVE STARS

Stars more massive than the Sun follow the same general course of events as the Sun, but with significant differences. The major similarity is that the majority of the star's lifetime is spent on the main sequence, while nuclear reactions in the core convert hydrogen

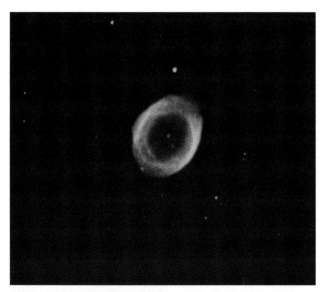

Figure 14.27   The Ring nebula. This is one of the best-known and most beautiful planetary nebulae, the Ring nebula in the constellation Lyra. (Lick Observatory photograph)

to helium, producing the energy necessary to balance gravity and supply the luminosity of the star. The principal contrasts with the evolution of a lower-mass star are that everything happens more rapidly in a massive star, and that many more stages of nuclear reactions can occur. The final result, or remnant, can be quite different as well.

Even though the first nuclear-burning stage is the same as in the Sun, the details are different. Whereas the low-mass stars convert hydrogen to helium by the proton-proton chain, in stars more than about twice as massive as the Sun, another reaction, the CNO cycle, is more efficient and is therefore dominant. (This reaction sequence is described in Chapter 13 and in Appendix 11.)

The luminosity and surface temperature of a star—and hence its location on the main sequence—depend on its initial mass. A star of 5 solar masses would have surface temperature roughly 12,000 K and spectral type B8, for example; a star of 10 solar masses would have surface temperature about 20,000 K and would be type B2; and a star of 20 solar masses would have surface temperature roughly 30,000 K and would be type O9.

The core temperature inside a star governs the rate of nuclear reactions, which is why massive stars evolve more rapidly than low-mass ones. The higher temperature in massive star cores (itself a direct result of the high mass; see the discussion in Chapter 13) results in rapid reaction rates and quick exhaustion of the nuclear fuel. A 5-solar-mass star will consume all the hydrogen in its core in only about 100 million years (about 1 percent of the main-sequence lifetime of the Sun), and a 20-solar-mass star will use up its hydrogen in even less time, perhaps 1 million years.

The high core temperatures of massive stars have another important impact on the reactions: they allow further reaction stages to occur. After each form of nuclear fuel is exhausted, the core contracts and the temperature increases, until they reach the point where a new reaction can begin. In the massive stars, helium will combine by the triple-alpha reaction to form carbon (while hydrogen continues to burn in a shell outside the core), and when a significant amount of carbon is present, it may react with helium to form oxygen through a process called an **alpha-capture reaction.** In alpha-capture reactions, helium nuclei (which are called alpha particles) are added to larger nuclei, increasing the mass of the nucleus by the addition of two neutrons and two protons and changing its identity. Carbon, with 6 neutrons and 6 protons, becomes oxygen, with 8 neutrons and 8 protons; oxygen can then add another alpha particle to become neon, with 10 neutrons and 10 protons, and so on (see Fig. 13.7). At sufficiently high temperatures, even more complex reactions can occur, in-

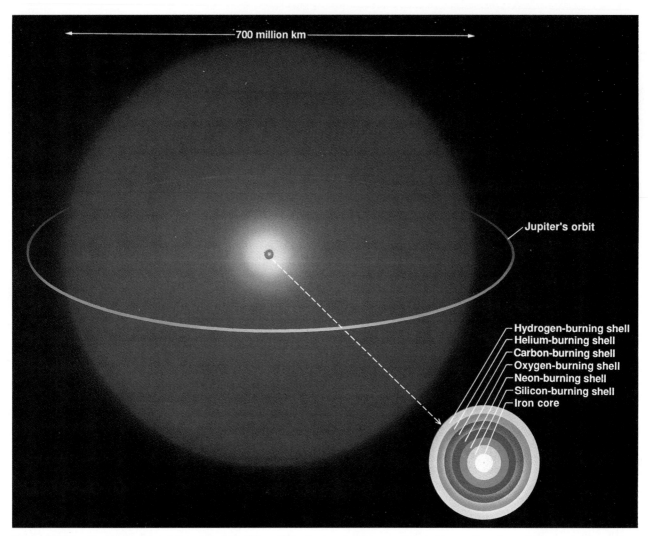

700 million km

Jupiter's orbit

Hydrogen-burning shell
Helium-burning shell
Carbon-burning shell
Oxygen-burning shell
Neon-burning shell
Silicon-burning shell
Iron core

Figure 14.28   A massive star at the end of its life.   This star has undergone all possible reaction stages in its core. The core consists of multiple layers of different composition, including six current or former shell sources and an inert iron core. The star is about to undergo collapse and supernova explosion.

volving the capture of free neutrons by large nuclei, which thus form ever-heavier nuclei. In this way, elements as heavy as iron can be produced in the core of a massive star.

After each nuclear reaction stage, the reaction continues in a shell around the core. Hence a star whose core has finished hydrogen burning will still have a shell outside the core where hydrogen continues to react. Later, when helium in the core is used up, there will be a helium-burning shell outside the core, and a hydrogen-burning shell will remain outside that. As successive reaction stages occur in the core, a series of concentric shells develops, each undergoing a different reaction (Fig. 14.28). The result is a star with an onionlike internal structure in which each layer has a different composition. Ultimately a core made of iron will form, after which the orderly progression

of reaction stages must end, with consequences as described in the next section.

As the successive nuclear reaction stages occur in the core of a star, the star's position in the H-R diagram changes in a fashion similar to that of a one-solar-mass star as it goes from main sequence to red giant. A massive star also makes this transition, although no helium flash occurs because the higher core temperature prevents the gas from becoming degenerate. Once the star becomes a red giant, it soon turns to the left and down in the H-R diagram as helium burning begins (Fig. 14.29). When the helium is used up, a second red giant phase follows, as the energy is again supplied primarily by a shell source and the outer layers expand and cool for a second time. Each time one form of fuel is exhausted, a new red giant phase begins, and each time a new fuel is ignited, the star will contract its outer layers and move to the left and down in the diagram. The number of these loops a star will undergo depends on how much mass it has, because this governs the internal temperature the star can ultimately reach, which this in turn determines how many reaction stages can occur.

Predicting exactly what happens to a star of a given mass is complicated by the fact that massive stars lose matter at several points in their lifetimes (see Chapter 13), and this can alter the internal conditions. Very massive stars eject matter in high-speed winds while on the main sequence, and stars of all masses lose matter through gradual winds when they are in the red giant stage. Through these processes, a significant fraction of the initial mass of a star can be lost during its lifetime.

We have seen that a star like the Sun will become a white dwarf when all possible nuclear reactions have run their course. The star becomes stable, with the pressure of degenerate electron gas supporting it against the force of gravity, and it remains a white dwarf perpetually (except under certain circumstances discussed in the next chapter). Some massive stars also become white dwarfs, although with different compositions from the white dwarfs left by low-mass stars (Fig. 14.30). If nuclear reactions proceed through helium burning, as is expected for the Sun, then the resulting white dwarf will be made of carbon; if the next stage occurs, in which the alpha-capture reaction of carbon with helium proceeds to completion, the resulting white dwarf will consist primarily of oxygen.

For stars above a certain mass, however, a white dwarf becomes an unattainable endpoint, because there is a firm limit on how massive a white dwarf can be. Above a certain mass, degenerate electron gas pressure is no longer sufficient to support the star

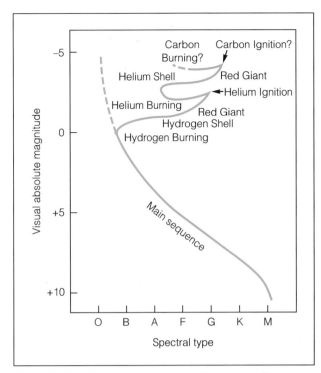

Figure 14.29   The evolution of a massive star.   This evolutionary track in the H-R diagram shows several steps followed by a star in the range of 5–10 solar masses. Each time the nuclear fuel in the core is exhausted, a shell burning source becomes dominant, causing the outer layers to expand and moving the star into the red giant region. When a new core source ignites, the star moves back to the left. The more massive the star, the greater the number of burning stages.

against gravity, and the star cannot exist as a stable white dwarf. The mass limit for white dwarfs, called the **Chandrasekhar limit,** is about 1.4 solar masses. Stars that start out with considerably more mass than this can end up as white dwarfs if they lose enough

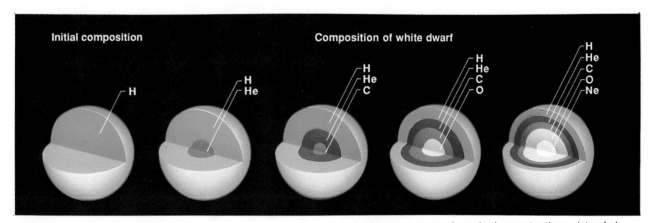

Figure 14.30   White dwarf composition.   The more nuclear burning stages a star goes through, the greater the variety of elements produced in its interior. The white dwarf that results, therefore, has a composition that depends on the number of reaction stages, and this, in turn, depends on the mass of the star.

mass during their lives; it is the *final* mass of a star that determines whether or not it can become a white dwarf. The calculations are somewhat uncertain (mostly due to poorly known rates of mass loss), but it is now thought that stars with initial masses up to 6 or 8 solar masses can end as white dwarfs. Stars that begin with more mass than this apparently do not lose enough to become white dwarfs, and something else must happen.

Stars that initially have between 8 and about 20 solar masses probably end up in another very compact state, even smaller and more dense than a white dwarf. Many of them reach this state by a very violent process called a **supernova** explosion (Fig. 14.31). Once iron has formed in the core of such a star, no further energy-producing reactions can take place as the core continues to contract and heat. Elements more massive than iron can be formed only by reactions that use up more energy than they create; these are called **endothermic** reactions. As reactions of this type begin under the incredibly high temperature (over 10^9K) and density (nearly 10^{10} grams/cm^3) conditions in the iron core, heat energy is removed from the surroundings by the reactions. This accelerates the contraction of the core, because the pressure is diminished, and the contraction becomes a free-fall collapse.

Very rapidly (within a few ten-thousandths of a second) the core reaches sufficient density that protons and electrons are forced together in a reaction called an **inverse beta decay,** forming a sea of neutrons (Fig. 14.32). A vast quantity of neutrinos, the tiny, massless particles discussed in Chapter 11, is released, carrying away enormous quantities of energy and further enhancing the collapse. The neutrons left behind, like electrons, cannot be packed closer together beyond a certain limiting density, and this creates a new kind of degenerate gas pressure. This **degenerate neutron gas pressure** is sufficient to halt the collapse of the core very abruptly. What was an effective vacuum into which the core was falling on itself suddenly becomes a very hard object, and material still falling in crashes onto its surface. A violent rebound, or shock wave, is formed and moves rapidly outward through the star. This shock may become stalled as it encounters the material of the outer layers, but the enormous burst of neutrinos from the core soon (within less than a second) pushes the

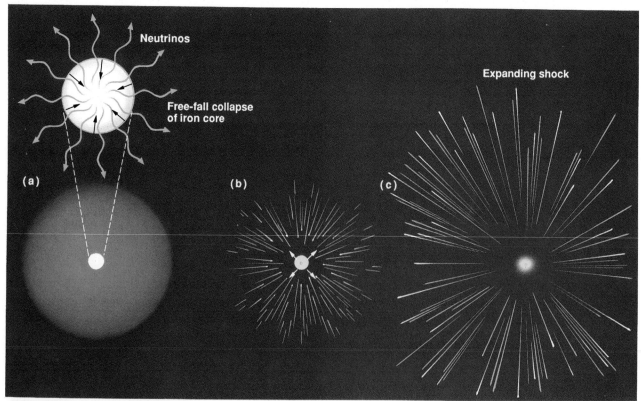

Figure 14.31   Type II supernova.   The tiny iron core at the star's center undergoes rapid collapse because the only nuclear reactions that can occur there remove heat energy (i.e., they are endothermic). The compression converts the core to pure neutrons, releasing copious quantities of neutrinos and becoming rigid in the process (a). The outer layer of the star falls in onto the neutron star core, and rebounds (b). This creates a shock wave which travels outward (c) and, with the help of the outflowing neutrinos, blows off the star's outer layers in a supernova explosion, leaving behind a neutron star remnant.

shock on its way again, and the entire outer portion of the star is blown away. This is the mechanism for a form of supernova called a **Type II supernova** and is basically what happened in the Large Magellanic Cloud in February 1987, when Supernova 1987A occurred (this is described in more detail in the next section). A Type II supernova is defined on the basis of the appearance of spectral emission lines due to hydrogen, which indicate that the exploded star still had an outer envelope of "normal" gas when its core collapsed. (In the next chapter we will discuss Type I supernovae, in which no hydrogen lines are present.)

The core that is left behind by the collapse and rebound of a star of 8 to 20 solar masses is made almost entirely of neutrons and is incredibly dense. A mass between 1 and 2 solar masses is compressed into a spherical volume whose diameter is only a few kilometers (typically 10 km, or about 6 miles), roughly the size of a modest city. This object, called a **neutron star,** is as dense as 10^{14} to 10^{15} grams/cm^3, comparable to the density of an atomic nucleus. Needless to say, matter in this form has bizarre properties that are not reproducible on the Earth. For years after the possibility of neutron stars was raised by theorists, no observational evidence for them was known, and they were seldom discussed. As we shall see in the next chapter, however, there is now compelling evidence that neutron stars are indeed very numerous in our galaxy.

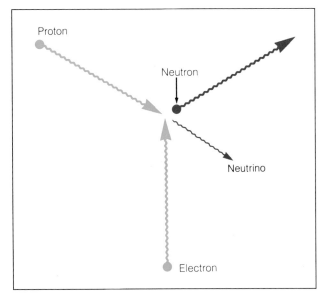

Figure 14.32   Inverse beta decay. In this process, which occurs at a significant rate only under conditions of very high pressure, a proton and an electron are forced together, forming a neutron and a neutrino. Energy is absorbed in the process, rather than being produced.

What of the stars that start life with more than 20 solar masses? These stars go through the same succession of nuclear reaction stages, although they do this so rapidly that their outer layers do not have sufficient time to respond to all the internal changes, and they may not actually go through the red giant part of the H-R diagram. In a matter of several hundred thousand to a million years, all the possible reactions may be completed and an iron core formed. In the meantime, these stars stay in the upper portion of the H-R diagram, remaining hot at their surfaces while losing their outer layers through dense, high-speed winds. A class of "peculiar" stars called **Wolf-Rayet stars,** characterized by high masses and unusual surface compositions along with rapid mass loss, are thought to represent an intermediate stage in the lives of very massive stars. In this stage products of the CNO cycle are exposed on the surface after the outermost layers have blown away.

Theorists believe that neutron stars form in the cores of the most massive stars just as they do in the intermediate-mass stars. The formation of a neutron star is a virtually automatic result of the rapid collapse of an iron core; as soon as the density is high enough, the core "hardens" and further infall is stopped by the outward shock that forms because of rebound. In the most massive stars, however, the shock becomes stalled and neutrino pressure from within is not sufficient to get it going again. Meanwhile, more material from the iron-producing layer in the star continues to fall slowly onto the core, adding to its mass. Soon the mass limit for a neutron star (not known exactly, but probably between 2 and 3 solar masses) is exceeded, and the stellar core collapses anew. This time nothing can stop it; there is no form of pressure that can counteract gravity, and the collapse continues without limit. Within less than one-thousandth of a second, the core becomes a **black hole.** The name derives from the fact that the force of gravity is so strong that even photons of light cannot escape; we have no direct means of communicating with or observing the inside of a black hole. Properties of these most bizarre objects are discussed in the next chapter.

Note that in the process of forming a black hole, a neutron star is created temporarily and a shock rebounds from its surface. The shock becomes stalled, however, by the large amount of matter above it, and the pressure of neutrinos streaming outward from the core is insufficient to push the shock outward. Therefore a supernova explosion probably does not take place in these circumstances. It appears that black hole formation by this mechanism is not accompanied by any spectacular, explosive event.

The mass limit dividing stars that become neutron stars from those that become black holes is not accurately known and may not even be a firm limit—it may depend instead on details such as the initial rotation rate of the star or minor aspects of its composition. Certainly the rate of mass loss during and after the star's main-sequence lifetime has an important effect on the end result, but as yet we cannot predict mass-loss rates with a high degree of accuracy. In this discussion, we have placed the cutoff between stars that form neutron stars and those that form black holes at about 20 solar masses; some think it may be as high as 30 or 40 solar masses.

We pointed out earlier in this discussion that the formation of iron is the natural endpoint of nuclear reactions in the core of even the most massive stars, because further reactions are endothermic and precipitate the collapse of the core. A variety of reactions can and do take place during the supernova outburst, however, and these are responsible for the formation of most of the elements heavier than iron. All of these reactions involve the capture of free neutrons by nuclei, thereby increasing the nuclear masses. There are two general classes of neutron-capture reactions: those that proceed slowly (these are called **s-process reactions**), and those that proceed rapidly (**r-process reactions**). The slow ones, s-process reactions, occur during the latter stages of evolution before the collapse of the iron core and produce some heavy elements then. The r-process reactions occur during the explosion and produce greater quantities of heavy elements. Whether formed before the supernova or during it, all of these heavy elements (as well as other products of stellar nuclear reactions, such as carbon, nitrogen, and oxygen) are dispersed into the interstellar medium as a result of the explosion and are thereby made available for subsequent generations of stars. Therefore we say that all the heavy elements crucial to life and technology are the result of supernova explosions.

Supernova 1987A

One of the most spectacular and scientifically beautiful events in the history of astronomy was observed on February 23, 1987 (Fig. 14.33). A supernova bright enough to be easily seen with the naked eye was discovered in the Large Magellanic Cloud, providing astronomers the first opportunity in nearly 400 years to study so bright a supernova. The new supernova, the first to be seen in 1987 and the first naked-eye one since the invention of the telescope, was named Supernova 1987A, following a tradition used by astronomers to designate supernovae in other galaxies.

Astronomers around the world were galvanized by the discovery as the news traveled through the International Astronomical Union's telegraph service and

Figure 14.33   The region of Supernova 1987A before and after the explosion.   Upper panel: At the left is a photograph made before the supernova occurred; the photograph at the right was made afterward. At this point, the supernova was a fourth-magnitude object, on its way to an eventual peak brightness of magnitude 2.9. Lower panel: This photograph shows both the supernova (lower right) and the Tarantula nebula. The large size of its image compared with normal stars is due to blurring on the photographic emulsion caused by the brilliance of the supernova; it was actually a point source, no bigger than the images of other stars. (© upper: 1987 REO/AAT Board; lower: David Dunlap Observatory)

by telephone. NASA's *International Ultraviolet Explorer* (IUE; see Chapter 4) was making ultraviolet spectroscopic observations within 14 hours, and ground-based observers everywhere in the Southern Hemisphere trained their telescopes on the supernova. Radio telescopes observed the supernova within hours of the discovery as well. Unfortunately the event occurred at a time when space astronomy was in a lull between major missions (the *Hubble Space Telescope,* for example, was still sitting on the ground because of delays in the shuttle program), so it was not possible to obtain all the data that would have been desirable. A Japanese X-ray telescope called

Ginga had just been launched, however, and was soon able to make X-ray observations. NASA moved quickly to mobilize a number of sounding-rocket and high-altitude aircraft and balloon experiments as well, so there was substantial coverage of the supernova at wavelengths that do not reach the ground. Perhaps the most spectacular observations of all did not involve any wavelength of electromagnetic radiation, however, but consisted instead of the detection of neutrinos from the explosion (this is discussed below).

Spectroscopic data quickly showed that the supernova was of Type II, because hydrogen lines were seen in its spectrum. The expansion velocity implied by the Doppler shifts of these lines was at least 20,000 km/sec. Recall that standard models assume that Type II supernovae result from the core collapse of massive stars that still have normal (that is, hydrogen-rich) gas in their outer layers.

Of the many unique opportunities the supernova has provided astronomers, one of the most important was the chance to analyze the star that blew up. Many photographic plates of the region exist in observatory files around the world, so it was possible to examine them to determine what kind of star the progenitor was. After some confusion caused by the crowded field of stars, the one that blew up was identified as a blue supergiant (type B3I) listed in catalogs as $Sk - 69°202$. This was a major surprise for many astronomers, because until then it had been thought that only red supergiants exploded in Type II supernovae (see the previous section).

The unusual nature of the progenitor, once it was established, helped explain some other peculiar features of the explosion. For example, the supernova never reached the luminosity normally expected, and it reached its peak brightness (just brighter than third magnitude) on May 20, some 85 days after its discovery, an unusually long time. Radio data, on the other hand, showed a rapid decline almost from the beginning, whereas other Type II supernovae typically do not reach their peak radio emission until hundreds of days after the outburst. The *Ginga* satellite showed no X rays coming from Supernova 1987A, although this may not be unusual. (There is very little information on X-ray emission in the early stages of other supernovae.)

All of these peculiarities fit well with supernova models involving core collapse in a massive star that is not as large as a red supergiant. The small size of the blue supergiant (compared with a red supergiant) accounts for the low luminosity, since relatively little surface area was available to radiate light. Another very important measure of the size of the star came from the detection of neutrinos, which escape directly from the core as the star collapses. The observed light is emitted later, when the shock front caused by the collapse reaches the star's surface. Thus the neutrinos reached the Earth before the supernova was visible, but the short delay (equivalent to the time it took for the shock to travel from the star's core to its surface) indicates that the star could not have been as large as a red supergiant. The best fit to all the observations suggests that $Sk - 69°202$ had a mass of about 20 solar masses and a diameter of roughly 43 times the diameter of the Sun.

As already mentioned, most astronomers were surprised that such a star would explode, although some model calculations had been made suggesting that it could happen. It is still not clear what stage of evolution the star was in, but most astronomers think it had been a red supergiant that then contracted and moved toward the left in the H-R diagram before it exploded. Whether this scenario is correct or not, it is generally believed that the explosion itself resulted from the collapse of an iron core through the mechanism described in the preceding sections. A complicating factor in determining the exact evolutionary stage the star was in is that the Large Magellanic Cloud has lower overall abundances of heavy elements than "normal" spiral galaxies like our own and most others where Type II supernovae are observed. Lower abundances of heavy elements can alter the evolution of a star.

Since Supernova 1987A failed (by about a factor of 5) to reach the magnitude expected for a Type II supernova, many at first assumed that this was a freak event and not any kind of common occurrence. Further analysis showed, however, that many supernovae of brightness comparable to this one could occur in other galaxies without being detected, because they would not stand out. Therefore, supernovae like 1987A might actually occur fairly frequently. There have been a small number of other "peculiar" supernovae with some similarities to 1987A.

The detection of neutrinos from Supernova 1987A was a spectacular event for astronomy and for physics. A burst of neutrinos was detected simultaneously by two separate experiments, one in Japan and one in Ohio. This marked the first time that extraterrestrial neutrinos had been detected except for those from the Sun (see Chapter 11), and some regard it as the beginning of a new branch of astronomy. As already explained, the timing of the arrival of the neutrinos provided valuable information on the size of the star. The neutrinos represent something far more important in terms of understanding supernovae, however; the energy the neutrinos carried away from the explosion was greater than that in all other forms by roughly a factor of 100. The visible light emitted over the first several months amounted to some 10^{42} joules of energy; the kinetic energy of the expanding gas cloud entailed perhaps a few times 10^{44} joules; and the energy released in neutrinos was about 3 or 4 times 10^{46} joules. These numbers are thought to be typical of Type II supernovae, according to model calculations, but Supernova 1987A provided the first confirmation of the important role of neutrinos. What we see of a supernova, as brilliant as it is, is only a small fraction of the energy produced. Furthermore, even though Supernova 1987A was dimmer than many other supernova in visible wavelengths of light, its total energy output was comparable to the brightest of supernovae.

Physicists who study elementary particles were as excited by the detection of neutrinos from Supernova 1987A as the astronomers were. Detailed analysis of the energy distribution of the neutrinos that were detected, along with the exact timing of their arrival, provided information about their basic properties. As mentioned elsewhere in this text, the question of neutrino mass has been the subject of some debate. The standard theories of neutrinos say that they are massless; yet controversial experimental data, along with a host of astronomical problems that would be solved, suggest that neutrinos might have mass. The data from the supernova neutrinos are not as complete as would have been desirable, but the information that was obtained still allows something to be learned about neutrino masses. If so, then they should travel through space with a range of speeds (unless they all have exactly the same energy), yet the neutrinos from Supernova 1987A all arrived very nearly at the same time, despite the large distance they had travelled, and despite the fact that they had a range of energies. The conclusion reached by physicists was that the

neutrino mass must be zero or very small—too small to solve the astronomical problems alluded to here (such as the total mass content of the universe and the low observed neutrino emission from the Sun).

After the initial outburst had diminished, during which the brightness of the supernova was caused by the high temperature of its expanding atmosphere, the main source of its luminosity was energy released by the radioactive decay of a form of nickel that was produced in large quantities (about 0.09 solar mass) during the explosion. This form of nickel decays to another radioactive substance, an isotope of cobalt, which then decays to iron. Thus the end product of the supernova is the production of great quantities of iron, but ironically this is not the iron from the core of the progenitor star, but rather iron produced indirectly by nuclear reactions resulting from the explosion.

One of the most important clues to the nature of the energy production in the supernova's expanding shell is the manner in which the brightness decreases with time. A graph showing magnitude versus time, called a *light curve*, is shown in Fig. 14.34. Radioactive decay causes the brightness to drop off in a specific fashion, and until recently, this is just what the observed light curve was doing. Then the brightness began to dip below the expected curve, probably because some of the energy from radioactive decay began to escape directly into space in the form of gamma rays rather than being absorbed and converted into visible light, an indication that the expanding cloud of debris is starting to thin.

It is considered a virtual certainty that a neutron star formed during the collapse and explosion of

Figure 14.34   The light curve of Supernova 1987A.   This graph shows the variation in magnitude of Supernova 1987A from its discovery until late fall 1987. The early part of the light curve is very unusual compared with other known supernovae, but once this one reached its peak, it began to behave very much like the light curves of other Type II supernovae. The reason for the difference in the early period is that the star that exploded was smaller than is thought typical, so it had less surface area to emit light. The later part of the light curve is consistent with energy production from the decay of radioactive nickel (^{56}Ni and then ^{57}Co). (J. Doggett and R. Fesen, University of Colorado)

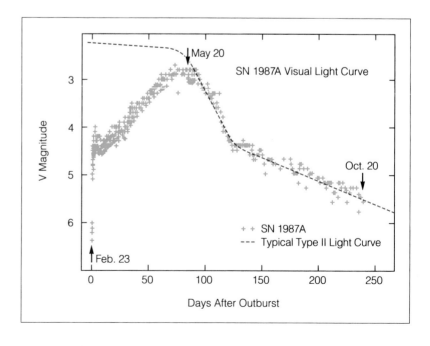

Sk − 69°202, because no other known mechanism could produce the huge quantity of neutrinos that was detected. If so, then eventually energy radiated by the neutron star should begin to dominate its luminosity, causing the light to stop diminishing so rapidly. Therefore astronomers are waiting for the first signs of energy from the neutron star, which could show up in X-ray wavelengths or in visible light, representing X-ray emission from the neutron star that has been absorbed and reemitted.

Some astronomers are surprised that the neutron star has not yet been detected, because it is thought that accretion of residual gas onto the neutron star should cause heating that would make it luminous enough to be seen. Evidently this accretion is not occurring, or perhaps the neutron star has acquired enough new mass to collapse, forming a black hole. Hence some astronomers believe a black hole now resides inside the cloud of debris from Supernova 1987A. We might not know whether this is true for some time, however; in about 300 years the debris should thin enough to reveal even a non-accreting neutron star, if it is there. If none is seen, then we may conclude that the remnant is a black hole instead.

The effects of the supernova on the interstellar gas and dust around it and along the line of sight between it and the Earth are also expected to be very interesting and informative. For example, astronomers suspect that much of the interstellar dust that pervades the space between the stars (see Chapter 16) forms through the condensation of gas in the material ejected from supernovae. Thus astronomers are monitoring the infrared emission from the vicinity of the supernova, looking for the glow that is expected from warm dust, if large quantities of dust are formed.

Previously existing interstellar dust has already played an important role in observations of the supernova. *Light echoes* from the supernova have been seen (Fig. 14.35). These are pale circles of light centered on the supernova, caused by scattering due to layers of dust in the foreground (Fig. 14.36). The dust particles reflect light most efficiently at a certain angle, so we see the reflected light only from the narrow ring of dust that lies at that angle relative to the direction of the supernova (a very similar effect due to ice particles in the Earth's upper atmosphere sometimes creates pale rings around the Moon). The light echoes from Supernova 1987A consist of light that was emitted during the initial outburst of the supernova, but was delayed in reaching the Earth, because it took an indirect route. The fact that two distinct echoes were seen indicates that there are two separate layers of interstellar dust lying on the near side of the supernova.

The detection of the light echoes from the supernova has fascinating implications. It is actually possible to "relive" the supernova—as though we were

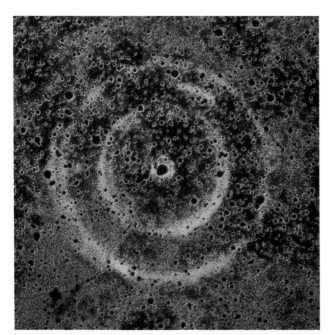

Figure 14.35   Light echoes from Supernova 1987A.   These rings of light from Supernova 1987A are reflected from sheets of gas and dust in the foreground. (© 1989 Anglo-Australian Telescope Board. Photo by David Malin)

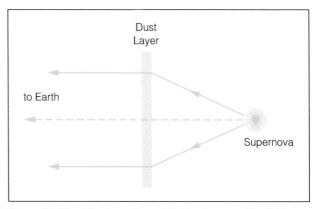

Figure 14.36   The formation of light echoes.   The light emitted straight toward the Earth got here first; its arrival first signaled the occurrence of the supernova. Light that struck the dust layer at just the right angle to be reflected (scattered) toward Earth arrived here several months after the initial outburst was observed, due to the longer path it traveled.

seeing a replay—by observing the light reaching us now from the dust layers along the way. The light from the outburst that is reflected from the dust particles is essentially unaltered, so in principle we can reobserve the supernova event in detail, perhaps picking up some information that we missed the first time. Unfortunately only a tiny fraction of the light is scattered in the direction of the Earth, however, so our "instant replay" is very, very dim, and making very precise measurements is difficult or impossible. It is interesting to note, though, that we can keep watching reruns of the supernova for a long time to come, as light being scattered farther and farther away from the direct path reaches us. In effect, the light echo rings are growing in diameter as light is scattered toward us from larger and larger angles (this growth of the rings has already been detected). The larger the angle of deflection, the longer the path the light has traveled to reach us, and the greater the delay until it arrives (refer to Fig. 14.36 to see this).

A ring of another sort was found when the *Hubble Space Telescope (HST)* obtained an image of Supernova 1987A (Fig. 14.37). Where ground-based photographs had barely suggested that the image of the supernova was not quite round, the superior image quality of the *HST* showed the ring easily. The ring is thought to be a remnant of the stellar wind stage that the star went through before it exploded. As the star blew off its outer layers in a steady wind, the ejected material was slowed down when it rammed into the surrounding interstellar gas and accumulated in a ring. When the intense flash of radiation from the supernova outburst reached this ring of gas, it was excited and began to glow. The size and shape of the ring have told astronomers a great deal about the stellar wind stage of the star's evolution (for example, that the outflow formed a ring instead of a spherical shell shows that the wind was concentrated in the equatorial plane of the star, an important bit of information for theories of stellar winds). Thus continued observations of the supernova are revealing new information about the nature of the star before it blew up.

The occurrence of Supernova 1987A has already had a major impact on nearly every field of astronomy. The technical journals are dominated by articles on theory or observations of the supernova, and we may expect this to continue for years. Eventually, as the expanding shell evolves into a normal supernova remnant, it is expected to become one of the brightest radio and X-ray sources in the sky, as the ionized gas disperses and emits energy by synchrotron and other nonthermal processes. In due course, perhaps in months or years, the gaseous remnant may become thin enough to allow astronomers to detect the neu-

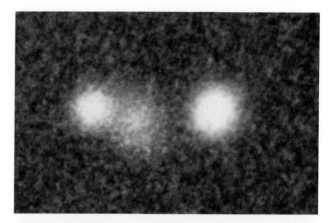

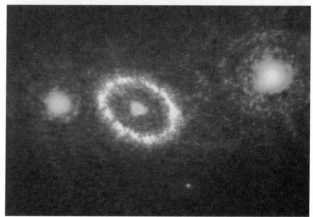

Figure 14.37   The ring around Supernova 1987A. The upper photo shows a recent image of the supernova, photographed from the ground, at visible wavelengths. Note the distinctly orange color, and note that the image appears slightly elongated. The lower photo, on approximately the same scale and with the same orientation, was obtained by the *Hubble Space Telescope.* This image reveals a ring around the supernova, consisting of gas lost before the explosion by the progenitor star. Radiation from the explosion is now causing this ring of gas to glow. (upper: © 1992 AAT Board; photo by D. Malin; lower: NASA/STSCI)

tron star, particularly if it is a pulsar (discussed in the next chapter). For decades or centuries, the gaseous remnant will be a strong source of visible and ultraviolet emission lines; for centuries it will remain an intense X-ray source; and for millennia it will be a strong radio emitter. We may expect many generations of astronomers to take part in the study of this supernova, and many future astronomy classes to learn about it.

EVOLUTION IN BINARY SYSTEMS

So far our discussion of stellar evolution has been concerned with single stars. About half the stars in

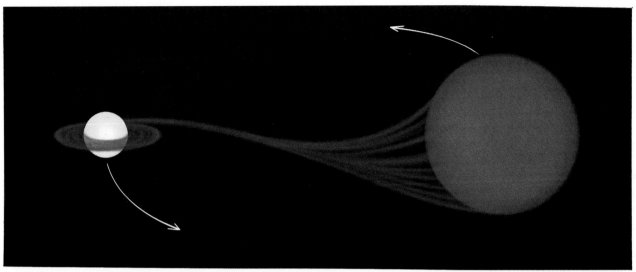

Figure 14.38   Mass transfer in a binary system.   When two stars in a binary system are close together, material from one may be transferred to the other. There is a point in space between the two stars at which material would feel equal gravitational forces from each star; if one star has a wind or expands as a red giant so that its outer layers reach this point, then matter can flow from one star to the other. This can alter the evolution of both.

the sky are double, however, and in some cases this can affect the lives of the stars. If the two stars are sufficiently far apart—perhaps several astronomical units or more—then most likely each evolves on its own, as though the companion were not there. If, however, they are very close together, significant interactions can drastically alter the normal course of events.

If either star is massive enough to have a strong wind, matter can be transferred to the other star (Fig. 14.38). Even if no strong wind is present initially, as soon as either star reaches the red giant stage, it can swell up sufficiently to dump material onto the other through the point between them where their gravitational forces are equal. If one of the two is a compact stellar remnant, such as a white dwarf, neutron star, or black hole, the result can be spectacular, as discussed in the next chapter.

Regardless of how it happens, once mass begins to transfer from one star to the other, the evolution of one or both is modified. The star that gains mass speeds up its evolution, while the star that loses mass may slow down. This process was discovered because certain binary systems were found in which the *less* massive star was the one that had evolved the farthest, quite contrary to what was expected. The most famous such system is the eclipsing binary known as Algol, or β Persei. This binary system consists of a red giant and a B main-sequence star. The B star is more massive, but is still on the main sequence, whereas the red giant has already left the main se-

quence. The "Algol paradox" was solved only when it was realized that the red giant must have been the more massive star initially but later lost mass to its companion (see the Astronomical Insight on the following page).

The most spectacular observational effects in mass-transfer binary systems occur when one star has gone all the way through its evolution and become a white dwarf, neutron star, or black hole and then gains new material from the companion star. Special effects, often including X-ray emission, are observed, and it is under these circumstances that stellar remnants are most easily detected. This will be discussed at some length in the next chapter.

PERSPECTIVE

We have now seen how stars are born, live their lives, and die, and we know how the sequence of events depends on mass. The least massive stars, those with much less than a solar mass, have been little discussed, simply because they evolve so slowly that none has yet had time, since the universe began, to complete its main-sequence lifetime. The evolution of these stars has had little effect on the rest of the galaxy, whereas the much more rare heavyweight stars are profoundly important because their rapid evolution has enriched the galaxy with heavy elements.

ASTRONOMICAL INSIGHT
The Mysteries of Algol

The bright double star known as Algol (β Persei) has played a leading role in the discovery of two important astronomical phenomena: the existence of binary stars and the process of mass transfer in close binaries. Both of these phenomena went unrecognized for some time because they seemed too radical for most astronomers to accept.

The first report that Algol's brightness varies came as long ago as 1667, when the Italian astronomer Geminiano Montanari noted that the star occasionally appeared fainter than normal. At that time a few variable stars were known, but no explanation for them had been found. It was not until 1782 that anyone rediscovered the variability of Algol; in that year the English astronomer John Goodricke found that the light variations were periodic, with a brief decrease in brightness occurring every 69 hours. Goodricke made the then-novel suggestion that the dimmings could be due to the presence of a companion star that orbited Algol, eclipsing it each time it passed in front. At that time there was controversy over earlier suggestions that some stars appeared to be double, and the prevailing view (including that of the eminent English astronomer William Herschel) was one of skepticism.

In independent studies of Goodricke's idea, the Swiss astronomer Daniel Huber and an obscure English clergyman named William Sewall soon analyzed the light variations of Algol to produce crude estimates of the properties of the suspected companion. The writings of both men, published originally in 1787 and 1791, were ignored by their contemporaries and then lost; they were not rediscovered until well into the twentieth century. Meanwhile, the notion that stars could be double gradually became accepted, as additional Algol-type stars were found, and observations of visual binaries became common. Interestingly, it was William Herschel who, in 1803, conclusively proved the existence of double stars when he was actually able to measure orbital motion in a visual binary system. Finally, in 1889, spectroscopic observations of Algol itself revealed the orbital motion of the bright star about its companion, establishing that stars can be double even though they present a single-point image when viewed in a telescope or photographed.

Much more recently, Algol played a similar leading role in the discovery of mass transfer in binary systems. During the twentieth century, Algol has been widely observed, using photometric and spectroscopic techniques, and a paradox was revealed as astronomers' understanding of stellar evolution developed. Analysis of the data on Algol using Kepler's third law and the shape of the light curve (a graph showing the brightness of the star versus time) showed that the visible star is a B main-sequence star having a mass 3.7 times that of the Sun, whereas the fainter star is a G giant star with a mass only 0.81 times the Sun's mass. From observations of cluster H-R diagrams and stellar evolution theory, it was known that more massive stars evolve faster than less massive ones. Thus the fact that the *less* massive member of the Algol system was already a red giant, while the *more* massive star was still on the main sequence presented a major mystery, which became known as the Algol paradox.

This puzzle has been solved only in the past 20 years or so, as ultraviolet observations have revealed massive stellar winds from some kinds of stars, and other evidence (mostly from telescopes in space) for mass transfer between stars in close binary systems has been found. Evidently the G star in the Algol system originally contained more mass and was once an upper main-sequence star, whereas its companion started out on the lower main sequence. As the first star, being more massive, became a red giant, it swelled up to such a large size that matter began to flow across to the companion. The first star became less massive and fainter than the companion, which in turn became hotter and brighter as its mass grew. Today the original situation is reversed, with the star that started as the less massive one now being the more massive. At some future time, there may be another reversal, as the B star becomes a red giant and swells up, spilling mass back onto its companion.

The existence of binary stars and the process of mass transfer in close binaries were both profound discoveries in astronomy. Ironically Algol provided hints of both, even though the evidence was found long before astronomers were ready to understand and accept the implications.

Before we finish discussing stars and turn our attention to larger-scale objects in the universe, we must consider the properties of the stellar remnants left behind as stars die.

SUMMARY

1. Steps in stellar evolution are deduced from observations of stars in various stages, particularly by comparing clusters of different ages, and from theoretical calculations.

2. The distances to clusters can be determined by plotting a color-magnitude diagram and comparing the location of the main sequence with that of a standard H-R diagram.

3. The fact that all the stars in a cluster have the same age and initial composition means that the observed differences from star to star within a cluster are due entirely to differences in stellar mass. This allows us to determine how stars of different mass evolve.

4. The age of a cluster can be inferred from the location of its main-sequence turnoff point.

5. Stars in the process of forming are often embedded within dense clouds and are best observed at infrared wavelengths.

6. A star forms from the gravitational collapse of an interstellar cloud, which condenses most rapidly at the center. The new star begins to heat up only after the gas in the core of the cloud becomes opaque, and the resulting protostar continues to shrink slowly until it becomes hot enough inside for nuclear reactions to begin.

7. A star with the mass of the Sun spends about 10^{10} years on the main sequence, producing its energy by nuclear reactions that convert hydrogen to helium in its core. Stars of higher mass have shorter lifetimes; those with less mass have longer lifetimes.

8. When the core hydrogen is gone, the reactions stop there but continue in a spherical shell outside the core. The outer layers expand and cool, and the star becomes a red giant.

9. The core continues to shrink and heat until it becomes degenerate. When it eventually gets hot enough a helium flash occurs, after which the star is powered by the triple-alpha reaction, converting helium into carbon.

10. Following the helium-burning phase, the star may undergo instabilities that cause the ejection of one or more planetary nebulae.

11. The star eventually becomes degenerate throughout and ends its evolution as a white dwarf.

12. A more massive star, in the range of 5 to 10 solar masses, evolves more quickly than the lower-mass stars, goes through multiple red giant stages, and ends up either as a white dwarf or a neutron star, depending on how much mass it loses during its evolution.

13. A very massive star evolves very quickly, going through many nuclear reaction stages before forming an iron core, which then collapses, leading to formation of a neutron star (accompanied by a Type II supernova explosion) or a black hole.

14. In close binary systems, mass can be transferred from one star to the other, speeding the evolution of the star that gains mass.

REVIEW QUESTIONS

1. If a cluster is observed to contain O stars, how do we know that it must be a young cluster? Is it possible for the same cluster also to contain M stars?

2. Suppose two stars in a cluster have identical $B - V$ color indices and apparent magnitudes. How do their masses and other properties compare? Could you make the same assumption for a pair of stars not in the same cluster, but which also have identical color indices and apparent magnitudes?

3. Comparison of the main sequence of a cluster color-magnitude diagram with a standard H-R diagram shows that the apparent magnitudes of the stars on the cluster diagram main sequence are 5 magnitudes larger than the absolute magnitudes of the corresponding stars on the standard H-R diagram. How far away is the cluster?

4. Is a young star in hydrostatic equilibrium when it is in the slow gravitational contraction stage? Explain.

5. Explain why the main sequence is somewhat broadened rather than being a thin strip in the H-R diagram. Where do you think the Sun lies within the broad strip of the main sequence?

6. We have learned that when helium-burning reactions start in a star like the Sun, they begin in an instantaneous "helium flash." Why is there not a "hydrogen flash" earlier in a star's lifetime, when it forms and nuclear reactions begin for the first time?

7. What is the wavelength of maximum emission for the central star of a planetary nebula whose surface temperature is 100,000 K? What kind of telescope would be best for observing this object?

8. If a neutron star has a radius that is 1/500 times that of a white dwarf of half the mass of the neutron star, compare the average densities of the two.

9. A supernova may have an absolute magnitude as small (bright) as -19. How much more luminous is such a supernova than the Sun, whose absolute magnitude is $+5$?

10. The relative abundances of the elements in the galaxy (Appendix 3) show that the abundances generally decline with increasing atomic weight, with hydrogen the most abundant, helium next, and so on. Contrary to this trend, iron is relatively abundant, compared with other elements near it in atomic weight. Based on what you have learned about stellar evolution, explain why this is so.

ADDITIONAL READINGS

Allen, D. A. 1987. Star formation and the IRAS galaxies. *Sky and Telescope* 73(4):372.

Bethe, H. A. 1985. How a supernova explodes. *Scientific American* 252(5):60.

Boss, A. P. 1991. The genesis of binary stars. *Astronomy* 19(6):34.

————. 1985. Collapse and formation of stars. *Scientific American* 252(1):40.

Fienberg, R. T. 1989. Brown dwarfs coming and going. *Sky and Telescope* 78:482.

Flannery, B. P. 1977. Stellar evolution in double stars. *American Scientist* 65:737.

Gehrz, R. D., D. C. Black, and P. M. Solomon. 1984. The formation of stellar systems from interstellar molecular clouds. *Science* 224(4651):823.

Graham, J. 1984. A new star becomes visible in Chile. *Mercury* 13(2):60.

Greenstein, G. 1985. Neutron stars and the discovery of pulsars. *Mercury* 14(2):34 (Part 1); 14(3):66 (Part 2).

Herbst, W., and G. E. Assousa. 1979. Supernovas and star formation. *Scientific American* 241(2):138.

Jones, B., D. Lin, and R. Hanson. 1984. T Tauri and TIRC: A star system in formation. *Mercury* 13(3):71.

Kaler, J. 1991. The brightest stars in the galaxy. *Astronomy* 19(5):30.

————. 1991. The faintest stars. *Astronomy* 19(8):26.

————. 1981. Planetary nebulae and stellar evolution. *Mercury* 10(4):114.

Kawaler, S. D., and D. E. Winget. 1987. White dwarfs: Fossil stars. *Sky and Telescope* 74(2):132.

Lada, C. J. 1986. A star in the making. *Sky and Telescope* 72(4):334.

————. 1982. Energetic outflows from young stars. *Scientific American* 247(1):82.

Loren, R. B., and F. J. Vrba. 1979. Starmaking with colliding molecular clouds. *Sky and Telescope* 57(6):521.

Malin, D., and D. Allen. 1990. Echoes of the Supernova. *Sky and Telescope* 79:22.

Rosa, M., and S. Harrington. 1988. Light echoes from Supernova 1987A. *Mercury* 17:63.

Schaefer, B. E. 1985. Gamma-ray bursters. *Scientific American* 252(2):52.

Schild, R. 1990. A star is born. *Sky and Telescope* 80(6):592.

Scoville, N., and J. S. Young. 1984. Molecular clouds, star formation, and galactic structure. *Scientific American* 250(4):42.

Seward, F. 1986. Neutron stars in supernova remnants. *Sky and Telescope* 71(1):6.

Soker, N. 1992. Planetary nebulae. *Scientific American* 266(5):78.

Spitzer, L. 1983. Interstellar matter and the birth and death of stars. *Mercury* 12(5):142.

Stahler, S. W. 1991. The early life of stars. *Scientific American* 265(1):48.

Sweigart, A. V. 1976. The evolution of red giant stars. *Physics Today* 29(1):25.

Tomkin, J., and D. L. Lambert. 1987. The strange case of Beta Lyrae. *Sky and Telescope* 74:354.

Waldrop, M. M. 1983. Stellar nurseries. *Science 83* 4(4):40.

Werner, M. W., E. E. Becklin, and G. Neugebauer. 1977. Infrared studies of star formation. *Science* 197:723.

Wheeler, J. C., and R. P. Harkness. 1987. Helium-rich supernovas. *Scientific American* 257(5):50.

White, R. E. 1991. Globular dusters: fads and fallacies. *Sky and Telescope* 81(1):24.

The following articles are all related to Supernova 1987A:

Helfand, D. 1987. Bang: The supernova of 1987. *Physics Today* 40(8):24.

Lattimer, J. M., and A. S. Burns. 1988. Neutrinos from Supernova 1987A. *Sky and Telescope* 76:348.

————. 1987. SN1987A: Watching and waiting. *Sky and Telescope* 74(1):14.

————. 1987. A supernova in our backyard. *Sky and Telescope* 73(4):382.

————. 1987. Supernova 1987A's fading glory. *Sky and Telescope* 74(3):258.

————. 1987. Supernova shines on. *Sky and Telescope* 73(5):470.

————. 1987. A surprising supernova. *Sky and Telescope* 73(6):582.

Marshall, L. A., and K. Brecher. 1992. Will Supernova 1987A shine again? *Astronomy* 20(2):30.

Time. March 23, 1987. 129(12):60.

Wheeler, J. C., and R. P. Harkness. 1987. Helium-rich supernovas. *Scientific American* 257(5):50.

ASTRONOMICAL ACTIVITY
The Pleiades

The galactic cluster known as the Pleiades provides a fine laboratory for the study of stellar evolution. It is also one of the most beautiful sights in the sky and easy to find. As part of this exercise, you should go out and have a look. The Pleiades cluster is up during the evening during late autumn and early winter. Find Orion, then follow the imaginary line through the three belt stars to the northwest. You will come to the bright star Aldebaran and then, a similar distance beyond Aldebaran in the same direction, you will see the Pleiades (see the accompanying sketch). The cluster is quite striking as seen by the naked eye and truly spectacular when viewed through binoculars or a small telescope (under very good conditions, you might even be able to see the wispy nebulosity that surrounds the stars).

Now you can use the Pleiades to learn about cluster H-R diagrams and stellar evolution. The table lists the visual magnitudes (V) and color indices ($B - V$) for stars in the Pleiades cluster. Make a color-magnitude diagram for the cluster, preferably on semitransparent graph paper, using the same scale as in Figure 12.24 (i.e., on the horizontal axis, plot $B - V$ with the same scale as in Figure 12.24; on the vertical axis, plot V with the same spacing per magnitude as in the figure). Now you can overlay your plot on Figure 12.24, and use the main-sequence fitting method to

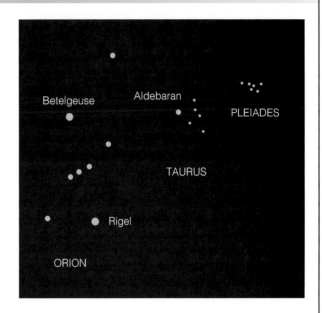

find the distance to the Pleiades by noting the difference, on the vertical axis, between the absolute and apparent magnitudes for the stars in the cluster. You can also see how far up the main sequence the Pleiades has stars, and from this you can comment on how old the cluster is.

Stars in the Pleiades*

Star Number	V	$B - V$	Star Number	V	$B - V$
3	8.2	.09	26	7.9	.04
5	8.1	.04	27	5.7	−.22
7	9.6	.33	28	6.4	−.20
8	8.1	.12	31	4.2	−.24
9	9.8	.41	32	10.4	.53
10	5.4	−.22	33	7.4	−.07
11	3.7	−.30	34	8.1	.21
13	10.4	.51	35	10.2	.62
15	8.6	.20	36	9.3	.34
16	5.6	−.26	39	10.5	.53
17	4.3	−.30	40	6.8	−.15
18	8.9	.31	41	8.4	.15
19	8.0	.07	42	9.4	.34
20	8.6	.21	43	6.9	−.14
21	10.1	.49	44	7.6	.05
22	7.2	.01	45	7.3	−.13
23	9.7	.43	46	7.7	.00
24	9.4	.37	47	6.8	−.10
25	3.9	−.25	56	6.9	−.03

*Data from H. L. Johnson and W. W. Morgan, 1951, *Astrophysical Journal*, 114:522.

Stellar Remnants

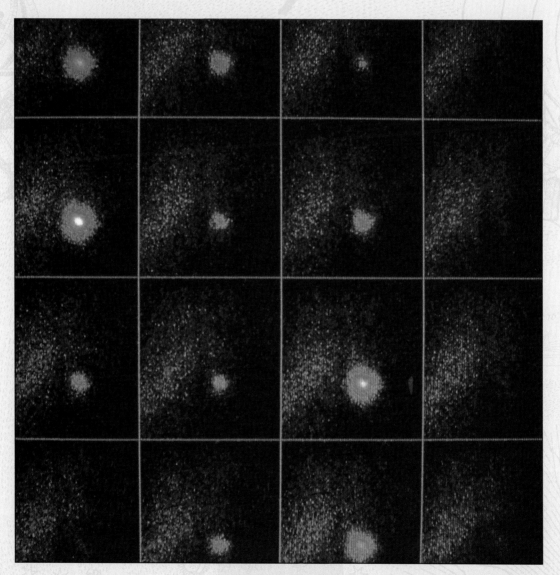

X rays from the Crab pulsar, turning on and off. (Harvard-Smithsonian Center for Astrophysics)

PREVIEW

The allure of astronomy for many people arises from the bizarre and exotic celestial objects that apparently are the source of some observed phenomena. Many of these objects, including black holes, neutron stars, pulsars, supernova remnants, white dwarfs, and X-ray binaries, are the result of stellar evolution and involve the remnants of once-ordinary stars. When a star completes its nuclear-burning lifetime, it must make its final accommodation to the relentless inward force of gravity, and stars of different masses find different final solutions. In this chapter you will learn how astronomers deduce the presence of the various types of stellar remnants, and you will find out what is known about their properties and their behavior. With this information in hand, we will be ready to move on to a discussion of the great stellar system to which our Sun belongs, the Milky Way Galaxy.

Our story of stellar evolution is almost complete. We have seen how the properties of stars are measured, and we have learned how they form, live, and die. The only missing link is what is left of a star after its life cycle is completed.

As we saw in the previous chapter, the form of remnant that remains at the end of a star's life depends on the mass the star had when it finally ran out of nuclear fuel. Three possibilities were mentioned: white dwarf, neutron star, and black hole. In this chapter, we discuss each of these bizarre objects, with emphasis on their observational properties.

WHITE DWARFS, BLACK DWARFS

Table 15.1 summarizes the properties of a typical white dwarf. An isolated white dwarf is not a very exciting object from an observational point of view. It is rather dim (Fig. 15.1), and its spectrum is nearly featureless, except for a few very broad absorption lines created by the limited range of chemical ele-

TABLE 15.1   Properties of a Typical White Dwarf

Mass: 1.0 $M_\odot$
Surface temperature: 10,000 K
Diameter: 0.008 $D_\odot$ (1.0 $D_\oplus$)
Density: 5×10^5 g/cm^3
Surface gravity: 1.3×10^5 g
Luminosity: 2×10^{31} erg/sec (0.005 $L_\odot$)
Visual absolute magnitude: 11

Figure 15.1   Sirius B.   The dim companion to Sirius (to the left of the large image of Sirius A) was the first white dwarf to be discovered; analysis of its mass, temperature, and luminosity led to the realization that it is very small and dense. (Lick Observatory)

ments it contains (Fig. 15.2). The great width of the lines is caused by the immense pressure in the star's outer atmosphere; pressure tends to smear out the atomic energy levels, because of collisions between atoms. In effect, the energy levels become broadened, and this results in broadened spectral lines when transitions of electrons occur between the levels.

Additional broadening or shifting of spectral lines may occur because of magnetic effects. If the star had a magnetic field before its contraction to the white dwarf state, that field is not only preserved but is intensified in the process of contraction. Thus a white dwarf may have a very strong magnetic field, and as we have seen previously (in Chapters 11 and 12), a magnetic field causes certain spectral lines to be split into two or more parts. In a white dwarf, where the lines are already very broad, this splitting (called the Zeeman effect) usually just makes the lines appear even broader.

The lines in the spectrum of a white dwarf are also shifted, always toward longer wavelengths. This shifting is not caused by motion of the star; if it were, it would be telling us that somehow all the white dwarfs in the sky are receding from us. Instead this type of redshift is caused by gravitational forces. One prediction of Einstein's theory of general relativity is that photons of light should be affected by gravitational fields. White dwarfs, with their immensely strong surface gravities, provide a confirmation of this predic-

ASTRONOMICAL INSIGHT
The Story of Sirius B

Sirius, also known as α Canis Majoris, is the brightest star visible in the heavens. In 1844 it was discovered to be a binary, because of its wobbling motion as it moves through space (recall the discussion of astrometric binaries in Chapter 12). Its period is about 50 years, and it has an orbit of about 20 astronomical units; analysis of these data using Kepler's third law leads to the conclusion that the companion must have a mass near that of the Sun. At the distance of Sirius (only 2.7 parsecs), a star like the Sun should be easily visible, yet the companion to Sirius defied detection for some time.

Part of the difficulty lay in the overwhelming brightness of Sirius itself—its light tended to drown out that of a lesser, very close star. Astronomers were so intrigued by the mysterious companion, however, that they made intensive efforts to find it, and finally, in 1863, it was seen with an exceptionally fine, new telescope.

Observations immediately confirmed that this star was unusually dim for its mass, having an apparent magnitude of 8.7 (the apparent magnitude of the primary star is −1.4) and an absolute magnitude of 11.5, nearly 7 magnitudes fainter than the Sun. Spectra obtained with great difficulty (again because of the brightness of the companion) showed that this object, called Sirius B, has a high temperature, similar to that of an O star.

When Henry Norris Russell constructed his first H-R diagram in 1913, he included Sirius B, which lay all by itself in the lower left-hand corner. The only way the star could have such a low luminosity with such a high temperature was for it to have a very small radius, and astronomers deduced that Sirius B was about the size of the Earth. The incredible density was estimated immediately, and astronomers realized that a totally new form of matter was involved. The term *white dwarf* was coined to describe Sirius B and other stars of this type, which were soon discovered in other binary systems.

The physics of this new kind of matter took some time to be understood, partly because the theory of

relativity had to be applied (the electrons in a degenerate gas move at speeds near that of light). The Indian astrophysicist S. Chandrasekhar was the first to solve the problem of the behavior of this kind of gas; by the 1930s he was able to construct realistic models of the structure of a white dwarf. From this work came the 1.4 solar-mass limit we have referred to in the text.

Modern observational techniques offer the possibility of detecting white dwarfs more easily. For example, even though Sirius B is about 10 magnitudes (or a factor of 10,000) fainter than Sirius A at visible wavelengths, its higher temperature makes it relatively brighter in the ultraviolet. There is even a wavelength, about 1100 Å, below which the white dwarf is actually brighter than its companion. Hence Sirius B and other white dwarfs have been detected with ultraviolet telescopes, and as more sophisticated space instruments are developed, it will become possible to study the properties of many white dwarfs in this way.

---

then the white dwarf can gradually gain mass without the infalling material causing any explosions. In this way, a white dwarf can approach the 1.4-solar-mass limit beyond which degenerate electron gas pressure can no longer support it. As a white dwarf gains mass, it becomes *smaller;* this is another peculiar property of degenerate matter. Thus, the white dwarf increases in density. What happens as the white dwarf approaches the mass limit depends on its composition; it may, more or less quietly, suddenly contract, becoming either a neutron star or black hole. If its composition is carbon, however, theory shows that nuclear reactions will begin instead, and the degeneracy will cause these reactions to very rapidly consume much of the white dwarf's mass, creating an immense explosion. This is another mechanism for producing a supernova explosion.

In the previous chapter, we discussed supernovae caused by the collapse and rebound of a massive star at the end of its nuclear lifetime; now we see that supernovae can also be caused by the addition of mass to a white dwarf. Supernovae created in both ways are generally similar in their maximum brightness and in the length of time they require to become dim again. But they differ in many details, such as the types of spectral lines observed and the shape of the **light curve,** a graph showing the brightness variations with time (Fig. 15.6). Supernovae caused by the explosion of a massive star are usually Type II supernovae, because hydrogen lines appear in the spectrum, while those produced in the explosion of a white dwarf are Type I supernovae with no hydrogen features.

White dwarfs, then, are most prominent in binary

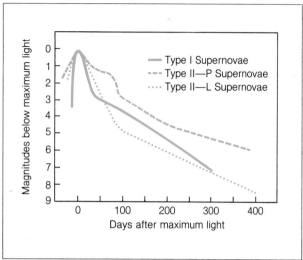

Figure 15.6 Supernova light curves.   The two types of supernovae are readily distinguished by their spectra and their light curves, which show marked differences. Type I supernovae usually have no hydrogen lines in their spectra and are thought to be explosions of carbon white dwarfs that have accreted matter. Type II supernovae have hydrogen lines in their spectra and are the explosions that occur in massive stars when all nuclear fuel is gone. Peak brightnesses for the two types are slightly different and, as seen here, the rates of decline are also quite different. Type II supernovae have been subclassified into types P and L. (J. Doggett and D. R. Branch)

TABLE 15.2   Prominent Supernova Remnants

Remnant	Age (yr)	Distance* (l.y.)
Vela X	10,000	1,600
Cygnus Loop	20,000	2,300
Lupus	975	4,000
IC 443	60,000	4,900
Crab nebula	936	6,520
Puppis A	4,000	7,500
Cassiopeia A	200	9,130
Tycho's supernova	400	7,800
Kepler's supernova	370	14,340

*Distance data are from a modern re-analysis by J. Saken, R. A. Fesen, and M. J. Shull, 1991. *Astrophys. J., Suppl.*

systems in which mass transfer from a normal companion takes place. Later in this chapter we will see that other forms of stellar remnants are also best observed in these circumstances.

SUPERNOVA REMNANTS

We have learned that stars can explode as supernovae in two different ways, in some cases leaving a stellar remnant (a neutron star or black hole). In all cases another form of matter will be left over as well—an expanding gaseous cloud called a **supernova remnant.**

Several supernova remnants are known to astronomers (Table 15.2). A few, like the prominent Crab nebula (Fig. 15.7), are detectable in visible light, but many are most easily observed at radio wavelengths. This is partly because visible light is affected by the interstellar dust that can hide our view of a remnant, whereas radio waves are largely unaffected. But it is primarily because, as a supernova remnant ages, its emission in visible light fades out long before it stops emitting radio waves. Thus there is an extended time

period (10,000 to 20,000 years) when the remnant emits radio radiation and little else. The radio emission is not a result of the temperature of the gas, as in thermal emission, but is produced instead by a different mechanism, called **synchrotron emission.**

Synchrotron emission, which derives its name from the particle accelerators in which it is produced on Earth, occurs when electrons move rapidly through a magnetic field. The electrons must have speeds near that of light; as they travel through the magnetic field, they are forced to move in a spiraling path, and they emit photons as they do so (Fig. 15.8). The emission occurs over a very broad range of wavelengths, including some visible, ultraviolet, and even X-ray radiation (Fig. 15.9), but these other wavelengths are often not as easily detected as radio, because the radio emission is the strongest and most persistent. Synchrotron radiation can be distinguished from thermal radiation (i.e., radiation due to the temperature of the object, as discussed in Chapter 4) in two ways: (1) the intensity of synchrotron radiation varies smoothly over a wide range of wavelengths (always increasing in brightness toward shorter wavelengths), rather than peaking sharply at a particular wavelength; and (2) synchrotron radiation is usually polarized (see Chapter 4).

Supernova remnants that are detected at visible wavelengths usually glow in the light of several strong emission lines, most notably the bright line of atomic hydrogen at a wavelength of 6563 Å, in the red portion of the spectrum. These lines often show large Doppler shifts, indicating that the gas is still moving rapidly as the entire remnant expands outward from the site of the explosion that gave it birth. Often a filamentary structure is seen, suggestive of turbulence, but probably modified in shape by magnetic fields.

Figure 15.7   Supernova remnants.   At the upper left is the Crab nebula, the remnant of a star that was seen to explode in A.D. 1054. At the upper right is a radio image of the remnant of Tycho's supernova. Below is a photograph of part of the Vela supernova remnant, showing its wispy, filamentary structure. (Top left: Lick Observatory photograph; top right: National Radio Astronomy Observatory; bottom: © 1992 AAT Board; photo by D. Malin)

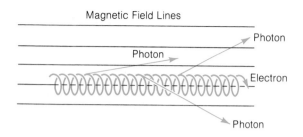

Magnetic Field Lines

Photon

Photon

Electron

Photon

Figure 15.8   The synchrotron process.   Rapidly moving electrons emit photons as they spiral along magnetic field lines. The radiation that results is polarized, and its spectrum is continuous but lacks the peaked shape of a thermal spectrum. The electrons must be moving very rapidly (at speeds near that of light), so whenever synchrotron radiation is detected, a source of large quantities of energy must be present.

Remnants are visible today at the locations of several famous historical supernova explosions. The Crab nebula is the most prominent, having been created in the supernova observed by Chinese astronomers in A.D. 1054. Other supernovae seen by Tycho Brahe (in 1572) and by Kepler and Galileo (1604) also left detectable remnants, although neither is as

bright in visible wavelengths as the Crab. Apparently a supernova remnant can persist for 10,000 years or more before becoming too dissipated to be recognizable any longer. One of the best-studied remnants, Cassiopeia A (see Fig. 15.9), is apparently only about 300 years old, but the supernova that created it was quite dim and not widely observed. This supernova may have been obscured by interstellar dust, or it may have been an unusually dim one intrinsically.

The energy of a supernova explosion is immense—comparable to the total amount of radiant energy the Sun will emit over its entire lifetime. Some of this energy takes the form of mass motions as the remnant expands. This kinetic energy heats the expanding gas to very high temperatures. The entire process has a profound effect on the interstellar gas and dust that permeates the galaxy, as we shall see in Chapter 16.

NEUTRON STARS

We have referred to a neutron star as a stellar remnant composed entirely of neutrons that are in a degenerate state, similar to that of the electrons in a white dwarf. The pressure created by the degenerate

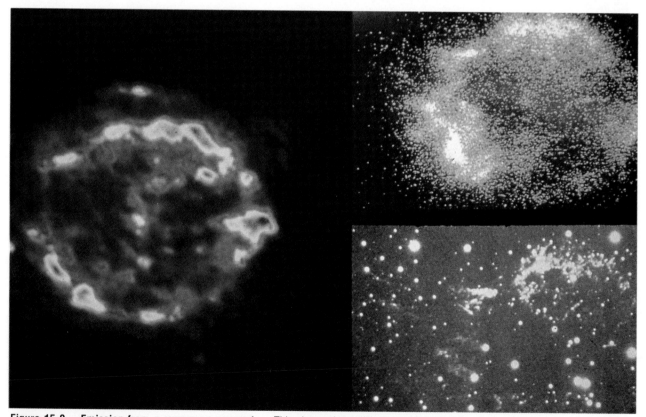

Figure 15.9   Emission from a supernova remnant.   This shows three different images of the same supernova remnant, Cassiopeia A. At the left is a radio map, at the upper right an X-ray image, and at the lower right a visible-light image. The remnant is less obvious in visible light because there are many other, brighter objects in the field of view. (Left: National Radio Astronomy Observatory; top right: Harvard-Smithsonian Center for Astrophysics; bottom right: K. Kamper and S. van den Bergh)

TABLE 15.3   Properties of a Typical Neutron Star

Mass: 1.5 $M_\odot$
Radius: 10 km
Density: 10^{14} g/cm^3
Surface gravity: 7×10^9 g
Magnetic field: 10^{12} Earth's field
Rotation period: 0.001–4 sec

---

neutron gas is greater than that of a degenerate electron gas, so a star slightly too massive to become a white dwarf can be supported by the neutrons. Again, however, there is a limit on the mass one of these objects can have. This limit, which depends on other factors such as the rate of rotation, is between 2 and 3 solar masses. Thus, there is a class of stars, whose masses at the end of their nuclear lifetimes are between 1.4 and about 2 or 3 solar masses, that become neutron stars.

The structure of a neutron star (Table 15.3 and Fig. 15.10) is even more extreme than that of a white dwarf. All of the mass is compressed into an even smaller volume (now the radius is about 10 km), and the gravitational field at the surface is immensely strong. The layer of normal gas that constitutes the atmosphere is only centimeters thick, and beneath it may be zones of different chemical composition resulting from previous shell-burning episodes. Each zone is only a few meters thick at the most. Inside these surface layers is the incredibly dense neutron gas core, which takes the form of a crystalline lattice. The temperature throughout is very high, but because the surface area is so small, a neutron star is very dim indeed. The compression of the star's original magnetic field may result in its having a magnetic field that is much more intense than even that of a white dwarf.

Pulsars: Cosmic Clocks

The properties of neutron stars were predicted theoretically several decades ago, but until 1967 it was not expected that they could be observed, because of their low luminosities. In that year, however, an accidental discovery by a radio astronomer established a new class of objects called **pulsars,** which are now believed to be neutron stars. As it turned out, pulsars are only one of two distinct ways in which neutron stars can be detected; the other is described in the next section.

Pulsars are radio sources that flash on and off very regularly and with a high frequency (Fig. 15.11). One of the most rapid—the one located in the Crab nebula—flashes 30 times a second, and several others are now known that repeat even more rapidly than the Crab pulsar; one of these pulses at the incredible frequency of 885 times per second! In contrast, the slowest known pulsars have cycle times of more than a second. In every case, the pulsar is only "on" for a small fraction of each cycle.

The discovery of pulsars generated a great deal of excitement and speculation, including the suggestion that they were beacons operated by an alien civilization. Once the initial shock of discovery wore off, however, a number of more natural explanations were offered by astronomers who sought to establish the identity of the pulsars. It was well known that a variety of stars pulsate regularly, alternately expanding and contracting, but none were known to do it so rapidly. Theoretical studies showed that such quick variations as those exhibited by the pulsars should occur only in very dense objects, denser even than white dwarfs. This led astronomers to think of neutron stars.

Enough was known from theoretical calculations, however, to rule out rapid expansion and contraction

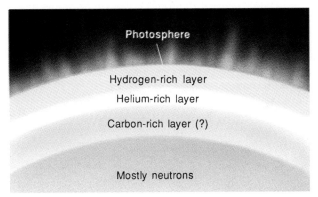

Figure 15.10   The structure of a neutron star.   As this diagram shows, the outer regions of a neutron star may consist of thin layers of various elements that were produced by nuclear reactions during the star's lifetime. These outer layers are thought to have a rigid crystalline structure because of the intense gravitational field of the neutron star.

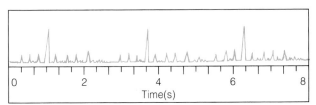

Figure 15.11   The radio emission from a pulsar.   Here, plotted on a time scale, is the radio intensity from a pulsar. The radio emission is weak or nonexistent except for a very brief flash once each cycle (in some cases, there is a weaker flash between each adjacent pair of strong flashes).

ASTRONOMICAL INSIGHT
Very Rapid Pulsars

Since the discovery of the first pulsars in 1967, many more have been found. The typical period of pulsation, thought to represent the rotation period of the neutron star responsible for the pulses, is 0.03 seconds or longer (the pulsar in the Crab supernova remnant, spinning 30 times per second, was for many years the most rapid known pulsar). Recently, however, some pulsars have been found that have periods far shorter than 0.03 seconds. These pulsars, dubbed "millisecond pulsars" because their periods are measured in units of thousandths of a second, probably have different origins from the majority of pulsars.

The frequency of pulses from the very rapid pulsars was at first difficult to understand. As a pulsar ages, its rotation rate should either stay constant or slow (slowing is caused by the drag between the magnetic field lines of the neutron star and the ionized gas that may exist around the pulsar). The pulsation rate of the Crab pulsar has been observed to be slowing, and similar slowing is expected for any neutron star formed in an environment where magnetic forces can act on surrounding gas. Yet the millisecond pulsars are apparently rotating at rates much faster than any reasonably old pulsar should be capable of. The contradiction between expectation and observation became far more severe recently, when a millisecond pulsar was discovered in a globular cluster. Recall (from Chapter 14) that globular clusters are very

old conglomerations of stars, not places where young, massive stars are expected to form and then die in supernova explosions creating neutron stars in the process. Therefore the existence of very young pulsars in globular clusters was difficult to understand.

Today the leading explanation of very rapid pulsars is that they are neutron stars whose spin rate has actually *increased* since they formed, quite in contradiction to the normal expectation. The increased spin rate is presumed to have come about because new matter has been added to these neutron stars, increasing their angular momentum and hence their spin rate. For example, in a binary system where one star is a neutron star and the other a normal star undergoing expansion to a red giant phase, it is likely that matter would transfer from the normal star to the neutron star (see the discussion of evolution in binary systems, Chapter 13). As matter is transferred to the neutron star, it follows a spiral path because of the relative orbital motion of the two stars, and this spiral path causes the new material to fall onto the neutron star with a large amount of angular momentum. This in turn causes the neutron star to spin more rapidly when it gains matter from its companion, turning the normal neutron star into a very rapidly spinning one. Thus, a once-slow pulsar can become a very rapid, millisecond pulsar.

Very recently a new millisecond pulsar was discovered in a binary system. Analysis of the orbital properties of the pulsar revealed that its companion star has a mass of only 2

percent of the mass of the Sun. This is far too little mass for a normal star, and astronomers have suggested that this star originally had much more mass, but has somehow lost most of its material. The leading hypothesis is that the neutron star has caused its companion to evaporate by heating the companion's outer layers. The neutron star in this system has been dubbed the "black widow" star, because it is "eating" its mate.

Since the discovery of the first millisecond pulsar, several more have been found, including a few in globular clusters. Too many have been discovered to allow an explanation in terms of residual rotation from the time of formation; it is now widely accepted that these rapid pulsars must have been spun up to high rotation rates by the addition of material from binary companions. It is quite possible that many of the rapid pulsars have in turn vaporized their companion stars, in a bizarre form of stellar cannibalism. The black widow star may not be the only one of its kind.

of neutron stars as the cause of the observed pulsations. The calculations showed that the vibration period of such an object would be even shorter than the observed periods of the pulsars.

A second possibility—that the rapid periods were produced by rotation of the objects—was considered. Like the physical contraction and expansion hypothesis, the rotation hypothesis ruled out ordinary stars and white dwarfs. To rotate several times per second, a normal star or white dwarf would have to have a surface speed in excess of the speed of light, a physical impossibility. The object would be torn apart by rotational forces before it approached such a rotational speed. A neutron star, on the other hand, could rotate several times per second and remain intact.

When the Crab pulsar was discovered, a great deal of additional information became available. Astronomers found that the pulse rate of this pulsar was gradually slowing, something that could best be explained if the pulses were linked to the rotation of the object. The rotation could slow as the pulsar gave up some of its energy to its surroundings.

This discovery cleared up another mystery. The source of energy that powers the synchrotron emission from the Crab nebula had been unknown. Now it was suggested that the slowdown of the rotating neutron star could provide the necessary energy, either by magnetic forces exerted on the surrounding ionized gas or by transfer of energy from the pulsar to the surrounding material through the emission of radio waves.

The remaining question was how the rotation created the pulses. Evidently a pulsar acts like a lighthouse, with a beam of radiation sweeping through space as it spins. But why should a rotating neutron star emit beams from just one or two points on its surface?

The most probable explanation has to do with the strong magnetic fields that neutron stars are likely to have. If a neutron star has a strong field, then electrons from the surrounding gas are forced to follow the lines of the field, hitting the surface only at the magnetic poles. The result is an intense beam of electrons traveling along field lines, especially concentrated near the magnetic poles of the neutron star, where the field lines are crowded together. The rapidly moving electrons emit synchrotron radiation as they travel along the field lines, creating narrow beams of radiation from both magnetic poles of the star. If the magnetic axis of the star is not aligned with the rotation axis (Fig. 15.12), these beams will sweep across the sky in a conical pattern as the star

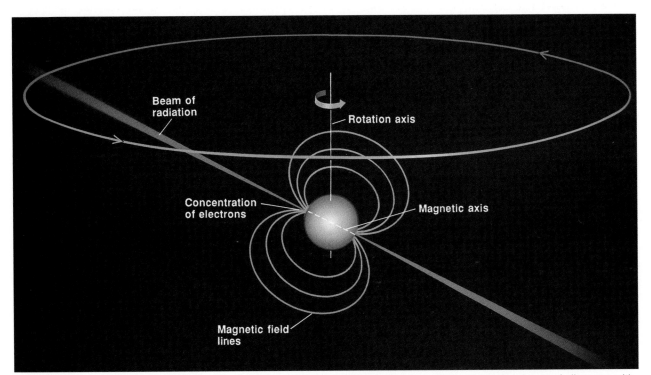

Figure 15.12   The pulsar mechanism.   Here we see a rapidly rotating neutron star, with its magnetic axis out of alignment with its rotation axis. Synchrotron radiation is emitted in narrow beams from above the magnetic poles, where charged particles, constrained to move along the magnetic field lines, are concentrated. These beams sweep the sky as the star rotates, and if the Earth happens to lie in a direction covered by one of the beams, we observe the star as a pulsar.

ASTRONOMICAL INSIGHT
Who Discovered the Pulsars?

Although the discovery of pulsars may be regarded as serendipitous, in fact it took a great deal of skilled detective work. It required the initial recognition that a series of squiggles on a lengthy, routine chart produced during a radio survey was significant, and not some random glitch in the machinery. It required substantial labor after that to go through the endless scrolls from previous observations to see if the squiggles were there at other times when the telescope scanned the same position on the sky (they were, but not every time). It required further searches to see whether other similar squiggles could be found (they were). And it required substantial analysis to rule out earthly sources of the mysterious, pulsing signals and to determine the possible properties of the objects that emitted them.

These steps were completed by two people working together in 1967. The initial discovery of a rapidly pulsating signal was made by Jocelyn Bell, a graduate student at Cambridge University. She had been assigned the task of studying the 400 or so feet of chart recorder paper that was being produced each week by an array of radio receivers designed to search for rapid variations (caused by interplanetary gas) in the intensity of distant, point radio sources such as quasars (see Chapter 20). When she discovered the signal, Bell showed it to her adviser, Antony Hewish, who was at first skeptical that it represented anything unusual. But Bell persisted, and she soon found that the signal reappeared from time to time, always at

the same position on the sky (which tended to rule out an earthly origin). By examining some three miles of charts from previous observations, she found three additional, similar signals. Hewish then realized that Bell had uncovered a significant new phenomenon, and he and Bell carefully analyzed the signals, while telling few others what they had found. In early 1968 they published a paper on the pulsating sources, arguing that they were truly celestial (as opposed to some kind of human-produced interference), and suggesting that pulsating white dwarfs or the then-hypothetical neutron stars could be responsible.

Once Hewish and Bell published their results, other radio astronomers immediately focused attention (and telescopes) on the mysterious pulsars, as they were being called. Soon it was realized that only spinning neutron stars could emit pulses so rapidly and so regularly (this conclusion is generally credited to astrophysicists Franco Pacini and Tommy Gold, who reached the idea independently of each other).

The discovery of pulsars is regarded as one of the most fundamental in recent astronomical history. In 1974 Hewish was awarded the Nobel Prize in Physics for his role in the discovery. Jocelyn Bell was not included in the award. Most members of the astronomical community regarded this as unfair, and Bell has generally been given at least equal credit for the discovery. It is not clear why the Nobel committee overlooked Bell. A likely possibility is that they assumed her role was minor because graduate students usually lack the maturity and authority to be leaders in original research. It was probably not a case of

prejudice against women (after all, women had been awarded Nobel prizes as long ago as 1930, when Marie Curie was honored). Possibly, the Nobel committee simply did not investigate the circumstances of the pulsar discovery and were not fully aware of Bell's role. For his part, Hewish gave Bell credit for bringing the strange signals to his attention, which allowed him to make the fateful discovery. But he has been quiet on the question of whether it was unfair that Bell did not receive equal recognition from the Nobel committee.

It often happens in science that one person makes a serendipitous discovery, and others work out the explanation. In the case of the pulsars, the person who made the initial discovery was ignored by the Nobel Prize committee, while the person who worked out the first suggested explanation was honored. Ironically, quite the opposite occurred when the cosmic background radiation (Chapter 19) was discovered. In that case, the co-discoverers, Arno Penzias and Robert Wilson, were awarded the Nobel Prize for having found an unexplained source of cosmic radio noise, while Robert Dicke, the person who explained it (and had in fact predicted it and was in the process of building a radio receiver to look for it), did not share the prize, as many thought he should.

rotates. If the Earth happens to lie in the direction intersected by one of these beams, then we see a flash of radiation every time the beam sweeps by us.

Synchrotron radiation is normally emitted over a very broad range of wavelengths, suggesting that pulsars may "pulse" in parts of the spectrum other than the radio. Indeed this is the case: the Crab pulsar has been identified as a visible-light and X-ray pulsar (Fig. 15.13). Although the details are still somewhat vague, the rotating neutron star model seems to be the best explanation of the pulsars. Since special conditions (i.e., nonalignment of the magnetic and rotation axes, and the need for the beam to cross the direction toward the Earth) are required for a neutron

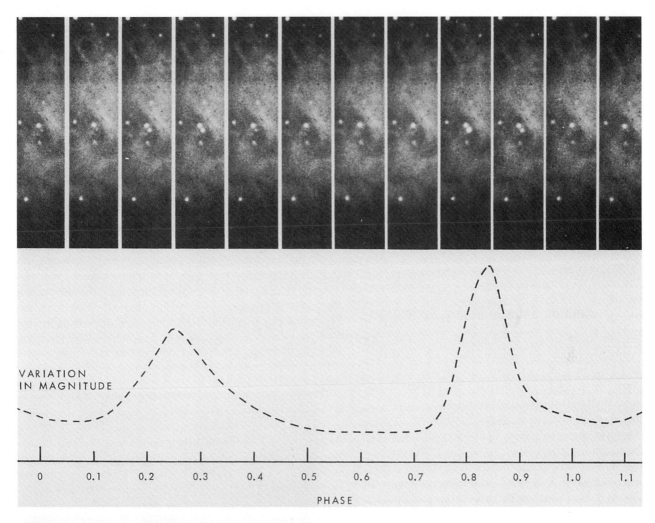

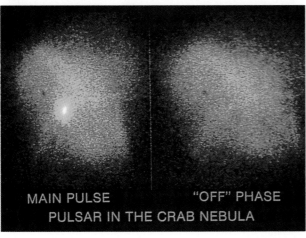

MAIN PULSE   "OFF" PHASE
PULSAR IN THE CRAB NEBULA

Figure 15.13   Visible and X-ray flashes from the Crab pulsar.   At the top is a sequence of visible-light photographs accompanied by a light curve, showing how the pulsar in the Crab nebula flashes on and off during its 0.03-second cycle. Below is a pair of X-ray images, showing the pulsar "on" and "off" in different parts of its cycle. (Top: National Optical Astronomy Observatories; bottom: Harvard-Smithsonian Center for Astrophysics)

Figure 15.14   An accretion disk.   Here a giant star, losing mass through a stellar wind, has a neutron star companion. Material that is trapped by the neutron star's gravitational field swirls around it in a disk, which is so hot from compression that it glows at X-ray wavelengths. A nearly identical situation can arise when the compact companion is a black hole.

star to be seen from Earth as a pulsar, it follows that there should be many neutron stars that do not manifest themselves as radio pulsars. In the next section, we will see how some of these neutron stars are detected.

Neutron Stars in Binary Systems

Earlier we made a general statement that stellar remnants are often most easily observed when they are in binary systems. We have already seen that a white dwarf in a binary can flare up violently if it receives new matter from its companion star. A neutron star reacts similarly in the same circumstances.

If a neutron star is in a binary system in which the companion object is either a hot star with a rapid wind or a cool giant losing matter because of its active chromosphere and low surface gravity, some of the ejected mass will reach the surface of the neutron star. As in the case of mass transfer involving white dwarfs, very little of it falls directly down onto the surface; instead, much of it swirls around the neutron star, forming an accretion disk (Fig. 15.14). The individual gas particles orbit the neutron star like microscopic planets and fall inward only as they lose energy in collisions with other particles. The accretion disk acts as a reservoir of material, slowly feeding it inward toward the neutron star.

The disk is very hot because of the immense gravitational field close to the neutron star. As gas in the accretion disk slowly falls in closer to the neutron star, gravitational potential energy is converted into heat. The gas in the disk reaches temperatures of several million degrees, hot enough to emit X rays. Hence a neutron star in a mass-exchange binary sys-

tem is likely to be an X-ray source, and a number of such systems have been found in the past 15 years (Fig. 15.15), since the advent of X-ray telescopes launched on rockets or satellites. Often we know that an X-ray object is part of a binary system, because the X rays are periodically eclipsed by the companion star. Eclipses are likely, because the two stars are close together (mass exchange would not occur unless it were a close binary) and because the mass-losing companion is likely to be a large star, either

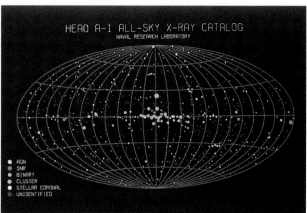

Figure 15.15   An X-ray map.   This shows the locations in our galaxy of X-ray sources cataloged by the *High Energy Astronomical Observatory 1,* an orbiting X-ray satellite that operated in the late 1970s. Most of the sources here are binary systems in which one member is a compact stellar remnant such as a neutron star or a black hole. The concentration of sources along the galactic equator indicates that the massive stars that can end up as neutron stars or black holes tend to form, evolve, and die close to the plane of our disk-like galaxy (see the discussion in Chapter 17). (Smithsonian Astrophysical Observatory)

TABLE 15.4   Selected Binary X-Ray Sources

Source	Period (days)	Mass of System ($M_\odot$)	Nature of Stars
Cygnus X-1	5.6	40	O9Ib supergiant; probable black hole
Centaurus X-3	2.09	20–25	B0Ib-III giant; pulsar (neutron star)
Small Magellanic Cloud X-1	3.89	15–25	B0Ib supergiant; pulsar (neutron star)
Hercules X-1	8.95	20–30	B0.5Ib supergiant; probable neutron star
Vela X-1	1.70	2–5	HZ Hercules (A star); pulsar (neutron star)
A 0620-00	0.32	11	K dwarf; probable black hole (companion mass $\simeq$ 8 $M_\odot$)
Large Magellanic Cloud X-3	1.70	15	B3 main-sequence star; probable black hole (mass $\simeq$ 9 $M_\odot$)

an upper main-sequence star or a supergiant.

The so-called **binary X-ray sources** (a few are listed in Table 15.4) are among the strongest X-ray–emitting objects known. Many of them are, most likely, neutron stars, although the possibility exists that some are black holes, which would similarly produce X-ray emission as material fell inward.

Neutron stars can also emit X rays in a slightly different way, which also occurs in mass-exchange binaries. If the infalling material trickles down onto the neutron star in a steady fashion, continuous radiation of X rays occurs, as we have just seen. If, however, the material falls in sporadically and arrives in substantial quantities every now and then, a major nuclear outburst occurs each time (Fig. 15.16), in close analogy to the nova process involving a white dwarf. This apparently happens in some cases, producing random but frequent X-ray outbursts. The intensity of the outburst, as in the case of a nova, depends on how much matter has fallen in and is consumed in the reactions. The neutron star binary systems where this occurs are called **bursters.** One important contrast with novae is that the flare-up of a burster occurs much more rapidly, lasting only a

few seconds. Another is that most of the emission occurs only in very energetic X rays, so these outbursts do not show up in visible light.

BLACK HOLES: GRAVITY'S FINAL VICTORY

In the last chapter we saw what happens to a massive star at the end of its life: It falls in on itself, and there is no barrier—neither electron degeneracy nor neutron degeneracy—that can stop it.

The gravitational field near the surface of a collapsing star grows in strength as the mass of the star becomes concentrated in an ever-smaller volume. This gravitational field has important effects in the near vicinity of the star, although at a distance it remains unchanged from what it was before the collapse. Close to the star, though, the structure of space itself is strongly distorted, according to Einstein's theory of general relativity. Einstein discovered that accelerations caused by changing motion and those caused by gravitational fields are equivalent. From this it follows that space must be curved in the presence of a gravitational field, so that moving particles follow the same path that they would follow if they were being accelerated. This applies to photons of light, as well as to other kinds of particles.

Normally the effects of the enhanced curvature of space are not noticeable except under very careful observation, or when considering very large distances. Later in this text, when we discuss the universe as a whole and its overall structure, we will consider the latter situation. For now, we confine ourselves to local regions where space can be distorted by very strong gravitational fields (Fig. 15.17).

Let us consider a photon emitted from the surface of a collapsing star (Fig. 15.18). If the photon is emitted at any angle away from the vertical, its path will be bent over farther. If the gravitational field is strong enough, the photon's path may be bent over so far

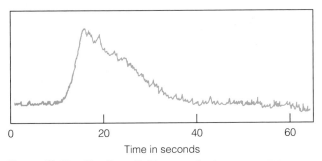

Figure 15.16   The X-ray light curve of a burster.   This schematic illustration of the X-ray intensity from a burster shows how rapidly the emission flares up and then drops off. In analogy with the nova process involving white dwarfs, these outbursts are thought to be caused by material falling onto the surface of a neutron star and igniting brief episodes of nuclear reactions.

Time in seconds
0   20   40   60

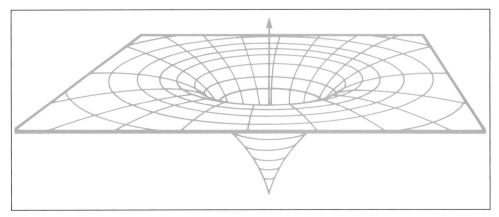

Figure 15.17   Geometry of space near a black hole.   Einstein's theory of general relativity may be interpreted in terms of a curvature of space in the presence of a gravitational field. Here we see a representation of how this curvature varies near a black hole (this is only a representation, not a realistic illustration, because the curvature of space near a black hole is actually a distortion of three-dimensional space, not two-dimensional space as shown here).

that it falls back onto the stellar surface. Photons emitted straight upward follow a straight path, but they lose energy to the gravitational field, causing their wavelengths to be redshifted (we discussed this phenomenon in the case of white dwarfs). When the gravitational field has become strong enough, the photon loses all of its energy and cannot escape. Another way of stating this is that the speed required for escape from the star exceeds that of light. To an outside observer, at the point when this happens the star becomes invisible, because no light from the star can reach the Earth.

The radius of the star at the time when its gravitational field becomes strong enough to trap photons

is called the **Schwarzschild radius,** after the German astrophysicist who first calculated its properties some 60 years ago. The Schwarzschild radius depends only on the mass of the star that has collapsed. For a star of 10 solar masses, it is 30 km; a star of twice that mass would have twice the Schwarzschild radius, and so on (Table 15.5).

When a collapsing star has shrunk inside its Schwarzschild radius, it is said to have crossed its **event horizon,** because an outside observer cannot see it or anything that happens to it after that point. We can have no hope of ever seeing what happens inside the event horizon, but since no force is known that could stop the collapse, we assume that it con-

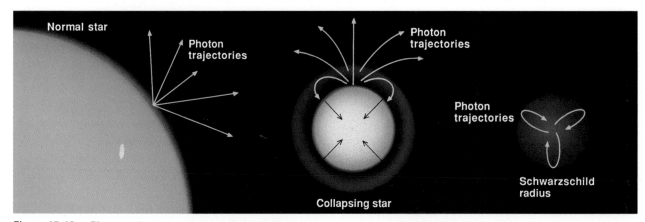

Figure 15.18   Photon trajectories from a collapsing star.   Light escapes in essentially straight lines in all directions from a normal star (left), whose gravitational field is not sufficient to cause large deflections. At an intermediate stage of collapse (center), photons emitted in a cone nearly perpendicular to the surface can escape, while others cannot. Those emitted at just the right angle go into orbit around the star, while those emitted at greater angles fall back onto the stellar surface. After collapse has proceeded to within the Schwarzschild radius, no photons can escape.

TABLE 15.5   Schwarzschild Radii for Various Objects

Object	Schwarzschild Radius
Sun	3 km
Earth	0.9 cm
150-lb person	1×10^{-23} cm
Jupiter	2.9 m
Star of 50 $M_\odot$	150 km
Typical globular cluster	5×10^4 km
Nucleus of a galaxy	$10^6 - 10^8$ km ($10^{-7} - 10^{-5}$ pc)
Massive cluster of galaxies	10^{15} km (100 pc)
The universe	10^{26} km (10^{13} pc)

tinues. The mass becomes concentrated in an infinitesimally small region at the center, which is called a **singularity,** because mathematically it is a single point.

What happens to the matter that falls into a singularity is a subject for speculation, but we can be more concrete about what happens just outside the event horizon. From the outsider's point of view, the rate of time is distorted by the extreme gravitational field and acceleration in such a way that the collapse would seem to slow gradually, taking an infinite time to reach the event horizon. This slowdown, however, is most significant in the last moments before the star's disappearance, and by then the star would be essentially invisible anyway, as most escaping photons would be redshifted into the infrared or beyond. The star would seem to disappear rather quickly, despite the stretching of the collapse in time caused by relativistic effects. To an observer unfortunate enough to be falling into the black hole, the fall would go very quickly; the rate of time would not be distorted by relativistic effects in the same way as that of the distant observer. (The infalling observer would suffer serious discomfort from being stretched out by tidal forces.)

Mathematically a black hole can be described completely by three quantities: its mass, its electrical charge, and its spin. The mass, of course, is determined by the amount of matter that collapsed to form the hole, plus any additional material that may have fallen in later. The electrical charge, similarly, depends on the charge of the material from which the black hole formed; if it contained more protons than electrons, for example, it would end up with a net positive charge. Because particles with opposite charges attract each other, and those with like charges repel, it is thought that electrical forces would maintain a fairly even mixture of particles during and after the formation of the black hole, so that

the overall charge would be nearly zero. To illustrate this, imagine that a black hole was formed with a net negative charge and that there was ionized gas around it afterwards (as there likely would be, with some of the matter from the original star still drifting inward). Then the negative charge of the hole would repel additional electrons, preventing them from falling in, while protons would be accelerated inward. In time, enough protons would be gobbled up to neutralize the negative charge.

The spin of a black hole is not so easily dismissed, however. It stands to reason that if the star were spinning before its collapse—and most likely it would have been—then conservation of angular momentum would cause the rotation to speed up greatly as the star shrank. A high spin rate actually shrinks the event horizon, allowing an outside observer to see closer in to the singularity residing at the center. It is even possible mathematically—although it presents a physical dilemma—to have sufficient spin that there is no event horizon and whatever is in the center is exposed to view. A **naked singularity,** as this has been dubbed, would not produce the usual gravitational effects of a black hole, and it would be possible to blunder into one without any forewarning.

For the most part, it is assumed that the spin rate is never so large that a naked singularity can form, and, in fact, black hole properties are usually specified by mass alone, neglecting the effects of charge and spin. Thus the assumption is that our main hope of detecting a black hole is by its gravitational effects, determined entirely by the mass.

Before turning to a discussion of how to find a black hole, we should mention that some black holes may be formed by processes other than the collapse of individual massive stars. These holes include the so-called mini black holes, very small ones postulated to have formed under conditions of extreme density early in the history of the universe; and supermassive black holes, thought possibly to inhabit the cores of large galaxies, where they formed from thousands or millions of stellar masses coalescing in the center. These other possibilities are discussed in later chapters; for now, we turn to the hunt for stellar black holes.

Do Black Holes Exist?

The mass that goes into a black hole during its formation still exists there, hiding inside its event horizon. Even though no light can escape, the gravitational effects persist. The gravitational force of the star is exactly the same as it was before the collapse, except at points so close that they would have been inside the original star.

Our best chance of detecting a black hole, then, is to search for an invisible object whose mass is too great to be anything else. Even if we do find such a thing, the evidence is really only circumstantial, because our conclusion that the object is a black hole relies on the theory that says neither a white dwarf nor a neutron star can survive if the mass is sufficiently great.

The best chance of determining the mass of an object is when it is in orbit around a companion star, where Kepler's third law can be applied to find the masses. The search for black holes, then, leads us to examine binary systems, looking for invisible, but massive, objects.

This search has been facilitated by another property of black holes: if they find new material to pull into themselves, the trapped matter forms an accretion disk, just as we described in the case of a neutron star in a mass-transfer binary. We have already seen that such a disk becomes so hot that it emits X rays. Probably some of the binary X-ray sources, therefore, are due to black holes in binary systems rather than to neutron stars. The best way to distinguish between the two possible types of remnants in these systems is to determine the mass of the invisible companion to see whether it is too great to be a neutron star.

The determination of the masses in a binary system is difficult, if not impossible, if only one of the two stars can be seen, because the information on their speeds and on the inclination angle of the orbit is incomplete. The fact that many X-ray binaries are eclipsing systems (at least the X-ray source is eclipsed; Fig. 15.19) helps to determine the tilt of the orbit in some cases. The orbital period is usually easily determined from the eclipse frequency as well, but the orbital velocities, needed to deduce the sizes of the orbits, are often very difficult or impossible to measure. The unseen star, of course, emits little or no light, so there is no hope of measuring Doppler shifts in its spectral lines. The normal companion star is often a hot giant or supergiant, and these stars usually have very broad spectral lines, whose shifts cannot be measured accurately. Nevertheless, sometimes enough information can be derived to at least place limits on the mass of the invisible companion. In one such case, the X-ray binary called Cygnus X-1 (Fig. 15.20), the collapsed object appears to contain at least 8 solar masses. This object is one of several leading black hole candidates (see Table 15.4).

Although the jury is still out on the question of the existence of black holes, the circumstantial evidence in favor is strong. Furthermore, it is easier to accept the existence of black holes than to find a way for a collapsing, massive star to avoid becoming one. As we have stressed, no force is known that can resist gravity in this circumstance. To believe that black holes do not form, one has to invent some way of preventing collapse, and also some way of explaining the observational evidence for invisible massive objects in some binary systems. It is far less complicated to accept that black holes do form and exist in these binaries, because doing so requires the fewest un-

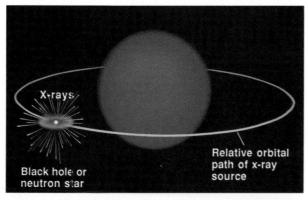

Figure 15.19   Eclipses of an X-ray source in a binary system.   Because the separation of the two stars in a mass-transfer system is always very small, and because the mass-losing star is usually very large (a red giant or a hot giant or supergiant with a stellar wind), eclipses occur easily. If the mass-receiving star is a neutron star or black hole, and therefore an X-ray emitter, the X rays are likely to be eclipsed periodically by the companion. Hence many such systems are eclipsing X-ray binaries.

Figure 15.20   Cygnus X-1.   This X-ray image obtained by the orbiting *Einstein Observatory* shows the intense source of X rays at the location of a dim, hot star that is thought to be a normal star that is losing mass to its invisible black hole companion. (Harvard-Smithsonian Center for Astrophysics)

proven assumptions. The principle called **Occam's razor**, a guiding philosophy for scientists, states that the explanation of any phenomenon that requires the fewest arbitrary assumptions is most likely to be the correct one.

Most astronomers today have adopted the concept of black holes and not only believe that they exist, but consider them an integral part of our universe, playing numerous important roles.

PERSPECTIVE

We have followed the stars into their graves. We have examined their corpses to see how they decay and have come away filled with wonder at the novel forms that matter can take. We have seen that the nature of stellar remnants depends entirely on how much mass is left when stars run out of nuclear fuel and collapse, the three possibilities being white dwarfs, neutron stars, and black holes. Stars that die alone, no matter what final form they take, are not likely to be detected again (except for the closest white dwarfs and the neutron stars that happen to appear to us as pulsars), while those that exist in binary systems may be reincarnated in spectacular fashion if mass exchange takes place.

Having learned all we can about stars as individuals, we are now ready to move on to larger scales in the universe, to examine galaxies and ultimately the universe itself.

SUMMARY

1. A white dwarf gradually cools off, taking billions of years to become a cold cinder.
2. If new matter falls onto the surface of a white dwarf—for example in a binary system in which the companion star loses mass—then the white dwarf may become a cataclysmic variable if the matter arrives in small amounts with high energy, or it may become a nova if matter builds up slowly to the point where a larger quantity is involved in the explosion. If it exceeds the white dwarf mass limit, it may become a neutron star or black hole, or it may explode as a Type I supernova.
3. Massive stars are likely to explode as Type II supernovae when all possible nuclear reaction stages have ceased. The supernova explosion creates an expanding cloud of hot, chemically enriched gas known as a supernova remnant.

4. In some cases, a remnant of 2 to 3 solar masses is left behind in the form of a neutron star, consisting of degenerate neutron gas.
5. A neutron star is too dim to be seen directly in most cases but may be observed as a pulsar (depending on the alignment of its magnetic and rotation axes, and our line of sight), and in a close binary system it may become a source of X-ray emission.
6. Some neutron stars that receive new material in clumps flare up occasionally as X-ray sources called bursters.
7. If the final mass of a star exceeds 2 or 3 solar masses, it will become a black hole at the end of its nuclear reaction lifetime.
8. The immensely strong gravitational field near a black hole traps photons of light, rendering the black hole invisible.
9. A black hole may be detected by its gravitational influence on a binary companion or by the X rays it emits if new matter falls in, as in a close binary system in which mass transfer takes place. A black hole binary X-ray source can be detected only by analysis of the orbits to determine the mass of the unseen object.

REVIEW QUESTIONS

1. Why is the final mass of a star, rather than the initial mass, the factor that determines what form of remnant the star will leave? How can the final mass be different from the initial mass?
2. Explain why white dwarfs, with no source of energy, can stay hot for as long as a billion years.
3. Describe the differences between a nova, a Type I supernova, and a Type II supernova.
4. How are neutron stars detected from the Earth?
5. If the rotation period of the Crab pulsar is 0.033 sec and its radius is 10 km, what is the speed of a point on its surface at the equator? What fraction of the speed of light is this?
6. Explain why the pulses from pulsars were finally judged to be the result of rapid rotation of neutron stars, rather than rapid expansion and contraction or rotation of some other type of star. What was the role of the Crab nebula and its pulsar in this discovery?
7. Both neutron stars and black holes can be X-ray sources in binary systems where mass exchange occurs. How can astronomers tell which kind of remnant is present in a given binary X-ray source?
8. Explain why some massive stars become black holes, while others become neutron stars or blow up completely, leaving no remnant.

9. Why would a neutron star or a black hole be expected to have a very rapid rotation rate?

10. Suppose an X-ray binary system is found in which the visible star is a red giant of a spectral type thought to have a mass of about 12 solar masses. The orbital period is 3.65 days, and the semimajor axis is 0.12 AU. Calculate the sum of the masses from Kepler's third law, then try to decide whether the compact companion star is a neutron star or a black hole.

ADDITIONAL READINGS

Backer, D. C., and S. Kulkarni. 1990. A new class of pulsars. *Physics Today* 43(3):26.

Bailyn, C. 1991. Problems with pulsars. *Mercury* 20(2):55.

Bethe, H. A. 1985. How a supernova explodes. *Scientific American* 252 (5):60.

————. 1990. Supernovae. *Physics Today* 43(9):24.

Cannizzo, J. K., and R. H. Kaitchnik. 1992. Accretion disks in interacting binary stars. *Scientific American* 266(1):92.

Croswell, K. 1992. The best black hole in the galaxy. *Astronomy* 20(3):30.

Graham-Smith, F. 1990. Pulsars today. *Sky and Telescope* 80(3):240.

Greenstein, G. 1985. Neutron stars and the discovery of pulsars. *Mercury* (2):34 (Part I); (3):66 (Part 2).

Gursky, H., and E. P. J. van den Heuvel. 1975. X-ray emitting double stars. *Scientific American* 232(3):24.

Helfand, D. J. 1978. Recent observations of pulsars. *American Scientist* 66:332.

Kafatos, M., and A. G. Michalitsianos. 1984. Symbiotic stars. *Scientific American* 251(1):84.

Kaler, J. 1991. The smallest stars in the universe. *Astronomy* 19(11):50.

Kaufmann, W. 1974. Black holes, worm holes, and white holes. *Mercury* 3(3):26.

Kawaler, S. D., and D. E. Winget. 1987. White dwarfs: fossil stars. *Sky and Telescope* 74:132.

Lewin, W. H. G. 1981. The sources of celestial X-ray bursts. *Scientific American* 244(5):72.

Margon, B. 1991. Exploring the high-energy universe. *Sky and Telescope* 82(6):607.

Marshall, L. A., and K. Brecher, 1992. Will Supernova 1987A shine again? *Astronomy* 20(2):30.

McClintock, J. 1988. Do black holes exist? *Sky and Telescope* 75:28.

Nather, R. E., and D. E. Winget. 1992. Taking the pulse of white dwarfs. *Sky and Telescope* 83(4):374.

Overbye, D. 1991. God's turnstile: the work of John Wheeler and Stephen Hawking. *Mercury* 20(4):98.

Price, R. H., and K. S. Thorne. 1988. The membrane paradigm for black holes. *Scientific American* 258(4):69.

Schaefer, B. E. 1985. Gamma-ray bursters. *Scientific American* 252(2):52.

Seward, F. 1986. Neutron stars in supernova remnants. *Sky and Telescope* 71(1):6.

Shaham, J. 1987. The oldest pulsars in the universe. *Scientific American* 256(2):50.

Smarr, L., and W. H. Press. 1978. Spacetime: Black holes and gravitational waves. *American Scientist* 66:72.

Van der Klis, M. 1988. Quasi-periodic oscillations in celestial X-ray sources. *Scientific American* 259(5):50.

Van Horn, H. M. 1973. The physics of white dwarfs. *Physics Today* 32(1):23.

Webbink, R. F. 1989. Cataclysmic variable stars. *American Scientist* 77:248.

Wheeler, J. C. 1973. After the supernova, what? *American Scientist* 61:42.

ASTRONOMICAL ACTIVITY
Black Hole or Neutron Star?

Suppose an eclipsing X-ray binary is discovered. Observations of its X-ray intensity produce a "light curve" as shown in the accompanying figure. The visible star (the companion to the X-ray source) is found to have spectral type B5V. Observations of its radial velocity produce the velocity curve shown in the figure. From this information and a few assumptions (described below), you can determine the masses of the two stars and decide whether the X-ray source is a black hole or a neutron star.

To do this, you may need to review some of the discussion of binary stars in Chapter 12. You will need to use Kepler's third law in the form $(m_1 + m_2)P^2 = a^3$, where m_1 and m_2 are the masses of the two stars (in units of the Sun's mass), P is the orbital period (in years), and a is the semimajor axis of the orbit (in AU).

To apply Kepler's third law, you need to determine the period P and the semimajor axis a, and you must convert the period to units of years and the semimajor axis to units of AU. You can find the period in days from either the light curve or the velocity curve; to convert to years, divide the period in days by 365.25.

To find a, you will have to use the information

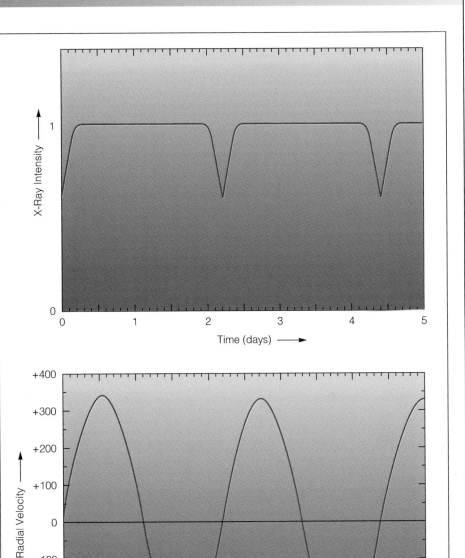

given on the orbital speed of the visible star, along with its period. The velocity curve shows how the observed radial velocity of the star varies as the star or-

continued on next page

bits the center of mass. The radial velocity reaches its extreme values when the star is moving directly toward or away from the Earth; the greatest positive value of the radial velocity occurs when the star is moving directly away from the Earth, and the greatest negative value of the radial velocity occurs when the star is moving directly toward the Earth. Thus, by finding these extreme values for radial velocity, you will find v, the orbital speed of the star.

The star travels a distance equal to the circumference of its orbit in a time equal to its orbital period. Since distance equals speed times time, you can write the equation

$$2\pi a = vP,$$

where $2\pi a$ is the circumference of the circular orbit having radius a, and vP is the distance the star travels in time P at speed v. Now you can solve this equation for a and substitute the values for v (from the velocity curve) and P (from the light curve—but you must convert P to seconds, since v is in units of km/sec). The value of a that you find will be in units of kilometers, but for Kepler's third law you need a in AU. To convert, divide your value in kilometers by the number of kilometers in one AU (see Appendix 2).

Once you have both a in AU and P in years, you can solve Kepler's third law for the sum of the masses of the two stars. Next, estimate the mass of the B5V star using standard tables of mass versus spectral type (see Table 12.4), and subtract. The remainder is the mass of the companion object, which is the source of the X rays. Finally, use the information given in this chapter to decide whether the mass you found falls into the range for neutron stars, or whether the object is so massive that it must be a black hole.

ASTRONOMICAL UPDATE

The Search for Planets and Failed Stars

In this section of the text, we learned that the most abundant stars are those of small mass. If we count the stars of each spectral type, we find that the number rises as the mass decreases, almost down to the lowest masses that are detectable. The observations become very difficult because the lowest-mass stars are also the dimmest. Below a certain limit (about 0.07 solar masses), the stellar interior never becomes hot enough to initiate nuclear reactions; an object falling below this limit is called a **brown dwarf.** Astronomers are intensely interested in the number of brown dwarfs because if they are very common, they may account for vast quantities of mass known to exist in unseen forms (this so-called **dark matter** is discussed in detail in the next section of the text). The general increase in the number of stars with decreasing mass suggests that brown dwarfs may be very common, but verifying this has proven difficult because the objects are hard to detect. Furthermore, there are indications, difficult to confirm, that the number of stars stops increasing with decreasing mass just at the limits of detectability; thus the trend is not clear enough to make us confident that brown dwarfs are numerous. Consequently, a great premium has been placed on detecting brown dwarfs directly.

One method for doing so is to look for their infrared glow, because brown dwarfs are expected to have surface temperatures of a few hundred degrees. But dim, low-mass stars also glow at infrared wavelengths, so it is not easy to distinguish between a cool, but normal star and a brown dwarf on the basis of the infrared emission alone. Several possible brown dwarf detections have been reported (some in binary

systems where mass estimates are possible, using Kepler's third law), but in each case the results are considered ambiguous. It has generally proven impossible in these cases to determine either the mass or the luminosity well enough to be certain that the object is a brown dwarf and not a dim star.

Very recently, however, a different observational approach seems to have yielded the first unambiguous detection of some brown dwarfs. And if this discovery is verified, it implies that the objects are very common indeed. As mentioned above, to be certain that a dim, infrared-emitting object is a brown dwarf, astronomers must determine that either its mass or its luminosity falls below the limit for normal stars. Determining mass with sufficient precision is very difficult, because many complexities are involved in applying Kepler's third law to a binary system in which the orbital tilt is unknown and the velocity variations of only one of the stars can be measured (see the discussion in Chapters 12 and 15). On the other hand, measuring the luminosity with accuracy is possible in cases where the distance is well established. This is the approach used in the recent, apparently successful study.

There are a few star clusters near the Sun whose distances are accurately known (see the discussion of clusters in Chapter 14). One such cluster is the Hyades, a familiar sight to amateur astronomers. In a recent study by a group of University of Minnesota astronomers headed by Claia Bryja, automated scans of photographic plates of the Hyades region revealed a number of faint, red objects. These appeared to be brown dwarf candidates, *if* they were at the distance of the cluster. Thus it became crucial to determine whether

these objects were cluster members or more distant, but more luminous objects.

To answer this question, Bryja and her colleagues compared photographs of the region taken at three different times over a 20-year interval. Careful measurements of the positions of the brown dwarf candidates showed that they are moving through space along with the cluster; this established that the objects are cluster members, rather than more distant stars that just happened to be seen in the same direction. Thus the distance to the objects was established quite accurately, and their luminosities could then be determined. Several of the objects proved to be too dim to be normal stars; instead their luminosities matched the expected values for brown dwarfs lying just below the 0.07 solar mass limit.

This successful detection of brown dwarfs has enormous implications for the mass content of galaxies like our own, for it means that much of the unseen matter may be in the form of brown dwarfs. The work of the Minnesota group will certainly stimulate further intensive searches for brown dwarfs, and you may find these obscure objects making news in the months to come.

Going even further in the direction of small-mass objects, several astronomers have been busy attempting to find planets orbiting stars other than the Sun. To establish the existence of other planetary systems is a prime goal because it will help to show how our own system was formed; furthermore, it will help to demonstrate whether other possible abodes for life exist in our galaxy.

Astronomers use several general techniques to try to detect planets

continued on next page

orbiting other stars, and all of them are difficult. Direct detection is perhaps the most problematic, because any planet will be so dim compared to its parent star that picking out the faint emission from the planet amid the glare from the star is virtually impossible. The best hope lies in detecting the subtle motions of the parent star as it orbits the center of mass of the system. As we learned in our discussions of binary stars, when two objects are in mutual orbit, both of them orbit the center of mass. Even when one is much more massive than the other, as in the case of a star and its planet, the massive object undergoes orbital motion. But this motion traverses a very limited range and is very slow. The Sun, for example, is most strongly affected by Jupiter, which causes the Sun to orbit about a point only a bit more than one solar radius from its center, with a period of nearly 12 years. The Sun's orbital velocity is only about 13 m/sec (note: *meters* per second, not kilometers). And from a typical interstellar distance, say, 10 pc, the angular motion of the Sun is only about 0.0005 arcsecond during the 12-year orbital period. Of course, the solar system has several planets in addition to Jupiter, so the Sun's motion is actually more complex than indicated. But in principle it would be possible to detect the presence of all the planets, given sufficiently accurate measurements of either the Sun's velocity variations or its positional shifts. Detection of such small velocities or angular motions is ex-

tremely difficult, and modern techniques are only now approaching the capability of achieving these levels of accuracy.

A very unexpected kind of planetary system was detected recently, however. Two planets were detected orbiting a neutron star! This discovery was achieved by taking advantage of the very stable pulse rate of a pulsar (a rotating, radio-emitting neutron star; see Chapter 15). Because the pulses from a pulsar are so incredibly steady in their frequency, it is possible to detect and accurately measure the slight shifts in pulse frequency that are caused by orbital motion. This is analogous to the Doppler effect, where each individual radio pulse acts like one of the light waves depicted in Figure 4.17 (except that the pulses are directed along a beam, rather than being emitted in all directions). When the neutron star approaches the observer, the pulses become bunched closer together, and when the neutron star recedes, the pulses become stretched farther apart in time. The high stability of the pulse rate, combined with the precise timing of detected pulses that can be achieved with radio telescopes, allows very small orbital velocities to be detected and measured.

Recently, radio astronomers Aleksander Wolszczan (National Astronomy and Ionosphere Center, Arecibo, Puerto Rico) and Dale Frail (National Radio Astronomy Observatory) reported that the pulsar PSR1257+12 has two companions,

with masses of at least 2.8 and 3.4 times the mass of the Earth, respectively. The nature of these objects is not well understood, of course, but clearly they are planetary in the sense that they are very much less massive than their parent "star." These planets are probably not originals; the parent star would have expanded to well beyond their orbital radii of 0.36 and 0.47 AU and engulfed them during its supergiant phases. Astronomers believe it is more likely that the planets formed from an accretion disk that surrounded the neutron star following the supernova that created it. Thus these planets probably bear little resemblance to a planetary system like our own.

The accuracy achievable by making radio timing measurements of neutron stars is far greater than can yet be accomplished for normal stars using optical techniques. Thus it will be some time before astronomers can detect Earth-like planets orbiting Sun-like stars. But sufficient accuracy to allow the detection of Jupiter-like planets is now being achieved, and several candidates have already been identified. Because the orbital periods are long (recall that Jupiter's period is nearly 12 years), several years of observations are required to be certain that a planet has been detected. But some of the ongoing searches may reach definitive results within the next few years, perhaps during your studies of astronomy.

SECTION

IV

A UNIVERSE OF GALAXIES

Having learned about the individual members of our great stellar system, the Milky Way, we now turn our attention to the larger picture. Having studied the trees, we now discuss the forest. Just as stars are the individual particles that make up a galaxy, galaxies are the building blocks of the universe as a whole (we note, however, that in view of the ample evidence for the existence of dark matter, it is possible that neither stars nor galaxies give us the true picture).

The first chapter of this section provides an overview of the structure and nature of the Milky Way, along with a description of the techniques used to learn about galactic properties. We will find that astronomers had difficulty determining the size and shape of our home galaxy, because our solar system is located at an interior position from which it is impossible to obtain an overall view. Once we have learned about the size, shape, and composition of the galaxy, we proceed in Chapter 17 to discuss its formation and evolution. Here we find some parallels with the formation of the solar system and of stars in general, in that the process began with a rotating cloud that flattened into a disk as it collapsed. The story of the evolution of the galaxy is the story of the formation and distribution of the elements, and elemental enrichment through stellar processing is a key theme of this chapter.

Along with our understanding of the nature of the Milky Way came the realization that our galaxy is only one among billions, and so the rest of this section is devoted to galaxies be-

yond our own. Chapter 18 provides an overview of the distribution and properties of galaxies and groupings of galaxies, thereby giving us a road map to the visible universe. The subject of an invisible universe comes up here as well, as we learn that perhaps 90 percent of all the mass is in some unknown, dark form.

In Chapter 19 we turn our attention from individual galaxies to two overall properties of the universe that have enormous implications: the general expansion of the universe, which shows that it had a beginning in a highly compressed state; and the cosmic background radiation, which is a relic of the early times when the universe was very hot and dense. These phenomena set the stage for our subsequence discussion of the formation and evolution of the cosmos.

Before discussing cosmology, however, we devote Chapter 20 to a study of violent, energetic phenomena such as radio galaxies, galaxies with active cores, and quasars. All of these objects appear to represent galaxies in upheaval. Because we see them primarily at great distances, they appear to us as they were long ago and thus represent early stages in the evolution of galaxies.

The final chapter in the section, Chapter 21, is dedicated to the grandest of all questions: How did the universe form? How did it reach its current state? What is its future? In attempting to answer these questions, we will be probing the very forefront of human knowledge of the cosmos.

CHAPTER
16

Structure and Organization of the Galaxy

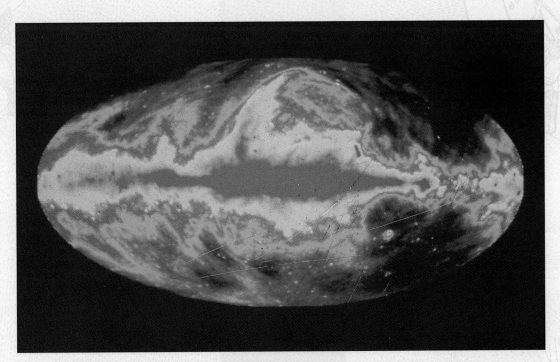

A radio map of the Milky Way (C. G. T. Haslam, Max Planck Institut fur Radioastronomie, Bonn)

PREVIEW

The Sun is just one of billions of stars in the Milky Way, a great pinwheel disk that is itself one of billions in the universe. In this chapter you will learn about the overall structure of our galaxy, starting from a historical context because the story of how the Milky Way came to be understood has much to tell us about the methods of astronomy. The system of stars is so vast, and the perspective from our own position within it so limited, that extensive detective work has been required in order for us to gain a clear overall view. Among the themes of this chapter will be the rich interplay between stars and the interstellar gas and dust from which they form and to which they return their enriched material as they evolve and die. You will also learn of two major mysteries confronting modern astronomy: the nature of the mysterious and immense source of energy that hides at the core of the Milky Way; and the form taken by the majority of the galactic mass, which is invisible. Both of these phenomena, whatever their explanation, will turn up again, as we discuss other galaxies in subsequent chapters.

We have discussed stars as individuals, as though they exist in a vacuum, isolated from the rest of the universe. For the purposes of analyzing the structure and evolution of stars, this is a suitable approach, but if we want to understand the full stellar ecology, the properties of individuals must be discussed in the context of the larger environment.

Even a casual glance at the nighttime sky shows that the stars tend to be grouped rather than randomly distributed. The most obvious concentration is the **Milky Way** (Fig. 16.1), a diffuse band of light stretching from horizon to horizon, clearly visible only in areas well away from city lights. To the ancients the Milky Way was merely a cloud; Galileo, with his primitive telescope, recognized it as a region with a great concentration of stars. In modern times we know the Milky Way as a galaxy, the great pinwheel of billions of stars to which the Sun belongs (Table 16.1). The hazy streak across our sky is a cross-sectional, or edge-on, view of the galaxy from a point within its disk.

The overall structure of the Milky Way resembles a phonograph record with a large central bulge, the center of which is called the **nucleus** (Figs. 16.2 and 16.3). Nearly all the visible light is emitted by stars in the plane of the disk, although the galaxy also has a **halo,** a distribution of stars and star clusters centered on the nucleus but extending well above and below the disk. The most prominent objects in the halo are the **globular clusters,** very old, dense clus-

TABLE 16.1 Properties of the Milky Way

Mass: 1.0×10^{11} M$_\odot$ (interior to Sun's position)
Diameter of disk: 30 kpc
Diameter of central bulge: 10 kpc
Sun's distance from center: 8.5 kpc
Diameter of halo: 100 kpc (very uncertain)
Thickness of disk: 1 kpc (at Sun's position)
Number of stars: 4×10^{11}
Typical density of stars: 0.1 stars/pc^3 (solar neighborhood)
Average density of interstellar matter: 10^{-24} g/cm^3 (roughly equal to one hydrogen atom/cm^3)
Luminosity: 2×10^{10} L$_\odot$
Absolute magnitude: -20.5
Orbital period at Sun's distance: 2.4×10^8 years

---

ters of stars characterized by their distinctly spherical shape.

To describe the size of the Milky Way, we must use a new unit of distance. In Chapter 12 we discussed stellar distances in terms of parsecs and found that the nearest star is about 1.3 parsecs (pc) from the Sun. To expand to the scale of the galaxy, we speak in terms of **kiloparsecs (kpc),** or thousands of parsecs. The visible disk of the Milky Way is roughly 30 kpc in diameter, and the disk is a few hundred parsecs thick. Light from one edge of the galaxy takes about 100,000 years to travel across to the far edge.

Our galaxy is so large that we must discuss new methods of measuring distance. Even the main-sequence fitting technique, the most powerful we have discussed so far, fails when star clusters are too distant and faint to allow determination of the individual stellar magnitudes and spectral types.

VARIABLE STARS AS DISTANCE INDICATORS

In Chapter 12 we stressed that it is always possible to determine the distance to an object if both its absolute and apparent magnitudes are known. We discussed spectroscopic parallax, where the absolute magnitude of a star is derived from its spectral class, and main-sequence fitting, where a cluster of stars is observed and its main sequence determined so that it can be fitted to the standard H-R diagram main sequence. Now we will learn about a special type of star whose absolute magnitude can be determined from its other properties, which allows it to be used to measure distances.

In Chapter 12 we briefly mentioned variable stars, including those that pulsate regularly (Table 16.2).

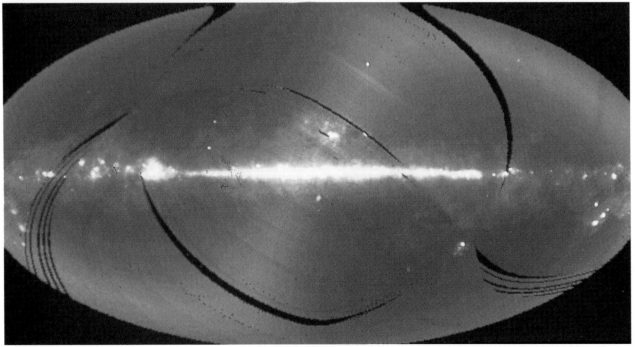

Figure 16.1   The Milky Way.   This illustration (top) shows the cross-sectional view of our galaxy that we see from Earth. The bottom image is an infrared view of the Milky Way, obtained by the *IRAS* satellite. (Top: Lund Observatory; bottom; NASA)

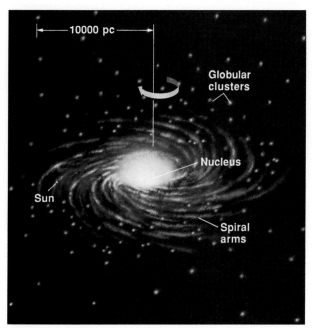

Figure 16.2   The structure of our galaxy.   This sketch illustrates the modern view of the Milky Way.

Figure 16.3   A spiral galaxy similar to the Milky Way.
This galaxy (NGC 2997) probably resembles our own, as seen from afar. (© 1980 AAT Board; photo by D. Malin)

These stars physically expand and contract, changing in brightness as they do so. One of the first such stars to be discovered was δ Cephei, which is sufficiently bright to be seen with the unaided eye. Following the discovery of δ Cephei in the mid-1700s, other similar stars were found, and these as a class became known as **Cepheid variables,** named after the prototype. Another type of pulsating variable, the **RR Lyrae stars,** were found to exist primarily in globular clusters and to have pulsation periods of less than a day, whereas the Cepheids have periods ranging from 1 to 100 days.

In 1912, through observations of variable stars in the Magellanic Clouds (two small, relatively nearby galaxies), American astronomer, Henrietta Leavitt,

discovered that there is a definite relationship between the pulsation period of these stars and their luminosities. For the Cepheids, the period increases with increasing luminosity (or decreasing absolute magnitude), while the RR Lyrae stars all have about the same luminosity. This relationship has a profound implication: When a star is recognized as a variable belonging to one of these classes, its absolute magnitude can be determined simply by measuring its

TABLE 16.2   Types of Pulsating Variables

Type of Variable	Spectral Type	Absolute Period (days)	Change of Magnitude	Magnitude
δ Cephei (Type I)	F, G supergiants	3–50	− 2 to − 6	0.1–2.0
W Virginis (Type II Cepheids)	F, G supergiants	5–30	0 to − 4	0.1–2.0
RR Lyrae	A, F giants	0.4–1	0.5 to 1.2	0.6–1.3
RV Tauri	G, K giants	30–150	− 2 to − 3	Up to 3
δ Scuti	F giants	0.1–0.2	2	0.1
β Cephei (β Canis Majoris)	B giants	0.1–0.2	− 3 to − 5	0.1
Long-period variables (Mira)	M supergiants	80–600	+ 2 to − 3	3–7

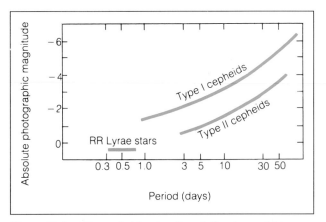

Figure 16.4 The period-luminosity relation for variable stars. This diagram shows how the pulsation periods for Cepheid and RR Lyrae variables are related to their absolute magnitudes. Initially astronomers did not recognize that these are two types of Cepheids with somewhat different relationships, and this led to some early confusion about distance scales.

Figure 16.5 Harlow Shapley. Shapley's work on the distances to globular clusters was a key step in the determination of the size of the Milky Way. Shapley also played a prominent role in the discovery of galaxies outside our own (see Chapter 18). Much of his observational work was done before 1920, when he was a staff member at the Mount Wilson Observatory. He later became director of the Harvard College Observatory. (Harvard College Observatory)

period and using the established period-luminosity relation (Fig. 16.4). Once the absolute magnitude is known, the distance can be found by comparing the absolute and apparent magnitudes. (Average values must be used, since the magnitudes actually vary with the pulsations.)

Because Cepheids are giant stars, they are very luminous and can be observed at great distances. This makes them very powerful tools for measuring distances beyond those reached by other techniques we have discussed. At first, astronomers did not realize that there are two types of Cepheids with slightly different period-luminosity relations (see Fig. 16.4), a fact that initially led to confusion in establishing the size of our galaxy.

THE STRUCTURE OF THE GALAXY AND THE LOCATION OF THE SUN

Because the solar system is located within the disk of the Milky Way, we have no easy way of getting a clear view of where we are in relation to the rest of the galaxy. All we see is a band of stars across the sky, which tells us that we are in the plane of the disk. It is not so easy to determine where we are with respect to the edge and center of the disk.

Early in this century no one knew that our galaxy is permeated with a very rarefied medium of gas and tiny solid particles in interstellar space, which contribute to the difficulty of determining the structure of the galaxy and our location in it. The haze of tiny solid particles, called **interstellar dust,** obscures our

view of distant stars, making it appear as though the number of stars in space decreases as we look farther and farther away. This illusion led one early researcher, the Dutch astronomer J. C. Kapteyn, to conclude that we are at the center of the galaxy because our Sun seemed to be in the densest region. But other techniques soon showed that we are far from the center of the galaxy.

One of these techniques, which was employed by Harlow Shapley (Fig. 16.5), made use of the globular clusters, the spherical star clusters found outside the confines of the galactic disk. These clusters tend to contain RR Lyrae variables, so Shapley could determine their distances and hence their locations with respect to the Sun and the disk of the Milky Way. He found that the globular clusters are distributed in a spherical arrangement centered on a point several thousand parsecs from the Sun (Fig. 16.6), and he argued that this point must represent the center of the galaxy. It would not make physical sense for the globular clusters to be concentrated around any location other than the center of the entire galactic system.

Shapley's conclusion, first published in 1917, was not widely accepted initially, but other supporting evidence was found in the 1920s. Two scientists, Jan Oort of Holland (already mentioned in connection with the origin of comets; see Chapter 9) and Bertil

ASTRONOMICAL INSIGHT
The Shape of the Milky Way

In many ways, studying our own galaxy is more difficult than observing other galaxies. The reason, of course, is that we can observe the Milky Way only from an interior position, where we see a cross-sectional view of the disk of the galaxy. We cannot step back and see the entire system of stars, and worse yet, interstellar gas and dust (discussed later in the chapter) limit the distance to which we can see even in our cross-sectional view. For these reasons, some of the most basic properties of our galaxy are still poorly known.

One of these properties is the shape of the galaxy. Galaxies outside our own have been classified according to their shapes (as discussed in Chapter 18), and it appears that shape has much to tell us about the overall properties and evolution of a galaxy. Thus it is of interest to know the shape of our own galaxy, so that we can compare it with others. Two recent developments have added to our knowledge of the shape of the Milky Way.

One development was the launch, in late 1989, of the *Cosmic Background Explorer (COBE)*, a satellite designed to measure infrared and short-wavelength radio emission from space (the reasons for doing this are discussed in Chapter 19). *COBE* has obtained data at the infrared wavelengths where the coolest stars emit, and a composite cross-sectional view of the Milky Way in these wavelengths has been assembled (see the photo above). In this view, we see much farther than in similar visible-light images, because the infrared radiation is not as strongly blocked out by interstellar dust. This image provides the best information yet on the size and shape of the galactic bulge, which contains mostly old, dim, red stars, in relation to the thickness of the galactic disk. For the first time, as-

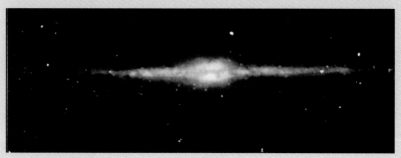

NASA: courtesy of the COBE Science Working Group

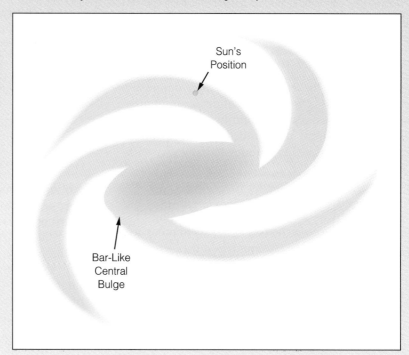

Sun's Position

Bar-Like Central Bulge

L. Blitz, University of Maryland, and D. Spergel, Princeton University

tronomers have a picture of our galaxy that can be compared with similar pictures of other spiral galaxies that are seen edge-on.

The second recent development involves the shape of the central bulge of the galaxy, as it would appear if we could see it face-on. Obviously we cannot do that, so we must use indirect methods to infer

the shape. One of the most common techniques is to measure the motions of hydrogen gas, observed through its 21-cm emission. As discussed in the text, this emission can be observed over enormous distances, and it allows astronomers to map the spiral structure of the galaxy. It also allows detailed studies of the motions of the gas at different distances from the galactic center.

Several observers have found evidence that the central bulge of the galaxy is elongated; that is, it is apparently not spherical, but instead has a barlike structure (see the sketch on the facing page). Most of the observations leading to this conclusion were made at positions close to the center of the galaxy, but now

even gas farther out has been shown to be moving in a way that is consistent with a central bar. A bar and a circular object have different gravitational effects, and the result is that gas orbiting the center follows ever-more-elliptical orbits, the closer in it is observed.

Many other spiral galaxies have barlike central bulges; about half of

all spirals are actually classified as barred spirals. Most are more highly elongated than the bar in our galaxy, but there seem to be many different degrees of elongation. As discussed in Chapter 18, the presence or absence of a bar may be linked to the mass and extent of the galaxy's halo.

Lindblad of Sweden, carried out careful studies of the motions of stars in the vicinity of the Sun. They found that these motions could best be understood if the Sun and the stars around it were assumed to be orbiting a distant point; that is, they found systematic, small velocity differences between stars, similar to those between runners on a track who are in the inside and outside lanes (Fig. 16.7). It appeared from these studies that the Sun is following a more-or-less circular path about a point several thousand parsecs away, which indicates that the center of the galaxy is located at that distant center of rotation. This supported Shapley's view of the galaxy, although the Sun's distance from the center was still uncertain.

The discovery of the general interstellar dust medium in 1930 laid to rest the apparent conflict between the findings of Kapteyn and those of Shapley, Oort, and Lindblad, and the only major question remaining was the true size of the galaxy. Determining

this required refinement of the period-luminosity relation for variable stars, which, as noted earlier, was confused for a while by the failure to recognize that there are two types of Cepheids. Eventually a consensus was reached that the Sun is about 10 kpc from the center of a disk-like galaxy whose total diameter is about 30 kpc.

More recently, evidence has emerged that indicates the Sun is a little closer to the center, perhaps only 7 or 8 kpc out (the current "official" distance, used by astronomers through international agreement, is 8.5 kpc). We see that the problem of determining

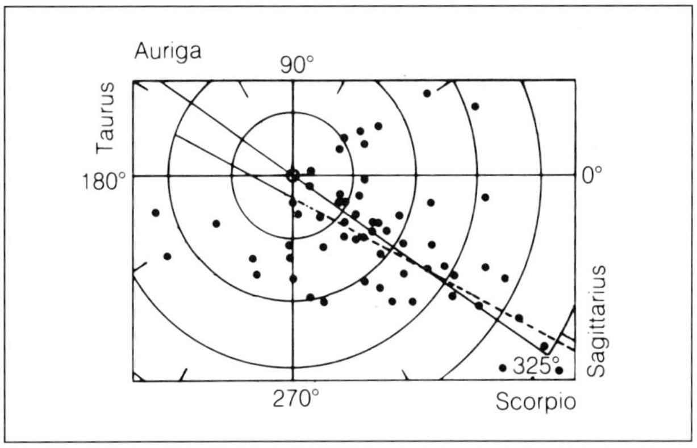

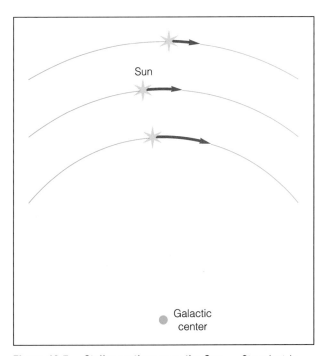

Figure 16.7   Stellar motions near the Sun.   Stars just inside the Sun's orbit move faster than the Sun, whereas those farther out move more slowly. Analysis of the relative speeds (as inferred from measurements of Doppler shifts) and distances of stars like these led to the realization that the Sun and stars near it are orbiting a distant galactic center.

Figure 16.6   Shapley's measurements of globular clusters.
This is one of Shapley's original figures illustrating the distribution of globular clusters in the galaxy. This is a projection down onto the galactic plane. The Sun is at the point where the straight lines intersect, and each circle centered on that point represents an increase in distance of 10,000 pc. Note that the distribution of globulars is centered some 10 to 20 kpc from the Sun. (Estate of H. Shapley, reprinted with permission.)

where we are in our galaxy is not necessarily completely solved even today. There are also significant uncertainties about the extent of the galaxy beyond the Sun's position, as we will see in later sections of this chapter.

GALACTIC ROTATION AND STELLAR MOTIONS

An important result of the work of Lindblad and Oort was an understanding of the overall motions in the galaxy. Oort's analysis was especially useful in this regard; he showed not only that the Sun and stars near it are orbiting the distant galactic center, but also that the rotation of the galaxy is differential. This means that the stars follow their own orbits at their own speeds, rather than moving together as though part of a fixed structure. Thus the disk of the galaxy is not rigid, but instead acts like a fluid, made of individual particles (stars) moving independently.

If the galaxy were a rigid disk, then the orbital speed would increase with increasing distance from the galactic center. Think about a record on a turntable; in order for the entire disk to rotate as one body, the outer edge, which has to travel much farther than the inner portions in order to make one revolution, must travel correspondingly faster. On the other hand, if the galaxy were made of individual stars that had no effect on each other, so that each one simply orbited the center as though all the mass were concentrated there, then we would expect the orbital speeds to decrease with distance from the center. To see this, recall Kepler's laws of planetary motion; the third law in particular, which relates orbital period to orbital radius (semimajor axis) shows that orbital speeds decrease with distance from the center (we know, for example, that each planet in succession outward from the Sun moves more slowly than the ones closer in).

So what do we actually find when we examine the orbital speeds of stars in the Milky Way? We can measure the **rotation curve** of the galaxy (Fig. 16.8); this is a graph showing how the orbital speed varies with distance from the center. What we find is an intermediate case; we see neither an increase nor a decrease of orbital speed as we go outward. This tells us that the galaxy is not a rigid disk, but on the other hand that the individual stars do not completely ignore the effects of other stars around them, so they do not act as completely free agents, either.

The rotation curve cannot be measured closer to the center than about 1 kpc, and has been measured no farther out than about 16 kpc. It is likely that there is a small inner disk, extending to perhaps 700 pc, where rigid-body rotation does occur, and the

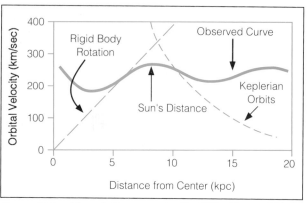

Figure 16.8   The rotation curve for the Milky Way.   This diagram shows how the stellar orbital velocities vary with distance from the center of the galaxy. The fact that the curve does not simply drop off to lower and lower velocities beyond the Sun's orbit, as it would if all the mass of the galaxy were concentrated at its center, indicates that there is a lot of mass in the outer portions of the galaxy. (Data from M. Fich)

orbital speeds rise quickly from zero at the very center to the steady level of above 200 km/sec, which extends as far out as the speeds can be measured.

The Sun's orbital speed is approximately 220 km/sec, and its orbital radius is about 8.5 kpc. Using these numbers to estimate the orbital period (by dividing the circumference of the orbit by the orbital speed, assuming the orbit to be a circle) yields a period of about 240 million years. Thus in the 4.5-billion year lifetime of the Sun, it has orbited the galaxy some 18 to 20 times.

Individual stellar motions do not necessarily follow precise circular orbits about the galactic center. Thus far we have described the overall picture that emerges from looking at the composite motions of large numbers of stars. If we look at the individual trees instead of the forest, we find that each star in the great disk has its own particular motion, which may deviate slightly from the ideal circular orbit. These individual motions are comparable to the paths of cars on a freeway, where the overall direction of motion is uniform, but a bit of lane-changing occurs here and there.

In the galaxy, the deviation in motion of a star from a perfect circular orbit is called the **peculiar velocity** of the star; in the case of the Sun, the peculiar velocity is called the **solar motion.** The Sun has a velocity of about 20 km per second with respect to a circular orbit, in a direction about 45° from the galactic center and slightly out of the plane of the disk. Most peculiar velocities of stars near the Sun are comparable, amounting to only minor departures from the overall orbital velocity of about 220 km per second. In Chapter 17 we will discuss the **high-velocity stars,** which deviate strongly from the circular orbits followed by most stars in the Sun's vicinity.

THE MASS OF THE GALAXY

Once the true size of the Milky Way was determined, it became possible to estimate its total mass. This could be achieved by measuring the star density in the vicinity of the Sun and then assuming that the entire galaxy has about the same average density. But a much simpler and more accurate technique is possible if we assume that the stars in our region of the galaxy approximately obey Kepler's laws (for this purpose this is a reasonable assumption).

Kepler's third law, in the more complete form developed by Newton, expresses a relationship among the period, the size of the orbit, and the sum of the masses of two objects in orbit about each other:

$$(m_1 + m_2)P^2 = a^3,$$

where m_1 and m_2 are the two masses (in units of the Sun's mass), P is the orbital period (in years), and a is the semimajor axis (in AU). If we consider the Sun to be one of the two objects and the galaxy itself to be the other, we can use this equation to determine the mass of the galaxy. As we have already seen, the orbital period of the Sun is roughly 240 million years: $P = 2.4 \times 10^8$ years. The orbit is nearly circular, with the galactic nucleus at the center, so the semimajor axis is approximately equal to the orbital radius; that is, $a = 8.5$ kpc $= 1.8 \times 10^9$ AU. Now we

can solve Kepler's third law for the sum of the masses:

$$m_1 + m_2 = a^3/P^2 = 1.0 \times 10^{11} \text{ solar masses.}$$

Since the mass of the Sun (1 solar mass) is inconsequential compared with this total, we can say that the mass of the galaxy itself is about 1.0×10^{11} solar masses. The Sun is slightly above average in terms of mass, so we conclude that the total number of stars in the galaxy must be a few hundred billion.

This method refers only to the mass inside the orbit of the Sun; the matter that is farther out has little effect on the Sun's orbit. Thus, when we estimate the mass of the galaxy by applying Kepler's third law in this way, we neglect all the mass that lies farther out. As we saw in the previous section, the orbital speeds do not drop off at greater distances, as they would if most of the galaxy's mass were concentrated in the inner regions. This tells us that a great deal of mass lies far from the center of the galaxy. It is now thought likely that most of the mass of our galaxy is hiding in the halo. This is discussed in a later section.

THE INTERSTELLAR MEDIUM

Astronomers have known for many decades that there is material in space between the stars. We have mentioned the haze of interstellar dust that permeates interstellar space, and photographs show very obvious concentrations of dark or glowing clouds here and there in the galaxy (Figs. 16.9 and 16.10). These

Figure 16.9   An emission nebula.   This dense cloud of interstellar gas and dust is being heated by hot stars embedded within it, which causes the gas to glow. Much of the emission is due to hydrogen atoms, which produce strong emission at 6563 Å, accounting for the red color seen here. (© 1984 ROE/AAT Board)

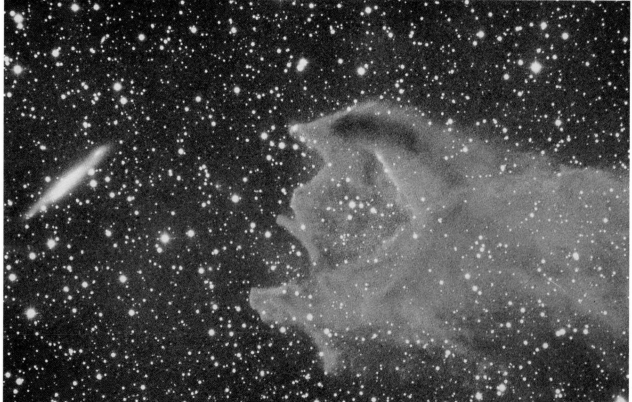

Figure 16.10   Dark clouds.   These dark, patchy regions are dense, cold interstellar clouds, the kind of regions where stars can form. The interior temperatures may be as low as 20 K, and the density roughly 10^5 particles per cubic centimeter, a high vacuum by earthly standards, but quite dense for the interstellar medium. The dark cloud in the lower photo, colored by gaseous emission and reddening due to dust absorption, is called a cometary globule because of its shape. (upper: © 1984 Anglo-Australian Telescope Board; lower: © 1992 AAT Board; photos by D. Malin)

TABLE 16.3   Interstellar Medium Conditions

Component	Temperature (K)	Density (per cm^3)	State of Gas	How Observed
Coronal gas (intercloud)	10^5–10^6	10^{-4}	Highly ionized	X-ray emission, UV absorption lines
Warm intercloud gas	1,000	0.01	Partially ionized	21-cm emission, UV absorption lines
Diffuse clouds	50–150	1–1,000	Hydrogen in atomic form, others ionized	Visible, UV absorption lines, 21-cm emission
Dark clouds	20–50	10^3–10^5	Molecular	Radio emission lines, IR emission and absorption lines

clouds are regions of relatively high density, whereas the lower-density material filling most of the volume of the galaxy is virtually transparent. In all, the interstellar gas and dust comprise some 10 to 15 percent of the mass of the galactic disk and play very important roles in its evolution. Stars form from interstellar clouds, as we have seen, and late in life, many stars return some of their substance to the interstellar medium. Consequently, the interstellar medium is studied in detail by astronomers interested in learning more about the lives and deaths of stars and of the galaxy itself.

The interstellar medium is a mixture of gas and dust, with a wide range of physical conditions (Table 16.3). The interstellar dust makes distant stars appear dimmer than they otherwise would and complicates distance determinations based on apparent magnitude measurements. The tendency of the dust to make stars appear dimmer is called **interstellar extinction.** Because the particles tend to block out short-wavelength light more effectively than the longer wavelengths, red light penetrates the interstellar medium more easily than blue light (Fig. 16.11). Thus distant stars appear not only dimmer but also redder than they otherwise would. This trend of in-creasing extinction with decreasing wavelength continues throughout the spectrum, so that the extinction at ultraviolet wavelengths is very severe, but it is minimal in the infrared portion of the spectrum (Fig. 16.12). Thus it is virtually impossible to observe the inside of even a moderately dense interstellar cloud with an ultraviolet telescope, but we can see deep inside the densest clouds at infrared and radio wavelengths. Infrared observations are a particularly effective technique for observing stars in the process of formation, as discussed in Chapter 14.

Interstellar grains are formed in the material ejected by various kinds of aged or dying stars. Red giant and supergiant stars, for example, are known to form grains in their outer atmospheres, and grain formation also seems to take place in planetary nebulae, novae, and supernova outbursts. Grains form through condensation, which occurs whenever the proper combination of pressure and temperature prevails. The formation of dust grains in the atmosphere of a red giant is analogous to the formation of frost on the lawn on a cool, humid morning.

Most (about 90 percent) of the mass in interstellar space is in the form of gas particles rather than grains. The gas is observed by means of absorption lines

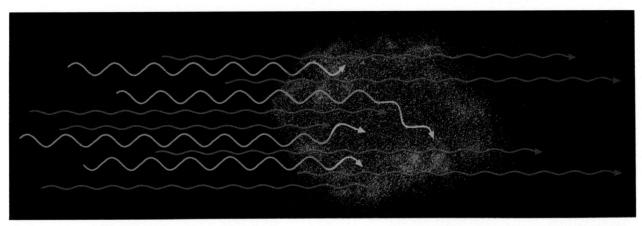

Figure 16.11   Scattering of light by interstellar grains.   The grains tend to absorb or deflect blue light more efficiently than red, so a star seen through interstellar material appears red.

ASTRONOMICAL INSIGHT
The Violent Interstellar Medium

Once it was realized (in the 1920s and 1930s) that there is a pervasive interstellar medium of gas and dust, the notion developed that the interstellar region was a cold, inactive place where nothing much ever changed. Now we know that picture to be false, because it was based on the very limited view of the interstellar medium that was available before the advent of space astronomy.

Because of the physical conditions that prevail in the interstellar medium, only a minute, insignificant portion of it can be observed in visible wavelengths of light. Before ultraviolet, radio, infrared, and X-ray technologies became available to astronomers, little was known about the material between the stars. The most active and important types of interstellar material were invisible.

Thus a great revolution in our knowledge of interstellar conditions has taken place in the past 20 years or so as telescopes have been developed for space observations. The first hints that the interstellar medium is actually a hot, violent place came in the early 1970s, when new ultraviolet and X-ray instruments were launched above the Earth's obscuring atmosphere. In the mid-1970s, the ultraviolet telescope/spectrometer *Copernicus* discovered that very highly ionized gases exist between the stars. The gases require temperatures of several hundred thousand degrees to achieve their high degree of ionization. (The most prominent example is oxygen with five of its eight electrons removed.) At about the same time, rocket-borne X-ray telescopes found that the interstellar medium emits a diffuse glow of X rays, indicating the presence of gas at a temperature of about 1 million degrees. Meanwhile spectroscopic observations in radio and ultraviolet wavelengths revealed clouds here and there in space that are moving at enormous velocities, well in excess of 100 km per second. It became clear that the old view of a placid, unchanging, cold, dark interstellar medium was incorrect.

It also became clear that a great deal of energy must be injected into interstellar space to create the observed high temperatures and velocities. It is now believed that the sources of this energy are supernova explosions, which occur on average only once every 30 to 50 years in our galaxy but release as much energy in an instant as our Sun will produce in its entire lifetime, and hot stars with rapid stellar winds, which produce similar quantities of energy in a few million years. These outbursts from massive stars create huge "bubbles" of superheated gas, which expand and join together, creating a galactic network of superhot, low-density gas that occupies much of the volume of the galactic disk. The high-velocity clouds that are observed are thought to be material swept up by the gas expanding outward from supernovae and hot stars with rapid winds.

We point out in the text that the interstellar medium is half of a cosmic recycling process. In this process material becomes concentrated in stars, where the chemical composition is altered by nuclear reactions, and then is dispersed into interstellar space through supernova explosions and stellar winds, where it forms the raw material for the formation of new stars. Now we find that the outpouring of matter from stars provides not only the material, but also the energy for the formation of stars. Compression of dense clouds by the expanding material from supernovae and stellar winds apparently is responsible for triggering the collapse of interstellar clouds to form new stars. Thus although the violence of the interstellar medium may have been a surprise when it was discovered, it is actually necessary for the vitality of our galaxy.

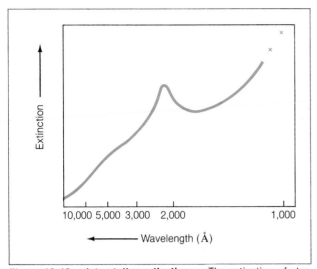

Figure 16.12   Interstellar extinction.   The extinction of starlight by interstellar dust is greater at short wavelengths, especially in the ultraviolet, and decreases toward longer wavelengths. It is very small in the infrared and virtually nonexistent at radio wavelengths.

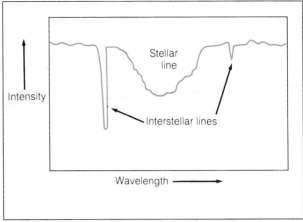

Figure 16.13   Interstellar absorption lines.   Gas atoms and ions in space absorb photons from distant stars, creating absorption lines in the spectra of the stars. Here we see some interstellar absorption lines at ultraviolet wavelengths, where most of these lines are found. The interstellar lines can be distinguished from the star's own spectral lines because they are narrower, usually have a different Doppler shift (because the star and the cloud producing the line move at different velocities), and represent different states of ionization.

formed in the spectra of distant stars (Fig. 16.13), mostly in the ultraviolet portion of the spectrum, or by means of various kinds of emission lines. Dense clouds that are heated by nearby hot stars glow by producing emission lines, primarily of hydrogen (Fig. 16.14). The strongest emission line of hydrogen lies at 6563 Å, in the red portion of the spectrum, so color photographs of these **emission nebulae** or **H II regions** show a vivid red appearance (Fig. 16.15).

Interstellar gas also produces radio emission lines. Again, hydrogen atoms play an important role, emitting at a wavelength of 21 centimeters. Since hydrogen is the most abundant element, observations of the 21-centimeter emission line are very useful for determining the structure of the galaxy (discussed in the next section). Molecules in the densest interstellar clouds also form radio emission lines, at wavelengths ranging from a few millimeters to several centimeters. Molecules, like atoms, have definite energy states, and photons are emitted or absorbed when the molecules change energy states. The energy levels that produce lines in the radio portion of the spectrum are related to the rotational energy of a molecule; being an extended object, a molecule can rotate, and the energy of rotation can have only certain values, just as the energy level of an electron orbiting an atom can have only certain values. Thus a spinning molecule can slow its spin only by making a sudden drop to a lower energy state, and in doing so it emits a photon whose energy corresponds to the loss of

rotational energy of the molecule. Each kind of molecule has its own unique set of rotational energy states and, therefore, its own unique spectrum of radio emission or absorption lines.

Over 100 different molecular species have been identified in dense, "dark" interstellar clouds (see Appendix 12), and more are being found frequently as radio telescope technology improves. By far the most abundant molecule is hydrogen (H_2), followed by carbon monoxide (CO). Ironically, hydrogen molecules do not have strong radio emission lines, and their presence must be inferred indirectly except in more rarefied clouds, where they can be observed through ultraviolet absorption lines. The less dense interstellar clouds have few molecules but are composed of a mixture of atoms and ions. Ionization in interstellar space occurs primarily through the absorption of ultraviolet photons rather than through collisions between atoms, as in the case of a star's interior or atmosphere.

The range of physical conditions in interstellar space is enormous. The densest interstellar clouds have densities of perhaps 10^4 to 10^6 particles per cubic centimeter. (Compare this with the Earth's atmospheric density of about 2×10^{19} particles per cubic centimeter!) These clouds are very cold, with temperatures typically between 20 and 50 K. The less dense "diffuse clouds" have densities of 1 to 1,000 particles per cubic centimeter and temperatures of 50 to 150 K. Most of the volume of space is filled by

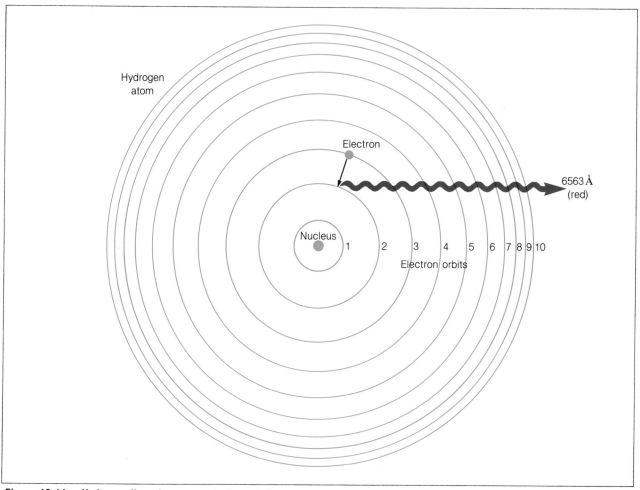

Figure 16.14   Hydrogen lines from an emission nebula.   In heated interstellar gas, electrons and protons combine to form hydrogen atoms, usually with the electron initially in an excited state. It then drops down to the lowest state, emitting photons at each step. The strongest emission at visible wavelengths corresponds to the jump from level 3 to level 2 and has a wavelength of 6563 Å. This is why emission nebulae appear red. (The energy level spacings are not drawn to scale.)

even more tenuous material, with a density as low as 10^{-4} particles per cubic centimeter and a temperature as high as 100,000 to 1 million K!

The enormously energized hot intercloud medium is heated by the blast waves from supernova explosions and the rapid outflows from hot stars with winds. The space between the obvious clouds is a violent place filled with superheated gas. The clouds themselves are often in motion, pushed around by the same forces that provide the energy for the intercloud heating. In the next section, we will learn more about how the interstellar material is distributed in the galaxy.

SPIRAL STRUCTURE AND THE 21-CENTIMETER LINE

So far we have spoken of the galactic disk as though it were a uniform, featureless object, but we know that this picture is not completely accurate. The Milky Way is a spiral galaxy, and if we could see it face-on, we would see the characteristic pinwheel shape normally found in galaxies of this type (Fig. 16.16).

Figure 16.15   The red glow of hydrogen gas in an H II region.   The red color of this cloud of hot gas is due to emission by hydrogen atoms. This is a nebula designated IC2944-48 (© 1992 Anglo-Australian Telescope Board)

Figure 16.16   Spiral structure.   Like our galaxy, this galaxy has prominent spiral arms, because hot, luminous young stars tend to be concentrated in the arms. This is the galaxy NGC 300, in Sculptor. (© 1992 AAT Board; photo by D. Malin)

A common misconception about the spiral structure in the Milky Way (and other spiral galaxies) is that there are few stars between the visible spiral arms. In reality, the density of stars between the arms is nearly the same as it is in the arms. The most luminous stars, however—the young, hot O and B stars—tend to be found almost exclusively in the arms. This is because the interstellar gas and dust tend to be concentrated in the arms, so young stars tend to be concentrated there also. Because these are the brightest stars, their presence in the arms makes the spiral structure stand out.

The fact that we live in a spiral galaxy was not easily discovered, again because we are located within it and see only a cross-sectional view. It was not until 1951 that investigations of the distribution of luminous stars revealed traces of spiral structure in the Milky Way, and even that technique was limited to a small portion of the galaxy. The obscuration caused by interstellar material limits our view of even the brightest stars to a local region, about 1,000 pc from the Sun at most.

A major advance in measuring the structure of the galaxy occurred in the same year, when the 21-centimeter radio emission of interstellar hydrogen atoms was first detected. It had been predicted that hydrogen should emit radiation at this wavelength, and the Americans E. M. Purcell and H. I. Ewen were the first to detect this emission, using a specially built radio telescope.

One great advantage of the hydrogen 21-centimeter emission for measuring galactic structure is its ability to penetrate the interstellar medium to great distances. Whereas the brightest stars can be detected at distances of at most one or two thousand parsecs, hydrogen clouds in space can be "seen" by radio telescopes from all the way across the galaxy, at distances of several thousand parsecs. Because hydrogen gas is the principal component of the interstellar medium and tends to be concentrated along

ASTRONOMICAL INSIGHT
21-Centimeter Emission from Hydrogen

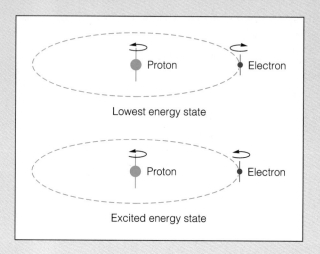

To understand how a hydrogen atom can emit radiation at a wavelength of 21 centimeters, we need to take a closer look at the structure of the atom.

A hydrogen atom consists only of a proton, which forms the nucleus, and a single electron in orbit about it. We have already learned that the electron can occupy a variety of possible orbits, each corresponding to a different energy state, and that a photon of light is absorbed by the electron if it moves from a low level to a higher one or is emitted if it drops from a high level to a lower one. The wavelength of the photon is related to the energy difference between the two electron energy levels; the greater the difference, the shorter the wavelength, and vice versa.

But the energy level structure of the electron is more complicated than previously described. The electron and the proton are both spinning, and the energy of the electron depends on whether it is spinning in the same or the opposite direction as the proton (see the figure). If both spin in the same (parallel) direction, the energy state is slightly greater than if they spin in opposite (antiparallel) directions.

As before, the electron can change from one state to the other by either emitting or absorbing a photon. The energy difference is so small, however, that the wavelength of the photon is very much longer than that of visible light. It is 21.1 centimeters.

Hydrogen atoms in space would normally tend to have the electron in the lowest possible energy state, with its spin antiparallel to that of the proton. Occasionally an atom will collide with another, however, causing the electron to jump to the higher state with the spin parallel to that of the proton. Following this the electron will spontaneously reverse itself, seeking to return to the lowest energy state, and when it does, it emits a photon of 21.1-centimeter wavelength. The probability that the electron will make the downward transition is very low, and the electron may remain in the upper state for as long as 10 million years before spontaneously dropping back down. There are so many hydrogen atoms in space, however, that at any given instant, many photons are being emitted. Hence radio telescopes capable of receiving this wavelength can trace the locations of hydrogen clouds throughout the galaxy and beyond.

the spiral arms, observations of the 21-centimeter radiation can be used to trace the spiral structure throughout the entire Milky Way (Fig. 16.17).

When a radio telescope is pointed in a given direction in the plane of the galaxy, it receives 21-centimeter emission from each segment of spiral arm in that direction. Because of differential rotation, each arm has a distinct velocity from the others that are closer to the center or farther out. Therefore, instead of a single emission peak at the laboratory wavelength of exactly 21.1 centimeters, what we see is a cluster of emission lines near this wavelength but separated from each other by the Doppler effect. By combining

measurements such as these with Oort's mathematical analysis of differential rotation, it was possible to reconstruct the spiral pattern of the entire galaxy. Radio emission from the carbon monoxide (CO) molecule has also been used to map the distribution and velocities of relatively dense interstellar clouds, providing further information on galactic rotation and spiral structure.

The pattern is much more complex than in some galaxies, where two arms elegantly spiral out from the nucleus. Instead the Milky Way consists of bits and pieces of a large number of arms, giving it a definite overall spiral form, but not a smoothly coherent one.

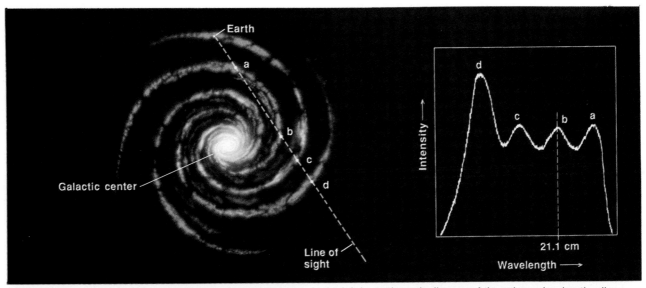

Figure 16.17   21-centimeter observations of spiral arms.   At the left is a schematic diagram of the galaxy, showing the direction in which a radio telescope might be pointed to record a 21-centimeter emission-line profile like the one at the right. The 21-centimeter line has many components, each corresponding to a distinct spiral arm, and each at a wavelength reflecting the Doppler shift between the velocity of that arm and the Earth's velocity.

As we will learn in Chapter 18, the differences between our type of spiral structure and the regular appearance of some other spiral galaxies may reflect different origins for the arms.

THE GALACTIC CENTER: WHERE THE ACTION IS

It is impossible to observe the central regions of our galaxy at visible wavelengths, because of obscuration by the intervening interstellar gas and dust (Fig. 16.18). From Shapley's work on the distribution of globular clusters, as well as Oort's analysis of galactic rotation, it was established by the 1920s that the center of the galaxy lies in the direction of the constellation Sagittarius. There we find immense concentrations of interstellar matter and a great concentration of stars. It is one of the richest regions of the sky to photograph, although the best views are seen only from the Southern Hemisphere.

The central region of the galaxy has different structures when examined on different scales. On the largest scale, the structure is dominated by a roughly spherical bulge consisting of old stars. This bulge is large enough (its radius is several hundred pc) that at some points we can see its outer regions, above and below the plane of the intervening disk (see Fig. 16.18). As we will see in the next chapter, this bulge

Figure 16.18   The galactic center.   The dark regions are dust clouds in nearby spiral arms. Because of obscuration, photographs such as this do not reveal the true galactic center, but only nearby stars and interstellar matter in the plane of the disk and the outer portions of the central bulge of the galaxy. (© Anglo-Australian Observatory photo by D. Malin)

probably represents the innermost, highly concentrated region of the galactic halo.

At the center of the bulge lies the nucleus of the galaxy. This region cannot be observed at visible wavelengths, but can be probed using infrared and radio telescopes, because the interstellar gas and dust do not block these wavelengths. Here again we see different structures when we look at different size scales. The use of interferometry in radio observations (see Chapter 4) allows us to observe details as small as a tiny fraction of a parsec, even though the galactic center is about 8,500 pc away from the Earth.

Radio data long ago revealed a central complex of glowing gas, which is called Sagittarius A. This was the first indication of highly energetic activity at the galaxy's core. Detailed observations subsequently showed that Sgr A consists of several smaller sources of emission, including one called Sgr A East, which has since been identified as a probable supernova remnant, and one called Sgr A West, which appears to be very close to the true center of the Milky Way. But Sgr A West itself consists of at least two distinct, tiny sources. One is called Sgr A*, which has a diameter of less than 0.001 parsec (that is, less than 100 AU, or roughly the size of the solar system). The other component of Sgr A West, lying very near to Sgr A*, is a bright infrared source, which has been identified as a dense cluster of massive stars, recently formed. Which of these two objects is the true center of the galaxy has been controversial, but most astronomers believe that Sgr A* is the more likely one. This, then, appears to be the point about which the entire galaxy rotates.

The nature of Sgr A* is not known. Studies of stellar motions in the vicinity indicate an increase in orbital velocity with decreasing distance from Sgr A; from these measurements it is possible to estimate the mass located at the central point. It is difficult to extend these observations close enough to the center to be certain of the mass, but the best indications are that there is a mass of about 1 million suns located within a very small volume. While some astronomers argue that this mass could exist in the form of a very dense cluster of stars, it is difficult to see how such a cluster could be present in such a small volume, and most believe that the object is instead a massive black hole. As we will see in later chapters (especially Chapter 20), many galaxies show evidence of having similar or more extreme activity at their nuclei, and in many cases it has been concluded that a supermassive black hole is the probable source of the energy.

Just outside of the tiny, energetic source that represents the galactic center lies a fantastic "mini spiral" structure, which shows up in both radio and infrared

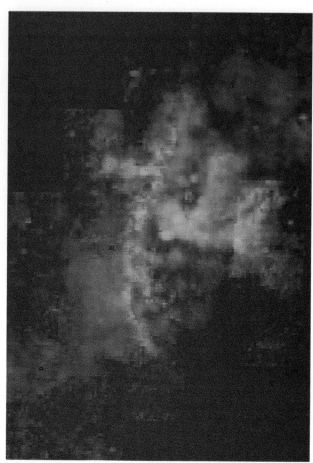

Figure 16.19   The central portion of the galactic nucleus.
This false-color infrared image shows the "mini spiral" in blue, ionized gas in green, and the dense, hot shell in red. The diameter of the shell is only about 5 pc. (AAT Board; courtesy of D. A. Allen)

observations (Fig. 16.19). The gas in this region is ionized, but the source of the heating is not well understood. The mini spiral is not part of the overall spiral structure of the galaxy; its size scale is very much smaller (its extent is only about 2 parsecs, as compared with thousands of parsecs for the spiral arms), and the mini spiral is tilted with respect to the plane of the galactic disk. Surrounding the mini spiral is a thin shell of gas which appears to form a boundary between the hot, ionized gas in the central region, and the colder, denser gas lying outside (this shell appears as a red glowing boundary in Fig. 16.19).

On a somewhat larger size scale, another very interesting structure is seen (Fig. 16.20). High-resolution radio maps reveal a large, arc-like structure which projects out of the plane of the galactic disk. This structure consists of filamentary arcs which lead astronomers to suspect that they are supported by magnetic forces, although it is not understood how

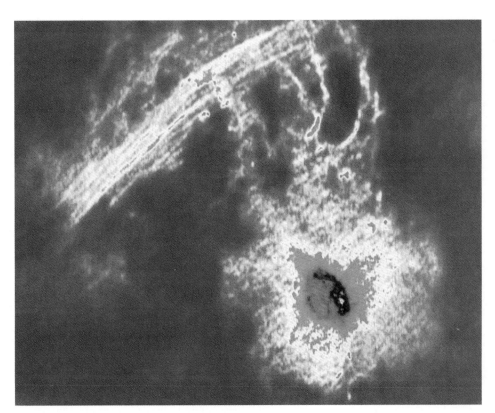

Figure 16.20   Wispy structure outside of the nucleus. This image, made at a wavelength of 6 centimeters (where thermal continuum emission is measured, rather than a spectral feature such as the 21-centimeter line), shows an arc of gas extending some 200 pc above the plane of the galactic disk. (F. Yusef-Zadeh, M. Morris, and D. Chance)

and why the galactic magnetic field would project nearly perpendicular to the plane of the disk.

The energetic activity at the core of the galaxy is reflected in structures much farther out as well. At a distance of about three kiloparsecs from the center lies a spiral arm which is moving outward at a speed of about 50 km/sec. This expansion, measured through the Doppler shift of the 21-cm emission line as well as absorption lines of the molecule CO, may be the result of a long-ago explosive event at the galactic center. Even farther out, some 5 or 6 kiloparsecs from the center, is a ring-like structure encircling the entire galaxy, which contains a high concentration of dense clouds. It is very difficult to get a clear picture of this "6-kpc ring," but one possible interpretation is that it is formed of swept-up material from an early outburst at the galactic center.

The overall view we get of the nucleus of the Milky Way is that it is a region of great turmoil, a place where star formation has taken place recently, in contrast with the rest of the central bulge, and a place where explosive events may have occurred in the past. In addition, there is a mysterious, very compact energy source at the very center which might be a million-solar mass black hole. As we will see in later chapters, it is possible that all of these forms of energetic activity are related to even more spectacular happenings at the cores of many other galaxies.

GLOBULAR CLUSTERS REVISITED

We have not yet said much about the globular clusters, the gigantic spherical conglomerations of stars that orbit the galaxy. A large globular cluster can be an impressive sight when viewed with a small telescope, and a number of them are popular objects for astronomical photography (Fig. 16.21). Well over 100 of these clusters have been cataloged; no doubt many more are obscured from our view by the disk of the Milky Way.

A single cluster may contain several hundreds of thousands of stars and have a diameter of 10 to 20 pc. The mass is usually in the range of several hundred thousand to a few million solar masses. To contain so many stars in a relatively small volume, globular clusters must be very dense compared with the galactic disk. The average distance between stars in a globular cluster is only about 0.1 pc (recall that the nearest star to the Sun is more than 1 pc away). If the Earth orbited a star in a globular cluster, the nighttime sky would be a spectacular sight, with hundreds of stars brighter than first magnitude.

An H-R diagram for a globular cluster (Fig. 16.22) looks rather peculiar compared with the familiar diagram for stars in the galactic disk. The main sequence

Figure 16.21   A globular cluster.   This cluster, designated NGC 6723, contains hundreds of thousands of stars. (© 1992 AAT Board; photo by D. Malin)

is almost nonexistent, having stars only on the extreme lower portion. On the other hand, many red giants and a number of blue stars lie on a horizontal sequence extending from the red giant region across to the left of the diagram. Together all this evidence points to a very great age for globular clusters. As we saw in Chapter 14, in a cluster H-R diagram the point where the main sequence turns off toward the red giant region indicates the age of the cluster; the lower the turnoff point, the older the cluster. Using this technique to date globular clusters leads typically to age estimates of 14 to 16 billion years, comparable to the accepted age of the galaxy itself. Thus globular clusters are among the oldest objects in the galaxy. As we shall see, their presence in the galactic halo provides important data on the early history of the galaxy.

The sequence of stars extending across the globular cluster H-R diagram from the red giant region to the left is called the **horizontal branch.** Most likely, these are stars that have completed their red giant stages (as discussed in Chapter 14) and are moving to the left on the diagram, perhaps on their way to becoming white dwarfs.

Some globular clusters have been found to contain X-ray sources in their centers (Fig. 16.23). In a few cases these are X-ray "bursters," or neutron stars that are slowly accreting new matter on their surfaces, probably in binary systems where the companion star is losing mass (see Chapter 15). The other globular cluster X-ray sources may be binaries with mass transfer onto a compact stellar remnant, but without the outbursts typical of bursters.

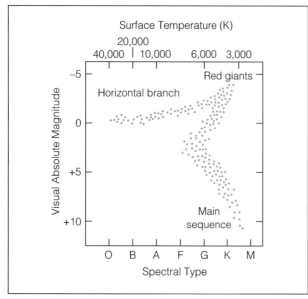

Figure 16.22   The H-R diagram for a typical globular cluster.   The main sequence branches off at a point far down from the top, indicating the great age of the cluster.

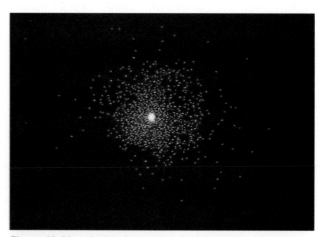

Figure 16.23   A globular cluster X-ray source.   Several globular clusters have been found to contain intense X-ray sources at their centers. This X-ray image obtained with the *Einstein Observatory* clearly shows the central source of a globular cluster. (Harvard-Smithsonian Center for Astrophysics)

A MASSIVE HALO?

The halo of the Milky Way galaxy has traditionally been envisioned as a very diffuse region, populated by a scattering of dim stars and the giant globular clusters. There was little or no evidence of any substantial amount of interstellar material, and the halo was assumed not to contain a significant fraction of the galaxy's mass.

Some of these ideas are changing as a result of very recent discoveries, and astronomers now believe that the halo of the galaxy may be much more extensive and massive than previously thought (Fig. 16.24). We mentioned in the earlier section of this chapter that radio observations have revealed an unexpectedly high amount of mass in the outer portions of the disk. This indicates that the galaxy is more extensive than previously believed. At the same time, other evidence (to be described in later chapters) led to a general expectation that many galaxies may have large quantities of matter in their halos. Finally, in the early 1980s, ultraviolet observations revealed a large amount of very hot, rarefied interstellar gas in the halo of our galaxy, probably extending several thousand parsecs above and below the plane of the disk. This gas is very turbulent, with clouds traveling at speeds of several hundred kilometers per second, and is highly ionized, indicating a temperature of 100,000 K or more.

The observed gas falls far short of supplying all the mass that is required to explain the rotation curve of the disk, however, and today astronomers speak of the "dark matter" in the halo. The form of this unseen material is the subject of intense speculation; suggestions include large numbers of low-mass stars (or sub-stellar objects known as brown dwarfs), or large numbers of black holes and neutron stars left over from a very early population of massive stars. Some even suggest that the halo of our galaxy contains a concentration of previously unknown subatomic particles; such particles have also been invoked to account for the vast quantities of dark matter that apparently permeate the entire universe (this is discussed in Chapters 18 and 21). Whatever the form of the unseen matter in the halo, it is intriguing to think that the vast conglomeration of visible stars and interstellar material in the Milky Way represent only a small fraction of its total mass.

PERSPECTIVE

We have discussed the anatomy of the galaxy, particularly its overall structure and motions. The solar system is located some 8 kpc from the center of a flattened rotating disk containing a few hundred billion stars. Some 10 to 15 percent of the mass of the disk is in the form of interstellar gas and dust, the interstellar medium serving as the source of new stars and the repository of material ejected by aging stars. The structure of the disk has been unraveled with the help of radio and infrared observations. It consists of spiral arms—delineated by bright, hot stars and interstellar gas—and a central nucleus that contains relatively dim red stars and a mysterious, energetic gremlin that stirs up the core region. Many questions remain about the structure and extent of our galaxy; the *Hubble Space Telescope* will help to answer some of them.

We have yet to discuss the workings of the galaxy; we have seen what it is like, but have said little about why. This will be the subject of the next chapter.

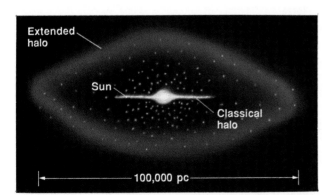

Figure 16.24   The extensive halo of the Milky Way.   This drawing illustrates the possible shape and scale of the very extended galactic halo, whose presence has been inferred from the shape of the galactic rotation curve (see Fig. 16.8) and from absorption lines formed by interstellar gas.

SUMMARY

1. The Milky Way is a spiral galaxy, consisting of a disk with a central nucleus and a spherical halo where the globular clusters reside. The disk is about 30 kpc in diameter.

2. Distances within the Milky Way can be determined by measuring variable star periods and applying the period-luminosity relation.

3. The true size of the Milky Way and the Sun's location within it were difficult to determine because the view from our location within the disk is obscured by interstellar extinction.

4. Star counts seem to indicate that the Sun is in the densest part of the Milky Way, but measurements of the distribution of globular clusters and analysis of stellar motions show that the Sun is very far (about 7–8 kpc) from the center of the galaxy. The discrepancy was resolved when it was discovered that interstellar extinction affects the star counts.

5. Stellar orbital speeds are nearly constant with distance from the galactic center, indicating that the disk is not rigid (except perhaps in a very small inner region), but at the same time that the stars do not act as completely free particles in accordance with Kepler's laws.

6. The mass of the galaxy, determined by the application of Kepler's third law to the Sun's orbit, is roughly 10^{11} solar masses. This technique ignores any mass that resides in the halo or in the galactic plane outside the Sun's position.

7. Between 10 and 15 percent of the mass of the disk is in the form of interstellar matter, including both dust particles and gas. This medium ranges from dark, cold clouds, where the gas is molecular and stars form, to more diffuse clouds, and to a very hot intercloud medium with temperatures as high as 1 million K. The interstellar material is stirred up by the supernovae, which account for the very hot regions and for motions of interstellar clouds.

8. The spiral structure of our galaxy is most easily and directly measured through radio observations of the 21-centimeter line of hydrogen atoms in space. The spiral pattern is complex, with many segments of spiral arms.

9. A variety of evidence indicates that there is chaotic, energetic activity associated with the central core of our galaxy. The data show that there is a compact, massive object there, and the best explanation is that it is a massive black hole.

10. The globular clusters that inhabit the halo of the galaxy are very old and therefore provide information on the early history of the galaxy.

11. The halo of the galaxy contains most of the total mass, in some invisible form referred to as dark matter. Speculations as to its nature include large numbers of low-mass stars or brown dwarfs, a quantity of black holes or neutron stars, or new kinds of subatomic particles.

REVIEW QUESTIONS

1. Compare the reasoning of Shapley in deciding that the globular clusters must be concentrated around the galactic center with that of Aristarchus who, more than 2,000 years earlier, deduced that the Sun is at the center of the solar system.

2. Explain why the 21-centimeter line of atomic hydrogen is a powerful tool for mapping the structure of the galaxy.

3. How can astronomers distinguish between a cool star that is intrinsically red and a hotter star that appears just as red because of interstellar dust?

4. How can absorption lines formed in the spectrum of a star by interstellar gas be distinguished from absorption lines formed in the atmosphere of the star itself?

5. Suppose that interstellar obscuration makes a star appear 5 magnitudes fainter than it would otherwise. If the star has apparent magnitude +17, what would its apparent magnitude be if it were not obscured by dust? If the star's absolute magnitude is +2, how far away is it? Compare your answer with what you would find if you did not correct the apparent magnitude for the effects of dust.

6. Summarize the evidence that a massive black hole might exist at the core of the galaxy.

7. How do we know that globular clusters are very old relative to the rest of the galaxy?

8. Given the great age of the globular clusters, do you think their chemical composition might differ from that of younger stars?

9. If there is an object at the galactic center with a mass equal to 1 million solar masses, what is the orbital period for a star 10 pc from this object? If this star's orbit is circular, what is its orbital speed? Comment on the relationship of your result to observations suggesting the presence of a massive black hole at the galactic core.

10. To appreciate what the nighttime sky might be like if the Earth orbited a globular cluster star, calculate the apparent magnitudes of the first five stars in the list of the brightest stars (Appendix 8) if each star were only 0.1 pc away (the average distance between stars in a globular cluster).

ADDITIONAL READINGS

Allen, D. A., and V. Meadows. 1992. The Huntsman nebula. *Astronomy* 20(9):38.

Blitz, L. 1982. Giant molecular cloud complexes in the galaxy. *Scientific American* 246(4):84.

Bok, B. J. 1984. A bigger and better Milky Way. *Astronomy* 12(1):6.

Bok, B. J., and P. Bok. 1981. *The Milky Way.* Cambridge, Mass.: Harvard University Press.

Chevalier, R. A., and C. L. Sarazin. 1987. Hot gas in the universe. *American Scientist* 75:609.

Clark, G. O. 1985. Ancients of the universe: Globular clusters. *Astronomy* 13(5):6.

Dame, T. M. 1988. The molecular Milky Way. *Sky and Telescope* 76:22.

de Boer, K. S., and B. D. Savage. 1982. The coronas of galaxies. *Scientific American* 247(2):54.

Eicher, D. J., and D. Higgins. 1987. The secret world of dark nebulae. *Astronomy* 15(9):46.

Geballe, T. R. 1979. The central parsec of the galaxy. *Scientific American*, 241(1):52.

Gehrz, R. D., D. C. Black, and P. M. Solomon. 1984. The formation of stellar systems from interstellar molecular clouds. *Science* 224(4651):823.

Gingerich, O., and B. Welther. 1985. Harlow Shapley and the Cepheids, *Sky and Telescope* 70(6):540.

Goldsmith, D., and N. Cohen. 1991. The great molecule factory in Orion. *Mercury* 20(5):148.

Greenberg, J. M. 1984. The structure and evolution of interstellar grains. *Scientific American* 250(6):124.

Hodge, P. 1986. *Galaxies.* Cambridge, Mass.: Harvard University Press.

————. 1987. The local group: Our galactic neighborhood. *Mercury* 6(1):2.

Kaler, J. B. 1982. Bubbles from dying stars. *Sky and Telescope* 63(2):129.

King, I. R. 1985. Globular clusters. *Scientific American* 252(6):78.

Malin, D. 1987. In the shadow of the horsehead. *Sky and Telescope* 74(3):253.

Mulholland, D. 1985. The beast at the center of the galaxy. *Science 85* 6(7):50.

Oort, J. 1992. Exploring the nuclei of galaxies (including our own). *Mercury* 21(2):57.

Smith, D. H. 1990. Seeking the origins of cosmic rays. *Sky and Telescope* 79:479.

Townes, C. H., and R. Genzel. 1990. What is happening in the center of our galaxy? *Scientific American* 262(4):46.

Tucker, W., and K. Tucker. 1989. Dark matter in our galaxy. *Mercury* 18(1):2 (Part 1); 18(2):51 (Part 2).

Verschuur, G. L. 1987. Molecules between the stars. *Mercury* 16(3):66.

————. 1992. Star dust. *Astronomy* 20(3):46.

————. 1992. Interstellar molecules. *Sky and Telescope* 83(4):379.

White, R. E. 1991. Globular clusters: fads and fallacies. *Sky and Telescope* 81(1):24.

ASTRONOMICAL ACTIVITY
Stellar Statistics

Taking a detailed and complete census of all the stars in the galaxy is impossible, but nevertheless it is not so difficult to get a good idea of the relative numbers of stars of different types. One way to do this is to count the stars in a representative volume of space and determine the relative numbers of different types. Then, by assuming that the galaxy as a whole has a similar distribution of types, you have learned something about the large-scale statistics of stars in the galaxy. This technique is very similar to the polling methods used to determine national opinions and preferences; a small sample of people is questioned, and then the results are applied to the overall population of the nation.

You can use information in Appendix 8 to carry out a limited census of stellar types in the galaxy. The second table in Appendix 8 is a list of nearby stars; it is a complete (or nearly so) compilation of all stars within 5 pc of the Sun. This is an extremely small volume of space to use as representative of the galaxy as a whole, but even so you will find some meaningful trends.

Draw an H-R diagram for the stars in the table, placing spectral type on the horizontal axis and absolute visual magnitude on the vertical (note that your scale will have to extend as far as +17 to include stars fainter than those shown on the H-R diagrams in the text). Omit white dwarfs and stars for which the spectral type is not given.

What does your diagram tell you about the relative numbers of different types of stars in the galaxy? Look at the first list in Appendix 8, which provides data on the 50 brightest stars in the sky. Would the H-R diagram for this sample of stars look the same as the one you just drew for the nearest stars? Why would the list of brightest stars not be a representative sample of the relative numbers of stars of different types in the galaxy?

The Formation and Evolution of the Galaxy

A galaxy similar to the Milky Way (© 1980 Anglo-Australian Observatory; photo by D. Malin)

PREVIEW

Just as individual stars evolve and change with time, so too does a galaxy like the Milky Way. In fact, it is the evolution of individual stars that causes the galaxy to evolve; as stars burn their nuclear fuel and return processed material to the interstellar medium, new generations of stars that form from this material start with enriched chemical compositions, which are then further altered as these stars evolve and die. As subsequent generations of stars form, live, and die, the overall composition and form of the Milky Way itself are altered. In this chapter you will learn how the galaxy is thought to have formed, how it has changed, and how it may continue to evolve. As in the previous chapter, we will examine this from a historical perspective, beginning with a discussion of the observational evidence before moving on to a treatment of the life story of the galaxy that is deduced from the data.

The galaxy we observe today is the end product of some 10 to 20 billion years of evolution. A variety of processes have shaped it, and many have left unmistakable evidence of their action. In this chapter we will discuss the major influences on the galaxy, along with the developments they have brought about.

We have laid the groundwork that will enable us to understand one of the most elegant concepts in the study of astronomy—the interplay between stars and interstellar matter and the long-term effects of this cosmic recycling on the evolution of the galaxy.

STELLAR POPULATIONS AND ELEMENTAL GRADIENTS

To begin this story, we focus first on the overall structure of the galaxy (Fig. 17.1) and the types of stars found in various regions. We have already seen that most of the young stars in the galaxy tend to lie along the spiral arms. Indeed, it is the brilliance of the hot, massive O and B stars and their associated H II regions that delineates the arms, making them stand out from the rest of the galactic disk. We have also noted that in the nuclear bulge of the galaxy, as well as in the globular clusters and the halo in general, there are few young stars and relatively little interstellar material.

Figure 17.1   A spiral galaxy similar to the Milky Way.   This vast conglomeration of stars and interstellar matter, designated M83, is following its own evolutionary course, just as individual stars do. Note that the spiral arms tend to be blue, while the central bulge is yellow. These colors indicate the types of stars that dominate in the different parts of the galaxy. The bar-shaped central region and somewhat filamentary spiral arm structure in this galaxy may bear a close resemblance to the Milky Way. (© 1977 AAT Board; photo by D. Malin)

Careful scrutiny has revealed a number of distinctions between the stars that lie in the disk, particularly those in the spiral arms, and the stars in the nucleus and halo. Analysis of the chemical compositions of the stars has shown that those in the halo and nucleus tend to have relatively low abundances of heavy elements such as metals, while in the vicinity of the Sun (and in spiral arms in general), these species are present in greater quantity. Nearly all stars are dominated by hydrogen, of course, but in the halo stars the heavy elements are present in even smaller traces than in the spiral arm stars. The relative abundance of heavy elements in a halo star may typically be a factor of 100 below that found in the Sun; if iron, for example, is 10^{-5} as abundant as hydrogen in the Sun, it may be only 10^{-7} as abundant as hydrogen in a halo star.

Such evidence has given us a picture of the galaxy with two distinct groups of stars, those in the halo on the one hand, and those in the spiral arms on the other (Fig. 17.2). Under the assumption that halo and disk stars represent truly distinct groupings, the concept of **stellar populations** was developed in the 1940s. In this scheme, halo stars are designated **Population II,** and those in the spiral arms **Population I** (Table 17.1). The Sun is thought to be typical of Population I, and its chemical composition is usually adopted as representative of the entire group.

Almost every time an attempt is made to classify astronomical objects into distinct groups, the boundaries between them turn out to be indistinct. There are usually intermediate objects whose properties fall between categories, and the stellar populations are no exception. For example, the stars in the disk of the galaxy that do not fall strictly within the spiral arms have properties intermediate between those of Population I and Population II and are often called simply the **disk population.** Furthermore, there are gradations within the two principal population groups, and we speak therefore of "extreme" or "intermediate"

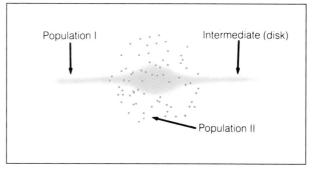

Figure 17.2   The distribution of Populations I and II.   This cross section of the galaxy shows that Population II stars lie in the halo and central bulge, whereas Population I stars inhabit the spiral arms in the plane of the disk. Intermediate population stars are distributed throughout the disk.

Population I or Population II objects. It is much more accurate to view the stellar populations as a smooth sequence of stellar properties represented by very low heavy element abundances at one end and Sun-like abundances at the other.

The differences in chemical composition from one part of the galaxy to another provide astronomers with very important clues in their efforts to piece together the history of the galaxy. Changes that occur gradually over substantial distances are referred to as **gradients,** and the variations of stellar composition from one part of the galaxy to another are therefore referred to as **abundance gradients.** There is a gradient of increasing heavy element abundances from the halo to the disk (Fig. 17.3). Within the disk itself, a similar gradient extends from the outer to the inner portions; thus, the central regions of the galaxy have relatively high abundances of heavy elements.

Most of the stars near the Sun belong to Population I; they are disk stars orbiting the galactic center in approximately circular paths, like the Sun. There are also a few Population II stars near the Sun, and they

TABLE 17.1   Stellar Populations

	Population I	Population II
Age	Young to intermediate	Old
Location	Disk, spiral arms	Halo, central bulge
Composition	Normal metals	Low in heavy elements
Constituents	Disk, arm stars	Stars low in heavy elements
	O, B stars	High-velocity stars
	Interstellar matter	Globular clusters
	Type I Cepheids	Type II Cepheids
	Type II supernovae	RR Lyrae stars
	The Sun	Type I supernovae
		Planetary nebulae

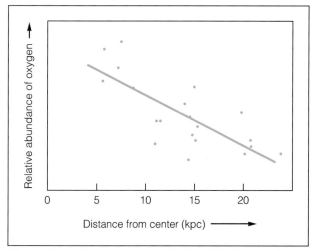

Figure 17.3   An abundance gradient.   This diagram shows how the abundance of oxygen (relative to hydrogen) varies with distance from the center of the Andromeda galaxy. More stellar generations have lived and died in the dense regions near the center, so nuclear processing is further advanced in the central region than it is farther out. (Data from W. P. Blair, R. P. Kirschner, and R. A. Chevalier, 1982, *Astrophysical Journal* 254:50)

are distinguished by several properties in addition to their compositions. Most easily recognized are their motions, which depart drastically from those of Population I stars. As a rule, Population II stars do not follow circular paths in the plane of the disk, but instead follow highly eccentric orbits that are randomly oriented (Fig. 17.4), much like cometary orbits in the solar system. Thus, most of the Population II stars that are seen near the Sun are just passing through the disk from above or below it. The Sun and the other Population I stars in its vicinity move in their orbits at speeds around 220 km/sec, so the Sun moves very rapidly with respect to the Population II stars that pass through its neighborhood in a perpendicular direction. From our perspective here on the Earth, these Population II objects therefore are classified as **high-velocity stars.**

Historically, high-velocity stars were discovered and recognized as a distinct class even before the differences between Population I and Population II were enumerated. It was only later that these objects were equated with Population II stars following randomly oriented orbits in the galactic halo.

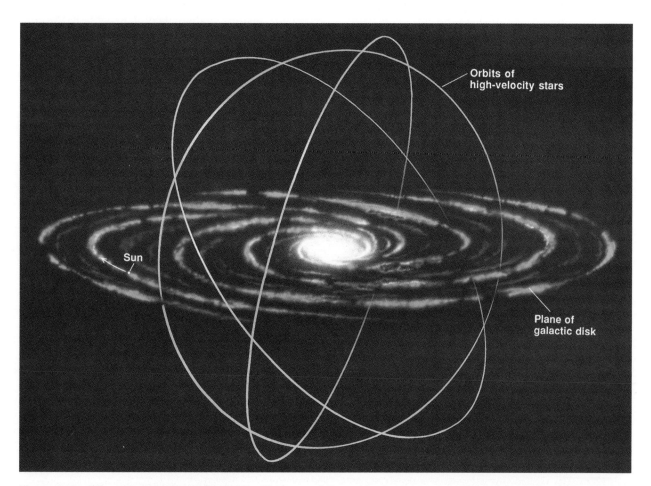

Figure 17.4   The orbits of high-velocity stars.   These stars follow orbits that intersect the plane of the galaxy. When such a star passes near the Sun, it has a high velocity relative to us, because of the Sun's rapid motion along its own orbit.

The systematic differences between the stellar populations arose from differing conditions at the times when they formed. In this regard it is important to recall that globular clusters, representative of Population II, are among the oldest objects in the galaxy, while Population I includes young, newly formed stars. We will follow this discussion further in a later section.

STELLAR CYCLES AND CHEMICAL ENRICHMENT

The fact that stars in different parts of the Milky Way have distinctly different chemical makeups is readily explained in terms of stellar evolution. Recall (from discussions in Chapter 14) that as a star lives its life, nuclear reactions gradually convert light elements into heavier ones. The first step occurs while the star is on the main sequence, when hydrogen nuclei deep in its interior are being fused into helium. Later stages, depending on the mass of the star, may include the fusion of helium into carbon and possibly the formation of even heavier elements by the addition of further helium nuclei. The most massive stars, which undergo the greatest number of reaction stages, also form heavy elements in the fiery instants of their deaths in supernova explosions.

All of these processes work in the same direction: They act together to gradually enrich the heavy element abundances in the galaxy. Material is cycled back and forth between stars and the interstellar medium, and with each passing generation, a greater supply of heavy elements is available. As a result, stars formed where a lot of stellar cycling has occurred in the past are born with higher quantities of heavy elements than those formed where little previous cycling has occurred. Stars formed before much

cycling occurred therefore tend to have low abundances of heavy elements. This explains why old stars belong to Population II; they formed out of material that had not yet been chemically enriched.

Population I stars, such as the Sun, are those that condensed from interstellar material that had previously been processed in stellar interiors and in supernova explosions. Therefore, as a rule, they formed at moderate to recent times in the history of the galaxy and are not as ancient as Population II objects.

Besides varying with age, elemental abundances can also vary with location. As noted in the preceding section, a distinct abundance gradient exists within the disk of the galaxy, with the stars nearer the center having higher abundances of heavy elements than those farther out. This distribution reflects enhanced stellar cycling near the center rather than differences in age. The density of stars and interstellar material is highest in the central portion of the disk, and therefore this region has been the site of relatively active stellar processing, with star formation proceeding at a greater rate than in the less dense outer portions of the disk. Over the lifetime of the galaxy, more generations of stars have lived and died in the central region than in the rarefied outer reaches.

THE MAINTENANCE OF SPIRAL ARMS

In Chapter 16 we described the structure of the galaxy and how its spiral nature was discovered and mapped, but we said nothing about how the spiral arms formed, nor why they persist. Both questions are important, for it is clear that if the arms were simply streamers trailing along behind the galaxy as it rotates, they should wrap themselves up tightly around the nucleus (Fig. 17.5), like string being wound into a ball. Because the galaxy is old enough

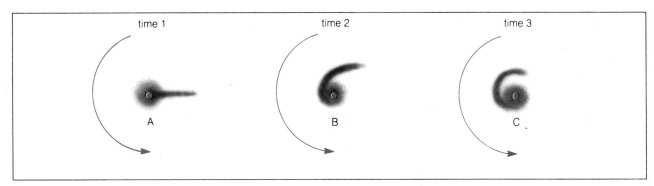

Figure 17.5   The windup of spiral arms.   If spiral arms were simply streamers of material attached to a rotating galaxy, they would wind tightly around the nucleus within a few hundred million years. The galaxy is much older than that, so some other explanation of the arms is needed.

ASTRONOMICAL INSIGHT
The Missing Population III

As the text explained, the two populations of stars should be viewed as having a range of stellar characteristics, such as age and metal content, rather than being separated into two completely distinct classes of stars. The most extreme Population II stars—those with the lowest metal abundances and the greatest ages—represent one end of the range, whereas the youngest stars in spiral arms, containing the highest metal abundances, represent the other end of the range.

The fact that the extreme Population II stars contain *some* heavy elements, even though the amount may be only 1 percent of the metal content of Population I stars, has proven difficult to explain. The picture of galactic formation and evolution presented in this chapter has portrayed the Population II stars as the first stars to form from the collapsing cloud of gas that was to become the galaxy. This cloud of gas is assumed to be primordial—that is, never to have been involved in stellar formation and processing. Therefore, the composition of the cloud should have consisted almost entirely of hydrogen and helium, with *no* heavy elements. The question, then, is why the oldest stars we observe, those belonging to extreme Population II, have the metal content that they do. Even these stars must have formed

from material that had previously been processed through stars, becoming somewhat enriched with heavy elements.

The missing population of stars formed from completely unprocessed material and having virtually zero heavy element abundances has become known as **Population III.** Several astronomers have spent considerable time and effort in the attempt to find such stars, with no success.

No widely accepted explanation for the lack of Population III stars has been devised yet, but there have been suggestions. One is that before the pregalactic cloud collapsed, there was a brief episode of star formation and nuclear processing, in which a number of very massive, short-lived stars were formed and quickly evolved to explode in supernovae, creating heavy elements. These heavy elements then became part of the gas that collapsed to form the disk, so that stars formed during and after the collapse—the oldest stars we now observe—contained some heavy elements and represent what we know today as Population II. None of the original Population III stars remain, because they were all too massive and short-lived to survive for long.

The chief difficulty in this and other pictures in which an early star-formation episode occurred is to ex-

plain how star formation became so active at a time when the pregalactic material was still widely dispersed in a giant, tenuous cloud. Despite our lack of understanding of *how* this occurred, it seems probable that it did, since we find no Population III stars. Furthermore, independent evidence indicates that galaxies do undergo periods of intense star formation and processing; such episodes are observed in some other galaxies. Therefore, it is quite possible that our galaxy went through such a time early in its history. Several studies are under way today in an effort to understand how and why this may have occurred.

Another possibility has recently been suggested, however. Contrary to long-standing assumptions about nuclear reactions in the earliest moments of the formation of the universe, it is now thought possible that some heavy elements were formed then. If so, the "primordial" material from which our galaxy formed may have contained some heavy elements, thus explaining how the oldest stars could have been born with the observed composition of Population II stars. In this view, there never was a Population III. The recent developments on the theory of the universe that lead to this suggestion are discussed in Chapter 21.

to have rotated at least 40 to 50 times since it formed, something must be preventing the arms from winding up, or they would have done so long ago.

Maintaining the spiral arms—a distinct question from how they formed in the first place—is a very complex business, and we do not yet fully understand how it is done. The first successful theory appeared in 1960 and is still being refined. The essence of this theory is that the large-scale organization of the galaxy is imposed on it by wave motions. We have understood waves as oscillatory motions created by disturbances, and we know that waves can be trans-

mitted through a medium over long distances while individual particles in the medium move very little. In the case of water waves, for example, a floating object simply bobs up and down as a wave passes by, whereas the wave itself may travel a great distance. Here we are distinguishing between the wave and the medium through which it moves.

The waves that apparently govern the spiral structure in our galaxy are not transverse waves, like those in water, but are compressional waves, similar to sound waves and to certain seismic waves (the P waves; see Chapter 6 for an elaboration on the differ-

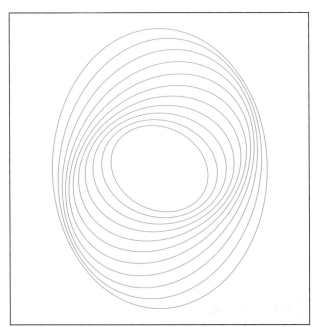

Figure 17.6   The density wave theory.   This drawing depicts the manner in which circular orbits are deformed into slightly elliptical ones by an outside gravitational force. The nested elliptical orbits are aligned in such a way that there are density enhancements in a spiral pattern. This pattern rotates at a steady rate and does not wind up more tightly.

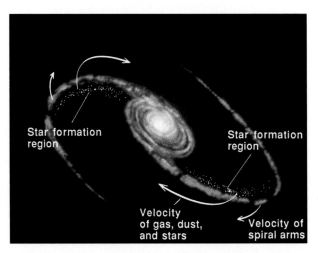

Figure 17.7   The effect of a spiral density wave on the interstellar medium.   As the wave moves through the medium, material is compressed, leading to enhanced star formation. The young stars, H II regions, and nebulosities along the density wave are what we see as a spiral arm.

ent types of waves). In this case the wave pattern consists of alternating regions of high and low density. When a compressional wave passes through a medium, the individual particles vibrate back and forth along the direction of the wave motion. In contrast to water waves, there is no motion perpendicular to the direction of wave motion.

The theory of galactic spiral structure that invokes waves as a means of maintaining the spiral arms is called the **density wave theory.** This theory supposes that there is a spiral-shaped wave pattern centered on the galactic nucleus, creating a pinwheel shape of alternating dense and relatively empty regions (Fig. 17.6). The density waves have more effect on the interstellar medium than on stars, so the spiral arms are characterized primarily by concentrations of gas and dust. These lead, in turn, to concentrations of young stars, because star formation is enhanced in regions where the interstellar material is compressed (Fig. 17.7).

In its simplest form, the wave pattern is double; that is, there are just two spiral arms emanating from opposite sides of the nucleus. Some galaxies have such a simple spiral structure, but many, including the Milky Way, are more complicated. The density wave theory allows the possibility of more arms, if the waves have shorter "wavelength," or distance between them (there is an alternative explanation for the somewhat disorganized spiral structure of galaxies like ours, which is discussed shortly).

The waves travel around the galaxy at a fixed rate that is constant from inner portions to the outer edge. Thus, while the outer portions of the arms appear to trail the rotation, in fact they move all the way around the galaxy in the same time period that the inner portions do. The waves are essentially rigid, in strong contrast with the motions of the stars. Just as in the case of water waves, the motion of the spiral density waves is quite distinct from the motions of individual particles (i.e., stars) in the medium through which the waves travel. In the Milky Way, individual stars orbit the galaxy at their own speed. In the inner part of the galactic disk, out to the Sun's position, the stars move faster than the density waves. This means that as a star circles the galaxy, every so often it will overtake and pass through a region of high density as it penetrates a spiral arm. Because their motion is slowed slightly when they are in a density wave, stars tend to become concentrated in the arms, just as cars on a highway become jammed together at a point where traffic flow is constricted. Farther out in the galaxy, the spiral density waves move faster than that of the individual stars.

The motion of an individual star about the galaxy is almost but not precisely circular. The spiral density waves impose slight oscillatory motions on the stars as they pass through the waves, with the result that the orbit of a star can be represented as an epicyclic motion (Fig. 17.8).

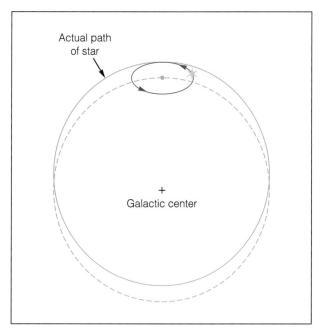

Actual path
of star

+
Galactic center

Figure 17.8   Epicyclic orbits.   The oscillatory motions created by spiral density waves cause individual stars to have orbits about the galaxy that are not perfect circles but are oval instead. The shape of the orbit may be viewed as a combination of a large circle and a small circle (epicycle). (Adapted from F. Shu, 1982, *The Physical Universe* [San Francisco: University Science Press], Fig. 12.12, p. 267.)

As we have noted, the main reason the spiral arms stand out is that they are the exclusive home of the hot, luminous young stars; these stars are found in the arms because star formation is enhanced there, where the interstellar gas is compressed. According to the density wave theory, the most active stellar nurseries should be located on the inside edge of the arms, where stars and gas catch up with and enter the compressed region. This is difficult to check observationally for our own galaxy, but it seems to be true of others.

The most luminous stars have such short lifetimes that they evolve and die before they have sufficient time to pull ahead of the arm where they were born; therefore, we find no bright, blue stars between the arms. Less luminous stars—those that live billions instead of only millions of years—can survive long enough to orbit the galaxy several times, and these stars become spread almost uniformly throughout the galactic disk, between the arms as well as in them. The Sun, for example, is old enough to have circled the galaxy some 15 to 20 times and therefore has passed through spiral arms and the intervening gaps on several occasions. It is only coincidence that the Sun is currently in a spiral arm; it is not necessarily the arm in which it formed.

The initial formation of the density wave that is the cause of the spiral arms is a separate question. A rotating disk of particles, such as the disk of a galaxy like ours, will naturally form spiral density waves if disturbed by an outside force (Fig. 17.9). This suggests that the spiral density waves in our galaxy were triggered by the gravitational influence of a neighbor galaxy, most likely the Magellanic Clouds, a pair of small galaxies that orbit the Milky Way.

Another possible mechanism for creating spiral density waves, with no help from neighboring galaxies, occurs in certain galaxies that have noncircular central bulges. In these **barred spiral galaxies** (see Chapter 18), the asymmetric shape of the central region can create the gravitational disturbance needed to initiate spiral density waves. Some calculations suggest that the barred spiral structure is the more natural form for a disklike galaxy to take, and that the presence of a massive, extended halo is what prevented our own galaxy from becoming a barred spiral instead of the "normal" spiral it is. If so, this suggests that other normal spiral galaxies may also have massive halos, whereas the barred spirals may not. In this context, it is very interesting that a recent study shows that the Milky Way may actually be a barred spiral (Fig. 17.10). A new analysis of gas motions in the outer galaxy indicates that the gas is orbiting a central bulge of stars that is elongated rather than round. The shape of the bulge has been likened to that of a watermelon. Thus our galaxy may be a mild form of barred spiral.

It is possible that in some spiral galaxies the structure is not governed by spiral density waves at all. The more chaotic and disorganized spirals, possibly including the Milky Way, consist of many fragments of spiral arms, and have no simple overall spiral pattern. The fragmentary spiral arms in these galaxies may be gigantic star-formation regions that have been stretched into arclike shapes by the differential rotation of the galaxy. If so, then in such galaxies the spiral structure must be regarded as a constantly changing and evolving pattern rather than the long-standing pattern that is maintained in other galaxies by density waves.

GALACTIC HISTORY

We are now in a position to tie together all the diverse information on the nature of the Milky Way and to develop a picture of its formation and evolution. Many important aspects are still uncertain, however, and some of these will be pointed out.

The pertinent facts that must be explained include the size and shape of the galaxy, its rotation, the

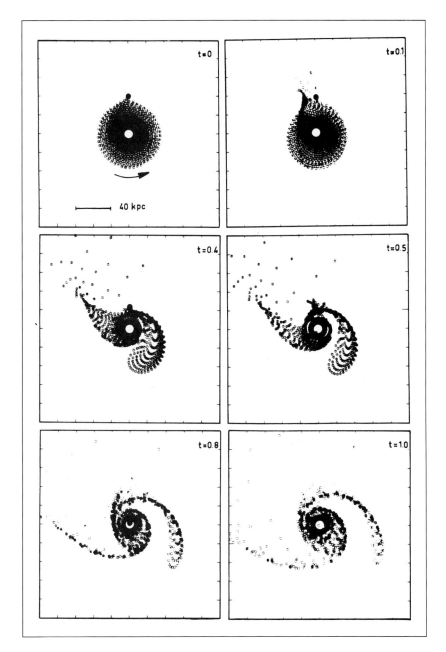

Figure 17.9   The creation of spiral structure.   This sequence of computer-generated models shows how a rotating disk forms spiral density waves when subjected to a gravitational force. (After T. M. Eneev, N. H. Kuzlov, and R. A. Sunyaev, 1973, *Astronomy and Astrophysics* 22:41)

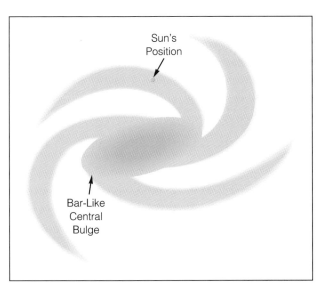

Figure 17.10   Is the Milky Way a barred spiral galaxy? This sketch illustrates a new picture of our home galaxy that has recently emerged as a result of the analysis of gas motions observed through the Doppler shifts in radio emission lines. As we will learn in the next chapter, about half of all spiral galaxies have barlike nuclei, so it may not be surprising that our own galaxy shows this trait. (data from L. Blitz, University of Maryland, and D. Spergel, Princeton University)

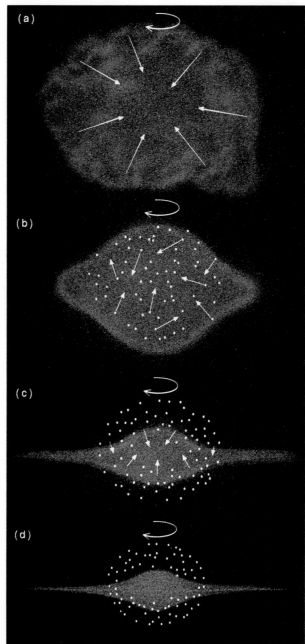

Figure 17.11   Formation of the galaxy.   The pregalactic cloud, rotating slowly, begins to collapse (a). The stars formed before and during the collapse have a spherical distribution and noncircular, randomly oriented orbits (b). The collapse leads to a disk with a large central bulge, surrounded by a spherical halo of old stars and clusters (c). The disk flattens further and eventually forms spiral arms (d). Stars in the disk have relatively high heavy element abundances because they formed from material that had been through stellar nuclear processing.

distribution of interstellar material, elemental abundance gradients, and the dichotomy between Population I and Population II stars in terms of composition, distribution, and motions. The task of fitting all the pieces of the puzzle together is made easier by the fact that we can reconstruct the time sequence—we know that stars with high abundances of heavy elements were probably formed more recently than those with lower abundances.

The oldest objects in the galaxy are in the halo, which is dominated by the globular clusters but contains a large (but unknown) number of isolated, dim, red stars as well. Estimates based on the main-sequence turnoff in the H-R diagrams for globular clusters indicate ages of between 14 and 16 billion years, and we conclude that the age of the galaxy itself is comparable (some globular clusters appear to be somewhat younger, as discussed in the next section).

The spherical distribution of the halo objects about the galactic center demonstrates that when they formed, the galaxy itself was round. Evidently the progenitor of the Milky Way was a gigantic spherical gas cloud, consisting almost exclusively of hydrogen and helium. Very early, perhaps even before the cloud began to contract (Fig. 17.11), the first stars and globular clusters formed in regions where localized condensations occurred. These stars contained few heavy elements and were distributed throughout a spherical volume, with randomly oriented orbits about the galactic center. Eventually the entire cloud began to fall in on itself, and as it did, star formation continued to occur, so that many stars were born with motions directed toward the galactic center. These stars assumed highly eccentric orbits, accounting for the motions observed today in Population II stars.

Apparently the pregalactic cloud was originally rotating, for we know that as the collapse proceeded, a disklike shape resulted. Rotation forced this to happen, just as it did in the contracting cloud that was to form the solar system (see Chapter 9). Rotational forces slowed the contraction in the equatorial plane, but not in the polar regions, so material continued to fall in there. The result was a highly flattened disk. Stars that had already formed before the disk took shape retained the orbits in which they were born; stars are unaffected by the fluid forces (e.g., viscosity) that caused the gas to continue collapsing to form a disk.

Stars that were formed after the disk had developed differed from their predecessors in at least two respects: they contained greater abundances of heavy elements, because by this time some stellar cycling had occurred, enriching the interstellar gas; and they were born with circular orbits lying in the plane of the disk. These are the primary traits of Population I stars.

Since the time of the formation of the disk, additional, but relatively gradual, changes have occurred. Further generations of stars have lived and died, continuing the chemical enrichment process, particularly in the inner region of the disk, and creating the chemical abundance gradient mentioned earlier. Apparently the enrichment process was once more rapid than it is today, because the present rate of star formation is too slow to have built up the quantities of heavy elements that are observed in Population I stars. At some time in the past, probably when the disk was forming, there must have been a period of intense star formation, during which the abundances of heavy elements in the galaxy jumped from almost none to nearly the present level. A large fraction of the galactic mass must have been cooked in stellar interiors and returned to space in a brief episode of stellar cycling whose intensity has not been matched since. (Today some galaxies are observed that seem to be in such a state of rapid stellar processing; these are called **starburst galaxies,** and they are discussed in Chapter 18).

Apparently, during this early phase of intense star formation, massive stars were created at a greater rate than they are today. Studies of star formation have shown that isolated dark clouds tend to form low-mass stars, while large star-formation regions, characterized by intense ultraviolet radiation and warm temperatures, tend to form stars of greater mass. Thus when intense star formation occurred early in the history of the galaxy, many massive stars were born, which quickly produced heavy elements and dispersed them into the interstellar medium.

If large numbers of massive stars did form early in the galaxy's history, then today there may be many compact stellar remnants such as neutron stars and black holes. As mentioned in Chapter 16, there could even be enough of these remnants left over to account for the large amount of invisible mass believed to exist in our galaxy.

We have now recounted almost all the events that led up to the present-day Milky Way, except for the formation of the spiral arms. It is not known when this took place, but it was probably soon after the formation of the disk itself. We see few gas-rich, disk-shaped galaxies without spiral structure, which leads to the conclusion that a disk galaxy containing large quantities of interstellar matter does not exist long in an armless state.

It is difficult to guess exactly what the future of the galaxy will be. The interstellar gas and dust are being consumed by star formation, but only gradually, so the general appearance of the galaxy may not change substantially for many billions of years. Perhaps in the very distant future, when most of the ingredients for new stars have been used up, our galaxy will become an armless disk containing only old stars and stellar remnants.

THE AGE OF THE GALAXY

The scenario we have described for the formation of the galaxy is thought to be generally correct, but some substantial uncertainties remain. Most of these have to do with the time scale on which the galaxy formed and the duration of certain stages along the way.

Estimates of the age of the galaxy using different techniques exhibit some discrepancies. We have already mentioned the use of globular cluster ages as an indicator of the age of the galaxy. The general idea is that the galaxy is probably comparable in age to the oldest systems within it—the globular clusters. This would not be correct, however, if there was some delay between the time the galaxy formed and the time the globular clusters developed. Recently we have received some hint that we do not fully understand the ages of the globular clusters.

The age of a cluster is determined from the main-sequence turnoff point, as described in Chapter 14. We did not emphasize, however, that while this method provides reliable *relative* ages, it relies on stellar evolution models to yield actual ages in years. The accuracy of the model calculations is always a bit uncertain, especially in cases of old clusters whose chemical composition is deficient in heavy elements.

Recent analyses of globular clusters in our galaxy reveal that they are probably not all the same age. Some appear to be as "young" as 11 billion years, while others are as old as 16 billion years. If the globular clusters really have such a wide range of ages, the formation of the galaxy was probably not as simple as outlined in the previous section. There must have been a rather extended period of time (perhaps 5 billion years) before the galaxy completed its contraction to a disklike shape. During this time the galaxy must have retained the roughly spherical shape of the halo, while the globular clusters formed here and there at different times.

It is not easy to explain why the galaxy would have undergone such a drawn-out early phase prior to collapsing to form a disk. One suggestion is that the galaxy formed from the conglomeration of several smaller systems of stars, which themselves formed earlier (and at different times, accounting for the spread in ages now observed). In this picture, our galaxy was created by the merger of a number of subgalaxies, systems of the stars much like globular clusters. Thus, rather than the entire galaxy forming from a single cloud of gas that collapsed, each of these subsystems would have formed separately from the collapse of a cloud. As we will see in the coming chapters, there is growing evidence that galaxies in general formed in this way; that the earliest stellar systems in the universe were smaller than galaxies, with galaxies forming later as the result of mergers of these smaller systems.

The age of the Milky Way can be estimated in at least two other ways, both of them independent of the globular cluster method. One method uses white dwarfs as age indicators. It is possible, through a statistical analysis of the luminosities of white dwarfs scattered around our part of the galaxy, to determine the age of the oldest (this depends on knowledge of the time required for white dwarfs to cool to the point of being too dim to be observed). This method yields a rather low estimate of the age of the galaxy—around 9 billion years. The significance is a bit unclear, however, since the white dwarfs near enough to be observed all lie in the local portion of the galactic disk, which is certainly not as old as the oldest portions of the galaxy.

The other method of age estimation also depends in part on theoretical calculations. From knowledge of the rates of the nuclear reactions that produce heavy elements, it is possible to estimate how old the galaxy must be in order to contain the quantities of these elements that exist today. This method involves substantial uncertainties and consequently does not provide very precise answers. It has been shown, however, that the current abundances of heavy elements are consistent with an age for the galactic disk of 11 to 15 billion years, in keeping with the age estimated from the oldest globular clusters.

It is not understood why the white dwarf and globular cluster methods of age determination provide such different answers. Certainly the picture we have developed of the formation and evolution of our galaxy is subject to further revision as this question is resolved and the wide range of globular cluster ages is further explored.

PERSPECTIVE

We have now seen our galaxy in a new light, as an active, dynamic entity. Individual stars orbit its center in individual paths, yet the overall machinery is systematically organized. The majestic spiral arms rotate at a stately pace, while stars and interstellar matter pass through them. We have come to understand the active processes of stellar cycling and chemical enrichment, which still occur today. In a constant turn-

over of stellar generations, the deaths of the old give rise to the births of the new, while the violence of their death throes energizes a chaotic interstellar medium.

The lessons we have learned by examining our own Milky Way will be remembered as we move out into the void, probing the distant galaxies.

SUMMARY

1. Stars in the galactic halo have low abundances of heavy elements and are referred to as Population II stars, whereas those in the spiral arms of the disk have "normal" compositions and are called Population I stars.

2. There are gradients of increasing heavy element abundance from the halo to the disk, and from the outer to the inner portions within the disk.

3. Population II stars have randomly oriented, highly eccentric orbits, whereas Population I stars have nearly circular orbits that lie in the plane of the disk. Population II stars, when passing through the disk near the Sun's location, are seen as high-velocity stars.

4. The variations in heavy element abundance from place to place within the galaxy reflect variations in age: Very old stars, formed before stellar nuclear reactions had produced a significant quantity of heavy elements, have low abundances of these elements, whereas younger stars were formed after the galactic composition had been enriched by stellar evolution.

5. Spiral arms are probably density enhancements produced by spiral density waves, which rotate about the galaxy while stars and interstellar material pass through them. Because the interstellar gas is compressed in these density waves, star formation tends to occur there, and this in turn explains why young stars are found predominantly in the spiral arms.

6. The existence and characteristics of Population I and II stars can be explained by a picture of galactic evolution that begins with a spherical cloud of gas that has little or no heavy elements at first. Population II stars formed while the cloud was still spherical or just beginning to collapse, but still had no significant quantities of heavy elements. Population I stars formed later, after the galaxy had collapsed to a disk, and after stellar evolution had produced some heavy elements.

7. The spiral arms formed after the disk had been created by the collapse of the original cloud. The spiral density waves that maintain the arms probably started as a result of gravitational disturbances, possibly from nearby galaxies.

8. Different methods for estimating the age of the galaxy provide contradictory values; furthermore, there is a wide spread in the age of globular clusters, suggesting that the galaxy may have formed from the merger of several subsystems of different ages.

REVIEW QUESTIONS

1. Summarize the properties of Population I and Population II objects in the galaxy.

2. How are high-velocity stars and comets similar?

3. The strong line of ionized calcium, whose rest wavelength is 3933.8 Å, is observed at a wavelength of 3932.0 Å in the spectrum of a star. To which population does this star belong? Explain how you decided.

4. How are the spiral arms of the galaxy similar to the rings of Saturn?

5. Explain the distinction between the orbital periods of stars in the galaxy and the rotation period of the spiral arms.

6. Why do stars form primarily in the spiral arms of a galaxy like ours?

7. Why do the globular clusters have the lowest heavy element abundances of any objects in the galaxy?

8. If the mass of the galaxy is 1.5×10^{11} solar masses, what is the orbital period of a globular cluster whose orbital semimajor axis is 20 kpc? (Remember to convert this into astronomical units.)

9. Explain why the evolution of massive stars, rather than that of the much more common low-mass stars, has been the principal contributor to the enrichment of heavy elements in the galaxy.

10. Summarize the various roles played by supernovae in the evolution of the galaxy.

ADDITIONAL READINGS

Bok, B. J. 1981. Our bigger and better galaxy. *Mercury* 10(5):130.

Bok, B. J., and P. Bok. 1981. *The Milky Way.* Cambridge, Mass.: Harvard University Press.

Burbidge, G., and E. M. Burbidge. 1958. Stellar populations. *Scientific American* 199(2):44.

Harris, W. E. 1991. Globular clusters in distant galaxies. *Sky and Telescope* 81(2):148.

Hodge, P. 1986. *Galaxies.* Cambridge, Mass.: Harvard University Press.

Iben, I. 1970. Globular cluster stars. *Scientific American* 223(1):26.

Larson, R. B. 1979. The formation of galaxies. *Mercury* 8(3):53.

Mathewson, D. 1985. The clouds of Magellan. *Scientific American* 252(4):106.

Nather, R. E., and D. E. Winget. 1992. Taking the pulse of white dwarfs. *Sky and Telescope* 83(4):374.

Scoville, N., and J. S. Young. 1984. Molecular clouds, star formation, and galactic structure. *Scientific American* 250(4):42.

Shu, F. H. 1973. Spiral structure, dust clouds, and star formation. *American Scientist* 61:524.

———. 1982. *The physical universe.* San Francisco: W. H. Freeman.

Silk, J. L. 1986. Formation of the galaxies. *Sky and Telescope* 72(6):582.

Weaver, H. 1975 and 1976. Steps towards understanding the spiral structure of the Milky Way. *Mercury* 4(5):18 (Part 1); 4(6):18 (Part 2); 5(1):19 (Part 3).

White, R. E. 1991. Globular clusters: fads and fallacies. *Sky and Telescope* 81(1):24.

Astronomical Activity
Finding the Age of the Galaxy

One method for measuring the age of the galaxy is to find the ages of the oldest objects in it, the globular clusters. Here you can use data on stars in a hypothetical globular cluster, along with methods described in Chapter 14, to make your own estimate of the age of the Milky Way.

The accompanying table lists values for the apparent visual magnitude (V) and the color index ($B-V$) for several stars in a globular cluster. To find the age of this cluster, you will need to construct an H-R diagram on which you plot V on the vertical axis (with the smallest values at the top) and $B-V$ on the horizontal axis (this is known as a color-magnitude diagram and is the normal form in which cluster H-R diagrams are plotted). When making your diagram, use graph paper that is thin enough to be used as tracing paper. Mark the spacings on the axes to match those in the H-R diagram in Figure 12.24. For the vertical axis, this means that the *scale,* or the number of magnitudes per centimeter, should be the same as in the figure, even though the values will be different because here you are plotting apparent magnitudes, whereas in the figure, absolute magnitudes are plotted on the vertical axis. On your horizontal axis, copy the $B-V$ scale directly from Figure 12.24; note that the spacings vary in size as you go from right to left.

Now make a point on your diagram for each star in the table. When you are done, you should be able to see the main sequence, and you should also see a number of stars that are off to the right. These are the giant stars that have evolved away from the main sequence as their core hydrogen has run out. To find the age of the cluster, determine from your diagram the main-sequence turnoff point; that is, find the value of $B-V$ for the highest point on the main sequence before the stars go off to the right.

You know that stars above the main-sequence turnoff point have had time to use up all of their core hydrogen, while those below that point have not yet used up their hydrogen. Therefore the age of the cluster is equal to the hydrogen-burning lifetime of stars at the turnoff point. Thus, to find the age, you need to determine the hydrogen-burning lifetime of a star whose $B-V$ value is equal to the turnoff value you have found. To do this, you will need to know both the mass and the luminosity of a star having this $B-V$

V	$B-V$	V	$B-V$
21.72	1.17	25.73	1.49
21.64	1.35	21.18	1.54
23.92	1.32	21.27	0.96
25.56	1.43	21.07	0.81
22.52	1.14	24.49	1.27
20.31	1.48	23.12	1.22
21.00	0.91	27.18	1.56
22.06	0.93	26.92	1.48
20.22	1.57	22.38	1.02
23.37	1.11	21.85	0.97
20.92	0.79	20.78	1.39
27.68	1.57	21.53	1.06
26.24	1.42	23.43	1.17
21.30	1.04	25.28	1.36
20.36	1.38	20.70	1.21
22.81	1.14	20.96	1.03
23.87	1.35	24.81	1.35
21.48	0.87	22.02	1.10
22.50	1.09	26.50	1.53
21.74	1.04	21.62	0.93
22.16	1.23	22.77	1.06
24.78	1.44		

value. You can determine these values from Table 12.4 (note that you may have to estimate the mass and luminosity if your value for $B-V$ falls between those listed in the table). Once you have found the mass and luminosity, then you can use the method described in Chapter 13 to find the lifetime of the star; essentially you compare the ratio of M/L for your star with the M/L value for the Sun, and multiply the result times the Sun's lifetime of 10 billion years.

As a by-product of making your color-magnitude diagram, you can find the distance to the globular cluster, using the main-sequence fitting technique. To do this, place your diagram on top of the H-R diagram in Figure 12.24, and line up the cluster main sequence with the main sequence in the diagram. Then look at the vertical axis at the left to find the difference ($m-M$) between the cluster apparent magnitude scale and the absolute magnitude scale of the H-R diagram. This difference, known as the distance modulus, corresponds to the distance of the cluster, as explained in Chapter 12.

Galaxies upon Galaxies

The M66 group of galaxies in Leo (© 1992 Anglo-Australian Observatory; photo by D. Malin)

PREVIEW

One of the great revolutionary concepts to develop in this century was the realization that our galaxy, this vast conglomeration of billions of stars, is not alone, but is instead one of many billions in the universe. The galaxies themselves are grouped and clustered on a hierarchy of scales that is still not fully known. In parallel with our earlier discussions of stars, in this chapter we first methodically discuss how galaxies are observed and what we can deduce about their physical properties; next we move on to describe their arrangement into groups and clusters, and then we examine the physical processes that shape them. Along the way we will confront a mystery that first arose in our discussion of the properties of the Milky Way: the riddle of the dark matter, the unknown substance that comprises as much as 99 percent of the matter in the universe, but is invisible to our telescopes. This will help set the stage for our subsequent discussions of the universe as a whole, its content and its evolution.

We have spoken of our galaxy as one of many, a single member of a vast population that fills the universe. Given all that we have learned about the ordinary position of human beings in the cosmos, this is no surprise. Having traced our painful progression from the geocentric view to the realization that we occupy an insignificant planet orbiting an ordinary star in an obscure corner of the galaxy, we should be surprised if we found that our galaxy held any kind of unique status in the larger environment of the universe as a whole. It doesn't.

Despite the seeming inevitability of this idea, the actual proof that our galaxy is not alone was some time in coming and arrived only after considerable debate and controversy. The so-called **nebulae**—dim, fuzzy objects scattered throughout the sky—have been known since the eighteenth century, but it was not until the mid-1920s that they were demonstrated to be galaxies, rather than closer objects such as gas clouds or star clusters. The proof of their extragalactic nature was announced in 1924, when the American astronomer Edwin Hubble (Fig. 18.1) reported that he had found Cepheid variables in the prominent Andromeda galaxy (until then known as the Andromeda nebula, since its true nature was not known) and used the period-luminosity relation to show that this nebula was entirely too distant to be within our own galaxy.

THE HUBBLE CLASSIFICATION SYSTEM

Having established that the nebulae were truly extragalactic objects, Hubble began a systematic study of their properties. The most obvious basis for classifying the nebulae was to categorize them according to shape. Hubble did this, designating the spheroidal nebulae as **elliptical galaxies** (Fig. 18.2), distinct from the **spiral galaxies** (Fig. 18.3). Within each of the

Figure 18.1   Edwin Hubble.   Hubble's discovery of Cepheid variables in the Andromeda nebula led to the unambiguous conclusion that this object lies well beyond the limits of the Milky Way and must therefore be a separate galaxy. Hubble later made important discoveries about the properties of galaxies and what they tell us about the universe as a whole (see Chapter 19). (Niels Bohr Library, California Institute of Technology)

Figure 18.2   An elliptical galaxy.   These smooth, featureless galaxies are probably more common than spirals. (National Optical Astronomy Observatories)

ASTRONOMICAL INSIGHT
The Shapley-Curtis Debate

Formal debate traditionally has been a respected means of determining right in cases of dispute. Debating is a time-honored practice, undertaken by some for entertainment and by others for more serious reasons. Our legal system, for example, is based on something akin to debates that take place in courtrooms, with a judge or jury to decide who is right. Occasionally, scientific questions are argued in a debate format, the judge being the test of time and the acquisition of new data to test the disputed theories, the jury being the community of scientists. A famous debate on the nature of the so-called nebulae was one of the singular events in early twentieth-century astronomy.

At the opening of this century, no one knew that there were galaxies beyond our own. (In fact, as we saw in Chapter 17, we did not even know where we were located in our own galaxy, nor did we have any good information on our galaxy's size and shape.) By 1920, large numbers of indistinct, hazy objects had been found sprinkled throughout the sky. Some astronomers believed these were other galaxies; others thought they were clouds of gas within our own galaxy. Both sides marshaled evidence to support their views.

Adherents of the "island universe" hypothesis, the idea that the so-called nebulae were distant galaxies like our own, tried to show that the nebulae lay outside our galaxy. In doing so, they argued that our galaxy was smaller than some had thought. This point in turn hinged on disagreements among astronomers over distance scales, particularly those based on pulsating variable stars in clusters. When the distribution of the nebulae was mapped, they were found to avoid the plane of the Milky Way, which indicated to the proponents of the island universe hypothesis that the nebulae lay far outside our galaxy, where the galactic dust layer could block our view of the ones lying near the plane of the disk. Observations of novae in some of the nebulae (particularly our neighbor, the Andromeda nebula) were used to show that they must be outside our galaxy, if they were as luminous as novae seen within the Milky Way.

Proponents of the local hypothesis for the nebulae argued that the Milky Way was large enough to include them. This is where the major disagreement on distance scales came into the argument, with one side believing the galaxy was some 10 times larger than did those on the other side of the issue. The problem was exacerbated by the assumption (on both sides) that there was no significant interstellar absorption, so that distances could be estimated simply from apparent magnitudes. Some "novae" were seen in other galaxies that, if comparable in luminosity to novae in our galaxy, would have to be close by, within the galaxy. (These were later identified as supernovae, which are far more luminous than ordinary novae.) The most telling argument favoring the local hypothesis was the observation, later shown to be incorrect, that some of the spiral nebulae were actually seen to be rotating. An object as large as the Milky Way would have to be rotating faster than the speed of light for the rotation to be observed from a great distance over a time of over several months. Consequently this was taken as proof that the nebulae could not be as large as galaxies and therefore must be close by.

The chief proponents of the opposing points of view met in formal debate before the National Academy of Sciences in Washington, D.C., on April 26, 1920. Representing the island universe hypothesis was H. D. Curtis, who had amassed most of the arguments cited above; opposing him was the formidable Harlow Shapley, who was already famous for his work on the size scale of the galaxy (see Chapter 17). The first part of the debate was devoted wholly to a discussion of the size of the Milky Way, and the second part of the debate dealt with the nature of the nebulae.

Shapley's arguments were considered more persuasive, and in fact he was more nearly correct on the question of the size scale of the galaxy (except for the omission of interstellar absorption). The judgment of time and new data found later by the jury eventually showed that Curtis was correct on the question of the nature of the nebulae, however. His arguments about the distribution of the nebulae, the existence of a class of "super" novae, and the invalidity of the rotation measurements for spiral nebulae were all correct, but all needed time and new information to be substantiated.

Today formal debate is rarely used as a tool for investigating scientific theories, but nevertheless the principles of debate often come into play. Usually the forum is the pages of a technical journal, where scientists publish their views and the evidence to support them. Verbal jousts often occur at scientific conferences, so the spirit of face-to-face debate is sometimes resurrected as well. Through these exchanges progress toward better understanding is made.

Figure 18.3   Spiral galaxies.   Here are four typical examples of spiral galaxies. At the top left is NGC 3627, at the top right is NGC 6946 (these designations refer to the Messier Catalog, see Appendix 15); at the bottom left is NGC 3623, at the bottom right is NGC 4321 (these designations are from the *New General Catalog*). (top left, bottom left: © 1992 AAT Board, photos by D. Malin; top right: U.S. Naval Observatory; bottom right: J. D. Wray, University of Texas)

two general types, Hubble established subcategories on the basis of less dramatic gradations in appearance (Table 18.1). The ellipticals displayed varying degrees of flattening and were sorted out according to the ratio of the long axis to the short axis (Fig. 18.4), with designations from E0 (spherical) to E7 (the most flattened). The number following the letter E is determined from the formula $10(1 - b/a)$, where a is the long axis and b is the short axis, as measured on a photograph.

TABLE 18.1   Types of Galaxies

Type	Designation	Characteristics
Elliptical	E0–E7	Spheroidal shape; subtype determined by the expression $10(1-b/a)$, where a is the long axis and b is the short axis of the galaxy.
Dwarf elliptical	dw E	Spheroidal shape; very low mass and luminosity.
S0	S0	Disk-like galaxies with no spiral structure.
Spiral	Sa–Sc	Disk-like galaxies with spiral arms; subtype determined by relative size of nucleus and openness of spiral structure; *Sa* refers to large nucleus and tightly wound arms, whereas *Sc* refers to small nucleus and open spiral arms.
Barred spiral	SBa–SBc	Spiral galaxies with elongated, barlike nuclei; subtypes determined the same way as in spiral galaxies.
Irregular I	Ir I	Disk-like galaxies with evidence of spiral structure, but not well organized.
Irregular II	Ir II	Galaxies that do not fit into any of the other types; some Ir II galaxies have been found to be normal spirals heavily obscured by interstellar gas and dust.

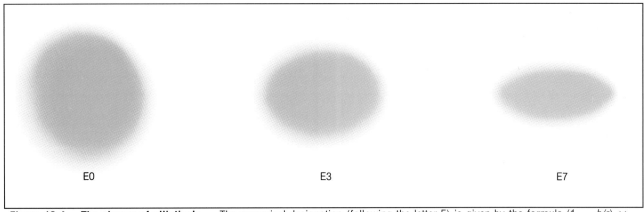

EO   E3   E7

Figure 18.4   The shapes of ellipticals.   The numerical designation (following the letter E) is given by the formula $(1 - b/a) \times 10$, where b is the short axis and a is the long axis of the galaxy image. Here are three examples: the EO galaxy has $a = b$ (that is, it is circular); the E3 galaxy has $a = 1.4b$; and the E7 has $a = 3.3b$. The E7 galaxy is the most highly elongated of the ellipticals.

Among the spirals, Hubble based his classification on the tightness of the arms and the compactness of the nucleus (Fig. 18.5). The types ranged from Sa (tight spiral, large nucleus) to Sc (open arms, small nucleus). The Milky Way is probably an Sb in this system, intermediate in both characteristics, although it is difficult to be sure, because we cannot get an outsider's view of what our galaxy looks like.

Hubble recognized a variation of the spiral galaxies in which the nucleus has extensions on opposing sides, with the spiral arms emanating from the ends of the extensions. He called these galaxies **barred spirals** (Fig. 18.6) and assigned them the designations SBa through SBc, using the same criteria as before in establishing the a, b, and c subclasses (Fig. 18.7).

Following Hubble's original work on galaxy classification, an intermediate class of S0 galaxies has been recognized. These appear to have a disk shape, but no trace of spiral arms.

Hubble arranged the types of galaxies in an organization chart that has become known as the "tuning-fork" diagram (Fig. 18.8). Because there are two types of spirals, Hubble chose not to force all the types into a single sequence but to split it into two branches. For a time astronomers thought that this diagram represented an evolutionary sequence, but when later studies showed that all types of galaxies contain old stars, it became clear that one type of galaxy does not evolve into another. The tuning-fork diagram is not an age sequence after all, and the differences in galactic type must be explained in some other way.

Most of the galaxies listed in catalogs are spirals, which are about evenly divided between normal and barred spirals. Only about 20 to 30 percent of the listed galaxies are ellipticals, and a comparable number are S0 galaxies. The remaining few percent are called irregular galaxies because they do not fit into the normal classification scheme. Because there are

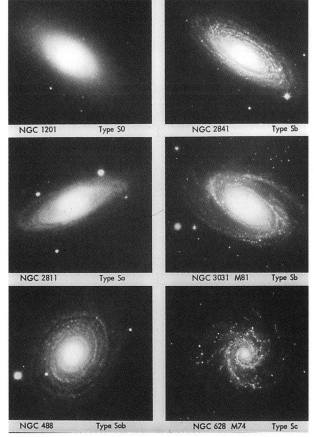

NGC 1201   Type S0   NGC 2841   Type Sb

NGC 2811   Type Sa   NGC 3031 M81   Type Sb

NGC 488   Type Sab   NGC 628 M74   Type Sc

Figure 18.5   An assortment of spirals.   This sequence shows spiral galaxies of several subclasses. (Palomar Observatory, California Institute of Technology)

probably many small, dim, elliptical galaxies that are usually not sufficiently prominent to appear in catalogs, it seems likely that ellipticals actually outnumber spirals in the universe. This certainly is the case

Figure 18.6   Barred spirals.   At the left is M83, a well-known example of a spiral galaxy with a barlike structure through the nucleus; at the right NGC 1365. (left: © 1977 Anglo-Australian Telescope Board; right: © 1992 AAT Board; photos by D. Malin)

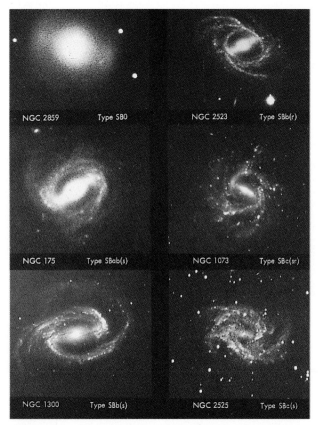

Figure 18.7   Barred spirals.   Roughly half of all spiral galaxies have central elongations, or bars, from which the spiral arms emanate. (Palomar Observatory, California Institute of Technology)

in dense clusters of galaxies, as we will see. Table 18.2 lists a few prominent galaxies whose properties indicate the typical values for galaxies of different types.

Although the irregular galaxies are, by definition, misfits, it has proven possible to find some systematic characteristics even in their case. Most have a hint of spiral structure, although they lack a clear overall pattern, and these have been designated as **Type I irregulars.** The rest, a small minority, simply do not conform in any way to the normal standards and are assigned the classification of **Type II irregulars** or peculiar galaxies (for example, see Fig. 18.9). The Magellanic Clouds are both Type I irregulars.

Recently, based largely on infrared observations, many Irr II galaxies have been found to be embedded in so much gas and dust that they are invisible except when observed at infrared wavelengths. Because very active star formation is taking place in these galaxies, as a class they are called *starburst galaxies* (see Fig. 18.9).

MEASURING THE PROPERTIES OF GALAXIES

To probe the physical nature of the galaxies, we must first determine their distances, because without this information, such fundamental parameters as masses and luminosities cannot be deduced.

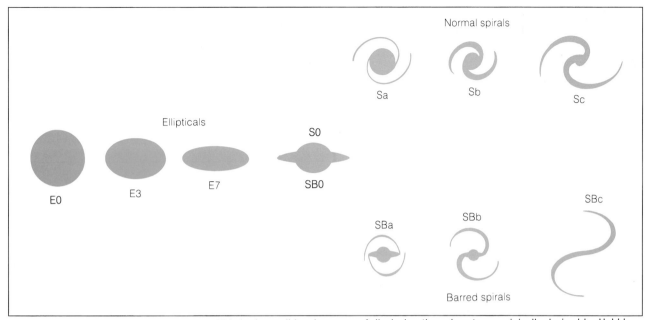

Figure 18.8   The tuning-fork diagram.   This is the traditional manner of displaying the galaxy types originally devised by Hubble. For quite some time this was thought to be an evolutionary sequence, although the imagined direction of evolution was reversed at least once. Now we know that in the normal course of events galaxies do not evolve from one type to another.

We have already mentioned one technique for distance determination that can be applied to some galaxies: the use of Cepheid variables. These stars are sufficiently luminous to be identified as far away as a few million parsecs (**megaparsecs,** abbreviated **Mpc**), which is sufficient to reach the Andromeda nebula and several other neighbors of the Milky Way.

It is not adequate, however, for probing the distances of most galaxies, so other techniques had to be developed.

Recall from our discussions of stellar distance determinations (Chapter 12) that we can always find the distance to an object if we know both its apparent and absolute magnitudes. This is what we do when

TABLE 18.2   Selected Bright Galaxies

Galaxy	Type	Angular Diameter (arcmin)	Diameter (kpc)	Distance (kpc)	Apparent Magnitude[a]	Absolute Magnitude[a]	Mass ($M_\odot$)
Large Magellanic Cloud	Ir I	460	7	55	0.1	−18.7	10^{10}
Small Magellanic Cloud	Ir I	150	3	63	2.4	−16.7	2×10^9
Andromeda (M31)	Sb	100	16	700	3.5	−21.1	3×10^{11}
M33	Sc	35	6	730	5.7	−18.8	1×10^{10}
M81	Sb	20	16	3,200	6.9	−20.9	2×10^{11}
Centaurus A	EOp	14	15	4,400	7	−20	2×10^{11}
Sculptor system	dw E	30	1	85	7	−12	3×10^6
Fornax system	dw E	40	2	170	7	−13	2×10^7
M83	SBc	10	12	3,200	7.2	−20.6	—
Pinwheel (M101)	Sc	20	23	3,800	7.5	−20.3	2×10^{11}
Sombrero (M104)	Sa	6	8	1,200	8.1	−22	5×10^{11}
M106	Sb	15	17	4,000	8.2	−20.1	1×10^{11}
M94	Sb	7	10	4,500	8.2	−20.4	1×10^{11}
M82	Ir II	8	7	3,000	8.2	−19.6	3×10^{10}
M32	E2	5	1	700	8.2	−16.3	3×10^9
Whirlpool (M51)	Sc	9	9	3,800	8.4	−19.7	8×10^{10}
Virgo A (M87)	E1	4	13	13,000	8.7	−21.7	4×10^{12}

[a]The apparent and absolute magnitudes represent the light from the entire galaxy. Because the light from a galaxy comes from a large area in the sky, a galaxy of a given apparent magnitude is not as easily seen by the eye as a star of the same magnitude.

Figure 18.9   Peculiar galaxies.   Not all galaxies fit the standard classifications. At left is M82, a well-known example of a galaxy with an unusual appearance. For a time, it was thought that the nucleus of this galaxy was exploding, but more recent analysis has indicated that the odd appearance is due instead to a very extensive region of interstellar gas and dust surrounding the center of a spiral galaxy that is undergoing an episode of intense star formation. At left is another galaxy shrouded in gas and dust, the starburst irregular galaxy AAT 64 (left: Lick Observatory photograph; right: © 1992 AAT Board; photo by D. Malin)

we use the spectroscopic parallax method, the main-sequence fitting technique, and even the Cepheid variable period-luminosity relation. A general term for any object whose absolute magnitude is known from its observed characteristics is **standard candle,** and an assortment of these are used in extending the distance scale to faraway galaxies.

The most luminous stars are the red and blue supergiants, which occupy the extreme upper regions of the H-R diagram. These can be seen at much greater distances than Cepheid variables and therefore are important links to distant galaxies. The absolute magnitudes of these stars are inferred from their spectral classes, just as in the spectroscopic parallax technique, although they are so rare that there is substantial uncertainty in assuming that they conform to a standard relationship between spectral class and luminosity. In a variation on this technique, it is simply assumed that there is a fundamental limit on how luminous a star can be, and that in a collection of stars as large as a galaxy, there will always be at least one star at this limit. Then to determine the distance to a galaxy, it is necessary only to measure the apparent magnitude of the brightest star in it and to assume that its absolute magnitude is at the limit, which is about $M = -8$. This technique extends the distance scale by about a factor of 10 beyond what is possible using Cepheid variables—that is, to 10 or more megaparsecs.

Other standard candles come into play at greater distances. These include supernovae (Fig. 18.10), which at peak brightness always reach the same absolute magnitude. (Care must be exercised in making measurements, however, because there are two distinct types of supernovae with different absolute magnitudes.) Supernovae can be observed at distances of hundreds of megaparsecs and are therefore very useful distance indicators, the major drawback being that distances can only be measured to galaxies that happen to have supernovae occur in them. The ultimate standard candle, useful to distances of thousands of megaparsecs, is the brightest galaxy in a cluster of galaxies. As in the case of the brightest star in a galaxy, astronomers assume that the brightest galaxy in a cluster always has about the same absolute magnitude. (Experience has shown that the brightest galaxy in a cluster may not be so standard from one cluster to another, and that the second- or third-ranked galaxy is a better standard candle.)

A new technique called the **Tully-Fisher method,** in which the luminosity of a galaxy is inferred from its mass, shows great promise for spiral galaxies. The mass is not measured directly; instead we measure the rotational velocity of the galaxy, which depends on its mass. The rotational velocity is estimated from the width of the 21-centimeter radio emission line from the galaxy. The Doppler effect broadens the line according to the maximum spread of velocities from

Figure 18.10   A Supernova in a Distant Galaxy.   Before the distinction between novae and supernovae became clear, these flare-ups (arrow) contributed to the controversy over the nature of the nebulae. Now that these occurrences in galaxies are known to be supernova explosions, comparisons of their apparent and assumed absolute magnitudes provide distance estimates. (Palomar Observatory, California Institute of Technology)

one side of the galaxy's disk to the other (Fig. 18.11). Thus the luminosity (hence the absolute magnitude) of a spiral galaxy can be inferred from the width of its 21-centimeter emission line, and its distance can then be found by measuring the apparent magnitude and comparing it with the absolute magnitude. This technique works best when infrared magnitudes are used, because infrared light is relatively unaffected by

interstellar dust within the galaxy being observed. The Tully-Fisher method is now being used in efforts to improve the intergalactic distance scale, which has very important implications for our understanding of the universe as a whole. A similar method has been developed for elliptical galaxies.

It is important to keep in mind how uncertain *all* galactic distance determinations are. To assume that all objects in a given class have identical basic properties such as luminosity is always a risky business, especially when such assumptions are applied to objects as distant as external galaxies or as rare as the brightest star in a galaxy. No other options are available, however, so we must simply recognize the inherent limitations in accuracy and take them into account. The uncertainties in distance determinations carry over to our measurements of other properties of galaxies.

The masses of nearby galaxies can be measured in the same manner as the mass of our own galaxy: by applying Kepler's third law to the orbital motions of stars or gas clouds in the outer portions. All that is required is to measure the orbital velocity at some point well out from the center and to determine how far from the center that point is (which in turn requires knowledge of the distance to the galaxy). Then Kepler's third law leads to

$$M = a^3/P^2 \ ,$$

where M is the mass of the galaxy (in solar masses), and a and P are the semimajor axis (in AU) and the

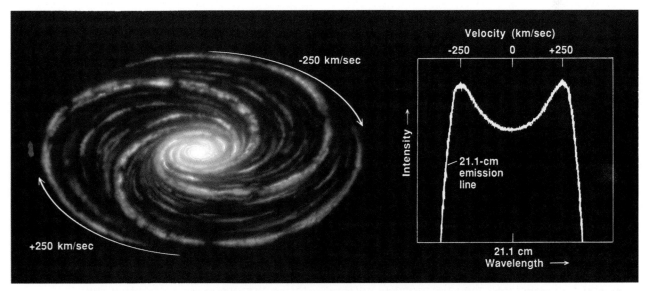

Figure 18.11   Rotation velocity and the width of the 21-centimeter line.   This shows how the rotation velocity of a spiral galaxy is related to the width of the 21-centimeter emission line from hydrogen gas in the galaxy. Because of the Doppler effect, 21-centimeter photons from one side of the galaxy are shifted toward shorter wavelengths, those from the other side are shifted toward longer wavelengths, and those from the central regions are not shifted. The result is a 21-centimeter line that is broadened according to how rapidly the galaxy is rotating. Because the rotation speed is governed by the central mass of a galaxy, the width of the line is related to the mass of the galaxy and can be used to estimate its absolute magnitude for distance determination.

ASTRONOMICAL INSIGHT
Infrared Galaxies

Until very recently, nearly all that we knew about the properties of galaxies and about their numbers was based on observations made in visible wavelengths of light. We now know that limiting ourselves to this small portion of the electromagnetic spectrum has severely limited our knowledge of galaxies and their evolution.

The *Infrared Astronomical Satellite* (*IRAS*) mapped the entire sky at four wavelengths far into the infrared portion of the spectrum. As noted in other chapters in this text, *IRAS* data have brought about major changes in our understanding of many objects in the universe, ranging from planets to newborn stars to interstellar clouds of gas and dust. The data have also produced new understandings of the galaxies that populate the universe.

The *IRAS* sky survey revealed huge numbers of faint, fuzzy sources of infrared emission not associated with any objects in our galaxy. These objects are now thought to be distant galaxies, many of which are invisible at all wavelengths except the infrared. This means that most of the energy that they emit comes from very cold material. These infrared objects are now believed to be spiral galaxies so obscured by dust

that they are seen only in the far-infrared wavelengths where the dust emits. Most of the light from the stars in these galaxies is absorbed by dust, which is heated in the process to temperatures of several tens of degrees and then emits infrared radiation.

The discovery of the infrared galaxies has revealed a whole new class of galaxies containing far more interstellar matter than normal spirals. The interpretation of these galaxies is uncertain: Are they aging galaxies undergoing a new phase of evolution in which stars emit vast quantities of material into space, or are they newly forming galaxies undergoing rapid star formation? The second point of view, that these are young galaxies undergoing massive star-formation episodes, is becoming more widely accepted. If this is correct, it has important implications for galactic evolution. It implies that intervals of very intense and rapid star formation may be normal stages in the early evolution of spiral galaxies. Visible-light observations, as well as infrared data, have already led astronomers to define a class of galaxies called **starburst galaxies**—galaxies thought to be in stages of rapid star formation. The new infrared data from *IRAS* appear to have estab-

lished the widespread existence of starburst galaxies.

If a starburst stage is normal in the early evolution of spiral galaxies, then one problem confounding studies of the history of our own galaxy may have been solved. As noted in Chapter 17, the abundance of heavy elements in the Milky Way is higher than would be expected if current star-formation rates had been the norm throughout the history of the galaxy. This has led to speculation that our galaxy was either formed from material that had already undergone elemental enrichment by nuclear reactions in some earlier generation of stars, or that the galaxy underwent a much more intense episode of star formation sometime in the past than it is undergoing now. The discovery by *IRAS* that starburst phases of galactic evolution may be common leads astronomers to believe that our own galaxy was once a starburst galaxy. If so, the Milky Way may once have been virtually invisible at all wavelengths except the far infrared. Perhaps alien astronomers several billion light-years away are at this moment discovering the Milky Way as a source of far-infrared radiation, to be included in their equivalent of our *IRAS* catalogs.

---

period (in years) of the orbiting material at the observed point, respectively. (See the more complete discussion in Chapter 16 to remind yourself how this equation was developed.) Note that the orbital period is far too long (typically hundreds of millions of years) to be observed directly. The period must be estimated by knowing the orbital speed and the distance from the center of the galaxy, and then computing how long it takes for a star to complete one orbit.

Using this technique presents two difficulties: (1) it is hard to isolate individual stars in distant galaxies and then measure their velocities using the Doppler shift, and (2) both the distance and the orientation

of the galaxy must be known before the true orbital velocity and semimajor axis can be determined. The orientation can usually be deduced for a spiral galaxy, since it has a disk shape whose tilt can be seen, and the distance can be estimated using one of the methods just outlined. In most cases the orbital velocities are measured at several points within a galaxy from the center out as far as possible, and the data are plotted as a **rotation curve** (Fig. 18.12), which is simply a diagram showing how orbital velocity varies with distance from the center. The most effective means of obtaining velocity data on the outer portions of a spiral galaxy is to measure the 21-centimeter emission from hydrogen, which can be detected at greater distances from the center than visible light from stars can be measured (Fig. 18.13).

The technique of measuring rotation curves can best be applied to spiral galaxies, where there is a disk with stars and interstellar gas orbiting in a coherent fashion. Elliptical galaxies have no such clearcut overall motion, and a slightly different technique must be used. The individual stellar orbits are randomly oriented in the outer portions of an elliptical galaxy, so that there is a significant range of velocities within any portion of the galaxy's volume. This range of velocities, called the **velocity dispersion,** is greatest near the center of the galaxy, where the stars move fastest in their orbits, and smaller in the outer regions. Furthermore, the greater the mass of the galaxy, the greater the velocity dispersion (at any distance from the center), so a measurement of this parameter can lead to an estimate of the mass of a galaxy. The velocity dispersion is deduced from the widths of spectral lines formed by groups of stars in different portions of a galaxy (Fig. 18.14): the greater the internal motion within the region observed, the

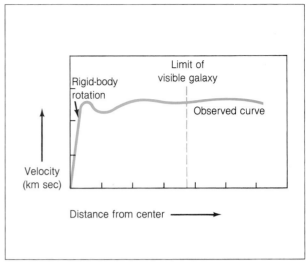

Figure 18.12   A schematic rotation curve for a galaxy.   As this diagram illustrates, in most spiral galaxies, the rotation velocity does not drop off in the outer regions, as it would if most of the mass were concentrated at the center of the galaxy. The data on rotation speed in the outermost regions come from radio observations.

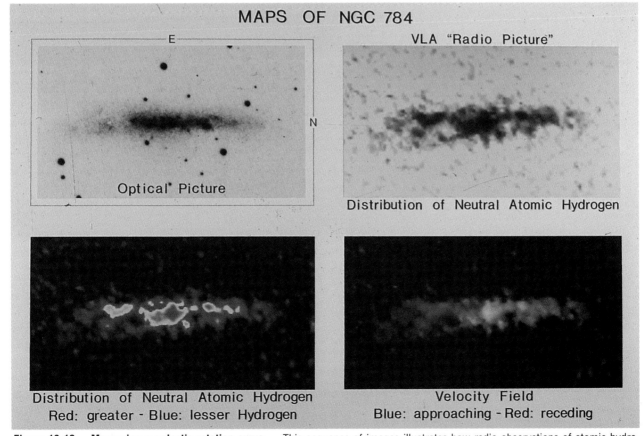

Figure 18.13   Measuring a galactic rotation curve.   This sequence of images illustrates how radio observations of atomic hydrogen are used to derive the rotation curve for a spiral galaxy. At the upper left is a visible-light photograph of the nearly edge-on galaxy; at the upper right is a radio map obtained in the 21-centimeter line of hydrogen. In the two lower panels, the radio data have been color-coded, on the left to indicate relative intensities of the hydrogen emission, and on the right to indicate relative line-of-sight velocities. In this last panel we can see that one side of the galaxy (red) is receding and the other side (blue) is approaching. Comparison of the velocities leads to a determination of the rotational speed as a function of distance from the galactic center. (National Radio Astronomy Observatory)

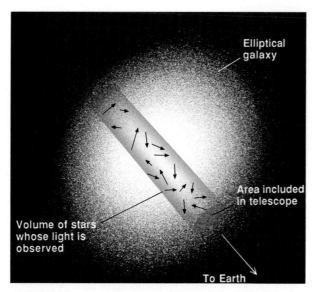

Elliptical
galaxy

Area included
in telescope

Volume of stars
whose light is
observed

To Earth

Figure 18.14   The velocity dispersion for an elliptical galaxy.   An elliptical galaxy has no easily measured overall rotation, so the mass cannot be estimated from a rotation curve. Instead the average velocities of stars at a known distance from the center are used. Light from a small area of the galaxy's image is measured spetroscopically. The random motions of the stars in the observed part of the galaxy broaden the spectral lines by the Doppler effect, and the amount of broadening is a measure of the average velocity of the stars in the observed region. At a given distance from the center of the galaxy, the higher the average velocity, the greater the mass contained inside that distance.

greater the widths of the spectral lines, due to the Doppler effect (caused by the fact that some stars are moving toward the Earth and others away from it).

Kepler's third law can sometimes be used in a different way entirely in determining galactic masses. Here and there in the cosmos are double galaxies, orbiting each other exactly like stars in binary systems. In these cases, Kepler's third law can be applied to derive an estimate of the combined mass of the two galaxies. The uncertainties are even more severe than in the case of a double star, however, because the orbital period of a pair of galaxies is measured in hundreds of millions of years. Thus the usual problems of not knowing the orbital inclination or the distance to the system are compounded by inaccuracies in estimating the orbital period, something that is usually well known for a double star. Still, this technique is useful and has one major advantage: it takes into account *all* the mass of a galaxy, including whatever part of it is in the outer portions, beyond the reach of the standard rotation curve or velocity dispersion measurements. Interestingly, galactic masses estimated from double systems are generally much larger than those based on measurements

of internal motions within galaxies, possibly indicating that most galaxies have extensive halos containing large quantities of matter. As noted in Chapter 16, independent evidence indicates that our own galaxy has a massive halo, perhaps containing as much as 90 percent of the total mass.

Once the distance to a galaxy has been established, its luminosity and size can be deduced directly from the apparent magnitude and apparent diameter. Both quantities are found to vary over wide ranges, with luminosities as low as 10^6 and as high as 10^{12} times that of the Sun, and diameters ranging from about 1 to 100 kiloparsecs (kpc).

Elliptical galaxies generally display a wider range of luminosities and sizes than the spirals. The latter tend to be more uniform, with luminosities usually between 10^{10} and 10^{12} solar luminosities and diameters between 10 and 100 kpc. The smallest elliptical galaxies are called **dwarf ellipticals.** These may be very common, but they are too dim to be seen at great distances, so we can only say for sure that there are many of them near our own galaxy and the Andromeda galaxy.

Galaxies of a given type seem to be fairly uniform in other properties, just as stars of a given spectral type are the same in other ways. One quantity often used by astronomers to characterize galaxies is the **mass-to-light ratio (M/L),** which is simply the mass of a galaxy divided by its luminosity, with both measured in solar units. Values of the mass-to-light ratio typically range from 5 to 200. Any value larger than 1 means that the galaxy emits less light per solar mass than the Sun; that is, the galaxy is dominated in mass by stars that are dimmer than the Sun. Even the smaller mass-to-light ratios that are observed for galaxies are much larger than 1; a value of M/L = 50, for example, means that 50 solar masses are required to produce the luminosity of one Sun. Interestingly, counts of stars in the Sun's part of our galaxy indicate a mass-to-light ratio near 1, whereas we expect (based on observations of similar galaxies) that it is much larger than 1 for our galaxy as a whole. This indicates that significant mass in galaxies like ours must be in forms that are very difficult to see, perhaps not in the form of normal stars at all. It is possible that the invisible mass resides in the halos of galaxies in some unseen form. Recall (from Chapter 16) that this appears to be true of our own galaxy. The whole question of the nature of the unseen "dark matter" in galaxies is currently an area of intense interest in astronomy.

Elliptical galaxies tend to have the largest mass-to-light ratios, consistent with our earlier statement that these galaxies are relatively deficient in hot, luminous stars. Spirals, on the other hand, which contain some of these stars, have lower mass-to-light values. It is worth stressing again, however, that even the low mass-to-light ratios for these galaxies are much greater than 1, indicating that they too are dominated in mass by very dim stars or dark matter (including, or course, interstellar gas and dust, but this usually accounts for less than 25 percent of the mass of spiral galaxies). Hence a spiral galaxy, with all its glorious bright blue disk stars, actually has far more dim red ones.

The colors of galaxies can also be measured by using filters to determine the brightness at different wavelengths. In general the spirals are not as red as the ellipticals, again indicating that the latter galaxies contain a higher percentage of cool, red stars. There are also color variations within galaxies: in a spiral, for example, the central bulge is usually redder than the outer portions of the disk where most of the young, hot stars reside.

Both the mass-to-light ratios and the colors of galaxies are indicators of the relative content of Population I and Population II stars. Recall that Population I stars tend to be younger and include all the bright, blue O and B stars. By contrast, Population II objects are old and include only cool, relatively dim main-sequence stars and red giants. Therefore a red overall color, along with a high mass-to-light ratio, implies that Population II stars are dominant, while a low mass-to-light ratio and a bluer color means that some Population I stars are included. Thus elliptical galaxies seem to contain almost entirely Population II objects, while spirals contain a mixture of the two populations.

This dichotomy between the two types of galaxies is also found when the interstellar matter content of spirals and ellipticals is compared. Photographs of spirals, especially those seen edge-on, often clearly show the presence of dark dust clouds, and face-on views typically reveal a number of bright H II regions. Neither shows up on photographs of elliptical galaxies. Radio observations of the 21-centimeter line of hydrogen bear this out; emission is usually present in the spectra of spirals but is weak or absent in the spectra of ellipticals.

THE ORIGINS OF SPIRALS AND ELLIPTICALS

In attempting to explain the differences between spiral and elliptical galaxies, astronomers have proposed several theories, none of which is fully developed at present.

The weight of all the evidence cited in the previous sections is that spiral galaxies are still dynamic, evolving entities, with active star formation and recycling

of material between stellar and interstellar forms, whereas elliptical galaxies have reached some sort of equilibrium in which these processes are not taking place at a significant rate. Because both types of galaxies contain old stars, we know that there are no significant age differences between the two types; we cannot simply conclude that the ellipticals are older and have run out of gas.

One of the earliest suggestions was that rotation is responsible for the differences between the two types. We learned in the previous chapter that our own galaxy is thought to have formed from a rotating cloud of gas that flattened into a disk as it contracted. The rotation of the cloud was the cause of the disk formation, so perhaps elliptical galaxies are the result of contracting gas clouds that did not rotate rapidly enough to form disks. It is not clear how this would account for the lack of interstellar material and star formation, however.

One problem with this idea is that elliptical galaxies can and do rotate, yet they have not formed disks. Therefore rotation cannot be the full explanation. Apparently the key is whether a disk forms before most of the gas in the contracting cloud has been consumed by star formation (Fig. 18.15). If all the gas is converted into stars before the collapse has proceeded very far, the result is an elliptical galaxy. If the initial rate of star formation is not so great, the cloud has time to form a disk while it still contains a large quantity of interstellar gas and dust. The reason the timing is so important is that stars, once formed, act as individual particles and continue to orbit the galaxy without forming a disk. In contrast, gas acts as a fluid and settles into a disk. (In essence, the question is whether there is viscosity that will dissipate energy and allow the material to sink into the plane of a disk; stars do not encounter each other frequently enough to create a fluid viscosity, but gas particles do.)

We may ask why star formation occurs quickly in some cases, consuming all the gas and leading to the formation of an elliptical galaxy, and more slowly in other cases, allowing time for the protogalactic cloud to collapse to a disk before being consumed by star formation. The most likely answer is that the initial rate of star formation depends on the average density of matter. In a dense region, star formation should proceed more rapidly than in a less dense region. Thus the type of galaxy that forms may be determined largely by the initial density of the material in the cloud.

A secondary question is why some spirals are barred and others are not. Theoretical studies suggest that the presence or absence of a massive halo may be the determining factor. Without a halo a disk of stars will be subject to an instability that naturally

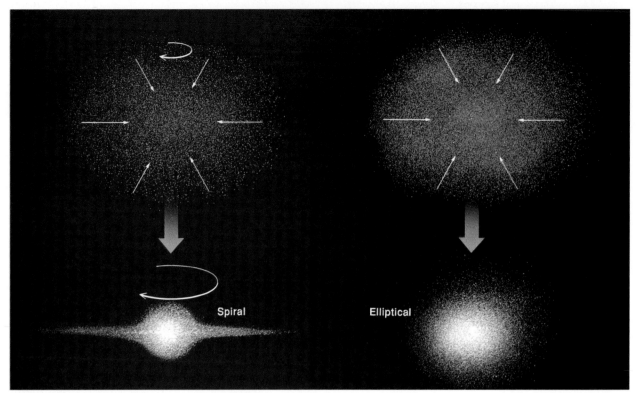

Figure 18.15   The origins of spirals and ellipticals.   Both types of galaxies begin as giant gas clouds that collapse gravitationally. If collapse to a disk occurs before all the gas is converted into stars, the result is a spiral galaxy. If the gas is entirely used up in star formation before a disk forms, the result is an elliptical. The rate of star formation relative to the rate of collapse to a disk is probably determined by the initial density of the cloud and may be influenced by the rotation rate.

causes its central region to become elongated and bar-like. Once a bar has formed, the asymmetry of the gravitational force it exerts on the surrounding disk is sufficient to create a spiral density wave, so that a barred spiral galaxy results. If, on the other hand, a massive halo is present, its gravitational influence inhibits the formation of a bar. Spiral density waves often form anyway, due to the gravitational influence of another galaxy nearby, and cause a normal spiral to form.

An entirely new explanation of the contrasts between spiral and elliptical galaxies is currently gaining favor. This theory supposes that spiral galaxies formed more readily during the era of galaxy formation, because most star or cloud systems that collapsed gravitationally would have had enough angular momentum to force the formation of a disk. Elliptical galaxies, in this view, formed subsequently, as the result of mergers of spiral galaxies. This theory is strongly supported by a growing body of evidence that most galaxies are the result of mergers of subordinate systems. When two or more spiral galaxies merged, their spin directions would have been randomly oriented, so that the resulting galaxy could easily end up with no systematic overall rotation, as is observed in elliptical galaxies. Thus, in this new

model, elliptical galaxies did indeed evolve from spirals, but not in the manner that Edwin Hubble originally envisioned. Detailed studies of the internal motions within elliptical galaxies now under way may eventually distinguish between the possible formation mechanisms for these galaxies.

CLUSTERS OF GALAXIES

Although galaxies may be considered the largest single objects in the universe (if indeed an assemblage of stars orbiting a common center can be viewed as a single object), there are yet larger scales on which matter is organized. Galaxies tend to be located in clusters (Fig. 18.16) rather than being distributed uniformly throughout the cosmos, and these in turn have an uneven distribution, with concentrations of clusters referred to as **superclusters.** It may be that superclusters are the largest entities in the universe.

Clusters of galaxies range in membership from a few (perhaps half a dozen) to many hundreds or even thousands (Appendix 13). Just as stars in a cluster orbit a common center, so galaxies in a group or cluster are gravitationally bound together and follow or-

Figure 18.16   A cluster of galaxies.   Several galaxies are visible in this photograph of a portion of a large cluster of galaxies, called the Fornax cluster. (© 1984 Royal Observatory, Edinburgh)

bital paths about a central point. In a large, rich cluster, frequent encounters between galaxies during the cluster-formation era cause the members to take on a smooth, spherical distribution; in a small group like the one to which the Milky Way belongs, the arrangement of individual galaxies is more haphazard, creating an amorphous overall appearance.

The Local Group

The Milky Way belongs to a small group of galaxies, known as the Local Group. This group consists of about 30 members arranged in a random distribution (Fig. 18.17). Despite the relative proximity of these galaxies to us, it has been difficult to ascertain their properties in some cases because of obscuration by our own galactic disk. We cannot even say with certainty how many members the Local Group has.

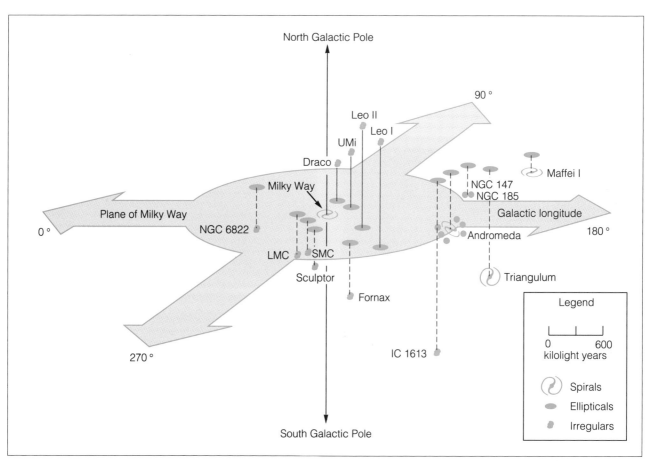

Figure 18.17   The Local Group.   This diagram schematically illustrates the arrangement of galaxies in the Local Group, with respect to the Milky Way.

Figure 18.18 The Andromeda galaxy.   At a distance of over 2 million light-years, this galaxy is the most distant object visible to the unaided eye. The naked eye sees only an extended, fuzzy patch of light, rather than the detailed view shown here, which was obtained with a telescope and a time exposure. (Palomar Observatory, California Institute of Technology)

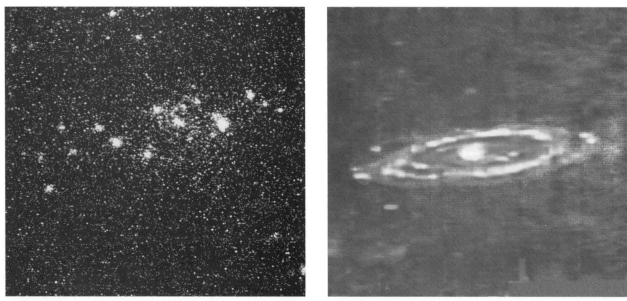

Figure 18.19 New views of the Andromeda galaxy.   Here are an X-ray image (left) and an infrared picture (right) of our neighbor spiral galaxy. The X-ray view reveals associations of stars with hot coronae and binary X-ray sources; the infrared image shows the locations of cold dust clouds. (Left: Harvard-Smithsonian Center for Astrophysics; right: NASA)

Among the member galaxies are three spirals, two of which—the Andromeda galaxy (Figs. 18.18 and 18.19) and the Milky Way—are rather large and luminous. These are probably the brightest and most massive galaxies in the entire cluster. Two large galaxies discovered in the 1970s through infrared observations (Fig. 18.20) were thought for a while to be members of the Local Group. They would have significantly altered the constitution of the cluster, but more recent evidence indicates that both are too distant to be members.

Most of the other members are ellipticals, many of them dwarf ellipticals (Fig. 18.21). There may be additional members of this type undetected so far because of their faintness. Four irregular galaxies (Fig. 18.22), two of them the Large and Small Magellanic Clouds (Fig. 18.23) are also found within the Local Group. In addition there are globular clusters, which are probably distant members of our galaxy, but they lie so far away from the main body of the Milky Way that they appear isolated. A list of known members of the Local Group appears in Appendix 13.

The Local Group is about 800 kpc in diameter and has a roughly disk-like overall shape, with the Milky

Figure 18.21   A dwarf galaxy.   This is NGC 6822, a faint, low-density spheroidal galaxy belonging to the Local Group. (© 1982. AAT Board; photo by D. Malin)

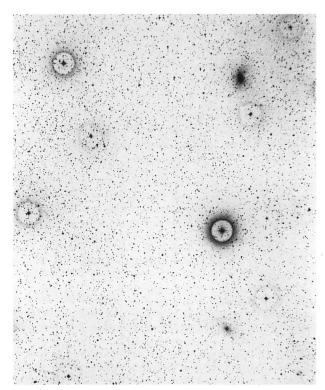

Figure 18.20   Maffei 1 and 2.   This infrared photograph reveals two large galaxies (the fuzzy images, upper right and lower right) that for a while were considered possible members of the Local Group. More extensive analysis has shown, however, that they probably are not. (The circular marks are overexposed images of closer stars.) (H. Spinrad)

Figure 18.22   IC1613.   This is an irregular galaxy, a member of the Local Group. (P. W. Hodge, photo taken at the Lick Observatory)

Figure 18.23   The Magellanic Clouds.   The Large Cloud is shown in the upper photo, and the Small Cloud in the lower picture. Our nearest neighbors, these galaxies orbit as satellites of the Milky Way. (Top: © 1984 Royal Observatory, Edinburgh; bottom: NOAO)

Way located a little off-center. Beyond the outermost portions of the Local Group, there are no conspicuous galaxies for a distance of some 1,100 kpc.

The Magellanic Clouds and the Andromeda galaxy (also known commonly as M31, its designation in the widely cited Messier Catalog; see Appendix 15) have been particularly well studied because of their proximity and prominence and because of what they can tell us about galactic evolution and stellar processing.

The Large and Small Magellanic Clouds appear to the unaided eye as fuzzy patches, easily visible only on dark, moonless nights. They lie near the south celestial pole and can therefore be seen only from the Southern Hemisphere. Their name originated from the fact that the first Europeans to see them were Ferdinand Magellan and his crew, who made the first voyage around the world in the early sixteenth century.

The Magellanic Clouds are considered to be satellites of the Milky Way. They have smaller masses and follow orbits about it, taking several hundred million years to make each circuit. Lying between 50 and 65 kpc from the Sun, both are Type I irregulars. They contain substantial quantities of interstellar matter and are quite obviously the sites of active star formation, with many bright nebulae and clusters of hot, young stars. Measurements of the colors of these galaxies, and of the spectra of some of their brighter stars, indicate that their heavy element abundances are 5 to 10 times lower than those of Population I stars in our galaxy. This seems to indicate that the Magellanic Clouds have not undergone as much stellar cycling and recycling as the Milky Way. These and other Type I irregular galaxies may generally be viewed as galaxies in extended adolescence that have not yet settled down into mature disks. In the case of the Magellanic Clouds, the Milky Way's tidal forces are probably the reason for the unrest.

The great spiral M31, the Andromeda galaxy, is the most distant object visible to the unaided eye, lying some 700 kpc from our position in the Milky Way. All that the eye can see is a fuzzy patch of light, even if a telescope is used, but when a time-exposure photograph is taken, the awesome disk stands out (although the spiral arms are still difficult to see because the disk is viewed from an oblique angle). The Andromeda galaxy is so large, extending over several degrees across the sky, that full portraits can be obtained only by using relatively wide-angle telescope optics. (Most large telescopes have extremely narrow fields of view.)

The Andromeda galaxy is probably very similar to our own galaxy and thus has taught us quite a bit about the nature of the Milky Way. The two stellar populations were first discovered through studies of stars in Andromeda, and the effects of stellar processing on the chemical makeup of different portions of the galaxy are better determined for Andromeda than for our galaxy. The study of this galaxy and other members of the Local Group has been very important in the development of distance-determination techniques for more distant galaxies and clusters.

Rich Clusters: Dominant Ellipticals and Galactic Mergers

Many clusters of galaxies are much larger than the Local Group and contain hundreds or even thousands of members (Fig. 18.24). The density of galaxies in such a cluster is relatively high, and therefore there are numerous close encounters between galaxies as they follow their individual orbits about the center. When two galaxies pass close by each other, they exert mutual gravitational forces that can have profound

Figure 18.24   A portion of a rich cluster of galaxies.   Most of the objects in this photograph are galaxies. This cluster lies in the constellation Hercules. (Palomar Observabory, California Institute of Technology)

lated galaxies. Near the center of a rich cluster, some 90 percent of the galaxies may be ellipticals or S0s, whereas about 60 percent of noncluster galaxies are spirals. This contrast is probably a direct result of the frequent near-collisions between galaxies in dense clusters. When two galaxies have a close encounter, the tidal forces they exert on each other stretch and distort them. Under some circumstances any interstellar matter in them can be pulled out and dispersed (Fig. 18.26). The outer regions, such as the halos, can also be stripped away. The net effect is analogous to what happens to rocks in a tumbler: the galaxies are gradually ground down into smooth remnants. A spiral galaxy subjected to these cosmic upheavals may assume the form of an elliptical. In time, most of the spirals in a cluster, particularly those in the dense central region, may be converted into ellipticals and S0s.

The frequent gravitational encounters between galaxies during cluster formation have another interesting effect: they cause a buildup of galaxies right at the center of the cluster. When two galaxies orbiting within a cluster come together, one always gains energy and moves to a larger orbit, while the other (the more massive of the two) loses energy and drops closer to the center. Thus a gradual sifting process—like the differentiation that occurs inside some planets as the heavy elements sink toward the core—gradually builds up a dense central conglomeration of galaxies at the heart of the cluster. There these galaxies may actually merge, the end result being

effects. These close encounters apparently were very frequent during the era when the rich clusters formed, which accounts for the smooth overall distribution of galaxies within the clusters, with the greatest density toward the center. By contrast small groups or clusters rarely reach this state and retain a less regular appearance.

The central regions of large, rich clusters have higher concentrations of elliptical and S0 galaxies (Fig. 18.25) than are found in small groups or iso-

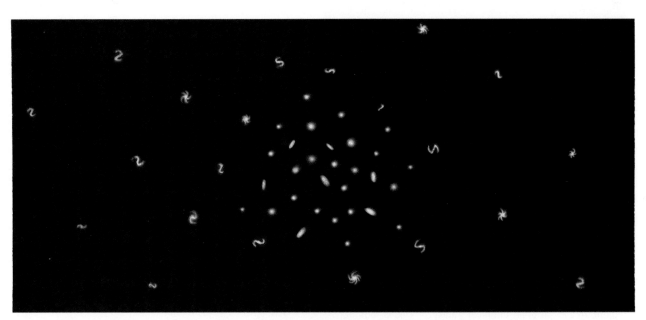

Figure 18.25   Galaxy types in a rich cluster.   In dense, highly populous clusters of galaxies, nearly all the members in the central portion are ellipticals, and a giant, dominant elliptical is often found at the center. Only in the outer portion are many spirals seen.

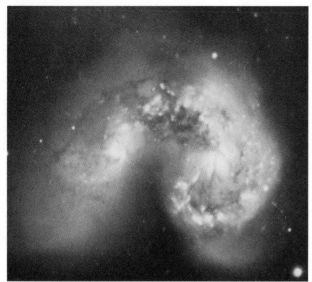

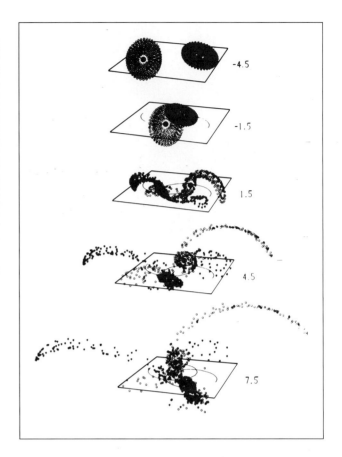

Figure 18.26   A collision between galaxies.   The photo above shows a pair of galaxies undergoing a near-collision. At the right is a computer simulation of their interaction, showing how the present appearance was created. Astronomers believe this type of interaction between galaxies is responsible in some cases for the removal of interstellar matter, converting spiral galaxies to elliptical or SO galaxies. (© 1992 AAT Board; photo by D. Malin; computer simulation from A. Toomre and J. Toomre)

Figure 18.27   A giant elliptical galaxy.   This is M87, a well-known example of a dominant central galaxy in a rich cluster. This galaxy is discussed further in Chapter 20. (© 1992 AAT Board; photo by D. Malin)

a single gigantic elliptical galaxy (Figs. 18.27 and 18.28), which continues to grow larger as new galaxies fall in.

Another distinctive characteristic of some rich clusters of galaxies is the existence of a very hot gaseous medium filling the spaces between the galaxies. This **intracluster gas** was discovered through X-ray observations (Fig. 18.29), which showed that the temperature of the gas is as high as a hundred million degrees, much hotter even than the highly ionized gas in the interstellar medium in our galaxy. This observation raised the possibility that the general intergalatic void is filled with such gas, although if it is, the density outside the clusters must still be rather low, or the X-ray data would have revealed its presence. (The significance of a possible general intergalactic medium is discussed in Chapter 21).

A different type of X-ray measurement indicates that the hot intracluster gas originates in the galaxies themselves, rather than entering the cluster from the intergalactic void. Spectroscopic measurements made at X-ray wavelengths have revealed that the gas contains iron, a heavy element, in nearly the same quantity (relative to hydrogen) as the Sun and other Population I stars. Such a high abundance of iron could only have been produced in nuclear reactions inside stars. Therefore this intracluster gas must once have

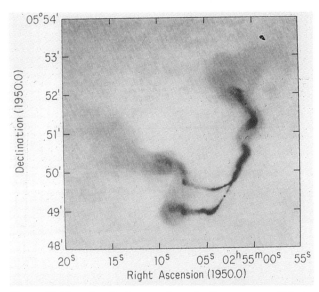

Figure 18.28   A radio image of a galactic merger.   This is a radio map of a giant elliptical galaxy with two nuclei (the bright points at lower center, each with a pair of wispy gaseous jets emanating from it; the jets are discussed in Chapter 20). Evidently this galaxy is in the process of forming from the merger of two galaxies whose centers have not yet completely merged. (National Radio Astronomy Observatory)

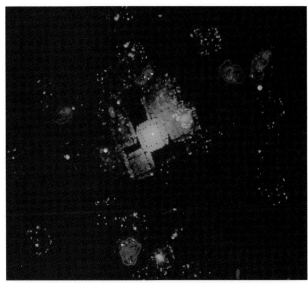

Figure 18.29   Intracluster gas in the Virgo cluster.   This is an X-ray image from the *Einstein Observatory,* showing emission from the entire central portion of this cluster of galaxies. The X rays are being emitted by very hot gas that fills the space between the galaxies. The X-ray emission is shown in blue; radio images of galaxies in the cluster are in red. (X-ray data from the Harvard-Smithsonian Center for Astrophysics; radio data and image processing by the National Radio Astronomy Observatory)

been involved in part of the cosmic recycling that goes on in galaxies, as stars gradually enrich matter with heavy elements before returning it to space. How the gas was expelled into the regions between galaxies is not clear, but it may have been swept out during near-collisions between galaxies, or it may have been ejected in galactic winds created by the cumulative effect of supernova explosions and stellar winds.

Cluster Masses

There are two distinct methods for measuring the masses of clusters of galaxies, and both are quite uncertain. This is a crucial problem, as we shall see in Chapter 21, because of the importance of knowing how much mass the universe contains.

The simpler and more straightforward of the two methods is to estimate the masses of the individual galaxies in a cluster, using techniques described earlier in this chapter, and add them up. In many cases, particularly for distant clusters where it is impossible to measure the rotation curves or velocity dispersions of individual galaxies, the only way to estimate a cluster's mass is to measure its brightness and then use a standard mass-to-light ratio to derive the mass. This is inaccurate, however, because it depends on know-

ing the cluster's distance from us and because it assumes that the galaxies adhere to the usual mass-to-light ratios for their types. It also neglects any matter in the cluster that may lie between the galaxies.

The second method (Fig. 18.30) is similar to the velocity dispersion technique used to estimate masses of elliptical galaxies. The mass of a cluster is estimated from the orbital speeds of galaxies in its outer portions; the faster they move, the greater the mass of the cluster. This method offers the advantage that it measures all the mass of the cluster, whether it is in galaxies or between them, but it has the disadvantage that the necessary velocity measurements, particularly for a very distant cluster, are difficult. Furthermore, the technique is valid only if the galaxies are in stable orbits about the cluster; if the cluster is expanding or some of the galaxies are not really gravitationally bound to it, the results will be incorrect. In rich clusters at least, the smooth overall shape and distribution give the appearance of a bound system, and this technique is probably valid. It may not be valid for small clusters such as the Local Group. This method always leads to an estimated cluster mass that is much greater than that derived by adding up the masses of the visible galaxies, probably because that technique neglects intracluster gas and other forms of dark matter (see the discussion in Chapter 21).

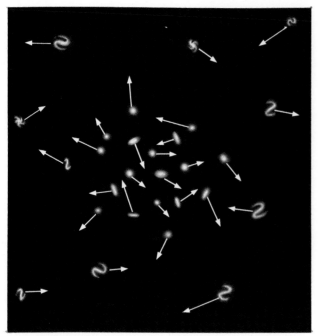

Figure 18.30   Galaxy motions within a cluster.   Galaxies move randomly within their parent cluster. In a large cluster, where the overall distribution of galaxies is uniform, analysis of the average galaxy velocities is used to estimate the total mass of the cluster, just as the velocity dispersion method is used to measure the masses of elliptical galaxies.

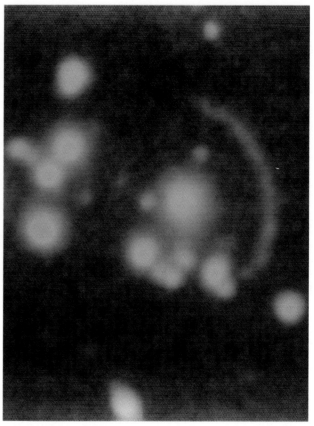

Figure 18.31   A luminous arc.   The blue, ringlike structure shown here is gigantic in scale, dwarfing the individual galaxies in this cluster. (F. Hammer and J. H. Jones, Canada-France-Hawaii Telescope)

GRAVITATIONAL LENSES AND DARK MATTER IN CLUSTERS

In our discussion of the derivation of masses of clusters of galaxies, we found that there seems to be much more mass present in many clusters than can be accounted for by the visible galaxies in those clusters. This implies that more mass is present in these clusters than we can see in the form of stars and galaxies. The invisible matter inferred to be present has been dubbed **dark matter.** Whatever this dark matter is, it has profound implications for the overall distribution of mass in the universe and plays a key role in determining the ultimate fate of the universe, as discussed in Chapter 21.

Very recent observations of clusters of galaxies have provided new insights into the distribution of the dark matter, while leaving its form and origin as mysterious as ever. Several clusters have been found with huge glowing arcs or rings of light around them (Fig. 18.31). These so-called **luminous arcs** are truly gigantic, if they lie at the same distance from us as the clusters in which they are seen. They appear to be millions of light-years in length, comparable in size to the largest structures associated with individ-

ual galaxies (such as the enormous radio-emitting regions surrounding radio galaxies; see Chapter 20).

Astronomers now think that the luminous arcs are not material structures at all, but extended, magnified, and distorted images of galaxies farther away. It was shown theoretically by Albert Einstein, and later confirmed by experiment, that a gravitational field deflects photons of light (this was discussed in Chapter 15). Because of this deflection, under certain circumstances a gravitational field can act like a lens, creating a magnified or distorted image of an object in the distance (Fig. 18.32). If the source of light, the massive object between, and the observer are perfectly aligned, the image formed by the gravitational lens will be a circle. If the alignment is not quite perfect, the image can consist of one or more arcs of light, or of multiple images. Several years ago, both members of a closely spaced pair of very distant objects called quasars (see Chapter 20) were observed to have virtually identical properties, and it was deduced that they were two images of the same quasar, formed by the gravitational lensing effect of a galaxy

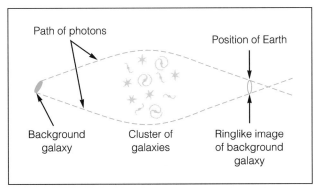

Figure 18.32   Gravitational lens creating luminous arcs.
The gravitational field of a foreground cluster of galaxies can deflect the light from a distant object, creating a ringlike image if the alignment is perfect as shown here or arclike images if the alignment is not quite perfect.

between the Earth and the quasar (Fig. 18.33). Since then many other examples of multiple images of quasars have been found. Now it is thought that the luminous arcs seen around some galaxy clusters are also the result of gravitational lenses.

If a distant galaxy lies behind a massive cluster of galaxies, then depending on the precision of the alignment as seen from the Earth, the gravitational field of the cluster can form one or more arclike, magnified images of the background galaxy. From the size and shape of the image, it is possible to calculate the mass contained in the cluster of galaxies, because the properties of the image depend on the strength of the gravitational field that forms the lens. Thus we have a new method of determining the masses of galaxy clusters, one that is quite independent of the usual assumptions that must made about the orbits of individual galaxies within the cluster.

This new method yields results that correspond very well with those obtained from other methods.

The masses of clusters are still found to be much larger than the sum of the masses of the visible galaxies. Thus this new tool is helping to confirm the growing belief among astronomers that the universe contains vast quantities of invisible, dark matter. At the same time, it is also providing new and dramatic confirmation of one of the predictions made by Einstein's theory of general relativity.

It has actually proven possible, through the analysis of the luminous arcs observed in and around a cluster, to determine the distribution of the dark matter in the cluster (Fig. 18.34). In some clusters, a large number of luminous arcs are seen. Using the equations that describe the deflection of light by a gravitational field, astronomers have been able to deduce the mass distribution that would be needed to create the observed pattern of arcs. In this way, they have actually been able to map the dark matter in several clusters of galaxies. The maps show that the dark matter traces the distribution of visible galaxies in each cluster, but without much fine-scale variation from place to place. So whatever the dark matter is, it tends to be associated with the glowing matter that makes up the visible stars and galaxies in the universe.

SUPERCLUSTERS, VOIDS, AND WALLS

We turn now to consider the overall organization of the matter in the universe. We noted earlier that clusters of galaxies may represent the largest scale on which matter in the universe is clumped, but that there is some evidence for a higher-order organization, even in the case of the Local Group (Fig. 18.35). It appears that clusters of galaxies are themselves concentrated in certain regions, commonly referred to as superclusters.

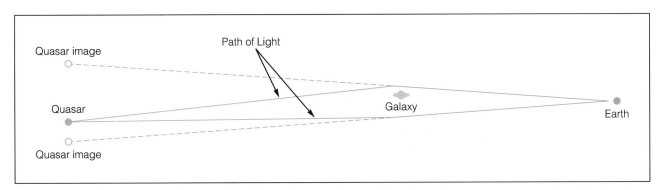

Figure 18.33   A double quasar image.   This diagram illustrates how an intervening galaxy can act as a gravitational lens forming two images of a distant quasar.

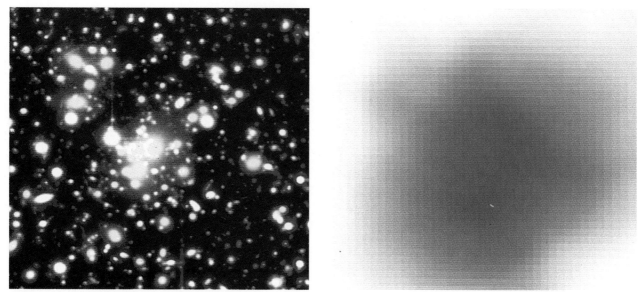

Figure 18.34   Dark matter in a cluster of galaxies.   It is possible to determine the distribution of dark matter in a cluster of galaxies by analyzing the luminous arcs created by the gravitational lensing effect of the dark matter. The left panel shows the optical image of a cluster of galaxies; close examination reveals a number of short, blue arcs. The right panel shows, on the same scale, the distribution of dark matter required to produce the blue arcs gravitationally. The dark matter is concentrated toward the center of the cluster but has a smoother distribution than the galaxies. (A. Tyson, Bell Laboratories)

Figure 18.35   The Local Supercluster.   This is an artist's concept of the cluster of galaxy clusters to which the Local Group belongs. The Local Group is at position (0,0) in this diagram. (Illustration from *Sky and Telescope Magazine* by Rob Hess. © 1982 by the Sky Publishing Corporation; also with the permission of R. B. Tully)

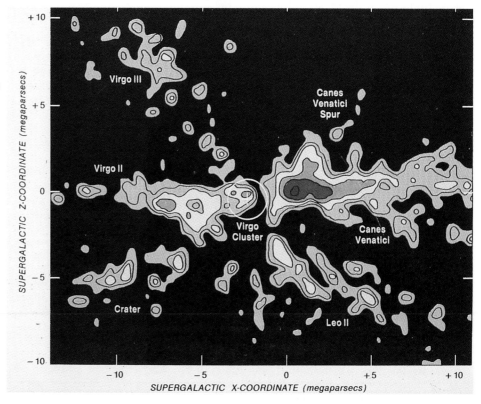

Astronomical Insight
Galaxies at the Edge of the Universe

Astronomers are forever probing the most distant reaches of the universe that can be reached with telescopes. These efforts are important because new kinds of objects may be revealed and new information on the overall nature of the universe and the distribution of matter in it may be found; furthermore, the farther away we look, the farther back in time we are seeing (because of light-travel time; this is discussed further in the next three chapters). One of the primary motivations for building the *Hubble Space Telescope* was to detect fainter objects than can be seen with telescopes that look up through the Earth's atmosphere.

It appears, however, that ground-based telescopes may have "scooped" the *Space Telescope* by observing the most distant objects it will ever be possible to see. Using modern electronic detectors in place of photographic film, astronomers have obtained images of regions of the sky that, on existing photographs, appear to be completely dark and devoid of objects. The superb sensitivity of the new detectors revealed a host of faint blue objects rather than a dark sky. These ethereal images represent objects as faint as magnitude 30, exceeding the limiting sensitivity of the *Space Telescope*.

It is not easy to demonstrate conclusively what the objects are, but the most likely explanation is that they are very distant galaxies. If so, the light from them has taken several billion years to reach us, which means that these galaxies must have been very young at the time the light was emitted.

It is possible that these objects are galaxies seen as they were during or shortly after their formation. If so, then a search for even fainter galaxies might reveal nothing. If there

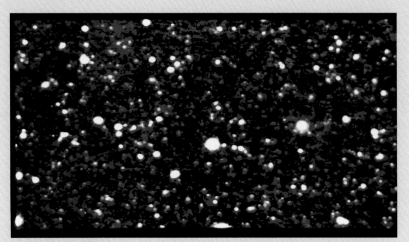

Very Faint Galaxies? The faint blue objects in this image, obtained by using a very sensitive electronic detector with a large telescope, are thought to be galaxies. If so, they are being seen as they were many billions of years ago, when the universe was young. (A. Tyson, Bell Laboratories).

were no galaxies at earlier times, as some theories of cosmology suggest, then a search for more distant galaxies would be the same as a search for galaxies at times earlier than the era when galaxies formed. If the theories are correct, there would have been no galaxies at these early times.

Most theories for the evolution of the universe (Chapter 21) postulate that at the earliest times no stars and galaxies existed. It is thought that all the material of the universe was in the form of atoms and ions and possibly huge quantities of "dark" matter, which may still be the dominant form of matter in the universe (see the discussions in this chapter and in Chapter 21). At some time, several billion years ago, stars and galaxies came into being through the gravitational condensation of atomic matter, and the universe began to look something like it does today. The very young galaxies now being detected may represent the time when galaxy formation was just beginning. If so, then no matter how

much farther we probe, nor how faint the objects we seek, we will find nothing but dark skies beyond these objects.

This finding, if confirmed, has profound implications. For one thing, it explains a long-standing riddle: If the universe were really infinite, then in any direction from the Earth our line of sight should eventually intersect the surface of a star, and the nighttime sky should be bright instead of dark. Now we see that in effect the universe is not infinite because there is a "horizon" in time before which no luminous objects such as stars and galaxies existed (see the discussion of Olbers's paradox in Chapter 21).

Even though no stars and galaxies can be seen farther away (i.e., longer ago) than some limiting distance (or some early time), we do have a way of observing the universe as it was even before there were stars and galaxies. The universe is filled with

continued on next page

radiation that is left over from the earliest moments of its creation, and this background radiation is observed today (this is discussed in the next chapter).

The conclusions discussed here concerning the faint objects seen in very sensitive images of the sky are quite new and must be regarded as tentative. When and if these results

are widely accepted, astronomers will still continue pushing toward ever-larger and more sensitive telescopes, because there is much yet to be learned about the universe. There is always more to be discovered, and making discoveries often requires gathering more light. For example, to determine definitely whether the newfound faint objects really are re-

cently formed galaxies, it will be necessary to obtain spectra of them, and this requires more light than is needed for simply detecting the objects. Obtaining their spectra is difficult or impossible with existing telescopes. It is safe to assume that astronomers will never outgrow their need for larger telescopes!

The reality of superclusters has been difficult to establish, and for a while many astronomers were not convinced. Today, however, few doubt that clusters of galaxies tend to be grouped, although there is still disagreement on the significance of the groups. The best evidence lies in the distribution of rich clusters (some 1,500 of these have been cataloged), which clearly tend to congregate in certain regions, with relatively empty space in between.

The uncertainty that remains has to do with whether the groupings are random occurrences or reflect a fundamental unevenness in the distribution of matter in the universe. This is usually tested by comparing the apparent grouping of clusters with what would be expected if they were actually distributed in a random fashion (after all, occasional accidental groupings will always appear in any random scattering of objects). Mathematical experiments of this type appear to show that the observed superclusters could be random concentrations of clusters and therefore may not have profound significance for the universe as a whole. On the other hand, very recent observations appear to show that a large number of clusters of galaxies may be linked into an enormous filamentary supercluster spanning a major portion of the sky. Previously unrecognized because it crosses the plane of the Milky Way, where our view of distant galaxies is obscured, this supercluster, if real, is far too large to have been the result of random groupings of clusters. The apparent existence of this supercluster may indicate that there is a basic unevenness in the distribution of matter in the universe.

Other signs of an uneven distribution of matter in the universe have begun to show up in the observations. As astronomers perform larger and larger surveys of galaxy positions and distances, probing more and more deeply into the far reaches of the universe, they find increasing evidence that the galaxies are not distributed randomly. The structures being discovered are larger even than superclusters, yet still much smaller than the overall size scale of the universe itself. Today it still is not clear what the term "supercluster" means. While the reality of enormous concentrations of clusters of galaxies is unquestioned, these structures appear to be segments of gigantic sheets and arcs rather than isolated groupings. Between the sheets and arcs are huge spaces that appear to be relatively devoid of visible matter.

Several years ago, a group of astronomers carrying out a large-scale survey of galaxies discovered a huge region that appeared to be empty, devoid of galaxies. This cavity or void was some hundreds of millions of parsecs across. More recently, this void and others like it have been found to be part of a cell-like structure in the universe, where sheets of galaxies and clusters of galaxies separate large, relatively empty regions from each other (Fig. 18.36). The structure of the universe may be likened to a gigantic Swiss cheese or bubble bath. Very recently, a huge wall-like structure was found, consisting of galaxies and clusters arranged in an enormous sheet so big that its true dimensions are yet to be uncovered.

The result of all these surveys and findings is that the universe appears to be far lumpier and more uneven than previously envisioned. The question now is how it got this way: Did the universe start out in a uniform state and then become lumpy, or was it somehow formed in an uneven, clumpy state? Possible answers to this question are discussed in the next section and in Chapter 21.

THE ORIGINS OF CLUSTERS AND SUPERCLUSTERS

The problem of explaining the formation of clusters of galaxies and larger structures involves fundamental questions about the universe itself. We must ask how and why (and when) the universe became clumpy enough for condensations of matter to collapse grav-

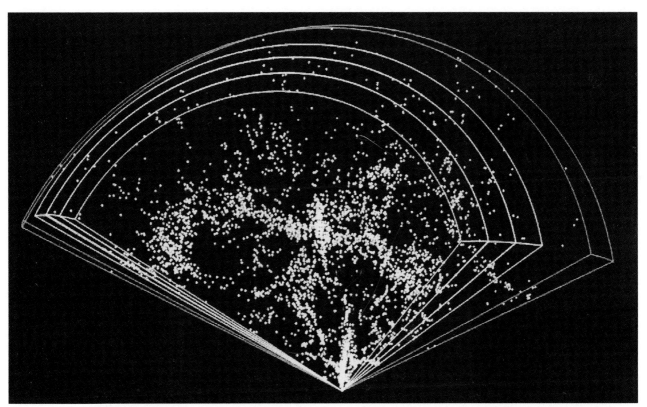

Figure 18.36   Arclike superclusters.   This shows the distribution of galaxies in a region of the sky that has recently been mapped completely, showing that superclusters are arranged in curved filamentary structures. Such groupings provide evidence that there is a large-scale structure in the universe and that matter is not distributed uniformly. (M. Geller, J. Huchra, M. Kurtz, and V. de Lapparent, Harvard-Smithsonian Center for Astrophysics)

itationally and form galaxies and clusters. In some respects, this brings up issues similar to those we confronted in explaining the formation of stars, because the probability that a cloud will collapse depends on its density and temperature (see the discussion of star formation in Chapter 14). For a given gas density, the cloud mass that will spontaneously collapse is determined by the temperature. The higher the temperature, the greater the internal energy of the gas, and the larger the region that must become self-gravitating if collapse is to occur. If we believe, as most astronomers now do, that 90 percent or more of the mass of the universe is in the form of dark matter rather than visible stars (see Chapter 21), then it is the temperature of the dark matter that determined how galaxy formation proceeded.

Thus, to understand galaxy and cluster formation, we need to invoke the properties of the mysterious, pervasive dark matter that apparently fills the universe but defies detection. The dark matter content of the universe is a central topic in modern discussions of cosmology; this is described in some detail in Chapter 21. For now, the essential point is that many lines of evidence suggest that the majority of

the mass in the universe is in forms that we cannot see. If this dark matter is hot, then the first condensations would have had masses comparable to or even greater than the masses of superclusters, around 10^{15} solar masses. If these were the first "objects" to form, then the resulting superclusters must have fragmented sequentially, forming clusters and then individual galaxies. This view, sometimes called the "top-down" theory of galaxy formation because it proposes that the largest structures formed first, would support the notion that superclusters represent a fundamental unevenness in the distribution of matter rather than a random aggregation of galaxies and clusters.

The opposing theory, called the "bottom-up" model, suggests that individual galaxies, or perhaps even smaller objects, formed first and then collected together to form clusters and superclusters. This would have happened if the dark matter was cold, because then the masses of the first condensations would have been around 10^5 or 10^6 solar masses, comparable to globular clusters. Some support for this view is found in the shapes of many galaxies, particularly those seen very far away and hence as they were very long ago, which appear to be the result of the merger of star clouds with masses smaller than a normal galaxy (as we pointed out in Chapter 17, it is possible that our own Milky Way Galaxy formed through the merger of smaller subsystems).

A crucial distinction between the top-down and bottom-up theories involves the time when galaxy formation occurred. In the cold dark matter (or bottom-up) picture, galaxy formation occurred relatively recently (that is, about 10 billion years ago), whereas in the hot dark matter (or top-down) theory, the first galaxies would have appeared much earlier, within the first billion years after the universe began. It is possible to estimate when galaxy formation took place, because we can look back into the past by looking sufficiently far away. The light from distant objects takes time to reach us; for example, if we look at a galaxy 1 billion light-years away, we see it as it was 1 billion years ago. If we can look back far enough to see galaxies in the process of formation, their distance from us will tell us how long ago the formation occurred. A possible protogalaxy, a huge amorphous gas and star-formation system some 12 billion light-years away, has been tentatively identified as a galaxy just in the process of formation. If this is correct, then galaxy formation apparently took place at about the right time to be consistent with the bottom-up theory.

The enormous extent of some superclusters presents a problem for the bottom-up theory, however. There appears to be an overall bubblelike network of filamentary superclusters (see Fig. 18.36), which is very difficult to explain as a result of the aggregation of clusters. Furthermore, indirect evidence (discussed in the next chapter) suggests the existence of huge concentrations of mass comparable to the 10^{15} solar masses expected from the top-down, or hot dark matter, theory. Thus, at present we do not have a clear basis for adopting either picture over the other; perhaps in the end we will find that both are wrong, or that some combination of the two prevails. In any event, the extent and nature of dark matter in the universe, as well as the manner in which it became lumpy and formed galaxies and clusters, are very important to our understanding of the universe as a whole.

Despite the apparent difficulty in explaining the enormous observed structures as due to random gravitational encounters among galaxies, recent model calculations show that this could have happened. Using larger computer models than previously available, a group of astronomers has recently shown that such structures as the huge filaments, voids, and "walls" of galaxies that have been observed could have formed by the gradual process of gravitational clumping, starting from an early universe in which the matter was uniformly and evenly distributed. If so, then the bottom-up theory, in which the dark matter is cold and therefore dominated by gravitational forces, may prove successful in explaining the observed overall distribution of matter in the universe.

PERSPECTIVE

We have at last completed our tour of the universe, having discussed nearly all the forms and types of organization that matter can take. We must still deal with a few specific kinds of objects that have not yet been described, but the overall picture of the universe in its present state is now more or less complete.

We are ready to tackle questions about the nature of the universe itself, its overall properties, and its dynamic nature, and we will do so before examining some of the peculiar objects that are clues to the past.

SUMMARY

1. Galaxies are categorized by shape in two general classes: spirals and ellipticals. There are also many S0 galaxies (disk-shaped, but with no spiral structure) and irregular galaxies.

2. Distances to galaxies are measured using a variety of standard candles, such as Cepheid variables, extremely luminous stars, supernovae, and galaxies of standard types. One useful technique uses a correlation between the width of the 21-centimeter line and luminosity to determine absolute magnitude and distance.

3. Masses of spiral galaxies are determined by applying Kepler's third law to the outer portions, where the orbital velocity and period are determined from a rotation curve. The internal velocity dispersion is used to determine the masses of elliptical galaxies. Galactic masses can also be determined by applying Kepler's third law to binary galaxies.

4. Galactic luminosities and diameters vary quite widely among elliptical galaxies but not so widely among spirals.

5. While all galaxies are dominated by Population II stars, spirals tend to contain a greater proportion of Population I stars, have substantial quantities of interstellar matter, and generally seem to be in a state of continuous evolution and stellar cycling. Ellipticals, on the other hand, have few or no Population I stars, contain little or no cold interstellar matter, and generally do not seem to have active stellar cycling at the present time.

6. Spiral galaxies appear to originate from rotating gas clouds that flatten into disks before all the gas is used up in star formation. Ellipticals seem to result when star formation consumes all of the gas before collapse to a disk occurs.

7. Many galaxies are members of clusters rather than being randomly distributed throughout the universe.

8. The Milky Way is a member of a cluster called the Local Group, which has about 30 members.

9. The nearest neighbors to the Milky Way are the Magellanic Clouds, both of which are Type I irregulars.

10. The Andromeda galaxy, a huge spiral of type Sb, is similar in size and general properties to the Milky Way, and the two are among the most prominent members of the Local Group.

11. The dominance of elliptical galaxies in rich clusters is probably the result of the conversion of spirals into ellipticals by collisions and tidal interactions with other galaxies. Many rich clusters have a giant elliptical galaxy at the center that probably formed from the merger of several galaxies that settled there as a result of collisions.

12. Masses of clusters can be determined from the sum of the masses of individual galaxies, or by the internal velocity dispersion of the galaxies in the clusters.

13. Observations of luminous arcs in clusters of galaxies confirm the presence of dark matter and allow the distribution of the dark matter to be mapped.

14. Although there is no evidence for a general intergalactic medium, there is a very hot, rarefied gas filling the space between galaxies in some rich clusters. Observed best at X-ray wavelengths, this gas appears to have been created by the galaxies themselves (through mass loss in galactic winds that result from stellar winds and supernova explosions).

15. Clusters of galaxies tend to be grouped into aggregates called superclusters, but these may be random concentrations rather than fundamental inhomogeneities of the universe. Superclusters appear to be sheet-like, with an overall distribution resembling a series of huge arclike structures, and this may tell us something about the early distribution of matter in the universe.

16. Clusters of galaxies formed because of an uneven distribution of matter at some point early in the history of the universe, but it is not known how this clumpiness originally developed. One of the two competing possibilities is that the early universe was hot, so that only very massive objects could collapse; this is the "top-down" theory, in which superclusters of galaxies formed first and then fragmented to form galaxies. The other scenario, called the "bottom-up" theory, assumes that the matter in the universe was cold, so that objects as small as galaxies could condense and then aggregate into clusters.

REVIEW QUESTIONS

1. How do we know that the tuning-fork diagram does not represent an evolutionary sequence?

2. A Type I supernova is observed in a distant galaxy. The absolute magnitude of such a supernova is assumed to be −19. If the supernova has an apparent magnitude of +24 (near the limit that can be observed with the largest telescopes), how far away is the galaxy in which it lies?

3. The mass-to-light ratio for spiral galaxies is usually much larger than 1, and for elliptical galaxies it is larger yet. What does this tell us about the types of stars that emit most of the light from such galaxies?

4. Elliptical galaxies are dominated by Population II stars, whereas spiral galaxies contain the younger Population I stars. Why are spiral galaxies thought to be as old as elliptical galaxies?

5. How does our location in a cluster (the Local Group) help us develop distance measurement techniques useful for determining distances to galaxies far away, outside our cluster?

6. Explain why the relative number of galaxies of different types may not be the same in a rich cluster of galaxies as it is among galaxies not belonging to such clusters.

7. If a dwarf elliptical galaxy has an absolute magnitude of $M = -15$ and could be detected with an apparent magnitude as faint as $m = +20$, how far away can this type of galaxy be found? Compare this distance with the diameter of the Local Group and with the distance to the Virgo cluster of galaxies, a moderately large cluster 15 Mpc away.

8. Why would it be surprising if the Local Group included a giant elliptical galaxy among its members?

9. Summarize the techniques that are used for finding the masses of clusters of galaxies. Explain any difficulties or shortcomings of each method.

10. Why is it thought that the intracluster gas observed in rich clusters of galaxies has been expelled from the galaxies in the cluster rather than being truly intergalactic gas?

11. Explain why there has been some question about the significance of superclusters, even though their existence is well established.

ADDITIONAL READINGS

Allen, D. A. 1987. Star formation and *IRAS* galaxies. *Sky and Telescope* 73:372.

Barnes, J., L. Hernquist, and F. Schweizer. 1991. What happens when galaxies collide. *Scientific American* 265(2):40.

Burns, J. O. 1986. Very large structures in the universe. *Scientific American* 255(1):38.

Chaffee, F. H. 1980. Discovery of a gravitational lens. *Scientific American* 243(5):60.

Chevalier, R. A., and C. L. Sarazin. 1987. Hot gas in the universe. *American Scientist* 75:609.

Davies, J., M. Disney, and S. Phillipps. 1990. Are spiral galaxies heavy smokers? *Sky and Telescope* 80:37.

de Boer, K. S., and B. D. Savage, 1982. The coronas of galaxies. *Scientific American* 247(2):54.

de Vaucouleurs, G. 1987. Discovering M31's spiral shape. *Sky and Telescope* 74:595.

Dressler, A. 1991. Observing galaxies through time. *Sky and Telescope* 82(2):126.

Geller, M. J. 1978. Large-scale structure of the universe. *American Scientist* 66:176.

Geller, M. J., and Huchra, J. P. 1991. Mapping the universe. *Sky and Telescope* 82(2):134.

Gorenstein, M. V. 1983. Charting paths through gravity's lens. *Sky and Telescope* 66(5):390.

Gorenstein, P., and W. Tucker. 1978. Rich clusters of galaxies. *Scientific American* 239(5):98.

Harris, W. E. 1991. Globular clusters in distant galaxies. *Sky and Telescope* 81(2):148.

Hirschfeld, A. 1980. Inside dwarf galaxies. *Sky and Telescope* 59(4):287.

Hodge, P. 1981. The Andromeda galaxy. *Scientific American* 244(1):88.

————. 1986. *Galaxies.* Cambridge, Mass.: Harvard University Press.

Kaufmann, M. 1987. Tracing M31's spiral arms. *Sky and Telescope* 73:135.

Lake, G. 1992. Understanding the Hubble sequence. *Sky and Telescope* 83(5):515.

Larson, R. B. 1977. The origin of galaxies. *American Scientist* 65:188.

Lawrence, J. 1980. Gravitational lenses and the double quasar. *Mercury* 9(3):66.

Levy, D. H. 1991. A grand gathering of galaxies. *Astronomy* 19(3):44.

Mathewson, D. 1984. The clouds of Magellan. *Scientific American* 252(4):106.

Oort, J. 1992. Exploring the nuclei of galaxies (including our own). *Mercury* 21(2):57.

Rubin, V. C. 1983. Dark matter in spiral galaxies. *Scientific American* 248(6):96.

Schramm, D. N. 1991. The origin of cosmic structure. *Sky and Telescope* 82(2):140.

Silk, J., A. S. Szaley, and Y. B. Zeldovich. 1983. The large-scale structure of the universe. *Scientific American* 249(4):56.

Strom, S. E., and K. M. Strom. 1977. The evolution of disk galaxies. *Scientific American* 240(4):56.

Struble, M. F., and H. J. Rood. 1988. Diversity among galaxy clusters. *Sky and Telescope* 75:16.

Talbot, R. J., E. B. Jensen, and R. J. Dufour. 1980. Anatomy of a spiral galaxy. *Sky and Telescope* 60(1):23.

Talcott, R. 1992. COBE's big bang. *Astronomy* 20(8):42.

Toomre, A., and J. Toomre. 1973. Violent tides between galaxies. *Scientific American* 229(6):38.

Tucker, W., and K. Tucker. 1989. Dark matter in our galaxy. *Mercury* 18(1):2 (Part 1): *Mercury* 18(2):51 (Part 2).

Turner, E. L. 1988. Gravitational lenses. *Scientific American* 259(1):54.

Tyson, A. 1992. Mapping dark matter with gravitational lenses. *Physics Today* 45(6):24.

ASTRONOMICAL ACTIVITY
Finding the Andromeda Galaxy

To see the Andromeda galaxy with the naked eye is an awesome experience and not too difficult to do. You will need a very clear, dark night with no Moon (i.e., the Moon should be at third quarter or new moon phase, if you want to view Andromeda during the early evening). And you will have better luck if you can get out in the country, away from city lights, although it is not impossible to see Andromeda from town, if you can find a location that is sheltered from direct illumination.

The constellation Andromeda is up during the evening in late summer and early autumn. The accompanying diagram shows where the galaxy is. To begin, find the large-scale pattern of Pegasus; this is a kite-shaped arrangement of bright stars. Then follow the diagram to the grouping near the Andromeda galaxy, and look for a faint, fuzzy patch of light at the indicated position.

The galaxy is large and dim and will not reveal its spiral structure or any other details to the unaided eye (only a time-exposure photograph taken through a telescope can do that). But nevertheless, the galaxy is an awesome sight, particularly when you consider that you are looking beyond the confines of our home galaxy, and that your eye is receiving photons of light that have been traveling for 2 million years. The Andromeda galaxy is about a thousand times more distant than any of the stars you can see and about six times farther away than the other neighbor galaxies that are naked-eye objects, the Magellanic Clouds (which are visible only from the Southern Hemisphere). To view the Andromeda galaxy is to look as far away (and as far back in time) as the human eye can see.

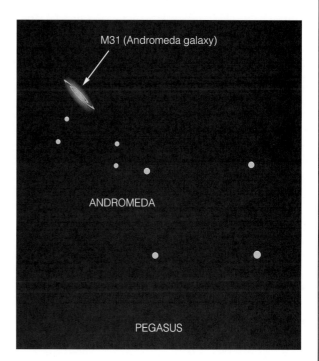

Universal Expansion and
the Cosmic Background

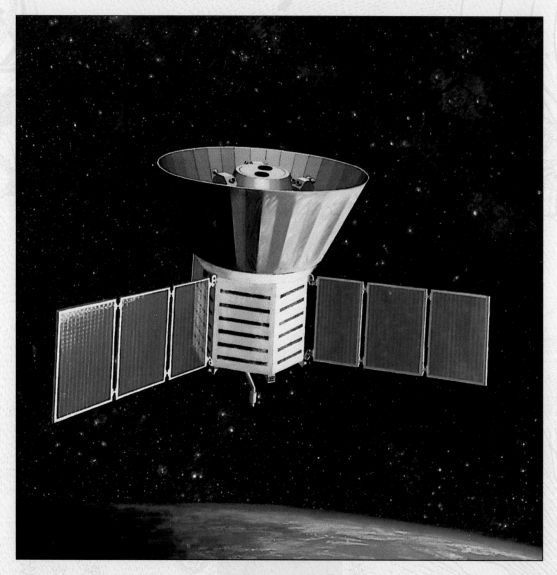

The *Cosmic Background Explorer* (NASA)

PREVIEW

This chapter presents you with two of the most significant phenomena ever discovered: the expansion of the universe and the radiation that fills it. The expansion, first discovered through the Doppler-shifted spectra of distant galaxies, is a cornerstone of modern cosmology, now so thoroughly embedded in our understanding that it is used to deduce the properties of faraway objects such as quasars (discussed in the next chapter). The cosmic background radiation, in turn, provides us with a powerful tool for probing the distant past of the universe. Taken together, the universal expansion and the background radiation have established beyond reasonable doubt in the minds of most astronomers that the universe itself has a life: it had a beginning, it evolves, and one day it may have an end. In this chapter we confine ourselves primarily to a description of the observational properties of the expansion and the background radiation, which remain two of the most pressing and active areas of research by modern astronomers; most discussion of the consequences of the observations is reserved for later.

A recurrent theme throughout our study of astronomy has been the dynamic nature of celestial objects. Everything we have studied in detail, from planets to stars to galaxies, has turned out to be in a constant state of change. Now that we are prepared to examine the entire universe as a single entity, we should not be surprised to see this theme maintained. That the universe itself is evolving, with a life story of its own, has been accepted by most astronomers, but until recently, some still doubted it. In this chapter, we will study the evidence that the universe is evolving.

HUBBLE'S GREAT DISCOVERY

Even before it was established unambiguously that the nebulae were really galaxies, a great deal of effort went into observing them. Their shapes were studied carefully, and their spectra were analyzed in detail. Typically the spectrum of a galaxy resembles that of a moderately cool star, with many absorption lines (Fig. 19.1). Because the spectrum represents the light from a huge number of stars, each moving at its own velocity, the Doppler effect causes the spectral lines to be rather broad and indistinct. Nevertheless, it is possible to analyze these lines in some detail, and during the decade before the Shapley-Curtis debate in 1920, one of the early workers in this field, V. M. Slipher, discovered that the nebulae tend to have large velocities away from the Earth. He found that the spectral lines nearly always were shifted toward

the red and that this tendency was most pronounced for the faintest nebulae. Slipher measured velocities as great as 1,800 km/sec.

Following Slipher's work, others continued to study spectra of nebulae. Their interest was heightened after Hubble's demonstration in 1924 that these objects were undoubtedly distant galaxies, comparable to the Milky Way in size and complexity. Hubble and others estimated distances to as many nebulae as possible, using the Cepheid variable period-luminosity relation for the nearby nebulae and other standard candles such as novae and bright stars for the more distant ones.

In 1929 Hubble made a dramatic announcement: The speed with which a galaxy moves away from the Earth is directly proportional to its distance. If one galaxy is twice as far away as another, for example, its velocity is twice as great. If it is 10 times farther away, it is moving 10 times faster.

The implications of this relationship between distance and velocity are enormous. It means that the universe itself is expanding, its contents becoming more widely dispersed in the process. All the galaxies in the cosmos are moving away from all the others (Fig. 19.2).

To envision why the velocity increases with increasing distance, it may be helpful to resort to a commonly used analogy: Imagine a loaf of bread that is rising. If the dough has raisins sprinkled uniformly through it, the raisins will move farther apart as the dough expands. Suppose that the raisins are 1 centimeter apart before the dough begins to rise, and after 1 hour, the dough has risen to the point where adjacent raisins are 2 cm apart. A given raisin is now 2 cm from its nearest neighbor, but 4 cm away from the next raisin over, 6 cm from the next one, and so on. The distance between any pair of raisins has doubled. From the point of view of any one raisin, its nearest neighbor had to move away from it at a speed of 1 cm/hr, the next raisin farther away had to move at 2 cm/hr, the next one had a speed of 3 cm/hr, and so on. If we let the dough continue to rise to the point where the adjacent raisins are 3 cm apart, then we find that all the distances between pairs have tripled over what they were initially, and the speeds needed to accomplish this are again directly proportional to the distance between a given pair. The farther away a raisin is to begin with, the farther it must move to maintain the regular spacing during the expansion, and therefore the greater is its velocity. In the same way, the galaxies in the universe must increase their separations from each other at a rate proportional to the distance between them (Fig. 19.3), or they would bunch up. By observing the velocities of galaxies, astronomers are keeping watch on the raisins to see how the loaf of bread is coming along.

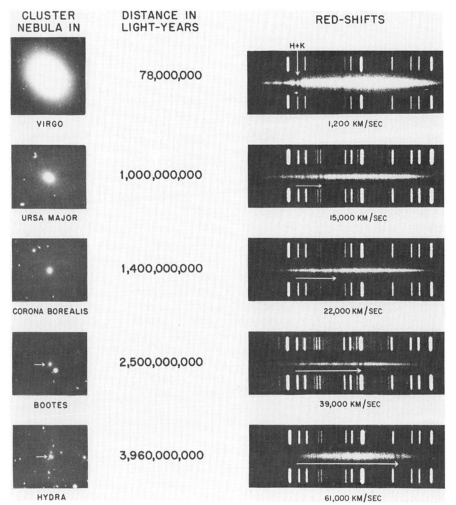

CLUSTER NEBULA IN	DISTANCE IN LIGHT-YEARS	RED-SHIFTS
VIRGO	78,000,000	1,200 KM/SEC
URSA MAJOR	1,000,000,000	15,000 KM/SEC
CORONA BOREALIS	1,400,000,000	22,000 KM/SEC
BOOTES	2,500,000,000	39,000 KM/SEC
HYDRA	3,960,000,000	61,000 KM/SEC

Figure 19.1   Spectra of galaxies. These examples illustrate the broad absorption lines characteristic of spectra of large groups of stars such as galaxies. Researchers noticed before 1920 that the spectral lines tend to be shifted toward the red in galaxy spectra, which indicates that galaxies as a rule are receding from us. (Palomar Observatory, California Institute of Technology)

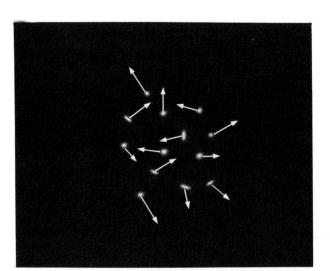
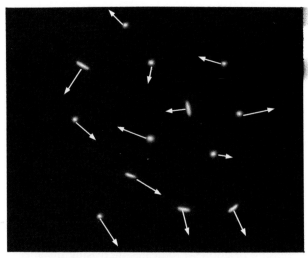

Figure 19.2   The expanding universe.   These drawings show the same group of galaxies at one time (left) and then at a later time (right). All of the spacings between the galaxies have increased because the universe is expanding.

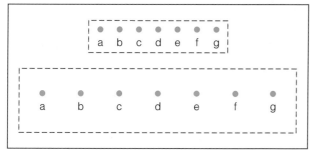

Figure 19.3   Velocity of expansion as a function of distance.   At the top is a row of galaxies and at the bottom is the same row 1 billion years later, when the universe (represented by the rectangular box) has expanded so that the distance between adjacent galaxies has doubled from 1 to 2 megaparsecs. From the viewpoint of an observer in any galaxy in the row, the recession velocity of its nearest neighbor is 1 Mpc per billion years; of its next nearest neighbor, 2 Mpc per billion years; of the next, 3 Mpc per billion years; and so on. The velocity is proportional to the distance in all cases, for any pair of galaxies.

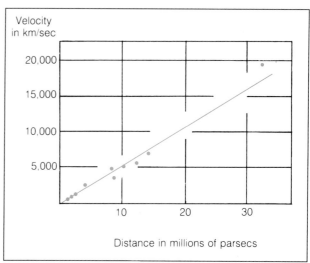

Figure 19.4   The Hubble law.   This early diagram prepared by E. Hubble and M. Humason shows the relationship between galaxy velocity and distance. The slope of the relation has since been altered as data for more galaxies have been added and the measurement of distances has improved, but the general appearance of this figure is the same today. (Estate of E. Hubble and M. Humason. Reprinted with permission)

It is important to realize that it doesn't matter which raisin we choose to watch; from any point in the loaf, all other raisins appear to move away with speeds proportional to their distances from that point. Thus we do not conclude that our galaxy is at the center of the universe; from *any* galaxy, it would appear that all others are receding. It is also important to realize that the raisin-bread analogy has serious shortcomings: it suggests that the universe can be viewed as a simple three-dimensional object, with an edge and a center, which is inappropriate (this is discussed further in Chapter 21). It is better to try to envision a universe with no boundaries and no center; space itself is getting bigger. There is no location within the universe that can today be identified as the center, nor is there an "outside" from which the universe can be viewed.

Following Hubble, several other astronomers extended the study of galaxy motions to greater and greater distances, with the same result: as far as the telescopes can probe, the galaxies are moving away from each other at speeds proportional to their distances. Expansion is a major feature of the universe we live in, one that must be taken into account as we seek to understand its origins and its fate.

It is often convenient to display the data showing the universal expansion in a graph of velocity versus distance (as Hubble originally did; see Fig. 19.4) or on a plot of redshift versus apparent magnitude (Fig. 19.5), since increasing apparent magnitude (i.e., decreasing brightness) indicates increasing distance.

The galaxies within a cluster have their own orbital motions, which are separate from the overall expansion (Fig. 19.6). As the universe expands, the clusters move apart from each other, while the individual galaxies within a cluster do not. The motions of galaxies within a cluster are relatively unimportant for very distant galaxies that are moving away from us at speeds of thousands of kilometers per second, but for those nearby, the orbital motion can rival or exceed the motion caused by the expansion of the universe. The Andromeda galaxy, for example, is actually moving *toward* the Milky Way at a speed of about 100 km/sec, whereas at its distance of 700 kpc from us, the universal expansion should give it a recession velocity of about 40 km/sec.

HUBBLE'S CONSTANT AND THE AGE OF THE UNIVERSE

The relation discovered by Hubble can be written in simple mathematical form:

$$v = Hd ,$$

where v is a galaxy's velocity of recession and d is its distance in megaparsecs. The Hubble constant, H, is then given in units of km/sec/Mpc, if v is in units of km/sec.

The value of H is difficult to establish (Table 19.1). It is done by collecting as large a body of data as

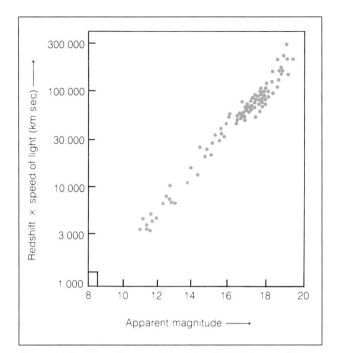

Figure 19.5   A modern version of the Hubble law. Often, apparent magnitude is used on the horizontal axis instead of distance, because the two quantities are related, particularly if the diagram is limited to galaxies of the same type, as this one is. The small rectangle at the lower left indicates the extent of the relationship as Hubble first discovered it; today many more, much dimmer galaxies have been included. The most rapid galaxies are not actually traveling at the speed of light as this diagram implies, because relativistic corrections have not been applied (see the discussion in the next chapter). (Adapted from J. Silk, 1980, *The Big Bang* [San Francisco: W. H. Freeman])

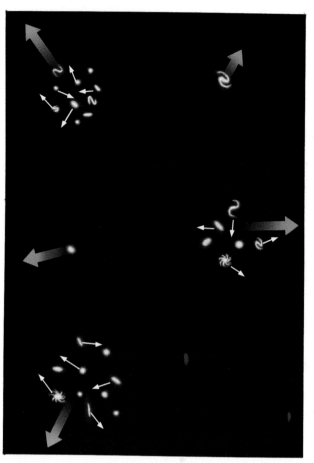

Figure 19.6   Local motions. Although the universe is undergoing a systematic overall expansion, individual galaxies within clusters, and even clusters within superclusters, have random individual motions. Hence within the Local Group the galaxies are not uniformly receding from each other.

TABLE 19.1   Measurements of the Hubble Constant

Value of H (km/sec/Mpc)	Observer	Date	Implied Age of Universe (years)
540	Hubble	1930	1.8×10^9
260	Baade	1949	3.8×10^9
180	Humason, et al.	1955	5.4×10^9
75	Sandage	1956	1.3×10^{10}
56	Sandage and Tammann	1974	1.8×10^{10}
100	van den Bergh	1975	1.0×10^{10}
80	Tully and Fisher	1977	1.2×10^{10}
100	de Vaucouleurs, et al.	1979	1.0×10^{10}
95	Aaronson	1980	1.0×10^{10}
65	Mould, et al.	1980	1.5×10^{10}
50	Sandage and Tammann	1982	2.0×10^{10}
82	Aaronson and Mould	1983	1.2×10^{10}

Note: Information from M. Rowan-Robinson, 1985, *The Cosmological Distance Ladder* (New York: W. H. Freeman).

possible on galactic velocities and distances, and then deducing the value of H that best represents the relationship between distance and velocity. Hubble did this first, finding that H = 540 km/sec/Mpc, meaning that for every megaparsec of distance, the velocity increased by 540 km/sec. The standard candles were not very well established in Hubble's day, when it was still news that the nebulae were distant galaxies, and this value for H turned out to be an overestimate. The best modern values for H are between 50 and 100 km/sec/Mpc. Until recently, a consensus had been developing that the best value is 55 km/sec/Mpc, but a new distance determination for a number of galaxies has now suggested a value close to 90 km/sec/Mpc. The precise value of H has extremely important implications for our understanding of the universe and its expansion, and intense research effort is being devoted to refining the estimates.

If the universe is expanding, it follows that all the matter in it used to be closer together than it is today (Fig. 19.7). Following this logic to its obvious conclusion, we find that the universe was once concentrated in a single point, from which it has been expanding ever since.

From the rate of expansion, we can calculate how long ago the galaxies were all together in a single point. From the simple expression

$$\text{time} = \frac{\text{distance}}{\text{velocity}} ,$$

we find

$$\text{age} = \frac{d}{Hd} = \frac{1}{H} .$$

The age of the universe is equal to $1/H$, if the expansion has been proceeding at a constant rate since it began. As we will see in Chapter 21, this assumption of a constant expansion rate is not strictly true (it was more rapid in the beginning), but it gives a useful estimate of the age for now.

If we adopt, for the sake of discussion, a value for H of 75 km/sec/Mpc, then the age of the universe is 4.1×10^{17} sec = 1.3×10^{10} years. (To do this calculation, you must first convert the value of a megaparsec into kilometers.)

A simple calculation based on the observed expansion rate has led to a profound conclusion: Our universe is about 13 billion years old. If this value is correct, it fits in with what we have learned about galactic ages. The Sun, for example, is thought to be about 4.5 billion years old, and the oldest globular clusters are apparently some 14 billion years old, acceptably close, given the uncertainties of the measurements. On the other hand, a much higher value of H would lead to an age for the universe that is much smaller than the estimated ages of globular clusters, and those who favor a value nearer to 100 km/sec/Mpc are facing just such a dilemma.

We can easily see what happens to our estimate of the age of the universe if we choose other values of

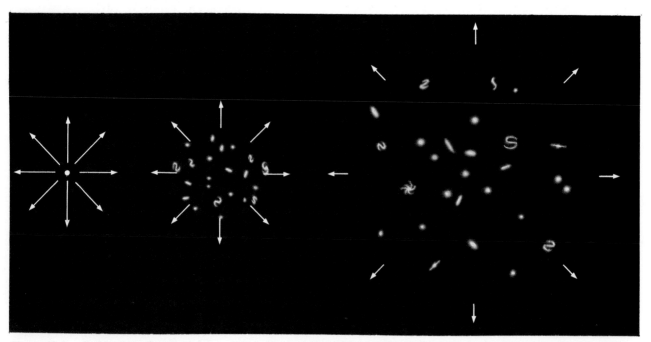

Figure 19.7   Backtracking the expansion.   If the universe began with a small size, the present rate of expansion provides information on how long ago the expansion began. With considerable uncertainty because the expansion rate has probably not been constant, this leads to an estimated age for the universe of between 10 and 20 billion years.

H. For example, if Hubble's value of 540 km/sec/Mpc had been correct, then the age would have been calculated as less than 2 billion years, and if the recently suggested value of about 90 km/sec/Mpc is found to be correct, then the age is 11 billion years. In either case, there is a conflict with the estimated ages of globular clusters; it will be interesting to see how this conflict is resolved if indeed a value for *H* of 90 km/sec/Mpc becomes widely accepted. The conflict worsens when we realize that age estimates based on 1/*H* are probably too large, because they do not take into account the fact that the expansion rate at the earliest times was somewhat more rapid than it is today. This further complicates things if a value of 90 km/sec/Mpc is found to be best. Whatever the correct ages are, this situation illustrates how uncertain some of the most important measurements in astronomy can be.

Because many other important questions about the universe also depend on the value of *H*, considerable effort is being expended to define its value as accurately as possible. One of the principal tasks of the *Hubble Space Telescope* is to probe the most distant galaxies possible to determine their distances and velocities so that this information can be used to help find the correct value of *H*.

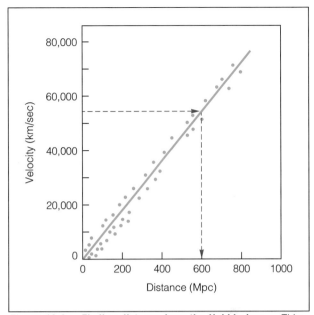

Figure 19.8   Finding distance from the Hubble law.   This diagram shows how the distance to a galaxy can be deduced directly from the Hubble law, if the redshift—and hence the velocity—is measured.

REDSHIFTS AS YARDSTICKS

The expansion of the universe has a very practical benefit for astronomers concerned with the properties of distant galaxies: it provides a means of determining their distances (Fig. 19.8). If we know the value of *H*, then we can find the distance to a galaxy simply by measuring its velocity as indicated by the Doppler shift of its spectral lines. The distance is given by

$$d = \frac{v}{H}.$$

Thus, if we adopt *H* = 75 km/sec/Mpc, then a galaxy found to be receding at 7,500 km/sec, for example, has the distance

$$d = \frac{7{,}500}{75} = 100 \text{ Mpc}.$$

This technique can be applied to the most distant and faint galaxies, where it is nearly impossible to observe any standard candles. It is important to remember, however, that the use of Hubble's relation to find distances is only as accurate as our value of *H*, and the value that we use is derived in turn from distance determinations that depend on standard candles. Furthermore, because the expansion rate of the universe has not been constant, for very distant galaxies it is

necessary to know that the rate was in the past in order to determine the distance accurately.

A COSMIC ARTIFACT: THE MICROWAVE BACKGROUND

Since Hubble's discovery of the universal expansion, many astronomers and physicists have explored its significance, both for what it may tell us about the future and for the information it provides about the past. A number of scientists, beginning in the 1940s with George Gamow and his associates, have done extensive studies of the nature of the universe in its earliest days. They quickly realized that if all the matter were originally compressed into a small volume, it must have been very hot, for the same reason that any gas heats up when compressed. The fiery conditions inferred for the early stages of the expansion have led to the term **big bang**, which is the commonly accepted name for all theories of the universal origin that start with an expansion from a state of high density and temperature.

Gamow's primary interest was in analyzing nuclear reactions and element production during the big bang (the modern understanding of this is described in Chapter 21). But in the course of his work, he came to another important realization: The universe must

ASTRONOMICAL INSIGHT
The Distance Pyramid

Having reached the ultimate distance scales, we can pause to look back at the steps we took to get here. As noted in the text, the use of Hubble's relation to estimate distances to the farthest galaxies depends critically on the value of H, which can be determined only from knowledge of the distances to a representative sample of galaxies. Those distances in turn depend on various standard candles, themselves calibrated through the use of other methods, such as the Cepheid variable period-luminosity relation, which are applicable only to the closest galaxies. Our knowledge of these close extragalactic distances rests on the known distances to objects within the Milky Way, such as star clusters containing variable stars, which are used to calibrate the period-luminosity relation. The distances to these clusters, in turn, are known from more local techniques, such as spectroscopic parallaxes and main-sequence fitting, and these ultimately depend on the distances to the nearest stars, derived from trigonometric parallaxes.

Thus the entire progression of distance scales, all the way out to the known limits of the universe, depends on our knowledge of the distance from the Earth to the Sun, the basis of trigonometric parallax measurements. At every step of the way outward, our ability to measure distances rests on the previous step. If we revise our measurement of local distances, we must accordingly alter all our estimates of the larger scales, affecting our perception of the universe as a whole. This elaborate and complex interdependence of

Distance (pc)	Key Object	Method
10^{-6}	Sun	Radar, planetary motions
10^{-5}		
10^{-4}		
10^{-3}		
10^{-2}		
10^{-1}		
1	Alpha Centauri	Trigonometric parallax
10^1		
10^2	Hyades cluster	Moving cluster method
10^3		
10^4	Limits of the Milky Way	Main-sequence fitting
10^5	Magellanic Couds	(spectroscopic parallax)
10^6	Andromeda Galaxy	Cepheid variables
10^7		Brightest stars
10^8	Virgo cluster	
10^9		
10^{10}		Brightest galaxies

distance-determination methods is known as the **distance pyramid,** and it is sometimes depicted in graphic form (see the table).

A very important step in the sequence shown here is represented by the Hyades cluster, a galactic star cluster some 43 parsecs away. The distance to this cluster is determined from the **moving cluster method,** in which a combination of Doppler shift measurements and apparent angular motion measurements reveal the true direction of cluster motion; then a comparison of angular motion and true velocity gives the distance to the cluster. Because much of the rest of the distance pyramid depends on the application of this technique to the Hyades cluster, great care has been taken in the analysis, and every so often it is redone. Whenever the distance to the Hyades is altered

slightly, the impact spreads as astronomers judge how this change affects other distance scales in the universe. One of the things the *Hubble Space Telescope* may accomplish is to make direct parallax measurements of stars in the Hyades, providing a new and important measurement of the distance to the cluster. If the *Space Telescope* succeeds, then once again we will have to revise our understanding of the distance scale of the universe.

have been filled with radiation when it was highly compressed and very hot, and this radiation should still be with us. At first, when the temperature was in the billions of degrees, the radiation was composed of gamma rays, but later, as the universe cooled, X rays and then ultraviolet light filled it. (Recall the relationship between temperature and wavelength of maximum emission, as stated in Wien's law; see Chapter 4). During the first million years or so after the expansion started, the radiation was constantly being absorbed and reemitted (i.e., scattered) by the matter in the universe, but eventually (when the temperature had dropped to around 3,000 K), protons and electrons combined to form hydrogen atoms, and from that time on there was little interaction between the matter and the radiation. The matter, in due course, organized itself into galaxies and stars, while the radiation has simply continued to cool as the universe has continued to expand (Fig. 19.9). The intensity and spectrum of the radiation are dependent only on the temperature of the universe and are described by the laws of thermal radiation (see Chapter 4).

Gamow and others made rough calculations showing that the present temperature of the radiation filling the universe should be very low indeed, about 25 K. More recently, R. H. Dicke and colleagues, quite independently, carried out new calculations, predicting an even lower temperature, about 5 K. Using Wien's law, it is easy to calculate the most intense wavelength of such radiation: If the temperature is 5 K, then the value of λ_{max} is 0.06 cm, in the microwave part of the radio spectrum.

Gamow was actually more interested in element production during the big bang than in the remnant radiation. Furthermore, in the late 1940s the technology needed to detect the radiation had not yet been developed. Hence no attempt was made in Gamow's time to search for the radiation, and the idea was forgotten until Dicke's work. The remnant radiation was eventually detected in 1965 by accident, when radio astronomers Arno Penzias and Robert Wilson found a persistent source of background noise while testing a new antenna (Fig. 19.10).

The implication soon became clear: the noise was the universal radiation that the theorists had predicted should exist. This was one of the most significant astronomical observations ever made, for it provided a direct link to the origins of the universe. More than a piece of circumstantial evidence, the radiation was a real artifact, the kind of hard evidence that carries weight in a court of law. Penzias and Wilson later were awarded the Nobel Prize for their discovery.

The Crucial Question of the Spectrum

Questions about the interpretation of the radiation remained, however, and some astronomers who did not accept the big bang theory of the origin of the universe were dubious about the radiation's origin. If the radiation was really the remnant of the primeval fireball, then it should fulfill certain conditions, and intensive efforts were made to find out whether it did.

One important prediction was that the spectrum

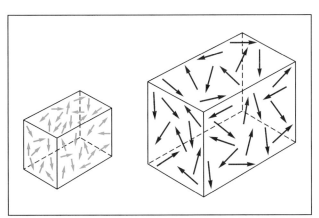

Figure 19.9   The cooling of the universal radiation.   At the left is a box representing the early universe filled with intense gamma-ray radiation. At the right is the same box, greatly expanded. The radiation is still there, but its wavelength has increased, and the temperature represented by its thermal spectrum has decreased.

Figure 19.10   The discoverers of the universal background. Robert Wilson, left, and Arno Penzias, in front of the horn-reflector antenna at Bell Laboratories with which they found a persistent noise that turned out to be the remnant radiation from the big bang. (Courtesy of AT&T)

should be that of a simple glowing object whose emission of radiation is the result only of temperature (that is, it should be a thermal spectrum, as explained in Chapter 4). If, on the other hand, the source of the radiation were something other than the remnant of the initial universe, such as distant galaxies or intergalactic gas, it could have a different spectrum. Thus the shape of the spectrum—its intensity as a function of wavelength—was a very important test of the interpretation of its origin.

Some of the early results were confusing due to unforeseen complications, but now the situation seems to have been resolved. The spectrum does indeed have the expected shape (Fig. 19.11), and the radiation is considered to be a relic of the big bang. The most difficult part of measuring the spectrum was observing the intensity at short wavelengths around 1 millimeter or less, where the Earth's atmosphere is especially impenetrable, but this was finally accomplished with high-altitude balloon payloads. Further confirmation, and a very precise temperature measurement, came from the *Cosmic Background Explorer (COBE)*, a NASA satellite launched in 1989; Fig. 19.12 (the *COBE* spectrum is shown in Fig. 19.11).

The net result of all the effort expended in determining the spectrum is that the temperature of the radiation is 2.730 K, and its peak emission occurs at a wavelength of 1.1 mm. The radiation is commonly referred to as the **microwave background,** or the **3-degree background radiation.**

The care that went into establishing the true nature of the background radiation reflects the importance of what it has to tell us about the universe. It is now widely accepted that the cosmic background radiation is indeed the result of a highly compressed and hot early universe, and this has proved to be very important in ruling out alternative theories about the earliest times.

Isotropy and Daily Variations

Another expectation of the background radiation, in addition to the nature of its spectrum, is that it should fill the universe uniformly, with no preference for any particular location or direction. A medium or radiation field that has no preferred orientation, but instead looks the same in all directions, is said to be **isotropic.** Some of the early observations of the 3-degree background were designed to test its isotropy, for again, failure to meet this expectation could imply some origin other than the big bang.

A test of isotropy, in simple terms, is just a measurement of the radiation's intensity in different directions to see whether it varies (Fig. 19.13). From the time it was discovered, the microwave back-

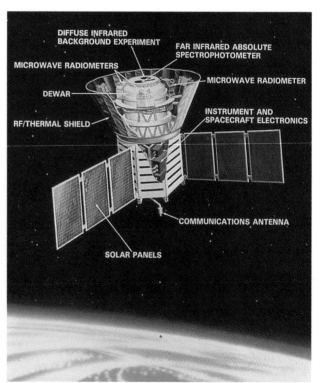

Figure 19.12   The Cosmic Background Explorer.   This satellite, launched in 1989, is making the most precise and extensive measurements yet of the microwave background radiation. (NASA)

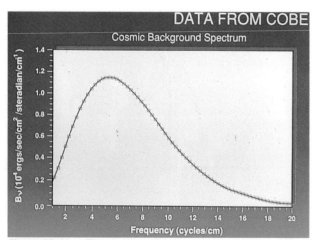

Figure 19.11   The measured background spectrum.   This shows the very accurate spectrum of the background radiation, as measured by the *Cosmic Background Explorer.* The solid line represents a purely thermal spectrum for an object having a temperature of 2.730 K; the dots represent actual measured intensities. (NASA/COBE Science Working Group)

ground showed a high degree of isotropy as expected. The issue has been pressed, however, for a variety of reasons. One possible alternative explanation of the radiation was that it arises from a vast number of individual objects, such as very distant galaxies, so closely spaced in the sky that their combined emission appears to come uniformly from all over. To test this idea, astronomers have attempted to find out whether the radiation is patchy on very fine scales, as it would be if it came from a number of point sources. So far, no evidence of any such fine-scale clumpiness has been found, and the big bang origin for the radiation has not been threatened. The big bang theory does lead to the expectation that very subtle large-scale variations of the background radiation might exist, as discussed in a later section.

24-Hour Variations

In addition to any subtle localized nonuniformities expected in the big bang theory, a broad asymmetry is expected, because of the motion of the Earth with respect to the frame of reference of the radiation. Because of the Doppler shift, the observed intensity of the radiation as seen by a moving observer will vary with direction, depending on whether the observer is looking toward or away from the direction of motion. If the observer looks ahead, toward the direction of motion, then a blueshift occurs because the peak of the spectrum is shifted slightly toward shorter wavelengths (Fig. 19.14). In terms of temperature, this means that the radiation looks a little hotter when

viewed in this direction. If the observer looks the other way, a slight redshift occurs, and the measured temperature is cooler.

The Earth is moving in its orbit about the Sun. In addition, the Sun is orbiting the galaxy, the galaxy is moving along its own path through the Local Group, and apparently the Local Group is moving with respect to the local supercluster. All of these motions combined represent a velocity of the Earth with respect to the frame of reference established by the background radiation, a velocity that should produce slight differences in the radiation temperature if it is viewed in different directions. Because of the Earth's rotation, if we simply point our radio telescope straight up, we should alternatively see high and low temperatures, as our telescope points toward and then (12 hours later) away from the direction of motion. Consequently there should be a daily variation cycle in the radiation, referred to as the **24-hour anisotropy** or the **dipole anisotropy** (*anisotropic* is the term describing a nonuniform medium).

The actual change in temperature from one direction to the other is very small and was not successfully measured until a few years ago (Fig. 19.15). The observed temperature difference from the forward to the rear direction is only a few thousandths of a degree and can only be measured with very sophisticated technology. The data show that the Earth is moving with respect to the background radiation at a speed of about 360 km/sec. When the Sun's known orbital velocity about the galaxy is taken into account, along with the galaxy's motion within the

Figure 19.13   A microwave receiver used to measure the cosmic background.   The pair of receiving horns were used to compare the radiation intensity in different directions. (G. Smoot, Lawrence Berkeley Laboratory)

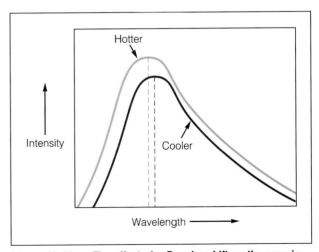

Figure 19.14   The effect of a Doppler shift on the cosmic background radiation.   If the observer on the Earth has a velocity with respect to the radiation, this velocity shifts the peak of the spectrum a little and affects the temperature that the observer deduces from the measurements. Motion of the Earth results in a daily cycle of tiny fluctuations in the observed temperature, known as the 24-hour anisotropy.

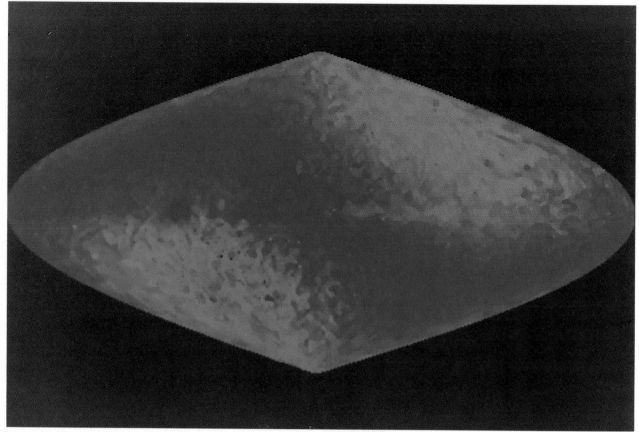

Figure 19.15   The 24-hour anisotropy as measured by *COBE*.   This global sky map shows slightly different background temperatures in opposite directions as different colors. (NASA/COBE Science Working Group)

Local Group, this implies that the Local Group is "falling" toward the very large Virgo cluster of galaxies with a speed of about 570 km/sec. Some of this motion is caused by the gravitational influence of the Virgo cluster, and apparently some is from the effects of distant superclusters. Very recently it has been found that the motion toward the Virgo cluster is a component of a much larger motion, a general streaming of all the clusters in our part of the universe. This streaming motion may be caused by the gravitational attraction of something extraordinarily massive—even compared with clusters of galaxies— yet invisible. This "Great Attractor," the subject of current excitement and controversy, may be related to the dark matter thought to contain most of the mass of the universe (see Chapter 21).

The use of the 3-degree background radiation as a tool for deducing the local motions with respect to the radiation is predicated upon the assumption that the radiation itself is fundamentally isotropic on large scales (as distinguished from the localized fluctuations discussed earlier). At present there is no reason

to doubt this, but the possibility remains that it is not—that the universe itself is somehow lopsided. If it is, it will be very difficult indeed to measure its asymmetry, because of the complexity of the motions that our observing platform, the Earth, undergoes.

Primordial Clumpiness

As mentioned earlier, the big bang model leads to the expectation that during the early expansion, the universe might have been clumpy, though perhaps only at a very subtle level. The reason for this expectation is that today's universe is populated by galaxies and groupings of galaxies which form a clumpy overall distribution (recall the discussion in Chapter 18 of superclusters and larger structures such as the so-called "Great Wall"). It is possible to explain today's large-scale structure if the early universe had an slightly uneven distribution of matter, so that the initial clumps of relatively high density could later pull themselves together gravitationally to form the large structures seen today.

ASTRONOMICAL INSIGHT
Falling Galaxies

In our discussion of the cosmic microwave background radiation, we pointed out the uniformity (that is, isotropy) of the radiation. At the same time, we observed that there is a daily variation in the radiation temperature that is measured from the Earth, because the Earth is moving through space relative to the reference frame of the radiation. Consequently, as the Earth rotates, we observe alternately in the forward and rearward directions. This daily variation, alternatively called the 24-hour anisotropy or the dipole anisotropy, has been analyzed to determine the direction and speed of the Earth, the Milky Way Galaxy, the Local Group of galaxies, and even the local supercluster of galaxies.

Carrying out this analysis is very complex, because all the different motions add together. To determine the overall motion of the Local Group or the local supercluster, one must first make allowance for the Earth's orbital motion around the Sun, the Sun's motion about the galaxy, and the galaxy's motion within the Local Group. When all this has been done, the result is not very definitive, and little can be said about the largest-scale motions.

There is, however, another approach to determining the overall motion of our galactic system within the framework established by the expanding universe. Careful statistical analyses of the observed motions of galaxies seen in different directions can indirectly reveal the motion of our own galaxy or group of galaxies. If a sample of galaxies is chosen for the best possible uniformity of properties and for the greatest possible extent in distance, then that set of galaxies forms its own reference frame against which the motion of the Local Group or even the local

supercluster can be measured. Distances for a large number of galaxies are measured, and spectra are obtained so that Doppler shifts can be measured, yielding line-of-sight velocities (i.e., radial velocities) for all the galaxies observed. If the Earth were at rest within the framework of this large collection of galaxies, then the average of all the velocities would be zero, because by random chance as many galaxies would be approaching us as receding from us. If, on the other hand, the Earth is moving in some steady direction with respect to the sample of galaxies, then we would see a trend toward redshifts in one direction and blueshifts in the other direction. This is analogous to measuring the anisotropy in the cosmic background radiation by determining its temperature (i.e., its Doppler shift) in opposite directions.

Many astronomers have attacked the problem of determining the Earth's motion relative to large populations of distant galaxies, with differing results. The difficulties lie in measuring the distances to the galaxies, which must be known in order to interpret the motions, and in obtaining sufficiently complete coverage of the sky to be sure that the distant galaxies really do represent a reference frame that is moving with the expansion of the universe (that is, a frame of reference against which we can measure our own motion with respect to the expansion).

Despite the difficulties, one of the most ambitious and complete of these studies has led to a fascinating conclusion: Our region of the universe, including the Local Group as well as other clusters of galaxies nearby, appears to be in motion with respect to the reference frame set by the expanding universe. That is, the

Local Group and the other galaxy clusters in our portion of the universe are all moving together with a speed of about 570 km per second. The motion is directed toward a large cluster of galaxies (known as the Virgo cluster), so it is tempting to conclude that this cluster is the cause. Its mass is not sufficient, however, and it appears that a far more massive object of some kind, lying about three times farther away (or about 50 Mpc from us) and having a mass of roughly 10^{16} solar masses (comparable to the largest known superclusters), is providing the gravitational attraction needed to explain the inferred motion of our cluster of galaxies. This unseen, mysterious object toward which we are falling has been dubbed the "Great Attractor."

When the presence of the Great Attractor was first proposed, astronomers were unable to find any galaxies in the region of the sky where it should be, and they speculated that perhaps the Great Attractor consists of dark matter. More diligent searches have since revealed some galaxies in the region, but it is still not known whether there are enough to explain the enormous gravitational attraction that is causing the motion of our part of the universe. Several astronomers are at work on this problem, and we may expect new developments on a continual basis for some time to come. Astronomers are especially intrigued by the Great Attractor because of what it may tell us about the structure of the universe, and because of the possibility that it is made of dark matter, the invisible material that may dominate the content of the universe.

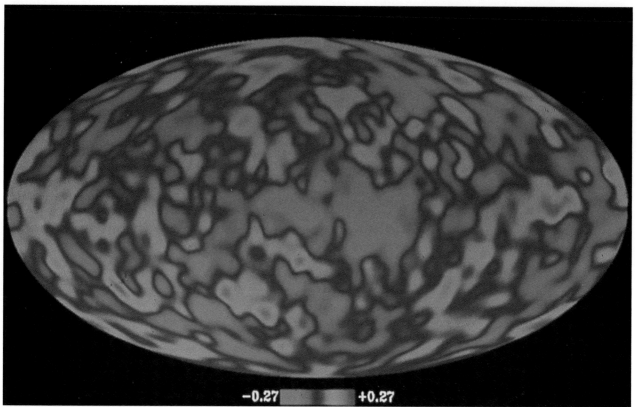

Figure 19.16   Clumpiness in the cosmic background radiation.   In this *COBE* sky map, the 24-hour anisotropy has been re-moved, and data from many months of observations are combined in order to produce maximum accuracy. The result, shown here in blue and pink, is that the radiation temperature displays very subtle variations with direction. These variations are interpreted as being caused by a slightly uneven distribution of matter in the early universe. (NASA/*COBE* Science Working Group)

If the early universe was a bit clumpy, this should show up as subtle variations or patchiness in the cosmic background radiation observed today. However, as mentioned earlier in this discussion, no such variations or anisotropies in the observed background radiation were seen by the early experimenters. The first results from the *COBE* satellite also failed to show any anisotropies, at a level of sensitivity where the lack of variations was beginning to become a concern for the big bang model. In short, for awhile it appeared that the early universe was too smooth to account for the formation of galaxies and groupings of galaxies.

Now, however, this problem appears to have been solved. Very recent results from *COBE*, gathered from more than two years' of observations, have finally revealed the expected anisotropies and confirmed that the early universe was indeed clumpy (Fig. 19.16). The variations show up as regions of slightly different radiation temperature; the differences from place to place are incredibly small, about 30 millionths of a degree absolute! Nevertheless, they indicate primordial density variations that were large enough to explain how galaxies, clusters, and superclusters of gal-axies could have formed by gravitational action. Thus the *COBE* results now have confirmed the two most important tests of the nature of the background radiation, that its spectrum is thermal and that it shows clumpiness due to the structure of the early universe. In the process *COBE* has substantially strengthened the big bang model of the origin of the universe. This model, and the implications of the *COBE* data, will be discussed further in Chapter 20.

PERSPECTIVE

We have learned now that the universe, like all the objects within it, is a dynamic, evolving entity. One of the grandest stories in astronomy has been the un-folding of the concept of universal expansion. An idea forced on astronomers by the evidence of the red-shifts, it has led directly to the big bang picture, which in turn tells us the age of the universe, explains the origin of some of the elements, and is verified through the presence of the microwave background radiation.

Before we finish the story by studying modern theories of the universe and their implications for its future, we must take a closer look at some of the objects that inhabit it. These include a breed of strange and bizarre entities whose very nature seems to be tied up with the evolution of the universe and which therefore will provide us with a few additional bits of evidence for our final discussion.

SUMMARY

1. Astronomers discovered early in this century that galaxies tend to have redshifted spectral lines, implying that they are moving away from us.

2. Hubble discovered in 1929 that the velocity of recession is proportional to distance, demonstrating that the universe is expanding.

3. The rate of expansion, expressed in terms of the Hubble constant H, is probably between 50 km/sec/Mpc and 100 km/sec/Mpc.

4. The fact that the universe is expanding implies that it originated from a single point, and the rate of expansion, if constant, tells us that it began some 10 to 20 billion years ago. This may be considered to be the age of the universe.

5. The relationship between speed and distance allows the distance to a galaxy to be estimated from its velocity, using the Doppler effect. This method of determining distance extends to the farthest limits of the observable universe.

6. The fact that the universe began in a single point implies that it was very hot and dense initially, and this in turn implies that the early universe was filled with radiation. As the expansion has proceeded, this radiation has been transformed into microwave radiation, with a thermal spectrum corresponding to a temperature of about 3 K. The cosmic background radiation was discovered in 1965.

7. To distinguish a primordial origin for the background radiation from other possible origins (such as many galaxies spread throughout the universe), it is necessary to measure the spectrum to see whether it truly is a thermal spectrum. Observations to date, including a very accurate and complete spectrum obtained by the *Cosmic Background Explorer,* show that it is.

8. Another important test is to determine whether the radiation is isotropic, or whether it is unevenly intense in different directions. Careful observations have revealed no evidence of fine-scale patchiness that might imply that the radiation arises in a large number of individual sources, thus confirming that the radiation is truly universal in origin.

9. There is a 24-hour (or dipole) anisotropy in the radiation, however, created by the Doppler effect resulting from the Earth's motion. By comparing the radiation intensity in the forward and backward directions, it has been possible to determine the Earth's velocity with respect to the reference frame of the radiation. The Earth's motion is a combination of its orbital motion about the Sun, the Sun's orbital motion about the galaxy, the galaxy's motion within the Local Group, and the Local Group's motion within the local supercluster.

10. Recent *COBE* results have revealed subtle, large-scale variations in the temperature of the background radiation, showing that the early universe had a slightly uneven distribution of matter. This "clumpiness" in the early universe helps to explain the origin of superclusters and clusters of galaxies.

REVIEW QUESTIONS

1. In your own words, explain how all galaxies can be receding from our own galaxy, even though our galaxy is not at the center of the universe.

2. Explain the distinction between the motions of galaxies within a cluster of galaxies and motions resulting from the expansion of the universe.

3. Suppose the current value of the Hubble constant is 75 km/sec/Mpc, but because the initial expansion was more rapid, the average value over the history of the universe is 80 km/sec/Mpc. What is the true age of the universe, and how does it compare with the value derived by assuming that expansion has been steady at the present rate?

4. How is the value of the Hubble constant measured? Why is it difficult to obtain a precise value?

5. Explain why the expansion of the universe is not as rapid today as it was long ago. What implications does the slowing of the expansion have for the future?

6. Suppose three galaxies are observed, and the position of the ionized calcium line in their spectra is observed to be 3936 Å for galaxy A, 4028 Å for galaxy B, and 3942 Å for galaxy C (the rest wavelength is 3933 Å). Find the velocities and distances to the galaxies, assuming a Hubble constant of $H = 75$ km/sec/Mpc.

7. Explain how the distances we measure to the most distant galaxies depend on how accurately we know the Sun-Earth distance.

8. Explain why the assumption that the early universe was hot and dense leads to the conclusion that it must have been filled with radiation.

9. Explain why the fine-scale isotropy of the background radiation is important in establishing whether it is remnant radiation from the early expansion of the universe.

10. If the background radiation is nearly isotropic, why are there small shifts in its temperature as observed in opposite directions from the Earth?

ADDITIONAL READINGS

Barrow, J. D., and J. Silk. 1980. The structure of the early universe. *Scientific American* 242(2):118.

Brush, S. G. 1992. How cosmology became a science. *Scientific American* 267(2):62.

Dressler, A. 1987. The large-scale streaming of galaxies. *Scientific American* 257(3):46.

Ferris, T. 1984. The radio sky and the echo of creation. *Mercury* 13(1):2.

Gulkis, S., P. M. Lubin, S. S. Meyer, and R. F. Silverberg. 1990. The *Cosmic Background Explorer. Scientific American* 262(1):132.

Hodge, P. 1984. The cosmic distance scale. *American Scientist* 72(5):474.

Kippenhahn, R. 1987. Light from the depths of time. *Sky and Telescope* 73(2):140.

Layzer, D. 1975. The arrow of time. *Scientific American* 233(6):56.

Muiller, R. A. 1978. The cosmic background radiation and the new aether drift. *Scientific American* 238(5):64.

Parker, B. 1986. The discovery of the expanding universe. *Sky and Telescope* 72(3):227.

Shu, F. H. 1982. *The physical universe.* San Francisco: W. H. Freeman.

Silk, J. 1980. *The big bang: The creation and evolution of the universe.* San Francisco: W. H. Freeman.

———. 1989. Probing the primeval fireball. *Sky and Telescope.* 79:600.

Smith, D. H. 1989. Cosmic fire, terrestrial ice. *Sky and Telescope* 78:471.

Webster, A. 1974. The cosmic background radiation. *Scientific American* 231(2):26.

Weinberg, S. 1977. *The first three minutes.* New York: Basic Books.

ASTRONOMICAL ACTIVITY
Measuring the Hubble Constant

The accompanying table provides you with data that can be used to determine the value of the Hubble constant. The table lists the measured wavelength, for several galaxies, of the spectral line of ionized calcium (designated Ca II), whose rest wavelength is 3933.663 Å Also given is the observed apparent magnitude of each galaxy, which can be used to derive its distance if the absolute magnitude is known (see Chapter 12 or Appendix 10 for a reminder of how to do this). Assume that all the galaxies are type Sb, and that all have the same absolute magnitude, $M = -21.0$.

To derive the value of the Hubble constant, you will need to compute the recession velocity of each galaxy, using the Doppler shift formula. You will have to use the relativistic formula for the Doppler shift if the redshift $\delta\lambda/\lambda$ is greater than about 0.01 (i.e., if the velocity is as large as 1 percent of the speed of light). The relativistic Doppler shift formula is given in Appendix 14. Once you have found the distances and the velocities of all the galaxies, convert the distances to units of megaparsecs, and make a graph (on graph paper with uniform-size squares in each direction) showing distance on the horizontal scale and recession velocity on the vertical scale (see Fig. 19.5).

The points representing the individual galaxies should form an approximately straight line running diagonally from the lower left toward the upper right. Using a ruler, draw the straight line that seems to fit best through the points (the points will tend to scatter a little around this line; this reflects observational uncertainty and local motions of galaxies). The *slope* of this line is the Hubble constant. To measure the slope, pick a point on the line, preferably near its upper end, and read off the values for distance (from the horizontal axis) and recession velocity (from the vertical axis) corresponding to that point. Then divide the recession velocity by the distance; the result will be the value of H, in units of km/sec/Mpc (note that for a straight

Galaxy	m	Wavelength (Å)
1	11.12	3961.032
2	10.70	3954.567
3	13.60	4023.457
4	13.40	4013.540
5	6.39	3936.418
6	12.96	4004.984
7	13.20	4010.998
8	12.27	3979.163
9	11.94	3975.318
10	10.05	3951.800
11	13.25	4005.384
12	12.39	3983.278
13	13.45	4018.496
14	11.64	3970.550
15	12.67	3987.264
16	13.32	4007.522
17	13.64	4019.836
18	12.79	3992.052
19	8.60	3939.568
20	12.21	3983.942
21	13.82	4033.938
22	13.17	4005.251
23	10.55	3955.754
24	9.19	3946.402
25	13.82	4030.038
26	12.91	3997.113
27	10.15	3949.429
28	11.83	3969.021
29	12.55	3993.383
30	13.49	4016.084

line, you will get the same value no matter where you pick your point).

How does your value of H compare with modern values as discussed in the text? Compute the age of the universe on the basis of your value (see the text). What do you think might be the major uncertainties in the value of H derived by this method?

Peculiar Galaxies,
Active Nuclei, and Quasars

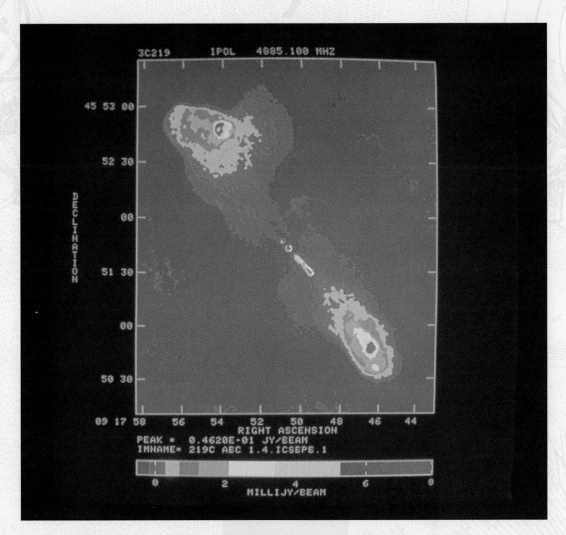

A VLA image of the radio galaxy 3C219 (NRAO)

PREVIEW

Earlier you learned that our galaxy harbors a compact yet massive object at its core, a mysterious beast that yields enormous quantities of energy. Now you will see that many galaxies contain similar, but larger, monsters at their centers that dwarf the object in the Milky Way in both mass and energy production. These energetic galactic nuclei are associated with several classes of "active" galaxies, which are discussed in this chapter. These include the radio galaxies, gigantic elliptical galaxies with enormous lobes of radio-emitting gas on opposite sides; the Seyfert galaxies, which are spirals with unusual bright blue nuclei; and the quasi-stellar objects or quasars, which appear to be the cores of young galaxies. The quasars are the most luminous objects in the universe and are seen at immense distances, so far from us that we know we are seeing them as they were billions of years ago. Thus the study of quasars gives us a glimpse into the early history of galaxies in the universe. In this chapter you will learn about the active galaxies and how their energy is thought to be created, and you will see how observations of these bizarre objects can tell us something about the early history of the universe.

In discussing the characteristics of galaxies in the preceding chapters, we have overlooked a variety of objects, some of them galaxies and some possibly not, that have unusual traits. As in many other situations in astronomy, the so-called peculiar objects, once understood, have quite a bit to tell us about the more normal ones.

To help you fully appreciate the bizarre nature of some of these astronomical oddities, we have delayed their introduction until this point, when we have essentially completed our survey of the universe. We know about the expansion and the big bang and are now in the process of tying it all together. In this chapter we will uncover a number of vital clues in the cosmic puzzle.

THE RADIO GALAXIES

Most of the first astronomical sources of radio emission to be discovered, other than the Sun, are galaxies. Most of these, when examined optically, have turned out to be large ellipticals, often with some unusual structure (Fig. 20.1). The first of these objects to be detected, and one of the brightest at radio wavelengths, is called Cygnus A (named under a preliminary cataloging system in which the ranking radio sources in a constellation are listed alphabetically). This galaxy has a strange double appearance, and it

was eventually found, after sufficient refinement of radio observing techniques (see the discussion of interferometry in Chapter 4) that the radio emission comes from two locations on opposite sides of the visible galaxy and well separated from it (Figs. 20.2 and 20.3). High-resolution observations with the *Very Large Array* have revealed remarkable detail in some radio galaxies (Fig. 20.4).

This double-lobed structure is a common feature of the so-called **radio galaxies,** those with unusually strong radio emission. Most, if not all, ordinary galaxies emit radio radiation, but the term *radio galaxy* is applied only to the special cases in which the radio intensity is many times greater than the norm. The core of the Milky Way is a strong radio source as viewed from the Earth, but it is not in a league with the true radio galaxies. Typical properties of radio galaxies are listed in Table 20.1.

Although the double-lobed structure is standard among elliptical radio galaxies, the visual appearance varies quite a bit. We have already noted the double appearance of Cygnus A; another bright source, the giant elliptical Centaurus A (the object shown in Fig. 20.1), appears to have a dense band of interstellar matter bisecting it, and it has not one, but two, pairs of radio lobes, one much farther out from the visible galaxy than the other. Other giant elliptical radio galaxies present other kinds of strange appearances. One of the most famous is M87, also known as Virgo A, which has a jet protruding from one side that is aligned with one of the radio lobes (Fig. 20.5). Careful examination shows that this jet contains a series of blobs or knots that appear to have been ejected from the core of the galaxy in sequence. Many galaxies have been discovered to have radio-emitting jets. These include galaxies already known to be peculiar, such as Centaurus A (see Fig. 20.1) and at least one spiral galaxy (Fig. 20.6). In some cases double jets are seen on opposite sides of a galaxy (Fig. 20.7).

The size scale of the radio galaxies can be enormous. In some cases, the radio lobes or jets extend as far as 5 million parsecs from the central galaxy.

TABLE 20.1   Properties of Radio Galaxies

Galaxy type: Most often elliptical or giant elliptical.

Radio luminosity compared with optical luminosity: 0.1 to 10.

Radio source shape: Either double-lobed or compact central source; often jets are seen that emit throughout the spectrum.

Variability: Intensity variations in times as short as days may occur in the compact radio sources.

Nature of spectrum: Usually synchrotron or inverse Compton spectrum, usually polarized.

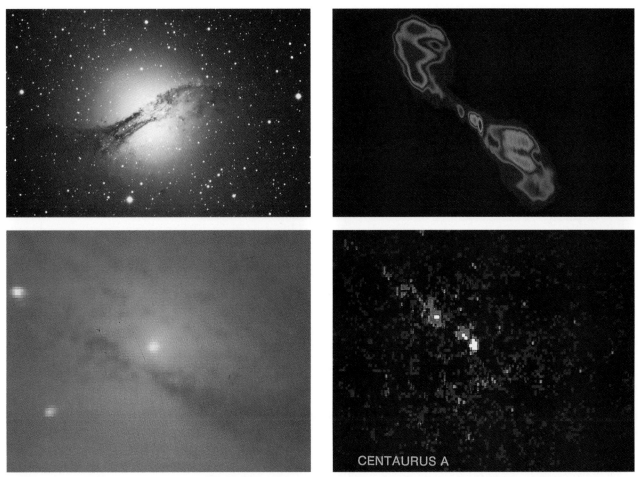

Figure 20.1 Centaurus A. This is a giant elliptical radio galaxy, showing a dense lane of interstellar gas and dust across the central region. At the upper left is a visible-light photograph; at lower left is an infrared image showing the galactic core; at upper right is a radio image; and at lower right is an X-ray image. Note that the radio and X-ray emission comes from the lobes above and below the visible galaxy. (Upper left: © 1980 AAT Board, photo by D. Malin; lower left: © 1992 AAT Board; image provided by D. A. Allen; upper left: NRAO; lower right: Harvard-Smithsonian Center for Astrophysics)

Figure 20.2 An image of a radio galaxy. This is the object 3C219. The visible galaxy is the small blue spot at center; the enormous radio lobes are shown in red, orange, and yellow. Near the center is a well-defined jet of ionized gas, apparently flowing outward from the core of the galaxy. (National Radio Astronomy Observatory)

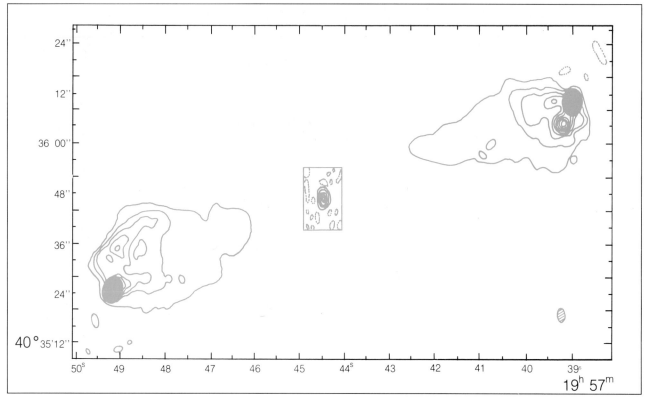

Figure 20.3   A map of a radio galaxy.   The intensity contours here represent radio emission from the double lobes of the radio galaxy Cygnus A. The blurred image at the center illustrates the size of the visible galaxy compared with the gigantic radio lobes. (Mullard Radio Astronomy Observatory, University of Cambridge)

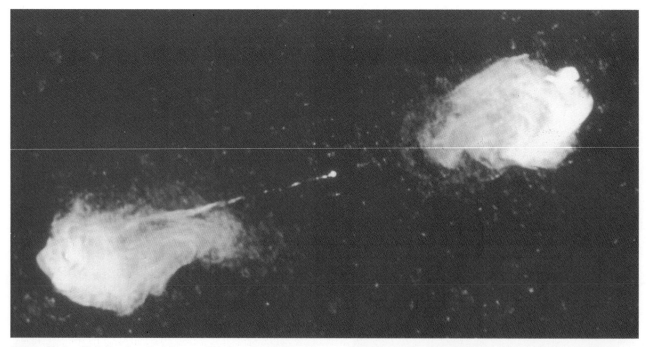

Figure 20.4   Details of radio lobes.   This image of Cygnus A was obtained with the *Very Large Array* and shows fine detail suggesting turbulence and flows in the lobes. (National Radio Astronomy Observatory)

Figure 20.5   A giant elliptical radio galaxy with a jet.   This is M87, also known as Virgo A, showing its remarkable linear jet. At left is a visible-light photograph taken with a long exposure time, showing the entire galaxy. At center is a short-exposure *Hubble Space Telescope* image, showing a bright concentration at the galaxy's core as well as the jet. At right is a radio montage, showing the radio lobes and jet on different scales. (Left: © 1992 AAT Board, photo by D. Malin, center: NASA/STScI; right: NRAO)

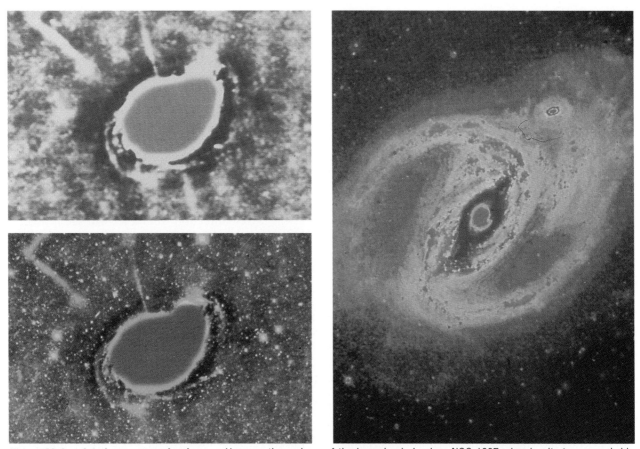

Figure 20.6   Jets in an unusual galaxy.   Here are three views of the barred spiral galaxy NGC 1097, showing its two remarkable linear jets. At the top left is a black-and-white photograph that has been processed to enhance the visibility of the jets. At the lower left is a false-color version. An enlarged view of the inner part of the galaxy is shown at the right. (H. C. Arp)

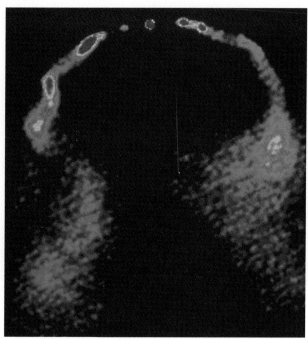

Figure 20.7   Dual radio jets.   While visible jets are rare, radio images often reveal jetlike structures. Here is a radio image of a galaxy with a pair of jets emanating from opposite sides and curving away from the galaxy, apparently because the galaxy is moving through an intergalactic gas medium. (National Radio Astronomy Observatory)

Recall that the diameter of a large galaxy is only about 2 percent of this distance, and that the Andromeda galaxy is less than 1 million pc from our position in the Milky Way.

Radio spectra show that radio galaxies emit by the synchrotron process (already mentioned in Chapter 15), which requires a strong magnetic field and a supply of rapidly moving electrons. The electrons are forced to follow spiral paths around the magnetic field lines, and as they do so, they emit radiation over a broad range of wavelengths. The characteristic signature of synchrotron radiation, in contrast with the thermal radiation from hot objects such as stars, is a sloped spectrum with no strong peak at any particular wavelength. The radiation from a synchrotron source is also polarized (see Chapter 4). Whenever an object is found to be emitting by the synchrotron process, astronomers immediately conclude that some highly energetic activity is taking place, producing the rapid electrons that are required to create the emission.

In some cases, it appears that a different process is responsible for the emission, particularly at wavelengths shorter than radio. If there is ordinary thermal radio emission from a galaxy, then the presence of electrons moving near the speed of light can alter the spectrum of the emission, as energy is transferred from the electrons to the radiation. This process is called **inverse Compton scattering,** and it has the same implication as synchrotron radiation: There must be a source of vast amounts of energy to produce the rapidly moving electrons required to create the radiation.

The source of this energy in radio galaxies is a mystery, although some interesting speculation has been fueled by certain characteristics of the galaxies. The double-lobed structure indicates that two clouds of fast electrons are located on either side of the visible galaxy, leading naturally to the impression that the clouds of hot gas may have been ejected from the galactic core, perhaps by an explosive event that expelled material symmetrically in opposite directions. This impression is heightened by the many cases in which jets have been seen protruding from the core (Figs. 20.5, 20.6 and 20.7) and by instances in which more than one pair of lobes, aligned in the same direction but with one farther out than the other, is seen. In these galaxies, it seems that superheated, highly ionized gas is continually being accelerated from the core in opposing directions.

But how can such prodigious amounts of energy be produced from a small volume such as the core of a galaxy? One possibility that has been suggested is that a giant black hole is at the core, surrounded by an accretion disk. The general idea is that superheated gas that has been compressed by the immense gravitational field of the black hole escapes along the rotation axes and is channeled into two opposing jets. It is probably better to view this process not as an explosion, but as a more or less steady expulsion of gas. Some have suggested that *all* radio galaxies have jets, and that this is the only mechanism for forming the double radio lobes. The fact that jets are not *observed* in all radio galaxies is thought to be an effect of how the jets are aligned with respect to our line of sight. The emission from rapidly moving particles is confined to a narrow angle; therefore, an observer in the wrong direction from a jet will see little of its radiation.

The theory that massive black holes are the energy sources in radio galaxies has received some observational support. Recently it was reported that evidence for such a central, massive object has been found in the heart of M87, the radio galaxy with the well-known visible jet. Based on measurements of the velocity dispersion near the center of the galaxy, researchers deduced that a large amount of mass must be confined in a very small volume at the core, and a black hole seemed the best explanation. Subsequent observations, however, have failed to confirm the reportedly high speeds of stars in the inner portions of the galaxy, and the case for a black hole in M87 has been weakened. Further efforts to find direct evidence of supermassive black holes are being made.

Figure 20.8   A Seyfert galaxy.   This photograph of NGC 4151, a well-studied example, shows the enormous intensity of light from the nucleus compared with the rest of the galaxy. (Palomar Observatory, California Institute of Technology)

TABLE 20.2   Properties of Seyfert Galaxies

Spiral galaxies.

Luminosities comparable to brightest normal spirals.

Bright, compact blue nuclei.

Nuclei show emission lines of highly ionized gas.

About 10% are radio sources.

Most are variable in times of days or weeks.

Emission from the nucleus is synchrotron radiation.

Some have radio jets.

Seyferts typically indicate higher-density gas that is more tightly confined in the galactic core. Evidently the cores of these galaxies are in extreme turmoil, with hot gas swirling about and tremendous amounts of energy being generated.

Few spirals show such effects, and it is not clear whether the Seyfert behavior is a phase they all pass through at some time, or whether only a few act up in this manner. Later in this chapter we will discuss evidence supporting the first of these possibilities.

SEYFERT GALAXIES AND ACTIVE NUCLEI

Although it is true that the strongest radio emitters among galaxies are giant ellipticals, they are by no means the only galaxies with evidence for explosive events in their cores. Some spiral galaxies also display such behavior. We pointed out in Chapter 16 that even the Milky Way is not immune and summarized the evidence for the existence of a compact, massive object at its center. There are other spiral galaxies with much more pronounced violence in their nuclei. These galaxies as a class are called **Seyfert galaxies** (Fig. 20.8 and Table 20.2), after Carl Seyfert, an American astronomer who discovered and cataloged many of them in the 1930s.

Seyfert galaxies have the appearance of ordinary spirals, except that the nucleus is unusually bright and blue, rather than the yellow of most normal spiral galaxy nuclei. About 10 percent of them are radio emitters, with spectra indicating that the synchrotron process is at work. The radio emission usually comes from the nucleus rather than from double lobes. The spectrum of the visible light from a Seyfert nucleus typically shows emission lines, something completely out of character for normal galaxies. For some nuclei, the emission lines, formed in an ionized gas, are broad, indicating speeds of several thousand km/sec in the gas that produces the emission, whereas for others the lines are narrower. The narrow-line Seyfert galaxy nuclei usually have emission lines characteristic of low-density regions, while the broad-line

THE DISCOVERY OF QUASARS

In 1960, spectra were obtained of two starlike, bluish objects that had been found to be sources of radio emission (Fig. 20.9). No radio stars were known, and astronomers were very interested in these two objects, called 3C 48 and 3C 273 (their designations in a catalog of radio sources that had recently been complied by the radio observatory of Cambridge University in England). The objects were called **quasars** (short for "quasi-stellar radio sources"), or **quasi-stellar objects,** because of their starlike appearance.

The spectrum of 3C 48 was obtained and found to consist of emission lines whose identities were unknown. Some three years later, the spectrum of 3C 273 was found to consist of emission lines also, and in this case they were recognized as the strong Balmer lines of hydrogen, but shifted toward far longer wavelengths than those measured in the laboratory. The lines in the spectrum of 3C 48 were then identified and also found to show very large shifts in wavelength. If these shifts are due to the Doppler effect, then tremendous speeds are implied. The redshift in the spectrum of 3C 273 is 16 percent, indicating that this object is moving away from us at 16 percent of the speed of light, or 48,000 km/sec! In 3C 48, the shift was even greater, about 37 percent, corresponding to a speed of 111,000 km/sec (these speeds actually have to be corrected for relativistic effects, as

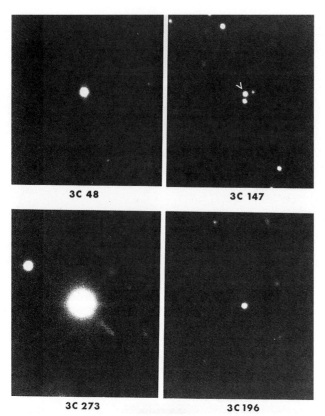

3C 48

3C 147

3C 273

3C 196

Figure 20.9   Quasi-stellar objects.   These starlike objects are quasars, whose spectra are quite unlike those of normal stars. As explained in the text, these objects are probably more distant, and therefore more luminous, than normal galaxies. (Palomar Observatory, California Institute of Technology)

described in Appendix 14; the correct values are 44,000 km/sec for 3C 273 and 91,000 km/sec for 3C 48).

Since the early 1960s, hundreds of additional quasars have been discovered (the current total is over 3000). Only a small fraction are radio sources, and in other ways they may differ from the first two, but they invariably have highly redshifted emission lines, and in very many cases weak absorption lines are also seen. In many quasars the redshift is so huge that spectral lines with rest wavelengths in the ultraviolet portion of the spectrum are shifted all the way into the visible region. The grand champion today is a quasar with redshift of 485 percent, so that the strongest of all the hydrogen lines, with a rest wavelength of 1216 Å, is shifted all the way to 7113 Å. This quasar is *not* traveling away from us at 4.85 times the speed of light, however; for very large velocities, the Doppler shift formula has to be modified in accordance with relativistic effects (Appendix 14). The velocity of this quasar is 94.3 percent of the speed of light, still an enormous speed.

THE ORIGIN OF THE REDSHIFTS

To understand the physical properties of the quasars, we must first discover the reason for the high redshifts. We have already tacitly adopted the most obvious explanation—that they are a result of the Doppler effect in objects that are moving very rapidly— but we still must ascertain the nature of the motions. Furthermore, at least one alternative explanation has been suggested that has nothing to do with motions at all.

One consequence of Einstein's theory of general relativity, verified by experiment, is that light can be redshifted by a gravitational field. Photons struggling to escape an intense field lose some of their energy in the process, and as this happens, their wavelengths are shifted toward the red. We have already been exposed to this concept in Chapter 15, when we discussed the behavior of light near black holes, which have such strong gravitational fields that no light can escape at all. For a while it was considered possible that quasars are stationary objects sufficiently massive and compact to have large gravitational redshifts.

This suggestion has been largely ruled out for two reasons. One is that there is no known way for an object to be compressed enough to produce such strong gravitational redshifts without falling in on itself and becoming a black hole. If enough matter were squeezed into such a small volume that its gravity produced redshifts as large as those in quasars, no known force could prevent this object from collapsing further. A neutron star does not have as large a gravitational redshift as those found in quasars, and we already know that the only possible object with a stronger gravitational field than that of a neutron star is a black hole.

The second objection is that, even if such a massive, yet compact object could exist, its spectral lines would be very much broader than those observed in quasar spectra. The high pressure would distort electron energy levels so that the spectral lines would be smeared out (as in a white dwarf, but more extreme), and light emitted from slightly different levels in the object would have different gravitational redshifts, again causing spectral lines to be broadened.

For both of these reasons, we are forced to accept the Doppler shift explanation for the redshifts. The problem then is to explain how such large velocities can arise. One possibility is that the quasars are relatively nearby (by intergalactic standards) and are simply moving away from us at very high speeds, perhaps as the result of some explosive event. This "local" explanation requires some care: To accept it, we must explain why no quasars have ever been found

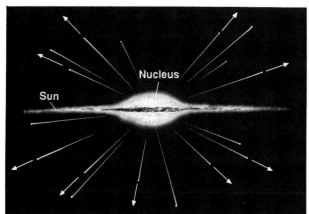

Figure 20.10 A "local" hypothesis for quasar redshifts. If quasars were very rapidly moving objects ejected from our galactic nucleus some time ago, they could all be receding from our position in the disk by now. This would explain why only redshifts are found, but there is no known mechanism for providing the energy required for such great speeds.

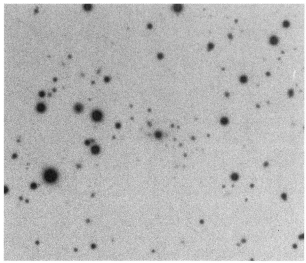

Figure 20.11 A quasar in a cluster of galaxies. The image at the center of this very long exposure photograph is a quasar. The much fainter objects around it are galaxies, forming a cluster of which the quasar is apparently a member. The galaxies and the quasar have the same redshift, indicating that they lie together at the same distance. (H. Spinrad)

to have blueshifts. In other words, if quasars are nearby objects moving very rapidly, it is not easy to see why they are all receding from us, with none approaching. It might be possible to argue that they originated in a nearby explosion (at the galactic center, perhaps) a long time ago, so any that happened to be aimed toward us have had time to pass by and are now receding (Fig. 20.10). There are serious difficulties with this picture, though, primarily in the amount of energy that would be required to get all these objects moving at the observed velocities, an amount that would dwarf the total light output of the galaxy over its entire lifetime.

We are left with one other alternative, one that still poses problems, but perhaps is more acceptable than the others. Let us consider the possibility that the quasars are very distant objects, moving away from us with the expansion of the universe. In this view they are said to be at "cosmological" distances, meaning that they obey Hubble's relation between distance and velocity, just as galaxies do. We find, if we adopt this assumption, that 3C 273 is some 586 Mpc away, and 3C 48 is more than 1,200 Mpc away. (These distances are based on the assumption that the value of the Hubble constant is $H = 75$ km/sec/Mpc and that the rate of expansion of the universe has been constant, something that is probably not true; this will be discussed in Chapter 21).

The best evidence that the quasars are at cosmological distances is the fact that some have been found in clusters of galaxies (Fig. 20.11), with the same redshift as the galaxies. This was an extremely difficult observation to make, because at the distances at which the quasars exist, the much fainter galaxies are very hard to see, even with the largest telescopes.

Supporting evidence for the cosmological redshift interpretation is provided by the **BL Lac objects.** Named for the prototype, an object called BL Lacertae that was once thought to be a variable star, these are galaxies with very bright central cores. The nucleus of a BL Lac object displays many of the properties of a quasar, with radio synchrotron emission, variable brightness, and enormous luminosity. The surrounding galaxy shows an almost featureless spectrum with a redshift that is consistent with its apparent distance and that is undoubtedly cosmological. Generally the redshifts of BL Lac objects are relatively small (by quasar standards), and these apparently are nearby objects very closely related to quasars. The implications of the fact that they are embedded in galaxies will be discussed later in this chapter.

Distances of hundreds or thousands of Mpc, inferred for the quasars with very large redshifts, have enormous implications. One is that the light that we receive from them has traveled on its way to us for many billions of years. When we look at quasars, we are looking far into the history of the universe and must keep this in mind as we attempt to interpret them. The fact that quasars are seen primarily at high redshifts, meaning that they exist primarily at great distances, indicates that they were more common in the early days of the universe than they are now. This will prove to be a very important clue to their true nature.

Another important implication of the great distances to quasars is that they are the most luminous

ASTRONOMICAL INSIGHT
The Redshift Controversy

In the text we present a standard set of arguments demonstrating that quasar redshifts are cosmological, resulting from the expansion of the universe; by implication, then, the quasars must be very distant objects. Although most astronomers accept this explanation, it has not been universally adopted, and some have proposed other explanations of the redshifts.

The evidence cited by the opponents of the cosmological interpretation consists primarily of cases where quasars are found apparently associated with galaxies or clusters of galaxies, but do not have the same redshift as the galaxies. The argument is that in these cases the quasar is physically associated with the galaxy or cluster of galaxies and is therefore at the same distance, so its redshift cannot be cosmological.

The evidence favoring these arguments can be quite striking. When a quasar is found within a cluster of galaxies, there is a natural tendency to think of it as a physical member of the group. It can be argued on statistical grounds that the chances of accidental alignments between galaxies and quasars are so low that the observed associations between these objects are unlikely to be coincidental. Long-exposure photographs have even appeared to show gaseous filaments connecting a galaxy with a quasar that has a different redshift. This would seem to prove that the redshift cannot be cosmological.

This radical view of quasars leaves many important questions open. The most difficult is how to explain the redshifts, if they really are not cosmological, since the arguments cited in the text against gravitational and local Doppler redshifts are still valid. No satisfactory explanation of the quasar redshifts has been offered by the opponents of the cosmological viewpoint.

The counterarguments center on statistical calculations. One point is that the calculations showing that the quasar-galaxy associations have a low probability of occurring by chance are made after the fact. That is, after a few such associations were found, arguments were made that the associations were very unlikely to occur by chance alignments of objects randomly distributed over the sky. This is somewhat akin to arguing that the chances of being dealt a certain combination of cards in a game are very low. This is true before the deal, but meaningless after the fact. Based on this and other analogies, many scientists think it incorrect to maintain that chance associations of quasars and galaxies are improbable. More to the point, they argue, quasars associated with galaxies are more likely to be discovered than those that are not; therefore, it is natural to find a disproportionately high number of quasars that happen to be aligned on the sky with galaxies. This is an example of a **selection effect,** because the process by which quasars are found is not perfectly random and thorough;

the attention of astronomers is naturally concentrated on regions where there are many galaxies, so those regions are more thoroughly searched for quasars.

Probably the most telling argument against the noncosmological redshifts for quasars arises from the fact (noted in the text) that a number of quasars are embedded in galaxies or within clusters of galaxies that share the same redshift. Recent observations show, in fact, that *every* quasar that is sufficiently nearby for an associated galaxy to be detected is indeed embedded within a galaxy that has the same redshift as the quasar. No exception to this rule has been found, and the data amount to virtual proof that quasars are the cores of galaxies and that their redshifts are cosmological.

The controversy is by no means over, however. Adherents of the noncosmological interpretation of the redshifts continue to find remarkable cases of apparent galaxy-quasar connections where the redshifts do not match. Perhaps observations made in the near future, with the *Hubble Space Telescope* or other planned instruments, will provide unequivocal evidence supporting one side or the other.

objects known. Their apparent magnitudes (the brightest are thirteenth-magnitude objects), combined with their tremendous distances, imply that their luminosities are far greater than even the brightest galaxies by factors of hundreds or even thousands. As we will see, explaining this astounding energy output is one of the central problems of modern astronomy.

THE PROPERTIES OF QUASARS

More than 5,000 quasars have been cataloged, and more are continually being discovered. Many are being studied with care, primarily by means of spectroscopic observations. Within the past few years, satellite observatories such as the *Infrared Astronomy Satellite (IRAS), International Ultraviolet Explorer (IUE),* and the *Einstein Observatory* have even made it possible to observe quasars at infrared, ultraviolet, and X-ray wavelengths. The faintness of the quasars makes these observations difficult, but their importance makes the effort worthwhile. The result of the intensive work being done on quasars is that a great deal is now known about their external properties (Table 20.3), even though their origin and especially their source of energy remain mysterious.

The blue color that characterized the first quasars discovered is a general property of all quasars (except for extremely redshifted ones, where the blue light has been shifted all the way into the red part of the spectrum), but the radio emission is not. Only about 10 percent are radio sources, contrary to the early impression that they all are. This type of misunderstanding is known as a **selection effect,** because at first quasars were selected based on a certain assumption about their characteristics. For a while, the only method used to look for quasars was to search for objects with radio emission, so naturally, all the quasars that were found were radio sources. The radio emission, and usually the continuous radiation of visible light as well, shows the characteristic synchrotron spectrum.

Apparently *all* quasars emit X rays, again by the synchrotron process, according to the data collected by the *Einstein Observatory* and by the U.S.-German X-ray observatory *ROSAT,* which has recently mapped the sky. Hence X-ray observations may be a reliable technique for finding new quasars, particularly those of relatively low redshift, which are relatively nearby and hence have stronger X-ray emission than the more distant ones.

In some cases, photographs of quasars reveal some evidence of structure instead of a single point of light.

TABLE 20.3 Properties of Quasars

Characterized by large redshifts.

Spectra dominated by emission lines of highly ionized gas.

Optical luminosities 100 to 1,000 times those of normal galaxies.[a]

About 10% are radio sources.

Appear as compact, blue objects.

Many are variable in times of days or weeks.

Emission due to synchrotron radiation.

Some have radio or optical jets.

[a]Assuming that quasars are at cosmological distances.

The most notable of these is 3C 273, which played such a key role in the initial discovery of quasars. This object shows a linear jet extending from one side (Fig. 20.12), closely resembling the jet emanating from the giant radio galaxy M87 (see Fig. 20.5). Perhaps this means that similar processes are occurring in these two rather different object.

Many quasars vary in brightness, usually over periods of several days to months or years (although, in at least one case, variations were seen over just a few hours). This variability is very important, for it provides information on the size of the region in the quasar that is emitting the light. This region cannot be any larger than the distance light travels in the time over which the intensity varies. There is no way

Figure 20.12 A quasar with a jet. This is 3C 273, one of the first two quasars discovered. This visible-light photograph reveals a linear jet very much like those seen in many radio galaxies. The radio structure of quasars is usually double-lobed, also similar to the structure found in radio galaxies. (National Optical Astronomy Observatories)

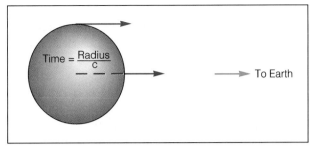

Figure 20.13 The implication of the time variability for the size of the emitting object. This sketch illustrates why an object cannot appear to vary in less time than it takes for light to travel across it. As a simple case, this spherical object is assumed to change its luminosity instantaneously. On Earth we observe the first hint of this change when light from the nearest part of the object reaches us, but we continue to see the brightness changing gradually as light from more distant portions reaches us.

an entire object that might be hundreds or thousands of light-years across can vary in brightness in a time of a month or so. The light-travel time across the object would guarantee that we would see changes only over times of hundreds or thousands of years, as the light from different parts of the object reached us (Fig. 20.13). Therefore the observed short-term variations in quasars tell us that the fantastic energy emitted by these objects is produced within a volume no more than a light-month, or about 0.03 pc, in diameter. This obviously places stringent limitations on the nature of the emitting object.

The spectra of quasars have already been described in broad outline, but they contain a great deal of detail as well (Fig. 20.14). All quasars have emission lines, generally of common elements such as hydrogen, helium, and often carbon, nitrogen, and oxygen (some of the latter elements only have strong lines in the ultraviolet and therefore are best observed when the redshift is sufficiently large to move these lines into the visible portion of the spectrum). In many

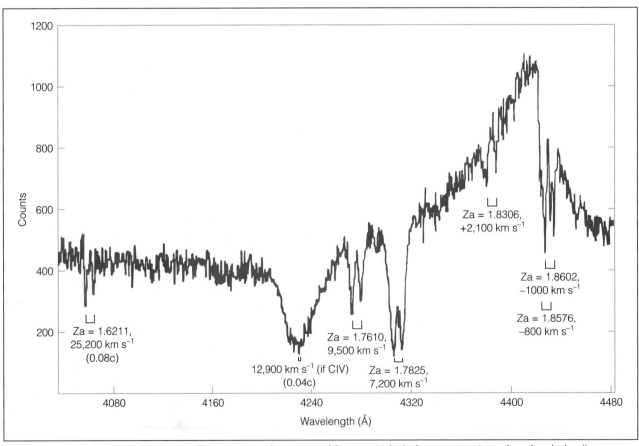

Figure 20.14 A quasar spectrum. This spectrum shows several features typical of quasar spectra: a broad emission line centered near 4400 Å, absorption near 4240 Å, and narrow absorption lines at several wavelengths. All features seen here are due to carbon that has lost three electrons (designated CIV). Redshifts are indicated (Z_a indicates the redshift of an absorption-line system). (Data from R. J. Weymann, R. F. Carswell, and M. G. Smith, 1981, *Annual Reviews of Astronomy and Astrophysics*, 19:41.)

cases, the emission lines are very broad, showing that the gas that forms them has internal motions of thousands of km/sec. The degree of ionization tells us that the gas is subjected to an intense radiation field, which continually ionizes the gas by the absorption of energetic photons.

Many quasars have absorption lines in addition to the emission features. Strangely enough, the absorption lines are usually at a different redshift (almost always smaller) than the emission lines, indicating that the gas that creates the absorption is not moving away from us as rapidly as the gas that produces the emission. Further confounding the issue is the fact that many quasars have multiple absorption redshifts; that is, they have several distinct sets of absorption lines, each with its own redshift, and each therefore representing a distinct velocity. This shows that there are several absorbing clouds in the line of sight. There are two possible origins for these clouds (Fig. 20.15). One is that the quasar itself, which is moving at the high speed indicated by the redshift of the emission lines, is ejecting clouds of gas, some of which happen to be aimed toward the Earth. The clouds therefore have lower velocities of recession than the quasar itself, so the redshift of the absorption lines they form is less than the redshift of the emission lines formed in the quasar.

The alternative view is that the absorption lines arise in the gaseous halos or disks of galaxies that happen to lie between us and the more distant quasar. Each galaxy has a recession velocity determined by the expansion of the universe, but because these intervening galaxies are not as far away as the quasars, they are not moving as fast and therefore have lower redshifts. If this interpretation of the absorption lines is correct, as is thought to be true in at least some cases, then their analysis should provide important information about the nature of the intervening galaxies. Of particular interest is the information quasar absorption lines might provide on the extensive halos that many galaxies are thought to have.

In some cases the absorption spectra of quasars show only lines formed by hydrogen and none due to heavier elements. These lines apparently arise in gas that has not been enriched at all by stellar processing, and it is thought that these might be intergalactic clouds that have never formed into galaxies, so that no stars have formed in them. These may be material remnants of the big bang; if so, they can tell us much about the exact amount of nuclear processing that occurred in the early stages of the expansion.

GALAXIES IN INFANCY?

A variety of theories, some of them rather fanciful, have been proposed to explain the quasars. One idea has gradually become widely accepted, however, and we will restrict ourselves to discussing that hypothesis, keeping in mind that there are other suggestions.

The prevailing interpretation of the nature of quasars was inspired by the fact that they existed only in the long-ago past, by their similarities to the BL Lac objects (which clearly are galaxies), and by their resemblance to the nuclei of Seyfert galaxies. The state-

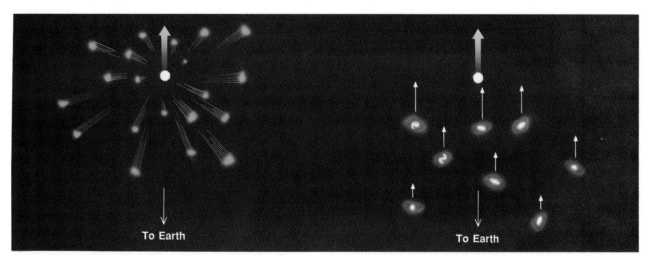

Figure 20.15   Two explanations of quasar absorption lines.   In one scenario (left), the absorption lines are formed by clouds of gas ejected from the quasar. Clouds ejected directly toward the Earth have lower velocities of recession than the quasar itself. In the other more likely view (right), the absorption lines are formed in the extended halos of galaxies that happen to lie between us and the quasar. The galaxies are closer to us than the quasar and therefore have lower velocities of recession and smaller redshifts.

ment that they existed only in the past is based on the fact that all are very far away, so that the light-travel time ensures that we are seeing things only as they were billions of years ago, not as they are today. The resemblance to Seyfert nuclei is striking: both are blue; both are radio sources in about 10 percent of the cases; both vary on similar time scales; and both have very similar emission-line spectra. Seyferts lack the complex absorption lines often seen in the spectra of quasars, but this would be expected if the quasar absorption lines are formed in the halos of intervening galaxies, since Seyfert galaxies are not so far away that there are likely to be many other galaxies along their lines of sight. The nuclei of Seyferts differ from quasars also in the amount of energy they emit, being considerably less luminous.

The picture that is developing is that quasars are young galaxies, with some sort of youthful activity taking place in their centers. In this view, the Seyfert galaxies are descendants of quasars, still showing activity in their nuclei, but with diminished intensity. If we carry this idea another step, we are led to the suggestion that normal galaxies like the Milky Way are later stages of the same phenomenon; recall the mildly energetic (by quasar standards) activity in the core of our galaxy.

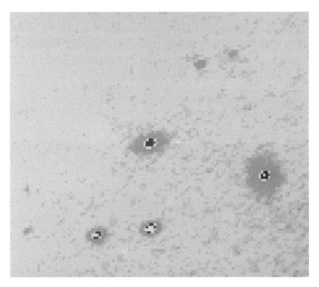

Figure 20.16   A quasar embedded in a galaxy.   This photograph, obtained with an electronic detector, contains several star images (top, sides), a galaxy (the slightly elongated object just to the right of center), and a quasar (lower center). The quasar image is extended, just as the galaxy image is, and the fuzzy region surrounding it is indeed a galaxy. In every case where a quasar is close enough to us that a galaxy would be bright enough to be detected, one has been, indicating that all quasars lie at the centers of galaxies. (M. Malkan)

The notion that quasars may be infant galaxies is supported by some rather direct evidence. As noted earlier, some quasars have been found associated with clusters of galaxies, showing that they can be physically located in the same region of space at the same time. In addition, a few examples are known where careful observation has revealed a fuzzy, dimly glowing region surrounding a quasar (Fig. 20.16). This is probably a galaxy with a quasar embedded at the center. It is likely that all quasars are located in the nuclei of galaxies, which are so much fainter that they usually cannot be seen. A short-exposure photograph of a Seyfert galaxy (Fig. 20.17) shows only the nucleus, which is very much brighter than the surrounding galaxy. Clearly, if we observed one of these galaxies from so far away that it was near the limit of detectability, all we would see would be a blue, starlike object resembling a quasar.

Although we can make a strong case that quasars are young galaxies, we are still far from understanding all of their properties. The main mystery remaining is the source of the tremendous energy of quasars and active galactic nuclei. Several possibilities have been raised, but today only one seems viable.

This possibility, which keeps coming up in situations where large amounts of energy seem to be coming from small volumes of space, is that quasars are powered by massive black holes (Fig. 20.18). In the chaotic early days of collapse, when a galaxy was just forming out of a condensing cloud of primordial gas, a great deal of material may have collected at the center. If many stars formed there, frequent collisions could have caused them to settle in more and more tightly, until they coalesced to form a black hole. The gravitational influence of such a massive object would have then stirred up the surroundings, creating the high electron velocities required to fuel the synchrotron emission. Infalling matter would have formed an accretion disk around the black hole, from which X rays would be emitted, by analogy with stellar black holes and neutron stars in binary systems. In this picture the source of energy in quasars is the same as in the radio galaxies described earlier in this chapter. Even the existence of jets is the same; such jets have been observed emanating from several quasars.

Infrared data from *IRAS* have led to a suggested mechanism for building up sufficient densities in the cores of young galaxies to form supermassive black holes. The properties of some of the very luminous infrared galaxies indicate that they may be pairs of galaxies that are colliding and merging. The infrared properties of quasars are very similar to those of luminous infrared galaxies, suggesting that quasars may form as the result of galactic mergers early in the history of the universe. Recent intensive observations

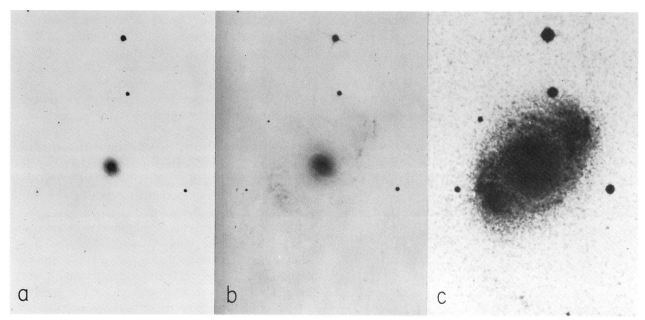

Figure 20.17   Three exposures of a Seyfert galaxy.   At the left is a short-exposure photograph of the Seyfert galaxy NGC 4151, in which only the nucleus is seen, resembling the image of a star. The center and right-hand images are longer exposures of the same object, revealing more of its fainter, outer structure. This sequence illustrates how a quasar, which is even more luminous than a Seyfert galaxy nucleus, can appear as a starlike image even though it is embedded in a galaxy. (Photographs from the Mt. Wilson Observatory, Carnegie Institution of Washington, composition by W. W. Morgan)

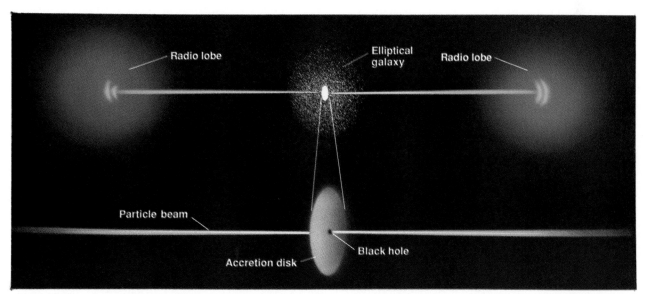

Figure 20.18   A geometrical model for a quasar, Seyfert galaxy, or a radio galaxy.   All of these objects have certain features in common that fit the picture shown here. A central object (most likely a supermassive black hole with an accretion disk) ejects opposing jets of energetic charged particles. These jets produce synchrotron radiation and build up double radio-emitting lobes on either side of the central object. These lobes extend well beyond the confines of the galaxy in which the central object is embedded.

of a number of quasars show that they may indeed be the result of mergers of galaxies that occurred at the time when the galaxies were just forming. There is growing evidence, in fact, that *all* galaxies formed through the merger of smaller systems of stars (this was discussed in Chapters 17 and 18).

Whatever the origin and power source of the quasars, we will certainly learn a lot about the nature of matter and the early history of the universe when we are able to answer all the questions about them. The *Hubble Space Telescope,* with its broad wavelength coverage and great sensitivity, will provide invaluable information on these fundamentally important objects.

PERSPECTIVE

In this chapter we have learned of new and wondrous things, the mighty radio galaxies and the enigmatic quasars. Along the way we have gained several important bits of information about the universe itself. It is time to tackle the fundamental question of the origin and fate of the comos.

SUMMARY

1. Radio galaxies emit vast amounts of energy in the radio portion of the spectrum. These galaxies, often giant ellipticals, commonly show structural peculiarities. The radio emission is nonthermal (usually synchrotron radiation) and in most cases is produced from two lobes on opposite sides of the visible galaxy.

2. The source of energy in a radio galaxy is unknown but appears to be concentrated at the core.

3. Seyfert galaxies are spiral galaxies with compact, bright blue nuclei that produce emission lines characteristic of high temperatures and rapid motions.

4. Several point sources of radio emission found in the early 1960s looked like blue stars and were therefore called quasi-stellar objects, or quasars.

5. Quasars have emission-line spectra with very large redshifts, corresponding to speeds that are a significant fraction of the speed of light. If these redshifts are cosmological, then quasars are on the frontier of the universe, are extremely luminous, and are seen as they were billions of years ago.

6. Quasars are sometimes radio sources, usually emit X rays, are compact and blue, and sometimes vary over times of only a few days. This implies that their tremendous energy output arises in a volume only a few light-days cross.

7. Many quasars have absorption lines at a variety of redshifts (almost always smaller than the redshift of the emission lines), which may be created by matter being ejected from the quasars or, more likely, by intervening galactic disks and halos.

8. The most satisfactory explanation of the quasars is that they are the cores of very young galaxies. This interpretation is supported by the fact that they are only seen at great distances (that is, they existed only long ago), and that they resemble the nuclei of Seyfert galaxies. In several cases, photographs have revealed galaxies surrounding quasars, supporting this suggestion.

9. The source of the energy that powers quasars is a premier mystery of modern astronomy. The most probable explanation is that a quasar has a massive black hole at its core; this can produce enough energy (from infalling matter) to account for the great luminosity of quasars, as well as their time variability. By inference, similar objects may exist in the nuclei of galaxies such as Seyfert galaxies, radio galaxies, and the Milky Way.

REVIEW QUESTIONS

1. Summarize the properties of radio galaxies, pointing out the ways in which they differ from normal galaxies.

2. Why do astronomers suggest that massive black holes may be responsible for the vast amounts of energy being emitted from the cores of radio galaxies, Seyfert galaxies, and quasars?

3. Suppose that the red hydrogen line (rest wavelength 6563 Å) is observed as a broad emission line from the nucleus of a Seyfert galaxy. The width of the line is 200 Å (that is, it extends for 100 Å in either direction from the rest wavelength). What is the maximum speed of the gas in the region where this line is emitted?

4. Explain, in your own words and using your own sketch, why the time scale of variations in light from a source such as a quasar places a limit on the diameter of the emitting region.

5. A quasar is found to have a redshift of $z = \Delta\lambda/\lambda = 2.45$. What is its recession velocity (use the relativistic Doppler shift formula given in Appendix 14)? How far away is it if the Hubble constant is $H = 75$ km/sec/Mpc? How long has the light from this quasar been on its way to us? How luminous is it if its apparent magnitude is $m = +18$?

6. Compile a list of the similarities and differences between quasars and Seyfert galaxies.

7. Why are quasar absorption lines always observed at smaller redshifts than the emission lines? Would this necessarily be true if the redshifts of the emission lines were not cosmological?

8. If quasars are young galaxies, how does this help demonstrate the difficulty of using faraway galaxies as standard candles in estimating large distances in the universe?

9. Explain why nearly all of the first quasars to be discovered are radio emitters, whereas only about 10 percent of all quasars are.

10. How can the material in intergalactic gas clouds give us information about conditions very early in the history of the universe?

Additional Readings

Barnes, J., L. Hernquist, and F. Schweizer, 1991. What happens when galaxies collide. *Scientific American* 265(2):40.

Bechtold, J. 1988. High-resolution spectroscopy of quasar absorption lines. *Mercury* 17:188.

Blandford, R. D., M. C. Begelman, and M. J. Rees. 1982. Cosmic jets. *Scientific American* 246(5):124.

Burbidge, G. 1988. Quasars in the balance. *Mercury* 17:136.

Burns, J. O., and R. Marcus. 1983. Centaurus A: The nearest active galaxy. *Scientific American* 249(5):50.

Courvoisier, T. J.-L., and E. I. Robson, 1991. The quasar 3C273. *Scientific American* 264(6):50.

Disney, M. J., and P. Veron. 1977. BL Lac objects. *Scientific American* 237(2):32.

Downes, A. 1986. Radio galaxies. *Mercury* 15(2):34.

Dressler, A. 1991. Observing galaxies through time. *Sky and Telescope* 82(2):126.

Feigelson, E. D., and E. J. Schreier. 1983. The X-ray jets of Centaurus A and M87. *Sky and Telescope* 65(1):6.

Ferris, T. 1984. The radio sky and the echo of creation. *Mercury* 13(1):2.

Finkbeiner, A. 1992. Active galactic nuclei: sorting out the mess. *Sky and Telescope* 84(2):138.

Gregory, S. 1988. Active galaxies and quasars: A unified view. *Mercury* 17:111.

Margon, B. 1991. Exploring the high-energy universe. *Sky and Telescope* 82(6):607.

McCarthy, P. 1988. Measuring distances to remote galaxies and quasars. *Mercury* 17:19.

Oort, J. 1992. Exploring the nuclei of galaxies (including our own). *Mercury* 21(2):57.

Osmer, P. S. 1982. Quasars as probes of the distant and early universe. *Scientific American* 246(2):126.

Overbye, D. 1982. Exploring the edge of the universe. *Discover* 3(12):22.

Posey, C. 1988. Three recent snapshots of how quasars can be triggered. *Mercury* 17:22.

Preston, R. 1988. Beacons in time: Maarten Schmidt and the discovery of quasars. *Mercury* 17:2.

Silk, J. 1980. *The big bang: The creation and evolution of the universe.* San Francisco: W. H. Freeman.

Strom, R. G., G. K. Miley, and J. Oort. 1975. Giant radio galaxies. *Scientific American* 233(2):26.

Tesch-Fienberg, R. 1987. *IRAS* and the Quasars. *Sky and Telescope* 73:13.

Tyson, A., and M. Gorenstein. 1985. Resolving the nearest gravitational lens. *Sky and Telescope* 70(4):319.

Weedman, D. 1988. Quasars: A progress report. *Mercury* 17:12.

Wilkes, B. J. 1991. The emerging picture of quasars. *Astronomy* 19(12):34.

Wyckoff, S., and P. Wehinger, 1981. Are quasars luminous nuclei of galaxies? *Sky and Telescope* 61(3):200.

ASTRONOMICAL ACTIVITY
Finding Quasar Redshifts

The accompanying figure is a sketch of the spectrum of a quasar, showing both emission lines and absorption lines. Also given is a table listing spectral lines (either emission or absorption) that might appear in the spectrum of a typical quasar. You can use this list to identify features in the spectrum and determine its redshift (hint: the lines represent more than one redshift; that is, this quasar has an emission redshift and at least two distinct absorption redshifts).

First, you will need to identify lines belonging to each redshift system, and then use the wavelengths you measure from the spectrum, along with the rest wavelengths given in the table, to determine the redshift $z = \Delta\lambda/\lambda$. You will have to identify the lines by pattern recognition, because the wavelengths are shifted so far that they are not even close to the laboratory wavelengths. A useful starting point is to assume that the 1216 Å line of hydrogen (H I) is normally the strongest line in each redshift system; note also that this line has the shortest wavelength of the lines typically seen. Once you have tentatively identified an H I line, compute its redshift and then see whether other lines from the list appear at approximately the same redshift. If you find several coinci-

Atom or Ion	Rest Wavelength (Å)
H I	1216
C III	1909
C IV	1549
N V	1240
Mg II	2800
Si II	1260
	1304
	1526
Si IV	1398

dences, then you have identified a redshift system. Note that to make things a little more complicated (but more realistic), not all of the lines in the table appear in each of the redshift systems in the quasar spectrum.

Once you have found the redshifts, you can determine the velocities of the quasar and of each absorbing cloud (using the relativistic Doppler shift formula; see Appendix 14); then you can use the Hubble law to find their distances (indicate what value of the Hubble constant H you chose).

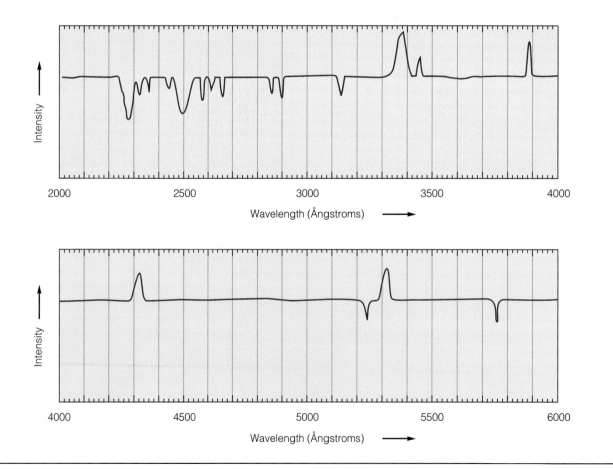

CHAPTER
21

Cosmology:
Past, Present, and Future
of the Universe

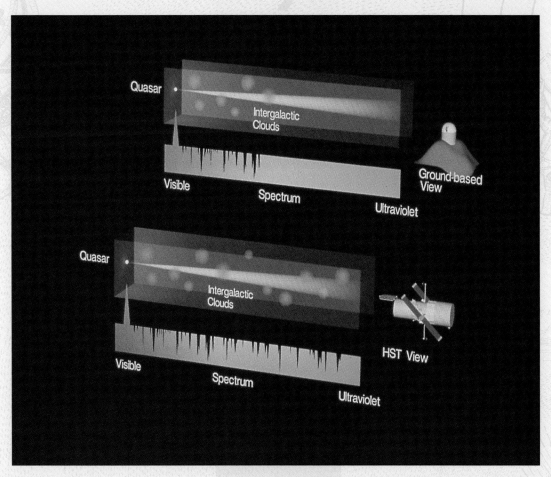

Quasar spectra showing redshifted hydrogen absorption lines
(NASA/STScI; data obtained by J. Bahcall)

PREVIEW

Astronomers are at a disadvantage relative to scientists in other fields in that we have only one universe to study, and we cannot perform experiments or alter it to see how it reacts. On the other hand, astronomers have the advantage of being able to study the universe at different ages by observing objects at various distances and thus at various times in the past, right back to the era when the background radiation was emitted. In this chapter you will see how astronomers deal with the biggest and most significant questions of all: How did the universe form, how has it evolved, and how will it end? The approach combines observation and theory, and we start by reviewing the assumptions of general relativity that are used to develop cosmological theories. The discussion then turns to observations and the ways in which they help refine the theory. You will find that a surprising amount of information is available on the earliest moments in the history of the universe, because physicists have learned much about the behavior of matter under conditions that simulate the early stages of the universal expansion. You will also learn that substantial problems remain, such as the nature of the dark matter that dominates the mass content of the universe and the process by which an initially smooth universe became lumpy enough to form galaxies and clusters and superclusters of galaxies. Finally, you will engage in some speculation about the future of the universe and the methods being used to predict its ultimate fate.

Given the time and opportunity, humans have, through the ages, devoted themselves to speculation on the grandest scale of all. Poets, philosophers, and theologians have approached the question of the origin and future of the universe in countless ways. So have astronomers, with the limitation that they must follow a somewhat restrictive set of rules: the answers they accept must not violate known laws of physics. By imposing this requirement, scientists attempt to approach problems in an objective, verifiable manner.

There are difficulties in maintaining this idealized posture, however, and we shall try to point them out. Because the universe in which we live is, as far as we know, unique, we have no opportunity to check our hypotheses by comparison with other examples. Furthermore, no matter how thoroughly and rigorously we can trace the evolution of the universe by application of known physical laws, there will always be fundamental questions that are beyond the scope of physics. As a result, even the most careful and objective scientists reach a point at which they have to make certain unverifiable assumptions, and at that point, they become philosophers or theologians. In this chapter we shall restrict ourselves to the questions that can in principle have objective, verifiable answers.

Technically the study of the universe as it now appears is **cosmology**; that is really what the preceding 20 chapters have all been about. The study of the origin of the universe is **cosmogony**, a word that applies to the big bang as well as to the earlier theories on the origin of the solar system, once thought to be the entire universe. In practice, the general subject of the nature of the universe and its evolution is lumped under the heading of cosmology, and so it is on a pursuit of this subject that we now embark.

UNDERLYING ASSUMPTIONS

To even begin to study the universe as a whole, we must make certain assumptions. These can, in principle, be tested, although it is not clear whether this can be done practically. One of the most important assumptions we make is that the known laws of physics can be applied to the universe as a whole. This is a bold assumption, since we are just now beginning to be able to observe significantly large distances (compared with the size of the universe), and it is possible that the laws that we know apply to our own times and our own region are merely subsets of some large-scale set of laws, as yet undiscovered. Nevertheless, for now the only way we can proceed is to assume that our known laws can be applied.

A central rule traditionally set forth by cosmologists is that the universe must look the same at all points within it. This does not mean that the appearance of the heavens should be literally identical everywhere; it means that the general structure, the density and distribution of galaxies and clusters of galaxies (Fig. 21.1) should be constant. This assumption states that the universe is **homogeneous.**

A related, but slightly different assumption has to do with the appearance of the universe when viewed in different directions. The assumed property in this case is **isotropy;** that is, the universe must look the same to an observer no matter in which direction the telescope points. We have already encountered this concept, in connection with the cosmic microwave background radiation. Again, the assumption of isotropy does not imply identical constellations or patterns of galaxies and stars in all directions, just comparable ones.

Both of these assumptions are thought to apply only on the largest scales, larger than any obvious clumpiness in the universe, such as clusters or superclusters of galaxies. It should also be pointed out here, although it is not discussed until later in the chapter, that both assumptions have been questioned in some modern theories and observations of the

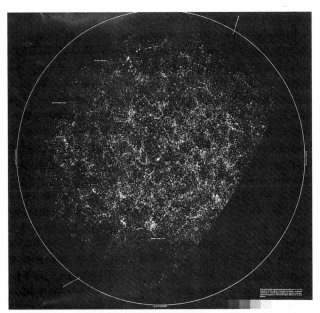

Figure 21.1   The cosmological principle.   The universe is assumed to be homogeneous and isotropic, meaning that it looks the same to all observers in all directions. Here is a segment of the universe, filled with galaxies. Their distribution, though not identical from place to place, is similar throughout. (© 1978, 1979 by The Co-Evolution Quarterly, Box 428, Sausalito, California)

Figure 21.2   Albert Einstein.   Among Einstein's great contributions was the development of a mathematical formalism to describe the interaction of gravity, matter, and energy in the universe with space-time. This framework, general relativity, has withstood all observational and experimental tests applied to it so far. (The Granger Collection)

structure of the universe. Some current theories even consider it surprising that the universe should be as homogeneous as it is. Hence the assumption of homogeneity and isotropy should now be viewed more as an observation that must be explained than as a postulate of cosmological theory.

The statement that the universe is homogeneous and isotropic is often referred to as the **Cosmological Principle** and is often stated in terms of how the universe looks to observers; that is, the universe looks the same to all observers everywhere. It is difficult to verify the Cosmological Principle. We can test the assumption of isotropy by looking in all directions from Earth, and indeed we do not find any deviations. On the other hand, we cannot test the homogeneity of the universe by traveling to various other locations to see whether things look any different. What we can do is count very dim, distant galaxies, to see whether their density appears to be any greater or less in some regions than in others, but even this is complicated by the fact that we are looking back in time to an era when the universe was more compressed and dense than it is now. The 3-degree background radiation gives us another tool for testing both homogeneity and isotropy, and it also appears to satisfy the Cosmological Principle. The best we can say for now is that the available evidence supports this principle.

EINSTEIN'S RELATIVITY: MATHEMATICAL DESCRIPTION OF THE UNIVERSE

The simple act of making assumptions about the general nature of the universe by itself tells us very little about its past or its future. It serves to elucidate certain properties of the universe as it is today, but to tie this into a quantitative description, to develop a theory capable of making predictions that can be tested, requires a mathematical framework. In developing such a mechanism for describing the universe, the astronomer seeks to reduce the universe to a set of equations and then to solve them, much as a stellar structure theorist studies the structure and evolution of stars by constructing numerical models.

The most powerful mathematical tool for describing the universe was developed by Albert Einstein (Fig. 21.2). His theory of general relativity represents

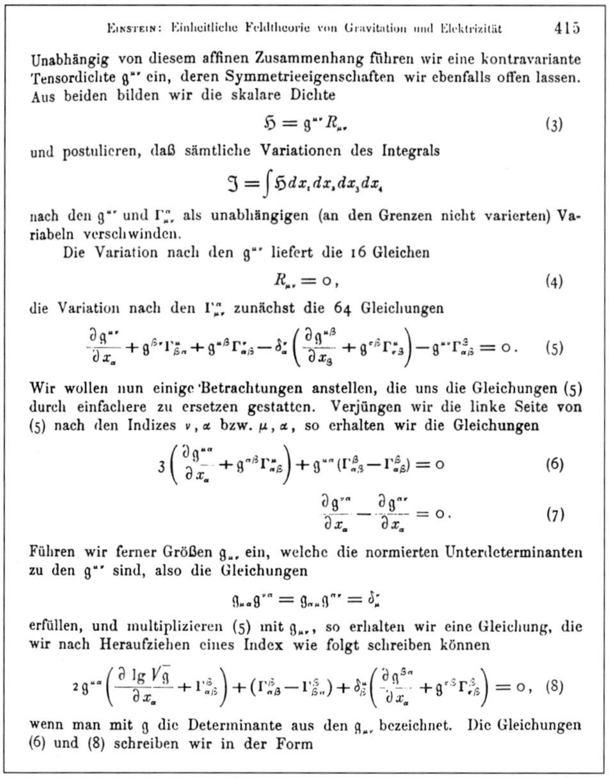

EINSTEIN: Einheitliche Feldtheorie von Gravitation und Elektrizität   415

Unabhängig von diesem affinen Zusammenhang führen wir eine kontravariante Tensordichte $\mathfrak{g}^{\mu\nu}$ ein, deren Symmetrieeigenschaften wir ebenfalls offen lassen. Aus beiden bilden wir die skalare Dichte

$$\mathfrak{H} = \mathfrak{g}^{\mu\nu} R_{\mu\nu}. \qquad (3)$$

und postulieren, daß sämtliche Variationen des Integrals

$$\mathfrak{J} = \int \mathfrak{H} dx_1 dx_2 dx_3 dx_4$$

nach den $\mathfrak{g}^{\mu\nu}$ und $\Gamma^{\alpha}_{\mu\nu}$ als unabhängigen (an den Grenzen nicht varierten) Variabeln verschwinden. Die Variation nach den $\mathfrak{g}^{\mu\nu}$ liefert die 16 Gleichungen

$$R_{\mu\nu} = 0, \qquad (4)$$

die Variation nach den $\Gamma^{\alpha}_{\mu\nu}$ zunächst die 64 Gleichungen

$$\frac{\partial \mathfrak{g}^{\mu\nu}}{\partial x_\alpha} + \mathfrak{g}^{\sigma\nu}\Gamma^{\mu}_{\beta\sigma} + \mathfrak{g}^{\mu\beta}\Gamma^{\nu}_{\alpha\beta} - \delta^{\nu}_{\alpha}\left(\frac{\partial \mathfrak{g}^{\mu\beta}}{\partial x_\beta} + \mathfrak{g}^{\epsilon\beta}\Gamma^{\mu}_{\epsilon\beta}\right) - \mathfrak{g}^{\mu\nu}\Gamma^{\beta}_{\alpha\beta} = 0. \qquad (5)$$

Wir wollen nun einige Betrachtungen anstellen, die uns die Gleichungen (5) durch einfachere zu ersetzen gestatten. Verjüngen wir die linke Seite von (5) nach den Indizes ν, α bzw. μ, α, so erhalten wir die Gleichungen

$$3\left(\frac{\partial \mathfrak{g}^{\mu\alpha}}{\partial x_\alpha} + \mathfrak{g}^{\epsilon\beta}\Gamma^{\mu}_{\alpha\beta}\right) + \mathfrak{g}^{\mu\alpha}(\Gamma^{\beta}_{\alpha\beta} - \Gamma^{\beta}_{\alpha\beta}) = 0 \qquad (6)$$

$$\frac{\partial \mathfrak{g}^{\nu\alpha}}{\partial x_\alpha} - \frac{\partial \mathfrak{g}^{\alpha\nu}}{\partial x_\alpha} = 0. \qquad (7)$$

Führen wir ferner Größen $\mathfrak{g}_{\mu\nu}$ ein, welche die normierten Unterdeterminanten zu den $\mathfrak{g}^{\mu\nu}$ sind, also die Gleichungen

$$\mathfrak{g}_{\mu\alpha}\mathfrak{g}^{\nu\alpha} = \mathfrak{g}_{\alpha\mu}\mathfrak{g}^{\alpha\nu} = \delta^{\nu}_{\mu}$$

erfüllen, und multiplizieren (5) mit $\mathfrak{g}_{\mu\nu}$, so erhalten wir eine Gleichung, die wir nach Heraufziehen eines Index wie folgt schreiben können

$$2\mathfrak{g}^{\mu\alpha}\left(\frac{\partial \lg \sqrt{\mathfrak{g}}}{\partial x_\alpha} + \Gamma^{\sigma}_{\alpha\sigma}\right) + (\Gamma^{\mu}_{\alpha\beta} - \Gamma^{\mu}_{\beta\alpha}) + \delta^{\nu}_{\beta}\left(\frac{\partial \mathfrak{g}^{\beta\mu}}{\partial x_\alpha} + \mathfrak{g}^{\epsilon\beta}\Gamma^{\mu}_{\epsilon\beta}\right) = 0, \quad (8)$$

wenn man mit $\mathfrak{g}$ die Determinante aus den $\mathfrak{g}_{\mu\nu}$ bezeichnet. Die Gleichungen (6) und (8) schreiben wir in der Form

Figure 21.3 The field equations. Here are portions of the equations that describe the physical state of the universe. To a large extent, the science of cosmology involves finding solutions to these equations. (The Granger Collection)

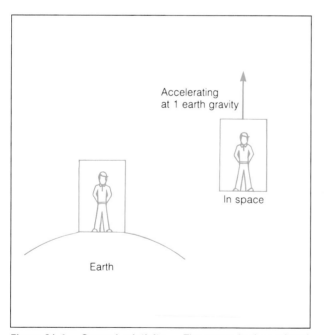

Accelerating at 1 earth gravity

In space

Earth

Figure 21.4 General relativity. The person in the enclosed room has no experimental or intuitive means of distinguishing whether he is motionless on the surface of the Earth or in space accelerating at a rate equivalent to 1 Earth gravity. The implication of this is that the acceleration due to motion and that due to gravity are equivalent, which in turn implies that space is curved in the presence of a gravitational field.

the properties of matter and its relationship to gravitational fields. Within the context of his theory, Einstein developed a set of relations, called the **field equations** (Fig. 21.3), that express in mathematical terms the interaction of matter, radiation, and gravitational forces in the universe. Although alternatives to general relativity have been developed and their consequences explored, most research in cosmology today involves finding solutions to Einstein's field equations and testing these solutions with observational data.

The basic premise of general relativity is that acceleration created by a gravitational field is indistinguishable from acceleration due to a changing rate or direction of motion. One way to visualize this is to imagine that you are inside a compartment with no windows (Fig. 21.4). If this compartment is on the Earth's surface, your weight feels normal because of the Earth's gravitational attraction. If, however, you are in space and the compartment is being accelerated at a rate equivalent to 1 Earth gravity, your weight will also feel normal. There is no experimental way to tell the difference between the two situations short of opening the door and looking out.

One consequence of the equivalence of gravity and acceleration is that an object passing near a source of gravitational field (that is, any other object having mass) undergoes acceleration and therefore follows a curved path. In a universe containing matter, this means that all trajectories of moving objects are curved, and it is often said that space itself is curved. The degree of curvature is especially high close to massive objects (Fig. 21.5), but the universe also has an overall curvature as a result of its total mass content. The solutions to the field equations specify, among other things, the degree and type of curvature. We will return to this point shortly.

Einstein's solution to the equations, developed in 1917, had a serious flaw in his view: it did not allow for a static, nonexpanding universe. In what he later admitted was the biggest mistake he ever made, Einstein added an arbitrary term called the **cosmological constant** to the field equations, solely for the purpose of allowing the universe to be stationary, neither expanding nor contracting.

Others developed different solutions, always by making certain assumptions about the universe. Some of these assumptions were necessary to simplify the field equations so that they could be solved. For example, W. de Sitter in 1917 developed a solution that corresponded to an empty universe, one with no matter in it.

By the 1920s, solutions for an expanding universe were found, primarily by the Soviet physicist A. Friedman and later by the Belgian G. LeMaître, who went so far as to propose an origin for the universe

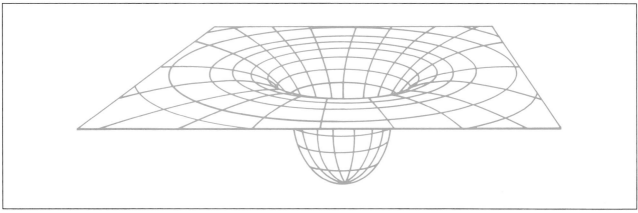

Figure 21.5   Curvature of space near a massive star.   This curved surface represents the shape of space very close to a massive star. The best way to envision what this curvature of space actually means is to imagine photons of light as marbles rolling on a surface of this shape.

in a hot, dense state from which it has been expanding ever since. This was the true beginning of the big bang idea, and it was developed some three years before Hubble's observational discovery of the expansion. Of course, once scientists found that the universe is not static but is actually in a dynamic state, the original need for Einstein's cosmological constant

disappeared. Nevertheless, modern cosmologists usually include it in the field equations, but its value is normally assumed to be zero.

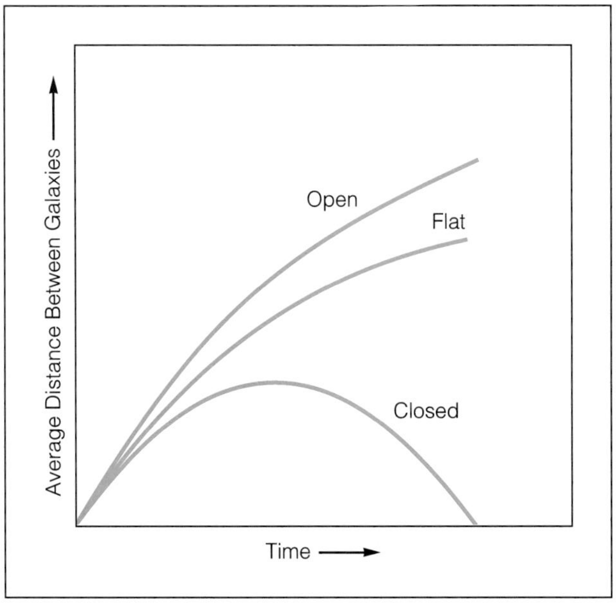

The general relativistic field equations allow for three possibilities regarding the curvature of the universe and three possible futures (Fig. 21.6). A central question of modern studies of cosmology involves deciding which possibility is correct. One is referred to as "negative curvature." An analogy to this is a saddle-shaped surface (Fig. 21.7), which is curved everywhere, has no boundaries, and is infinite in extent. This type of curvature corresponds to what is called an **open universe,** a solution to the field equations in which the expansion continues forever, never stopping. (It does slow down, however, because the gravitational pull of all the matter in it tends to hold back the expansion.)

Figure 21.6   The three possible fates of the universe. This diagram shows the manner in which the average distance between galaxies will change with time for the open, flat, and closed universe. In the first case, galaxies will continue to separate forever, although the rate of separation will slow. In the second case, the rate of separation will, in an infinite time, slow to a halt but will not reverse. In the third case, the galaxies eventually begin to approach each other, and the universe returns to a single point.

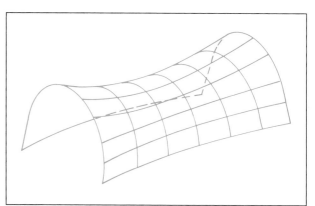

Figure 21.7   A saddle surface.   This is a representation of the geometry of an open universe, which has negative curvature, is infinite in extent, has no center in space, and has no boundaries.

ASTRONOMICAL INSIGHT
The Mystery of the Nighttime Sky

A very simple question, raised over two centuries ago and discussed at length in the early 1800s by the German astronomer Wilhelm Olbers, leads to profound consequences. The question is: Why is the sky dark at night?

Olbers argued that in an infinite universe filled with stars, every line of sight, regardless of direction, should intersect a stellar surface. The nighttime sky should therefore be uniformly bright, not dark, as the star images in the sky would literally overlap each other. The fact that this is not so presents a paradox, one worth pondering. The dilemma created by asking why the sky is dark is called **Olbers' paradox.**

One suggested solution is that interstellar extinction can diminish the light from distant stars sufficiently to make the sky appear black. Careful consideration shows, however, that this is not so. Light that is absorbed by dust in space causes the grains to heat up. If the universe were really filled with starlight, as suggested by Olbers's paradox, then the grains

would become so hot that they would glow, and we would still have a uniformly bright nighttime sky.

Another suggested explanation of Olbers's paradox involves the expansion of the universe. The redshift due to the expansion, if we look sufficiently far away, becomes so great that the light never gets here in visible form. It becomes redshifted to extremely large wavelengths, losing nearly all of its energy. This solution to Olbers's paradox also works in an infinitely old universe, so that even if there is no beginning to the universe, there is still a horizon.

While the expansion of the universe partially explains Olbers's paradox, it does not completely account for the darkness of the sky between galaxies. Yet another solution to the paradox turns out to be more important. To understand this solution, we must realize that the universe is not really infinite, at least from a practical point of view. At sufficiently great distances, we expect to see no galaxies or stars, because we will be looking back to a time before the

universe began. Thus, even though the universe has no edge, there is a horizon beyond which stars and galaxies cannot be seen. Therefore, this argument goes, all lines of sight do not have to intersect stellar surfaces; many reach the darkness beyond the horizon of the universe, and we have a dark nighttime sky.

Another factor that accounts for the darkness of the nighttime sky is also related to the finite size of the universe. The source of energy available to produce light—that is, all the stars in the universe—is limited. If we calculate the total energy that possibly can be produced by all stars everywhere, we find that it is too little to make our sky appear bright. The universe simply does not contain enough energy.

The simple question of why the sky is dark at night, if analyzed fully, could have led to the conclusion that the universe has a finite age or a finite size a full century before this was learned from other observational and theoretical developments.

A second possibility is that the curvature is "positive," corresponding to the surface of a sphere (Fig. 21.8), which is curved everywhere, has no boundaries, but is finite in extent. This possible solution to the field equations is called the **closed universe,** and it implies that the expansion will eventually be halted by gravitational forces and will reverse itself, leading to a contraction back to a single point.

The third and last possibility is that the universe is a **flat universe,** with no curvature. In this possibility, outward expansion is precisely balanced by the inward gravitational pull of the matter in the universe, so that the expansion will eventually come to a stop but will not reverse itself. This balanced, static state requires a perfect coincidence between the energy of expansion and the inward gravitational pull of the combined total mass of the universe and may therefore seem very unlikely. But we do not know

what set of rules governed the amount of matter the universe was born with, so there is no logical reason to rule out this possibility. Most astronomers, however, generally pose the question in terms of whether the universe is open or closed, without specifically mentioning the third possibility that it is flat. As we will see in the next section, if a perfect balance has been achieved, it will be very difficult to determine this from the observational evidence, which has large uncertainties.

OPEN OR CLOSED: THE OBSERVATIONAL EVIDENCE

Substantial intellectual and technological resources are devoted today to the question of the fate of the

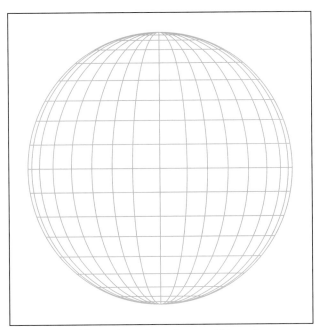

Figure 21.8   The surface of a sphere.   This surface represents a closed universe, which has positive curvature and a finite extent but no boundaries and no center.

universe. Theorists are at work developing and refining the solutions to the field equations or seeking alternatives to general relativity that might provide equally or more valid representations. Observers are busily attempting to test the theoretical possibilities by finding situations in which competing theories should lead to different observational consequences. This is a difficult job, because most of the differences show up only on the largest scales, and therefore detecting them requires observations of the farthest reaches of the universe.

Within the context of general relativity, the premier question traditionally has been whether the universe is open or closed. Historically, two general observational approaches have been used in answering this: determining whether there is enough matter in the universe to produce sufficient gravitational attraction to close it, or measuring the rate of deceleration of the expansion to see whether it is slowing rapidly enough to eventually stop and reverse itself.

Total Mass Content

The total mass in the universe is the quantity that determines whether or not gravity will halt the expansion. The field equations express the mass content of the universe in terms of the density, the amount of mass per cubic centimeter. This is convenient for

observers, because it is obviously simpler to measure the density in our vicinity than to try to observe the total mass everywhere in the universe. The field equations can be solved for the value of the density that would produce an exact balance between expansion and gravitational attraction (the density corresponding to a flat universe). If the actual density is greater than this critical density, then there is sufficient mass to close the universe, and the expansion will stop. The critical density depends on the value of Hubble's constant H; for a value near 55 km/sec/Mpc, it is calculated to be roughly 6×10^{-30} g/cm^3, or about 3 protons per cubic meter, a very low value by any earthly standards.

The most straightforward way to measure the density of the universe is to simply count galaxies in some randomly selected volume of space, add up their masses, and divide the total by the volume. Care must be taken to choose a very large sample volume, so that clumpiness from clusters of galaxies is not important. This technique yields values for the density between 10^{-31} to 10^{-30} g/cm^3, only a small percentage of the critical density. It may seem that we have already answered the question, and that the universe is open, but this method may overlook substantial quantities of mass.

One clue to this arises from determinations of cluster masses based on the velocity dispersion of the galaxies in them (see the discussion in Chapter 18). These mass measurements, for reasons still not entirely clear, always yield much higher values than the estimated total mass of the visible galaxies in the cluster. The disparity can be as great as a factor of 10, so if the larger values are correct, the average density of the universe comes closer to the critical value. Even the larger values based on velocity dispersion measurements, however, fall short of the amount needed to close the universe. If the mass density is to exceed the critical value, there must be large quantities of matter in some form that has not yet been detected (Fig. 21.9). The hidden matter cannot be concentrated inside clusters of galaxies, because its presence would have been detected by the velocity dispersion measurements.

If there is a lot of dark matter in the universe, it could take several forms. One possibility that is obvious to us by now is that there may be many black holes in the space between clusters. The idea that there may be black holes in intergalactic space has been seriously suggested by the astrophysicist Stephen Hawking, who has hypothesized the existence of countless numbers of "mini" black holes, formed during the early stages of the big bang. These would have very small masses, much smaller even than the mass of the Earth, but would be so numerous that they could easily exceed the critical density. Current

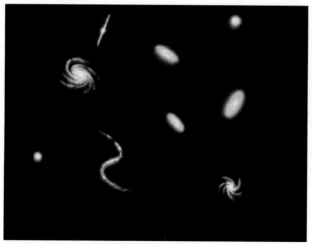 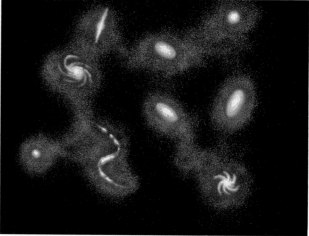

Figure 21.9   Dark matter in the universe.   Are the isolated galaxies that we see all that there is in the universe? Or are they merely points that happen to glow, embedded in a universal sea of unknown substance, whose mass density overwhelms that represented by the galaxies?

observational evidence argues against the existence of such objects in great numbers, however.

Another possibility is that neutrinos, the elusive subatomic particles produced in nuclear reactions, have mass. Recall from Chapter 11 that these particles permeate space, traveling freely through matter and vacuum alike, but that standard theory says they are massless. A recent controversial experiment indicates that this last assertion may not be correct and that they may contain minuscule quantities of mass after all. If so, they are sufficiently plentiful to provide more than the critical density, thereby closing the universe. Further experiments designed to determine whether the neutrino has mass have been performed, but so far all of their results have been more consistent with the standard notion that these particles are massless.

Significant information on the possibility that neutrinos have mass came from the recent supernova (Supernova 1987A) in the Large Magellanic Cloud. Neutrinos from reactions that took place during the explosion were detected at the Earth, and the time of arrival and the energies of these neutrinos allowed scientists to place a limit on how massive they can be. It was found that neutrinos probably cannot be massive enough to bring the average density of the universe up to the point where it is closed. Therefore, if the universe is closed, the dark, unseen matter must be in some form other than neutrinos.

Other kinds of elementary particles have been postulated to exist on the basis of elementary particle theory. These particles have not been detected experimentally (the energies required far exceed those possible in any existing particle accelerators), but it is

possible that they are sufficiently numerous and massive to close the universe. We must await new generations of particle accelerators before we will have an answer to the question of the significance of such particles.

If we base our current thinking only on the density of matter that has actually been observed, we are forced to conclude from the available observational evidence that the mass density of the universe is less than the critical density, and therefore that the universe is open. It is worth stressing, however, that many scientists believe there are substantial quantities of dark matter in the universe, and that it might be closed or flat.

The Deceleration of the Expansion

Another approach to answering the question of whether the universe is open or closed is perhaps being pursued more vigorously today. The objective is to compare the present expansion rate with what it was early in the history of the big bang to see how much slowing, or deceleration, has occurred (Fig. 21.10). The expansion has certainly slowed; the question is how much. If it has only decelerated a little, then we infer that is not going to slow down enough to stop and reverse itself; but if it has decelerated quite a bit, then we conclude that the expansion is coming to a halt, and that the universe is closed.

Establishing the deceleration rate requires knowing both the present expansion rate and the expansion rate at a time long ago, shortly after the big bang began. Neither quantity is easy to determine: To begin with, we have already learned that the value of

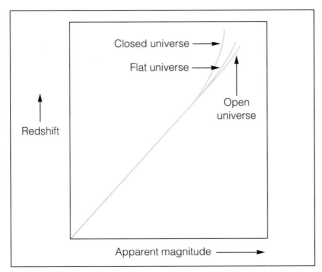

Figure 21.10   The effect of deceleration in the Hubble diagram.   This figure shows the relationship between velocity of recession and apparent magnitude (an indicator of distance) for the three possible cases: closed (upper curve), flat (middle), or open (lower). Present data seem to favor the open universe, although there is substantial uncertainty, largely because of problems in applying standard candle techniques to galaxies so far away that they are being seen as they were at a young age. The distinction is also made difficult by the subtlety of the difference between the shapes of the curves.

the Hubble constant H, which tells us the present expansion rate, is uncertain. To measure the expansion rate at early times in the history of the universe is even more difficult, for it involves determining the distances and velocities of very faraway galaxies, so distant that we see them as they were when the universe was young. Such objects are now at the very limits of detectability, and researchers hope that the *Hubble Space Telescope* will push the frontier back far enough to permit observations of velocities in the early days of the expansion.

We still must rely on standard candles to establish the distance scale, and for such distant objects this procedure becomes even more uncertain than usual. When we look so far back in time, we are seeing galaxies that are much younger than those near us, and therefore they may have different properties from mature galaxies. For example, their content of stellar populations may be rather different from those in nearby galaxies, their brightest stars and nebulae may be different, and even their total galactic luminosities may be different. It may not be valid to assume that the brightest galaxy in a cluster of galaxies has the same absolute magnitude as in more nearby clusters.

The present evidence on deceleration is considered

ambiguous by most astronomers, for the reasons just mentioned. In effect, to measure the deceleration requires determining the curvature at the extreme upper right-hand end of the Hubble diagram (see Fig. 21.10), which is very difficult in view of the many uncertainties.

An indirect technique that has been employed recently to measure the deceleration avoids many of the difficulties of observing extremely distant galaxies. Some light elements were created in the early stages of the expansion, and the amounts that were produced depend on the fraction of the universal energy that was in the form of mass rather than radiation at the time of element formation. The primordial mass density is in turn related to the present-day deceleration of the expansion, so if the primordial density is deduced from the abundances of elements that were created, we can infer the deceleration rate. The strategy, therefore, is to measure the abundance of some element that was produced in the big bang, and from that to derive the deceleration (but this method says nothing about mass that might be present in the form of elementary particles other than those that constitute ordinary atoms; therefore use of deceleration measures to infer the mass density of the universe does not help in determining the dark matter content).

The most abundant element (other than hydrogen) produced in the big bang is helium, so if we could measure how much helium was created, this might be a good indicator of the deceleration. One problem with this element, however, is that it is also produced in stellar interiors, so it is difficult to determine how much of what we see in the universe today is really left over from the big bang. Another problem with using helium as a probe of the early universal expansion rate is that the quantity produced is not expected to vary much with different expansion rates; that is, the primordial helium abundance, even if we knew it well, is not a very sensitive indicator of the early expansion rate.

Better candidates are ^{7}Li, an isotope of the element lithium, and deuterium, the form of hydrogen that has one proton and one neutron in the nucleus. These species were also produced in the big bang and, as far as is currently known, are not made in any other way. To date, most efforts have been concentrated on measuring the present-day abundance of deuterium, although in principle ^{7}Li can be measured as well. While deuterium can be destroyed by nuclear processing in stars, it is not produced in that way. (Even if it were, it would not survive the high temperatures of stellar interiors without undergoing further reactions.) The present deuterium abundance in the universe should therefore represent an upper limit on the

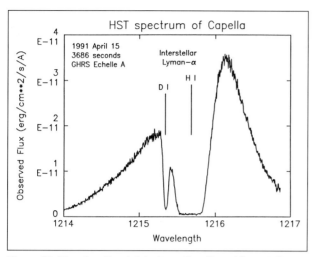

Figure 21.11   An ultraviolet absorption line of interstellar deuterium.   The abundance of deuterium in space is determined from the analysis of absorption lines it forms in the spectra of background stars. This is a portion of a spectrum obtained by the *Hubble Space Telescope,* showing a broad ultraviolet emission line of hydrogen, formed in the chromosphere of the red giant star Capella. There are two interstellar absorption lines superimposed on the broad emission; the stronger one, at right, is due to hydrogen atoms along the line of sight to Capella; the weaker one, at left is due to deuterium atoms. Analysis of the strengths and shapes of these absorption lines yields the relative abundance of deuterium and hydrogen in space. (NASA/STScI; data provided by J. L. Linsky)

quantity created in the big bang. Direct measurements of the amount of deuterium in space are possible, through observations of ultraviolet absorption lines of interstellar deuterium atoms (Fig. 21.11). The abundance of deuterium that is found is sufficiently high that it implies a low density, in turn pointing to a small amount of deceleration. Thus this test, in agreement with others, indicates that the universe is open. There is some uncertainty, though, because the observations seem to indicate an uneven distribution of interstellar deuterium throughout the galaxy, and this would not be expected if the deuterium formed exclusively in the big bang. A clumpy distribution of deuterium in the galaxy could indicate that some of it is somehow produced by stars, in which case its abundance is not a strict test of deceleration. A recent measurement of the ^{7}Li abundance in the galaxy is consistent with the deuterium results; that is, the observed quantity of ^{7}Li is also large enough to imply that the universe is open.

Another technique, the direct measurement of the expansion rate at different times, based on counting galaxies with different redshifts, may be viewed as a modification of the attempts to infer the deceleration of the expansion. The result of the first application

of this technique is that the universe appears to be flat; that is, the expansion appears to have just the rate needed to eventually stop (in an infinite time) but never to contract again. Certainly this technique will be further exploited, and the results scrutinized in the coming years.

It is worth noting that all the observational techniques used thus far to discover whether the universe is open or closed have pointed toward a fairly close balance; that is, the data do not point to a universe that is closed or open by a wide margin. Whatever the correct answer is, we can say that the universe is close to being flat. This is very significant, for such a close balance is not necessarily expected in the big bang theory and must be regarded in this theory as a coincidence. There are models, however, in which a flat or nearly flat universe is expected; one of these is described in the next section.

THE INFLATIONARY UNIVERSE

The big bang cosmology is widely accepted, but some nagging problems remain. One is the improbability that a universe which started from a singularity should turn out to be as symmetric as the observed universe. It is unlikely that the universe would appear as homogeneous and isotropic as observations tell us it is. A great deal of effort has been put into testing this property of the universe, because any departures from homogeneity would provide information on imbalances or asymmetries in the early epochs of the expansion. It is also surprising that the universe is so closely balanced between open and closed.

A new kind of expansion model has been developed in which the close balance and the homogeneity of the universe are easily understood. Calculations have shown that, at a certain point very early in the history of the universe, when the temperature was around 10^{27}K, conditions were right for small regions to separate themselves from the rest of the universe and then to expand very rapidly to much larger sizes. A reasonable analogy would be the creation of bubbles in a liquid, with the bubbles then growing much larger almost instantaneously. In this cosmological model, each "bubble" becomes a universe in its own right, with no possibility of communication with other bubbles.

The major advantage of this so-called **inflationary universe** model is that the universe born of a tiny cell in the early expansion would be expected to remain symmetric and homogeneous as it grew. The reasons for this are somewhat abstract but have to do with the fact that the various portions of the tiny region that was to expand into the present universe were so close together originally that they were in a form of equilibrium, with little or no variation in physical conditions being possible. In a larger region, such as in the standard big bang cosmology, no such equilibrium would have existed, and there would be no reason to expect such uniformity throughout the resulting universe.

Another advantage of the inflationary model is that in such a universe, the expansion would be expected to go on forever but would continue to slow, approaching a stationary state; that is, a flat universe is the natural result of rapid expansion from a tiny bubble as envisioned in this model. As we have seen, observations tell us that our universe is nearly flat and may be precisely flat. If it were not so closely balanced, it would not be so difficult to determine whether it is open or closed.

The very success of the inflationary model in explaining the uniformity of the universe has also proven to be a problem: it fails to explain how the matter in the universe became clumpy enough to form galaxies, clusters, and superclusters. This has been a problem for the standard big bang model as well, and has led some astronomers to question the basic validity of the big bang theory. The recent discovery of subtle non-uniformities in the cosmic background radiation has alleviated these concerns, however (see discussion in Chapter 19), and has increased the general level of confidence in the big bang/inflationary models.

The inflationary universe theory has reached a high degree of acceptance among cosmologists. This does not mean that the big bang model is wrong, but rather that it is a subordinate part of a much bigger picture, much as Newton's concept of gravity is now seen as a subset of Einstein's more complete theory, general relativity.

One of the most pleasing aspects of the inflationary model is also one that is very difficult to understand intuitively. In the standard big bang, we are forced to accept that the state of the universe as it began to expand was a given, something that just was. In other words, the big bang theory can only explain what happened after the beginning. In the inflationary model, the entire universe (and the infinite number of others that are possible) was literally created out of a vacuum. This means that no untidy, inexplicable initial conditions have to be invoked.

In a vacuum, energy may be present even though mass is not, and quantum mechanics tells us that particles can appear out of nothing. These occurrences may be viewed as random events, reflecting the fact that on the smallest of scales the fabric of the universe undergoes fluctuations. The spontaneous appearance

of particles out of a vacuum has been observed in laboratory experiments. General relativity theory tells us that if the energy of the vacuum is high enough, a strange state called "false vacuum" is created, in which a peculiar form of "negative" pressure acts as a repulsive force, causing the local region of space-time to expand if matter is present. Another strange property of false vacuum is that during expansion the density remains constant, which means that additional matter is created out of the energy of the vacuum. In a very brief moment (about 10^{-30} seconds), the early universe would have expanded in size by a factor of 10^{25} or more and would have grown enormously in mass. This was the inflation phase already mentioned, which forms the basis of the inflationary model of the universe. The cooling that occurred during the expansion eventually (that is, after 10^{-30} seconds) would have broken the false vacuum, and the universe would then have been transformed into the more normal form of space we are familiar with, containing matter and having enough residual energy to power the expansion that has been proceeding ever since.

In the inflationary theory, then, the origin of our universe was a random quantum mechanical event, something now believed to occur naturally and inevitably, requiring nothing more than a vacuum with sufficiently high energy. It is now being speculated that perhaps new universes can begin within existing ones, that generation after generation of new bubbles of false vacuum occur randomly and expand. In this view it seems likely that our universe was born of another earlier one; it would be pretentious to think that, of all the possible universes, ours was the first.

If a new universe formed from a random fluctuation in our own, there might be no way for us to know it, because the expansion could take place in different dimensions of space-time from those that we occupy and can sense. In effect, a tiny bit of space-time, if given enough energy, could pinch off and disappear from our universe, spawning a whole new universe that we cannot perceive. An enormous concentration of energy in a tiny space would be required for this to happen, so it is a very low-probability event. But in a sufficiently large universe, even a low-probability event can occur somewhere at some time.

As already noted, once the false vacuum was broken and the universe reverted to normal vacuum filled with radiation and matter, the rest of the expansion is described very well by the standard big bang theory. Hence, the long legacy of big bang cosmologies, beginning in the 1940s with the work of George Gamow and his associates, still provides a useful description of the evolution of our universe. This description is summarized in the next section.

ASTRONOMICAL INSIGHT
Particle Physics and Cosmology

There is some irony in the fact that the science of the smallest scales, particle physics, is finding its greatest opportunities in cosmology, the science of the largest scales. One of the most rapidly advancing areas in modern particle physics coincides with one of the brightest fields in modern astrophysics, bringing full circle one of the themes of astronomy, the relationship between the science of the atom and the science of the cosmos.

In particle physics, scientists are concerned with discovering the nature of matter and energy at its most fundamental level. We now know that the atom and the major subatomic particles such as protons and neutrons are actually composed of more basic particles, known in general as elementary particles. The most modern theories of elementary particles postulate that all matter is made up of two types of basic particles: **leptons,** which include electrons, and **quarks,** which can have several different combinations of properties. Quarks and other particles known only from theory are sought experimentally in particle accelerators, enormous machines that can accelerate particles to such high kinetic energies that when they collide they are broken down into their constituent elementary particles. Because there is an equivalence between matter and energy (as ex-

pressed in Einstein's theory of relativity), elementary particles can be created from pure energy in sufficient quantities. Thus the objective in building particle accelerators to probe elementary particles is to inject enough energy into the accelerators to create the particles being sought. Progress is made every time new levels of energy are reached.

The planned Superconducting Super Collider (SSC), recently approved for construction by the U.S. government, will provide higher energies than any existing accelerators and will give scientists new capabilities for producing elementary particles. This instrument will consist of a huge circular tunnel (53 miles in circumference) in which particle beams will be accelerated by gigantic magnets and then allowed to collide. The collision energies will be comparable to particle energies in the early universe, only a small fraction of a second after the expansion began.

Even the SSC, however, will not answer questions about the very earliest times. Some of the particles that are predicted by theory will require energies so high that it may never be practical to create them in accelerators. This is where astrophysics, specifically cosmology, may come to the rescue. Energies high enough to create *all* possible particles certainly existed during the earliest stages of the universal expansion. Thus elementary particle physicists may find some of the answers they seek by studying what is known about the big bang.

We cannot actually observe the composition of the universe at the time of the early big bang, but we know that today's universe is the product of that era. Thus, by examining the detailed atomic composition of the universe today, particle physicists hope to deduce the mixture of particles that inhabited the universe during its initial moments. A remarkable degree of success has already been achieved in this way, as particle theorists have been able to derive, with considerable certainty, the physical conditions that must have prevailed in early times. We are confident in our knowledge about the formation of the elements in the early universe because the theory of elementary particles is capable of reproducing the observed abundances of the elements believed to be present a fraction of a second after the start of the expansion. Now the hope is that the process can be reversed—that particle physicists can learn about elementary particle processes by observing the state of today's universe.

THE HISTORY OF EVERYTHING

It is impossible to give a physical description of the universe at the precise moment that the expansion (or the inflation of our "bubble") began; it is physical nonsense to deal with infinitely high temperature and density. We can, however, calculate the conditions immediately after the expansion started, and at any later time. Many of the most interesting events in the early history of the universe, and the ones currently under the most active investigation, occurred very early, before even a ten-thousandth of a second had passed. Under the conditions of density and temperature that existed then, matter and the forces that act on it were quite different from anything we can experience, even in the most advanced laboratory experiments. Even the familiar subatomic particles such as protons and neutrons could not exist, but were replaced by *their* constituent particles.

Current particle physics theory holds that the most fundamental particles are **leptons,** which include electrons and positrons, and **quarks.** Modern theory also provides a basis for believing that all four of the fundamental forces in nature (see the Astronomical Insight, opposite) may really be manifestations of the same phenomenon. So far it has been possible to show that three of the fundamental forces (the electromagnetic force and the weak and strong nuclear forces) are manifestations of the same basic interaction. Under physical conditions that we are used to, these forces behave very differently, but when the density and temperature are very high, as they were early in the expansion of the universe, the forces are indistinguishable. Some aspects of the theory that show this have been confirmed by laboratory experiments. The theoretical framework connecting the three forces is called **Grand Unified Theory** (**GUT** for short). This name may be somewhat exaggerated, because the fourth force, gravity, has not yet been shown to be unified with the other three, and some theories of gravity indicate that it may not be.

The unification of the other three forces implies that, in the very early moments following the beginning of the expansion, until 10^{-35} seconds had passed, the universe contained only leptons and quarks and related particles and radiation, and the only forces operating in it were gravity and the unified force that was later to become recognizable as the electromagnetic, strong and weak forces.

As the universe expanded and cooled, the three forces sequentially separated and became distinguishable from each other. Meanwhile a rich stew of particles was brewing, because particles can appear spontaneously from photons of radiation, and in its earliest times the universe was filled with very high energy photons. Recall that mass and energy are interchangeable according to the special relativity relation $E = mc^2$; particles (mass) can appear from photons (energy) if sufficient energy is present in the photons. This was the case for a brief time during the early expansion of the universe.

When particles appear in this way, they do so in pairs with opposing properties. Thus, instead of a single electron, an electron and its oppositely charged counterpart, the positron, would appear. Protons appeared paired with "antiprotons," particles identical in mass but having opposite charges. For each type of particle, there is an antiparticle, its opposite in key properties. (In some cases of neutral particles, the antiparticle and the normal particle are the same.) Pair production of particles from energetic photons (or from high-energy particle collisions) has been observed in particle accelerators.

Once a particle-antiparticle pair formed, each of the two particles moved freely through the young universe, until by random chance it encountered its antiparticle. When this happened, the two particles annihilated each other, converting their combined mass into energy and creating a new photon. Thus an equilibrium existed in which energy and matter were continually being converted back and forth, as photons created particle pairs and particle pairs created photons.

This process could have continued forever if not for the expansion and cooling of the universe. The cooling meant that the energy of the photons decreased steadily, according to Wien's law (Chapter 4). Soon the photon energies were insufficient to produce particle pairs, and pair production ceased. This occurred at different times for particles of different masses; thus as the universe expanded, each type of particle passed a threshold after which that particle was no longer produced.

Because particles and their antiparticles were both created, eventually they found each other and annihilated. If particles and their antiparticles had formed in exactly equal numbers, then in due time all would have annihilated, and there would be no matter in the universe today. The fact that there is matter tells us that particles and antiparticles must have been produced in slightly unequal numbers. The current understanding of how this came to be is that at the time when the four forces were separating from each other, there were very slight asymmetries that favored normal particles over their antiparticles. Theorists estimate that normal particles had to outnumber antiparticles by only one part in 10^{16} to leave all the matter in the universe that we see today. It is thought possible that in other universes that may exist (see the preceding section), different physical laws could prevail, and perhaps antimatter could dominate over normal matter.

Let us now consider an epoch in the early universe when matter was beginning to take on familiar forms, and the four fundamental forces were already acting as four distinct forces. In the inflationary theory, this followed the phase of rapid expansion, which was finished by 10^{-30} seconds after the beginning. The temperature dropped to 10 billion degrees (10^{10} K) by 1.09 seconds after the start, and by then protons and neutrons were appearing (Fig. 21.12). At this point, most of the energy of the universe was still in the form of radiation. Within a few minutes, conditions became better suited for nuclear reactions to take place efficiently, and the most active stage of element creation began. The principal products, in addition to the helium and deuterium mentioned in an earlier section, were tritium (another form of hydrogen, with one proton and two neutrons in the nucleus) and the elements lithium (three protons and four neutrons) and to a minor extent beryllium (four protons and five neutrons). Nearly all the available neutrons combined with protons to form helium nuclei, a process that was complete within 4 minutes of the beginning

of the expansion. At this point, with some 22 to 28 percent of the mass in the form of helium, the reactions were essentially over, except for some production of lithium and beryllium over the next half-hour (see Fig. 21.12).

The expansion and cooling continued, but nothing significant happened for a long time after the nuclear reactions stopped. Eventually (at least 500,000 years after the beginning), the density of the radiation had dropped sufficiently that its energy was less than that contained in the mass (that is, the energy derived from $E = mc^2$ became greater than that contained in the radiation). At that point, it is said, the universe became matter-dominated rather than radiation-dominated.

The matter and radiation continued to interact, however, because free electrons scatter photons of light very efficiently. The strong interplay between matter and radiation finally ended nearly a million years after the start of the big bang, when the temperature became low enough to allow electrons and protons to combine into hydrogen atoms. The atoms still absorbed and reemitted light, but much less effectively, because they could do so only at a few specific wavelengths. From this time on, the matter and the radiation went separate ways. The radiation simply continued to cool as the universe expanded,

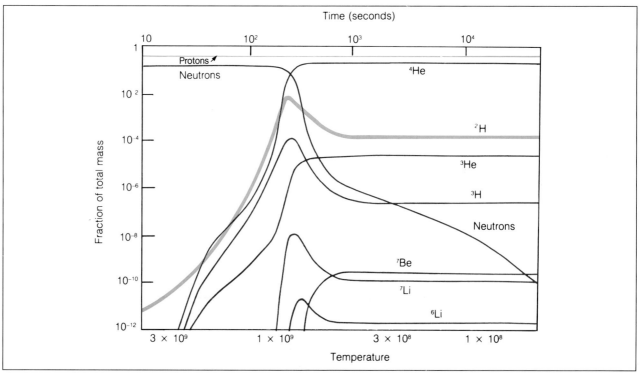

Figure 21.12   Element formation in the big bang.   This diagram illustrates the relative abundances of various light elements formed by nuclear reactions during the early stages of the big bang expansion. The relative abundances are shown as functions of time and temperature. (Data from R. V. Wagoner, 1973, *The Astrophysical Journal* 179:343)

reaching its present temperature of 2.7 K some 10 to 20 billion years after the expansion began.

Up to this point, our story fits well with observations, including the observed nature of the cosmic background radiation. But now we face a very difficult challenge: how to explain the formation of galaxies and groupings of galaxies (Fig. 21.13). As discussed in Chapter 18, there are two competing scenarios for this: the "hot dark matter" and "cold dark matter" theories.

Both of these rest on the assumption that the formation of galaxies and clusters of galaxies was controlled by dark matter, since the vast majority of all the mass in the universe is dark. The hot dark matter theory invokes very rapidly moving (thus "hot") particles such as neutrinos as the basic form of matter. A hot medium can only condense to form enormous structures, and in this theory the first concentrations of mass that formed would have been comparable to today's superclusters. Smaller systems, such as clusters of galaxies and individual galaxies, would have formed through subsequent fragmentation of the larger ones. The hot dark matter theory has been in disfavor with most theorists in recent years, primarily because there is no convincing evidence that neutrinos have mass. In addition, the time required for a universal sea of hot dark matter to settle down and form concentrations of mass is far too long. Therefore the cold dark matter model has been favored.

In the cold dark matter scenario, the nonglowing particles responsible for most of the mass are exotic subatomic particles whose existence has been suggested by some theories (such as the Grand Unified Theory mentioned earlier in this chapter). So far scientists have been unable to confirm the existence of these particles, because it is not possible to build particle accelerators with sufficient energy to produce them (see the Astronomical Insight on p. 496). Nevertheless, the cold dark matter model has appeared promising, particularly because computer simulations have shown that a uniform medium of cold dark matter can develop into a lumpy one through random gravitational interactions and collisions. In other words, a universe made of cold dark matter could, in time, develop into a universe of galaxies, clusters, and superclusters such as we observe today (in saying this, theorists are assuming that the visible matter follows the distribution of the dark matter, an idea that seems to be supported by observations, as pointed out in Chapter 18).

A major requirement of the cold dark matter theory is that the early universe should have been clumpy. Some unevenness in the distribution of matter is needed in order to start the process of gravitational condensation which led to the formation of galaxies, clusters, and superclusters. Until very recently,

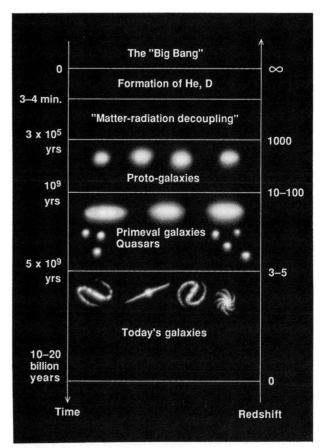

Figure 21.13   The evolution of the universe. This diagram shows, as a function of time since the big bang, significant stages in the evolution of the universe. It also shows the types of objects that existed at each stage and the redshift at which they are (or would be) observed at the present time. (J. M. Shull)

it appeared that observations were close to ruling out the cold dark matter model, because no evidence of the needed clumpiness had been found. Clumpiness in the early universe would show up today in the form of patchiness, or anisotropy, in the cosmic background radiation. As noted in Chapter 19, for some time no such anisotropy was found, even in the early results from the *COBE* satellite, which had been designed specifically to search for it. Recently, however, in a dramatic announcement, the *COBE* scientific team has reported that very low-level variations in the temperature of the background radiation have been discovered. The new *COBE* results show wispy, cloud-like structures corresponding to enormous size scales (see Fig. 19.16). Assuming that these structures are some 15 billion light years away (the "distance" to the time when recombination occurred and the background radiation was released from its close ties to matter in the universe), the largest of the clumps correspond to structures roughly a quarter of a billion light years in length.

ASTRONOMICAL INSIGHT
Dark Matter in the Universe

Throughout the latter portions of this book, we have made many references to "dark matter," in our galaxy, in the halos of other galaxies, or in intergalactic space. In all these environments, astronomers are finding evidence that the majority of the matter that is present is invisible. Perhaps it is useful at this point to summarize the evidence for dark matter in the universe and to review some of the suggestions that have been made for its composition.

Several lines of argument favor the conclusion that a substantial fraction, perhaps as much as 90 percent, of the mass in our galaxy is in some form other than normal stars. The motions of star clusters and gas clouds in the outer portions of the galaxy, as evidenced by the flat rotation curve beyond the Sun's position (Chapter 16), shows that there must be large amounts of mass in the outer galaxy. Ninety percent or more of the mass of the Milky Way Galaxy may reside in the form of invisible matter, most of it in the halo.

Suggestions for this galactic dark matter include dim stars, which may be evolved stellar remnants such as black holes, neutron stars, and white dwarfs; and very low-mass normal stars that are simply too underluminous to be seen at any substantial distance. Both of these possibilities have run into difficulties—the former because an enormous number of stars would have had to evolve to stellar remnants already, and the latter because a huge quantity of low-mass stars would have had to form, in contradiction to what is known of the distribution of stellar masses that normally forms today. Even the so-called brown dwarfs, objects too low in mass to undergo core nuclear reactions, are probably not abundant enough to account for the dark matter in our galaxy. If all these stellar (or prestellar) forms of normal matter can be ruled out, then the best remaining suggestions are those that invoke elementary particles whose properties are not yet known even theoretically or from laboratory ex-

periments.

On scales beyond our galaxy, one of the most direct lines of evidence for large quantities of dark matter arises in clusters of galaxies. Measurements of the overall gravitational effect of such a cluster, either from the orbital motions of the individual galaxies or from the masses required to create the observed gravitational lens effect (Chapter 18), invariably show the presence of much more mass than can be accounted for by the visible galaxies in the cluster. In this case, no quantity of dim stars within galaxies can explain the discrepancy, because such forms of matter are included in the estimates of the masses of individual galaxies. The most striking evidence that the dark matter is not within the galaxies in clusters comes from the analysis of the so-called luminous blue arcs that have been observed in many clusters (see Chapter 18). These arcs are distorted images of background galaxies, and the analysis of their shapes and distribution

---

The new *COBE* results help explain not only how galaxies and clusters and superclusters came into being, but they also explain the recent discovery, by the *ROSAT* x-ray mission (Chapter 4), that clustering had occurred by the time quasars were forming. As discussed in the previous chapter, quasars are thought to be young galaxies, seen as they were long ago, when galaxy formation was taking place. The new *ROSAT* observations show that quasars can appear in clusters, another difficult fact to assimilate into the cold dark matter model, unless the early universe was already clumpy, even before the time when quasars formed. Now this appears to be the case, and it is not difficult to understand how quasars came to form in clusters.

The discovery of early anisotropies in the universe has placed the modern theory of big-bang cosmology on a very firm footing. For the time being, most astronomers are not questioning the basic foundation

of the model, as they were before the new *COBE* results were announced. Instead the focus is now being placed on more detailed questions having to do with how the early irregularities might have formed under the inflation hypothesis, how element enrichment has proceeded since the era of star formation began, and what the future of the universe will be.

Possible Heavy Element Formation in the Big Bang

The standard model of the expansion of the universe shows that only a few of the lightest-weight elements could have formed, because the expansion and cooling were so rapid that there was insufficient time for heavier elements to build up. Furthermore, the next heavier elements after helium and lithium tend to be unstable, immediately decaying by radioactive processes after being formed; thus, even if the reactions can yield information on the distribution of mass within the foreground cluster of galaxies; that is, by studying the images of background galaxies, it is possible to ascertain the shape of the lens that is forming them. In the clusters where this analysis has been carried out, astronomers have found that the matter creating the gravitational lens is more spread out and smoothly distributed than the galaxies themselves. Furthermore, the mass required to create the lens effect is far greater than the mass known to be present in the form of hot, intracluster gas.

On scales beyond clusters of galaxies, the evidence for dark matter becomes more indirect. Other than the local streaming motion of our galaxy and many others toward the "Great Attractor" (Chapter 19), there is little opportunity to analyze motions of clusters of galaxies to look for anomalous effects that might require the presence of large amounts of unseen mass. The composition of the Great Attractor itself is still unknown, but recently some have argued that it may be an ordinary (but very massive) supercluster containing the usual allotment of luminous galaxies. On the other hand, some

have suggested that it is indeed a concentration of dark matter. We may soon know more about this in view of the intensive observational studies of the Great Attractor now taking place.

Perhaps the most compelling argument for a universal medium of dark matter comes from studies, both observational and theoretical, of the evolution of the universe itself. In this chapter, we have explored the current status of these theories and observations and have found strong theoretical arguments that the universe is flat. That is, some modern theories call for a balance between the momentum of expansion of the universe and the self-gravitation that is slowing the expansion. Furthermore, observational estimates of the masses of galaxy clusters based on mass-to-light ratios (see Chapter 18) always fall short of the masses inferred from the motions of galaxies within the clusters, implying that at least 90 percent of cluster masses are invisible.

In order for the universe to be flat, there must be at least 10 times more mass in it than is observed, and the actual discrepancy is probably more like a factor of 100. This means that in the universe as a

whole, dark matter may constitute as much as 80 to 99 percent of all the mass (the uncertainty arises because it is so difficult to measure precisely how much mass is present in the form of visible matter; current estimates are that the greatest quantity of visible mass allowed by observations is only one-fifth of the amount needed to make the universe flat, and that the most likely amount of visible matter present is only about 1 percent of the amount needed).

Clearly the issue of dark matter in the universe represents one of the most important questions astronomers are asking today. Much of the effort in observational astronomy will be focused on this problem, making use of the biggest telescopes available today and new instruments such as the *Hubble Space Telescope*. Similarly, particle physicists are working through theory and experiment to discover the existence and nature of exotic particles that may account for the dark matter. In the coming years, we may expect news from many fronts, but it remains to be seen how much light will be shed on this dark and dim problem.

had formed elements beyond lithium, they would not have lasted. For these reasons, theorists have generally assumed that the universe contained essentially only hydrogen, helium, and a little lithium by the time it was a few minutes old, and that all other elements present today were formed by reactions in stars. One difficulty, however, has been that no "pure" hydrogen and helium stars have ever been found; even the oldest stars observed today contain some traces of heavy elements. The origin of these traces has been difficult to unravel.

Now it has been suggested that perhaps some heavy elements were created in the early big bang after all. There is a family of reactions, called **neutron-capture reactions,** that are capable of producing heavy elements if a supply of free neutrons is available to combine with existing nuclei. Under conditions of sufficiently high density, these reactions could conceivably have created some heavy elements

during the early big bang, but for this to happen, neutron-rich regions would have had to exist. Recent theories indicate that quantum variations during the earliest moments of the universal expansion might have produced clumps of dense material surrounded by lower-density regions that were neutron rich. If so, then neutrons could have penetrated the dense regions, creating some heavy elements by neutron-capture reactions. Detailed models of this process, which depend in turn on the quantum models of early fluctuations in the universe, have been successful in producing enough heavy elements to match what is observed in the oldest stars.

This new picture of element formation during the big bang is not yet well established, but it is generating considerable interest because of its potential for explaining a long-standing problem with the composition of the oldest stars. We may expect further developments in coming years.

WHAT NEXT?

Any discussion of the future of the universe is obviously a speculative venture. We cannot even answer, with absolute certainty, the basic question of whether it is open or closed. Nevertheless, just as theorists have attempted to describe the early stages of the universe, as discussed in the last section, efforts have been made to calculate future conditions in the universe. The uncertainties are much greater in describing the future, however, so the following discussion should be regarded as very speculative.

If the universe is open or flat, as in the inflationary model, then it will have no definite end; it will just gradually run down. The radiation background will continue to decline in temperature, approaching absolute zero. As stellar processing continues in galaxies, the fraction of matter that is in the form of heavy elements will continue to grow, and the supply of hydrogen, the basic nuclear fuel, will diminish. It is predicted that all the hydrogen will be gone by about 10^{14} years after the birth of the universe, so the universe in which stars dominate has now lived approximately one ten-thousandth of its lifetime. The recycling process between stars and the interstellar medium will continue until this time, but gradually the matter will become locked up in black holes, neutron stars, white dwarfs, and black dwarfs. Dead and dying stars will continue to interact gravitationally, eventually colliding often enough in their wanderings that all planets will be lost (by about 10^{17} years) and galaxies will dissipate as their constituent stars are lost to intergalactic space (by about 10^{18} years). At later times, new physical processes will take over. The new Grand Unified Theory tells us that the proton, a basically stable particle, may disintegrate in a very low probability process that occurs, on average, once in 10^{32} years for a given proton. When the universe reaches an age of about 10^{20} years, enough protons will begin to evaporate here and there that the energy produced will keep the remnant stars heated, although only to the modest temperature of perhaps 100 K. At 10^{32} years, most protons will have decayed, and the universe will consist largely of free positrons, electrons, black holes, and radiation (the extremely cold remnant of the big bang). The final stage that has been foreseen occurs at an age of 10^{100} years, when sufficient time has passed for all black holes to evaporate, leaving nothing but a sea of positrons, electrons, and radiation (theory says that black holes can disintegrate with very low probability, meaning that if we wait long enough, eventually they will do so).

If the universe is closed, then someday, perhaps some 50 billion years from now, the expansion will stop and will be replaced by contraction. The deterioration described above will still take place, until the final moments when the universe once again becomes hot and compressed, entering a new singularity. In some views, purely conjectural and without possibility of verification, such a contraction would be followed by a new big bang, and the universe would be reborn. This concept of an oscillating universe, pleasing to the minds of many, will not occur unless the present weight of the evidence favoring an open or flat universe is found to be in error.

PERSPECTIVE

Our story is essentially complete. We have now discussed the universe and all of the various objects and structures in it, and we have described what is known of its evolution and its fate.

We have, however, omitted consideration of what is perhaps the most important ingredient of all: life. Although much of what we might say about this is beyond the scope of astronomy, it is appropriate to assess what can be learned from objective scientific consideration. We will do so in the next chapter.

SUMMARY

1. In cosmological studies, astronomers usually adopt the Cosmological Principle, which states that the universe is both homogeneous and isotropic. Existing observational data tend to support this assumption.

2. Einstein's theory of general relativity, which describes gravitation and its equivalence to acceleration, is used to mathematically describe the universe as a whole. In the context of this theory, field equations are written that represent the interaction of matter, radiation, and energy in the universe. Solutions of the field equations define the properties of the universe.

3. There are three general solutions to the field equations that are considered by modern cosmologists. They correspond to a closed universe (positive curvature), an open universe (negative curvature), and a flat universe (no curvature).

4. A major question in modern astrophysics is whether the universe is closed (expansion eventually to be reversed) or open (expansion never to stop).

5. There are two general types of tests of whether the universe is open or closed: (1) ascertaining whether there is enough mass density in the universe to halt the expansion gravitationally, or (2) determining the rate of deceleration of the expansion, to see whether it is slowing enough to eventually stop.

6. If visible galaxies are used to determine the mass density of the universe, it appears that there is not enough mass to close the universe. Various suggestions have been made concerning other forms in which the necessary mass could exist.

7. The deceleration is measured in two ways: (1) by comparing past and present expansion rates, through observations of very distant galaxies, and (2) by inferring the early expansion rate from the present-day abundances of elements that formed only during the big bang.

8. Both the mass density that is observed and the inferred deceleration of the expansion indicate that the universe is open or flat. This conclusion is not universally accepted, and observational tests are continuing.

9. Recent inflationary models, in which our universe formed through rapid expansion of an initially tiny bubble, show great promise. In these models a homogeneous, flat universe is expected, consistent with observations.

10. The early stages of the universal expansion, up to the time when matter and radiation decoupled, can be described quite precisely and certainly by modern physics. Following the initial moment of infinite density and temperature, the first atomic nuclei formed just over 1 second later, and all of the early element production was finished within a few minutes. Matter and radiation decoupled almost a million years later; the manner in which the universe subsequently organized itself into stars and galaxies is not so well understood, but recently it was discovered through *COBE* measurements of anisotropies in the cosmic background radiation that the early universe was clumpy.

11. Apparently two possibilities exist for the future of the universe. If it is open or flat, it will gradually become cold and disorganized. If it is closed, it will eventually contract, perhaps to a new beginning in another big bang.

Review Questions

1. Explain how the universe can be homogenous even though galaxies are concentrated in clusters and superclusters.

2. Recall from the previous chapter that there is a daily anisotropy in the cosmic background radiation. Does this violate the assumption that the universe is isotropic? Explain.

3. Explain in your own words why general relativity implies that space is curved near massive objects.

4. Summarize the future of the universe under the three possibilities discussed in the text; that is, what will happen if the universe is open, closed, or flat?

5. Explain why it is difficult to determine the average mass density of the universe from observations.

6. The interstellar medium in the disk of our galaxy has an average density of about one hydrogen atom for every 10 cm^3 of volume. Convert this density into units of g/cm^3, assuming that the mass of a hydrogen atom is 1.7×10^{-24} g. How does your answer compare with the critical density for closing the universe? What implications does this have for whether the universe is open or closed?

7. It has been suggested (as discussed in Chapter 20) that most or perhaps all galaxies have very massive black holes in the cores. If so, would this change observational estimates of the mass density of the universe and therefore help determine whether the universe is open or closed? Explain.

8. Why is the deuterium abundance a better indicator of the early expansion rate of the universe than helium?

9. What are the advantages of the inflationary universe theory over the conventional big bang model?

10. Why did element production occur only during the first few minutes of the universal expansion?

Additional Readings

Abbott, L. 1988. The mystery of the cosmological constant. *Scientific American* 258(5):106.

Adair, R. K. 1988. A flaw in a universal mirror. *Scientific American* 258(2):50.

Barrow, J. D., and J. Silk. 1980. The structure of the early universe. *Scientific American* 242(4):118.

Bartusiak, M. 1985. Sensing the ripples in spacetime. *Science 85* 6(3):58.

————. 1986. *Thursday's universe*. New York: Times Books.

Boslough, J. 1981. The unfettered mind: Stephen Hawking. *Science 81* 2(9):66.

Brush, S. G. 1992. How cosmology became a science. *Scientific American* 267(2):62.

Chevalier, R. A., and C. L. Sarazin. 1987. Hot gas in the universe. *American Scientist* 75:609.

Davies, P. 1987. Particle physics for everybody. *Sky and Telescope* 74:582.

———. 1990. Matter-antimatter. Sky and Telescope 79:257.

———. 1991. Everybody's guide to cosmology. *Sky and Telescope* 81(3):250.

———. 1992. The first are second of the universe. *Mercury* 21(3):82.

Dicus, D. A., J. R. Letaw, D. C. Teplitz, and V. L. Teplitz. 1983. The future of the universe. *Scientific American* 248(3):90.

Dressler, A. 1991. Observing galaxies through time. *Sky and Telescope* 82(2):126.

Ferris, T. 1984. The radio sky and the echo of creation. *Mercury* 13(1):2.

Field, G. B. 1982. The hidden mass in galaxies. *Mercury* 11(3):74.

Gaillard, M. K. 1982. Toward a unified picture of elementary particle interactions. *American Scientist* 70(5):506.

Geller, M. J., and J. P. Huchra, 1991. Mapping the universe. *Sky and Telescope* 82(2):134.

Guth, A. H., and P. J. Steinhardt. 1984. The inflationary universe. *Scientific American* 250(5):90.

Halliwell, J. J. 1991. Quantum cosmology and the creation of the universe. *Scientific American* 265(6):76.

Hawking, S. W. 1984. The edge of spacetime. *American Scientist* 72(4):355.

Kanipe, J. 1992. Beyond the big bang. *Astronomy* 20(4):30.

Malaney, R. A., and W. A. Fowler. 1988. The transformation of matter after the big bang. *American Scientist* 76:472.

Mallove, E. F. 1988. The self-reproducing universe. *Sky and Telescope* 76:253.

Manda, R. 1992. Shedding light on dark matter. *Astronomy* 20(2):44.

McCrea, A. W. 1991. Arthur Stanley Eddington. *Scientific American* 264(6):92.

Nather, R. E., and D. E. Winget, 1992. Taking the pulse of white dwarfs. *Sky and Telescope* 83(4):374.

Odenwald, S. 1991. Einstein's fudge factor. *Sky and Telescope* 81(4):362.

Overbye, D. 1983. The universe according to Guth. *Discover* 4(6):92.

———. 1985. The shadow universe: Dark matter. *Discover* 6(5):12.

———. 1991. God's turnstile: the work of John Wheeler and Stephen Hawking. *Mercury* 20(4):98.

Page, D. N., and M. R. McKee, 1983. The future of the universe. *Mercury* 12(1):17.

Parker, B. 1988. The cosmic cookbook: The discovery of how the elements came to be. *Mercury* 17:171.

Penzias, A. A. 1978. The riddle of cosmic deuterium. *American Scientist* 66:291.

Sadoulet, B., and J. W. Cronin. 1992. Subatomic astronomy. *Sky and Telescope* 83(1):25.

Schramm, D. N. 1974. The age of the elements. *Scientific American* 230(1):69.

———. 1991. The origin of cosmic structure. *Sky and Telescope* 82(2):140.

Schwinger, J. 1986. *Einstein's legacy*. New York: W. H. Freeman.

Shu, F. 1983. The expanding universe and the large-scale geometry of spacetime. *Mercury* 12(6):162.

Silbar, M. L. 1982. Neutrinos: Rulers of the universe? *Griffith Observer* 46(1):9.

Silk, J., A. S. Szalay, and Y. B. Zel'dovich. The large-scale structure of the universe. *Scientific American* 249(4):56.

Spergel, D. N., and N. G. Turk. 1992. Textures and cosmic structure. *Scientific American* 266(3):52.

Talcott, R. 1992. COBE's big bang. *Astronomy* 20(8):42.

Trefil, J. S. 1978. Einstein's theory of general relativity is put to the test. *Smithsonian* 11(1):74.

———. 1983a. How the universe began. *Smithsonian* 14(2):32.

———. 1983b. How the universe will end. *Smithsonian* 14(3):72.

———. 1983. *The moment of creation*. New York: Scribners.

Tucker, W., and K. Tucker, 1982. A question of galaxies. *Mercury* 11(5):151.

———. 1989. Dark matter in our galaxy. *Mercury* 18(1):2 (Part 1); 18(2):51 (Part 2).

ASTRONOMICAL ACTIVITY
A Model for the Expanding Universe

An ordinary rubber balloon can be used as an aid in understanding the properties of a uniformly expanding universe. The balloon is a three-dimensional analogy to a four-dimensional reality, but the main points can be made. In the balloon model, the universe consists of the two-dimensional surface of the balloon; there is no universe, no perceptible dimension, outside of this surface, to a being that might live in it (see A. E. Abbott's *Flatland* for a fascinating insight into what life in such a two-dimensional world might be like).

To do this experiment, you will need a balloon that has approximately a spherical shape when inflated—so don't choose one that is elongated (this would represent a different universe, one that seems to be in conflict with modern observations). Blow the balloon up partially, just enough so that the surface becomes smooth. Mark dots on it at random locations, using a felt-tip marker. Label several of the dots with numbers or letters, so that you can keep track of them as the balloon is inflated further. Now measure the distances between a few dot pairs, some close together and others more widely separated (to do this, use a tape measure that can follow the curvature of the balloon's surface, or, if using a rigid ruler, "roll" the ruler around the balloon so that you are measuring the distance along its curved surface). Make a list of the distances you have measured.

Now inflate the balloon further, as far as you can without danger of its bursting. Measure the same dot-to-dot distances as before, and record the results in a column adjacent to the list of previous measurements. Finally, compute the ratio of the new distance divided by the old distance for each of your measured pairs, and write the ratio in another column.

You should find that the ratio of new distance to old is roughly the same for all dot pairs. That is the same as saying that the distance between each pair of dots increased in proportion to the original separation. You may find some deviations from this, because the balloon is probably not perfectly spherical (which would mean that it does not expand uniformly in all directions) and it may also have variations in the elasticity of its surface (which would correspond to an anisotropic universe, in which the expansion rate depended on direction). But generally speaking, to a creature living in the two-dimensional universe of the balloon's surface, the dots would appear to be receding at a rate that is proportional to their distance.

Carrying the analogy a bit further, if we inflated the balloon at a uniform rate, then this hypothetical observer would be able to derive the equivalent of the Hubble law and the Hubble constant, and determine the age of his universe (i.e. the time since the balloon started inflating from an initial single point). If the elasticity of the balloon began to fail at some point, forcing the expansion to slow, this would be analogous to the deceleration of the expansion of our universe due to the self-gravitation of the matter in it. Our tiny resident astronomer might conclude that his universe was flat, as the expansion rate slowed and approached zero (one can only guess what he might think when the balloon subsequently exploded; let us hope that an equivalent failure in the fabric of space-time does not occur in our universe!).

The Elusive Hubble Parameter

It is both fitting and ironic that observations made with the *Hubble Space Telescope* have become a major factor in a current controversy over the expansion rate and age of the universe. As befits its name, the *Hubble Space Telescope* is being employed to measure the value of the Hubble constant.

The general relativistic theory that describes the universe is expressed in terms of five basic parameters: H, the Hubble constant, which represents the rate of expansion; q, the deceleration parameter, which measures the extent to which the expansion has slowed; T, the age of the universe; Ω, the ratio of the average density of the universe to the critical density required to close it; and Λ, Einstein's "cosmological constant," introduced originally in order to allow for a static universe.

The values for the five basic parameters are not entirely independent of each other; if we can pinpoint two or three of them, then the others are determined as well. But measuring even one of them with certainty is very difficult. The Hubble constant has been discussed in Chapter 19, where we noted that current estimates of its value range between 50 and 100 km/sec/Mpc. The value of q, the deceleration parameter, would be ½ if the universe were just closed (and if $\Lambda = 0$, which is normally assumed); in this case, $\Omega = 1$, meaning that the density is equal to the critical density. The age of the universe, T, is closely related to H, as described in Chapter 19; if the rate of expansion has been constant since it began, then we have the relation $T = 1/H$.

If we try to measure H and T separately, then, we must hope that the two values are consistent with each other; that is, the age implied by our value of H should be similar to the age we estimate from independent techniques. This was a problem for some of the early measurements of H, which were large and therefore implied a young age for the universe. Recall (from Chapter 19) that Hubble's first estimate of the value of H led to a universal age that was only about 2 billion years, much smaller than the known age of the Earth. Subsequent measurements have reduced the value of H, but some still run into conflict with the estimated age of the universe. For example, if $H = 100$ km/sec/Mpc, then $T = 9.8$ billion years, whereas current estimates of globular cluster ages range as high as 15 billion years (see Chapters 14 and 17). From this point of view, a value for H closer to 50 km/sec/Mpc would be more "comfortable," in that it would not be in conflict with such age indicators as the globular clusters.

Thus, intensive efforts are being made to refine the measured value of H. But rather than improving, the situation seems to be getting worse, because leading researchers are finding increasingly conflicting results. In principle, it would seem as though H could be readily determined. All we need to do is measure the expansion velocity of faraway galaxies whose distances are well known. But, as discussed in Chapter 19, this has proven difficult to do with much accuracy. The main problem is that distance determination methods are not very accurate for

faraway galaxies; all of the standard candles (see Chapter 18) rely on assumptions that are fairly imprecise and become less and less reliable as we go to great distances and thus to long-ago times.

Despite the difficulties, modern measurements have generally converged toward a value of H of around 75 km/sec/Mpc. The recent measurements are based on a variety of distance indicators, such as the Tully-Fisher relation (see Chapter 18), along with various standard candles, including the absolute magnitudes of Type I supernovae. Many different authors have reached reasonably consistent results that point toward a value of H of close to 75 km/sec/Mpc. This was shown recently by Sidney van den Bergh (Dominion Astophysical Observatory, Canada), who in late 1992 made a comprehensive analysis and intercomparison of virtually all modern distance measurements for galaxies. His recommended value is $H = 76 \pm 9$ km/sec/Mpc (the "± 9" means that the uncertainty in the final value, when all the various measurement uncertainties are taken into account, is 9 km/sec/Mpc).

One of the most important steps in deriving the value of H is the use of supernovae as standard candles. These are considered the best indicator of distances for galaxies beyond 30 Mpc or so, which makes them extremely important. To determine H reliably, it is necessary to use galaxies at great distances, so local motions do not affect the result. Thus a great deal hinges on the assumed calibration of supernovae absolute magnitudes. One particular type of supernova, Type Ia, has proven to be especially critical. And it is in calibrating the Type Ia supernova that the current conflict arises.

In late 1992, one of the pioneers of modern observational cosmology, Allan Sandage (Observatories of the Carnegie Institution of Washington), and several colleagues announced the results of a new study. Sandage and his group used observations of Cepheid variables in a galaxy where a Type Ia supernova occurred in 1937. This galaxy, known as IC4182, lies at a distance of around 5 Mpc, and previous determinations of its distance were based on the use of red supergiants as standard candles. Now Sandage and his coworkers have been able to discover and determine the periods of several (27) Cepheid variables in this galaxy, using images obtained by the *Hubble Space Telescope*. By using the period-luminosity relation for Cepheids (see Chapter 16), they established a new distance measurement for the galaxy. This distance (4.94 Mpc), combined with the peak apparent magnitude of the 1937 supernova that occurred in IC4182, allowed Sandage and his colleagues to determine the absolute magnitude of the supernova. Finally, by assuming that this peak absolute magnitude is standard for all supernovae of Type Ia, Sandage and his colleagues derived a new distance scale for more remote galaxies, and hence a new value for H. The result is a major surprise, for they found $H = 45 \pm 9$ km/sec/Mpc.

It is not clear how to explain the discrepancy between the new value found by Sandage and his colleagues and the higher one recommended by van den Bergh, since both rely on supernovae of Type Ia. The argument comes down to how much variation there is in the absolute magnitudes of supernovae of this type: Sandage and his group argue that there is little intrinsic variation, and that these supernovae are reliable standard candles. Van den Bergh argues, to the contrary, that there may be substantial differences from one supernova of this type to the next (at the time, van den Bergh was unaware of the result obtained by Sandage and his group; it is not clear that he thinks the range in absolute magnitudes for Ia supernovae can be large enough to explain the new discrepancy). No doubt, the months to come will see intensive efforts to reconcile the differing results; perhaps a consensus will be reached soon. In the meantime we must accept that this most essential of all universal parameters is not universally understood.

The value of H found by Sandage and his colleagues implies a comfortably old age for the universe, one that is not in conflict with the ages of globular clusters. By using the relation $T = 1/H$, we find $T = 22$ billion years. Correcting for the fact that the expansion was faster at early times (i.e., correcting for deceleration) leads to an estimated age of 15 billion years.

If Sandage should turn out to be wrong and the consensus favored by van den Bergh right, then we will have to face the issue of how to reconcile the relatively young age implied by the larger value of H with the estimated ages of the globular clusters. One possibility, of course, is that the globular cluster ages are incorrect, and that even the most ancient of these systems is no more than about 10 billion years old. This would be very upsetting for current understandings of stellar evolution, however, and such a revision of globular cluster ages will be strongly resisted.

Another way to resolve such a crisis will be to assume that the value of Ω, the density parameter, is not 1. In most modern cosmologies, it is assumed that $\Omega = 1$, which means that we have a flat universe (see Chapter 21). Observations show that the visible matter in the universe only provides about one-tenth of the critical density, but there are many indications that there may be enough dark matter to reach the critical density and bring the value of Ω up to 1. In the inflationary models that are now highly favored, it is expected that $\Omega = 1$. But if in fact this is not the case, the Ω is much less than 1, then the age of the universe is greater than $1/H$. In that event, a value of H near 76 km/sec/Mpc can be consistent with the estimated ages for globular clusters. So one way out of the dilemma posed by a high value of H is to upset current thinking about Ω and the inflationary universe model.

Meanwhile, it is obviously very important to extend the work of Sandage and his colleagues, using the *Hubble Space Telescope* to find Cepheid variables in other galaxies where supernovae have occurred. Plans are under way to do this, so we may expect new information on the value of H to be forthcoming soon.

LIFE IN THE UNIVERSE

We finish our survey of the dynamic universe with a discussion of living organisms and the prospects for finding that we are not alone in the cosmos. To many, this is the most central question we can ask. In the sole chapter of this section, we will examine the astronomer's attempt to answer it.

In discussing the question of life elsewhere in the universe, we will begin by looking into the origins of life here on Earth. In doing so, we will learn of biological evolution and the evidence that has been unearthed to tell us of our own beginnings. To do this, we will step outside the normal arena of astronomy, but we will then return to it in assessing whether the conditions that led to life on Earth might exist on other planets in our solar system and in the galaxy at large. We will discuss the probability that technological civilizations might be thriving here and there in the Milky Way.

We will complete our brief treatment of life in the universe by discussing strategies for searching for other civilizations. It seems most likely that radio communications will be our initial means of contact, and the choice of the wavelength at which to search for or to send signals involves an interesting exercise in logic.

The Chances of Companionship

The global distribution of plant life, on the continents and in the seas (NASA)

PREVIEW

To many people, a question even more interesting and significant than how the universe came to be is whether life exists elsewhere. We cannot yet answer that question, but in this chapter you will learn something about how life is thought to have formed on the Earth, which provides some insight into the probability that it could have happened on some other planet circling a faraway star. You will also follow the reasoning of astronomers who attempt to estimate the probability that we may someday detect other intelligent and technological beings in our galaxy, and you will learn how the search is being carried out. You will find that despite the vast array of possible means of communication, logical arguments can be used to limit the likely possibilities to a relatively small range of radio frequencies. Given this and the advances made recently in the technology of radio telescopes and signal-processing computers, we may have some optimism that if other technological civilizations exist in the galaxy, we may soon find them. If we do not, then the lack of detection itself will become a significant argument that we are, after all, alone.

Figure 22.1   Charles Darwin.   Scientific inquires by Darwin led to an understanding of evolution, one of the most profound concepts of human intellectual development. (The Granger Collection)

We have attempted to answer all of the fundamental questions about the physical universe that can be treated scientifically. Having done this, we know our place in the cosmos: we know something of its scale and age, and we realize how insignificant our habitat is.

In this chapter we contemplate whether we as living creatures are unique in the universe, or whether even that distinction must be shared. Nothing that we have learned so far leads us to rule out the possibility that other life-forms, some of them intelligent, may exist. We believe that the Earth and the other planets in our solar system are a natural by-product of the formation of the Sun, and we have evidence that some of the essential ingredients for life were present on the Earth from the time it formed. Similar conditions must have developed countless times in the history of the universe and will occur countless more times in the future.

Science cannot yet tell the full story of how life began, however, and we have not found any evidence that it actually does exist elsewhere. The mystery remains.

LIFE ON EARTH

We start our discussion of possible extraterrestrial life by discussing the origins of the only life we know. Besides giving us some insight into the processes thought to have been at work on the Earth, this will help us later, when we are speculating about whether the same processes have occurred elsewhere.

Before the time of Charles Darwin (Fig. 22.1) in the mid-1800s, the view was widely held that life could arise spontaneously from nonliving matter. Darwin's work in the study of evolution, showing how species develop gradually as a result of environmental pressures, made such an idea seem improbable.

An alternative to the spontaneous formation of life was proposed in 1907 by the Swedish chemist S. Arrhenius, who suggested that life on Earth was introduced billions of years ago from space, arriving originally in the form of microscopic spores that float through the cosmos landing here and there to act as seeds for new biological systems. This idea, called the *panspermia* hypothesis, cannot be ruled out, but several arguments make it seem unlikely. Such spores would take a very long time to permeate the galaxy, and their density in space would have to be a very high for one or more of them to reach the Earth by chance. More importantly, it seems very unlikely that the spores could survive the hazards of space, such as ultraviolet light and cosmic rays. Even if the panspermia concept is correct, the question of the ulti-

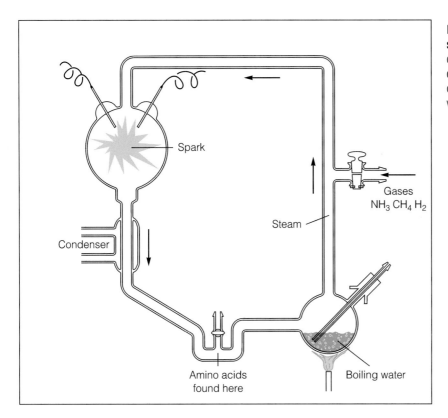

Figure 22.2 Simulating the atmosphere of early Earth. Urey and Miller constructed this apparatus to reproduce conditions on the primitive Earth, in hopes of learning how life-forms could have developed. (© West Publishing Co.)

mate origin of life remains, although it is transferred to some other location. In view of what is known today about the evolution of life and the early conditions on the Earth, scientists generally agree that life arose through natural processes occurring here and was not introduced from elsewhere.

It is believed that the early atmosphere of the Earth was composed in part of hydrogen and hydrogen-bearing molecules such as ammonia (NH_3) and methane (CH_4), as well as water (H_2O). Therefore, the first organisms must have developed in the presence of these ingredients.

In the 1950s, scientists began to perform experiments in which they attempted to reproduce the conditions of the early Earth. The starting point of these experiments was to place water in containers filled with the type of atmosphere just described. Water was introduced because it is apparent that the Earth had oceans from very early times, and because it is thought that life started in the oceans, where the liquid environment provided a medium in which complex chemical reactions could take place. Reactions occur much more slowly in solids and in gases—in solids because the atoms are not free to move about easily and interact, and in gases because the density is low, and particles are relatively unlikely to encounter each other. Water is the most stable and abundant liquid that can form from the common ele-

ments thought to be present when the Earth was young.

The first of these experiments (Fig. 22.2) was performed in 1953 by the American scientists H. Urey and S. Miller, who concocted a mixture of methane, ammonia, water, and hydrogen and exposed it to electrical discharges, a possible source of energy on the primitive Earth. (Ultraviolet light from the Sun is another, but it was more difficult to work with in the lab.) After a week, the mixture turned dark brown, and Urey and Miller analyzed its composition. They found large quantities of amino acids, complex molecules that form the basis of proteins, which are the fundamental substance of living matter. Other experimenters later showed that exposure to ultraviolet light produced the same results. These experiments demonstrated that at least some of the precursors of life probably existed in the primitive oceans almost immediately after the Earth had cooled enough to support liquid water. Other similar experiments have produced more complex molecules, including sugars and larger fragments of proteins.

As noted in Chapter 9, amino acids may have been present in the solar system even before the Earth formed. Traces of them have been found in some meteorites (Fig. 22.3), and we know that meteorites are very old, representing the first solid material in the solar system. We also know that several kinds of

Figure 22.3   Primordial amino acids.   This is a section of the Murchison meteorite, which fell in Australia. Amino acids found in this carbonaceous chondrite were apparently present in it when it fell. (Photo by John Fields, the Trustees, the Australian Museum)

complex molecules, including organic (carbon-bearing) molecules, exist inside dense interstellar clouds (see Chapter 16), and we may speculate that perhaps amino acids may also have formed in these regions. To have survived on the Earth, these primordial amino acids would have to reach our planet sometime *after* its molten period.

The direction things took once amino acids and other organic molecules existed is no so clear. Somehow these building blocks had to combine to form **ribonucleic acid (RNA)** and **deoxyribonucleic acid (DNA;** Fig. 22.4). These very complex molecules carry the genetic codes that allow living creatures to reproduce. Experiments have successfully produced molecules that are fragments of RNA and DNA from conditions like those that prevailed on the early Earth, but not the complete forms required. Maybe it is simply a matter of time; if such experiments could be performed for years or millennia, perhaps the vital forms of these proteins would appear. This is one of the areas of greatest uncertainty in our present knowledge of how life began.

Fossil records (Fig. 22.5) tell us that the first micro-organisms appeared some 3 to 3.5 billion years ago, when the Earth was barely 1 billion years old. Following their appearance, the evolution of increasingly complex species seems to have followed naturally (Table 22.1). At first the development was very slow, only reaching the level of simple plants such as algae 1 billion years later. Increasingly elaborate multicellular plant forms followed and gradually altered the Earth's atmosphere by introducing free oxygen. Meanwhile, the gases hydrogen and helium, light and fast-moving enough to escape the Earth's gravity, essentially disappeared. Nitrogen, always present from outgassing and volcanic activity, became more pre-

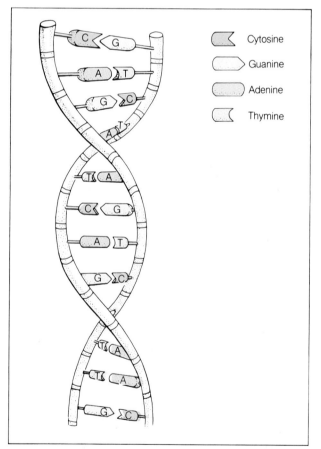

Cytosine

Guanine

Adenine

Thymine

Figure 22.4   The DNA molecule.   This is a schematic diagram of a section of a DNA molecule. DNA carries the genetic code that allows organisms to reproduce themselves. A critical question in understanding the development of life on Earth is to learn how DNA arose. (© West Publishing Co.)

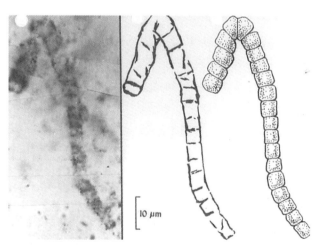

10 μm

Figure 22.5   A fossil microorganism.   This fossil alga is evidence that primitive life-forms existed on the Earth billions of years ago. (Photo courtesy of J. William Schopf, University of California, Los Angeles)

TABLE 22.1 Steps in the Evolution of Life on Earth

Era	Period	Age (yr)	Biological Developments
Archeozoic		3.5×10^9	
	Archean		No life-forms
Proteozoic		$1.5\text{–}3.5 \times 10^9$	
	Algonkin		Radiolaria; marine algae
Paleozoic		$0.25\text{–}1.5 \times 10^9$	
	Cambrian		Marine faunas, primitive vegetation
	Ordovician		Fishlike vertebrates
	Silurian		Air-breathing invertebrates, first land plants
	Devonian		Fishes and amphibians; primitive ferns
	Mississippian		Ancient sharks, mosses, ferns
	Pennsylvanian		Amphibians, insects
	Permian		Primitive reptiles, mosses, ferns
Mesozoic		$70\text{–}250 \times 10^6$	
	Triassic		Reptiles, dinosaurs, ferns
	Jurassic		Toothed birds, primitive mammals, palms
	Cretaceous		Decline of dinosaurs, modern insects, birds, snakes
Cenozoic		$1.5\text{–}70 \times 10^6$	
	Paleocene		Large land mammals
	Eocene		Primates, first horse, modern plants
	Oligocene		Larger horse, modern plants
	Miocene		Proliferation of mammals, larger horse, modern plants
	Pliocene		Grassland mammals, earliest hominids
Modern		1.5×10^6 to present	
	Pleistocene		Hominids, modern-size horse, modern plants
	Holocene		*Homo sapiens*, present-day vegetation

dominant through the decay of dead organisms. By about 1 or 2 billion years ago, the Earth's atmosphere had reached its present composition.

The first broad proliferation of animal life occurred about 600 million years ago, and the great reptiles arose some 350 million years later. The dinosaurs died out after about 200 million years, and mammals came to dominance about 65 million years ago. Our primitive ancestors appeared only in the last 3 or 4 million years. Once the development of intelligence had provided the ability to control the environment, the entire world became our ecological home, and our physical evolution essentially stopped. It remains to be seen whether future ecological pressures will lead to future evolution of the human species.

COULD LIFE DEVELOP ELSEWHERE?

The scenario just described, if at all accurate, seemingly should occur almost inevitably, given the proper conditions. If this is so, then the question of whether life exists elsewhere amounts to asking whether the conditions that existed on the primitive Earth could have arisen elsewhere. (For a dissenting view—that life was still improbable—see the Astronomical Insight on p. 512.)

It is clear that no other planet in the solar system could have provided an environment exactly like that on the early Earth. It therefore seems unlikely that Earth-like organisms will be found anywhere else in the solar system. Mars is the only planet where life-forms have been sought so far (Fig. 22.6), but even there the conditions are quite unlike those on the Earth. Given the billions of stars in the galaxy, however, and the vast number that are very similar to the Sun, it seems highly probable that the proper conditions must have been reproduced many times in the history of the galaxy.

So far we have worked from the tacit assumption that life on other planets, if it exists, must be similar to life on Earth, and we have considered only the question of whether other Earth-like environments may exist. We may question the premise that life could have developed only in the form that we are familiar with, however. Here, obviously, we must indulge in speculation, having no examples of other types of life at hand for examination.

Life as we know it is based on carbon-bearing molecules, and some have argued that only carbon is capable of combining chemically with other common elements in a sufficiently wide variety of ways to produce the complexity of molecules thought to be necessary. After all, it is clear that the basic elements available to begin with must be the same everywhere,

ASTRONOMICAL INSIGHT
The Case for a Small Value of N

Although it is an interesting exercise to write down an equation for N, the number of technological civilizations in the galaxy, no two astronomers can agree on the values of the various terms.

One very pessimistic argument concludes that we are alone in the galaxy. This conclusion is based on the likelihood that if there were other civilizations, a significant fraction of them would have arisen long before ours. This is a straightforward consequence of the great age of the galaxy, some 5 to 10 billion years greater than the age of the Earth. The argument postulates further that if technological civilizations, once begun, live a long time, then the galaxy must be inhabited by a number of very advanced races, much older and more mature than our own. Up to this point, none of these assertions is particularly controversial.

The next logical step in the argument, however, is debatable; it says that any advanced civilization that has survived for millions or even billions of years will necessarily have done so by perfecting some form of interstellar travel and colonizing other planets. Once this process had started, the argument goes, the spread of a given civilization

throughout the galaxy would accelerate rapidly, and by this time in galactic history, no habitable planet such as the Earth would remain uncolonized. The first successful galactic civilization, in this view, would quickly rise to complete dominance.

The only logical corollary, if the argument is accepted up to this point, is that no other civilizations exist, because if they did, the Earth would have been colonized long ago. The absence of interstellar visitors on the Earth is taken as proof that there are no other civilizations in the galaxy.

Those who adhere to this line of reasoning believe that $N = 1$. Therefore, at least one of the terms in Drake's equation must be very much smaller than the more optimistic values discussed in the text. One suggestion is that the temperate zone around a Sun-like star where planets can have moderate conditions is really very much narrower than usually supposed, perhaps because a Venus-like greenhouse effect develops more readily than is normally thought to be the case. If the temperate zone is very small, then the term n_e, representing the number of habitable, Earth-like planets per star, would be very much smaller

than the value of 0.1 adopted in the text.

Another term that has recently been singled out is f_l, the fraction of Earth-like planets on which life begins. As we learned in the text, somehow the amino acids that were present on the primitive Earth had to arrange themselves into special combinations to produce the long, chainlike protein molecules RNA and DNA. From a purely statistical point of view, the probability that the proper combination would happen to come together by chance is very small. This apparently happened on the Earth, but it may be so unlikely that it has not occurred anywhere else in the galaxy (actually it is thought that these molecules formed in steps, so that a random encounter of all the requisite atoms, a very unlikely event indeed, was not needed).

Whatever the correct values of the terms in the equation, the debate will rage on until our civilization reaches its life expectancy L and dies, until we discover another civilization, or until we ourselves expand to colonize the galaxy and find no one else out there.

given the homogeneous composition of the universe. This may seem to rule out life-forms based on anything other than carbon. However, it has been pointed out that another common element, silicon, also has a very complex chemistry and therefore might provide a basis for a radically different type of life. If so, we cannot begin to speculate on the conditions necessary for such life-forms to arise.

Another assumption that might be subject to question is that water is a necessary medium. As noted earlier, it is the most abundant liquid that can form under the temperature and pressure conditions of the early Earth, and it is thought that only a liquid medium could support the required level of chemical

activity. Other liquids can exist under other conditions, however, and it is interesting to consider whether life-forms of a wholly different type might arise in oceans of strange composition. It is interesting to speculate, for example, about what goes on in the lakes of liquid methane or liquid nitrogen that are thought possibly to exist on Titan, the mysterious giant satellite of Saturn.

If we are satisfied that life probably has formed naturally in many places in the universe, we can address a related, and to many, a more important, question: Given the existence of life, how likely is it that intelligence will follow? Here we have no means of answering, except to reiterate that as far as we know,

Figure 22.6   Searching for life in the solar system.   Mars was long thought to be the most likely home for extraterrestrial life in the solar system. Here we see a Viking lander in a simulated Martian environment. The *Viking* missions reached Mars in 1976 but found no evidence for life-forms. (NASA)

the evolution of our species on Earth was the natural product of environmental pressures.

THE PROBABILITY OF DETECTION

In view of the limitations that prohibit faster-than-light travel, it is exceedingly unlikely that we will ever be able to visit other solar systems, seeking out life-forms that may live there. We will continue to explore our own system, so there is a reasonable chance that if life exists on any of the other planets of our Sun, we will someday discover it. It seems, however, that our best hope of finding other intelligent races in the galaxy will be to make long-range contact with them through radio or light signals. Since this requires both a transmitter and a receiver, we can hope to contact only civilizations as advanced as ours, with the capability of constructing the necessary devices for interstellar communication.

A mathematical exercise in probabilities has been used for some years as a means of assessing, as objectively as possible, the chances for making contact with an extraterrestrial civilization. The aim is to separate the question into several distinct steps, each of which can then be treated independently. The underlying assumption is that the number of technological civilizations in our galaxy today with the capability for interstellar communication is the product of the number of planets that exist with appropriate conditions, the probability that life developed on those planets, the probability that such life has developed intelligence that gave rise to a technological

civilization, and, finally, the likelihood that the civilization has not killed itself off, through evolution or catastrophe.

Mathematically, the so-called Drake equation, (after Frank Drake, who has been its best-known advocate) is written

$$N = R_* f_p n_e f_l f_i f_c L,$$

where N is the number of technological civilizations currently in existence in our galaxy, R_* is the number of stars of appropriate spectral type formed per year in the galaxy, f_p is the fraction of these that have planets, n_e is the number of Earth-like planets per star, f_l is the fraction of these on which life arises, f_i is the fraction of those planets on which intelligence has developed, f_c is the fraction of planets with intelligence on which a technological civilization has evolved to the point at which interstellar communications would be possible, and, finally, L is the average lifetime of such a civilization.

By expressing the number of civilizations in this way, we can isolate the factors about which we can make educated guesses from those about which we are more ignorant. It is an interesting exercise to go through the terms in the equation one by one, to see what conclusions we reach under various assumptions. People who do this have to make sheer guesses for some of the terms, and the result is a variety of answers ranging from very optimistic to very pessimistic. In the following, we will adopt middle-of-the-road numbers for most of the unknown terms.

The first two factors, R_* and f_p, are, in principle, quantities that can be known with some certainty from observations. The rate of star formation in the equation refers to stars similar to the Sun. A much cooler star would not have a temperate zone around it where a planet could have the moderate temperatures needed for life to begin (we are, throughout this exercise, limiting ourselves to life-forms similar to our own), and a very hot star would be short-lived, so there would be insufficient time for life to develop on its planets before the parent star blew up in a supernova explosion. Taking these considerations into account, some estimate that up to 10 suitable stars form in our galaxy per year; for the sake of discussion, we will be more cautious and adopt $R_* = 1/\text{year}$. This may even be a little high for the present epoch in galactic history, but the star-formation rate was surely much higher early in the lifetime of the galaxy, and low-mass stars of solar type formed that long ago are still in their prime. Thus the adoption of an average formation rate of one Sun-like star per year is probably reasonable.

From what we know of the formation of our solar system, it seems that the formation of planets is almost inevitable, except perhaps in double- or multiple-star systems. For the sake of discussion, let us assume that $f_p = 1$; that is, that all stars of solar type have planets. Observations made with instruments such as the *Hubble Space Telescope* may eventually provide real information on this term.

The number of Earth-like planets, n_e, is highly uncertain and depends on the width of the zone where the appropriate temperature conditions could exist. Recent studies show that this zone might be rather small—that is, that the Earth would not have been able to support life if it were only about 5 percent closer to or farther from the Sun than it is. Estimates of n_e vary from 10^{-6} to 1. Let us be moderate and assume $n_e = 0.1$; that is, that in 1 out of 10 planetary systems around solar-type stars, there is a planet within the temperature zone where life can arise.

Now we get to the *really* speculative terms in the equation. We have no way of estimating how likely it is that life should begin, given the right conditions. From the seeming naturalness of its development on Earth, it can be argued that life would always begin if given the chance. Let us be optimistic here and agree with this, adopting $f_l = 1$.

Again, the chance of this life developing intelligence and then advancing to the point of being capable of interstellar communication is completely unknown. All we know is that in the only example that has been observed, both things happened. For the sake of argument, we therefore set both f_i and f_c equal to 1.

At this point, it is instructive to put the values adopted so far into the equation. We find

$$N = (1/\text{year})(1)(0.1)(1)(1)(1)L$$
$$= 0.1L$$

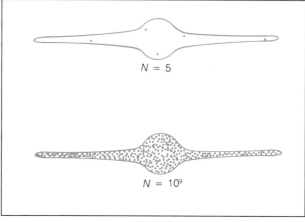

Figure 22.7   Possible values of N.   The number of technological civilizations in the galaxy (N) may be very small (upper), in which case the average distance between them is very large; or N may be large, so that the distance between civilizations is relatively small (lower). The chances for communication with alien races are much higher in the latter case, where the distances are only a few tens of light-years.

Having taken our chances and guessed at the values for all the other terms, we now face a critical question: How long can a technological civilization last? Our has been sufficiently advanced to send and receive interstellar radio signals for only about 50 years, and our society is sufficiently unstable to lead some pessimists to think that we will not last many more decades. If we take this viewpoint and adopt 50 years as the average lifetime, then we find

$$N = 5 \; ,$$

meaning that we should expect the total number of technological civilizations present in the galaxy at one moment to be very small, about five. If this is correct, then the average distance between these outposts of civilization is nearly 20,000 light-years (Fig. 22.7). The time it would take for communications to travel between civilizations would therefore be very much longer than their lifetimes, and there would be no hope of establishing a dialogue with anyone out there. If this estimate is correct, it is not surprising that we haven't heard from anybody yet.

We can be more optimistic, though, and assume that a technological civilization solves its internal problems and lives much longer than 50 years. Extremely optimistic people would argue that a civilization is immortal; that it colonizes star systems other than its own, so that it is immune to any local crises such as planetary wars or suns expanding to become red giants. In that case, allowing a few billion years for the development of such civilizations, we can set $L = 10^{10}$ years (i.e., nearly equal to the age of the

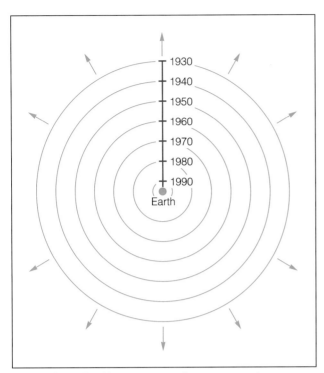

Figure 22.8 Earth's message to the cosmos.   As our entertainment and communications broadcasts travel out into space, they provide a history of our culture for anyone who may be receiving the signals. At the present time, the growing sphere that is filled with our broadcasts has a radius of over 50 light-years.

Figure 22.9 Another message from Earth.   This recording of a message from Earth is traveling beyond the solar system aboard the *Voyager* spacecraft. Only an advanced race of beings would have the technological skills necessary to learn how to listen and to decode the message of peace that it contains. (NASA)

galaxy), and we find

$$N = 10^9 ,$$

where the average distance between civilizations is only about 15 light-years (see Fig. 22.7), coincidentally comparable to the distance our own radio signals have traveled (Fig. 22.8) since the early days of radio and television. If this estimate of N is correct, we should be hearing from somebody very soon.

We have presented two extreme views of the likelihood that other civilizations exist in our part of the galaxy. As we mentioned earlier, opinions among scientists who seriously study this question vary throughout this large range. Those who favor the optimistic viewpoint advocate the idea that we should make deliberate attempts to seek out other civilizations.

THE STRATEGY FOR SEARCHING

The probability arguments outlined in the preceding section are amusing and perhaps somewhat instructive, but obviously not very accurate. There are entirely too many unknowns in the equation for us to

develop a reliable estimate of the chances for galactic companionship. Perhaps we will not know for certain what the answer is until we make contact with another civilization.

The problem of developing an experiment to search for or to send interstellar signals is that we do not know the ground rules. There are an infinite number of ways in which a distant civilization might choose to communicate, and we cannot search for all of them. We must guide our attempts by making our coverage of the possibilities as broad as possible, and by making reasoned guesses as to the best methods to use.

In view of the power that is transmitted by radio signals and the relative lack of natural noise in the galaxy in that part of the spectrum, it has normally been assumed that radio communications are most likely to succeed, although other techniques have been tried (Fig. 22.9). Several searches for extraterrestrial radio communications have been carried out, starting with *Project Ozma* in the 1960s, in which a large radio telescope at the U.S. National Radio Astronomy Observatory was used to search for signals from the directions of nearby Sun-like stars. Other searches have been conducted, primarily in the U.S. and the former U.S.S.R., but so far without success.

ASTRONOMICAL INSIGHT
The Complexity of UFOs

In this chapter we have discussed the strategies for detecting extraterrestrial civilizations using radio communications, with little or no mention of the possibility of direct physical contact. Yet there is a 40-year history of "sightings" of objects thought to be alien spacecraft visiting Earth that suggests that the first contact might be in the form of face-to-face meetings. Since 1947, there have been countless reports of strange objects or lights in the sky, and some have attributed these "Unidentified Flying Objects" or "UFOs" to alien visitors.

To help decide what to think about UFOs, it is very useful to apply Occam's razor, the principle that the simplest explanation for an observation is most likely the correct one. By simplest we mean the explanation that requires the fewest unsupported assumptions. Let's see how the alien spacecraft interpretation of UFOs fares under this criterion.

The laws of physics dictate that a spacecraft cannot travel faster than the speed of light. Therefore, allowing for acceleration time, it must take decades or centuries for a ship to travel the distance from even the nearest star to the Earth (recall that the nearest star is over 4 light-years away; calculations based on known propulsion systems indicate that it would take roughly 40 years to get there from here, and changing the propulsion system to some as-yet uninvented type would still require some acceleration time). Furthermore, the energy required for such a journey is immense; it has been estimated that the energy needed to boost a colonizing spaceship from Earth to a nearby star is enough to power the entire United States at its present rate of consumption for one hundred years! Is it likely that Congress would ever approve such an expenditure? This enormous energy requirement would be faced by any civilization planning interstellar travel and cannot be lessened by advanced technology (unless there are ways of violating the known laws of physics, but there is no evidence that this is possible, and there is plentiful evidence that it is not).

Given the time and energy constraints imposed on any civilization thinking about interstellar travel, it is very difficult indeed to accept the notion that UFOs are sightseeing or spying spacecraft from another star system or that an alien civilization would expend the required time and energy routinely and often. It is far simpler to accept the alternative theory of UFOs, that they are phenomena that occur in the Earth's atmosphere and are not spacecraft. This explanation is made all the more likely by the fact that virtually every UFO sighting that has been carefully studied has been found to be due to something normal such as an airplane, a balloon of some sort, or a bright planet seen near the horizon.

Many people are tempted to believe that UFOs represent visiting aliens, but objective scientific analysis makes this belief very difficult to accept. Occam's razor tells us that a much more mundane explanation of UFOs is likely to be correct, and that our best hope of contacting alien civilizations lies in long-range radio communications, which travel at the speed of light and require very little energy. Of course, the fact that no one really *knows* what alien technology and psychology might be like makes the speculation all the more interesting.

One of the problems faced by anyone wanting to search for extraterrestrial radio communications is obtaining observing time on a radio telescope. Most major radio facilities are already in high demand by astronomers for a wide range of research projects that are more likely to produce significant results; thus it is difficult to persuade the proposal review panels that large blocks of time should be devoted to the search for extraterrestrial civilizations. This problem has generally prevented any large-scale, systematic searches from being conducted, with a couple of exceptions. One project called *Serendip* has worked in a parasitic fashion, in which a special receiver has been attached to radio telescopes carrying out routine astronomical research. The extra receiver searches for intelligent signals at selected frequencies, in whatever direction the telescope happens to be pointed. This does not cover any kind of planned search pattern, but instead takes advantage of the "free" observing time, in the expectation that at least some of the time the telescope will be pointed toward interesting stars.

The broadest dedicated search for extraterrestrial civilizations yet undertaken has been under way for some time at Harvard University. There a small radio telescope has been equipped with a special receiver and a sophisticated computer data-analysis system, so that a wide range of frequencies can be searched. The telescope is devoted completely to the search, and it is expected that thousands of nearby stars will be scrutinized for non-natural radio signals in years to

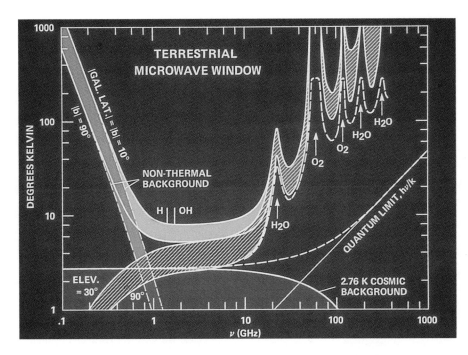

Figure 22.10   The galactic noise spectrum. This graph illustrates the relative intensities of various sources of natural background noise, from both astronomical sources and the earth's atmosphere. As seen here, the region between the lines of hydrogen and the hydroxyl radical (OH) has the lowest background noise. For this reason, the so-called "water hole" has been proposed as a likely spectral region in which interstellar communications might be carried out. (NASA/SETI Institute)

come. Interestingly, major funding for this project has been provided by movie producer Steven Spielberg, who has been responsible for a number of films about contact with extraterrestrial civilizations.

Most of the searches so far have not had the luxury of covering a very wide range of frequencies, so difficult choices had to be made. On the assumption (or hope) that there might be civilizations that are trying to send signals deliberately, some guesswork led to the hypothesis that the signals might be sent at or near the same wavelengths most often observed by radio astronomers. For this reason, many of the early searches have concentrated on the 21-cm wavelength of atomic hydrogen, reasoning that alien scientists would realize the significance of this wavelength, and therefore might use it in attempts to send signals that would be noticed by distant colleagues. Another good argument for searching at this wavelength is that many of the world's radio telescopes are already equipped with receivers designed to operate at 21 cm.

Another wavelength that has been considered, but only tried in a limited search, lies between the 21-cm line of hydrogen and a series of lines due to the hydroxyl radical (OH), lying between 17.4 and 18.6 cm. The region in between, dubbed the "water hole" because both hydrogen and OH are constituents of water, lies just in the wavelength range where natural galactic noise is minimized (Fig. 22.10), and this adds to the argument favoring a search there.

One pessimistic viewpoint is that we are all listening, and no one is sending signals. In that case, we can only hope to pick up accidental emissions, such as entertainment broadcasts on radio or television, and these would be much weaker and more difficult to detect. Deliberately sent signals can be detected over much greater distances than accidental transmissions, because more power would be put into a deliberate signal, and it could be directed specifically toward candidate stars, whereas our accidental radio and television signals are broadcast indiscriminately in all directions. Therefore it makes a big difference whether someone out there is trying to send a message or not.

Humans have attempted to send messages. In 1974, a message from Earth was sent from the giant Arecibo radio telescope (see Fig. 4.28) toward the globular cluster M13, about 25,000 light years away. The reason for choosing a globular cluster was that it contains hundreds of thousands of stars, many of which are similar to the Sun in spectral type, and all of them would be within the beam of the radio telescope's transmission. The message consists of a stream of numbers which, when arranged into a two-dimensional array, form a pattern that illustrates schematically such things as the structure of DNA, the form of the human body, Earth's population, and the location of the Earth in the solar system. If this message is received and understood by anyone in M13, it will be some 25,000 years from now, and any answer that they may send back will arrive here about 50,000 years from now.

There have been many proposals for large-scale searches for extraterrestrial signals, but it has been difficult to obtain the necessary resources. Not only is telescope time difficult to obtain, but a truly major, systematic search will take a long time and will cost

Figure 22.11   A radio antenna used in the search.   This is the NASA Deep Space Network antenna near Canberra, Australia. This 70-m dish is one of three being used in the High Resolution Microwave Survey project, the most ambitious search for extraterrestrial civilizations yet undertaken. This one provides the first broad coverage of skies in the southern hemisphere. (NASA/SETI Institute)

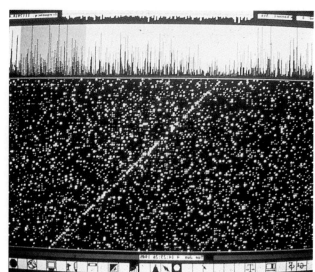

Figure 22.12   A successful detection of an extraterrestrial signal.   The clearly-visible bright streak across this display of radio intensities represents the signal received from the *Pioneer 10* spacecraft, over 3.5 billion km from the Earth and emitting with a power of only about 1 watt. Each row of dots in the lower-central position of this display represents an individual spectrum, taken at successive times 2 seconds apart. The reason the signal from *Pioneer 10* appears as a slanted line is that the Earth's velocity relative to the spacecraft was changing due to the Earth's orbital motion, so that the signal was Doppler-shifted to a slightly different frequency in each successive spectrum. (NASA/SETI Institute)

a lot of money. Various projects have been proposed over the years, including a grandiose one involving the construction of hundreds of radio dishes in a giant array called *Project Cyclops,* but until recently none was funded. One major new development has been the emergence of compact, highly powerful (and relatively inexpensive) computer systems that are capable of analyzing a broad range of frequencies automatically, searching for patterns that might indicate unnatural origins.

In 1992 the U.S. space agency, NASA, initiated a project called the *High Resolution Microwave Survey,* which will use a combination of dedicated telescopes and shared time on other facilities (Fig. 22.11) to carry out a search for over a million stars, covering a very broad range in frequencies. Begun in 1992, this project is expected to take about 10 years to be completed. Initial tests showed that the detection technology is very efficient; the project had no difficulty in detecting signals from the *Pioneer 10* spacecraft, which is radiating with a power of only about 1 watt, from a distance of 5.3 billion km (Fig. 22.12). The project will make use of several existing telescopes,

including three antennas in NASA's Deep Space Network, used for communications with distant spacecraft; the 43-m dish at Green Bank, West Virginia (part of the National Radio Astronomy Observatory); and the giant 300-m dish at Arecibo. Scientists around the world are watching with anticipation this newest and broadest attempt to detect signals from extraterrestrial civilizations, knowing that success will be perhaps the most significant discovery of all time.

PERSPECTIVE

In this chapter we have introduced a bit of speculation, well founded perhaps in the discussion of life on Earth, but less so in the sections on the possibilities of life elsewhere.

We are now prepared to understand and appreciate new advances in astronomy. Having completed our study of the present knowledge of the universe, we will be able to put into perspective the major new discoveries that are sure to come as technology and theory continue to improve. The *Hubble Space Telescope,* the large optical telescopes now being designed, the comparable advances in radio, infrared, and X-ray techniques—all of these will inevitably lead to novel and unforeseen breakthroughs in our view of the universe. The next few decades will be exciting times for astronomy, and we can anticipate their coming with a sense of excitement and wonder.

restrial Intelligence (SETI), has been funded by NASA and is now in full operation. *SETI* will search about 1 million stars in 8 million radio wavelengths and should result in a detection if the number of technological civilizations in the galaxy is large.

SUMMARY

1. Although some scientists have suggested that life started on Earth spontaneously or by primitive spores from space, most scientists today accept the theory that life began through natural, evolutionary processes.

2. Amino acids, fundamental components of living organisms, are formed readily in experiments designed to simulate early conditions on the Earth.

3. The steps that led to the development of the necessary forms of RNA and DNA are not yet fully understood and probably occurred over a very long period of time.

4. Fossil evidence provides a record of the evolutionary steps leading from the first primitive lifeforms to modern life.

5. The conditions that prevailed on the early Earth have probably been duplicated on other planets in the galaxy, though probably not on other planets in the solar system.

6. It is often assumed that only Earth-like life could develop, because carbon is nearly unique in the complexity of its chemistry, but some have suggested that at least one other element (silicon) may have the necessary properties.

7. Estimates of the number N of technological civilizations now in the galaxy can be made, with great uncertainty, based on what is known of the formation rate of Sun-like stars, what is guessed for the probability that such stars have planets with the proper conditions, and the probability that life, leading to intelligence and technology, will develop on these planets. Estimates range from $N = 1$ to $N = 10^9$.

8. Some attempts have been made to search for radio signals at 21 cm and other wavelengths, but with no success.

9. The chances for detecting or being detected by other civilizations depend strongly on whether or not deliberate attempts are made to send signals.

10. Future projects to search for and to send signals have been planned, and one, the *Search for Extrater-*

REVIEW QUESTIONS

1. To what extent has life on Earth modified its atmosphere? Does this suggest a method for determining, from remote spectroscopic observations, whether a distant planet might have life?

2. If "I Love Lucy" broadcasts started in 1953, in what year did they first reach Alpha Centauri, 1.3 pc away? When will they reach Canopus, 30 pc away? When might we expect an answer from Canopus, if anyone is there to receive our signals?

3. Suppose a planet is orbiting a star exactly like the Sun, and the planet's semimajor axis is 0.9 AU. How would the intensity of sunlight on that planet compare with the intensity of sunlight on the Earth? What effect might this have on the possibility that life could develop on this planet?

4. Summarize the arguments that lead some scientists to expect life elsewhere to be similar in chemistry to life on Earth.

5. Based on what is assumed about how life started on Earth, do you think it likely that life has developed in the atmosphere of Jupiter? Do you think it could develop in an interstellar cloud?

6. Recall, from Chapter 17, the distinction between Population I and Population II stars, and the location of these stars in the galaxy. Do you think planets orbiting Population II stars are as likely to have lifeforms as planets orbiting Population I stars like the Sun? Explain.

7. Explain in your own words the meaning of the various terms in the Drake equation, and how this equation represents the number of technological civilizations in the galaxy at any one time.

8. Repeat the calculation of N, the number of technological civilizations in the galaxy, if the probability of life starting on an Earth-like planet (the term f_l in the Drake equation) is only 10^{-4} instead of 1, as assumed in the text. Assume that the lifetime L is 10^{10} years, then redo the calculation for $L = 10^5$ years.

9. Why are radio communications thought to be the best method for detecting evidence for other civilizations?

10. What are the factors that govern the choice of wavelength at which to search for signals?

ADDITIONAL READINGS

Abt, H. A. 1979. The companions of sun-like stars. *Scientific American* 236(4):96.

Ball, J. A. 1980. Extraterrestrial intelligence: Where is everybody? *American Scientist* 68:656.

Barber, V. 1974. Theories of the chemical origin of life on Earth. *Mercury* 3(5):20.

Beatty, J. K. 1991. Killer crater in the Yucatan? *Sky and Telescope* 82(1):38.

Black, D. C. 1991. Worlds around other stars. *Scientific American* 264(1):76.

Burt, D. M. 1989. Mining the Moon. *American Scientist* 77:574.

Comins, N. F. 1991. Life near the center of the galaxy. *Astronomy* 19(4):46.

Crosswell, K. 1991. Does Alpha Centauri have intelligent life? *Astronomy* 19(4):28.

Drake, F. 1988. The *Pioneer* message plaques. *Mercury* 17:88.

Goldsmith, D., ed. 1980. *The quest for extraterrestrial life*. Mill Valley, Calif.: University Science Books.

Guin, J. 1980. In the beginning. *Science 80* 1(5):45.

Horgan, J. 1991. In the beginning . . . *Scientific American* 264(2):116.

Horowitz, P. 1986. *To utopia and back: The search for life in the solar system*. New York: W. H. Freeman.

Naeye, R. 1992. SETI at the crossroads. *Sky and Telescope* 84(5):507.

Olson, E. C. 1985. Intelligent life in space. *Astronomy* 13(7):6.

O'Neill, G. K. 1974. The colonization of space. *Physics Today* 27(9):32.

Overbye, D. 1982. Is anyone out there? *Discover* 3(3):20.

Papagiannis, M. D. 1982. The search for extraterrestrial civilizations—A new approach. *Griffith Observer* 11(1):112.

————. 1984. Bioastronomy: The search for extraterrestrial life. *Sky and Telescope* 67(6):508.

Pelligrino, C. R. 1979. Organic clues in carbonaceous meteorites. *Sky and Telescope* 57(4):330.

Pollard, W. G. 1979. The prevalence of Earth-like planets. *American Scientist* 67:653.

Rood, R. T., and J. S. Trefil. 1981. *Are we alone?* New York: Scribner.

Sagan, C. 1982. The search for who we are. *Discover* 3(3):30.

Sagan, C., and F. Drake. 1975. The search for extraterrestrial intelligence. *Scientific American* 232(5):80.

Sagan, C., I. S. Shklovskii. *Intelligent Life in the universe*. San Francisco: Holden-Day.

Shneartzman, D., and L. J. Rickard, 1988. Being optimistic about the search for extraterrestrial intelligence. *American Scientist* 76:364.

Scheaffer, R. 1986. *The UFO verdict*. Buffalo, N.Y.: Prometheus Books.

Special Report on Life in Space. 1983. *Discover* 4(3).

Tipler, F. J. 1982. The most advanced civilization in the galaxy is ours. *Mercury* 11(1):5.

Weissman, P. 1990. Are periodic bombardments real? *Sky and Telescope* 79:266.

Wetherill, C., and W. T. Sullivan. 1979. Eavesdropping on the Earth. *Mercury* 8(2):23.

ASTRONOMICAL ACTIVITY
Constructing an Alien

It is clear that the forms life has taken here on the Earth have been influenced very strongly by local conditions. Our eyes can see only the portion of the electromagnetic spectrum that penetrates the atmosphere; our skeletal systems are of appropriate size and strength to allow motion under one Earth gravity; we breathe the gases that are most common in the Earth's atmosphere; and we derive our energy from food sources that are abundant on the Earth. All of these characteristics of humans and other earthly species would be very different if the Earth itself were very different.

We have shown in this chapter that Earth-like conditions do not exist anywhere else in the solar system (although conditions similar to the early Earth might once have prevailed on Mars), so it is not likely that the solar system harbors any other life-forms similar to us. But this does not rule out the possibility that life of a rather different kind than we are familiar with may exist elsewhere in the solar system.

To stimulate your thinking about this, try writing a description of a possible life-form that might exist on another planet (or a satellite or an interplanetary body) in the solar system. Try to think of realistic ways in which this creature would get food (and what kind of food it might use), and think about how its body chemistry might work. Such factors as the physical state of matter in its environment will come into play; for example, if you choose to invoke a life-form that lives in the Sun's outer layers, you should explain how it remains in solid form even though all matter in that environment is vaporized; conversely, if you choose a very cold environment, you need to think about how the alien's body might transport nutrients internally when everything around it is frozen solid. You might want to think about motion and how the creature could travel around to catch food or pursue changing environments as the seasons change. What kind of sensory organs might be useful?

Perhaps in doing this exercise you will hit upon some new methods that astronomers might try as they search for extraterrestrial life. In any event, you will gain a better appreciation for the special circumstances that allowed us to evolve here on the Earth.

ASTRONOMICAL UPDATE

The Modern Search for Companionship

On October 12, 1992, switches were turned on at the great Arecibo radio telescope in Puerto Rico and at one of NASA's Deep Space Network antennas (in California), officially marking the beginning of the most ambitious and comprehensive search for extraterrestrial civilizations yet undertaken. The project, called the *High Resolution Microwave Survey (HRMS)*, is funded by NASA and involves a 10-year radio search for nonnatural signals from beyond the solar system. The former name for this search was *Project SETI* (the acronym SETI stood for "Search for Extra-Terrestrial Intelligence"), but the name was changed in 1992 when NASA assigned the project to its program *Toward Other Planetary Systems*, or *TOPS*.

The total cost of the *HRMS* project is expected to be around $100 million, averaging about $10 million per year from now through the year 2001. This may sound like a lot to spend on such a speculative program, but as is the case in any research project, it is very difficult to assign cost-effectiveness figures when the ultimate outcome will not be known for a long time (and in any case, the quest for knowledge is very difficult to appraise in terms of dollars and cents). It is noteworthy that the cost of *HRMS* is only about one-

tenth of 1 percent of the projected NASA budget for the next decade, and that a positive result (i.e., the discovery of an extraterrestrial civilization) would surely rank as one of the most important discoveries ever made—perhaps the most important.

Funding for *HRMS* (or *Project SETI*) has been controversial. Several times, Congress has stopped attempts to get the project started; at one point *SETI* even won the dubious distinction of a "Golden Fleece" award (these "awards" were announced occasionally by former Senator William Proxmire of Wisconsin, based on his opinion of certain congressional expenditures as wasteful). As recently as 1992, Congress considered canceling funding that had been budgeted, but this threat was averted when the project was transferred to the *TOPS* office at NASA.

In some ways, the many delays in starting the search may have been beneficial. Vast economies have been achieved in the project design from the time (more than a decade ago) when an earlier version called *Project Cyclops* was envisioned. *Cyclops,* which would have cost billions of dollars today, called for the construction of a huge array of radio antennas designed to search very sensitively for artificial signals from

space. Now, in place of the hundreds of antennas of *Cyclops,* we have sophisticated computers capable of deciphering complex signals from just one receiver, greatly reducing the expense and complexity of the search, but preserving its breadth and sensitivity.

As in most previous searches for extraterrestrial civilizations, the *HRMS* strategy is based on the assumption that radio communications are the most likely means of detecting alien societies. There are several reasons for this assumption: far less energy is needed to send electromagnetic communications than to send spacecraft or probes over interstellar distances; radio frequencies require less energy than higher-frequency bands such as visible light; and the natural background of astronomical noise is minimized in the microwave portion of the radio spectrum.

As mentioned above, technology advances in receivers and in computer data-processing techniques now make it possible to search a broad range of frequencies and automatically detect any unnatural signals that may be present. The capability of searching a broad range of frequencies means that scientists no longer have to guess at what wavelength (or frequency) alien signals might be sent. As discussed in Chapter 22, in previous searches it was often necessary to choose a limited frequency range (centered on the 21-cm line of hydrogen, for example) and hope that the alien societies would be sending messages in that range.

Another advantage of the *HRMS* over previous *SETI* projects is that the data will be analyzed instantly, as they are received, so that if a possible signal is detected, astronomers will be alerted immediately. Then the telescope can quickly be pointed toward the location of the possible detection, so that it can be verified. Past searches have found several possible signals, but in each case the discovery came some time after the observation had actually been made, and when the position was scanned again later, nothing was seen.

The modern receiver and data analysis technologies to be applied to the *HRMS* are so efficient that in the first few minutes after the systems were activated on October 12, more data were acquired than in all the previous *SETI* searches dating back to the first in 1960.

The *HRMS* project has two distinct phases: (1) a targeted search, in which some 1,000 nearby solar-type stars will be observed intensively; and (2) a sky survey, covering the entire sky at lower sensitivity. The targeted search is using the enormous 305-m Arecibo radio telescope (see Fig. 4.28), as well as the three 70-m dishes of NASA's Deep Space Network, to seek artificial radio signals from solar-type stars within about 100 light-years of the Sun. This search is sufficiently sensitive that accidental transmissions, such as our own entertainment broadcasts on radio and television, should be detectable if we have neighbors with similar technologies.

The sky survey, which will utilize several 34-m radio antennas that are also part of NASA's Deep Space Network, will include the entire galaxy within its coverage, but will be able to detect only strong signals that might be deliberately sent by faraway civilizations. The kind of accidental transmissions sought in the targeted search will be too weak for detection by the sky survey, except perhaps from very nearby stars. The sky survey will cover 10,000 times more range in frequency than previous *SETI* efforts and will have a sensitivity at least 300 times as great.

None of the radio telescopes being used in the *HRMS* is dedicated full-time to the project; this is the main reason it will take some 10 years to complete the search. The timing is very good in one sense, because human-generated radio noise is becoming an ever more serious problem for radio astronomy and affects *SETI* efforts as well as basic astronomical research being conducted at radio wavelengths. In a few years, it may become impossible to conduct a search as sensitive as this one, simply because our own radio noise will drown out any signals that might reach us from the stars. Thus the opportunity we have now to search for other civilizations may never be available again. It is fortunate that we are taking advantage of this opportunity. Now we can only wait with anticipation for the results.

Appendices

APPENDIX 1:   Symbols Commonly Used in this Text

Symbol	Meaning	Symbol	Meaning
Å	Angstrom, a unit of length often used to measure wavelengths of light; $1\text{ Å} = 10^{-10}$ m	$\Delta\lambda$	The Greek letters delta lambda, used to designate a shift in wavelength, as in the Doppler effect
c	Standard symbol for the speed of light	γ	The Greek letter gamma, sometimes used to designate a gamma-ray photon
G	Standard symbol for the gravitational constant		
H	Symbol for the constant in the Hubble expansion law	λ	The Greek letter lambda, usually used to designate wavelength
h	Standard symbol for the Planck constant		
K	Kelvin, the unit of temperature in the absolute scale	ν	The Greek letter nu, the standard symbol for frequency, also used to designate a neutrino in nuclear reactions
z	Symbol commonly used to designate the Doppler shift ($z = \Delta\lambda/\lambda = v/c$ for velocities much less than the speed of light)	π	The Greek letter pi, usually used to designate the parallax angle; also used for the ratio of the circumference of a circle to its diameter
α	The Greek letter alpha, sometimes used to designate an alpha particle		

APPENDIX 2:   Physical and Mathematical Constants

Constant	Symbol	Value
Speed of light	c	2.9979249×10^8 m/sec
Gravitation constant	G	6.6720×10^{-11} N·m²/kg²
Planck constant	h	6.62618×10^{-34} J·sec
Electron mass	m_e	9.10953×10^{-31} kg
Proton mass	m_p	1.67265×10^{-27} kg
Stefan-Boltzmann constant	σ	5.67032×10^{-8} W/m²kg⁴
Wien constant	W	0.00289776 m·deg
Boltzmann constant	k	1.38066×10^{-23} J/deg
Astronomical unit	AU	1.49599×10^8 km
Parsec	pc	3.085678×10^{13} km = 3.261633 light-years
Light-year	ly	9.460530×10^{12} km
Solar mass	$M_\odot$	1.9891×10^{30} kg
Solar radius	$R_\odot$	6.9600×10^8 m
Solar luminosity	$L_\odot$	3.827×10^{26} W
Earth mass	$M_\oplus$	5.9742×10^{24} kg
Earth radius	$R_\oplus$	6378.140 km
Tropical year (equinox to equinox)		365.241219878 days
Sidereal year (with respect to stars)		365.256366 days = 3.155815×10^7 sec

APPENDIX 3: The Elements and Their Abundances

Element	Symbol	Atomic No.	Atomic Weight[a]	Abundance[b]	Element	Symbol	Atomic No.	Atomic Weight[a]	Abundance[b]
Hydrogen	H	1	1.0080	1.00	Iodine	I	53	126.9045	4.07×10^{-11}
Helium	He	2	4.0026	0.085	Xenon	Xe	54	131.30	1×10^{-10}
Lithium	Li	3	6.941	1.55×10^{-9}	Cesium	Cs	55	132.905	1.26×10^{-11}
Beryllium	Be	4	9.0122	1.41×10^{-11}	Barium	Ba	56	137.34	6.31×10^{-11}
Boron	B	5	10.811	2.00×10^{-10}	Lanthanum	La	57	138.906	6.46×10^{-11}
Carbon	C	6	12.0111	0.000372	Cerium	Ce	58	140.12	4.37×10^{-11}
Nitrogen	N	7	14.0067	0.000115	Praseodymium	Pr	59	140.908	4.27×10^{-11}
Oxygen	O	8	15.9994	0.000676	Neodymium	Nd	60	144.24	6.61×10^{11}
Fluorine	F	9	18.9984	3.63×10^{-8}	Promethium	Pm	61	146	—
Neon	Ne	10	20.179	3.72×10^{-5}	Samarium	Sm	62	150.4	4.57×10^{-11}
Sodium	Na	11	22.9898	1.74×10^{-6}	Europium	Eu	63	151.96	3.09×10^{-12}
Magnesium	Mg	12	24.305	3.47×10^{-5}	Gadolinium	Gd	64	157.25	1.32×10^{-11}
Aluminum	Al	13	26.9815	2.51×10^{-6}	Terbium	Tb	65	158.925	2.63×10^{-12}
Silicon	Si	14	28.086	3.55×10^{-5}	Dysprosium	Dy	66	162.50	1.29×10^{-11}
Phosphorus	P	15	30.9738	3.16×10^{-7}	Holmium	Ho	67	164.930	3.1×10^{-12}
Sulfur	S	16	32.06	1.62×10^{-5}	Erbium	Er	68	167.26	5.75×10^{-12}
Chlorine	Cl	17	35.453	2×10^{-7}	Thulium	Tm	69	168.934	2.69×10^{-11}
Argon	Ar	18	39.948	4.47×10^{-6}	Ytterbium	Yb	70	170.04	6.46×10^{-12}
Potassium	K	19	39.102	1.12×10^{-7}	Lutetium	Lu	71	174.97	6.92×10^{-12}
Calcium	Ca	20	40.08	2.14×10^{-6}	Hafnium	Hf	72	178.49	6.3×10^{-12}
Scandium	Sc	21	44.956	1.17×10^{-9}	Tantalum	Ta	73	180.948	2×10^{-12}
Titanium	Ti	22	47.90	5.50×10^{-8}	Tungsten	W	74	183.85	3.72×10^{-10}
Vanadium	V	23	50.9414	1.26×10^{-8}	Rhenium	Re	75	186.2	1.8×10^{-12}
Chromium	Cr	24	51.996	5.01×10^{-7}	Osmium	Os	76	190.2	5.62×10^{-12}
Manganese	Mn	25	54.9380	2.63×10^{-7}	Iridium	Ir	77	192.2	1.62×10^{-10}
Iron	Fe	26	55.847	2.51×10^{-5}	Platinum	Pt	78	195.09	5.62×10^{-11}
Cobalt	Co	27	58.9332	3.16×10^{-8}	Gold	Au	79	196.967	2.09×10^{-12}
Nickel	Ni	28	58.71	1.91×10^{-6}	Mercury	Hg	80	200.59	1×10^{-9}
Copper	Cu	29	63.546	2.82×10^{-8}	Thallium	Tl	81	204.37	1.6×10^{-12}
Zinc	Zn	30	65.37	2.63×10^{-8}	Lead	Pb	82	207.19	7.41×10^{-11}
Gallium	Ga	31	69.72	6.92×10^{-10}	Bismuth	Bi	83	208.981	6.3×10^{-12}
Germanium	Ge	32	72.59	2.09×10^{-9}	Polonium	Po	84	210	—
Arsenic	As	33	74.9216	2×10^{-10}	Astatine	At	85	210	—
Selenium	Se	34	78.96	3.16×10^{-9}	Radon	Rn	86	222	—
Bromine	Br	35	79.904	6.03×10^{-10}	Francium	Fr	87	223	—
Krypton	Kr	36	83.80	1.6×10^{-9}	Radium	Ra	88	226.025	—
Rubidium	Rb	37	85.4678	4.27×10^{-10}	Actinium	Ac	89	227	—
Strontium	Sr	38	87.62	6.61×10^{-10}	Thorium	Th	90	232.038	6.61×10^{-12}
Yttrium	Y	39	88.9059	4.17×10^{-11}	Protactinium	Pa	91	230.040	—
Zirconium	Zr	40	91.22	2.63×10^{-10}	Uranium	U	92	238.029	4.0×10^{-12}
Niobium	Nb	41	92.906	2.0×10^{-10}	Neptunium	Np	93	237.048	—
Molybdenum	Mo	42	95.94	7.94×10^{-11}	Plutonium	Pu	94	242	—
Technetium	Tc	43	98.906	—	Americium	Am	95	242	—
Ruthenium	Ru	44	101.07	3.72×10^{-11}	Curium	Cm	96	245	—
Rhodium	Rh	45	102.905	3.55×10^{-11}	Berkelium	Bk	97	248	—
Palladium	Pd	46	106.4	3.72×10^{-11}	Californium	Cf	98	252	—
Silver	Ag	47	107.868	4.68×10^{-12}	Einsteinium	Es	99	253	—
Cadmium	Cd	48	112.40	9.33×10^{-12}	Fermium	Fm	100	257	—
Indium	In	49	114.82	5.13×10^{-11}	Mendelevium	Md	101	257	—
Tin	Sn	50	118.69	5.13×10^{-11}	Nobelium	No	102	255	—
Antimony	Sb	51	121.75	5.62×10^{-12}	Lawrencium	Lr	103	256	—
Tellurium	Te	52	127.60	1×10^{-10}					

[a]The atomic weight of an element is its mass in *atomic mass units*. An atomic mass unit is defined as one-twelfth of the mass of the most common isotope of carbon, and has the value 1.660531×10^{-27} kg. In general, the atomic weight of an element is approximately equal to the total number of protons and neutrons in its nucleus.

[b]The abundances are given in terms of the number of atoms of each element compared to hydrogen, and are based on the composition of the Sun. For very rare elements, particularly those toward the end of the list, the abundances can be quite uncertain, and the values given should not be considered exact.

APPENDIX 4:   Temperature Scales

At the most basic level, temperature can be defined in terms of the motion of particles in a gas (or a solid, or a liquid). We all have an intuitive idea of what heat is, and we are all familiar with at least one scale for measuring temperature.

The most commonly used scales are somewhat arbitrarily defined, with zero points not representing any truly fundamental physical basis. The popular *Fahrenheit* scale, for example, has water freezing at a temperature of 32°F and boiling at 212°F. On this scale absolute zero, the lowest possible temperature (where all molecular motions cease), is −459°F.

The centigrade (or Celsius) scale is perhaps better founded, although it is based on the freezing and boiling points of water, rather than the more fundamental absolute zero. In this system, the freezing point is defined as 0°C, and 100°C is the boiling point. This scale has the advantage over the Fahrenheit scale that there are exactly 100° between the freezing and boiling points, rather than 180°, as in the Fahrenheit system. To convert from Fahrenheit to centigrade, we must subtract 32° first, and then multiply the remainder by 100/180, or 5/9. For example, 50°F is equal to 5/9 × (50 − 32) = 10°C. To convert from centigrade to Fahrenheit, first multiply by 9/5 and then add 32°. Thus, −10°C = 9/5 × (−10 + 32) = 14°F. On the centigrade scale, absolute zero occurs at −273°C.

The temperature scale preferred by scientists is a modification of the centigrade system. In this system, named after its founder, the British physicist Lord Kelvin, the same degree is used as in the centigrade scale; that is, one degree is equal to one one-hundredth of the difference between the freezing and boiling points of water. The zero point is different from the zero point on the centigrade scale, however; it is equal to absolute zero. Hence on this scale water freezes at 273°K and boils at 373°K. Comfortable room temperature is around 300°K. In modern usage the degree symbol (°) is dropped, and we speak simply of temperatures in units of Kelvins (273 K, for example).

APPENDIX 5:   Radiation Laws

Several laws described in Chapter 4 apply to continuous radiation from hot objects such as stars. In the text these laws were discussed in general terms, and a few simple applications were explained. Here the same laws are given in more precise mathematical form, and their use in that form is illustrated.

Wien's Law

In general terms, Wien's law says that the wavelength of maximum emission from a glowing object is inversely proportional to its temperature. Mathematically this can be written as

$$\lambda_{max} \propto 1/T,$$

where λ_{max} is the wavelength of strongest emission, T is the surface temperature of the object, and $\propto$ is a special mathematical symbol meaning "is proportional to."

Experimentation can determine the *proportionality constant*, specifying the exact relationship between λ_{max} and T, and Wien did this, finding

$$\lambda_{max} = W/T,$$

where W has the value 0.0029 if λ_{max} is measured in meters and T in degrees absolute. With this equation it is possible to calculate λ_{max}, given T, or vice versa. Thus, if we measure the spectrum of a star and find that it emits most strongly at a wavelength of 2,000 Å = 2 × 10^{-7} m, then we can solve for the temperature

$$T = W/\lambda_{max} = .0029/2 \times 10^{-7} = 14{,}500 \text{ K.}$$

This is a relatively hot star, and it would appear blue-white to the eye. Note that it was necessary to measure the spectrum in ultraviolet wavelengths in order to find λ_{max}.

When solving problems using Wien's law, it is always possible to use the equation form, as we have just done in this example. Often, however, it is more convenient to compare the properties of two objects by considering the ratio of the temperatures or of the wavelengths of maximum emission. In effect this is what we did in the text when we compared two objects of different temperatures in order to determine how their λ_{max} values compared. For example, we said that if one object is twice as hot as another, its value of λ_{max} is half that of the other.

We can see how this ratio technique works by writing the equation for Wien's law separately for object 1 and object 2:

$$\lambda_{max_1} = W/T_1, \quad \text{and} \quad \lambda_{max_2} = W/T_2.$$

Now we can divide one equation by the other:

$$\frac{\lambda_{max1}}{\lambda_{max2}} = \frac{W/T_1}{W/T_2}$$

or

$$\frac{\lambda_{max1}}{\lambda_{max2}} = \frac{T_2}{T_1}.$$

The numerical factor W has canceled out, and we are left with a simple expression relating the values of λ_{max} and T for the two objects. Now we see that if $T_1 = 2T_2$ (object 1 is twice as hot as object 2), then

$$\frac{\lambda_{max1}}{\lambda_{max2}} = \frac{T_2}{2T_2} = \frac{1}{2},$$

or λ_{max} for object 1 is one-half that for object 2. In this extremely simple example, it probably would have been easier just to work it out in our heads, but what if we have a case where one object is 3.368 times hotter than the other, for example?

A great deal can be learned from making comparisons in this way. Astronomers often use the Sun as the standard for comparison, expressing various quantities in terms of the solar values. Another reason for using comparisons occurs when the numerical constants (such as Wien's constant in the foregoing examples) are not known. If the trick of comparing is kept in mind, it is often possible to work out answers to astronomical questions simply by carrying around in one's head a few numbers describing the Sun.

The Stefan-Boltzmann Law

As discussed in the text, the energy emitted by a glowing object is proportional to T^4, where T is the surface temperature. This can be written mathematically as

$$E = \sigma T^4,$$

where σ stands for a proportionality constant that has the value 5.7×10^{-8} in the SI system.

If we now consider how much surface area a star has, then the total energy it emits, called its *luminosity* and usually denoted L, is

$$\begin{aligned} L &= \text{surface area} \times E \\ &= 4\pi R^2 \sigma T^4, \end{aligned}$$

where R is the radius of the star. This equation is called the Stefan-Boltzmann law.

As in other cases, the law can be used directly in this form, or we can choose to compare properties of stars by writing the equation separately for two stars

and then dividing. If we do this, we find

$$\frac{L_1}{L_2} = \frac{R_1^2 T_1^4}{R_2^2 T_2^4} = \left(\frac{R_1}{R_2}\right)^2 \left(\frac{T_1}{T_2}\right)^4.$$

The constant factors 4π and σ have canceled out.

As an example of how to use this expression, suppose we determine that a particular star has twice the temperature of the Sun but only one-half the radius, and we wish to know how this star's luminosity compares to that of the Sun. If we designate the star as object 1 and the Sun as object 2, then

$$\frac{L_1}{L_2} = \left(\frac{1}{2}\right)^2 (2)^4 = \frac{1}{4} \times 16 = 4$$

This star has 4 times the luminosity of the Sun.

The Stefan-Boltzmann law is particularly useful because it relates three of the most important properties of stars to each other.

The Planck Function

The radiation laws described above and in the text are actually specific forms of a much more general law, discovered by the great German physicist Max Planck. Wien's law, the Stefan-Boltzmann law, and some others not mentioned in the text bear the same kind of relation to Planck's law as the laws of planetary motion discovered by Kepler do to Newton's mechanics. Kepler's laws were first discovered by observation, but with Newton's laws of motion it is possible to derive Kepler's laws theoretically. In the same fashion the radiation laws discussed so far were found experimentally but can be derived mathematically from the much more general and powerful Planck's law.

Planck's law is usually referred to as the Planck function, a mathematical relationship between intensity and wavelength (or frequency) that describes the spectrum of any glowing object at a given temperature. The Planck function specifically applies only to objects that radiate solely because of their temperature and do not reflect any light or have any spectral lines. The popular term for such an object is *blackbody*, and we often refer to radiation from such an object as *blackbody radiation* or, more commonly, *thermal radiation*, the term used in the text. Stars are not perfect radiators, but to a good approximation can be treated as such; hence in the text we apply Wien's law and the Stefan-Boltzmann law to stars (and even planets in certain circumstances) without pointing out the fact that to do so is only approximately correct.

The form of the Planck function for the radiation intensity B as a function of wavelength is

$$B = \frac{2hc^2}{\lambda^5} \frac{1}{e^{hc/\lambda kT} - 1},$$

where h is the Planck constant, c is the speed of light, and k is the Boltzmann constant (the values of all three are tabulated in Appendix 2), λ is the wavelength (in m), and T is the temperature of the object (on the absolute scale). The symbol e represents the base of the natural logarithm, something not used elsewhere in this text; for the present purpose, this may be regarded simply as a mathematical constant with the value 2.718.

In terms of frequency ν rather than wavelength, the expression is

$$B = \frac{2h\nu^3}{c^2} \frac{1}{e^{h\nu/kT} - 1}.$$

Either expression may be used to calculate the spectrum of continuous radiation from a glowing object at a specific temperature. In practice the Planck function is used in a wide assortment of theoretical calculations that call for knowledge of the intensity of radiation so that its effects can be assessed on physical conditions such as ionization.

APPENDIX 6:   Major Optical Telescopes of the World

Observatory	Location[a]	Telescope
European Southern Observatory	(Cerro La Silla, Chile)	16-m *Very Large Telescope* (four 8-m instruments)
Keck Observatory (California Institute of Technology and the University of California)	Mauna Kea, Hawaii Mauna Kea, Hawaii	10.0-m Keck I Telescope 10.0-m Keck II Telescope
National Optical Astronomy Observatories	(Mauna Kea, Hawaii)	8-m
National Optical Astronomy Observatories	(Cerro Tololo, Chile)	8-m
Japanese National Observatory	(Mauna Kea, Hawaii)	7-m
Smithsonian Observatory	Mount Hopkins, Arizona	6.5-m (former 4.5-m Multiple Mirror Telescope)
Special Astrophysical Observatory	Mount Pastukhov, USSR	6-m Bol'shoi Teleskop Azimutal'nyi
Hale Observatories	Palomar Mountain, California	5.08-m George Ellery Hale Telescope
Royal Greenwich Observatory	La Palma, Canary Islands	4.2-m William Herschel Telescope
Cerro Tololo Observatory	Cerro Tololo, Chile	4.0-m
University of North Carolina and Columbia University	(Cerro Pachon, Chile)	4.0-m SOAR Telescope
Anglo-Australian Observatory	Siding Spring Mountain, Australia	4.0-m Anglo-Australian Telescope
Kitt Peak National Observatory	Kitt Peak, Arizona	3.8-m Nicholas U. Mayall Telescope
Royal Observatory Edinburgh	Mauna Kea, Hawaii	3.8-m United Kingdom Infrared Telescope
Canada-France-Hawaii Observatory	Mauna Kea, Hawaii	3.6-m Canada-France-Hawaii Telescope
European Southern Observatory	Cerro La Silla, Chile	3.57-m
Max Planck Institute (Bonn)	Calar Alto, Spain	3.5-m
Astronomical Research Corporation	Apache Point, New Mexico	3.5-m
University of Wisconsin, Indiana University, Yale University, and the National Optical Astronomy Observatories	(Kitt Peak, Arizona)	3.5-m WIYN Telescope
European Southern Observatory	Cerro La Silla, Chile	3.5-m New Technology Telescope
Lick Observatory	Mount Hamilton, California	3.05-m C. Donald Shane Telescope
Mauna Kea Observatory	Mauna Kea, Hawaii	3.0-m NASA Infrared Telescope
McDonald Observatory	Mount Locke, Texas	2.7-m
Haute Provence Observatory	Saint Michele, France	2.6-m
Crimean Astrophysical Observatory	Simferopol, USSR	2.6-m
Byurakan Astrophysical Observatory	Byurakan, USSR	2.6-m
Mount Wilson and Las Campanas Observatories	Las Campanas, Chile	2.5-m Irenée du Pont Telescope
Royal Greenwich Observatory	Canary Islands	2.5-m Isaac Newton Telescope
Dartmouth College and University of Michigan	Kitt Peak, Arizona	2.4-m McGraw-Hill Telescope
Australian National University	Siding Spring Mountain, Australia	2.4-m
Wyoming Infrared Observatory	Mount Jelm, Wyoming	2.3-m Wyoming Infrared Telescope

[a]Telescopes planned or under construction have their locations indicated in parentheses.

APPENDIX 6:   Major Optical Telescopes of the World, *Continued*

Observatory	Location[a]	Telescope
Steward Observatory	Kitt Peak, Arizona	2.3-m Wyoming Infrared Telescope
Mauna Kea Observatory	Mauna Kea, Hawaii	2.2-m
Max Planck Institute (Bonn)	Calar Alto, Spain	2.2-m
Max Planck Institute	(South West Africa)	2.2-m
University of Mexico Observatory	(Mexico)	2.16-m
La Plata Observatory	La Plata Argentina	2.15-m
Kitt Peak National Observatory	Kitt Peak, Arizona	2.1-m
McDonald Observatory	Mount Locke, Texas	2.1-m Otto Struve Telescope
Karl Schwarzschild Observatory	Tautenberg, East Germany	2.0-m (largest Schmidt telescope)

[a]Telescopes planned or under construction have their locations indicated in parentheses.

APPENDIX 7:   Planetary and Satellite Data

Orbital Data for the Planets

Planet	Sidereal Period	Semimajor Axis	Orbital Eccentricity[a]	Inclination of Orbital Plane	Rotation Period	Tilt of Axis
Mercury	0.241 yr	0.387 AU	0.2056	7°0′15″	58^d.65	28°
Venus	0.615	0.723	0.068	3°23′40″	243	3°
Earth	1.000	1.000	0.0167	0°0′0″	23^{h}56^m	23°27′
Mars	1.881	1.524	0.0934	1°51′0″	24^{h}37^m	23°59′
Jupiter	11.86	5.203	0.0485	1°18′17″	9^{h}55^m.5	3°5′
Saturn	29.46	9.555	0.0556	2°29′33″	10^{h}39^m.4	26°44′
Uranus	84.01	19.22	0.0472	0°46′23″	17^{h}14^m.4	97°52′
Neptune	164.79	30.11	0.0086	1°46′22″	1606.6	29°34′
Pluto	248.5	39.44	0.250	17°10′12″	6^d.387	122.5°

[a]The eccentricity of an orbit is defined as the ratio of the distance between foci to the semimajor axis. In practice it is related to the perihelion distance P and the semimajor axis a by $P = a(1 - e)$, where e is the eccentricity; and to the aphelion distance A by $A = a(1 + e)$.

Physical Data for the Planets

Planet	Mass[a]	Average, 1 Bar Diameter[a]	Density	Surface Gravity	Escape Speed	1 Bar Temperature	Albedo
Mercury	0.0558	0.381	5.50 g/cm^3	0.38 g	4.3 km/sec	100–700 K	0.106
Venus	0.815	0.951	5.3	0.90	10.3	730	0.65
Earth	1.000	1.000	5.518	1.00	11.2	200–300	0.37
Mars	0.107	0.531	3.96	0.38	5.0	130–290	0.15
Jupiter	317.89	10.85	1.327	2.64	60.0	165	0.52
Saturn	95.184	8.99	0.688	1.13	36.0	134	0.47
Uranus	14.536	3.96	1.272	0.89	21.2	76	0.50
Neptune	17.148	3.85	1.640	1.13	23.5	74	0.5:
Pluto	0.0022	0.18	1.7	0.06	1.2	40	0.06:
							0.44–0.61

[a]The masses and diameters are given in units of the Earth's mass and diameter, which are 5.974×10^{24} kg and 12,734 km, respectively.

continued on next page

APPENDIX 7:   Planetary and Satellite Data, *Continued*
Satellites

Planet	Satellite	Semimajor Axis	Period	Diameter	Mass (kg)	Density
Earth	Moon	3.84×10^5 km	27^d322	3476 km	7.35×10^{22}	3.34 g/cm^3
Mars	Phobos	9.38×10^3	0.3189	$27 \times 22 \times 19$	9.6×10^{15}	2:
	Deimos	2.35×10^4	1.2624	$15 \times 12 \times 11$	1.9×10^{15}	2:
Jupiter	Metis	1.280×10^5	0.295	20:	9.5×10^{16}	
	Adrastea	1.290×10^5	0.298	$20 \times 20 \times 15$	1.9×19^{16}	
	Amalthea	1.81×10^5	0.498	$270 \times 170 \times 150$	7.2×10^{18}	
	Thebe	2.22×10^5	0.675	110×90	7.6×10^{17}	
	Io	4.22×10^5	1.769	3630	8.92×10^{22}	3.53
	Europa	6.71×10^5	3.551	3138	4.87×10^{22}	3.03
	Ganymede	1.07×10^6	7.155	5262	1.49×10^{23}	1.93
	Callisto	1.88×10^6	16.689	4800	1.08×10^{23}	1.70
	Leda	1.11×10^7	238.7	16:	5.7×10^{15}	
	Himalia	1.15×10^7	250.6	186:	9.5×10^{18}	
	Lysithea	1.17×10^7	259.2	36:	7.6×10^{16}	
	Elara	1.18×10^7	259.7	76:	7.6×10^{17}	
	Ananke	2.12×10^7	631	30:	3.8×10^{16}	
	Carme	2.26×10^7	692	40:	9.5×10^{16}	
	Pasiphae	2.35×10^7	735	50:	1.9×10^{17}	
	Sinope	2.37×10^7	758	36:	7.6×10^{16}	
Saturn	1981 S13 (Pan)	1.33×10^5	0.573	20:		
	Atlas	1.377×10^5	0.602	40×20		
	Prometheus	1.394×10^5	0.613	$140 \times 100 \times 80$		
	Pandora	1.417×10^5	0.629	$110 \times 90 \times 80$		
	Epimetheus	1.514×10^5	0.694	$140 \times 120 \times 100$		
	Janus	1.514×10^5	0.695	$220 \times 200 \times 160$		
	Mimas	1.855×10^5	0.942	392	4.5×10^{19}	1.43
	Enceladus	2.381×10^5	1.370	500	7.4×10^{19}	1.13
	Telesto	2.947×10^5	1.888	$34 \times 28 \times 26$		
	Calypso	2.947×10^5	1.888	$34 \times 22 \times 22$		
	Tethys	2.947×10^5	1.888	1060	7.4×10^{20}	1.19
	Dione	3.774×10^5	2.737	1120	1.05×10^{21}	1.43
	Helene	3.781×10^5	2.737	$36 \times 32 \times 30$		
	Rhea	5.271×10^5	4.518	1530	2.50×10^{21}	1.33
	Titan	1.222×10^6	15.945	5150	1.35×10^{23}	1.89
	Hyperion	1.481×10^6	21.277	$410 \times 260 \times 220$	1.71×10^{19}	
	Iapetus	3.561×10^6	79.330	1460	1.88×10^{21}	1.15
	Phoebe	1.295×10^7	550.5	220		

APPENDIX 7: **Planetary and Satellite Data,** *Continued*
Satellites

Planet	Satellite	Semimajor Axis	Period	Diameter	Mass (kg)	Density
Uranus	Cordelia	4.97×10^4	0.336	40		
	Ophelia	5.38×10^4	0.377	50		
	Bianca	5.92×10^4	0.435	50		
	Cressidia	6.18×10^4	0.465	60		
	Nesdimona	6.27×10^4	0.476	60		
	Juliet	6.46×10^4	0.494	80		
	Portia	6.61×10^4	0.515	80		
	Rosalind	6.99×10^4	0.560	60		
	Belinda	7.53×10^4	0.624	60		
	Puck	8.60×10^4	0.764	170		
	Miranda	1.298×10^5	1.413	484	7.5×10^{19}	1.26
	Ariel	1.912×10^5	2.520	1160	1.4×10^{21}	1.65
	Umbriel	2.660×10^5	4.144	1190	1.3×10^{21}	1.44
	Titania	4.358×10^5	8.706	1610	3.5×10^{21}	1.59
	Oberon	5.826×10^5	13.463	1550	2.9×10^{21}	1.50
Neptune	1989N6	4.80×10^4	0.296	54		
	1989N5	5.00×10^4	0.313	80		
	1989N3	5.25×10^4	0.333	150		
	1989N4	6.20×10^4	0.429	180		
	1989N2	7.36×10^4	0.554	190		
	1989N1	1.176×10^5	1.121	400		
	Triton	3.548×10^5	5.875	2705	2.21×10^{22}	2.07
	Nereid	5.5134×10^6	360.129	340	2.1×10^{19}	
Pluto	Charon	1.964×10^4	6.387	1186		

APPENDIX 8: Stellar Data
The Fifty Brightest Stars

Star		Spectral Type	Apparent Magnitude	Distance	Position (1980)	
					Right Ascension	Declination
α Eri	Achernar	B3 V	0.51	36 pc	01^{h}37^{m}0	−57°20′
α UMi	Polaris	F8 Ib	1.99	208	02 12.5	+89 11
α Per	Mirfak	F5 Ib	1.80	175	03 22.9	+49 47
α Tau	Aldebaran	K5 III	0.86	21	04 34.8	+16 28
β Ori	Rigel	B8 Ia	0.14	276	05 13.6	−08 13
α Aur	Capella	G8 III	0.05	14	05 15.2	+45 59
γ Ori	Bellatrix	B2 III	1.64	144	05 24.0	+06 20
β Tau	Elnath	B7 III	1.65	92	05 25.0	+28 36
ε Ori	Alnilam	B0 Ia	1.70	490	05 35.2	−01 13
ζ Ori	Alnitak	09.5 Ib	1.79	490	05 39.7	−01 57
α Ori	Betelgeuse	M2 Iab	0.41	159	05 54.0	+07 24
β Aur	Menkalinan	A2 V	1.86	27	05 58.0	+44 57
β CMa		B1 II-III	1.96	230	06 21.8	−17 56
α Car	Canopus	F0 Ib-II	−0.72	30	06 23.5	−52 41
γ Gem	Alhena	A0 IV	1.93	32	06 36.6	+16 25
α CMa	Sirius	A1 V	−1.47	2.7	06 44.2	−16 42
ε CMa	Adhara	B2 II	1.48	209	06 57.8	−28 57
δ CMa		F8 Ia	1.85	644	07 07.6	−26 22
α Gem	Castor	A1 V	1.97	14	07 33.3	+31 56
α CMi	Procyon	F5 IV-V	0.37	3.5	07 38.2	+05 17
β Gem	Pollux	K0 III	1.16	11	07 44.1	+28 05
γ Vel		WC8	1.83	160	08 08.9	−47 18
ε Car	Avior	K3 III?	1.90	104	08 22.1	−59 26
ζ Vel		A2 V	1.95	23	08 44.2	−54 38
β Car	Miaplacidus	A1 III	1.67	26	09 13.0	−69 38
α Hya	Alphard	K4 III	1.98	29	09 26.6	−08 35
α Leo	Regulus	B7 V	1.36	26	10 07.3	+12 04
γ Leo		K0 III	1.99	28	10 18.8	+19 57
α UMa	Dubhe	K0 III	1.81	32	11 02.5	+61 52
α Cru A	Acrux	B0.5 IV	1.39	114	12 25.4	−62 59
α Cru B		B1 V	1.86	114	12 25.4	−62 59
γ Cru	Gacrux	M4 III	1.69	67	12 30.1	−57 00
β Cru		B0.5 III	1.28	150	12 46.6	−59 35
ε UMa	Alioth	A0p	1.79	21	12 53.2	+56 04
α Vir	Spica	B1 V	0.91	67	13 24.1	−11 03
η UMa	Alkaid	B3 V	1.87	64	13 46.8	+49 25
β Cen	Hadar	B1 III	0.63	150	14 02.4	−60 16
α Boo	Arcturus	K2 III	−0.06	11	14 14.8	+19 17
α Cen A	Rigil Kentaurus	G2 V	0.01	1.3	14 38.4	−60 46
α Cen B		K4 V	1.40	1.3	14 38.4	−60 46
α Sco	Antares	M1 Ib	0.92	160	16 28.2	−26 23
α TrA	Atria	K2 Ib	1.93	25	16 46.5	−68 60
λ Sco	Shaula	B1 V	1.60	95	17 32.3	−37 05
θ Sco		F0 Ib	1.86	199	17 35.9	−42 59
ε Sgr	Kaus Australis	B9.5 III	1.81	38	18 22.9	−34 24
α Lyr	Vega	A0 V	0.04	8	18 36.2	+38 46
α Aql	Altair	A7 IV-V	0.77	5	19 49.8	+08 49
α Pav	Peacock	B2.5 V	1.95	95	20 24.1	−56 48
α Cyg	Deneb	A2 Ia	1.26	491	20 40.7	+45 12
α Gru	Al Na'ir	B7 IV	1.76	20	22 06.9	−47 04
α PsA	Fomalhaut	A3 V	1.15	7	22 56.5	−29 44

APPENDIX 8: Stellar Data, *Continued*
Nearby Stars (Within 5 Parsecs of the Sun)[a]

Star	Spectral Type	Visual Apparent Magnitude	Visual Absolute Magnitude	Parallax	Distance	Proper Motion
α Centauri	G2V	−0.1	4.8	0.753"	1.33 pc	3.68"/yr
Barnard's star	M5V	9.5	13.2	0.544	1.84	10.31
Wolf 359	M8V3	13.5	16.7	0.432	2.31	4.71
BD + 36°2147	M2V	7.5	10.5	0.400	2.50	4.78
Luyten 726-8	M6Ve	12.5	15.4	0.385	2.60	3.36
Sirius	A1V	−1.5	1.4	0.377	2.65	1.33
Ross 154	M5Ve	10.6	13.3	0.345	2.90	0.72
Ross 248	M6Ve	12.3	14.8	0.319	3.13	1.58
ε Eridani	K2V	3.7	6.1	0.305	3.28	0.98
Ross 128	M5V	11.1	13.5	0.302	3.31	1.37
Luyten 789-6	M6V	12.2	14.6	0.302	3.31	3.26
61 Cygni	K5Ve	5.2	7.5	0.292	3.42	5.22
ε Indi	K5Ve	4.7	7.0	0.291	3.44	4.69
τ Ceti	G8V	3.5	5.9	0.289	3.46	1.92
Procyon	F5V	0.4	2.7	0.285	3.51	1.25
Σ 2398	M4V	8.9	11.2	0.284	3.52	2.28
BD + 43°44	M1Ve	8.1	10.4	0.282	3.55	2.89
CD − 36°15693	M2Ve	7.4	9.6	0.279	3.58	6.90
G51-15		14.8	17.0	0.273	3.66	1.26
L725-32	M5Ve	11.5	13.6	0.264	3.79	1.22
BD + 5°1668	M4V	9.8	12.0	0.264	3.79	3.73
CD − 39°14192	M0Ve	6.7	8.8	0.260	3.85	3.46
Kapteyn's star	M0V	8.8	10.8	0.256	3.91	8.89
Kruger 60	M4V	9.7	11.7	0.254	3.94	0.86
Ross 614	M5Ve	11.3	13.3	0.243	4.12	0.99
BD − 12°4523	M5V	10.0	11.9	0.238	4.20	1.18
Wolf 424	M6Ve	13.2	15.0	0.234	4.27	1.75
van Maanen's star	W.D.	12.4	14.2	0.232	4.31	2.95
CD − 37°15492	M3V	8.6	10.4	0.225	4.44	6.08
Luyten 1159-16	M8V	12.3	14.0	0.221	4.52	2.08
BD + 50°1725	K7V	6.6	8.3	0.217	4.61	1.45
CD − 46°11540	M4V	9.4	11.1	0.216	4.63	1.13
CD − 49°13515	M3V	8.7	10.4	0.214	4.67	0.81
CD − 44°11909	M5V	11.2	12.8	0.213	4.69	1.16
BD + 68°946	M3.5V	9.1	10.8	0.213	4.69	1.33
G158-27		13.7	15.5	0.212	4.72	2.06
G208-44/45		13.4	15.0	0.210	4.76	0.75
BD − 15°6290	M5V	10.2	11.8	0.209	4.78	1.16
40 Eridani	K0V	4.4	6.0	0.207	4.83	4.08
L145-141	W.D.	11.4	12.6	0.206	4.85	2.68
BD + 20°2465	M4.5V	9.4	10.9	0.203	4.93	0.49
70 Ophiuchi	K1V	4.2	5.7	0.203	4.93	1.13
BD + 43°4305	M4.5Ve	10.2	11.7	0.200	5.00	0.83

[a]Data from Lippencott, L. S., 1978, *Space Science Reviews* 22:153. Many of the stars listed here are multiple systems; in these cases, the data in the table refer only to the brightest member of the system.

APPENDIX 9:   The Constellations

Name	Genitive	Abbreviation	Position Right Ascension	Declination
Andromeda	Andromedae	And	01^h	+40°
Antlia	Antliae	Ant	10	−35
Apus	Apodis	Aps	16	−75
Aquarius	Aquarii	Aqr	23	−15
Aquila	Aquilae	Aql	20	+05
Ara	Arae	Ara	17	−55
Aries	Arietis	Ari	03	+20
Auriga	Aurigae	Aur	06	+40
Bootes	Bootis	Boo	15	+30
Caelum	Caeli	Cae	05	−40
Camelopardalis	Camelopardalis	Cam	06	−70
Cancer	Cancri	Cnc	09	+20
Canes Venatici	Canum Venaticorum	CVn	13	+40
Canis Major	Canis Majoris	CMa	07	−20
Canis Minor	Canis Minoris	CMi	08	+05
Capricornus	Capricorni	Cap	21	−20
Carina	Carinae	Car	09	−60
Cassiopeia	Cassiopeiae	Cas	01	+60
Centaurus	Centauri	Cen	13	−50
Cepheus	Cephei	Cep	22	+70
Cetus	Ceti	Cet	02	−10
Chamaeleon	Chamaeleonis	Cha	11	−80
Circinis	Circini	Cir	15	−60
Columba	Columbae	Col	06	−35
Coma Berenices	Comae Berenices	Com	13	+20
Corona Australis	Coronae Australis	CrA	19	−40
Coronoa Borealis	Coronae Borealis	CrB	16	+30
Corvus	Corvi	Crv	12	−20
Crater	Crateris	Crt	11	−15
Crux	Crucis	Cru	12	−60
Cygnus	Cygni	Cyg	21	+40
Delphinus	Delphini	Del	21	+10
Dorado	Doradus	Dor	05	−65
Draco	Draconis	Dra	17	+65
Equuleus	Equulei	Equ	21	+10
Eridanus	Eridani	Eri	03	−20
Fornax	Fornacis	For	03	−30
Gemini	Geminorum	Gem	07	+20
Grus	Gruis	Gru	22	−45
Hercules	Herculis	Her	17	+30
Horologium	Horologii	Hor	03	−60
Hydra	Hydrae	Hya	10	−20
Hydrus	Hydri	Hyi	02	−75
Indus	Indi	Ind	21	−55

APPENDIX 9:   The Constellations, *Continued*

Name	Genitive	Abbreviation	Position Right Ascension	Position Declination
Lacerta	Lacertae	Lac	22	+45
Leo	Leonis	Leo	11	+15
Leo Minor	Leonis Minoris	LMi	10	+35
Lepus	Leporis	Lep	06	−20
Libra	Librae	Lib	15	−15
Lupus	Lupi	Lup	15	−45
Lynx	Lincis	Lyn	08	+45
Lyra	Lyrae	Lyr	19	+40
Mensa	Mensae	Men	05	−80
Microscopium	Microscopii	Mic	21	−35
Monoceros	Monocerotis	Mon	07	−05
Musca	Muscae	Mus	12	−70
Norma	Normae	Nor	16	−50
Octans	Octantis	Oct	22	−85
Ophiuchus	Ophiuchi	Oph	17	00
Orion	Orionis	Ori	05	+05
Pavo	Pavonis	Pav	20	−65
Pegasus	Pegasi	Peg	22	+20
Perseus	Persei	Per	03	+45
Phoenix	Phoenicis	Phe	01	−50
Pictor	Pictoris	Pic	06	−55
Pisces	Piscium	Psc	01	+15
Piscis Austrinus	Piscis Austrini	PsA	22	−30
Puppis	Puppis	Pup	08	−40
Pyxis	Pyxidis	Pyx	09	−30
Reticulum	Reticuli	Ret	04	−60
Sagitta	Sagittae	Sge	20	+10
Sagittarius	Sagittarii	Sgr	19	−25
Scorpius	Scorpii	Sco	17	−40
Sculptor	Sculptoris	Scl	00	−30
Scutum	Scuti	Sct	19	−10
Serpens	Serpentis	Ser	17	00
Sextans	Sextantis	Sex	10	00
Taurus	Tauri	Tau	04	+15
Telescopium	Telescopii	Tel	19	−50
Triangulum	Trianguli	Tri	02	+30
Triangulum Australe	Trianguli Australi	TrA	16	−65
Tucana	Tucanae	Tuc	00	−65
Ursa Major	Ursae Majoris	UMa	11	+50
Ursa Minor	Ursae Minoris	UMi	15	+70
Vela	Velorum	Vel	09	−50
Virgo	Virginis	Vir	13	00
Volans	Volantis	Vol	08	−70
Vulpecula	Vulpeculae	Vul	20	+25

APPENDIX 10:   Mathematical Treatment of Stellar Magnitudes

Logarithmic Representation

In the text, magnitudes are discussed in terms of the brightness ratios between stars of different magnitudes. We generally avoided discussion of cases where two stars differ by a fraction of a magnitude, because in such cases it is no longer simple to calculate the brightness ratio corresponding to the magnitude difference. If star 1 is 0.5 magnitudes brighter than star 2, for example, what is the brightness ratio? Or if star 1 is a factor of 48.76 fainter than star 2, what is the difference in magnitudes?

Astronomers use an exact mathematical relationship between magnitude differences and brightness ratios, written as

$$m_1 - m_2 = 2.5 \log (b_2/b_1),$$

where m_1 and m_2 are the magnitudes of two stars, and b_2/b_1 is the ratio of their brightnesses. The notation "$\log (b_2/b_1)$" means the **logarithm** of this ratio; a logarithm is the power to which 10 must be raised to give this ratio. Hence, for example, if $b_2/b_1 = 100$, $\log (b_2/b_1) = \log (10^2) = 2$, because 10 must be raised to the second power to give 100. The magnitude difference is 2.5 log (100) = 2.5 × 2 = 5. Similarly, if $b_2/b_1 = 0.001$, then $\log (b_2/b_1) = \log (0.001) = \log (10^{-3}) = -3$, and in this case the magnitude difference is 2.5 × -3 = -7.5 (the minus sign indicates that star 1 is brighter than star 2 in this example).

The method works equally well in cases where the power of 10 is not a whole number, as in the example where $b_2/b_1 = 48.76$. Here $\log (b_2/b_1) = \log (48.76) = 1.69$ (this is usually found by consulting tables of logarithms or by using a scientific calculator). In this example, the magnitude difference is 2.5 log (b_2/b_1) = 2.5 log(48.76) = 2.5 × 1.69 = 4.23, so star 1 is 4.23 magnitudes fainter than star 2.

The equation can be used in other ways as well; solving for b_2/b_1 yields

$$\frac{b_2}{b_1} = 10^{(m_1 - m_2)/2.5}$$
$$= 10^{0.4(m_1 - m_2)}.$$

Thus, if we know that the magnitudes of two stars differ by $m_1 - m_2$, we multiply this difference by 0.4 and raise 10 to the power $0.4(m_1 - m_2)$, again using a calculator or tables, to get the brightness ratio b_2/b_1. As a simple example, suppose $m_1 - m_2 = 5$; then $0.4(m_1 - m_2) = 0.4 \times 5 = 2$, and $10^{0.4(m_1 - m_2)} = 10^2 = 100$, as we knew it should. As a more complex example, consider the stars Betelgeuse (magnitude +0.41) and Deneb (magnitude +1.26). From the equation above, we see that Betelgeuse is

$10^{0.4(1.26 - 0.41)} = 10^{0.34} = 2.19$ times brighter than Deneb.

While in most cases it is possible to follow the discussions of magnitudes and brightness ratios in the text without using this exact mathematical technique, it is still useful to be familiar with it.

The Distance Modulus

Whenever we know both the apparent and absolute magnitudes of a star, a comparison of the two will give its distance. In the text we have seen how to make this calculation by the following several steps:

1. Convert the difference $m - M$ between apparent and absolute magnitude into a brightness ratio, that is, a numerical factor indicating how much brighter or fainter the star would appear at 10 parsecs distance than at its actual distance.
2. Using the inverse square law, determine the change in distance required to produce this change in brightness.
3. Multiply this distance factor by 10 parsecs to find the distance to the star.

To do calculations mentally in this way can be laborious, especially in cases where the magnitude difference does not correspond neatly to a simple numerical factor, as it did in the examples given in the text. Hence astronomers use a mathematical equation expressing the relationship between distance and the distance modulus $m - M$. This equation, which works equally well for all cases, is

$$d = 10^{1 + .2(m - M)},$$

where d is the distance in parsecs to a star whose apparent magnitude is m and whose absolute magnitude is M.

In a simple example, where $m = 9$ and $M = -6$, we have

$$d = 10^{1 + .2(15)}$$
$$= 10^{1 + 3}$$
$$= 10^4 \text{ parsecs.}$$

Now let's try a more complex case. Suppose the star is an M2 main-sequence star, so that $M = 13$, as found from the H-R diagram. The apparent magnitude is $m = 16$. Our equation tells us that

$$d = 10^{1 + .2(16 - 13)}$$
$$= 10^{1 + .6}$$
$$= 10^{1.6}$$
$$= 39.8 \text{ parsecs.}$$

It is necessary to use a slide rule, calculator, or mathematical table to do this, of course, but it is still relatively straightforward compared with following the mental steps outlined above.

The Effect of Extinction On the Distance Modulus

When we discussed distance determination techniques (in Chapter 12), we ignored the effects of interstellar extinction. In any method that depends on the apparent brightness of a star, however, extinction can be important, particularly for very distant stars. Because the effect of extinction is to make a star appear fainter than it otherwise would, the tendency is to overestimate distances if no allowance is made for it.

Recall that in the spectroscopic parallax technique, the distance modulus $m - M$ is used to find the distance to a star from the equation given above:

$$d = 10^{1 + .2(m - M)},$$

where m is the apparent magnitude, M is the absolute magnitude, and d is the distance to the star in parsecs.

To correct this equation for extinction, we add a term, A_v, which refers to the extinction (in magnitudes) in visual light. Thus if the extinction toward a particular star makes that star appear 2 magnitudes fainter than it otherwise would, $A_v = 2$. If we insert this into the equation, we find:

$$d = 10^{1 + .2(m - M - A_v)}$$

Let us consider a simple example, a star whose apparent magnitude is $m = 12.4$ and whose absolute magnitude is $M = 2.4$. First let us consider its distance if extinction is ignored:

$$d = 10^{1 + .2(12.4 - 2.4)} = 10^3 = 1,000 \text{ parsecs.}$$

Now suppose it is determined that the extinction in the direction of this star amounts to 1 magnitude. Now the distance is

$$d = 10^{1 + .2(12.4 - 2.4 - 1)} = 10^{2.8} = 631 \text{ parsecs.}$$

One magnitude is only a modest amount of extinction, yet by neglecting it, we overestimated the distance to this star by almost 60 percent. We can see from this example that extinction can have a drastic effect on distance estimates.

It is worthwhile to add a note about how the extinction A_v is determined. It is possible to determine how much redder, in terms of the $B - V$ color index, a star appears because of extinction. To carry that a step further, astronomers define a *color excess* called $E(B - V)$, which is the difference between the observed and intrinsic values of $B - V$:

$$E(B - V) = (B - V)_{\text{observed}} - (B - V)_{\text{intrinsic}}$$

Studies of the variation of interstellar extinction with wavelength show that the extinction at the visual wavelength is approximately three times the color excess; that is:

$$A_v = 3E(B - V).$$

Hence determination of excess reddening leads to an estimate of A_v, and this in turn can be used in the modified equation for the distance.

APPENDIX 11: Nuclear Reactions in Stars

In the text we did not spell out the details of the reactions that occur in stellar cores, although they are quite simple. To do so here, we will use the notation of nuclear physics. This is basically a shorthand in symbols. For example, a helium nucleus, containing two protons and two neutrons, is designated ^4_2He, the subscript indicating the **atomic number** (the number of protons) and the superscript the **atomic weight** (the total number of protons and neutrons). Similarly a hydrogen nucleus is ^1_1H, and deuterium, a form of hydrogen with an extra neutron in the nucleus, is ^2_1H. A special symbol (ν) is used for the **neutrino,** a massless subatomic particle emitted in some reactions, and e^+ indicates a **positron,** which is equivalent to an electron but has a positive electrical charge.

The symbol γ indicates a gamma ray, a very short wavelength photon of light emitted in some reactions.

The Proton-Proton Chain

Using this notation system, we can now spell out the proton-proton chain:

$$^1_1\text{H} + {}^1_1\text{H} \rightarrow {}^2_1\text{H} + e^+ + \nu$$
$$^2_1\text{H} + {}^1_1\text{H} \rightarrow {}^3_2\text{He} + \gamma.$$

The ^3_2He particle is a form of helium, but not the common form. Once we have this particle it will combine with another:

$$^3_2\text{He} + {}^3_2\text{He} \rightarrow {}^4_2\text{He} + {}^1_1\text{H} + {}^1_1\text{H}.$$

We end up with a normal helium nucleus. A total of six hydrogen nuclei went into the reaction (remember, the first two steps had to occur twice, in order to produce two $_2^3\text{He}$ particles for the final reaction), while there were two left at the end, so the net result is the conversion of four hydrogen nuclei into one helium nucleus.

The CNO Cycle

The CNO cycle, which dominates at higher temperatures, is more complex, involving not only carbon but also nitrogen and oxygen. Each of these elements has more than one form, which differ in their numbers of neutrons. Some of these **isotopes** are unstable and spontaneously emit positrons, decaying into other species in the process. Here is the CNO cycle:

$$_6^{12}\text{C} + _1^1\text{H} \rightarrow _7^{13}\text{N} + \gamma$$
$$_7^{13}\text{N} \rightarrow _6^{13}\text{C} + e^+ + \nu$$
$$_6^{13}\text{C} + _1^1\text{H} \rightarrow _7^{14}\text{N} + \gamma$$
$$_7^{14}\text{N} + _1^1\text{H} \rightarrow _8^{15}\text{O} + \gamma$$
$$_8^{15}\text{O} \rightarrow _7^{15}\text{N} + e^+ + \nu$$
$$_7^{15}\text{N} + _1^1\text{H} \rightarrow _6^{12}\text{C} + _2^4\text{He}.$$

Here we end up with a helium nucleus and a carbon nucleus, while the particles going into the reaction were four hydrogen nuclei and a carbon nucleus. Along the way, three isotopes of nitrogen and one of oxygen were created and then converted into something else, leaving neither element at the end. As in the proton-proton chain, the net result is the conversion of four hydrogen nuclei into one helium nucleus.

The Triple-Alpha Reaction

Stars that have used up all the available hydrogen in their cores may become hot enough to undergo a new reaction in which helium is converted into carbon. Helium nuclei, consisting of two protons and two neutrons, are called alpha particles, and the reaction is called the triple-alpha reaction, since three of these particles are involved:

$$_2^4\text{He} + _2^4\text{He} \rightarrow _4^8\text{Be}$$
$$_4^8\text{Be} + _2^4\text{He} \rightarrow _6^{12}\text{C} + \gamma.$$

In this reaction sequence, $_4^8\text{Be}$ is a form of the element beryllium, and the end product, $_6^{12}\text{C}$, is the most common form of carbon. The second step must follow very quickly after the first because $_4^8\text{Be}$ is unstable and will break apart into two $_2^4\text{He}$ particles in a very short time. Therefore the third $_2^4\text{He}$ particle must react with the $_4^8\text{Be}$ particle almost immediately; thus this reaction sequence can be viewed as a three-particle reaction.

Other Reactions

Following helium burning in the triple-alpha reaction, other reactions can take place if the stellar core becomes hot enough. The first of these reactions are **alpha-capture reactions,** in which one form of nucleus adds an alpha particle to become a new form with two more protons and two additional neutrons. One example is the carbon alpha capture:

$$_6^{12}\text{C} + _2^4\text{He} \rightarrow _8^{16}\text{O} + \gamma,$$

in which $_8^{16}\text{O}$, the most common form of oxygen, is produced. In another sequence of alpha captures, nitrogen is converted into another isotope of oxygen, which can then be converted into neon:

$$_7^{14}\text{N} + _2^4\text{He} \rightarrow _8^{18}\text{O} + e^+ + \nu,$$

and

$$_8^{18}\text{O} + _2^4\text{He} \rightarrow _{10}^{22}\text{Ne} + \gamma.$$

Additional alpha-capture reactions can occur, but at high enough temperatures other, more complex, reactions also take place. For example, two carbon nuclei can react, forming a number of different products, including sodium, neon, and magnesium. Two oxygen nuclei can also react, creating such species as sulfur, phosphorus, silicon, and magnesium. As a massive star evolves and its core contracts and heats following each reaction stage, a wide variety of elements is created with generally increasing atomic numbers. As explained in the text, the heaviest and most complex element produced in stable reactions in stellar cores is iron ($_{26}^{52}\text{Fe}$). Once a star has an iron core, it cannot undergo any further reaction stages without major disruption by a supernova explosion or collapse to a neutron star or black hole.

APPENDIX 12:   Detected Interstellar Molecules[a]

Number of Atoms	Symbol	Name	Number of Atoms	Symbol	Name
2	H_2	Molecular hydrogen	4	$NCNH^+$	Protonated hydrogen cyanide
	C_2	Diatomic carbon		$HNCS$	Isothiocyanic acid
	CH	Methylidyne		C_3N	Cyanoethynyl
	CH^+	Methylidyne ion		C_3O	Tricarbon monoxide
	CN	Cyanogen		H_2CS	Thioformaldehyde
	CO	Carbon monoxide	5	C_4H	Butadiynyl
	CS	Carbon monosulfide		C_3H_2	Cyclopropenylidene
	OH	Hydroxyl		HCO_2H	Formic acid
	NO	Nitric oxide		CH_2CO	Ketene
	NS	Nitrogen sulfide		HC_3N	Cyanoacetylene
	SiC	Silicon carbide			
	SiO	Silicon monoxide		NH_2CN	Cyanamide
	SiS	Silicon sulfide		CH_2NH	Methanimine
	SO	Sulfur monoxide		CH_4	Methane
3	H_2D^+	Protonated hydrogen deuteride		SiH_4	Silane[b]
	C_2H	Ethynyl	6	C_5H	Pentynylidyne[b]
	HCN	Hydrogen cyanide		CH_3OH	Methanol
	HNC	Hydrogen isocyanide		CH_3CN	Methyl cyanide
	HCO	Formyl		CH_3SH	Methyl mercaptan
	HCO^+	Formyl ion		NH_2CHO	Formamide
	N_2H^+	Protonated nitrogen	7	CH_2CHCN	Vinyl cyanide
	HNO	Nitroxyl		CH_3C_2H	Methylacetylene
	H_2O	Water		CH_3CHO	Acetaldehyde
	HCS^+	Thioformyl ion		CH_3NH_2	Methylamine
	H_2S	Hydrogen sulfide		HC_5N	Cyanodiacetylene
	OCS	Carbonyl sulfide	8	$HCOOCH_3$	Methyl formate
	SO_2	Sulfur dioxide		CH_3C_3N	Methylcyanoacetylene
	SiC_2	Silicon dicarbide[b]	9	CH_3C_4H	Methyldiacetylene
4	C_2H_2	Acetylene		CH_3CH_3O	Dimethyl ether
	C_3H	Propynylidyne		CH_3CH_2CN	Ethyl cyanide
	H_2CO	Formaldehyde		CH_3CH_2OH	Ethanol
	NH_3	Ammonia		HC_7N	Cyano-hexa-tri-yne
	$HNCO$	Isocyanic acid	11	HC_9N	Cyano-octa-tetra-yne
	$HOCO^+$	Protonated carbon monoxide	13	$HC_{11}N$	Cyano-deca-penta-yne

[a]This list does not include isotopic variations (identical molecules except that one or more atoms are in rare isotopic forms, such as deuterium in place of hydrogen, or ^{13}C instead of the much more common ^{12}C).
[b]These species have been detected only in the dense circumstellar clouds surrounding red giant or supergiant stars.

APPENDIX 13:   Clusters of Galaxies
Galaxies of the Local Group

Galaxy[a]	Type[b]	Absolute Magnitude	Position Right Ascension	Position Declination
M31 (Andromeda)	Sb	−21.1	$00^h 40.^m 0$	+41° 00′
Milky Way	Sbc	−20.5	17 42.5	−28 59
M33 = NGC 598	Sc	−18.9	01 31.1	−30 24
Large Magellanic Cloud	Irr	−18.5	05 24	−69 50
IC 10	Irr	−17.6	00 17.6	+59 02
Small Magellanic Cloud	Irr	−16.8	00 51	−73 10
M32 = NGC 221	E2	−16.4	00 40.0	+40 36
NGC 205	E6	−16.4	00 37.6	+41 25
NCG 6822	Irr	−15.7	19 42.1	−14 53
NCG 185	Dwarf E	−15.2	00 36.1	+48 04
NGC 147	Dwarf E	−14.9	00 30.4	+48 14
IC 1613	Irr	−14.8	01 02.3	+01 51
WLM	Irr	−14.7	23 59.4	−15 44
Fornax	Dwarf sph	−13.6	02 37.5	−34 44
Leo A	Irr	−13.6	09 56.5	+30 59
IC 5152	Irr	−13.5	21 59.6	−51 32
Pegasus	Irr	−13.4	23 26.1	+14 28
Sculptor	Dwarf sph	−11.7	00 57.5	−33 58
And I	Dwarf sph	−11	00 42.8	+37 46
And II	Dwarf sph	−11	01 13.6	+33 11
And III	Dwarf sph	−11	00 32.7	+36 14
Aquarius	Irr	−11	20 44.1	−13 02
Leo I	Dwarf sph	−11	10 0.58	+12 33
Sagittarius	Irr	−10	19 27.1	−17 47
Leo II	Dwarf sph	− 9.4	11 10.8	+22 26
Ursa Minor	Dwarf sph	− 8.8	15 08.2	+67 18
Draco	Dwarf sph	− 8.6	17 19.4	+57 58
Carina	Dwarf sph		06 40.4	−50 55
Pisces	Irr	− 8.5	00 01.2	+21 37

[a]Galaxy names are derived from a variety of sources, including several catalogs (such as those designated M, NGC, and IC) and colloquial names bestowed by discoverers. Many in this list are simply named after the constellation where they are found.
[b]The galaxy types listed here are described in the text, except for the *Dwarf sph* designation, which stands for "dwarf spheroidal" and refers to dwarf galaxies that do not fit easily into the designation of dwarf ellipticals. Note that the absolute magnitudes of some of these are comparable to those of the brightest individual stars in our galaxy.

APPENDIX 13: Clusters of Galaxies, *Continued*
Clusters of Galaxies within 1,000 Mpc[a]

Cluster	Distance (Mpc)	Radial Velocity (km/sec)	Diameter (Mpc)	Number of Galaxies	Density of Galaxies(Mpc^{-3})
Virgo	19	1,180	4	2,500	500
Pegasus I	65	3,700	1	100	1,100
Pisces	66	250	12	100	250
Cancer	80	4,800	4	150	500
Perseus	97	5,400	7	500	300
Coma	113	6,700	8	800	40
UMa III	132		2	90	200
Hercules	175	10,300	0.3	300	
Cluster A	240	15,800	4	400	200
Centaurus	250		9	300	10
UMa I	270	15,400	3	300	100
Leo	310	19,500	3	300	200
Cluster B	330		4	300	200
Gemini	350	23,300	3	200	100
CrB	350	21,600	3	400	250
Bootes	650	39,400	3	150	100
UMa II	680	41,000	2		400
Hydra	1,000	60,600			

[a]Data from Allen C. W., 1973, *Astrophysical Quantities*, 3d ed. (London: Athlone Press).

APPENDIX 14: The Relativistic Doppler Effect

The simple formula given in the text relating the Doppler shift of an object's spectrum to its velocity is accurate only when the velocity is much less than the speed of light. That simple relation is

$$v = (\Delta\lambda/\lambda)c,$$

where v is the object's velocity, $\Delta\lambda$ is the shift in wavelength of a spectral line whose rest wavelength is λ, and c is the speed of light.

When v is a significant fraction of c, the correct equation must be used. The error in determining v caused by using the approximate formula is 1 percent of v when $\Delta\lambda/\lambda$ is only 0.02; and the error becomes 5 percent when $\Delta\lambda/\lambda$ is 0.1. Thus failure to use the correct, relativistic formula becomes important for speeds of only a few percent of the speed of light.

For simplicity of notation, astronomers usually use the symbol z to represent the Doppler shift; that is,

$$z = \Delta\lambda/\lambda.$$

From Einstein's theory of special relativity, it can be shown that the correct relationship between the shift in wavelength and the speed of the emitting object is

$$z = \Delta\lambda/\lambda = \sqrt{\frac{1 + v/c}{1 - v/c}} - 1,$$

which leads to the following solution for the velocity:

$$v = c\left[\frac{(z + 1)^2 - 1}{(z + 1)^2 + 1}\right]$$

Let us apply this to the redshifts of quasars. One of the first quasars discovered, 3C273, has $z = 0.16$, which would imply a velocity of $0.16c = 48,000$ km/sec, if we used the simple, nonrelativistic equation. Use of the relativistic formula leads instead to $v = 0.147c = 44,200$ km/sec. In this case, use of the approximate formula leads to an error of almost 9 percent. The quasars with the highest known redshifts have z greater than 4.0, which would lead to the nonsensical conclusion that their speeds are more than 4 times that of light, if the nonrelativistic equation were used. Use of the correct equation for $z = 4.0$ gives $v = 0.923c = 277,000$ km/sec, still an enormous velocity.

APPENDIX 15:   The Messier Catalog

Number	Right Ascension	Declination	Magnitude	Description
M1	05^{h}33.^{m}3	+22°01'	11.3	Crab Nebula in Taurus
M2	21 32.4	−00 54	6.3	Globular cluster in Aquarius
M3	13 41.3	+28 29	6.2	Globular cluster in Canes Venatici
M4	16 22.4	−26 27	6.1	Globular cluster in Scorpio
M5	15 17.5	+02 07	6	Globular cluster in Serpens
M6	17 38.9	−32 11	6	Open cluster in Scorpio
M7	17 52.6	−34 48	5	Open cluster in Scorpio
M8	18 02.4	−24 23		Lagoon Nebula in Sagittarius
M9	17 18.1	−18 30	7.6	Globular cluster in Ophiuchus
M10	16 56.0	−04 05	6.4	Globular cluster in Ophiuchus
M11	18 50.0	−06 18	7	Open cluster in Scutum
M12	16 46.1	−01 55	6.7	Globular cluster in Ophiuchus
M13	16 41.0	+36 30	5.8	Globular cluster in Hercules
M14	17 36.5	−03 14	7.8	Globular cluster in Ophiuchus
M15	21 29.1	+12 05	6.3	Globular cluster in Pegasus
M16	18 17.8	−13 48	7	Open cluster in Serpens
M17	18 19.7	−16 12	7	Omega Nebula in Sagittarius
M18	18 18.8	−17 09	7	Open cluster in Sagittarius
M19	17 01.3	−26 14	6.9	Globular cluster in Ophiuchus
M20	18 01.2	−23 02		Trifid Nebula in Sagittarius
M21	18 03.4	−22 30	7	Open cluster in Sagittarius
M22	18 35.2	−23 55	5.2	Globular cluster in Sagittarius
M23	17 55.7	−19 00	6	Open cluster in Sagittarius
M24	18 17.3	−18 27	6	Open cluster in Sagittarius
M25	18 30.5	−19 16	6	Open cluster in Sagittarius
M26	18 44.1	−09 25	9	Open cluster in Scutum
M27	19 58.8	+22 40	8.2	Dumbbell Nebula; planetary nebula in Vulpecula
M28	18 23.2	−24 52	7.1	Globular cluster in Sagittarius
M29	20 23.3	+38 27	8	Open cluster in Cygnus
M30	21 39.2	−23 15	7.6	Globular cluster in Capricornus
M31	00 41.6	+41 09	3.7	Andromeda galaxy
M32	00 41.6	+40 45	8.5	Elliptical galaxy, companion of M31
M33	01 32.8	+30 33	5.9	Spiral galaxy in Triangulum
M34	02 40.7	+42 43	6	Open cluster in Perseus
M35	06 07.6	+24 21	6	Open cluster in Gemini
M36	05 35.0	+34 05	6	Open cluster in Auriga
M37	05 51.5	+32 33	6	Open cluster in Auriga
M38	05 27.3	+35 48	6	Open cluster in Auriga
M39	21 31.5	+48 21	6	Open cluster in Cygnus
M40	12 20	+59		Double star cluster in Ursa Major
M41	06 46.2	−20 43	6	Open cluster in Canis Major
M42	05 34.4	−05 24		Orion Nebula
M43	05 34.6	−05 18		Small extension of the Orion nebula
M44	08 38.8	+20 04	4	Praesepe; open cluster in Cancer
M45	03 46.3	+24 03	2	The Pleiades; open cluster in Taurus
M46	07 40.9	−14 46	7	Open cluster in Puppis
M47	07 35.6	−14 27	5	Open cluster in Puppis
M48	08 12.5	−05 43	6	Open cluster in Hydra
M49	12 28.8	+08 07	8.9	Elliptical galaxy in Virgo
M50	07 02.0	−08 19	7	Open cluster in Monocerotis
M51	13 29.0	+47 18	8.4	Whirlpool galaxy; spiral galaxy in Canes Venatici
M52	23 23.3	+61 29	7	Open cluster in Cassiopeia
M53	13 12.0	+18 17	7.7	Globular cluster in Coma Berenices
M54	18 53.8	−30 30	7.7	Globular cluster in Sagittarius
M55	19 38.7	−31 00	6.1	Globular cluster in Sagittarius

APPENDIX 15:   The Messier Catalog, *Continued*

Number	Right Ascension	Declination	Magnitude	Description
M56	19 15.8	+30 08	8.3	Globular cluster in Lyra
M57	18 52.9	+33 01	9.0	Ring Nebula; planetary nebula in Lyra
M58	12 36.7	+11 56	9.9	Spiral galaxy in Virgo
M59	12 41.0	+11 47	10.3	Elliptical galaxy in Virgo
M60	12 42.6	+11 41	9.3	Elliptical galaxy in Virgo
M61	12 20.8	+04 36	9.7	Spiral galaxy in Virgo
M62	16 59.9	−30 05	7.2	Globular cluster in Scorpio
M63	13 14.8	+42 08	8.8	Spiral galaxy in Canes Venatici
M64	12 55.7	+21 48	8.7	Spiral galaxy in Coma Berenices
M65	11 17.8	+13 13	9.6	Spiral galaxy in Leo
M66	11 19.1	+13 07	9.2	Spiral galaxy, companion of M65
M67	08 50.0	+11 54	7	Open cluster in Cancer
M68	12 38.3	−26 38	8	Globulalr cluster in Hydra
M69	18 30.1	−32 23	7.7	Globular cluster in Sagittarius
M70	18 42.0	−32 18	8.2	Globular cluster in Sagittarius
M71	19 52.8	+18 44	6.9	Globular cluster in Sagitta
M72	20 52.3	−12 39	9.2	Globular cluster in Aquarius
M73	20 57.8	−12 44		Open cluster in Aquarius
M74	01 35.6	+15 41	9.5	Spiral galaxy in Pisces
M75	20 04.9	−21 59	8.3	Globular cluster in Sagittarius
M76	01 40.9	+51 28	11.4	Planetary nebula in Perseus
M77	02 41.6	−00 04	9.1	Spiral galaxy in Cetus
M78	05 45.8	+00 02		Emission nebula in Orion
M79	05 23.3	−24 32	7.3	Globular cluster in Lepus
M80	16 15.8	−22 56	7.2	Globular cluster in Scorpio
M81	09 54.2	+69 09	6.9	Spiral galaxy in Ursa Major
M82	09 54.4	+69 47	8.7	Irregular galaxy in Ursa Major
M83	13 35.9	−29 46	7.5	Spiral galaxy in Hydra
M84	12 24.1	+13 00	9.8	Elliptical galaxy in Virgo
M85	12 24.3	+18 18	9.5	Elliptical galaxy in Coma Berenices
M86	12 25.1	+13 03	9.8	Elliptical galaxy in Virgo
M87	12 29.7	+12 30	9.3	Giant elliptical galaxy in Virgo
M88	12 30.9	+14 32	9.7	Spiral galaxy in Coma Berenices
M89	12 34.6	+12 40	10.3	Elliptical galaxy in Virgo
M90	12 35.8	+13 16	9.7	Spiral galaxy in Virgo
M91				Not identified; possibly M58
M92	17 16.5	+43 10	6.3	Globular cluster in Hercules
M93	07 43.6	−23 49	6	Open cluster in Puppis
M94	12 50.1	+41 14	8.1	Spiral galaxy in Canes Venatici
M95	10 42.8	+11 49	9.9	Barred spiral galaxy in Leo
M96	10 45.6	+11 56	9.4	Spiral galaxy in Leo
M97	11 13.7	+55 08	11.1	Owl Nebula; planetary nebula in Ursa Major
M98	12 12.7	+15 01	10.4	Spiral galaxy in Coma Berenices
M99	12 17.8	+14 32	9.9	Spiral galaxy in Coma Berenices
M100	12 21.9	+15 56	9.6	Spiral galaxy in Coma Berenices
M101	14 02.5	+54 27	8.1	Spiral galaxy in Ursa Major
M102				Not identified; possibly M101
M103	01 31.9	+60 35	7	Open cluster in Cassiopeia
M104	12 39.0	−11 35	8	Sombrero galaxy; spiral galaxy in Virgo
M105	10 46.8	+12 51	9.5	Elliptical galaxy in Leo
M106	12 18.0	+47 25	9	Spiral galaxy in Canes Venatici
M107	16 31.8	−13 01	9	Globular cluster in Ophiuchus
M108	11 10.5	+55 47	10.5	Spiral galaxy in Ursa Major
M109	11 56.6	+53 29	10.6	Barred spiral galaxy in Ursa Major

Glossary

Absolute magnitude   The magnitude a star would have if it were precisely 10 parsecs away from the Sun.

Absolute zero   The temperature where all molecular or atomic motion stops, equal to $-273°C$ or $-459°F$.

Absorption line   A wavelength at which light is absorbed, producing a dark feature in the spectrum.

Abundance gradient   A systematic change in the relative abundances of elements over a distance; particularly in a galaxy.

Acceleration   Any change—either of speed or direction—in the state of rest or motion of a body.

Accretion disk   A rotating disk of gas surrounding a compact object (such as a neutron star or black hole), formed by material falling in.

Aether   The substance of which the heavens and all bodies in the sky were thought to be composed, in ancient Greek cosmology.

Albedo   The fraction of incident light that is reflected from a surface, such as that of a planet.

Alpha-capture reaction   A nuclear fusion reaction in which an alpha particle merges with an atomic nucleus. A typical example is the formation of ^{16}O by the fusion of an alpha particle with ^{12}C.

Alpha particle   A nucleus of ordinary helium containing two protons and two neutrons.

Altitude   The angular distance of an object above the horizon.

Amino acid   A complex organic molecule of the type that forms proteins. Amino acids are fundamental constituents of all living matter.

Andromeda galaxy   The large spiral galaxy located some 700,000 parsecs from the Sun; the most distant object visible to the unaided eye.

Angstrom   The unit normally used in measuring wavelengths of visible and ultraviolet light; one angstrom is equal to 10^{-8} centimeter.

Angular diameter   The diameter of an object as seen on the sky, measured in units of angle.

Angular momentum   A measure of the mass, radius, and rotational velocity of a rotating or orbiting body. In the simple case of an object in circular orbit, the angular momentum is equal to the mass of the object, times its distance from the center of the orbit, times its orbital speed.

Annual motions   Motions in the sky caused by the Earth's orbital motion about the Sun. These include the seasonal variations of the Sun's declination and the Sun's motion through the zodiac.

Annular eclipse   A solar eclipse that occurs when the Moon is near its greatest distance from the Earth, so that its angular diameter is slightly smaller than that of the Sun, and a ring, or annulus, of the Sun's disk is visible surrounding the disk of the Moon.

Anticyclone   A rotating wind system around a high-pressure area. On the Earth, an anticyclone rotates clockwise in the Northern Hemisphere and counterclockwise in the Southern Hemisphere.

Antimatter   Matter composed of the antiparticles of ordinary matter. For each subatomic particle, there is an antiparticle that is its opposite in such properties as electrical charge, but its equivalent in mass. Matter and antimatter, if combined, annihilate each other, producing energy in the form of gamma rays according to the formula $E = mc^2$.

Aphelion   The point in the orbit of a solar system object where it is farthest from the Sun.

Apollo asteroid   An asteroid whose orbit brings it closer to the Sun than 1 astronomical unit (AU).

Apparent magnitude   The observed magnitude of a star or other celestial object, as seen from the Earth. (*see magnitude*)

Arctic Circle   The region extending from the North Pole to 66.5° N latitude, where the Sun stays below the horizon for the full 24-hour day at the time of the winter solstice.

Asteroid   (*see minor planet*)

Asthenosphere   The upper portions of the Earth's mantle, below the zone (the lithosphere) where convection currents are thought to operate. The term is also applied to similar zones in the interiors of the Moon and other planets.

Astrology   The ancient belief that earthly affairs and human lives are influenced by the positions of the Sun, Moon, and planets with respect to the zodiac.

Astrometric binary   A double star recognized as such because the visible star or stars undergo periodic motion that is detected by astrometric measurements.

Astrometry   The science of accurately measuring stellar positions.

Astronomical unit (AU)   A unit of distance used in astronomy, equal to the average distance between the Sun and the Earth; 1 AU is equal to 1.4959787×10^8 kilometer.

Astronomy   The science that deals with the universe beyond the Earth's atmosphere.

Astrophysics   The application of physical laws to astronomical phenomena; interchangeable with astronomy in modern usage.

Atomic number   The number of protons in the nucleus of an element. The atomic number defines the identity of an element.

Atomic weight   The mass of an atomic nucleus in atomic mass units [one atomic mass unit (amu) is defined as the average mass of the protons and neutrons in a nucleus of ordinary carbon, ^{12}C]. For most atoms, the atomic weight is approximately equal to the total number of protons and neutrons in the nucleus.

Aurorae australis   "Southern lights"; the visual emission from the Earth's upper atmosphere, caused by charged particles from space.

Aurorae borealis   "Northern lights"; (*see Aurorae australis*)

Autumnal equinox   The name used in the Northern Hemisphere for the point where the Sun crosses the celestial equator from north to south, around September 21. See also **equinox** and **vernal equinox.**

Bailey's beads   Small glowing regions seen at the edges of the Moon's disk during a solar eclipse, due to irregularities in the shape of the lunar disk, which allows sunlight to pass through to the Earth.

Barred spiral galaxy   A spiral galaxy whose nucleus has linear extensions on opposing sides, giving it a barlike shape. The spiral arms usually appear to emanate from the ends of the bar.

Basalt   An igneous silicate rock common in regions formed by lava flows on the Earth, the Moon, and probably on the other terrestrial planets.

Beta decay   A spontaneous nuclear reaction in which a neutron decays into a proton and an electron, with a neutrino being emitted also. The term has been generalized to mean any spontaneous reaction in which an electron and a neutrino (or their antimatter equivalents) are emitted.

Big bang   A term referring to any theory of cosmogony in which the universe began at a single point, was very hot initially, and has been expanding from that state since.

Binary star   A double star system in which the two stars orbit a common center of mass.

Binary X-ray source   A close binary system containing a compact stellar remnant which emits X rays as it accretes matter from its companion.

BL Lac object   A class of active galactic nuclei in which it is thought that the accretion disk is viewed nearly face-on, so that one of the axial jets is directed nearly along the line of sight toward the Earth.

Black hole   An object that has collapsed under its own gravitation to such a small radius that its gravitational force traps photons of light.

Bode's law   Also known as the Titius-Bode relation, a simple numerical sequence that approximately represents the relative distances of the inner seven planets from the Sun. The relation is derived by adding 4 to each number in the sequence 0, 3, 6, 12, . . ., and then dividing each sum by 10, resulting in the sequence 0.4, 0.7, 1.0, 1.6, . . ., which is approximately the sequence of planetary distances from the Sun in astronomical units.

Bolide   An extremely bright meteor that explodes in the upper atmosphere.

Bolometric correction   The difference between the bolometric magnitude and the visual magnitude for a star; the bolometric correction is always negative because the star is brighter when all wavelengths are included.

Bolometric magnitude   A magnitude in which all wavelengths of light are included.

Breccia   Lunar rocks consisting of pebbles and soil fused together by meteorite impacts.

Burster   A sporadic source of intense X-rays, probably consisting of a neutron star onto which new matter falls at irregular intervals.

Carbonaceous chondrite   A meteorite containing chondrules having a high abundance of carbon and other volatile elements. Carbonaceous chondrites, thought to be very old, have apparently been unaltered since the formation of the solar system.

Cassegrain focus   A focal arrangement for a reflecting telescope in which a convex secondary mirror reflects the image through a hole in the primary mirror to a focus as the bottom of the telescope tube. This arrangement is commonly used in situations where relatively lightweight instruments for analyzing the light are attached directly to the telescope.

Cataclysmic variable   A binary system in which mass exchange causes irregular outbursts, usually involving a white dwarf which accretes mass from its companion.

Catastrophic theory   Any theory in which observed phenomena are attributed to sudden changes in conditions or to the intervention of an outside force or body.

CCD   Charge-coupled device; an electronic detector that receives photons of light and converts them into electrical charge patterns which are then converted into digital data. (*see detector*)

Celestial equator   The imaginary circle formed by the intersection of the Earth's equatorial plane with the celestial sphere. The celestial equator is the reference line for north-south (declination) measurements in the standard equatorial coordinate system.

Celestial pole   The point on the celestial sphere directly overhead at either of the Earth's poles.

Celestial sphere   The imaginary sphere formed by the sky. It is a convenient device for discussing and measuring the positions of astronomical objects.

Center of mass   In a binary star system, or any system consisting of several objects, the point about which the mass is "balanced;" that is, the point that moves with a constant velocity through space while the individual bodies in the system move about it.

Cepheid variable   A pulsating variable star, of a class named after the prototype δ Cephei. Cepheid variables obey a period-luminosity relationship and are therefore useful as distance indicators. There are two classes of Cepheid variables, the so-called classical Cepheids, which belong to Population I, and the Population II Cepheids, also known as W Virginis stars.

Chondrite   A stony meteorite containing chondrules.

Chondrule   A spherical inclusion in certain meteorites, usually composed of silicates and always of very great age.

Chromatic aberration   The creation of images that are dispersed according to wavelength perpendicular to the focal plane, caused by the use of lenses in refracting telescopes; different wavelengths of light are brought to a focus at different positions.

Chromosphere   A thin layer of hot gas just outside the photosphere in the Sun and other cool stars. The temperature in the chromosphere rises from about 4,000 K at its inner edge to 10,000 or 20,000 K at its outer boundary. The chromosphere is characterized by the strong red emission line of hydrogen.

Closed universe   A possible state of the universe in which the expansion will eventually be reversed and which is characterized by positive curvature, being finite in extent but having no boundaries.

Cluster variable   (see *RR Lyrae variable*)

CNO cycle   A nuclear fusion reaction sequence in which hydrogen nuclei are combined to form helium nuclei, and other nuclei, such as isotopes of carbon, oxygen, and nitrogen, appear as catalysts or by-products. The CNO cycle is dominant in the cores of stars on the upper main sequence.

Color-magnitude diagram   The equivalent of an H-R diagram, for a cluster of stars. Because the stars are at a common distance, the apparent magnitude may be plotted on the vertical axis in place of the absolute magnitude; the horizontal axis is usually the color index $(B-V)$, rather than spectral type.

Color excess   The amount, measured in magnitude units, by which a star appears redder in color due to interstellar dust than it would without dust in the line of sight. Technically the definition is $E(B-V) = (B-V)$ observed $- (B-V)$ intrinsic, where $(B-V)$ is the color index.

Color index   The difference $B-V$ between the blue (B) and visual (V) magnitudes of a star. If B is less than V (that is, if the star is brighter in blue than in visual light), the star has a negative color index and is a relatively hot star. If B is greater than V, the color index is positive, and the star is relatively cool.

Coma   The extended glowing region that surrounds the nucleus of a comet.

Comet   An interplanetary body, composed of loosely bound rocky and icy material, which forms a glowing head and extended tail when it enters the inner solar system.

Compressional waves   Waves in which the oscillations are in the direction of wave motion. (*see also P wave*)

Configuration   The position of a planet or the Moon relative to the Sun-Earth line.

Conjunction   The alignment of two celestial bodies on the sky. In connection with the planets, a conjunction is the alignment of a planet with the Sun, an inferior conjunction being the occasion when an inferior planet is directly between the Sun and the Earth, and a superior conjunction being the occasion when any planet is directly behind the Sun as seen from the Earth.

Constellation   A prominent pattern of bright stars, historically associated with mythological figures. In modern usage each constellation incorporates a precisely defined region of the sky.

Constructive interference   The overlapping of waves (including light waves) in which the peaks and valleys of separate waves match each other, adding to their strength (or intensity, in the case of light).

Continental drift   The slow motion of the continental masses over the surface of the Earth, caused by the motions of the Earth's tectonic plates, which in turn are probably caused by convection in the underlying asthenosphere.

Continuous radiation   Electromagnetic radiation that is emitted in a smooth distribution with wavelength, without spectral features such as emission and absorption lines.

Convection   The transport of energy by fluid motions occurring in gases, liquids, or semirigid material such as the Earth's mantle. These motions are usually driven by the buoyancy of heated material, which tends to rise while cooler material descends.

Co-orbital satellites   Satellites that share the same orbit; usually one or more small satellites orbiting 60° ahead or 60° behind a larger one, kept in place by the combined gravitation effects of the large satellite and the parent planet.

Coriolis force   The apparent force felt by a particle moving away from the equator on a spinning body such as a planet; the particle veers in the direction of planetary rotation due to the fact the surface rotational speed decreases with distance from the equator.

Coronal holes   Regions of relatively low density in the solar corona, from which the solar wind emanates.

Corona   The very hot, extended outer atmosphere of the Sun and other cool, main-sequence stars. The high temperature in the corona $(1 - 2 \times 10^6$ K) is probably caused by the dissipation of mechanical energy from the convective zone just below the photosphere.

Cosmic background radiation   The primordial radiation field that fills the universe, having been created in the form of gamma rays at the time of the big bang, but having cooled since so that today its temperature is 2.73 K, and its peak wavelength is near 1.1 millimeters, in the microwave portion of the spectrum. Also known as the 3° background radiation.

Cosmic ray   A rapidly moving atomic nucleus from space. Some cosmic rays are produced in the Sun, while others come from interstellar space and probably originate in supernova explosions.

Cosmogony   The study of the origins of the universe.

Cosmological constant   A term added to the field equations by Einstein to allow solutions in which the universe was static (neither expanding nor contracting). Although the need for the term disappeared when it was discovered that the universe is expanding, the cosmological constant is retained in the field equations by modern cosmologists but is usually assigned the value zero.

Cosmological principle   The postulate, made by most cosmologists, that the universe is both homogeneous and isotropic. It is sometimes stated as, "The universe looks the same to all observers everywhere."

Cosmological redshift   A Doppler shift toward longer wavelengths that is caused by a galaxy's motion of recession due to the expansion of the universe.

Cosmology   The study of the universe as a whole.

Coudé focus   A focal arrangement for a reflecting telescope, in which the image is reflected by a series of mirrors to a remote, fixed location where a massive, immovable instrument can be used to analyze it.

Crater   A depression (usually circular) in the surface of a planet or satellite, caused by either an impact from space or volcanic activity.

Cyclone   A rotating wind system about a low-pressure center, often associated with storms on the Earth. On the Earth, cyclones rotate counterclockwise in the Northern Hemisphere and clockwise in the Southern Hemisphere.

Dark matter   The invisible material postulated to account for the majority of mass in clusters of galaxies and in the universe as a whole.

Declination   The coordinate in the equatorial system that measures positions in the north-south direction, with the celestial equator as the reference line. Declinations are measured in units of degrees, minutes, and seconds of arc.

Deferent   The large circle centered on or near the Earth on which the epicycle for a given planet moved, in the geocentric theory of the solar system developed by ancient Greek astronomers such as Hipparchus and Ptolemy.

Degenerate gas   A gas in which either free electrons or free neutrons are as densely spaced as allowed by laws of quantum mechanics. Such a gas has extraordinarily high density, and its pressure is not dependent on temperature as it is in an ordinary gas. Degenerate electron gas provides the pressure that supports white dwarfs against collapse, and degenerate neutron gas similarly supports neutron stars.

Density wave   A stable pattern of alternating dense and rarified regions, usually in a rotating fluid disk such as the ring system of Saturn or a spiral galaxy.

Deoxyribonucleic acid (DNA)   A complex protein consisting of a double helix structure composed of amino acid bases, responsible for carrying the genetic code in the nuclei of cells.

Destructive interference   The overlapping of waves (including light waves) in which the peaks of one wave coincide with the valleys of the other, diminishing both (and reducing the intensity, in light).

Detector   The general term for any device that receives and records the intensity of light at the focus of a telescope. Photographic film has been used traditionally in astronomy, but modern telescopes often use electronic devices, such as CCDs, instead.

Deuterium   An isotope of hydrogen containing in its nucleus one proton and one neutron.

Differential gravitational force   A gravitational force acting on an extended object, so that the portions of the object closer to the source of gravitation feel a stronger force than the portions that are farther away. Such a force, also known as a tidal force, acts to deform or disrupt the object and is responsible for many phenomena, ranging from synchronous rotation of moons or double stars to planetary ring systems and the disruption of galaxies in clusters.

Differentiation   The sinking of relatively heavy elements into the core of a planet or other body. Differentiation can only occur in fluid bodies, so any planet that has undergone this process must once have been at least partially molten.

Diffraction   The process in which waves (including light waves) bend as they pass an obstacle.

Dipole anisotropy   (see **24 hour anisotropy**)

Distance modulus   The difference $m - M$ between the apparent and absolute magnitudes for a given star. This difference, which must be corrected for the effects of interstellar extinction, is a direct measure of the distance to the star.

Diurnal motion   Any motion related to the rotation of the Earth. Diurnal motions include the daily risings and settings of all celestial objects.

DNA   (see **deoxyribonucleic acid**)

Doppler effect   The shift in wavelength of light or sound caused by relative motion between the source (or sound) and the observer.

Doppler shift   The observed shift in wavelength (and frequency) of a wave due to relative motion between the source of the wave and the observer.

Dwarf elliptical galaxy   A member of a class of small spheroidal galaxies similar to standard elliptical galaxies except for their small size and low luminosity. Dwarf galaxies are probably the most common in the universe but cannot be detected at distances beyond the Local Group of galaxies.

Dwarf nova   A close binary system containing a white dwarf in which material from the companion star falls onto the other at sporadic intervals, creating brief nuclear outbursts.

Eccentricity   A measure of the degree to which an elliptical orbit is elongated; technically, the eccentricity is equal to the ratio of the distance between the foci to the length of the major axis.

Eclipse   An occurrence in which one object is partially or totally blocked from view by another or passes through the shadow of another.

Eclipsing binary   A double star system in which one or both stars are periodically eclipsed by the other as seen from Earth. This situation can occur only when the orbital plane of the binary is viewed edge-on from the Earth.

Ecliptic   The plane of the Earth's orbit about the Sun, which is approximately the plane of the solar system as a whole. The apparent path of the Sun across the sky is the projection of the ecliptic onto the celestial sphere.

Ejecta   Material blasted out of the ground by a meteorite impact.

Electromagnetic force   The force created by the interaction of electric and magnetic fields. The electromagnetic force can be either attractive or repulsive and is important in countless situations in astrophysics.

Electromagnetic radiation   Waves consisting of alternating electric and magnetic fields. Depending on the wavelength, these waves may be known as gamma rays, X-rays, ultraviolet radiation, visible light, infrared radiation, or radio radiation.

Electromagnetic spectrum   The entire array of electromagnetic radiation arranged according to wavelength.

Electron   A tiny, negatively charged particle that orbits the nucleus of an atom. The charge is equal and opposite to that of a proton in the nucleus, and in a normal atom the number of electrons and protons is equal so that the overall electrical charge is zero. The electrons emit and absorb electromagnetic radiation by making transitions between fixed energy levels.

Ellipse   A geometrical shape such that the sum of the distances from any point on it to two fixed points called foci is constant. In any bound system where two objects orbit a common center of mass, their orbits are ellipses with the center of mass at one focus.

Elliptical galaxy   One of a class of galaxies characterized by smooth spheroidal forms, few young stars, and little interstellar matter.

Emission line   A wavelength at which radiation is emitted, creating a bright line in the spectrum.

Emission nebula   A cloud of interstellar gas that glows by the light of emission lines. The source of excitation that causes the gas to emit may be radiation from a nearby star or heating by any of a variety of mechanisms.

Endothermic reaction   Any nuclear or chemical reaction that requires more energy to occur than it produces.

Energy   The ability to do work. Energy can be in either kinetic form, when it is a measure of the motion of an object, or potential form, when it is stored but capable of being released into kinetic form.

Epicycle   A small circle on which a planet revolves, which in turn orbits another, distant body. Epicycles were used in ancient theories of the solar system to devise a cosmology that placed the Earth at the center but accurately accounted for the observed planetary motions.

Equatorial coordinates   The astronomical coordinate system in which positions are measured with respect to the celestial equator (in the north–south direction) and a fixed direction (in the east–west dimension). The coordinates used are declination (north–south, in units of angle) and right ascension (east–west, in units of time).

Equinox   Either of two points on the sky where the planes of the ecliptic and the Earth's equator intersect. When the Sun is at one of these two points, the lengths of night and day on the Earth are equal. (see also *autumnal equinox* and *vernal equinox*)

Erg   A unit of energy equal to the kinetic energy of an object of 2 grams mass moving at a speed of 1 centimeter per second, but defined technically as the work required to move a mass of 1 gram through a distance of 1 centimeter at an acceleration of 1 cm/sec^2.

Escape speed   The upward speed required for an object to escape the gravitational field of a body such as a planet. In a more technical sense, the escape speed is the speed at which the kinetic energy of the object equals its gravitational potential energy; if the object moves any faster, its kinetic energy exceeds its potential energy, and it can escape the gravitational field.

Event horizon   The "surface" of a black hole; the boundary of the region from within which no light can escape.

Evolutionary theory   Any theory in which observed phenomena are thought to have arisen as a result of natural processes requiring no outside intervention or sudden changes.

Excitation   A process by which one or more electrons of an atom or ion are raised to energy levels above the lowest possible one.

Excited state   A state in which an atom, ion, or molecule has more energy than the lowest possible state (known as the ground state). For an atom or ion, this usually refers to the energy state of one or more electrons; for a molecule, the excited state may refer to an electron energy state, a vibrational energy state, or a rotational energy state.

Field equations   A set of equations in general relativity theory that describe the relationship of matter, energy, and gravitation in the universe.

Fission reaction   A nuclear reaction in which a large nucleus is split into one or more smaller nuclei.

Flat universe   A possible state of the universe in which the momentum of expansion is exactly balanced by self-gravitation, so that the expansion will slow to a stop in an infinite time, but will not reverse itself and become a contraction. This state is predicted to exist by inflationary universe models.

Fluorescence   The emission of light at a particular wavelength following excitation of the electron by absorption of light at another, shorter wavelength.

Focus   (1) The point at which light collected by a telescope is brought together to form an image; (2) one of two fixed points that define an ellipse. (*see also ellipse*)

Force   Any agent or phenomenon that produces acceleration of a mass.

Fraunhofer lines   The series of prominent solar absorption lines identified and cataloged in the early nineteenth century by Josef Fraunhofer, based on visual observations made with a spectroscope.

Frequency   The rate (in units of hertz, or cycles per second) at which electromagnetic waves pass a fixed point. The frequency, usually designated ν, is related to the wavelength λ and the speed of light c by $\nu = c/\lambda$.

Fusion reaction   A nuclear reaction in which atomic nuclei combine to form more massive nuclei.

Galactic cluster   A loose cluster of stars located in the disk or spiral arms of the galaxy.

Gamma ray   A photon of electromagnetic radiation whose wavelength is very short and whose energy is very high. Radiation whose wavelength is less than one angstrom is usually considered to be gamma-ray radiation.

Gegenschein   The diffuse glowing spot, seen on the ecliptic opposite the Sun's direction, created by sunlight reflected off interplanetary dust.

Globular cluster   A large spherical cluster of stars located in the halo of the galaxy. These clusters, containing up to several hundred thousand members, are thought to be among the oldest objects in the galaxy.

Gram   A unit of mass, equal to the quantity of mass contained in 1 cubic centimeter of water.

Grand Unified Theory   A theory being sought by particle physicists in which three of the four fundamental forces (the weak and strong nuclear forces, and the electromagnetic force) are shown to be manifestations of the same phenomenon, and indistinguishable from each other under conditions of sufficiently high temperature and pressure.

Granulation   The spotty appearance of the solar surface (the photosphere) caused by convection in the layers just below.

Gravitational redshift   A Doppler shift toward long wavelengths caused by the effect of a gravitational field on photons of light. Photons escaping a gravitational field lose energy to the field, which results in the redshift.

Greatest elongation   The greatest angular distance from the Sun that an inferior planet can reach, as seen from Earth.

Great Red Spot   An oval-shaped, reddish feature on Jupiter's surface, thought to be a long-lived storm system.

Greenhouse effect   The trapping of heat near the surface of a planet by atmospheric molecules (such as carbon dioxide) that absorb infrared radiation emitted by the surface.

H-R diagram   (*see Hertzsprung-Russell diagram*)

H II region   A volume of ionized gas surrounding a hot star. (*see also emission nebula*)

Half-life   The time required for half of the nuclei of an unstable (radioactive) isotope to decay.

Halo   (1) The extended outer portions of a galaxy (thought to contain a large fraction of the total mass of the galaxy, mostly in the form of dim stars and interstellar gas); (2) the extensive cloud of gas surrounding the head of a comet.

Helium flash   A rapid burst of nuclear reactions in the degenerate core of a moderate mass star in the hydrogen shell-burning phase. The flash occurs when the core temperature reaches a sufficiently high temperature to trigger the triple-alpha reaction.

Herbig-Haro object   A bright object often associated with young stars, thought to be a region of ionization caused by the jets of high-speed gas emanating from the polar regions of newly-formed stars.

Hertzsprung-Russell diagram   A diagram on which stars are represented according to their absolute magnitudes (on the vertical axis) and spectral types (on the horizontal axis). Because the physical properties of stars are interrelated, they do not fall randomly on such a diagram but lie in well-defined regions according to their state of evolution. Very similar diagrams can be constructed using luminosity instead of absolute magnitude and temperature or color index in place of spectral type.

Hertz   A unit of frequency used in describing any oscillation phenomena, such as electromagnetic radiation; 1 hertz (1 Hz) is equal to one cycle or wave per second.

High-velocity star   A star whose velocity relative to the solar system is large. As a rule, high-velocity stars are Population II objects following orbital paths that are highly inclined to the plane of the galactic disk.

Horizontal branch   A sequence of stars in the H-R diagram of a globular cluster, extending horizontally across the diagram to the left from the red giant region. These are probably stars undergoing helium burning in their cores by the triple-alpha reaction.

Hubble constant   The numerical factor, usually denoted H, which describes the rate of expansion of the universe. It is the proportionality constant in the Hubble law $v = Hd$, which relates the speed of recession of a galaxy (v) to its distance (d). The present value of H is not well known, with estimates ranging between 55 and 90 kilometers per second per megaparsec.

Hydrostatic equilibrium   The state of balance between gravitational and pressure forces that exists at all points inside any stable object such as a star or planet.

Igneous rock   A rock that was formed by cooling and hardening from a molten state.

Impact crater   A crater formed on the surface of a terrestrial planet or a satellite by the impact of a meteoroid or planetesimal.

Inertia   The tendency of an object to remain in its state of rest or of uniform motion. This tendency is directly related to the mass of the object.

Inferior planet   One of the planets whose orbits lie closer to the Sun than that of the Earth (Mercury or Venus).

Inflationary universe   A theory of cosmology in which the universe expanded rapidly from a very compact and homogeneous initial state. This early rapid expansion was followed by the big bang expansion, which still continues.

Infrared radiation   Electromagnetic radiation in the wavelength region just longer than that of visible light; that is, radiation whose wavelength lies roughly between 7,000 Å and 0.01 centimeter.

Interferometry   The use of interference phenomena in electromagnetic waves to measure positions precisely or to achieve gains in resolution. Interferometry in radio astronomy entails the use of two or more antennae to overcome the normally very coarse resolution of a single radio telescope; in visible-light observations, the object is to eliminate the distorting effects of the Earth's atmosphere.

Interplanetary dust   Tiny solid particles in interplanetary space, concentrated in the ecliptic plane.

Interstellar cloud   A region of relatively high density in the interstellar medium. Interstellar clouds have densities ranging between 1 and 10^6 particles per cubic centimeter and, in aggregate, contain most of the mass in interstellar space.

Interstellar dust   The diffuse medium of tiny solid particles that permeates the space between the stars in a galaxy.

Interstellar extinction   The obscuration of starlight by interstellar dust. Light is scattered off dust grains, so that a distant star appears dimmer than it otherwise would. The scattering process is most effective at short (blue) wavelengths, so that stars seen through interstellar dust are reddened and dimmed.

Inverse Compton scattering   Scattering of photons by relativistic electrons, in which photon energies are increased.

Inverse square law   In general, any law describing a force or other phenomenon that decreases in strength as the square of the distance from some central reference point. In particular the term *inverse square law* is often used by itself to mean the law stating that the intensity of light emitted by a source such as a star diminishes as the square of the distance from the source.

Ionization   Any process by which an electron or electrons are freed from an atom or ion. Ionization generally occurs in two ways: by the absorption of a photon with sufficient energy or by collision with another particle.

Ionosphere   The zone of the Earth's upper atmosphere, between 80 and 500 kilometers altitude, where charged subatomic particles (chiefly protons and electrons) are trapped by the Earth's magnetic field. (*see also* **Van Allen belts**)

Ion   Any subatomic particle with a nonzero electrical charge. In standard practice, the term *ion* is usually applied only to positively charged particles, such as atoms missing one or more electrons.

Io torus   A zone of gas particles concentrated in the orbit of Io (the innermost major satellite of Jupiter); the gas originates in volcanic eruptions on Io, and is dispersed throughout the satellite's orbit by magnetic forces, once it becomes ionized.

Isotope   Any form of a given chemical element. Different isotopes of the same element have the same number of protons in their nuclei, but different numbers of neutrons.

Isotropic   Having the property of appearing the same in all directions. In astronomy, this term is often postulated to apply to the universe as a whole.

Jeans criterion   The condition required for a cloud of gas to spontaneously collapse due to its own self-gravitation, usually expressed in terms of the size (or mass) of the cloud relative to its density and temperature.

Joule   A unit of energy defined as the work required to move a mass of 1 kilogram through a distance of 1 meter at an acceleration of 1 m/sec^2.

Kelvin   A unit of temperature, equal to 0.01 of the difference between the freezing and boiling points of water, used in a scale whose zero point is absolute zero. A Kelvin is usually denoted simply by K.

Kilogram   A unit of mass, equal to 1,000 grams.

Kiloparsec   A unit of distance, equal to 1,000 parsecs.

Kinetic energy   The energy of motion. The kinetic energy of a moving object is equal to ½ times its mass times the square of its velocity.

Kirkwood's gaps   Narrow gaps in the asteroid belt created by orbital resonance with Jupiter.

Kuiper belt   A disk-like collection of Sun-orbiting bodies thought to be concentrated between 30 and 50 AU from the Sun; the source of most periodic comets.

Latitude   The distance north or south of the Earth's equator, measured in units of angle.

Lepton   One of the two fundamental classes of elementary particles; includes electrons.

Light curve   A graph showing the intensity of light (or magnitude) of a celestial object versus time. Light curves are often useful in diagnosing the properties of variable stars and other objects.

Light-gathering power   The ability of a telescope to collect light from an astronomical source; the light-gathering power is directly related to the area of the primary mirror or lens.

Limb darkening   The dark region around the edge of the visible disk of the Sun or of a planet caused by a decrease in temperature with height in the atmosphere.

Liquid metallic hydrogen   Hydrogen in a state of semirigidity that can exist only under conditions of extremely high pressure, as in the interiors of Jupiter and Saturn.

Lithosphere   The layer in the Earth, Moon, and terrestrial planets that includes the crust and the outer part of the mantle.

Local Group   The cluster of about thirty galaxies to which the Milky Way belongs.

Logarithm   The logarithm of a number is the power to which 10 must be raised to equal that number. For example, $100 = 10^2$; so the logarithm of 100 is 2.

Luminosity class   One of several classes to which a star can be assigned on the basis of certain luminosity indicators in its spectrum. The classes range from I for supergiants to V for main-sequence stars (also known as dwarfs).

Luminosity   The total energy emitted by an object per second (the power of the object). For stars the luminosity is usually measured in units of ergs per second.

Luminous arcs   Enormous arc-like glowing structures seen in some dense clusters of galaxies. These are thought to be distorted images of background galaxies or quasars, formed by the bending of light due to the cluster gravitation field. (*see **gravitational lens***)

Lunar eclipse   An eclipse of the Moon, caused by its passage through the shadow of the earth.

Lunar month   The synodic period of the Moon, equal to 29 days, 12 hours, 44 minutes, 11 seconds.

L wave   A type of seismic wave that travels only over the surface of the Earth.

Magellanic Clouds   The two irregular galaxies that are the nearest neighbors to the Milky Way and are visible to the unaided eye in the Southern Hemisphere.

Magma   Molten rock from the Earth's interior.

Magnetic braking   The slowing of the spin of a young star such as the early Sun by magnetic forces exerted on the surrounding ionized gas.

Magnetic dynamo   A rotating internal zone inside the Sun or a planet, thought to carry the electrical currents that create the solar or planetary magnetic field.

Magnetosphere   The region surrounding a star or planet that is permeated by the magnetic field of that body.

Magnitude   A measure of the brightness of a star, based on a system established by Hipparchus in which stars were ranked according to how bright they appeared to the unaided eye. In the modern system, a difference of 5 magnitudes corresponds exactly to a brightness ratio of 100, so that a star of a given magnitude has a brightness that is $100^{1/5} = 2.512$ times that of a star 1 magnitude fainter.

Main sequence   The strip in the H-R diagram, running from upper left to lower right, where most stars that are converting hydrogen to helium by nuclear reactions in their cores are found.

Main-sequence fitting   A distance-determination technique in which an H-R diagram for a cluster of stars is compared with a standard H-R diagram to establish the absolute magnitude scale for the cluster H-R diagram.

Main-sequence turnoff   In an H-R diagram for a cluster of stars the point where the main-sequence turns off toward the upper right. The main-sequence turnoff, showing which stars in the cluster have evolved to become red giants, is an indicator of the age of the cluster.

Mantle   The semirigid outer portion of the Earth's interior extending from roughly the midway point nearly to the surface and consisting of the mesosphere (the lower portion) and the asthenosphere.

Mare (*pl.* maria)   Any of several extensive, smooth lowland areas on the surface of the Moon or Mercury that were created by extensive lava flows early in the history of the solar system.

Mass   A measure of the quantity of matter contained in an object.

Mass-luminosity relation   The correspondence between the masses of stars and their luminosities; first found empirically, this relation has since been duplicated by theoretical models of stellar structure.

Mass-to-light ratio   The mass of a galaxy, in units of solar masses, divided by its luminosity, in units of the

Sun's luminosity. The mass-to-light ratio is an indicator of the relative quantities of Population I and Population II stars in a galaxy.

Maunder minimum   An interval during the latter half of the seventeenth century when the number of sunspots was abnormally low.

Maxwell distribution   A mathematical representation of the distribution of particle speeds in a gas of a certain temperature.

Mean solar day   The average length of the solar day as measured throughout the year; the mean solar day is precisely 24 hours.

Megaparsec (Mpc)   A unit of distance, equal to 10^6 parsecs.

Meridian   The great circle on the celestial sphere that passes through both poles and directly overhead; that is, the north–south line directly overhead.

Mesosphere   (1) The layer of the Earth's atmosphere between roughly 50 and 80 kilometers in altitude, where the temperature decreases with height; (2) the layer below the asthenosphere in the Earth's mantle.

Metamorphic rock   A rock formed by heat and pressure in the Earth's interior.

Meteor   A bright streak or flash of light created when a meteoroid enters the Earth's atmosphere from space.

Meteorite   The remnant of a meteoroid that survives a fall through the Earth's atmosphere and reaches the ground.

Meteoroid   A small, interplanetary body.

Meteor shower   A period during which meteors are seen with high frequency, occurring when the Earth passes through a swarm of meteoroids.

Micrometeorite   A microscopically small meteorite.

Microwave background   (see **cosmic background radiation**)

Milky Way   Historically, the diffuse band of light stretching across the sky; our cross-sectional view of the disk of our galaxy. In modern usage, the term *Milky Way* refers to our galaxy as a whole.

Minor planet   One of thousands of large (up to 1000 km in diameter) non-planetary bodies orbiting the Sun; most of the orbits lie between Mars and Jupiter.

Moving cluster method   A distance-determination technique in which the radial velocities of stars in a cluster are combined with knowledge of the direction of motion gained from observed proper motions, so that the true space velocity can be used in combination with the proper motion to derive the distance.

Neutrino   A subatomic particle without mass or electrical charge that is emitted in certain nuclear reactions.

Neutron   A subatomic particle with no electrical charge and a mass nearly equal to that of the proton. Neutrons and protons are the chief components of the atomic nucleus.

Neutron-capture reactions   Nuclear reactions in which neutrons are captured by nuclei. (*see also* ***r-process reactions*** *and* ***s-process reactions***)

Neutron star   A very compact, dense stellar remnant whose interior consists entirely of neutrons, and which is supported against collapse by degenerate neutron gas pressure.

Newtonian focus   A focal arrangement for reflecting telescopes in which a flat mirror is used to reflect the image through a hole in the side of the telescope tube.

Nonthermal radiation   Radiation not due only to the temperature of an object. The term is most often applied to sources of continuous radiation such as synchrotron radiation.

North celestial pole   The projection of the Earth's North Pole onto the sky.

Nova   A star that temporarily flares up in brightness, most likely as the result of nuclear reactions caused by the deposition of new nuclear fuel on the surface of a white dwarf in a binary system. (*see also* ***recurrent nova***)

Nucleus   The central, dense concentration in an atom, a comet, or a galaxy.

OB association   A group of young stars whose luminosity is dominated by O and B stars.

Obliquity   A measure of the tilt of a planet's rotation axis; technically, the angle between the rotation axis and the perpendicular to the orbital plane.

Occam's razor   The principle that the simplest explanation of any natural phenomenon is most likely the correct one. *Simple* in this case usually means requiring few assumptions or unverifiable postulates.

Olbers' paradox   The apparent conflict created by the fact that in a universe that is infinite in extent and in age, every line of sight should intersect a stellar surface, producing a uniformly bright nighttime sky; the fact that the sky is dark demonstrates that the universe is not infinite in both extent and age.

Oort cloud   The cloud of bodies, hypothesized to be orbiting the Sun at a great distance, from which comets originate.

Open cluster   A loosely-bound association of stars in the galactic disk. (*see* ***galactic cluster***)

Open universe   A possible state of the universe in which its expansion will never stop, and it is characterized by negative curvature, being infinite in extent and having no boundaries.

Opposition   A planetary configuration in which a superior planet is positioned exactly in the opposite direction from the Sun as seen from Earth.

Optical double   A pair of stars that happen to appear near each other on the sky but are not in orbit; not a true binary.

Orbital resonance   A situation in which the periods of two orbiting bodies are simple multiples of each other so that they are frequently aligned and gravitational forces due to the outer body may move the inner body out of its original orbit. This is one mechanism thought responsible for creating the gaps in the rings of Saturn and Kirkwood's gaps in the asteroid belt.

Organic molecule   Any of a large class of carbon-bearing molecules found in living matter.

Outgassing   The process by which gases escape from a planetary interior, often through volcanic venting or eruptions.

Ozone   A form of oxygen containing three oxygen atoms bonded together.

Paleomagnetism   Vestigial traces or artifacts in rocks of ancient magnetic fields.

Parallax   Any apparent shift in position caused by an actual motion or displacement of the observer. (*see also stellar parallax*)

Parsec   A unit of distance, equal to the distance to a star whose stellar parallax is 1 arcsecond. A parsec is equal to 206, 265 AU, 3.03×10^{13} kilometers, or 3.26 light-years.

Peculiar velocity   The deviation in a star's velocity from perfect circular motion about the galactic center.

Penumbra   (1) The light, outer portion of a shadow, such as the portion of the Earth's shadow where the Moon is not totally obscured during a lunar eclipse; (2) the light, outer portion of a sunspot.

Perihelion   The point in the orbit of any Sun-orbiting body where it most closely approaches the Sun.

Permafrost   A permanent layer of ice just below the surface of certain regions on the Earth and probably on Mars.

Photometer   A device, usually using a photoelectric cell, for measuring the brightnesses of astronomical objects.

Photon   A particle of light having wave properties but also acting as a discrete unit.

Photosphere   The visible surface layer of the Sun and stars; the layer from which continuous radiation escapes and where absorption lines form.

Planck constant   The numerical factor h relating the frequency v of a photon to its energy E in the expression $E = hv$. The Planck constant has the value $h = 6.62620 \times 10^{-27}$ erg second.

Planck function (also known as the **Planck law**)   The mathematical expression describing the continuous thermal spectrum of a glowing object. For a given temperature, the Planck function specifies the intensity of radiation as a function of either frequency or wavelength.

Planetary nebula   A cloud of glowing, ionized gas, usually taking the form of a hollow sphere or shell, ejected by a star in the late stages of its evolution.

Planetesimal   A small (diameter up to several hundred kilometers) solar system body of the type that first condensed from the solar nebula. Planetesimals are thought to have been the principal bodies that combined to form the planets.

Plate tectonics   A general term referring to the motions of lithospheric plates over the surface of the Earth or other terrestrial planets. (*see also continental drift*)

Polarization   A preferential orientation of electromagnetic waves from a source; that is, the magnetic (or electric) fields show a preference for a certain orientation, rather than having random orientations.

Population I   The class of stars in a galaxy with relatively high abundances of heavy elements. These stars are generally found in the disk and spiral arms of spiral galaxies and are relatively young. The term *Population I* is also commonly applied to other components of galaxies associated with the star formation, such as the interstellar material.

Population II   The class of stars in a galaxy with relatively low abundances of heavy elements. These stars are generally found in a spheroidal distribution about the galactic center and throughout the halo and are relatively old.

Population III: A population of ancient stars thought to have formed before any enrichment of heavy elements had taken place, so that they contained only hydrogen and helium. No Population III stars have been found.

Positron   A subatomic particle with the same mass as the electron but with a positive electrical charge; the antiparticle of the electron.

Potential energy   Energy that is stored and may be converted into kinetic energy under certain circumstances. In astronomy the most common form of potential energy is gravitational potential energy.

Power   The rate of energy expenditure per second; for stars the world *luminosity* is used instead.

Precession   The slow shifting of star positions on the celestial sphere caused by the 26,000-year periodic wobble of the Earth's rotational axis.

Primary mirror   The principal light-gathering mirror in a reflecting telescope.

Prime focus   The focal arrangement in a reflecting telescope in which the image is allowed to form inside the telescope structure at the focal point of the primary mirror, so that no secondary mirror is needed.

Prograde motion   Orbital or spin motion in the "normal" direction; in the solar system, this is counterclockwise as viewed from above the North Pole.

Prominence   A cloud or column of heated, glowing gas extending from the chromosphere into the corona of the Sun, its structure controlled by magnetic fields.

Proper motion   The motion of a star across the sky, usually measured in units of arcseconds per year.

Proton-proton chain   The sequence of nuclear reactions in which four hydrogen nuclei combine, through intermediate steps involving deuterium and ^{3}He, to form one helium nucleus. The proton-proton chain is responsible for energy production in the cores of stars on the lower main sequence.

Protostar   A star in the process of formation, specifically one that has entered the slow gravitational contraction phase.

Pulsar   A rapidly rotating neutron star that emits periodic pulses of electromagnetic radiation, probably by the emission of beams of radiation from the magnetic poles, which sweep across the sky as the star rotates.

P wave   A seismic wave that is a compressional, or density, wave. P waves can travel through both solid and liquid portions of the Earth and are the first to reach any remote location from an earthquake site. (*see also* **Compressional waves**)

QSO   (*see* **quasistellar object**)

Quadrature   The configuration where a superior planet or the Moon is 90 degrees away from the Sun, as seen from the Earth.

Quantum mechanics   The physics of atomic structure and the behavior of subatomic particles, based on the principle of the quantum.

Quantum   The amount of energy associated with a photon, equal to $h\nu$, where h is the Planck constant, and ν is the frequency. The quantum is the smallest amount of energy that can exist at a given frequency.

Quark   One of the two fundamental classes of elementary particles; quarks are the constituent subparticles of protons and neutrons.

Quasar   (*see* **quasistellar object**)

Quasistellar object   Any of a class of extragalactic objects characterized by emission lines with very large redshifts. The quasistellar objects are thought to lie at great distances, in which case they existed only at earlier times in the history of the universe; they may be young galaxies.

Radial velocity   The component of motion of a star or other body along the line of sight; i.e. the portion of the relative velocity that is directed straight toward or away from the observer.

Radiation pressure   Pressure created by the forces exerted by photons of light when they are absorbed or reflected.

Radiative transport   The transport of energy, inside a star or in other situations, by radiation.

Radioactive dating   A technique for estimating the age of material such as rock, based on the known initial isotopic composition and the known rate of radioactive decay for unstable isotopes initially present.

Radioactivity   The spontaneous emission of subatomic particles (alpha rays or beta rays) or high-energy photons (gamma rays) by unstable nuclei. The identity of the emitting substance is changed in the process.

Radio galaxy   Any of a class of galaxies whose luminosity is greatest in radio wavelengths. Radio galaxies are usually large elliptical galaxies, with synchrotron radiation emitted from one or more pairs of lobes located on opposite sides of the visible galaxy.

Ray   A bright streak of ejecta emanating from an impact crater, especially on the Moon or on Mercury.

Recurrent nova   A star known to flare up in nova outbursts more than once. A recurrent nova is thought to be a binary system containing a white dwarf and a mass-losing star, in which the white dwarf sporadically flares up when material falls onto it from the companion.

Red giant   A star that has completed its core hydrogen-burning stage and has begun hydrogen shell burning, which causes its outer layers to become very extended and cool.

Reflecting telescope   A telescope that brings light to a focus by using mirrors.

Reflection nebula   An interstellar cloud containing dust that shines by light reflected from a nearby star.

Refracting telescope   A telescope that uses lenses to bring light to a focus.

Refractory   The property of being able to exist in solid form under conditions of very high temperature. A refractory element is one that is characterized by a high temperature of vaporization; refractory elements are the first to condense into solid form when a gas cools, as in the solar nebula.

Regolith   The layer of debris on the surface of the Moon created by the impact of meteorites; the lunar surface layer.

Resolution   In an image, the ability to separate closely spaced features, that is, the clarity or fineness of the image. In a spectrum, the ability to separate features that are close together in wavelength.

Retrograde motion   Orbital or spin motion in the opposite direction from prograde motion; in the solar system retrograde motions are clockwise as seen from above the North Pole.

Ribonucleic acid (RNA)   A complex protein consisting of amino acid bases, responsible for transferring the genetic code during reproduction of life forms.

Right ascension   The east–west coordinate in the equatorial coordinate system. The right ascension is measured in units of hours, minutes, and seconds to the east from a fixed direction on the sky, which itself is defined as the line of intersection of the ecliptic and the celestial equator.

Rille   A type of winding, sinuous valley commonly found on the Moon.

RNA   (*see* **Ribonucleic acid**)

Roche limit   The point near a massive body such as a planet or star inside which the tidal forces acting on an orbiting body exceed the gravitational force holding it together. The location of the Roche limit depends on the size of the orbiting body.

Rotation curve   A plot showing the orbital velocity of stars in a spiral galaxy versus distance from the galactic center.

r-process reactions   Nuclear fusion reactions in which neutrons are captured by nuclei in rapid succession, so that there is insufficient time for beta-decay to take place between captures. These reactions are important only in very high-energy situations, such as in supernova explosions.

RR Lyrae variable   A member of a class of pulsational variable stars named after the prototype star, RR Lyrae. These stars are blue-white giants with pulsational periods of less than one day and are Population II objects found primarily in globular clusters.

Russell-Vogt theorem   The statement that the properties of a star are fully determined by its mass and its composition.

Saros cycle   An eighteen-year, eleven-day repeating pattern of solar and lunar eclipses caused by a combination of the tilt of the lunar orbit with respect to the ecliptic and the precession of the plane of the Moon's orbit.

Scarp   A long cliff or series of cliffs, usually created by shrinking or settling of a planetary crust.

Scattering   The random reflection of photons by particles such as atoms or ions in a gas or dust particles in interstellar space.

Schwarzschild radius   The radius within which an object has collapsed to the stage when light can no longer escape the gravitational field as the object becomes a black hole.

Secondary mirror   The second mirror in a reflecting telescope (after the primary mirror), usually either convex in shape, to reflect the image out a hole in the bottom of the telescope to the cassegrain focus, or flat, to reflect the image out of the side of the telescope to the Newtonian focus or along the telescope mount axis to the coudé focus.

Sedimentary rock   A rock formed by the deposition and hardening of layers of sediment, usually either underwater or in an area subject to flooding.

Seeing   The blurring and distortion of point sources of light, such as stars, caused by turbulent motions in the Earth's atmosphere.

Seismic wave   A wave created in a planetary or satellite interior, usually caused by an earthquake.

Selection effect   The tendency for a conclusion based on observations to be influenced by the method used to select the objects for observation. An example was the early belief that all quasars are radio sources, when the principal method used to discover quasars was to look for radio sources and then see if they had other properties associated with quasars.

Semimajor axis   One-half of the major, or long, axis of an ellipse.

Sextant   A device consisting of a pair of pointers and a segment (one-sixth) of a circle inscribed with angle markings, for the measurement of altitudes or angular separations between objects on the sky.

Seyfert galaxy   Any of a class of spiral galaxies, first recognized by Carl Seyfert, with unusually bright blue nuclei.

Shear wave   A wave that consists of transverse motions, that is, motions perpendicular to the direction of wave travel.

Sidereal day   The rotation period of the Earth with respect to the stars, or as seen by a distant observer, equal to 23 hours, 56 minutes, 4.091 seconds.

Sidereal period   The orbital or rotational period of any object with respect to the fixed stars or as seen by a distant observer.

Singularity   A structure that is defined mathematically as a single point, and which therefore cannot be described in physical terms. In astronomy the center of a black hole is described as a singularity.

SNC meteorites   A class of meteorites whose distribution and composition suggests that they are ejecta from one or more impacts that occurred on Mars.

Solar active region   A region on the Sun, usually associated with sunspots, where activity such as solar flares originates.

Solar constant   The intensity of sunlight (in units of ergs per square centimeter) striking the earth above the atmosphere. The value is approximately 1.4×10^6 erg/sec.

Solar day   The synodic rotation period of the Earth with respect to the Sun; that is, the length of time from one local noon, when the Sun is on the meridian, to the next local noon.

Solar eclipse   An eclipse of the Sun, occurring when the Moon's disk blocks some or all of the Sun's disk, as seen from the Earth.

Solar flare   An explosive outburst of ionized gas from the Sun, usually accompanied by X-ray emission and the injection of large quantities of charged particles into the solar wind.

Solar motion   The deviation of the Sun's velocity from perfect circular motion about the center of the galaxy; that is, the Sun's peculiar velocity.

Solar nebula   The primordial gas and dust cloud from which the Sun and the planets condensed.

Solar wind   The stream of charged subatomic particles flowing steadily outward from the Sun.

Solstice   The occasion when the Sun, as viewed from the Earth, reaches its farthest northern (summer solstice to Northern Hemisphere observers) or southern point (winter solstice).

South celestial pole   The projection of the Earth's south pole onto the sky.

s-process reactions   Nuclear reactions in which nuclei capture neutrons at a rate slow enough that beta-decay can occur before an additional neutron is captured. These reactions are responsible for the creation of many heavy elements in massive stars before they explode as supernovae.

Spacetime   The term for four-dimensional space, consisting of the three spatial dimensions plus the time dimension, as described in general relativity theory.

Spectrogram   A photograph of a spectrum.

Spectrograph   An instrument for recording the spectra of astronomical bodies or other sources of light.

Spectroscope   An instrument allowing an observer to view the spectrum of a source of light.

Spectroscopic binary   A binary system recognized as a binary because its spectral lines undergo periodic Doppler shifts as the orbital motions of the two stars cause them to move toward and away from the Earth. If lines of only one star are seen, it is a single-lined spectroscopic binary; if lines of both stars are seen, it is a double-lined spectroscopic binary.

Spectroscopic parallax   The technique of distance determination for stars in which the absolute magnitude is inferred from the H-R diagram and then compared with the observed apparent magnitude to yield the distance.

Spectroscopy   The science of analyzing the spectra of stars or other sources of light.

Spectrum   An arrangement of electromagnetic radiation according to wavelength.

Spectrum binary   A binary system recognized as a binary because its spectrum contains lines of two stars of different spectral types.

Spicules   Narrow, short-lived jets of hot gas extending upward from the solar chromosphere.

Spin-orbit coupling   A simple relationship between the orbital and spin periods of a satellite or planet, caused by tidal forces that have slowed the rate of rotation of the orbiting body. Synchronous rotation is the simplest and most common form of spin-orbit coupling.

Spiral density wave   A spiral wave pattern in a rotating, thin disk, such as the rings of Saturn or the plane of a spiral galaxy like the Milky Way. (*see also density wave*)

Spiral galaxy   Any of a large class of galaxies exhibiting a disk with spiral arms.

Standard candle   A general term for any astronomical object, the absolute magnitude of which can be inferred from its other observed characteristics, and which is therefore useful as a distance indicator.

Starburst galaxy   A galaxy, usually a spiral or an irregular, undergoing a phase of intense star formation. Many are detected only in infrared wavelengths because of obscuration by dense clouds of interstellar dust associated with the star formation.

Stefan-Boltzmann law   The law of continuous radiation stating that for a spherical glowing object such as a star the luminosity is proportional to the square of the radius and the fourth power of the temperature.

Stefan's law   An experimentally derived law of continuous radiation stating that the energy emitted by a glowing body per square centimeter of surface area is proportional to the fourth power of the absolute temperature.

Stellar occultation   The passage of a foreground object in front of a background star. Such occurrences are sometimes useful in analyzing the properties of the occulting body, such as in the *Voyager* spacecraft observations of fine detail in the ring systems of the outer planets, or in the analysis of planetary atmospheres observed as they occult background stars.

Stellar parallax   The apparent annual shifting of position of a nearby star with respect to more distant background stars. The term *stellar parallax* is often assumed to mean the parallax angle, which is one-half of the total angular motion a star undergoes. (*see also parallax and parsec*)

Stellar wind   Any stream of gas flowing outward from a star, including the very rapid winds from hot, luminous stars; the intermediate-velocity, rarefied winds from stars like the Sun; and the slow, dense winds from cool supergiant stars.

Stratosphere   The layer of the Earth's atmosphere, between 10 and 50 kilometers in altitude, where the temperature increases with height.

Strong nuclear force   The force that holds the nucleons (protons and neutrons) together in the nucleus of an atom. This is the strongest of the four known natural forces, but it acts only over very short (nuclear) distances.

Subduction   The process in which one tectonic plate is submerged below another along a line where two plates collide. A subduction zone is usually characterized by a deep trench and an adjoining mountain range.

Sublimation   Evaporation of a compound directly from the solid to the gaseous state.

Supercluster   A cluster of clusters of galaxies.

Supergiant   A star in its late stages of evolution that is undergoing shell burning and is therefore very extended in size and very cool; an extremely large giant.

Supergranulation   The pattern of large cells seen in the Sun's chromosphere when viewed in the light of the strong emission line of ionized hydrogen.

Superior planet   Any planet whose orbit lies beyond the Earth's orbit around the Sun.

Supernova   The explosive destruction of a massive star that occurs when all sources of nuclear fuel have been consumed, and the star collapses catastrophically.

Synchronous rotation   A situation in which the rotational and orbital periods of an orbiting body are equal so that the same side is always facing the companion object.

Synchrotron radiation   Electromagnetic radiation emitted by rapidly-moving charged particles (usually electrons) spiralling around magnetic field lines; the radiation is polarized and is emitted in a conical pattern in the direction of particle motion.

Synodic period   The orbital or rotational period of an object as seen by an observer on the Earth; for the Moon or a planet the synodic period is the interval between repetitions of the same phase or configuration.

Système Internationale   The system of units, based on the meter (for distances), kilogram (for masses), and second (for time), which has been adopted by international agreement of physicists.

S wave   A type of seismic wave that is a transverse, or shear, wave and can travel only through rigid materials. (*see* **shear wave**)

Tectonic activity   Geophysical processes involving motions of tectonic plates and associated volcanic and earthquake activity. (*see also* **plate tectonics** *and* **continental drift**)

Temperature   A measure of the internal energy of a substance. At the atomic level, temperature is related to the motions of atoms and molecules that make up the substance.

Thermal radiation   Continuous radiation emitted by any body whose temperature is above absolute zero.

Thermal spectrum   The spectrum of continuous radiation from a body that glows because it has a temperature above absolute zero. A thermal spectrum is one that is described mathematically by the Planck function.

Thermosphere   The layer of the Earth's atmosphere above the mesosphere, extending upward from a height of 80 kilometers, where the temperature rises with altitude.

Tidal force   A gravitational force that tends to stretch or distort an extended object. (*see* **differential gravitational force**)

Total eclipse   Any eclipse in which the eclipsed body is totally blocked from view or totally immersed in shadow.

Transit telescope   A telescope designed to point straight overhead and accurately measure the times at which stars cross the meridian.

Transverse wave   (*see* **shear wave**)

Triple-alpha reaction   A nuclear fusion reaction in which three helium nuclei (or alpha particles) combine to form a carbon nucleus.

Trojan asteroid   Any of several asteroids orbiting the Sun at stable positions in the orbit of Jupiter, either 60 degrees ahead of the planet or 60 degrees behind it.

Tropical year   The year as defined by the length of time required for the Sun to move through the celestial sphere once, as defined by its passage through the vernal equinox.

Troposphere   The lowest temperature zone in the Earth's atmosphere, extending from the surface to a height of about 10 kilometers, in which the temperature decreases with altitude.

Tully-Fisher method   A technique for estimating the distances of spiral galaxies, based on a correlation between the width of the 21-cm emission line and the luminosity of a galaxy.

Tuning fork diagram   A diagram developed by Edwin Hubble in which the morphological types of galaxies are displayed in a sequence having two branches (representing spiral galaxies and barred spiral galaxies), resembling the shape of a tuning fork.

Type I irregular   A galaxy not conforming to any of the standard types, but appearing to have some spiral structure.

Type II irregular   A galaxy not conforming to any of the standard types, and showing no evidence of spiral structure.

Type I Supernova   A violent stellar explosion whose spectrum does not include lines of hydrogen, thought to be caused by the nuclear destruction of a white dwarf star that has gained mass from a companion star.

Type II Supernova   A violent stellar explosion whose spectrum contains lines of hydrogen, thought to be caused by the collapse and rebound of a massive star that has formed an iron core at the end of its nuclear-burning lifetime.

T Tauri star   A young star still associated with the interstellar material from which it formed, typically exhibiting brightness variations and a stellar wind.

Ultraviolet   The portion of the electromagnetic spectrum between roughly 100 and 3,000 angstroms.

Umbra   (1) The dark inner portion of a shadow, such as the part of the Earth's shadow where the Moon is in total eclipse during a lunar eclipse; (2) the dark central portion of a sunspot.

Van Allen belts   Zones in the Earth's magnetosphere where charged particles are confined by the Earth's magnetic field. There are two main belts, one centered at an altitude of roughly 1.5 times the Earth's radius, and the other between 4.5 and 6.0 times the Earth's radius.

Velocity dispersion   A measure of the range of speeds of particles or stars in a group or cluster with random internal motions. In globular clusters and elliptical galaxies, the velocity dispersion can be used to infer the central mass.

Vernal equinox   The Northern Hemisphere name for the occasion when the Sun crosses the celestial equator from south to north, usually occurring around March 21. (*see also* **autumnal equinox** *and* **equinox**)

Visual binary   Any binary system in which both stars can be seen through a telescope or on photographs.

Volatile   The property of being easily vaporized. Volatile elements stay in gaseous form except at very low temperatures and did not condense into solid form during the formation of the solar system.

Watt   A unit of power, equal to 1 joule/sec.

Wavelength   The distance between wavecrests in any type of wave.

White dwarf   The compact remnant of a low-mass star, supported against further gravitational collapse by the pressure of the degenerate electron gas that fills its interior.

Wien's law   An experimentally discovered law applicable to thermal continuum radiation, which states that the wavelength of maximum emission intensity is inversely proportional to the absolute temperature.

X-ray   A photon of electromagnetic radiation in the wavelength interval between about 1 and 100 angstroms.

Zeeman effect   The broadening or splitting of spectral lines due to the presence of a magnetic field in the gas where the lines are formed.

Zenith   The point directly overhead.

Zero-age main sequence   The main sequence in the H-R diagram formed by stars that have just begun their hydrogen-burning lifetimes and have not yet converted any significant fraction of their core mass into helium. The zero-age main sequence forms the lower left boundary of the broader band representing the general main sequence.

Zodiac   A band circling the celestial sphere along the ecliptic that is broad enough to encompass the paths of all the planets visible to the naked eye; in some usages the sequence of constellations lying along the ecliptic.

Zodiacal light   A diffuse band of light visible along the ecliptic near sunrise and sunset, created by sunlight scattered off interplanetary dust.

21-centimeter line   The emission line of atomic hydrogen, whose wavelength is 21.11 centimeters, emitted when the spin of the electron reverses itself with respect to that of the proton. The 21-centimeter line is the most widely used and effective means of tracing the distribution of interstellar gas in the Milky Way and other galaxies.

24-hour anisotropy   The daily fluctuation in the observed temperature of the cosmic background radiation caused by a combination of the Earth's motion with respect to the background and its rotation.

3° background   The radiation field filling the universe, left over from the big bang. (*see also* **cosmic background radiation**)

Index

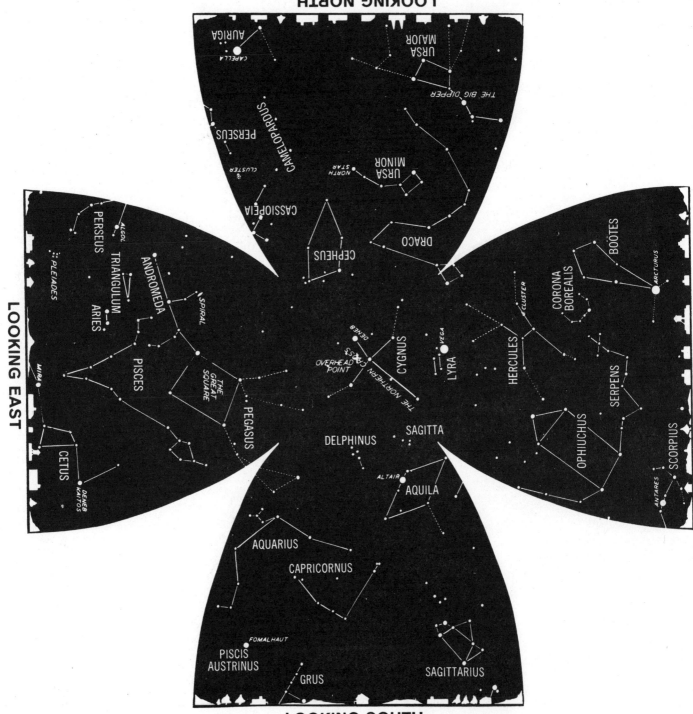

LOOKING NORTH

LOOKING EAST

LOOKING SOUTH

This map represents the sky   SEPTEMBER   1 at 10 p.m.
at the following standard times   SEPTEMBER 16 at   9 p.m.
(for daylight saving time, add one hour):   OCTOBER   1 at   8 p.m.